Carpentry Framing and Finishing

Level Two

Trainee Guide
Fourth Edition

PEARSON
Prentice
Hall

Upper Saddle River, New Jersey
Columbus, Ohio

NCCER
President: Don Whyte
Director of Product Development: Daniele Stacey
Carpentry Project Manager: Daniele Stacey
Production Manager: Jessica Martin
Product Maintenance Supervisor: Debie Ness
Editors: Brendan Coote and Bethany Harvey
Desktop Publishers: Jessica Martin and James McKay

Writing and development services provided by Topaz Publications, Liverpool, New York.

Pearson Education, Inc.
Product Manager: Lori Cowen
Project Manager: Stephen C. Robb
Design Coordinator: Diane Y. Ernsberger
Text Designer: Kristina D. Holmes
Cover Designer: Kristina D. Holmes
Copy Editor: Sheryl Rose
Scanning Technician: Janet Portisch
Operations Supervisor: Pat Tonneman
Marketing Manager: Derril Trakalo

This book was set in Palatino and Helvetica by Carlisle Communications, Ltd. It was printed and bound by Courier Kendallville, Inc. The cover was printed by Phoenix Color Corp.

This information is general in nature and intended for training purposes only. Actual performance of activities described in this manual requires compliance with all applicable operating, service, maintenance, and safety procedures under the direction of qualified personnel. References in this manual to patented or proprietary devices do not constitute a recommendation of their use.

Pearson Prentice Hall™ is a trademark of Pearson Education, Inc.
Pearson® is a registered trademark of Pearson plc
Prentice Hall® is a registered trademark of Pearson Education, Inc.

Pearson Education Ltd.
Pearson Education Singapore Pte. Ltd.
Pearson Education Canada, Ltd.
Pearson Education—Japan

Pearson Education Australia Pty. Limited
Pearson Education North Asia Ltd.
Pearson Educación de Mexico, S.A. de C.V.
Pearson Education Malaysia Pte. Ltd.

10 9 8
ISBN-13: 978-0-13-614410-6
ISBN-10:   0-13-614410-1

Preface

TO THE TRAINEE

Congratulations! If you're training under an NCCER-Accredited Training Sponsor, you have successfully completed *Carpentry Fundamentals: Level One*. Now you are well on your way to more specific skills training in the carpentry trade. As you are aware, there are many choices to be made in life—your career is just one. As you continue with your training, the skills you gain will help to craft your career in the carpentry trade.

NCCER has developed *Carpentry Framing and Finishing: Level Two* to support your skill development, no matter which career track you decide upon. The competencies taught in this level support training in both residential and commercial carpentry. With the help of your instructor, guidance counselor, or other mentor, you may choose to focus on one track over the other or decide to complete the entire program. Ultimately, that choice will be up to you.

We wish you success on your journey through this training program. Should you have any comments on how NCCER might improve upon this text, please complete the user feedback form located at the back of each module and send it to us. We will always consider and respond to input from our customers.

NEW WITH *CARPENTRY FRAMING AND FINISHING: LEVEL TWO*

NCCER and Prentice Hall are pleased to present the fourth edition of *Carpentry Framing and Finishing: Level Two*. This edition presents a new design and includes expanded coverage of commercial applications, as well as new modules on commercial drawings and cold-formed steel framing.

Check out the opening pages of each of the twelve modules in this textbook to see award-winning construction projects from the two largest construction trade associations in the nation, Associated Builders and Contractors (ABC) and Associated General Contractors (AGC). ABC's Awards of Excellence Program and AGC's Build America Program have become the premier competitions within the construction

industry to recognize outstanding construction projects across the nation.

Associated Builders and Contractors is a national association representing 23,000 merit shop construction and construction-related firms in 79 chapters across the United States. ABC's membership represents all specialties within the U.S. construction industry and is comprised primarily of firms that perform work in the industrial and commercial sectors of the industry. To learn more about ABC, visit **www.abc.org.**

Associated General Contractors is the nation's oldest construction trade association, established in 1918 after a request by President Woodrow Wilson. President Wilson recognized the construction industry's national importance and desired a partner with which the government could discuss and plan for the advancement of the nation. AGC has been fulfilling that mission for almost 90 years. To learn more about AGC, visit **www.agc.org.**

We also invite you to visit the NCCER website at **www.nccer.org** for the latest releases, training information, newsletter, Contren® product catalog, and much more. Your feedback is welcome. You may email your comments to **curriculum@nccer.org** or send general comments and inquiries to **info@nccer.org.**

NEW MULTI-TIERED CREDENTIALING

Opportunities for multiple credentials now exist due to the reorganization of the Carpentry program. (See Figures 1 and 2.) Are you training for a career in the commercial sector? Then follow the commercial carpentry course map within the new *Carpentry Level Two*. Successful completion of this level (in combination with *Core Curriculum* and *Carpentry Level One*) will lead to a certificate in Commercial Carpentry from NCCER's National Registry. Or are you training to go to work in the residential field? Then follow the alternate course map in Carpentry Level Two. Successful completion will lead to a certificate in Residential Carpentry.

If you're training through a traditional registered apprenticeship program, NCCER still meets

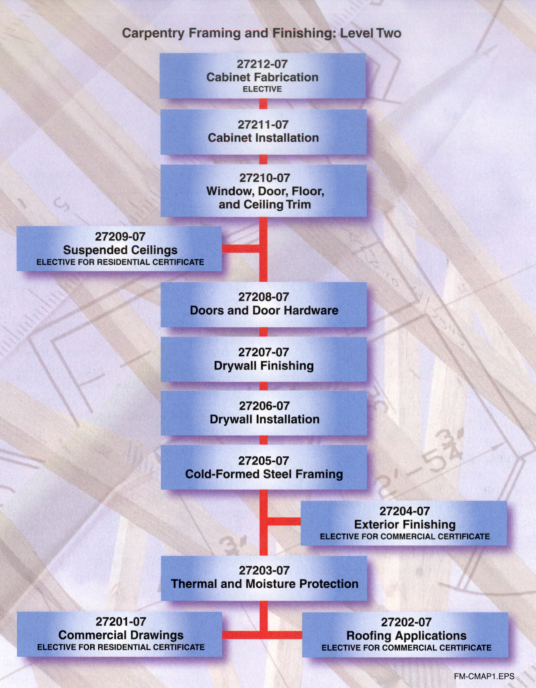

Carpentry Framing and Finishing: Level Two

27212-07
Cabinet Fabrication
ELECTIVE

27211-07
Cabinet Installation

27210-07
Window, Door, Floor, and Ceiling Trim

27209-07
Suspended Ceilings
ELECTIVE FOR RESIDENTIAL CERTIFICATE

27208-07
Doors and Door Hardware

27207-07
Drywall Finishing

27206-07
Drywall Installation

27205-07
Cold-Formed Steel Framing

27204-07
Exterior Finishing
ELECTIVE FOR COMMERCIAL CERTIFICATE

27203-07
Thermal and Moisture Protection

27201-07
Commercial Drawings
ELECTIVE FOR RESIDENTIAL CERTIFICATE

27202-07
Roofing Applications
ELECTIVE FOR COMMERCIAL CERTIFICATE

FM-CMAP1.EPS

Figure 1 ◆ Course maps are included at the beginning of each module. This course map shows modules common to both Commercial and Residential, as well as modules required for each tier.

the time-based requirements set forth by the Department of Labor's Office of Apprenticeship. A credential for a pure Carpentry Level Two completion is available through NCCER as well.

Now there's even more reason to train under an NCCER-accredited training sponsor—multi-tiered

credentialing! And nothing gives the job-seeker more of an edge than an industry-recognized credential. If you're not training through an NCCER-accredited training sponsor, find out where one exists near you. Call NCCER at 1-888-NCCER20 or visit **www.nccer.org**.

27211-07 Cabinet Installation	27211-07 Cabinet Installation
27210-07 Window, Door, Floor, and Ceiling Trim	27210-07 Window, Door, Floor, and Ceiling Trim
27209-07 Suspended Ceilings	27208-07 Doors and Door Hardware
27208-07 Doors and Door Hardware	27207-07 Drywall Finishing
27207-07 Drywall Finishing	27206-07 Drywall Installation
27206-07 Drywall Installation	27205-07 Cold-Formed Steel Framing
27205-07 Cold-Formed Steel Framing	27204-07 Exterior Finishing
27203-07 Thermal and Moisture Protection	27203-07 Thermal and Moisture Protection
27201-07 Commercial Drawings	27202-07 Roofing Applications
Mandatory Modules for Commercial Carpentry Certificate	**Mandatory Modules for Residential Carpentry Certificate**

FM-CMAP2.EPS

Figure 2 ◆ These simplified maps extracted from Figure 1 show the modules required for Commercial Carpentry (left) and Residential Carpentry (right) certificates.

CONTREN® LEARNING SERIES

The National Center for Construction Education and Research (NCCER) is a not-for-profit 501(c)(3) education foundation established in 1995 by the world's largest and most progressive construction companies and national construction associations. It was founded to address the severe workforce shortage facing the industry and to develop a standardized training process and curricula. Today, NCCER is supported by hundreds of leading construction and maintenance companies, manufacturers, and national associations. The Contren® Learning Series was developed by NCCER in partnership with Prentice Hall, the world's largest educational publisher.

Some features of NCCER's Contren® Learning Series are as follows:

- An industry-proven record of success
- Curricula developed by the industry for the industry
- National standardization providing portability of learned job skills and educational credits
- Compliance with Department of Labor's Office of Apprenticeship requirements for related classroom training (CFR 29:29)
- Well-illustrated, up-to-date, and practical information

NCCER also maintains a National Registry that provides transcripts, certificates, and wallet cards to individuals who have successfully completed modules of NCCER's Contren® Learning Series. *Training programs must be delivered by an NCCER Accredited Training Sponsor in order to receive these credentials.*

Contren® Curricula

NCCER's training programs comprise more than 40 construction, maintenance, and pipeline areas and include skills assessments, safety training, and management education.

Boilermaking
Carpentry
Cabinetmaking
Careers in Construction
Concrete Finishing
Construction Craft Laborer
Construction Technology
Core Curriculum: Introductory Craft Skills
Currículo Básico
Electrical
Electronic Systems Technician
Heating, Ventilating, and Air Conditioning
Heavy Equipment Operations
Hydroblasting
Highway/Heavy Construction
Instrumentation
Insulating
Ironworking
Maintenance, Industrial
Masonry
Millwright
Mobile Crane Operations
Painting
Painting, Industrial
Pipefitting

Pipelayer
Plumbing
Reinforcing Ironwork
Rigging
Scaffolding
Sheet Metal
Site Layout
Sprinkler Fitting
Welding

Pipeline

Control Center Operations, Liquid
Corrosion Control
Electrical and Instrumentation
Field Operations, Liquid
Field Operations, Gas
Maintenance
Mechanical

Safety

Field Safety
Orientación de Seguridad
Safety Orientation
Safety Technology

Management

Introductory Skills for the Crew Leader
Project Management
Project Supervision

Special Features of This Book

In an effort to provide a comprehensive user-friendly training resource, we have incorporated many different features for your use. Whether you are a visual or hands-on learner, this book will provide you with the proper tools to get started in the construction industry.

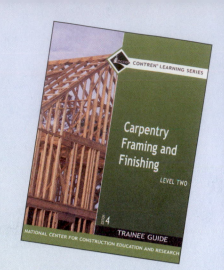

Introduction Page

This page is found at the beginning of each module and lists the Objectives, Trade Terms, Required Trainee Materials, Prerequisites, and Course Map for that module. The Objectives list the skills and knowledge you will need in order to complete the module successfully. The list of Trade Terms identifies important terms you will need to know by the end of the module. Required Trainee Materials list the materials and supplies needed for the module. The Prerequisites for the module are listed and illustrated in the Course Map. The Course Map also gives a visual overview of the entire course and a suggested learning sequence for you to follow.

Notes, Cautions, and Warnings

Safety features are set off from the main text in high-lighted boxes and organized into three categories based on the potential danger of the issue being addressed. Notes simply provide additional information on the topic area. Cautions alert you of a danger that does not present potential injury but may cause damage to equipment. Warnings stress a potentially dangerous situation that may cause injury to you or a co-worker.

Did You Know?

The Did You Know? features introduce historical tidbits or modern information about the carpentry industry. Interesting and sometimes surprising facts about carpentry are also presented.

Inside Track

Inside Track features provide a head start for those entering the carpentry field by presenting technical tips and professional practices from master carpenters in a variety of disciplines. Inside Tracks often include real-life scenarios similar to those you might encounter on the job site.

Color Illustrations and Photographs

Full-color illustrations and photographs are used throughout each module to provide vivid detail. These figures highlight important concepts from the text and provide clarity for complex instructions. Each figure is denoted in the text in *italic type* for easy reference.

Trade Terms

Each module presents a list of Trade Terms that are discussed within the text, defined in the Glossary at the end of the module. These terms are denoted in the text with **blue bold type** upon their first occurrence. To make searches for key information easier, a comprehensive Glossary of Trade Terms from all modules is found at the back of this book.

Step-by-Step Instructions

Step-by-step instructions are used throughout to guide you through technical procedures and tasks from start to finish. These steps show you not only how to perform a task but how to do it safely and efficiently.

Review Questions

Review Questions are provided to reinforce the knowledge you have gained. This makes them a useful tool for measuring what you have learned.

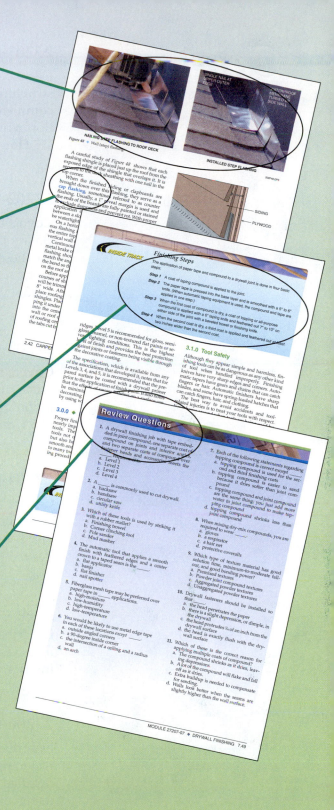

Contents

Acknowledgments

This curriculum was revised as a result of the farsightedness and leadership of the following sponsors:

ABC National
Associated Training Services–ABC Wisconsin
Brasfield & Gorrie
Cianbro Corporation
Greenville Technical College
Guilford Technical Community College

Onslow-Sheffield, Inc.
River Valley Technical Center
SC Department of Education
Steel Framing Alliance
The Shaw Group

This curriculum would not exist were it not for the dedication and unselfish energy of those volunteers who served on the Authoring Team. A sincere thanks is extended to the following:

Howard Davis
Shane Harvey
Curtis Haskins
Erin Hunter
Peter Klapperich
Dan McNally

Perry Moore
Mike Noble
Mark Onslow
Jay A. Pearson
Maribeth Rizzuto
Todd Staub

A final note: This book is the result of a collaborative effort involving the production, editorial, and development staff at Prentice Hall and the National Center for Construction Education and Research. Thanks to all of the dedicated people involved in the many stages of this project.

NCCER PARTNERING ASSOCIATIONS

American Fire Sprinkler Association
American Society for Training & Development
API
Associated Builders & Contractors, Inc.
Associated General Contractors of America
Association for Career and Technical Education
Carolinas AGC, Inc.
Carolinas Electrical Contractors Association
Center for Improvement of Construction
    Management and Processes
Construction Industry Institute
Construction Users Roundtable
Design-Build Institute of America
Electronic Systems Industry Consortium
Merit Contractors Association–of Canada
Metal Building Manufacturers Association
National Association of Minority Contractors
National Association of State Supervisors for
    Trade and Industrial Education

National Association of Women in Construction
National Insulation Association
National Ready Mixed Concrete Association
National Systems Contractors Association
National Utility Contractors Association
National Technical Honor Society
National Utility Contractors Association
North American Crane Bureau
North American Technician Excellence
Painting & Decorating Contractors of America
Portland Cement Association
SkillsUSA
Steel Erectors Association of America
Texas Gulf Coast Chapter ABC
U.S. Army Corps of Engineers
University of Florida
Women Construction Owners & Executives, USA

Commercial Drawings
27201-07

27201-07
Commercial Drawings

Overview

Reading and interpreting drawings and specifications are essential to construction work. The project drawings will tell you everything from where to place the building on the site to how to construct an exterior wall. The specifications provide details on materials and construction methods. If you can't read and interpret construction drawings, you can't be an effective carpenter.

Objectives

When you have completed this module, you will be able to do the following:

1. Recognize the difference between commercial and residential construction drawings.
2. Identify the basic keys, abbreviations, and other references contained in a set of commercial drawings.
3. Accurately read a set of commercial drawings.
4. Identify and document specific items from a door and window schedule.
5. Explain basic construction details and concepts employed in commercial construction.
6. Calculate the floor area of each room in a floor plan.

Trade Terms

Beams
Callouts
Civil drawings
Contour lines
Elevation view
Girders
Isometric drawing
Joists
Landscape drawings
Plan view
Riser diagram

Required Trainee Materials

1. Pencil and paper
2. Appropriate personal protective equipment

Prerequisites

Before you begin this module, it is recommended that you successfully complete *Core Curriculum*; and *Carpentry Fundamentals Level One*.

This course map shows all of the modules in the second level of the *Carpentry* curriculum. The suggested training order begins at the bottom and proceeds up. Skill levels increase as you advance on the course map. The local Training Program Sponsor may adjust the training order.

27212-07
Cabinet Fabrication
ELECTIVE

27211-07
Cabinet Installation

27210-07
Window, Door, Floor, and Ceiling Trim

27209-07
Suspended Ceilings
ELECTIVE FOR RESIDENTIAL CERTIFICATE

27208-07
Doors and Door Hardware

27207-07
Drywall Finishing

27206-07
Drywall Installation

27205-07
Cold-Formed Steel Framing

27204-07
Exterior Finishing
ELECTIVE FOR COMMERCIAL CERTIFICATE

27203-07
Thermal and Moisture Protection

27202-07
Roofing Applications
ELECTIVE FOR COMMERCIAL CERTIFICATE

27201-07
Commercial Drawings
ELECTIVE FOR RESIDENTIAL CERTIFICATE

FRAMING AND FINISHING

CARPENTRY FUNDAMENTALS

CORE CURRICULUM:
Introductory Craft Skills

201CMAP.EPS

1.0.0 ◆ INTRODUCTION

A large part of the construction market in the United States is commercial and industrial construction. Commercial buildings come in various shapes and sizes (*Figure 1*). In order to build commercial structures, you must be able to understand and interpret the architect's plans and drawings. Commercial plans and drawings are usually more complex than residential plans. This module will introduce you to commercial drawings. The basic principles, such as scaling and dimensioning, are the same as those that apply to residential drawings. However, commercial structures will have more dimensions to interpret. Additional section and detail drawings mean there will be more changes in scale.

201F01.EPS

Figure 1 ◆ Modern buildings.

2.0.0 ◆ REQUIREMENTS AND CONTENTS

Commercial or industrial construction work is often called heavy construction. This is because heavy equipment and heavy materials are used for these jobs. Cranes, hoists, and graders are heavy equipment. Heavy materials include steel and concrete. Most commercial projects are larger and more complicated than residential projects. They require a greater variety of construction techniques, equipment, and materials.

Consequently, plans and drawings for a commercial project are also more complicated than residential plans and drawings. Commercial structures have many different uses. Safety and environmental requirements must also be considered. The complexity of the commercial project is reflected in the complexity of the plans.

2.1.0 Requirements for Commercial Plans

There are several reasons why commercial construction plans are more detailed than residential plans. First, the structures are usually larger and more expensive to build. Another major consideration is legal liability. The contractor's legal liability is far greater in commercial construction because there are more applicable codes, ordinances, and regulations. Other reasons that commercial plans are more complex include the following:

- The architectural plans, drawings, schedules, and specifications are legal documents. Many state, local, and federal agencies demand greater detail in commercial construction drawings to substantiate any legal disputes.
- Code restrictions and safety requirements for commercial and industrial buildings are far more complicated than for residential construction. More detailed drawings are used to make certain that all codes and local ordinances are met.
- The size of a commercial building requires a greater number of drawings, sections, details, and schedules, with more detail required to correlate the various parts of the structure.
- The materials used in commercial construction call for more detailed information on construction techniques, especially for structural steel.

A major commercial project may have fifty or sixty drawings in its plan set, plus associated schedules. Specialty subcontractors will generally get a partial set of the plans relating to their work. At least one complete set of plans will be kept in the field office for reference.

DID YOU KNOW?

Brick upon Brick

For nearly 40 years, the Empire State Building was the tallest building in the world. Approximately 10 million bricks were used in its construction. It was completed in 1931. Today it is likely that curtain walls containing brick facades would be used in place of individual bricks.

Despite the enormity of the project, construction of the Empire State Building was completed in about 15 months. One of the methods used to speed up construction was to have trucks dump the bricks down a chute, instead of dumping them in the street. The chute led to a large hopper from which the bricks were then dumped into carts, and hoisted to the location where they were needed. The innovative technique eliminated the backbreaking work of moving bricks from the pile to the bricklayer using a wheelbarrow.

201SA01.EPS

2.2.0 Commercial Plan Contents

Construction drawings consist of several different kinds of drawings assembled into a set (*Figure 2*). Each type of drawing is assigned a letter. For each type, there may be several drawings, which are then numbered. For example, the first few electrical drawings would be numbered E1, E2, and E3. A complete set of commercial construction plans typically includes the following drawing types:

- Architectural–A
- Structural–S
- Mechanical–M
- Plumbing–P
- Electrical–E

The exact content of the plan set will vary, depending on the type and size of the job and local code requirements. For example, a drawing set for a commercial office building may include landscape drawings. These would be denoted with the letter L. In some commercial drawing sets the site plans are called civil drawings. They are marked C1, C2, C3, and so forth. Specifications and schedules may be referenced or included with each type of drawing. These two components are similar in content, if not in quantity, to architectural drawings for residences. *Figure 3* shows common drawing lines and *Figure 4* shows common symbols for various materials.

Object lines show the main outline of the structure. Dimension lines and extension lines show the size of an object. Leader lines are used to connect the notes to features on the drawing. Arrows indicate the extent of the measurement. The measurement is written above the dimension line or in a break.

Section lines, or cutting plane lines, show the location of a section view of that particular part of the structure. Designers use various formats to indicate sections. Usually, they are in the form of a U with arrows at the tips. A letter identifies the section. The section drawing is marked with a corresponding letter. The section view is placed either on the same page as the main drawing or on a separate page for all sections. In the latter case, the section is referenced.

Symbols provide another graphic indication for different types of materials. There are many commonly used symbols. However, there are no standardized symbols for specific materials. The best known reference for standard symbols, lines, and abbreviations is *Architectural Graphic Standards* prepared by the American Institute of Architects.

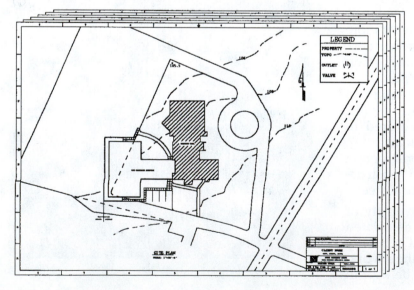

TITLE SHEET(S)
ARCHITECTURAL DRAWINGS
- SITE (PILOT) PLAN
- FOUNDATION PLAN
- FLOOR PLANS
- INTERIOR/EXTERIOR ELEVATIONS
- SECTIONS
- DETAILS
- SCHEDULES

STRUCTURAL DRAWINGS

PLUMBING DRAWINGS

MECHANICAL DRAWINGS

ELECTRICAL DRAWINGS

201F02.EPS

Figure 2 ◆ Drawing set.

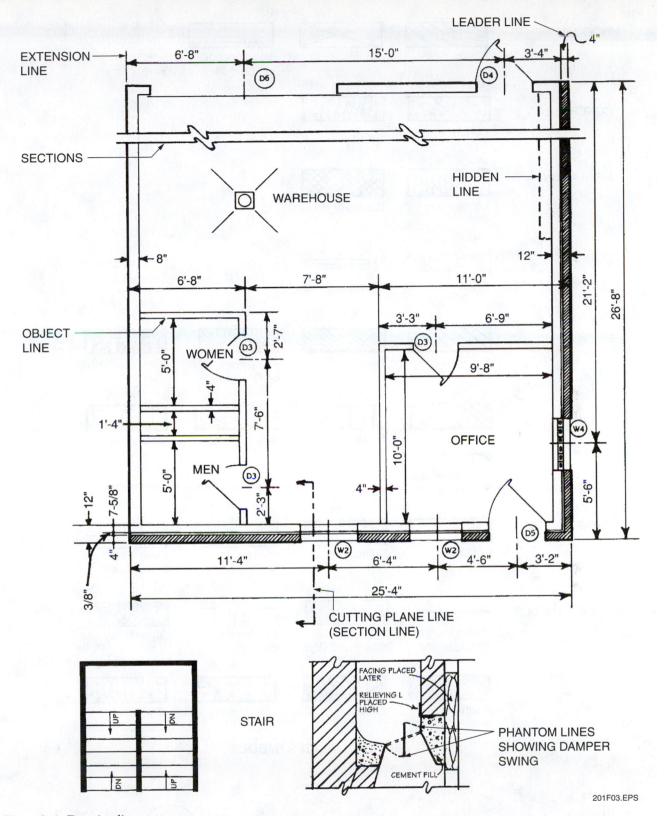

Figure 3 ◆ Drawing lines.

201F03.EPS

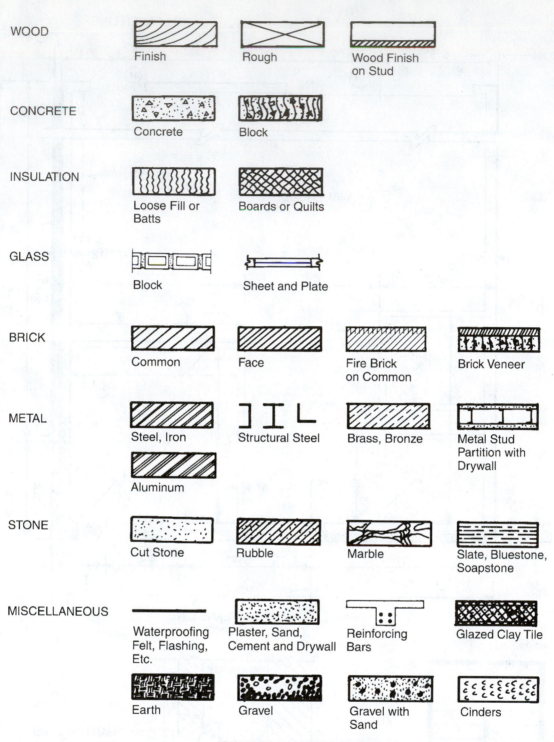

WOOD
- Finish
- Rough
- Wood Finish on Stud

CONCRETE
- Concrete
- Block

INSULATION
- Loose Fill or Batts
- Boards or Quilts

GLASS
- Block
- Sheet and Plate

BRICK
- Common
- Face
- Fire Brick on Common
- Brick Veneer

METAL
- Steel, Iron
- Structural Steel
- Brass, Bronze
- Metal Stud Partition with Drywall
- Aluminum

STONE
- Cut Stone
- Rubble
- Marble
- Slate, Bluestone, Soapstone

MISCELLANEOUS
- Waterproofing Felt, Flashing, Etc.
- Plaster, Sand, Cement and Drywall
- Reinforcing Bars
- Glazed Clay Tile
- Earth
- Gravel
- Gravel with Sand
- Cinders

GENERAL PLAN SYMBOLS

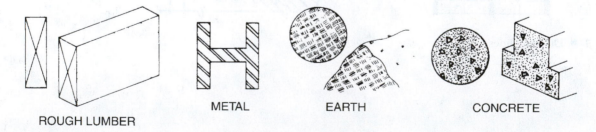

ROUGH LUMBER

METAL

EARTH

CONCRETE

SECTION VIEW SYMBOLS

201F04.EPS

Figure 4 ◆ Material symbols.

Large commercial projects will have a key block or legend. It may be on the index page or a separate page. It lists the symbols and abbreviations used throughout the plan set.

2.2.1 Architectural Drawings

The architectural drawings are usually labeled with page numbers beginning with the letter A. These contain general design features of the building, room layouts, construction details, and materials requirements. The architectural drawings will include the following:

- A site plan and/or other location plans
- Floor plans
- Wall sections
- Door and window details and schedules
- Elevations
- Special application details, including finish details and schedules

The site plan shows topographic features including trees, bodies of water, and ground cover. It will also show man-made features such as roads, railroad tracks, and utility lines. Typical topographic symbols are shown in *Figure 5*.

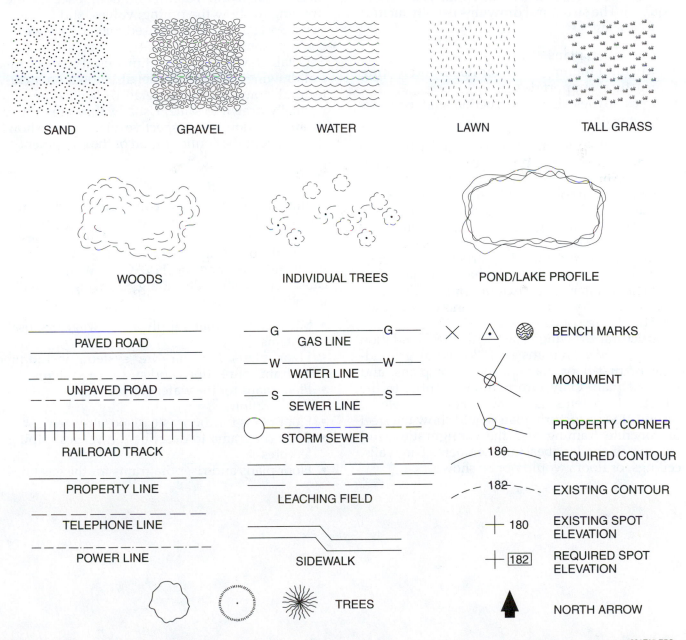

Figure 5 ◆ Topographic symbols.

201F05.EPS

2.2.2 Structural Drawings

Structural drawings provide a view of the structural members of the building and how they will support and transmit those loads to the ground. Structural drawings are numbered sequentially and designated by the letter S. They are normally located after the architectural drawings in a plan set.

A structural engineer prepares the structural drawings. They must calculate the forces on the building and the load that each structural member must withstand. The structural support information includes the foundation, size, and reinforcing requirements; the structural frame type and size of each member; and details on all connections required. The structural drawings usually include the following:

- Foundation plans
- Structural framing plans for floors and roofing
- Structural support details
- Notes to describe construction and code requirements

Structural support for a commercial building may be steel framing, precast concrete structural elements, or cast-in-place concrete. Unlike residences, modern commercial buildings do not have wooden frames, except in unusual circumstances.

Structural drawings provide useful information. They can stand alone for craftworkers such as framers and erectors. The structural drawings show the main building members and how they relate to the interior and exterior finishes. They do not include information that is unnecessary at the structural stage of construction.

Structural drawings start with the foundation plans. Foundation plans are followed by ground-floor or first-floor plans, upper-floor plans, and the roof plan. Only information essential to the structural systems is shown. For example, a second-floor structural plan would show the steel or concrete framing and the configuration and spacing of the loadbearing members. The walls, ceilings, or floors would not be shown.

2.2.3 Mechanical Drawings

Mechanical drawings show the different mechanical systems of the building. Specifically, they include the heating, ventilating, and air conditioning; plumbing; and fire protection drawings. These drawings have a prefix of M, P, and FP respectively. The heating, ventilating, and air-conditioning systems are commonly referred to as HVAC systems. The **plan view** format is commonly used on these drawings. This view offers the best illustration of the location and configuration of the work. The drawings serve as a diagram of the system layout.

A large amount of information is required for mechanical work. There is limited space on the drawing to show the piping, valves, and connections. Special symbols are used for clarity. *Figure 6* contains some common HVAC symbols.

Detail drawings are sometimes used on mechanical drawings. Unlike the details shown on architectural drawings, these detail drawings are not normally drawn to scale. Usually, they are drawn as an elevation or perspective view. They show details about the configuration of the equipment.

2.2.4 Plumbing Drawings

Plumbing drawings are considered part of the mechanical plans. However, they are usually placed on their own set of drawings for clarity. Unless the building is very basic, placing them on the same sheets as the HVAC drawings would cause confusion. Plumbing drawings usually include the following:

- Site plan for water supply and sewage disposal systems
- Floor plans for the fire system, including hydrant connections and sprinkler systems
- Floor plans for the water supply system and fixture location
- Floor plans for the waste disposal system
- **Riser diagrams** to describe the vertical piping features
- Floor plans and riser diagrams for the gas lines

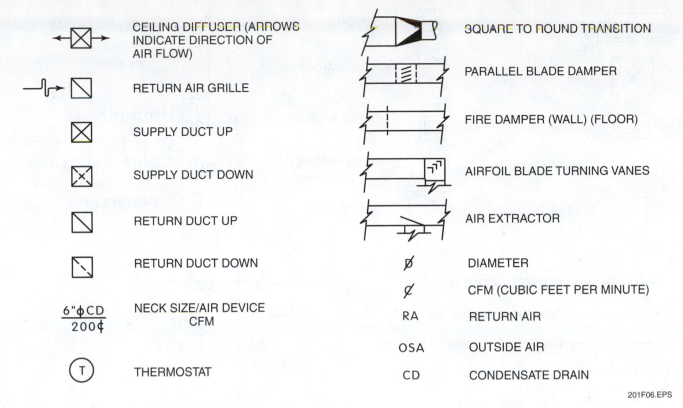

Figure 6 ◆ HVAC symbols.

The most common symbols used to designate the water and gas systems on plumbing drawings are shown in *Figure 7*. In addition to these symbols, there will be specific graphic symbols for the layout of such items as toilets, sinks, water heaters, and sump pumps. These will vary from structure to structure, depending on the specific design.

2.2.5 Electrical Drawings

The last part of the plan set usually contains the electrical drawings. They show the various electrical and communications systems of the building.

The electrical drawings are labeled with page numbers beginning with the letter E. They contain information on the electrical service requirements for the building. They also show the location of all outlets, switches, and fixtures. In addition to schematics of the branch circuits for the building, electrical drawings will include the following:

- Site plan for electrical service requirements
- Floor plans for the outlet and switch locations and the branch circuit requirements
- Lighting plans
- Emergency power and lighting systems
- Life safety systems
- Any backup power generation facilities
- Notes and details to describe other parts of the electrical system

Like mechanical drawings, electrical drawings use the plan view to show system layout. Details and schedules provide clarification. One drawing may include power, lighting, and telecommunications layouts. In more complex structures the systems are shown separately. There are many different symbols for electrical connections and fixtures. Commonly used symbols are shown in *Figure 8*. Special symbols for components such as power supplies, security systems, and circuit boards are usually designated by the manufacturer.

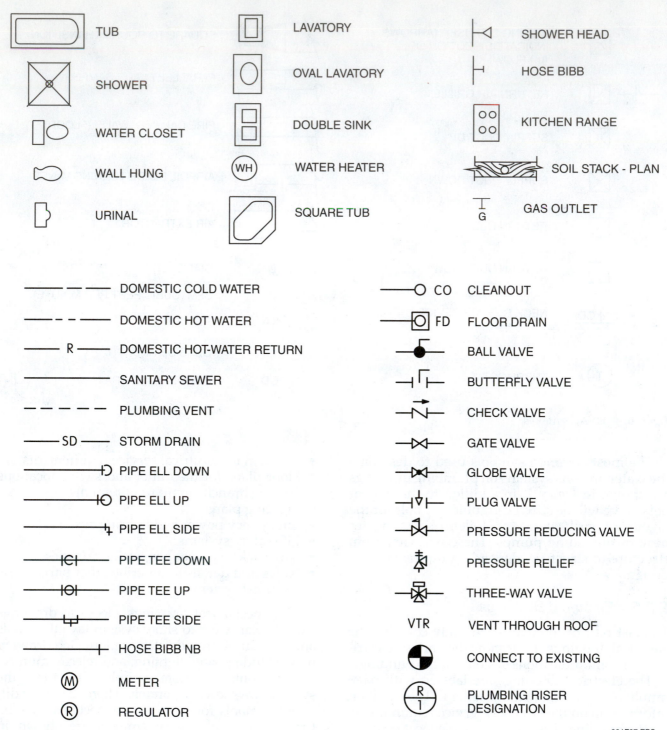

Figure 7 ◆ Plumbing symbols.

201F07.EPS

GENERAL OUTLETS

Junction Box, Ceiling

Fan, Ceiling

Recessed Incandescent, Wall

Surface Incandescent, Ceiling

Surface or Pendant Single
Fluorescent Fixture

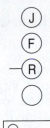

SWITCH OUTLETS

Single-Pole Switch

Double-Pole Switch

Three-Way Switch

Four-Way Switch

Key-Operated Switch

Switch w/Pilot

Low-Voltage Switch

Door Switch

Momentary Contact Switch

Weatherproof Switch

Fused Switch

Circuit Breaker Switch

S

S_2

S_3

S_4

S_K

S_P

S_L

S_D

S_{MC}

S_{WP}

S_F

S_{CB}

RECEPTACLE OUTLETS

Single Receptacle

Duplex Receptacle

Triplex Receptacle

Split-Wired Duplex Recep.

Single Special Purpose Recep.

Duplex Special Purpose Recep.

Range Receptacle

Switch & Single Receptacle

Grounded Duplex Receptacle

Duplex Weatherproof Receptacle

GFCI

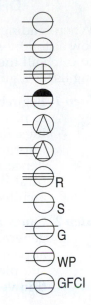

AUXILIARY SYSTEMS

Telephone Jack

Meter

Vacuum Outlet

Electric Door Opener

Chime

Pushbutton (Doorbell)

Bell and Buzzer Combination

Kitchen Ventilating Fan

Lighting Panel

Power Panel

Television Outlet

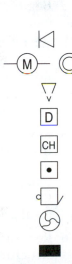

201F08.EPS

Figure 8 ◆ Electrical symbols.

3.0.0 ◆ READING AND UNDERSTANDING DRAWINGS

When reading commercial plans, it is best to follow a step-by-step process to avoid confusion and to catch all the important details. Use the following list as a guide:

Step 1 Begin by reading the project specifications to pick up details not found on the drawings. Remember that in cases of conflicting information, the specifications take precedence over the drawings, and the structural drawings take precedence over the architectural drawings.

Step 2 Quickly review all drawings that give a general impression of the shape, size, and appearance of the structure, including the site plan, floor plans, and exterior **elevation views.**

Step 3 Begin correlating the floor plans with the exterior elevations, making certain that the parts appear to fit together logically.

Step 4 Next, look at the wall sections and determine the wall types (materials; loadbearing or nonbearing) and construction details or procedures.

Step 5 Review the structural plans. Determine the foundation requirements and the type of structural system. Turn back to the site plan, floor plans, wall sections, and elevations as often as needed to relate the structural and architectural drawings.

Step 6 Review all details on the architectural and structural drawings. Carefully consider all items that may require special construction procedures.

Step 7 Review all interior elevations and try to get a clear picture of what the interior of the building will look like.

Step 8 Review the finish schedule.

Step 9 Review the mechanical and electrical plans.

If some part of a drawing is not clear, check with your supervisor. There may be a logical explanation for the item in question, or the plans may be incorrect. It is best to find an answer before construction begins. The architect should clarify and resolve all conflicts in writing so there will not be any confusion later in the project.

3.1.0 Architectural Drawings

Architectural drawings are the core drawings of any plan set. They are sequentially numbered, usually starting with the site plan or the basement. In some cases, the exterior drawings, such as the site plan and landscaping plans, are numbered separately from the architectural drawings. If this is the case, the architectural drawings are numbered in order of basement or ground-floor plans; upper-level floor plans; exterior elevations; sections; interior elevations; details; and window, door, and room finish schedules.

3.1.1   Site Plans

The main purpose of the site plan is to locate the structure within the confines of the building lot. The site plan clearly shows the building's dimensions (*Figure 9*). The building is usually shown by the size of the foundation and the distances to the respective property lines.

A commercial building needs a site that is appealing, convenient for customer traffic, and in harmony with other commercial structures in the

Drawing Revisions

When a set of drawings has been revised, always make certain that the most up-to-date set is used for all future work. Either destroy the old, obsolete drawing or else clearly mark on the affected sheets *Obsolete Drawing–Do Not Use.* A good practice is to remove the obsolete drawings from the set and file them as history copies for possible future reference.

Also, when working with a set of construction drawings and written specifications for the first time, thoroughly check each page to see if any revisions or modifications have been made to the original. Doing so can save time and expense for all concerned.

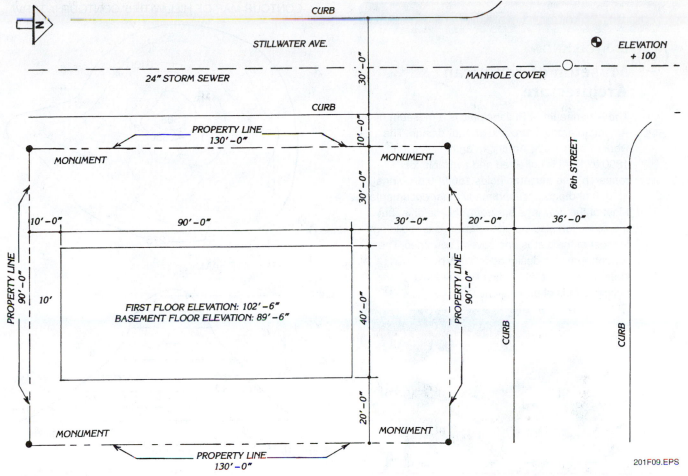

Figure 9 ◆ Site plan.

area. Commercial site drawings show many of the same features as residential drawings, but are far more detailed. They include details on site improvement features, existing and finish **contour lines**, paving areas, and site access. Since a commercial structure serves the public, the architect, owner, and zoning board should note and specify all site changes. Commercial site plans will also show the location of public utilities.

When looking at a commercial site plan for the first time, notice the features that also appear on residential location plot plans, including the following:

- Survey data such as directional arrow, property lines, and the structure's relationship to other features
- Geographic data such as lot corner elevations, existing and proposed contours, and landscaping features
- Building features such as floor elevations, exterior wall positioning, roof overhang, and drainage

Features on the commercial site plans that may *not* be found on residential plans include the following:

- Details of paved areas, including walkways, driveways, and parking areas
- Details on grading and other site work
- Notations that refer to details on the site plan or elsewhere in the drawings
- Position of existing and new utility lines
- Dimensions are usually specified in feet and tenths of feet.
- Referenced details, sections, and elevations that pertain to site preparation features
- Symbols or notations detailing materials types, sizes, and positions

Another primary purpose of the site plan is to show the unique surface conditions, or topography, of the lot. The topography of a particular lot may be shown right on the site plan. For projects in which the topography must be shown separately for clarity, a grading plan is used.

DID YOU KNOW?

Museum of American Architecture

The Athenaeum of Philadelphia is a museum of American architecture and interior design. The work of over 1,000 American architects from 1800 to 1945 is collected and available for research. The museum holds 150,000 drawings, 50,000 photographs, and many other documents. Most of the drawings are from Philadelphia, but the collection also includes drawings for buildings in most of the states and several countries. The Athenaeum was designed in 1845 by John Notman. It is one of the first Philadelphia buildings built of brownstone.

201SA02.EPS

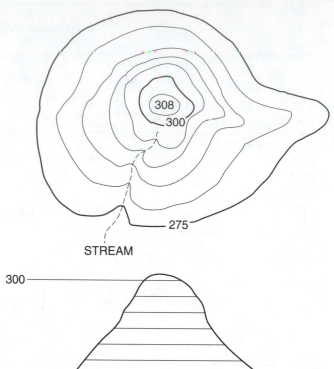

201F10.EPS

Figure 10 ◆ Contour map of a hill.

The topographical information includes changes in the elevation of the lot such as slopes, hills, valleys, and other variations in the surface. *Figure 10* shows the contour map of a hill. These changes in the surface conditions are shown on a site plan by means of a contour, which is a line connecting points of equal elevation. An elevation is a distance above or below a known point of reference, called a datum. This datum could be sea level, or may be an arbitrary plane of reference established for the particular building.

Two important characteristics of the contour need to be observed when reading a site plan:

- Contours are continuous and frequently enclose large areas in comparison to the size of the building lot. For this reason, contours are often drawn from one edge to the other edge of the site plan.
- Contours do not intersect or merge together. The only exception to this rule is in the case of a vertical wall or plane. For example, a retaining wall shown in plan view would show two contours touching, and a cliff that overhangs would be the intersection of two contours.

The existing contour is shown as a dashed line; the new or proposed contour is shown as a solid line (*Figure 11*). Both are labeled with the elevation of the contour in the form of a whole number. The spacing between the contour lines is at a constant vertical increment, or interval. The typical interval is five feet, but intervals of one foot are not uncommon for site plans requiring greater detail, or where the change in elevation is gradual.

A known elevation on the site used as a reference point during construction is called a bench mark. The bench mark is established in reference

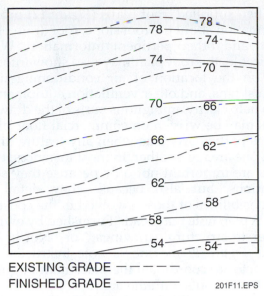

EXISTING GRADE — — — — —
FINISHED GRADE ——————————

201F11.EPS

Figure 11 ◆ Contour lines used for grading.

to the datum and is commonly noted on the site print with a physical description and its elevation relative to the datum. For example: "Northeast corner of catch basin rim—Elev. 102.34'" might be a typical bench mark found on a site plan. When individual elevations, or grades, are required for other site features, they are noted with a "+" and the grade. Grades vary from contours in that a grade has accuracy to two decimal places, whereas a contour is expressed as a whole number.

Sophisticated site plans showing utilities and drainage services often require a legend. The legend is similar to that on the architectural drawings, listing the different symbols and abbreviations found in the particular group of site plans.

As always, you should review materials symbols, dimensioning and scaling, and fundamental construction techniques before attempting to understand a full set of plans. Carefully review the site plan and get an overall concept of the work required. Look at the contours to determine where excavations will be required. For example, a point is at an existing elevation of about 68 feet and a finish elevation of 70 feet. At this location, 2 feet of fill will be required. It is often helpful to divide the site plan into sections or by grid lines to fully understand the amount of work required on the site.

Site plans are drawn using any convenient scale. This may be ⅛ inch to 1 foot, or it may be an engineering scale, such as 1 inch to 20 feet.

Larger projects have several site plans showing different scopes of related or similar work. One such plan is the drainage and utility plan. Utility drawings show locations of the water, gas, sanitary sewer, and electric utilities that will service the building. Drainage plans detail how surface water will be collected, channeled, and dispersed on- or off-site. Drainage and utility plans illustrate in plan view the size and type of pipes, their length, and the special connections or terminations of the various piping.

The elevation of a particular pipe below the surface is given with respect to its invert. The invert is the bottom of the pipe that the liquid flows through. This is typically noted with the abbreviation for invert and an elevation, for example: "INV. 12.34'." The inverts are shown at the intersections of pipes or other changes in the continuous run of piping, such as a manhole, catch basin, sewer manholes, and so on. Inverts are usually given for piping that has a gravity flow or pitch. Using bench marks, contours, or spot elevations, you can quickly calculate the distance of the piping below the surface and the direction of the flow.

With projects of a more sophisticated nature, separate drawings showing various site improvements may be needed for clarification. Site improvements may include such items as curbing, walks, retaining walls, paving, fences, steps, benches, and flagpoles.

Paving and curbing plans show the various types of bituminous, concrete, and brick paving and curbing, as well as the limits of each. Use this information to calculate the area and measurements of paving and curbing. Review the legend symbols to understand where one material ends and another begins. Do not make assumptions. Details show subsurface sections. They show the thickness of the paving and the substrate.

3.1.2 *Floor Plans*

Floor plans of both commercial and residential structures show various floor levels as if a horizontal plane had been cut through the structure.

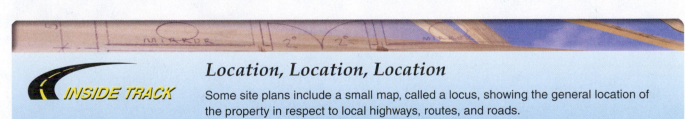

Location, Location, Location

INSIDE TRACK

Some site plans include a small map, called a locus, showing the general location of the property in respect to local highways, routes, and roads.

This results in an overhead view of each floor. One common use of the floor plan is to calculate the area of each room and other areas represented on the floor plan. This information is needed to determine the amount of floor covering, wallboard, and other material required. For most rooms, it is simply a matter of multiplying the length and width dimensions to determine the square footage of the room. If the room is not a perfect rectangle, however, the calculation must be done by breaking the room into pieces, then adding the square footage of the pieces.

The most noticeable difference between commercial and residential floor plans is the amount of detail. Commercial plans contain more details of room use, finishes, wall types, sound transmission, and fire retardation. In fact, most commercial floor plans will incorporate a legend or chart to specify the various interior wall types shown on the plan. The drawing scale is normally ¼ inch to 1 foot for the plan views. The detailing instructions usually include the following types of information:

- Numerous **callouts** specifying sectional views and details
- Room assignment designations by function or number
- Detailed dimensioning of all visible parts of the structure
- Finish designations referenced to schedules

On commercial plans, little is left to chance for several reasons. First, the construction itself is varied and complex. Second, construction must meet all code specifications. Finally, different contractors will be working on the same job. Since there are many ways to accomplish the same task, the requirements are specified in detail to achieve consistency throughout the structure.

Each door or window on a floor plan for a commercial building is typically accompanied by a number, letter, or both. This number/letter is an identifier that refers to a door or window schedule that describes the corresponding door by size, type of materials, or model number for the specific door. Schedules are discussed in more detail later in this section.

When supplied, roof plans (*Figure 12*) provide information about the roof slope, roof drain placement, and other pertinent information. Where applicable, the roof plan may also show information on the location of air conditioning units, exhaust fans, and other ventilation equipment.

To help clarify ambiguous parts of the drawing, notes may be written on commercial floor plans. This is particularly true when any feature differs from one area to another. In most instances, these notes are important not only because they show variations, but also because they detail the responsibilities of those involved in the construction. A plan note may read "Furnished by owner," "Refer to structural drawings," or "Not in contract." *Figure 12* includes a note instructing the contractor to coordinate the placement of gutters and downspouts with the architect.

For large buildings, the architect often will divide the floor plan into sections by grid lines. The grid is the same for all floors of the building. Using the grid allows the architect and engineer to locate features very specifically anywhere in the building. The grid lines are useful for locating features that are repeated on one floor or from one floor to another. The grid markings on the floor plan also reappear on the structural drawings where they locate footings and columns.

3.1.3 Schedules and Details

Since a commercial building has so many varied parts and functions, it is impossible to draw all of the features on one drawing. Many of the smaller detail features are in notes on the actual drawings. Even then, the plans may become cluttered if too many notes appear. To prevent such a problem, schedules, detail drawings, and written specifications are used to describe the construction materials and procedures required.

In most cases, the architectural drawings will include schedules for doors, windows, and interior wall finishes (*Figure 13*). A schedule is actually a chart or table that provides detailed information corresponding to various parts of the drawings. Common features include:

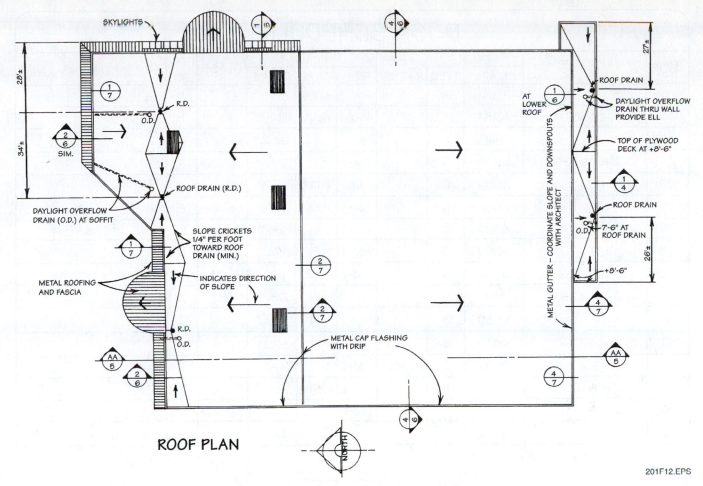

ROOF PLAN

Figure 12 ◆ Roof plan.

201F12.EPS

- A reference mark, letter, or number, which corresponds to the markings on the drawings
- The desired manufacturer for a specific item
- Information on item part numbers, sizes, special finishes, and hardware requirements

Door and window schedules are designated by a number or letter on the drawings. The same letter or number is duplicated in the schedule, with a brief description of the item. Typical door and window schedules must show their relationship to the floor plan.

A finish schedule (*Figure 14*) references a location by a room number noted on the schedule; the drawing usually references the schedule with a symbol or note. Finish schedules for commercial buildings will provide detailed lists of finishing materials and their application, plus installation procedures for floors, walls, base trim, ceilings, and molding.

Certain areas of a floor plan, elevation, or other drawing may be enlarged for greater clarity. These enlargements are drawn to a larger scale. They are

DOOR SCHEDULE

DOOR	DOOR SIZE (1¾" THICK UNLESS OTHERWISE NOTED)	DOOR		FIRE RATING	FRAME			GLASS	HDW GRP.	REMARKS
		MAT'L	TYPE		ELEV.	TYPE	DETAIL			
1011	3'-0" × 7'-0"	ALUM	A		6/A2.1	ALUM	22/A5.1		1	
1012	3'-0" × 7'-0"	ALUM	A		4/A2.1	ALUM	23/A5.1		1	
1031	3'-0" × 7'-0"	HM	B		HM-1	F1	2/A2.1		3	COORDINATE DOOR UNDERCUT WITH SPEC'D THRESHOLD
1041	3'-0" × 7'-0"	WOOD	C		HM-1	F1	1/A2.1	1/4" TEMP	4	
1051	3'-0" × 7'-0"	WOOD	C		HM-1	F1	1/A2.1	1/4" TEMP	4	
1052	3'-0" × 7'-0"	WOOD	C		HM-1	F1	1/A2.1	1/4" TEMP	4	
1071	3'-0" × 7'-0"	WOOD	C		HM-1	F1	1/A2.1	1/4" TEMP	5	
1081	3'-0" × 7'-0"	WOOD	C		HM-1	F1	1/A2.1	1/4" TEMP	5	
1091	3'-0" × 7'-0"	WOOD	B		HM-1	F2	1/A2.1		5	
1092	3'-0" × 7'-0"	HM	B		HM-1	F1	2/A2.1		2	COORDINATE DOOR UNDERCUT WITH SPEC'D THRESHOLD
1101	3'-0" × 7'-0"	WOOD	B		HM-1	F1	1/A2.1		6	
1111	3'-0" × 7'-0"	WOOD	B		HM-1	F1	1/A2.1		6	
1121	3'-0" × 7'-0"	HM	B	3 HR	HM-2	F1	3/A2.1		7	
1131	3'-0" × 7'-0"	HM	B		HM-1	F1	2/A2.1		2	COORDINATE DOOR UNDERCUT WITH SPEC'D THRESHOLD
1141	3'-0" × 7'-0"	WOOD	C		HM-1	F1	1/A2.1	1/4" TEMP	4	
1161	2'-4" × 7'-0" PAIR	WOOD	B		HM-1	F1	1/A2.1		8	4'-8" × 7' × 0" FRAME

WINDOW SCHEDULE

SYMBOL	WIDTH	HEIGHT	MAT'L	TYPE	SCREEN	QUANTITY	REMARKS	MANUFACTURER	CATALOG NUMBER
A	3'-8"	3'-0"	ALUM.	DOUBLE HUNG	YES	2	4 LIGHTS, 4 HIGH	LBJ Window Co.	141 PW
B	3'-8"	5'-0"	ALUM.	DOUBLE HUNG	YES	1	4 LIGHTS, 4 HIGH	LBJ Window Co.	145 PW
C	3'-0"	5'-0"	ALUM.	STATIONARY	STORM ONLY	2	SINGLE LIGHTS	H & J Glass Co.	59 PY
D	2'-0"	3'-0"	ALUM.	DOUBLE HUNG	YES	1	4 LIGHTS, 4 HIGH	LBJ Window Co.	142 PW
E	2'-0"	6'-0"	ALUM.	STATIONARY	STORM ONLY	2	20 LIGHTS	H & J Glass Co.	37 TS
F	3'-6"	5'-0"	ALUM.	DOUBLE HUNG	YES	1	16 LIGHTS, 4 HIGH	LBJ Window Co.	143 PW

201F13.EPS

Figure 13 ◆ Door and window schedule.

called details (*Figure 15*). Details can be found either on the sheet where they are first referenced, or grouped together on a separate detail sheet included in the set of drawings. These drawings are important sources of information for the contractor and craftsperson.

3.1.4 Elevations and Sections

Elevation views on commercial construction drawings are similar to residential elevations, but provide more information. Exterior elevations (*Figure 16*) provide views of the building from each major orientation, as well as references for

ROOM FINISH SCHEDULE

ROOMS	FLOOR				CEILING				WALL				BASE				TRIM			REMARKS
	CARPET	CERAMIC TILE	RUBBER TILE	CONCRETE	ACOUSTIC TILE	DRYWALL	PAINT	CERAMIC TILE	DRYWALL	PAINT	WALLPAPER	CERAMIC TILE	WOOD	RUBBER	CERAMIC TILE	STAIN	WOOD	STAIN	PAINT	
ENTRY		✓			✓				✓	✓	✓		✓			✓	✓	✓		See owner for all painting
HALL	✓				✓				✓	✓			✓			✓	✓	✓		
BEDROOM 1	✓				✓				✓	✓	✓		✓			✓	✓	✓		See owner for grade of carpet
BEDROOM 2	✓				✓				✓	✓			✓			✓	✓	✓		See owner for grade of carpet
BEDROOM 3	✓				✓				✓	✓			✓			✓	✓	✓		See owner for grade of carpet
BATH 1	✓	✓			✓			✓	✓	✓	✓	✓	✓			✓	✓	✓		Wallpaper 3 walls around vanity
BATH 2		✓			✓			✓	✓	✓	✓	✓			✓		✓	✓		Water-seal tile / Wallpaper w/wall
UTIL + CLOSETS	✓		✓			✓	✓		✓	✓			✓			✓	✓	✓	✓	Use off-white flat latex
KITCHEN			✓		✓				✓	✓				✓			✓	✓		
DINING	✓				✓				✓	✓	✓		✓			✓	✓	✓		
LIVING	✓				✓				✓	✓			✓			✓	✓	✓		See owner for grade of carpet
GARAGE				✓		✓	✓					✓	✓				✓	✓		

201F14.EPS

Figure 14 ◆ Finish schedule.

section views. Elevations are normally drawn to the same scale as the floor plans.

Some interior elevations provide vertical dimensioning for interior work, materials lists, and construction details. They are important for built-in cabinets, shelving, finish carpentry, or millwork items. The scale of the drawing depends on the detail required. This may be as small as ¼ inch to 1 foot, or as large as ¾ inch to 1 foot.

There are many wall sections to show the different types of exterior and interior walls. The interior sections will detail the construction of each wall type, such as curtain walls, partitions, loadbearing walls, fire-resistant walls, and noise-reduction walls. The floor plans incorporate a legend that specifies the interior wall type. The wall section drawing provides the necessary detail. Wall sections usually provide the following details:

- Construction techniques and materials types
- Stud types and placement
- Fire ratings of various materials, which is measured in terms of hours of resistance
- Sound barrier placement or materials
- Insulation applications and materials
- In-wall features such as recesses or chases

Elevations and sections are also drawn for landings, stairways, and backfilled retaining walls.

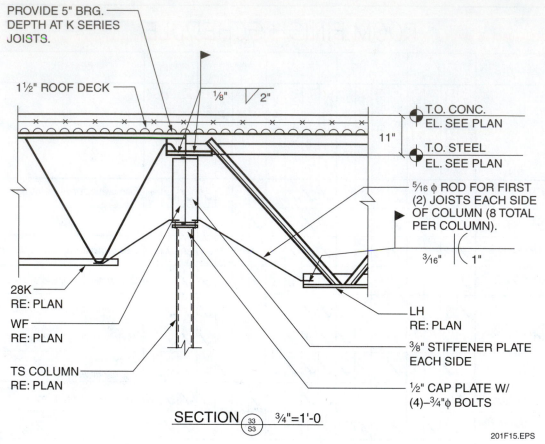

PROVIDE 5" BRG. DEPTH AT K SERIES JOISTS.

1½" ROOF DECK

⅛"   2"

T.O. CONC. EL. SEE PLAN

11"

T.O. STEEL EL. SEE PLAN

⁵⁄₁₆ φ ROD FOR FIRST (2) JOISTS EACH SIDE OF COLUMN (8 TOTAL PER COLUMN).

³⁄₁₆"   1"

28K RE: PLAN

WF RE: PLAN

TS COLUMN RE: PLAN

LH RE: PLAN

⅜" STIFFENER PLATE EACH SIDE

½" CAP PLATE W/ (4)–¾"φ BOLTS

SECTION 33 S3   ¾"=1'-0

201F15.EPS

Figure 15 ◆ Detail drawing.

The section drawings detail construction techniques or materials. For instance, a stairway will have details, sections, and elevations with information on tread, landing, and handrail construction. The amount of detail will depend on the complexity of the stairway and on the various building code specifications.

Longitudinal and transverse wall sections show construction features that are expanded on the structural plans. These sections will show the following:

- Relationships of all wall features from the footings through the roof
- Footing and foundation placement in relation to other elevations
- Exterior materials symbols and notations
- Framework type and placement

As with the elevations, the sections provide an overall view of the proposed structure rather than the detail required for construction. Detail notations or callouts will refer directly to the structural plan sheets. Most callouts will refer to details on roofing **beams** and trusses, foundations, framing features, or other structural components.

3.2.0 Structural Drawings

Structural drawings (*Figure 17*) provide detailed information on the structural features of the building. This includes information on the loadbearing design and materials, such as masonry, reinforced concrete, steel framing, or oversize timber. The structural drawings include plan views, sections, details, schedules, and notes. They provide information on the size and placement of loadbearing elements. They also show how they are connected to each other and to other parts of the structure.

Typical structural drawings include a foundation plan, floor framing plans, and a roof framing plan (*Figure 18*). The plan view will have sections, details, schedules, and notes located in any available space on the drawing sheet. Each plan view should have a north directional arrow to maintain a consistent orientation. Plan views are typically drawn to the scale of ⅛ or ¼ inch to 1 foot; sections are ½ or ¾ inch to 1 foot; details are 1 or 1½ inch to 1 foot.

Each plan is referenced to a grid or checkerboard identifying the placement of columns and/or footings. The grid is the same on all plans.

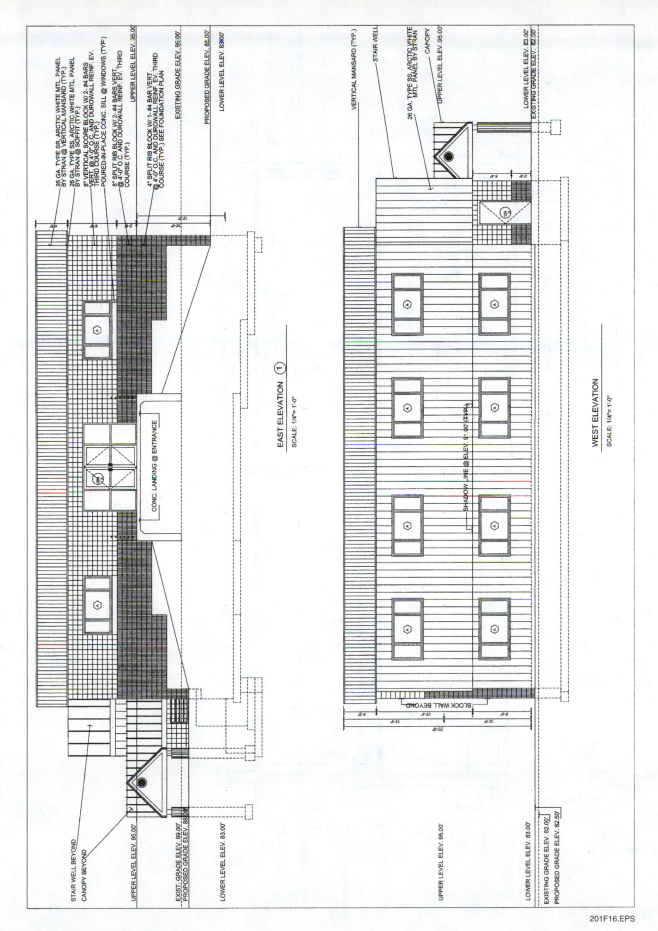

Figure 16 ◆ Elevation drawings.

201F16.EPS

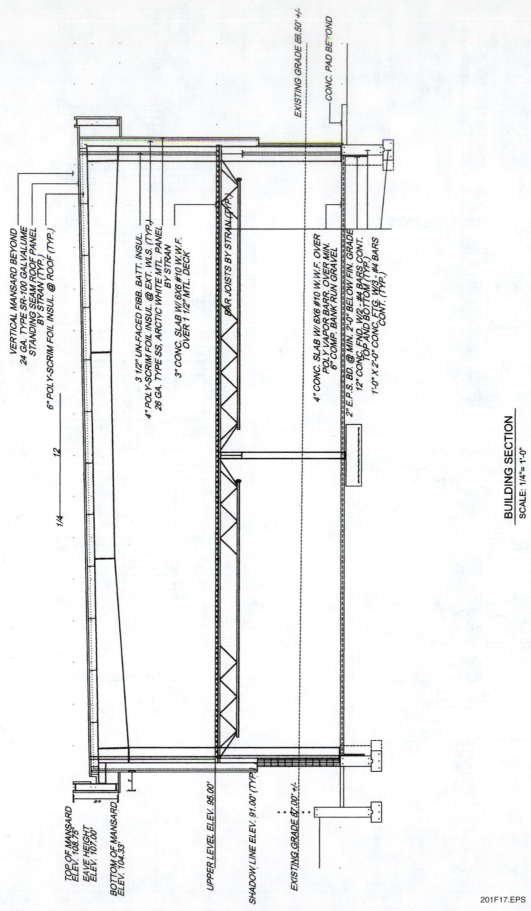

VERTICAL MANSARD BEYOND
24 GA. TYPE SR-100 GALVALUME STANDING SEAM ROOF PANEL BY STRAN (TYP.)
6" POLY-SCRIM FOIL INSUL. @ ROOF (TYP.)

3 1/2" UN-FACED FIBB. BATT. INSUL.
4" POLY-SCRIM FOIL INSUL. @ EXT. WLS. (TYP.)
26 GA. TYPE SS, ARCTIC WHITE MTL. PANEL BY STRAN
3" CONC. SLAB W/ 6X6 #10 W.W.F. OVER 1 1/2" MTL. DECK

BAR JOISTS BY STRAN (TYP.)

4" CONC. SLAB W/ 6X6 #10 W.W.F. OVER POLY VAPOR BARR. OVER MIN. 6" COMP. BANK RUN GRAVEL
2" E.P.S. BD. @ MIN. 2'-0" BELOW FIN. GRADE
12" CONC. FND. W/2 - #4 BARS CONT. TOP AND BOTTOM (TYP.)
1'-0" X 2'-0" CONC. FTG. W/3 - #4 BARS CONT. (TYP.)

EXISTING GRADE 85.50' +/-
CONC. PAD BEYOND

TOP OF MANSARD ELEV. 108.75'
EAVE HEIGHT ELEV. 107.00'
BOTTOM OF MANSARD ELEV. 104.33'

UPPER LEVEL ELEV. 95.00'

SHADOW LINE ELEV. 91.00' (TYP.)

EXISTING GRADE 87.00' +/-

12
1/4

BUILDING SECTION
SCALE: 1/4"= 1'-0"

201F17.EPS

Figure 17 ◆ Structural drawing.

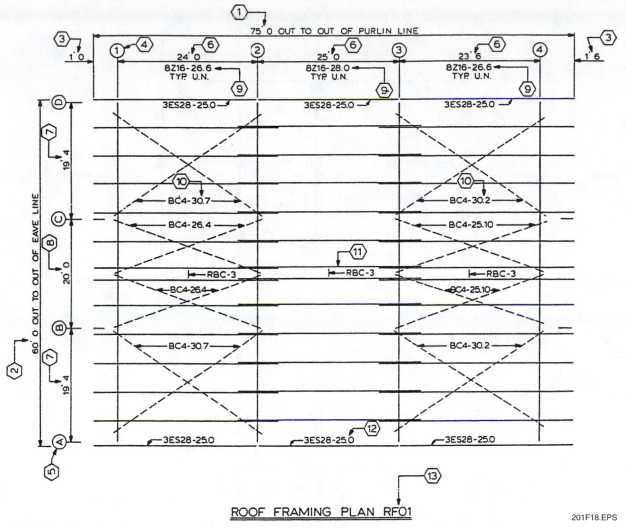

Figure 18 ◆ Roof framing plan.

As noted in the section on floor plans, the grid also shows dimensioning lines. In the structural drawings, a callout sequence number or mark will identify and show the placement of columns on the grid. Some project plans do not use a grid, but use a callout sequence marking. In either case, the identification marks will be referenced to schedules or notes. For example, in *Figure 19* the notation B2 may be referencing a footing, pier, or column on the appropriate schedule.

The structural drawings will show the type of framing and loadbearing for the building. For example, if the floor, roof beams, or trusses place their weight directly on the wall materials, the structure has loadbearing walls. This means that the exterior walls are constructed of materials with high compressive strength such as brick, block, or cast concrete. The walls support their own weight as well as that of the various floor and roof elements. The plans for a loadbearing wall will show no structural beams or columns along the wall. This type of construction is typical in smaller commercial buildings with spans of about 40 feet and height of no more than three stories.

In reinforced concrete construction, the load-bearing elements usually include reinforced concrete footings, foundations, piers, columns, and pillars. Exterior masonry is typically nonbearing curtain or panel walls. The structural drawings will show the size, type, and placement of reinforcing materials and of various jointing techniques. For instance, sections may provide information for the placement of reinforcing bar or wire mesh reinforcement, while details may show the types of saddles, chairs, stirrups, or joints to use. A reinforcing bar notation will usually include the bar size and the bar spacing. As shown in *Figure 20*, the notation for "footing B2" refers to the schedule, where the reinforcing is detailed. This is the same callout referenced in *Figure 19*.

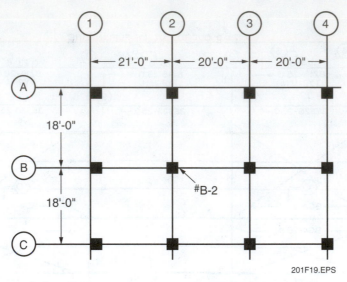

Figure 19 ◆ Grid lines for structural plan.

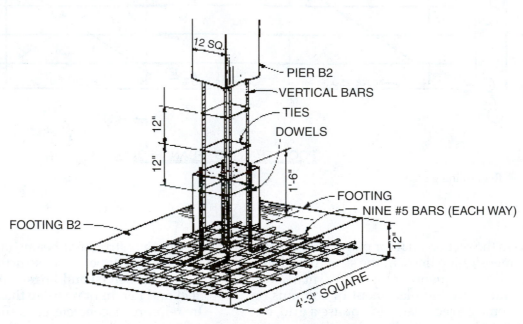

FOOTING/PIER SCHEDULE								
MARK	PIER SIZE	PIER REINF.	FOOTING SIZE	FOOTING DEPTH	REINF. EA. WAY	COLUMN BASE PL.	ANCHOR BOLTS	REMARKS
B1	14" × 14"	FOUR #5 VERT. #3 TIES @12"	3'-2" □	12"	SIX #4	4" × 10" × 11½"	TWO 10" × ⅜"	USE FOUR #5 DOWELS EXT. 1'-6" SBOVR FTG.
B2	12" × 12"	FOUR #6 VERT. #3 TIES @12"	4'-3" □	12"	NINE #5	4" × 8" × ½"	TWO 10" × ½"	USE FOUR #6 DOWELS EXT. 1'-6" SBOVR FTG.
B3	DO	DO	2'-4" □	10"	SIX #5	DO	DO	DO

201F20.EPS

Figure 20 ◆ Footing/pier drawing and schedule.

High-rise buildings are typically steel frame construction with masonry curtain or panel walls. The framework for the entire structure is formed by bolting or welding various steel elements together. The loads from floors and roofing are transferred to beams and **girders** and down columns to the footings. When the joints are bolted together, the schedules will designate the number, size, and material requirements for the bolts. *Figure 21* shows the most common shapes for steel frame elements and lists the typical plan designations.

Timber construction is still used in some commercial roofing. The beams usually have 6-inch nominal dimensions. They may be solid or laminated. The drawings and schedules will show manufacturer's designations and notations for connecting hardware. The metal parts, such as strap hangers, brackets, base plates, and lag screws, are listed on the schedule. They are used to connect the wood to a concrete or steel support.

The standard structural steel notation gives the type or shape of the beam, the depth of the web, and the weight per foot. For example, the notation "W18 × 77" refers to a wide-flange shape with a nominal depth of 18 inches and a weight of 77 pounds per linear foot. The length of the beam is found on the plan view or the shop drawings. The fabricator will cut these beams to the specified length, label them, and predrill connection holes.

3.2.1 Foundation Plans

Commercial buildings that carry heavy loads receive a great deal of design attention at the foundation and footing level. Soil sampling, laboratory tests, and engineering analysis determine the foundation type and size. The foundation size is determined by the load to be carried. The type of foundation is determined by a combination of load and soil capacity. Foundations are categorized as shallow, intermediate, or deep.

Descriptive Name	Shape	Identifying Symbol	Typical Designation height/wt/ft in lb	Nominal Size height width
WIDE FLANGE SHAPES	I	W	W21 × 132	21 × 13
MISCELLANEOUS SHAPES	I	M	M8 × 65	8 × 2¼
AMERICAN STANDARD BEAMS	I	S	S8 × 23	8 × 4
AMERICAN STANDARD CHANNELS	[	C	C6 × 13	6 × 2
MISCELLANEOUS CHANNELS	[	MC	MC8 × 20	8 × 3
ANGLES EQUAL LEGS	L	L	L 6 × 6 ×½	6 × 6
ANGLES UNEQUAL LEGS	L	L	L 8 × 6 × ½	8 × 6
STRUCTURAL TEES (cut from wide flange)	T	WT	WT12 × 73	9
STRUCTURAL TEES (cut from am. std. beams)	T	ST	ST9 × 35	9

201F21.EPS

Figure 21 ◆ Structural steel notations.

Shallow foundations are set to a depth just below the frost line or slightly lower to reach soil with adequate bearing capacity. Shallow foundations take these forms:

- A continuous reinforced concrete footing around the entire building perimeter carrying wall loads directly
- Isolated reinforced concrete footings located under loadbearing columns
- A concrete slab that is placed in a single operation and can carry wall loads directly, or through columns, or both
- Grade beams of reinforced concrete set below grade level and supported by other foundation elements

Intermediate foundations are set to a depth generally not exceeding 15 to 20 feet. Intermediate foundations take these forms:

- Mat foundations are large, heavily reinforced concrete mats under the complete building area; sometimes called raft foundations.
- Drilled piers are concrete or reinforced concrete piers formed by placing concrete in deep columnar holes drilled in the earth; these are designed to carry column loads or grade beams.

Deep foundations are set to a depth over 20 to 30 feet. These are used where surface soil is not adequate for building loads. Steel or concrete pilings are driven into the ground until they reach a strong, stable soil layer, or until they generate enough frictional resistance to compensate for the building load. Piles are typically capped with concrete that supports column loads or grade beams.

More detailed information is needed if the foundation goes well below the earth's surface. However, any foundation plan should provide the following information:

- Plan views for footings, piers, and/or columns with notations on position and size
- Schedules with dimensional notations, shapes, reinforcing, and construction requirements
- Sections showing the smaller construction details of footings, columns, connections, and callout marks, which refer to the schedules

3.2.2 Framing Plans

The structural engineer draws a framing plan or diagram for the roof and for each floor level that will be framed. On these drawings, the exterior walls or bearing walls are often drawn in lightly while heavier lines represent the framing. The resulting plan looks very much like a graph or diagram, as shown in *Figure 22*.

Looking at this diagram, or at any framing plan, you should see the following:

- Notations identifying beams, *joists*, and girders by size, shape, and material
- Column, pier, and support locations and their relationship to joists or framing
- Notes or callouts identifying corresponding sections or detail drawings

You may also see details for locations of stairs, recesses, and chimney placements. Details show additional unique framing around these areas. The dimensions on the drawings are centerline dimensions, not actual member dimensions. The member size must be less than the centerline dimensions to allow for construction tolerances.

The columns on a framing plan are shown from the top. Lines running between columns are

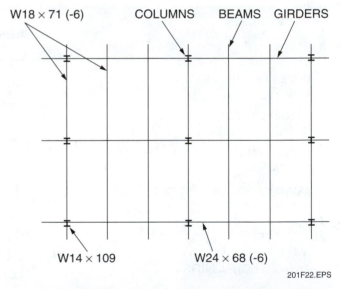

Figure 22 ◆ Structural steel framing diagram.

beams. Beams fasten directly to the columns. Joists fasten between beams or between beams and walls. To keep long spans from swaying or twisting in the center, bridging or support members are placed between joists. All of these members will either be referenced to a schedule or to notes directly written on the plans.

Structural drawings include sheets of details, schedules, and notes with the framing plans. They help workers to understand and follow the specifications for the structure. They provide information on the following areas:

- Reinforcing information for all areas where steel rods or wire mesh will be used
- Information for each type of connection made in framing members
- Bearing plate information detailing the features of all members that will bear directly on other members
- Information for positioning ties, stirrups, or saddles
- Placing and constructing information for any unique features that cannot be adequately described with a drawing

The details also identify load limits, test strength, fastener types, and uniform specifications, which must be applied where specific information is not given.

There is a great deal of information on the framing plans. It is easier to read such plans by isolating the separate bays or spans between columns. Read the details for that area before moving to other areas of the plan.

3.3.0 Mechanical Drawings

Mechanical drawings provide information about the HVAC systems, plumbing systems, and fire protection systems. The work required to install these systems in commercial structures is typically performed by trades or firms that specialize in the particular craft.

While carpenters generally do not get involved with working on these systems, there are many situations where you will be required to make passage for or work around mechanical items. One example would be the through-wall ductwork for a forced-air HVAC system. Because of these types of requirements, you must carefully review all the mechanical drawings for information on dimensions and measurements. Locate services entering the building and passing through interior walls and coordinate with the HVAC workers.

3.3.1 HVAC Systems

Mechanical drawings that contain HVAC systems should be carefully studied for the layout, locations, and sizes of ductwork and fin tube radiation baseboard. Special mechanical drawings that illustrate the components of boiler and rooftop- or grade-mounted HVAC units may also be included.

Figure 23 shows a typical mechanical drawing for an HVAC system. This involves laying out all the piping, controllers, and air handlers to scale on a basic floor plan. Most of the details that are normally on a floor plan have been removed so that the details of the HVAC system can be seen.

INSIDE TRACK

Coordination Drawings

Coordination drawings are produced by the individual contractors for each trade in order to prevent a conflict in the installation of their materials and equipment. Coordination drawings are produced prior to finalizing shop drawings, cut lists, and other drawings and before the installation begins. Development of these drawings evolves through a series of review and coordination meetings held by the various contractors.

Some contracts require coordination drawings, while others only recommend them. In the case where one contractor elects to make coordination drawings and another does not, the contractor who made the drawings may be given the installation right-of-way by the presiding authority. As a result, the other contractor may have to bear the expense of removing and reinstalling equipment if the equipment was installed in a space designated for use by the contractor who produced the coordination drawings.

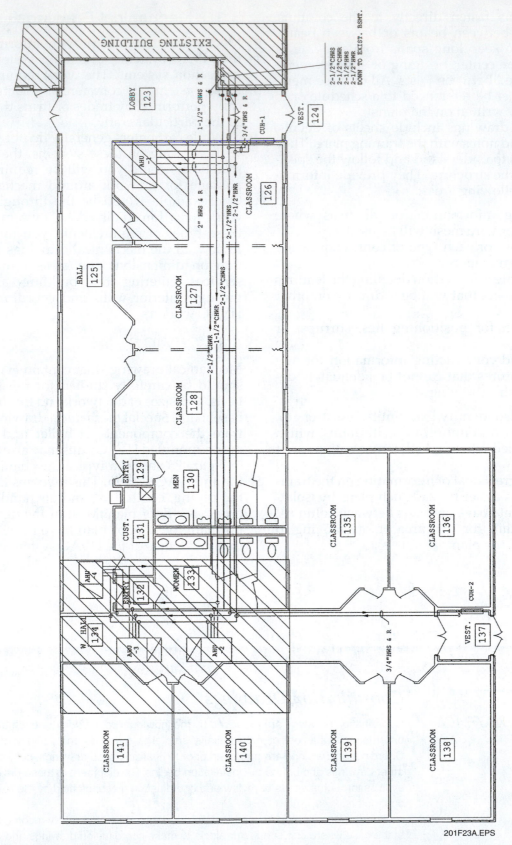

Figure 23 ◆ HVAC mechanical plan (1 of 2).

201F23A.EPS

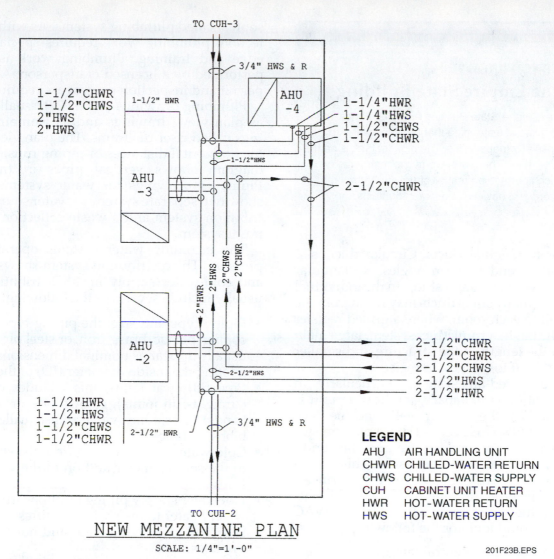

TO CUH-3

3/4" HWS & R

1-1/2"CHWR
1-1/2"CHWS
2"HWS
2"HWR

1-1/2" HWR

AHU
-4

1-1/4"HWR
1-1/4"HWS
1-1/2"CHWS
1-1/2"CHWR

1-1/2"HWS

AHU
-3

2-1/2"CHWR

2"HWR
2"HWS
2"CHWS
2"CHWR

AHU
-2

2-1/2"CHWR
1-1/2"CHWR
2-1/2"CHWS
2-1/2"HWS
2-1/2"HWR

2-1/2"HWS

1-1/2"HWR
1-1/2"HWS
1-1/2"CHWS
1-1/2"CHWR

2-1/2" HWR

3/4" HWS & R

TO CUH-2

NEW MEZZANINE PLAN
SCALE: 1/4"=1'-0"

LEGEND

AHU	AIR HANDLING UNIT
CHWR	CHILLED-WATER RETURN
CHWS	CHILLED-WATER SUPPLY
CUH	CABINET UNIT HEATER
HWR	HOT-WATER RETURN
HWS	HOT-WATER SUPPLY

201F23B.EPS

Figure 23 ◆ HVAC mechanical plan (2 of 2).

Information shown in *Figure 23* includes the following:

- Layout of all of the water supply and return and air handling units
- Piping diagrams for the hot-water and chilled-water coils
- Legend that identifies the abbreviations used in the drawings

Mechanical plans typically contain schedules that identify the different types of HVAC equipment. As appropriate, the plans will include a detailed view describing the installation of the HVAC equipment. Depending on the nature of the project, these views can include a refrigeration piping schematic, chilled-water coil and hot-water coil piping schematics, and piping runs for other HVAC equipment.

A basic understanding of the air handling system is a primary requirement for reading an HVAC drawing. In a forced-air system, fans move the heated or cooled air through ductwork into the working areas, offices, and public spaces. Air return ducts collect air from these areas and channel it back through the system or to the outside atmosphere.

The ductwork for these systems is usually one of the following types:

- Individual duct systems
- Trunk duct systems
- Crawl space plenum systems

The first two delivery systems may be used either from the floor or ceiling levels. The ductwork begins at the central unit and moves out to each area of work space. The plans may specify circular or rectangular ducts. Circular ducts slip together and bend to form angles. Rectangular ducts fit between joists and studs. In the individual duct system, many small ducts may reach from the central unit to each room, where a grilled register delivers the air. In a trunk duct system, large main ducts run the length of the structure and smaller ducts branch off to each room register.

In some forced-air systems, particularly in buildings with a crawl space, the heating unit is designed to heat the exterior walls and the floor above the crawl space. A crawl space plenum system heats the entire crawl space and the conditioned air rises through floor registers. This system requires no ductwork to transfer the warm air since it is transferred through the structural frame.

To gain the most information from HVAC plans, you should look for the following:

- The direction of the central fan
- Duct supply lines; these deliver the conditioned air to all parts of the structure
- Return ducts; some HVAC systems will have return ducts to return air to the central unit for more heating or cooling

3.3.2 Plumbing Systems

All buildings that will be occupied require some type of plumbing. Warehouses have simple toilets, whereas hospitals and restaurants have sophisticated plumbing systems. As with fire protection, plumbing work requires special knowledge and training. Plumbing work is usually performed by a licensed craftsperson. A separate permit and inspection process are required.

Plumbing drawings (*Figure 24*) usually appear as plan view drawings and as isometric drawings called riser diagrams. The plan views show the horizontal distances or piping runs. The riser diagram shows vertical pipes in the walls. Plumbing drawings for water systems usually show two separate systems: a water source or distribution system and a waste collection and disposal system.

The incoming water systems operate under pressure. The distribution system shows the hot- and cold-water supply lines. Distribution plans usually include several of the following:

- Piping type and size; the piping types include copper, plastic, brass, iron, or steel piping; sizes usually appear as nominal dimensions approximating the inside diameter (ID) of the pipe
- Pipe fitting at joints; this includes couplings (straight-run joints), elbows (45- or 90-degree bends), tees, and valves (also called cocks, bibbs, or faucets)
- Cold water supply piping from the street main to the service meter and on to the various fixtures
- Hot water supply piping locations from the hot water heater to the various fixtures
- A legend identifying cold- and hot-water supply lines

The waste system works using gravity and must be vented. The waste system is known as drain, waste, and vent, or DWV, piping. Reading waste disposal plans, you can expect to see a combination of several notations. You will usually find the following information on a plumbing drawing for a waste disposal system:

- Waste disposal line size
- Location of fixture branches (horizontal disposal pipes)

INSIDE TRACK

Isometric Drawings

Isometric means equal measurement. A designer uses the true dimension of an object to construct the drawing. An isometric drawing shows a three-dimensional view of where the pipes should be installed. The piping may be drawn to scale or to dimension, or both. In an isometric drawing, vertical pipes are drawn vertically on the sketch, and horizontal pipes are drawn at an angle to the vertical lines.

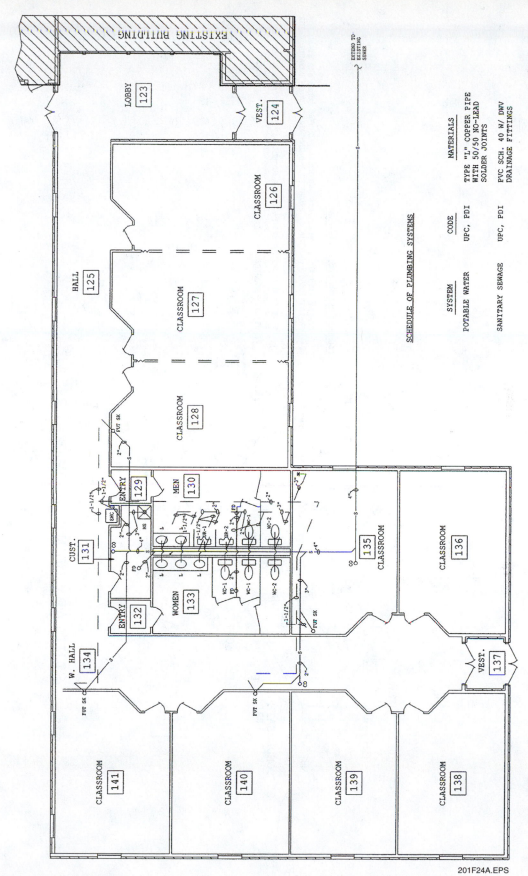

Figure 24 ◆ Sanitary plumbing plan (1 of 2).

201F24A.EPS

PLUMBING SYSTEM SPECIFICATIONS

COMPLY WITH APPLICABLE STATE AND LOCAL PLUMBING CODES AND STANDARDS
PERTAINING TO MATERIALS, PRODUCTS AND INSTALLATION OF POTABLE WATER
AND SANITARY SEWAGE SYSTEMS.

TEST EACH PLUMBING SYSTEM IN ACCORDANCE WITH APPLICABLE CODES AND
STANDARDS, STERILIZE POTABLE WATER SYSTEMS PER STATE AND LOCAL
UTILITY REQUIREMENTS.

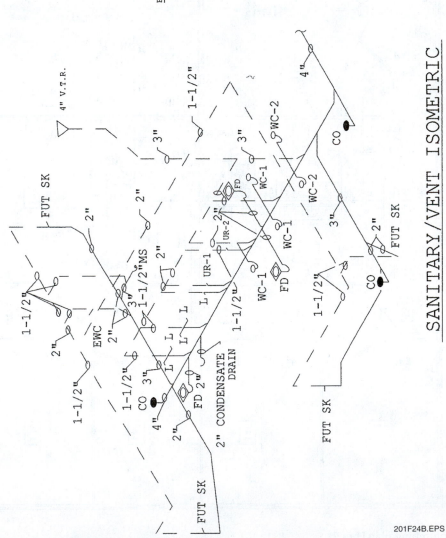

SANITARY/VENT ISOMETRIC

201F24B.EPS

Figure 24 ◆ Sanitary plumbing plan (2 of 2).

- Stack locations; vertical disposal pipes are called waste stacks if they carry toilet waste, soil stacks if they carry waste materials other than toilet waste, and vent stacks if they extend through the roof to release gases within the stacks
- Locations for drains, sewer lines, cleanouts, traps, meters, and valves
- Because waste disposal occurs using the force of gravity, fixture branches are sloped between ⅓ and ½ inch per foot away from the fixtures

Piping for distribution of natural gas within the structure is often considered part of the plumbing work. The utility company installs a meter where natural gas enters the building. The piping starts at the meter. Natural gas is distributed to the gas-fueled appliances within the building such as boilers, furnaces, water heaters, and rooftop HVAC units.

Typically, black steel pipe with threaded fittings is used to distribute natural gas. The fittings are similar to those of other piping systems. They include elbows, bends, unions, and tees. Valves to control the flow of gas within the pipe are typically brass. Gas systems may require openings through masonry walls to accommodate piping, valves, and regulators. Through-wall piping must not be rigidly connected to the wall due to the different rates of expansion and contraction between the piping and the masonry.

3.4.0 Electrical Drawings

Electrical drawings identify the layout of the electrical distribution system, the lighting requirements, and the telecommunications and computer connections. This information is provided in plan view on a copy of the floor plan.

Although you do not need to understand the details of electrical circuitry to read an electrical plan, you should understand the basic design of the system. All systems begin at the service source. Like water and natural gas, it begins with a meter installed by the utility company. From the meter, service moves through a main cutoff switch to the service entry panel. The panel separates service into branch circuits. These circuits are protected from overload by circuit breakers with various capacities. The branches then feed out to specific fixtures and outlets throughout the building.

Electrical systems, like plumbing systems, may be very basic for structures like general warehouses and office buildings. They can also be very complicated for research laboratories and hospitals. Some complex facilities have backup generators and parallel wiring systems.

The power plan illustrates the power requirements of the structure. It shows the panels, receptacles, and the circuitry of power-utilizing equipment. Some buildings need multiple panels. The power plan would then include panel schedules. Panel schedules list the circuits in the panel, the individual power required for each panel, and the total power requirement for the system. The schedules total the power requirements to help the electrical contractor size the panel. The power company sizes the overall service requirements.

The lighting plan locates the various lighting fixtures in the building. It is complemented by a fixture schedule that lists the types of light fixtures to be used. The fixture schedule is organized by number or letter. These refer to the manufacturer and model, the wattage of the lamps, voltage, and any special remarks concerning the fixture. Outlets vary according to their electrical capacity and type. Different symbols are used for outlets according to their voltage requirements. This is also true of most other types of electrical items. Details such as type, size, and voltage are listed on the schedules.

The lighting plan can also include smoke- and fire-detection equipment, emergency lighting, cable TV, and telephone outlets. Buildings with more complex electrical systems may include separate plans for fire prevention and telecommunications systems.

Figure 25 contains two drawings for electrical installations. The top drawing shows the wiring

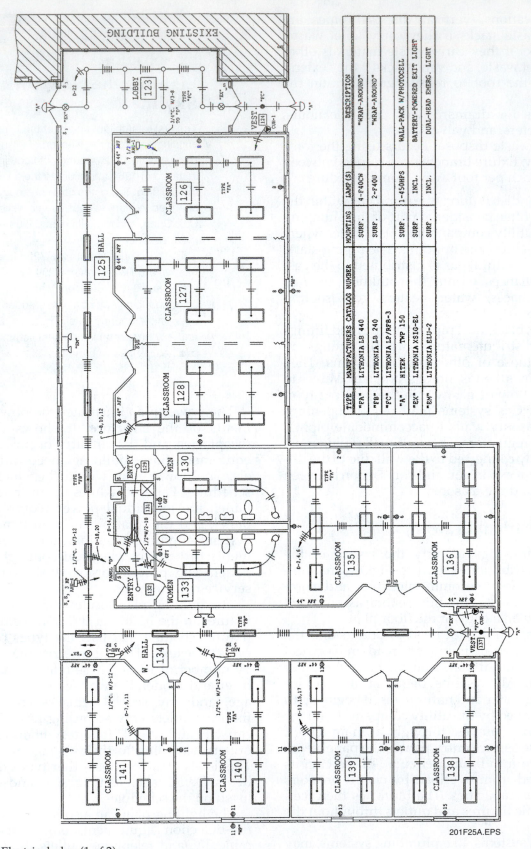

TYPE	MANUFACTURERS CATALOG NUMBER	MOUNTING	LAMP (S)	DESCRIPTION
"FA"	LITHONIA LB 440	SURF.	4-F40CN	"WRAP-AROUND"
"FB"	LITHONIA LB 240	SURF.	2-F40U	"WRAP-AROUND"
"FC"	LITHONIA LP/RFB-3	SURF.	1-450HPS	WALL-PACK W/PHOTOCELL
"A"	HITEK TWP 150	SURF.	INCL.	BATTERY-POWERED EXIT LIGHT
"EX"	LITHONIA XS1G-EL	SURF.	INCL.	DUAL-HEAD EMERG. LIGHT
"EM"	LITHONIA ELU-2	SURF.		

Figure 25 ◆ Electrical plan (1 of 2).

201F25A.EPS

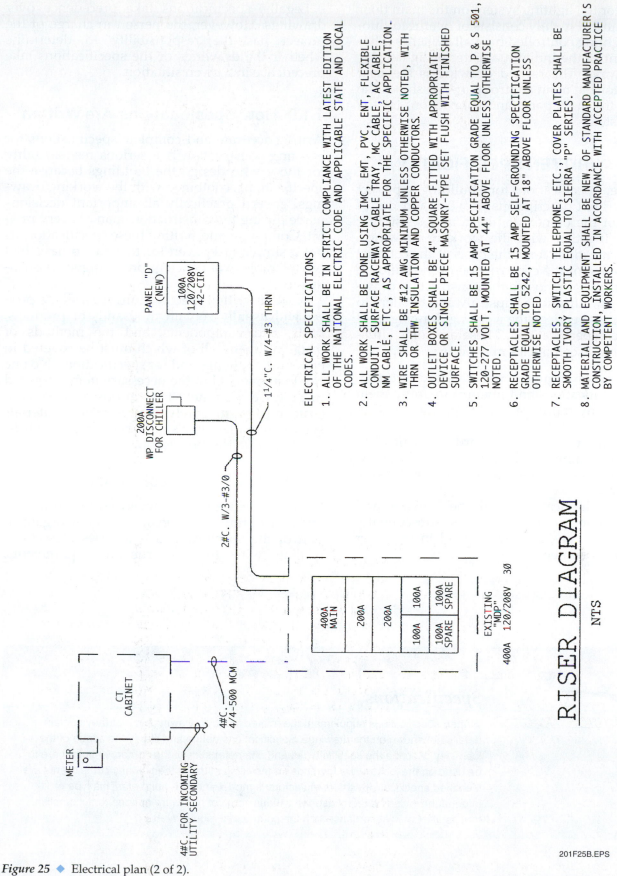

PANEL "D"
(NEW)
100A
120/208V
42-CIR

200A
WP DISCONNECT
FOR CHILLER

1¼"C. W/4-#3 THRN

2#C. W/3-#3/0

CT
CABINET

METER

4#C.
4/4-500 MCM

4#C. FOR INCOMING
UTILITY SECONDARY

400A MAIN		
200A		
200A		
100A	100A	
100A SPARE	100A SPARE	

EXISTING
"MDP"
400A   120/208V   30

ELECTRICAL SPECIFICATIONS

1. ALL WORK SHALL BE IN STRICT COMPLIANCE WITH LATEST EDITION OF THE NATIONAL ELECTRIC CODE AND APPLICABLE STATE AND LOCAL CODES.

2. ALL WORK SHALL BE DONE USING IMC, EMT, PVC, ENT, FLEXIBLE CONDUIT, SURFACE RACEWAY, CABLE TRAY, MC CABLE, AC CABLE, NM CABLE, ETC., AS APPROPRIATE FOR THE SPECIFIC APPLICATION.

3. WIRE SHALL BE #12 AWG MINIMUM UNLESS OTHERWISE NOTED, WITH THRN OR THW INSULATION AND COPPER CONDUCTORS.

4. OUTLET BOXES SHALL BE 4" SQUARE FITTED WITH APPROPRIATE DEVICE OR SINGLE PIECE MASONRY-TYPE SET FLUSH WITH FINISHED SURFACE.

5. SWITCHES SHALL BE 15 AMP SPECIFICATION GRADE EQUAL TO P & S 501, 120-277 VOLT, MOUNTED AT 44" ABOVE FLOOR UNLESS OTHERWISE NOTED.

6. RECEPTACLES SHALL BE 15 AMP SELF-GROUNDING SPECIFICATION GRADE EQUAL TO 5242, MOUNTED AT 18" ABOVE FLOOR UNLESS OTHERWISE NOTED.

7. RECEPTACLES, SWITCH, TELEPHONE, ETC., COVER PLATES SHALL BE SMOOTH IVORY PLASTIC EQUAL TO SIERRA "P" SERIES.

8. MATERIAL AND EQUIPMENT SHALL BE NEW, OF STANDARD MANUFACTURER'S CONSTRUCTION, INSTALLED IN ACCORDANCE WITH ACCEPTED PRACTICE BY COMPETENT WORKERS.

RISER DIAGRAM
NTS

201F25B.EPS

Figure 25 ◆ Electrical plan (2 of 2).

diagram for the lighting system on the main floor of a building. It mainly consists of overhead fluorescent lighting controlled by wall switches. Photocells control the individual exterior lights over the doorways. The second drawing is the riser diagram for the main control panel. It shows the layout of the main panel and connections for the HVAC system.

4.0.0 ◆ WRITTEN SPECIFICATIONS

The written specifications for a building or project are the written descriptions of work and duties required of the owner, architect, and consulting engineer. Together with the working drawings, these specifications form the basis of the contract requirements for the construction of the building or project. Those who use the construction drawings and specifications must always be alert to discrepancies between the working drawings and the written specifications. These are some situations where discrepancies may occur:

- Architects or engineers use standard or prototype specifications and attempt to apply them without any modification to specific working drawings.
- Previously prepared standard drawings are changed or amended by reference in the specifications only, and the drawings themselves are not changed.
- Items are duplicated in both the drawings and specifications, but an item is subsequently amended in one and overlooked in the other contract document.

Legally, specifications take precedence over drawings. However, the person in charge of the project has the responsibility to determine whether the drawings or the specifications take precedence in a given situation.

4.1.0 How Specifications Are Written

Writing accurate and complete specifications for building construction is a serious responsibility for those who design the buildings because the specifications, combined with the working drawings, govern practically all important decisions made during the construction span of every project. Compiling and writing these specifications is not a simple task, even for those who have had considerable experience in preparing such documents.

A set of written specifications for a single project will usually contain thousands of products, parts, and components, and the methods of installing them, all of which must be covered in either the drawings and/or specifications. No one can memorize all of the necessary items required to describe accurately the various areas of construction. One must rely upon reference materials such as manufacturer's data, catalogs, checklists, and high-quality specifications.

4.2.0 Format of Specifications

For convenience in writing, speed in estimating, and ease of reference, the most suitable organization of the specifications is a series of sections dealing with the construction requirements,

Specifications

Written specifications supplement the related working drawings in that they contain details not shown on the drawings. Specifications define and clarify the scope of the job. They describe the specific types and characteristics of the components that are to be used on the job and the methods for installing some of them. Many components are identified specifically by the manufacturer's model and part numbers. This type of information is used to purchase the various items of hardware needed to accomplish the installation in accordance with the contractual requirements.

products, and activities. They should be easily understandable by the different trades. Those people who use the specifications must be able to find all information needed without spending too much time looking for it.

The most commonly used specification format in North America is *MasterFormat*™. This standard was developed jointly by the Construction Specifications Institute (CSI) and Construction Specifications Canada (CSC). For many years prior to 2004, the organization of construction specifications and supplier's catalogs has been based on a standard with 16 sections, known as divisions. The divisions and their subsections were individually identified by a five-digit numbering system. The first two digits represented the division number and the next three individual numbers represented successively lower levels of breakdown. For example, the number 13213 represents division 13, subsection 2, sub-subsection 1 and sub-sub-subsection 3.

In 2004, the *MasterFormat*™ standard underwent a major change. What had been 16 divisions was expanded to four major groupings and 49 divisions with some divisions reserved for future expansion (*Figure 26*). The first 14 divisions are essentially the same as the old format. Subjects under the old Division 15 – *Mechanical* have been relocated to new divisions 22 and 23. The basic subjects under old Division 16 – *Electrical* have been relocated to new divisions 26 and 27.

In addition, the numbering system was changed to 6 digits to allow for more subsections in each division for finer definition. In the new numbering system, the first two digits represent the division number. The next two digits represent subsections of the division, and the two remaining digits represent the third level sub-subsection numbers. The fourth level, if required, is a decimal and number added to the end of the last two digits. For example, the number 132013.04 represents division 13, subsection 20, sub-subsection 13, and sub-sub-subsection 04.

5.0.0 ◆ REBAR DRAWINGS

The structural drawings identify the types and locations of rebar. See the Appendices for examples of drawings created for large rebar installation projects.

Using the structural drawings as a guide, the rebar detailer prepares placing drawings that give the size, location, spacing, and other information needed to develop bar lists. These drawings are also used by the ironworker to install the rebar.

The detailer then uses the placing drawings to prepare a bar list that contains quantities, sizes, lengths, bending dimensions, and placement locations of all the rebar.

The rebar fabricator uses the bar list to cut, bend, and tag rebar. Installers use the bar lists to check shipments, sort rebar, and place rebar in forms.

Each bundle of rebar contains a tag that shows the quantity, size, and bend configuration of a bundle of bars. This tag is developed by the fabricator. In many cases, the tag is computer-generated from data that went into the development of the bar list. The tag also shows the location in which the bars are to be placed. Bars are usually sorted into bundles of the same type and building location. However, this may vary, depending on the quantity and sizes of the bars.

MasterFormat™ 2004 Edition – Numbers & Titles | Division Numbers & Titles

6/8/04

Division Numbers and Titles

PROCUREMENT AND CONTRACTING REQUIREMENTS GROUP

Division 00 Procurement and Contracting Requirements

SPECIFICATIONS GROUP

GENERAL REQUIREMENTS SUBGROUP

Division 01 General Requirements

FACILITY CONSTRUCTION SUBGROUP

Division 02 Existing Conditions
Division 03 Concrete
Division 04 Masonry
Division 05 Metals
Division 06 Wood, Plastics, and Composites
Division 07 Thermal and Moisture Protection
Division 08 Openings
Division 09 Finishes
Division 10 Specialties
Division 11 Equipment
Division 12 Furnishings
Division 13 Special Construction
Division 14 Conveying Equipment
Division 15 Reserved
Division 16 Reserved
Division 17 Reserved
Division 18 Reserved
Division 19 Reserved

FACILITY SERVICES SUBGROUP

Division 20 Reserved
Division 21 Fire Suppression
Division 22 Plumbing
Division 23 Heating, Ventilating, and Air Conditioning
Division 24 Reserved
Division 25 Integrated Automation
Division 26 Electrical
Division 27 Communications
Division 28 Electronic Safety and Security
Division 29 Reserved

SITE AND INFRASTRUCTURE SUBGROUP

Division 30 Reserved
Division 31 Earthwork
Division 32 Exterior Improvements
Division 33 Utilities
Division 34 Transportation
Division 35 Waterway and Marine Construction
Division 36 Reserved
Division 37 Reserved
Division 38 Reserved
Division 39 Reserved

PROCESS EQUIPMENT SUBGROUP

Division 40 Process Integration
Division 41 Material Processing and Handling Equipment
Division 42 Process Heating, Cooling, and Drying Equipment
Division 43 Process Gas and Liquid Handling, Purification, and Storage Equipment
Division 44 Pollution Control Equipment
Division 45 Industry-Specific Manufacturing Equipment
Division 46 Reserved
Division 47 Reserved
Division 48 Electrical Power Generation
Division 49 Reserved

Div Numbers - 1

201F26.EPS

Figure 26 ◆ 2004 *MasterFormat*™.

Summary

Commercial and industrial construction make up a large part of the construction market across the United States. These structures are usually more difficult to build than residential buildings. They also involve many code, safety, and environmental requirements.

In order to build commercial structures, you must be able to understand and interpret the plans and drawings. The basic principles such as scaling and dimensioning apply to both residential and commercial drawings. However, commercial drawings are much more detailed. They can contain additional structural, mechanical, and electrical systems. These require close study and review. Understanding the plan will help keep construction activities well-organized and true to the design.

A typical commercial plan set includes architectural, structural, mechanical, plumbing, and electrical drawings. There is a letter designation for each section. Each section may include more than one sheet, depending on the size and complexity of the project. Architectural drawings include the site plan, floor plans, wall sections, door and window details, and schedules. Structural drawings include foundation plans, framing plans, support details, schedules, and notes. Mechanical drawings include HVAC plans in both plan view and isometric view. Plumbing drawings and mechanical drawings are similar in format. They provide a plan view and isometric view of the fresh water system and the waste water system. Electrical drawings usually include a power plan and a lighting plan. These may be on the same or separate sheets, depending on the complexity of the project.

When reading commercial plans, follow a step-by-step process to avoid confusion and see all the details involved. Begin by reading the project specifications to pick up details not found on the drawings. Quickly review all the drawings in order to get a general impression of the shape, size, and appearance of the structure. Begin correlating the floor plans with the exterior elevations. Look at the wall sections to determine the wall types and construction details.

Review the structural plans to determine the foundation requirements and the type of structural system. Carefully consider all items that may require special construction procedures. Review all interior elevations. Try to get a clear picture of what the interior of the building will look like. Review the finish schedule and the mechanical and electrical plans. In particular, look for work items to be performed by carpenters or features that will affect the work of carpenters. If some part of the drawing is not clear, check with your supervisor. There may be a logical explanation, or there may be a mistake. The architect should clarify and resolve all conflicts in writing.

Specifications provide written instructions for the owner, architect, and engineer. A set of written specifications for a large commercial project will usually contain thousands of products, parts, components, and the methods of installing them. A standard format has been voluntarily adopted for commercial construction throughout North America. Manufacturers and suppliers key their catalogs and data to this standard format.

The standard format makes understanding the specifications much easier. However, discrepancies can arise if the standard specifications are not modified to match the drawings of a particular project. All discrepancies must be resolved by the person in charge before work begins.

1. In some commercial drawing sets the site plans are called _____ drawings.
   a. civil
   b. commercial
   c. layout
   d. elevation

2. Architectural drawings are identified with the letter _____.
   a. E
   b. A
   c. M
   d. P

3. Structural drawings usually include _____ plans.
   a. HVAC
   b. riser
   c. piping
   d. foundation

4. Which of the following would be seen on a second-floor structural drawing?
   a. The placement of partition walls
   b. The placement of electrical service drops
   c. The floor layout
   d. The configuration of loadbearing members

5. Mechanical drawings for HVAC systems are typically shown in a(n)_____ view.
   a. section
   b. plan
   c. elevation
   d. cutaway

6. The emergency lighting systems are part of the _____ drawings.
   a. electrical
   b. civil
   c. mechanical
   d. architectural

7. When reading commercial plans, use a _____ to avoid missing details.
   a. step-by-step process
   b. schedule
   c. local code
   d. plan handbook

8. Unlike residential plans, commercial site plans usually include _____.
   a. survey data
   b. paving details
   c. geographic data
   d. elevations

9. On a site plan, the typical interval between contour lines is _____.
   a. one foot
   b. two feet
   c. five feet
   d. ten feet

10. A known elevation on the site that is used as a reference point during construction is known as a(n)_____.
    a. elevation
    b. contour
    c. datum
    d. bench mark

11. To help clarify ambiguous parts of the drawing, _____ may be written on commercial floor plans.
    a. notes
    b. details
    c. symbols
    d. schedules

12. Of the following, the _____ would *not* be seen in a door schedule.
    a. manufacturer's name
    b. riser diagram
    c. hardware requirements
    d. door dimensions

13. An enlargement of an area on the floor plan is called a(n) _____.
    a. elevation
    b. schedule
    c. section
    d. detail

14. Interior elevation drawings may have a scale as small as _____ inch to 1 foot, or as large as _____ inch to 1 foot.
    a. ¼; ¾
    b. ⅛; ¼
    c. ⅜; ⅝
    d. ½; ¾

15. Stairways will usually be shown as elevations and _____ on the drawings.
    a. sections
    b. grids
    c. materials
    d. schedules

16. Floor framing plans are found in the _____ drawings.
    a. architectural
    b. structural
    c. mechanical
    d. electrical

17. The placement of a column in a structural drawing will be shown by a(n) _____.
    a. circle with a smaller circle inside it
    b. circle with an X through it
    c. callout sequence number or mark
    d. open grid square

18. Shallow, intermediate, and deep are typical categories for _____.
    a. girders
    b. columns
    c. beams
    d. foundations

19. Drilled piers are designed to carry _____.
    a. footings
    b. grade beams
    c. mats
    d. girders

20. Deep foundations are usually constructed using _____.
    a. mats
    b. piers
    c. pilings
    d. footings

21. The structural engineer draws a framing plan for _____.
    a. each floor and the roof
    b. the plumbing, electrical, and HVAC systems
    c. landscaping and grading
    d. the foundation and exterior walls

22. The arrangement of beams, joists, and girders is shown on _____.
    a. elevations
    b. sections
    c. schedules
    d. framing plans

23. It is easier to read framing plans by _____.
    a. starting at the bottom and working upwards
    b. starting at the top and working downwards
    c. isolating the separate bays or spans between columns
    d. reading the foundation plans first

24. Mechanical plans typically contain schedules that identify the different types of _____ equipment.
    a. heavy
    b. grading
    c. safety
    d. HVAC

25. A plumbing riser diagram shows _____.
    a. vertical pipes in the wall
    b. horizontal distance or pipe runs
    c. the hot- and cold-water supply lines
    d. the couplings, elbows, tees, and valves needed

26. Piping for distribution of natural gas is included in the _____ plans.
    a. HVAC
    b. plumbing
    c. structural
    d. mechanical

27. The electrical system begins with a _____.
    a. service entry panel
    b. main cut-off switch
    c. circuit breaker
    d. meter

28. There may be discrepancies between the written specifications and the working drawings because _____.
    a. the specifications were amended and the drawings were not
    b. they refer to different parts of the project
    c. specifications are only used by the engineer and architect
    d. specifications are no longer needed once the drawings are prepared

29. Specifications follow a standard format _____.
    a. as outlined in the building codes
    b. for convenience in writing and ease of reference
    c. because it is required by law
    d. specified by ANSI

30. The most commonly used specification is developed by the _____ and Construction Specifications Canada.
    a. Brick Industry Association (BIA)
    b. American National Standards Institute (ANSI)
    c. Construction Specifications Institute (CSI)
    d. International Code Council (ICC)

Trade Terms Introduced in This Module

Beams: Loadbearing horizontal framing elements supported by walls or columns and girders.

Callouts: Markings or identifying tags describing parts of a drawing; callouts may refer to detail drawings, schedules, or other drawings.

Civil drawings: A drawing that shows the overall shape of the building site. It is also called a site plan.

Contour lines: Imaginary lines on a site plan/plot plan that connect points of the same elevation. Contour lines never cross each other.

Elevation view: A drawing giving a view from the front, rear, or side of a structure.

Girders: Large steel or wooden beams supporting a building, usually around the perimeter.

Isometric drawing: A three-dimensional type drawing in which the object is tilted so that all three faces are equally inclined to the picture plane.

Joists: Horizontal members of wood or steel supported by beams and holding up the planks of floors or the lathes of ceilings. Joists are laid edgewise to form the floor support.

Landscape drawings: A drawing that shows proposed plantings and other landscape features.

Plan view: A drawing that represents a view looking down on an object.

Riser diagram: A type of isometric drawing that depicts the layout, components, and connections of a piping system.

Rebar Drawings for a Mining Conveyor Base

CONDOR rebar consultants inc.

EXITOSO TRANSITION PROJECT

1403 rev. 0

FLAG 2

FLAG 1

Condor Job #: FD232
Client Job #: 03264400-CTS-002

Release Location:
CONVEYOR 115-CV-205
Tunel transportador 115-CV-205
CONVEYOR FOOTING FOR PEDESTALS P2
Fundacion para pedestales P2 del tunel transportador

Release Weight: 9398 kg

Client:
FLUOR DANIEL

Release color code: Dark Green
Detailer: LD
Placing Sheets: 4
Cutting Sheets: 1

Reference Drawings:
115-C-DW-225 rev.0, 115-C-DW-226 rev.0

Reference Releases:
1402, 1404

Reason for Revision:	To:	Date:
Function:		
E-mail		
E-mail		
E-mail		
FTP server		
Fax		

Notes:
All steel on this release to be grade INN A63-42H

1 All identifying project data clearly presented, including:
- Job numbers (Client and Condor's)
- Description of concrete element
- Job Site Location
- Release Weight
- Multiple languages

2 All technical data clearly presented, including:
- Release number
- Quantity of sheets in this release package
- Reference Drawings
- Reference Releases
- Reviewer's Block
- Miscellaneous Notes

201A01.EPS

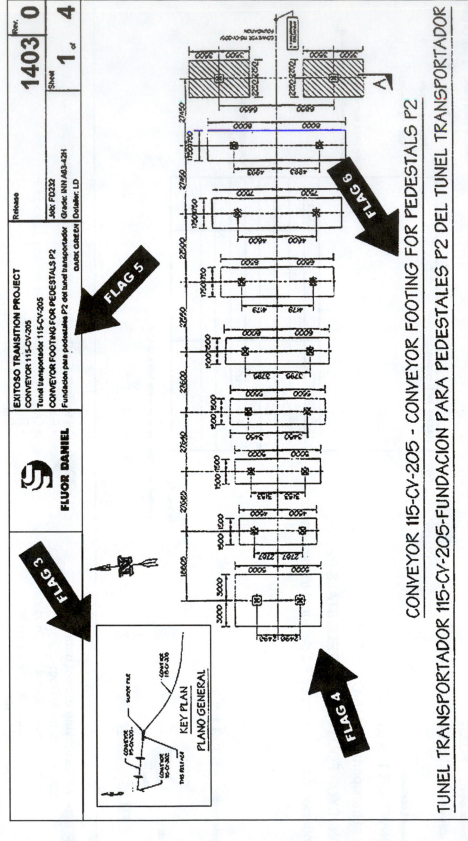

CONVEYOR 115-CV-205 - CONVEYOR FOOTING FOR PEDESTALS P2

TUNEL TRANSPORTADOR 115-CV-205-FUNDACION PARA PEDESTALES P2 DEL TUNEL TRANSPORTADOR

3 ◄ Site key plan to locate area of project covered by this release

4 ◄ Key Plan to locate items covered by this release

5 ◄ Title Block includes all pertinent data as detailed on the Cover Page (FLAG 1&2)

6 ◄ Repeats the clear description of the items covered in this release

201A02.EPS

3 | **PLAN SHEET**

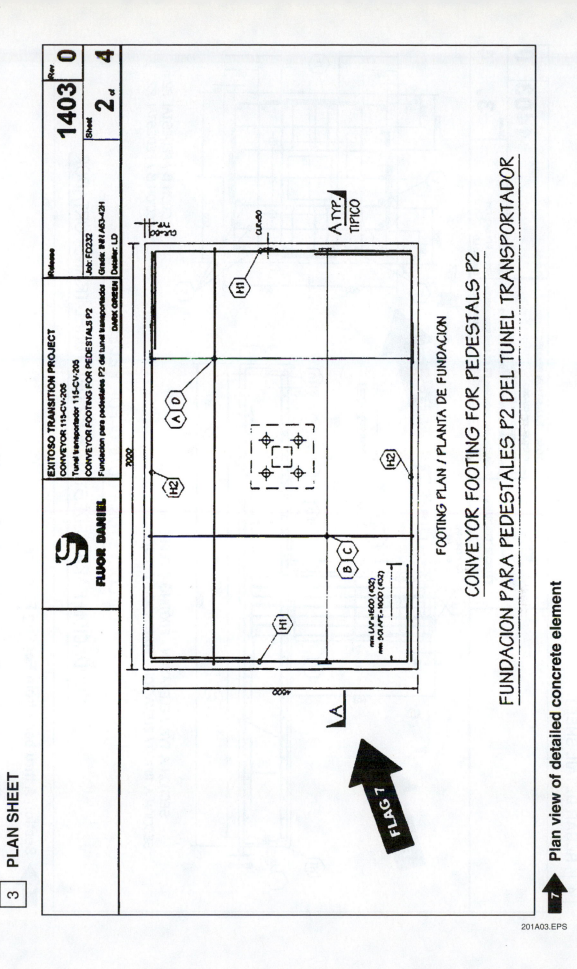

FOOTING PLAN / PLANTA DE FUNDACION

CONVEYOR FOOTING FOR PEDESTALS P2

FUNDACION PARA PEDESTALES P2 DEL TUNEL TRANSPORTADOR

7 **Plan view of detailed concrete element**

201A03.EPS

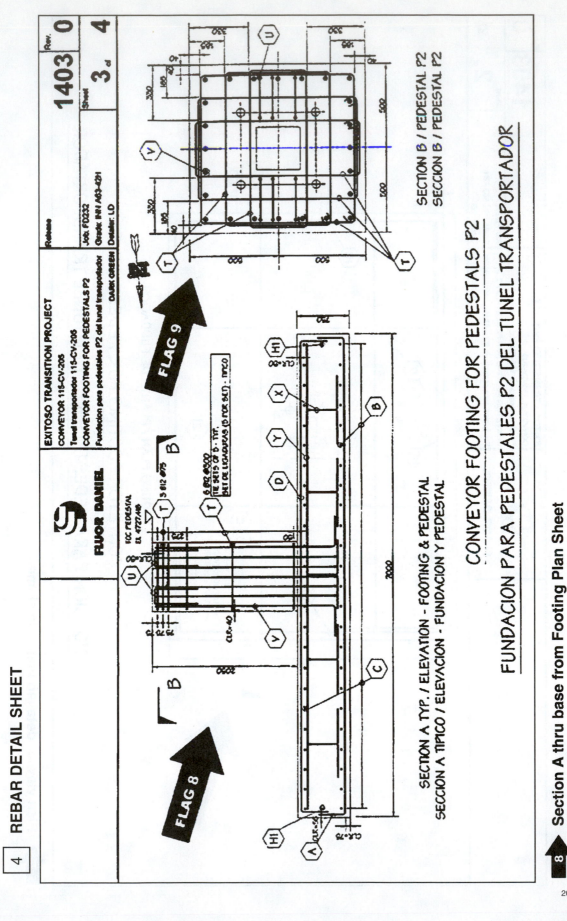

REBAR DETAIL SHEET

CONVEYOR FOOTING FOR PEDESTALS P2

FUNDACION PARA PEDESTALES P2 DEL TUNEL TRANSPORTADOR

SECTION A TYP. / ELEVATION - FOOTING & PEDESTAL
SECCION A TIPICO / ELEVACION - FUNDACION Y PEDESTAL

SECTION B / PEDESTAL P2
SECCION B / PEDESTAL P2

Section A thru base from Footing Plan Sheet

Section B thru base from Footing & Pedestal Section Sheet

201A04.EPS

FLUOR DANIEL

EXITOSO TRANSITION PROJECT
CONVEYOR 115-CV-205
Tunel transportador 115-CV-205
CONVEYOR FOOTING FOR PEDESTALS P2
Fundacion para pedestales P2 del tunel transportador
DARK GREEN

Release **1403** Rev. **0**
Sheet **4** of **4**

Job: FD032
Grade: INN A63-42H
Detailer: LD

No	Location	Qty	Size	Length	Mark	Type	A	B	C	D	E	F	G	H	K	O	Remarks	Weight
1	FOOTING (2RQD) / FUNDACION (2RQD)																	
2																		
3	BOTTOM STEEL / ACERO INFERIOR																	
4	A	46	Ø32	8000	32001	17		550	6900	550							(2x23) @200 LOWER LAYER / capa inferior	2332
5	B	46	Ø32	5500	32002	17		550	4400	550							(2x24) @300 UPPER LAYER / capa superior	1656
6																		
7	EDGE STEEL / ACERO DE BORDES																	
8	H1	4	Ø32	6000	32003	2	1700	4300									(2x2) @200 HORIZONTALS / horizontales	151
9	H2	4	Ø32	8500	32004	2	1700	8800									(2x2) @300 HORIZONTALS / horizontales	215
10																		
11	SUPPORT STEEL / SOPORTES																	
12	X	40	Ø16	2180	16001	28		500	480	200	480	500					(2x4x5) @1200 CHAIRS / soportes	137
13	Y	8	Ø16	6900		STR		6900									(2x4) @1200 SUPPORT / soportes	87
14																		
15	TOP STEEL / ACERO SUPERIOR																	
16	C	48	Ø32	5500	32002	17		550	4400	550							(2x24) @300 LOWER LAYER / capa inferior	1656
17	D	46	Ø32	7950	32005	17		550	6900	500							(2x23) @200 UPPER LAYER / capa superior	2308
18																		
19	PEDESTAL STEEL (2 RQD) / ACERO DE PEDESTAL (2 RQD)							13(1,12)/9.4										
20	V	48	Ø25	3040	25001	2	430	2810									(2x24) @200 VERTICALS / verticales	562
21																		
22	T	18	Ø12	3910	12001	T1	115	920	920	920	920		115				(2x9) TIES (SET OF 5)	62
23		36	Ø12	2750	12002	T1	115	340	920	340	920		115				(2x9x2) ligaduras (5 por set)	88
24		36	Ø12	3330	12003	T1	115	630	920	630	920		115				(2x9x2)	106
25																		
26	U	12	Ø16	1505	16002	17		820	285	620							(2x2x3) U-BARS / U-barras	29
27																		
28																		
29																		
30																	TOTAL WEIGHT THIS RELEASE:	9392 kg

FLAG 10 **FLAG 11**

10 All bend shapes for this release shown for Placer's information

11 Bars listed generally in the sequence of placing:
- Bars grouped in easily identifiable components;
- Includes labels (in hexagons) which correspond with labels on sketches;
- Qty, Size, Length, Mark, and Bend Type in conventional presentation;
- All dimensions shown for Placer's information;
- "Remarks" column gives placing instructions;
- "Weight" column gives the weight of each item.

201A05.EPS

CUTTING AND BENDING LIST

FLUOR DANIEL

EXITOSO TRANSITION PROJECT
CONVEYOR 115-CV-205
Tunel transportador 115-CV-205
CONVEYOR FOOTING FOR PEDESTALS P2
Fundacion para pedestales P2 del tunel transportador
DARK GREEN

Release:
Job: FD232   Grade: BRN A63-42H   Detail: LD

1403   Rev: 0
Sheet 1 of 1

Qty	Size	Length	Mark	Type	A	B	C	D	E	F	G	HM1	H2	J	K	R	O	Weight
8	Ø16	6900		STR		6900												87
4	Ø32	8600	32004	2	1700	6800												215
46	Ø32	8000	32001	17		550	6900	650										2322
46	Ø32	7950	32005	17		550	6900	500										2308
4	Ø32	6000	32003	2	1700	4300												151
96	Ø32	5500	32002	17		550	4400	550										3332
48	Ø25	3040	25001	2	430	2610												662
40	Ø16	2160	16001	26		600	480	200	480	500								137
12	Ø16	1505	16002	17		620	265	620										29
16	Ø12	3910	12001	T1	115	920	920	920	920		115							62
36	Ø12	3330	12003	T1	115	630	920	630	920		115							106
36	Ø12	2760	12002	T1	115	340	920	340	920		115							88

TOTAL WEIGHT THIS RELEASE: 9398 kg

12 All bend shapes on this release shown for quick reference

13 Bars listed in optimum sequence for the Fabricator:
- Straight bars are separated from bent bars;
- Bars are listed from largest size to smallest size;
- Bars are listed from longest length to shortest;
- All dimensions and weights are shown;
- This sheet can be digitized for download into rebar management software.

14 Weights are listed for each bar type. They are totalled and shown on both the Placing list and the Cutting and bending lists.

201A06.EPS

Rebar Drawings for a Steam Turbine Generator Foundation

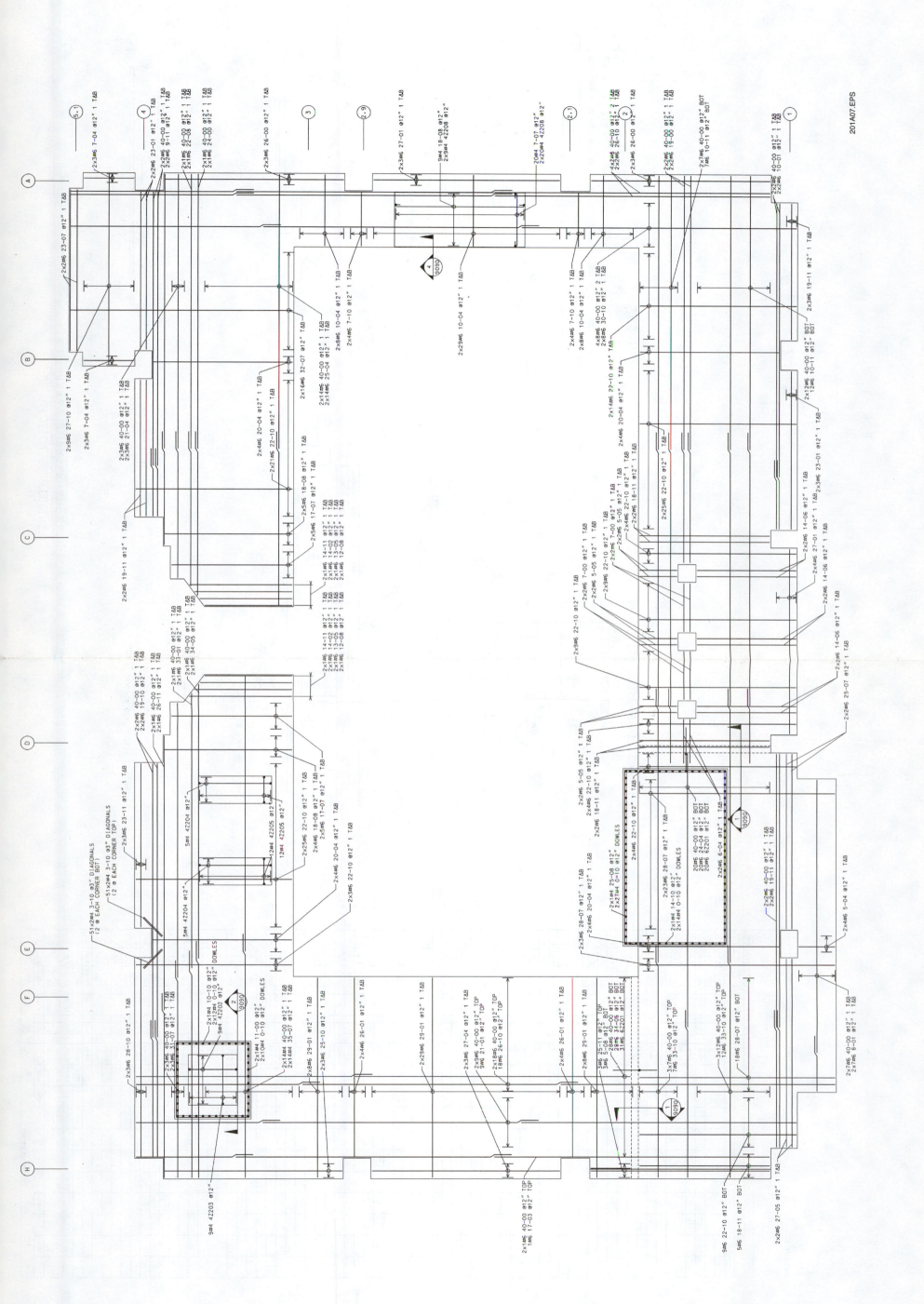

201A07.EPS

Bar Bending Schedule (Main)

QTY	SIZE	WGT	BAR MARK EP	BND V TYP CD	A	B	C	D	E F/R	G	H	J	K	O
51	#6	504	62201	31	0-10	2-07	1-05	2-07		0-10	1-00		1-00	
9	#4	57	42202	2	0-10	7-10				0-10	1-00			
10	#4	58	42203	2	0-10	8-00				0-10	1-00			
9	#4	80	42204	2	0-10	10-04				0-10	1-00			
24	#4	86	42205	2	0-10	3-08				0-10	1-00			
323	#6	19406			40-00									
30	#6	1551			34-05									
19	#6	966			33-10									
2	#6	99			33-01									
32	#6	1566			32-07									
16	#6	285			31-07									
16	#6	741			30-10									
90	#6	3931			29-01									
6	#6	260			28-10									
70	#6	3005			28-07									
18	#6	753			27-10									
4	#6	165			27-05									
6	#6	246			27-04									
14	#6	570			27-01									
2	#6	81			26-11									
22	#6	887			26-10									
16	#6	627			26-01									
12	#6	469			25-11									
3	#6	117			25-10									
6	#6	233			25-07									
28	#6	1065			25-04									
20	#6	731			24-04									
2	#6	72			24-00									
4	#6	216			23-11									
4	#6	142			23-07									
10	#6	347			23-01									
245	#6	8402			22-10									
2	#6	68			22-08									
6	#6	192			21-04									
9	#6	285			21-01									
32	#6	977			20-04									
14	#6	419			19-11									
4	#6	119			19-10									
13	#6	369			19-00									
18	#6	505			18-11									
20	#6	528			18-08									
1	#6	26			17-07									
					17-03									
12	#6	261			14-11									
4	#6	90			14-06									
12	#6	261			14-02									
4	#6	85			13-05									
4	#6	81			12-08									
19	#6	312			11-00									
90	#6	1397			10-11									
4	#6	61			10-04									
14	#6	170			10-01									
16	#6	188			9-11									
12	#6	132			8-01									
8	#6	84			7-10									
4	#6	38			7-04									
3	#6	26			7-00									
12	#6	98			6-04									
8	#4	64			5-08									
28	#4	748			5-05									
28	#4	666			5-04									
2	#4	34			40-00									
2	#4	20			35-07									
2	#4	15			25-08									
2	#4	14			14-10									
20	#4	129			11-00									
204	#4	522			10-10									
126	#4	70			9-08									
					8-08									
					3-10									
					0-10									

TOTAL: 57042
Total for size 6: 54416
Total for size 4: 2626

TOP & BOTTOM REINFORCING
STEAM TURBINE AREA

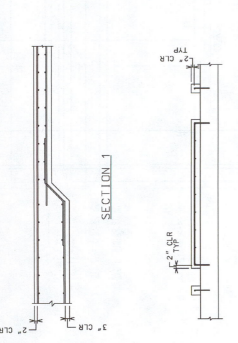

6" CLR FOR DOWELS THIS SIDE ONLY
2" CLR (TYP)
2" CLR

DRILL & GROUT USING HILTI HIT HY-150 TECHNIQUE WITH 6" EMBEDMENT

SECTION 4
NO SCALE
SEE THIS DWG

QTY	SIZE	WGT	BAR MARK EP	BND V TYP CD	A	B	C	D	E F/R	G	H	J	K	O
58	#4	58	42208	17	18-08	0-10	0-08							
9	#4	112			7-07									
20	#4	101												

TOTAL: 271
Total for size 4: 271

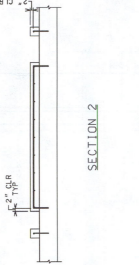

SECTION 1
2" CLR
3" CLR
2" CLR TYP

SECTION 2
2" CLR TYP

SECTION 3
2" CLR TYP
2" CLR

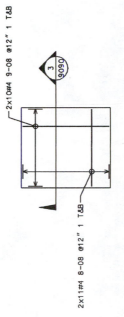

2x11#4 8-08 @12" 1 T&B
2x10#4 9-08 @12" 1 T&B
9090

TOP & BOTTOM REINFORCING
STEAM TURBINE GENERATOR SURGE & VT COMPARTMENT

NOTES

1. SEE DRAWING S5950 FOR GENERAL NOTES, LEGEND AND TYPICAL DETAILS.
2. FOR CONCRETE OUTLINE SEE S5000 SERIES DRAWINGS (UNO).
3. LAP SPLICE IS 30" FOR #6 BAR.

Additional Resources and References

This module is intended to be a thorough resource for task training. The following reference works are suggested for further study. These are optional materials for continued education rather than for task training.

Architectural Graphic Standards, Eighth Edition, The American Institute of Architects, John Wiley & Sons, New York, NY, 1988.

Basics for Builders: Plan Reading & Material Takeoff, Wayne J. DelPico, R.S. Means Company, Inc., Kingston, MA, 1994.

NCCER CURRICULA — USER UPDATE

NCCER makes every effort to keep its textbooks up-to-date and free of technical errors. We appreciate your help in this process. If you find an error, a typographical mistake, or an inaccuracy in NCCER's curricula, please fill out this form (or a photocopy), or complete the online form at **www.nccer.org/olf**. Be sure to include the exact module ID number, page number, a detailed description, and your recommended correction. Your input will be brought to the attention of the Authoring Team. Thank you for your assistance.

Instructors – If you have an idea for improving this textbook, or have found that additional materials were necessary to teach this module effectively, please let us know so that we may present your suggestions to the Authoring Team.

NCCER Product Development and Revision

13614 Progress Blvd., Alachua, FL 32615

Email: curriculum@nccer.org
Online: www.nccer.org/olf

❏ Trainee Guide ❏ AIG ❏ Exam ❏ PowerPoints Other _____

Craft / Level: _____ Copyright Date: _____

Module ID Number / Title: _____

Section Number(s): _____

Description: _____

_____

_____

_____

_____

Recommended Correction: _____

_____

_____

_____

_____

Your Name: _____

Address: _____

_____

Email: _____ Phone: _____

Roofing Applications
27202-07

27202-07
Roofing Applications

Topics to be presented in this module include:

Overview

As you travel around, take note of the different types of residential and commercial structures. Note how many different kinds of roof construction there are, and note how many different types of roofing materials are used. Part of your work as a carpenter will involve preparing roof decks to receive the finish roofing material, and you may even find yourself installing roofing material.

The roof is the most vulnerable part of a building. If it is not properly installed, it will leak. In some situations, it could even collapse. Safety is always a major consideration when working on a roof.

Objectives

When you have completed this module, you will be able to do the following:

1. Identify the materials and methods used in roofing.
2. Explain the safety requirements for roof jobs.
3. Install fiberglass shingles on gable and hip roofs.
4. Close up a valley using fiberglass shingles.
5. Explain how to make various roof projections watertight when using fiberglass shingles.
6. Complete the proper cuts and install the main and hip ridge caps using fiberglass shingles.
7. Lay out, cut, and install a cricket or saddle.
8. Install wood shingles and shakes on roofs.
9. Describe how to close up a valley using wood shingles and shakes.
10. Explain how to make roof projections watertight when using wood shakes and shingles.
11. Complete the cuts and install the main and hip ridge caps using wood shakes/shingles.
12. Demonstrate the techniques for installing other selected types of roofing materials.

Trade Terms

Asphalt roofing cement	Scrim
Base flashing	Selvage
Bundle	Side lap
Cap flashing	Slope
Exposure	Square
Head lap	Top lap
Overhang	Underlayment
Pitch	Valley
Ridge	Valley flashing
Roof sheathing	Vent stack flashing
Saddle	Wall flashing

Required Trainee Materials

1. Pencil and paper
2. Appropriate personal protective equipment

Prerequisites

Before you begin this module, it is recommended that you successfully complete *Core Curriculum*; *Carpentry Fundamentals Level One*; and *Carpentry Framing and Finishing Level Two*, Module 27201-07.

FRAMING AND FINISHING

27212-07
Cabinet Fabrication
ELECTIVE

27211-07
Cabinet Installation

27210-07
Window, Door, Floor, and Ceiling Trim

27209-07
Suspended Ceilings
ELECTIVE FOR RESIDENTIAL CERTIFICATE

27208-07
Doors and Door Hardware

27207-07
Drywall Finishing

27206-07
Drywall Installation

27205-07
Cold-Formed Steel Framing

27204-07
Exterior Finishing
ELECTIVE FOR COMMERCIAL CERTIFICATE

27203-07
Thermal and Moisture Protection

27202-07
Roofing Applications
ELECTIVE FOR COMMERCIAL CERTIFICATE

27201-07
Commercial Drawings
ELECTIVE FOR RESIDENTIAL CERTIFICATE

CARPENTRY FUNDAMENTALS

CORE CURRICULUM:
Introductory Craft Skills

202CMAP.EPS

This course map shows all of the modules in *Carpentry Framing and Finishing Level Two*. The suggested training order begins at the bottom and proceeds up. Skill levels increase as you advance on the course map. The local Training Program Sponsor may adjust the training order.

1.0.0 ◆ INTRODUCTION

Roofing materials are used to protect a structure and its contents from the elements. Besides rain protection, some materials are especially suitable for use in areas where fire, high wind, or extreme heat problems exist or in areas where cold weather, snow, and ice are problems. Materials can contribute to the attractiveness of the structure due to the careful selection of texture, color, and pattern. However, the design of the structure as well as local building codes may limit the choice of materials because of the **pitch** of the roof or because of other considerations at a particular location. In any case, the project specifications must be checked to determine the type of roofing materials to be used.

2.0.0 ◆ TYPICAL ROOFING MATERIALS

Many types of roofing materials as well as commercial roofing systems are available today. This section will cover the most common materials used on residential and small commercial structures. Commercial roofing systems used on larger or more expensive structures will be described in more detail in a later level.

2.1.0 Composition Shingle Roofing

Composition shingles (*Figure 1*) are the most common roofing material in North America. They are available in a wide variety of colors, textures, types, and weights (thicknesses). The standard shingle is made of a fiber or fiber-mat material, coated or impregnated with asphalt, and then coated with various mineral granules to provide color, fire resistance, and ultraviolet protection.

In the past, composition shingles were made using asbestos fiber or organic fiber and were commonly referred to as asphalt shingles. The manufacture of asbestos-fiber shingles has been prohibited because the asbestos poses a cancer risk and environmental disposal hazard.

Organic fiber shingles, which have a lifespan of 15 to 20 years, have been largely replaced by asphalt-coated, fiberglass mat shingles, simply called fiberglass shingles, which have a lifespan of 20 to 25 years.

More expensive architectural shingles with lifespans of 25 to 40 years or more are also available. Architectural shingles are constructed of multiple layers of fiberglass that are laminated

INSIDE TRACK

Metric Shingles

Most shingles manufactured today are available in foot-pound (English) and metric dimensions. The standard foot-pound, three-tab shingle or laminated architectural shingle is 36" long. The metric versions are 39⅜" long.

INSIDE TRACK

Special Architectural Shingles

These shingles are very expensive and are used primarily in elegant residential applications. Various styles of special architectural shingles that provide different pattern effects on a roof are available.

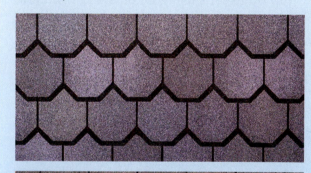

202SA01.EPS

together when manufactured or job-applied in layers to create a heavy shadow effect (*Figure 2*). Manufacturers may also provide their shingles in fungus/algae-resistant versions for use in damp locations where unsightly fungus growth or black streaks caused by algae tend to be a problem. The fungus/algae resistance is provided by copper granules incorporated with the mineral aggregates that are bonded to the surface of the shingle.

Composition shingles generally suit every climate in North America and can normally be applied to any roof with a **slope** of 4 in 12 (*Figure* 3) up to 21 in 12, provided they have factory-applied, seal-down adhesive strips. By applying double-lap **underlayment**, they can be

ARCHITECTURAL SHINGLE

THREE-TAB SHINGLE

202F01.EPS

Figure 1 ◆ Typical composition shingles.

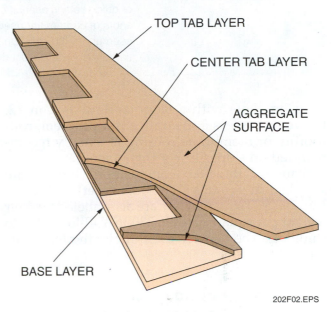

TOP TAB LAYER

CENTER TAB LAYER

AGGREGATE SURFACE

BASE LAYER

202F02.EPS

Figure 2 ◆ Typical architectural shingle.

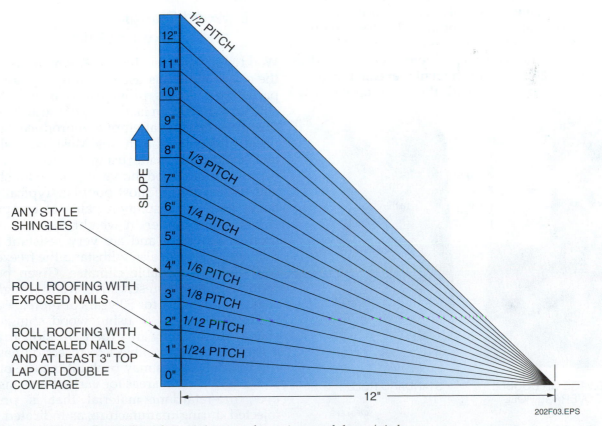

12"
11"
10"
9"
8"
7"
6"
5"
4"
3"
2"
1"
0"

1/2 PITCH

1/3 PITCH

1/4 PITCH

1/6 PITCH

1/8 PITCH

1/12 PITCH

1/24 PITCH

SLOPE

ANY STYLE SHINGLES

ROLL ROOFING WITH EXPOSED NAILS

ROLL ROOFING WITH CONCEALED NAILS AND AT LEAST 3" TOP LAP OR DOUBLE COVERAGE

12"

202F03.EPS

Figure 3 ◆ Typical shingle or roll roofing applications for various roof slopes/pitches.

used on roofs with a slope as low as 2 in 12. However, unless appearance is a problem, roll-roofing or membrane roofing is usually recommended on slopes lower than 4 in 12.

Standard three-tab composition shingles are supplied in **squares** (100 square feet) with three **bundles** per square. They are also labeled by their weight per square (210-lb, 230-lb, 240-lb, or 250-lb shingles). The heavier the shingle, the longer its life.

2.2.0 Roll Roofing

Roll roofing (*Figure 4*) is available in 50-lb to 90-lb weights with the same materials and colors as composition shingles. While it is inexpensive and quick to apply, the life of rolled roofing is typically only 5 to 12 years. It is recommended for use on shallow-slope roofs with slopes of less than 4 in 12 where appearance is not a problem. It can be used on slopes of 2 in 12 with the nails revealed or on slopes as low as 1 in 12 with the nails concealed.

SELVAGE OR DOUBLE
COVERAGE ROLL

STANDARD ROLL

202F04.EPS

Figure 4 ◆ Typical roll roofing.

Roll roofing can also be used as **valley flashing** to match the color of the roof shingles. It is supplied in three types. One type is a smooth surface roofing not covered with any granules. It is used on roofs that are subsequently covered with a hot or cold asphalt and separately applied aggregate. The other two types are for finish roofs. One is completely covered with granules and the other type, called **selvage** or double-coverage roll roofing, is only half covered with granules and is designed to be applied with a cemented-down half lap, with concealed fasteners used on very low slopes. The rolls are normally 36" wide. However, granule-covered, half-width, starter-strip rolls are also available.

2.3.0 Wood Shingle and Shake Roofing

Wood shingles and shakes (*Figure 5*) are among the oldest materials used in shingling, and both, particularly shakes, provide a pleasing and desirable visual effect. In fact, many of the architectural fiberglass shingles attempt to reproduce the rustic visual effect of wood shakes. Most wood shingles and shakes are made using western red cedar, cypress, or redwood; however, red cedar shingles and shakes are the most popular. Typically, they have twice the insulation value of composition shingles, are lighter in weight than most other roofing materials, and are very resistant to hail damage. They can also withstand the freeze-thaw conditions of variable climates. Given periodic coatings of wood preservative, wood shingles and shakes should last for 50 years or more.

The drawbacks of using wood shingles and shakes are that they are expensive, slow to install, a fire hazard, and subject to insect damage and rot. Building codes may prohibit the use of wood shingles in certain areas for various reasons; however, fire-retardant material that is pressure-injected during manufacture, as indicated by the industry designation Certi-Guard™, is supposed

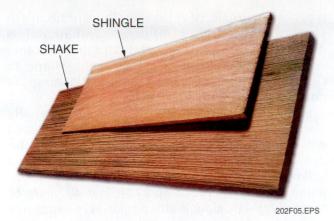

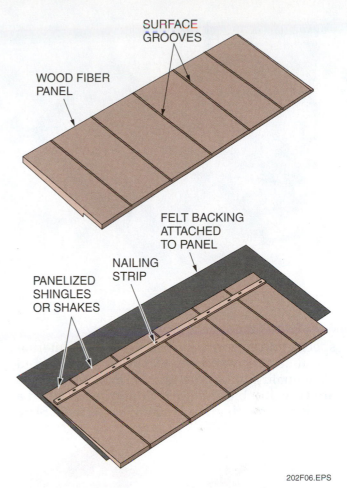

SHAKE

SHINGLE

202F05.EPS

Figure 5 ◆ Typical wood shake and shingle.

SURFACE GROOVES

WOOD FIBER PANEL

FELT BACKING ATTACHED TO PANEL

NAILING STRIP

PANELIZED SHINGLES OR SHAKES

202F06.EPS

Figure 6 ◆ Panelized shakes or shingles and wood fiber hardboard panel.

to conform to all state and local building codes for use in fire hazard regions.

Wood shingles are machine-cut and smoothed on both sides. They are uniform in thickness and length but vary in width. Hand-split wood shakes, which are more expensive, are either hand-split on one side and machine smoothed on the other side, or hand-split on both sides. The hand splitting produces a rough, rustic surface. Hand-split shakes vary in thickness and width, but are uniform in length. Because of the varying thickness, as well as the rough surface of at least one face of the shakes, an underlayment is used between each course of the shakes to prevent wind-driven rain from being forced back under the shakes. Machine sawn and grooved shakes that simulate hand-split shingles are also available.

Wood shingles and shakes are also available as 4' to 8' pre-bonded panels that are in two-ply and three-ply sections with **exposures** of 5½" to 9" (*Figure 6*). These panelized shingles and shakes, with course-separating underlayment pre-installed, are more expensive than traditional shakes and shingles. However, they are claimed to be up to two times faster to install than composition shingles and up to four times faster to install than traditional wood shakes and shingles. A simulated wood shingle made of wood fiber composition hardboard is also available in panel form. The panels are 12" × 48" and are applied lengthwise across the roof. They are embossed with deep shadow lines and random-cut grooves that mimic the look of shakes. The panels overlap with a shiplap joint between courses and between panels in the same course. After exposure, they weather to a gray that is similar to cedar shingles. Simulated shingle and shake metal roofing panels that are 4' long and completely fireproof are also available. These metal panels are very much like metal siding. They are applied horizontally and have an interlocking joint at the top and bottom edges.

2.4.0 Slate Roofing

Today, real slate roofing (*Figure 7*) is probably the most expensive roofing option in terms of roof framing materials, roofing materials, and increased installation time. Even though it has a very long service life of 60 to 100 years or more, it is usually not financially practical to use slate unless it is required or desired strictly for architectural purposes. No other material can match the high-quality look of slate. Unfortunately, besides being expensive, slate is heavy—about 7 to 10 lbs per square foot. As a result, the roof framing must be engineered to be substantially stronger to support the slate load, as well as any anticipated snow and ice loads.

Slate is completely fireproof and rot-proof. It is available in a wide variety of grades, thicknesses, and colors ranging from gray or black to shades of green, purple, and red. The colors are qualified as unfading or, if subject to some change, weathering. The industry often uses the old federal grading system of A (best) through C to indicate the quality of slate; however, architectural

Figure 7 ◆ Slate roofing.

types with streaks of color usually contain intrusions of sand or other impurities and are not as durable. Premium quality slate in various colors can be obtained from quarries from Maine to Georgia. Copper slater's nails and flashing are normally used with slate roofs.

Synthetic slate is made of fiber mat (usually fiberglass) that has been impregnated and coated with cement. It looks like real slate, but is lighter in weight. Even though it is lighter than slate, it still requires strong roof framing for support. Synthetic slate is fireproof and can last 40 years or more. Synthetic slate is still a relatively expensive material and is typically used on structures where historically appropriate materials are required. The synthetic slate material is difficult to cut and must be carefully fastened to avoid cracking the shingles.

specifications normally use the ASTM International testing numbers to specify the desired slate quality. New York and Vermont slate types are very durable and have a uniform color and a straight, smooth grain running lengthwise. Slate

2.5.0 Tile Roofing

Like slate, glazed and unglazed clay and ceramic tile roofing products (*Figure 8*) are fireproof and rot-proof, and last from 50 to 100 years. Tile roofing is expensive and heavy (7 to 10 pounds per

VILLA	ROMA	CLASSIC	HOMESTEAD
FRENCH	ENGLISH	MISSION S-STYLE	SHAKE AND SLATE
MISSION	RIDGE AND HIP	HIP STARTER	RAKE
SHINGLE	SPANISH	THREE-WAY APEX	FOUR-WAY APEX

Figure 8 ◆ Typical tile roofing styles.

Synthetic Tiles, Shakes, and Shingles

One alternative to clay and ceramic tiles is synthetic tile made of fiber-reinforced concrete or shredded tire material. These tiles are very durable and lighter than clay or ceramic tiles. In some cases, they may be light enough to install on roofs intended to support composition shingles. Synthetic tiles are available in many colors and styles. Like clay and ceramic tiles, synthetic tiles are fireproof and rot-proof. They last almost as long as clay and ceramic tiles. Other alternatives to clay, ceramic, or concrete tiles are interlocking baked-enamel steel and vinyl-coated aluminum tiles and shingles. These types of tiles and shingles have a life expectancy exceeded only by real slate or tile. They are available in styles similar to shakes, slate, and tile roofing and are light as well as less expensive.

SLATE

SHAKE

202SA02.EPS

square foot) and requires appropriate roof framing to support the tile weight and any other anticipated loads. It is used in the South and to a great extent in the southwest areas of the country because it is fireproof and impervious to damage caused by intense sunlight. It is available in Spanish, Mission (barrel or S-type), and other styles known by various names such as French, English, Roma, and Villa. Other styles may also include flat tiles that may resemble slate or wood shakes. Matching hip and **ridge** tiles, rake/barge tiles, and hip starter tiles are also available. Every tile for a particular style furnished by a manufacturer is a uniform size in width, thickness, and length.

Tiles are fastened with non-corrosive nails (copper, galvanized, or stainless steel). Copper nails have the advantage of being soft enough to allow expansion and contraction without causing the tiles to crack if they are fastened too tight. They also allow easier replacement of any broken tiles. Because of its durability and resistance to any alkaline corrosion, copper is generally used as flashing.

2.6.0 Metal Roofing

Metal roofing is available in a great variety of materials and styles. These materials can be purchased with a baked enamel, ceramic, or plastic coating. They can also be purchased without a coating so that a separate roof coating can be applied after installation. The following are some common metal roofing materials:

- Aluminum (plain or coated)
- Galvanized steel (plain or coated)
- Terne metal (heat-treated, copper-bearing steel hot-dipped in terne metal comprised of 80 percent lead and 20 percent tin)
- Aluminum/zinc-coated steel (plain or coated)
- Stainless steel

Besides the common residential panel roofing styles (*Figure 9*), many new engineered/preformed, architectural metal fascia/roofing systems (*Figure 10*) are available for commercial or residential use in a wide selection of colors and in styles that include shingles, panels, and tiles.

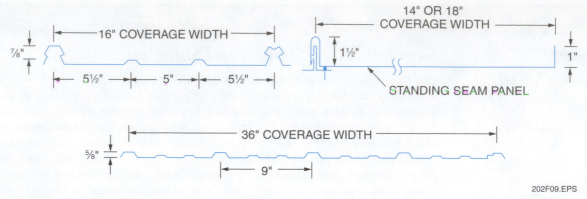

Figure 9 ◆ Common metal roofing styles.

Figure 10 ◆ Example of architectural metal fascia/roofing system.

2.7.0 Built-Up Roofing Membrane

Conventional built-up roofing (BUR) membrane (*Figure 11*) has been used for over 100 years on very low-slope roofs of residential and commercial structures. While it is still a viable form of roofing, it is gradually being replaced by premanufactured membrane roofing systems.

Built-up roofing, which is field-fabricated, consists of three to five layers of heavy, asphalt-coated polyester/fiberglass felt embedded in alternate layers of hot-applied or cold-applied bitumens (coal tar or asphalt based) that are the waterproofing material. Most commercial applications use hot asphalt as the waterproofing material. The top surface layer of bitumen is sometimes left smooth but is most often covered with either a mineral-coated cap sheet or embedded with a loose mineral surface consisting of small aggregates such as washed gravel, pea rock, or crushed stone.

Hot asphalt applications require special heating equipment along with some method of transporting the hot asphalt to the roof unless the heating equipment is lifted and positioned onto the roof. Installing this type of roof is labor-intensive and requires experienced roofers. Because the membrane is created in the field, its quality is subject to many variables, including the weather, application techniques, and the experience of the roofers. Most BUR is placed over one or more insulation boards that are bonded or fastened to the roof substrate.

The lifespan of a correctly applied, built-up roof is about 10 to 20 years, depending on the number of layers. Generally, specific damage to this type of roof can be easily repaired. However, the normal, gradual deterioration of the roof, which may result in deep splits over much of the surface or delamination of the layers, will generally require that the roof be completely removed and a new roof applied.

2.8.0 Premanufactured Membrane Roofing

There are two general categories of premanufactured membrane roofing systems: modified bitumen systems or single-ply systems. The single-ply membrane systems are wholly synthetic roofing materials that exhibit elastomeric (rubber-like) properties to various degrees. There are a number of types within these two general systems. Today, there are many manufacturers of both systems, and it is important to note that each manufacturer requires specific, compatible materials and accessories for their versions of each type within each system. These materials and accessories include flashing, fasteners, drain and vent boots, inside/outside corners, cant strips (coving), cleaners, solvents, adhesives, caulking, sealers, and tapes. To obtain the maximum performance from the system

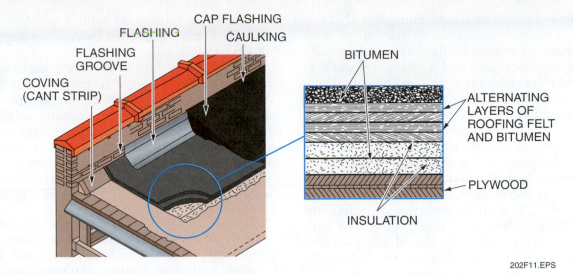

Figure 11 ◆ Conventional built-up roofing membrane.

and for warranty purposes, compatible materials and accessories must be used and manufacturer-specified application procedures must be strictly followed. Normally, these systems are installed over insulation boards by roofers experienced with a specific manufacturer's product.

2.8.1 Modified Bitumen Membrane Roofing Systems

Modified bitumen roofing systems can be classified as either styrene butadiene styrene (SBS) or atactic polypropylene (APP) modified bitumen products. The SBS products are usually a composite of polyester or glass fiber and modified asphalts coated with an elastomeric blend of asphalt and SBS rubber. The APP products are coated with an elastomeric blend of asphalt and atactic polypropylene. Both products are the weatherproofing medium and are used in a hybrid BUR as a cap sheet over one or more base/felt plies that have been secured with hot asphalt or cold adhesive to a deck board covering the insulation or directly to the insulation.

Depending on the manufacturer, SBS products are either supplied with a preapplied adhesive and the product is rolled for adhesion, or the product is secured with hot asphalt or a cold adhesive applied separately. APP products, called torch-down roofing, are usually secured to the layers below by heat welding. Using flame heating equipment, the back of the product roll is heated as the product is unrolled and pressed down. The back coating of the product is heated to the point where the bitumen coating acts as an adhesive, bonding the product to the layer below and the overlapped edges of the adjacent sheet.

Flame heating equipment (*Figure 12*) can be used for BUR and torch-down roofing systems. Both types of systems require a mineral covering or protective coating to prevent ultraviolet destruction of the modified bitumen materials. In some cases, the product is available with various colors of ceramic roofing granules preapplied to the exposed surface.

2.8.2 Single-Ply Membrane Roofing Systems

Single-ply roofing systems can be classified as either thermoplastic (plastic polymer) or thermoset (rubber polymer) systems. Within these classifications, a number of types exist. One of the most common thermoset polymer membranes is

Figure 12 ◆ Flame heating equipment for BUR and torch-down roofing.

an ethylene propylene diene monomer (EPDM) product. This polymer, usually reinforced with polyester scrim, retains its flexibility over a wide range of temperatures and is very resistant to ozone and ultraviolet ray damage. The membrane is usually supplied in large sheets or rolls and is spliced together with compatible adhesives or tapes.

Thermoplastic single-ply membranes have become very popular for commercial and industrial roofing. Two of the most common types of thermoplastic membranes available are polyvinyl chloride (PVC) and thermoplastic polyolefin (TPO). PVC membrane is usually reinforced with a polyester scrim and can be joined using solvent or hot-air welded seams that are extremely durable (*Figure 13*). PVC membrane is lightweight and aesthetically pleasing. It is very resistant to ozone and ultraviolet ray damage. It is also puncture- and tear-resistant. TPO membrane combines the advantages of the solvent or hot-air seam welding capability of PVC with the greater weatherability and flexibility benefits of the more traditional EPDM membranes. TPO membranes

may also be reinforced with a polyester scrim. Both PVC and TPO membranes are supplied in sheets or wide rolls.

Single-ply roofing systems are clean and economical to install because they do not use hot-asphalt installation techniques common to BUR and modified bitumen membrane roofing systems. Most single-ply membranes are underlaid with insulation boards or protective mats along with a fireproof slipsheet, if necessary.

Single-ply membranes are usually anchored to a roof structure in one of four ways (*Figure 14*).

- *Loose laid/ballasted* – The perimeter is anchored with adhesives or mechanical fasteners, and the entire surface is weighted down with a round stone ballast or walking pavers (thin concrete blocks). This method is used only on very low-slope roofs capable of supporting the ballast load. It is fast and economical.
- *Partially adhered* – The entire area of the membrane is spot-adhered to the roof with mechanical fasteners and/or adhesives. This method produces a membrane with a dimpled, wind-resistant surface.
- *Mechanically adhered* – The entire area of the membrane is spot-adhered to the roof with mechanical fasteners. This method produces a dimpled, wind-resistant installation and allows easy removal of the membrane during future roof replacement.
- *Fully adhered* – The entire area of the membrane is completely cemented down with an adhesive. This method produces a smooth, windproof surface.

2.9.0 Drip Edge and Flashing

Figure 15 shows various types of drip edges that are available. Some are used in reroofing applications only. Drip edges and any other flashing must be made of materials that are compatible with the roofing and any items that are being flashed. They must last as long as the finish roof

202F13.EPS

Figure 13 ♦ Single-ply membrane installation.

Open-Flame Heat Welding

INSIDE TRACK

Today, most seam-sealing methods and equipment recommended by manufacturers of single-ply membrane roofing systems make use of hot-air welding methods that are quite safe. However, some contractors still use open-flame heating equipment for seam welding. Open-flame seam sealing can be hazardous and may cause damage to the membrane. For these reasons, open-flame seam sealing of polymer membrane roofing should be avoided.

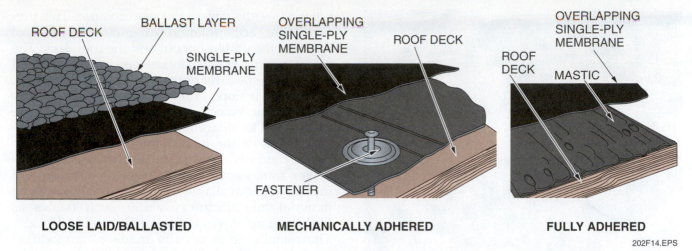

LOOSE LAID/BALLASTED   MECHANICALLY ADHERED   FULLY ADHERED

202F14.EPS

Figure 14 ◆ Typical methods of anchoring single-ply membrane to a roof.

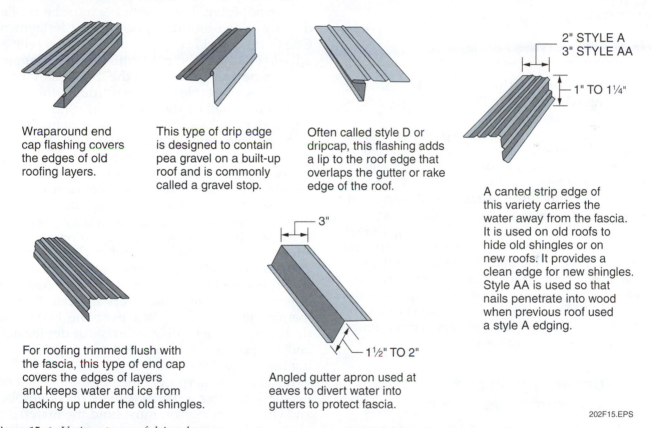

Wraparound end cap flashing covers the edges of old roofing layers.

This type of drip edge is designed to contain pea gravel on a built-up roof and is commonly called a gravel stop.

Often called style D or dripcap, this flashing adds a lip to the roof edge that overlaps the gutter or rake edge of the roof.

2" STYLE A
3" STYLE AA
1" TO 1¼"

A canted strip edge of this variety carries the water away from the fascia. It is used on old roofs to hide old shingles or on new roofs. It provides a clean edge for new shingles. Style AA is used so that nails penetrate into wood when previous roof used a style A edging.

For roofing trimmed flush with the fascia, this type of end cap covers the edges of layers and keeps water and ice from backing up under the old shingles.

3"
1½" TO 2"

Angled gutter apron used at eaves to divert water into gutters to protect fascia.

202F15.EPS

Figure 15 ◆ Various types of drip edges.

material. Today, aluminum, galvanized steel, copper, vinyl, and stainless steel are the most common materials used for drip edges and flashing. Galvanized steel, copper, or stainless steel must be used with any cement-based roofing material or against masonry materials due to the corrosive nature of cement. Copper is normally used with slate roofing because of its long life and resistance to corrosion.

For best results, any galvanized drip edge or flashing should be coated with an appropriate primer before installation. All drip edges and flashing must be fastened with nails or staples made of a compatible material to prevent electrolytic corrosion between the fasteners and the flashing. The roofing material manufacturer's recommendations for flashing must be followed.

Figure 16 shows a W-metal (so named because its end profile looks like a letter W) or standing-seam **valley** flashing. This type of valley flashing is available in 8' to 10' lengths in widths of 20" to 24". If desired, it can be field-fabricated using flat

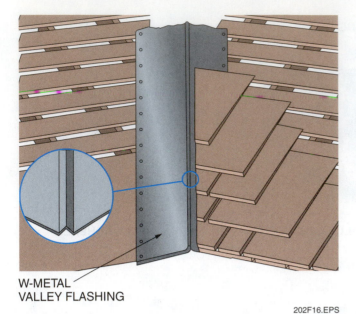

W-METAL VALLEY FLASHING

202F16.EPS

Figure 16 ◆ W-metal valley flashing.

roll flashing and a metal brake. In any case, it must be wide enough so that the finish roofing overlaps the metal by more than 6".

The W-metal valley flashing is preferred if open valleys will be used. Open valleys are defined as valleys where the valley flashing material will be visible after the finish roofing is applied. Open valleys are more difficult to install but accommodate higher rainfall rates than closed valleys. They can also be used with any type of finish roofing. Closed valleys can be used only on composition shingle roofs. The ridge in the middle of the W-metal valley (about ¾" to 1" high) prevents water rushing down the slope of one roof from washing under the shingles of the intersecting roof. On short valleys or relatively low-slope roofs, ordinary flat-roll flashing can be used in the valleys.

2.10.0 Underlayment and Waterproof Membrane

Underlayment is available as 60-lb rolls of nonperforated, 15-lb, 30-lb, and 60-lb asphalt-saturated felt (*Table 1*). The felt used under roofing materials

must allow the passage of water vapor. This prevents the accumulation of moisture or frost between the underlayment and the roof deck. The correct weight of felt to be used is usually specified by the manufacturer of the finish roofing material. Roofs with a slope of more than 4 in 12 normally use 15-lb felt. Wood shakes usually require the use of 30-lb felt to separate the courses.

In areas of the country where water backup under the finish roof is a problem due to wind-driven rain, ice, and snow buildup, a waterproof membrane is available from a number of roofing material manufacturers under such names as Storm-Guard™, Water-Guard™, and Dri-Deck™. The membrane is usually made of a modified asphalt-impregnated fiberglass mat, coated on the bottom side with an elastic-polymer sealer that is also an adhesive. The membrane must be applied directly to the roof deck, not the underlayment. One of the top side edges is also covered with an adhesive so that the next course of overlapping membrane will adhere to the previous course. Any nails or staples driven through the membrane are sealed by the membrane, thus preventing water leakage. Because the membrane is impermeable, any moisture from inside the structure condensing under the membrane will eventually damage any wood directly under the membrane. As a result, some manufacturers suggest that the membrane only be applied along the bottom edges of the eaves and up the roof to at least 24" above the outside wall, along the rake edges, up any valleys, around skylights, and on any **saddles** or other problem areas. The rest of the roof is covered with conventional underlayment. However, other manufacturers indicate that the membrane can be applied over the entire roof deck, if adequate ventilation exists under the deck and a vapor barrier is installed on any inside ceiling under the deck.

Installing waterproof membrane is a two-person job. After a manageable strip is unrolled and cut off, one person must hold the material in place while another peels away a protective sheet covering the bottom adhesive. The membrane adhesive remains tacky during installation and

Table 1	Sizes, Weights, and Coverage of Asphalt-Saturated Felt							
Approximate Weight per Roll	Approximate Weight per Square	Squares per Roll	Roll Length	Roll Width	Side or End Laps	Top Lap	Exposure
60#	15#	4	144'	36"	4" to 6"	2"	34"
60#	30#	2	72'	36"	4" to 6"	2"	34"
60#	60#	1	36'	36"	4" to 6"	2"	34"

Waterproof Membranes

Waterproof membranes used on roof edges and valleys are self-healing. This means that if the membrane is intentionally or accidentally penetrated by a screw or nail, a modified asphalt coating will flow to seal the penetration when the roof and membrane are heated by the sun. Because they have an adhesive that secures them to the roof deck, waterproof membranes are virtually impossible to remove once the adhesive is heat-set by the sun. This makes reroofing of a structure very expensive if the membrane must be removed. This is because the roof deck must be replaced if it is wood or a wood product. Newer versions of membranes are available with a granular surface that allows the overlying shingles to be removed without damaging the membrane.

the membrane can be lifted off the roof deck and repositioned, if necessary. However, after a short time it will set up and cannot be removed without damaging the membrane.

2.11.0 Roofing Nails

Common roofing materials must be fastened with nails of the proper length and made of a material that is compatible or the same as the drip edges and flashing. *Figures 17* and *18* show the most common nails used for composition shingles and wood shakes or shingles. These nails are usually available in galvanized steel; some other types are aluminum, stainless steel, or copper. Normally, copper slater's nails are used for slate roofs, and stainless steel nails are used for tile roofs. About 1 pound of nails per square (100 square feet) is required to fasten composition shingles. For wood shakes and shingles, 2 to 4 pounds of nails will be required. For other types of nailed roofing, follow the

manufacturer's recommendations. Check your local code for nail penetration requirements.

For composition roofs or underlayment, nails can be installed using pneumatic-powered or electric-powered nailing guns.

2.12.0 Cold Asphalt Roofing Cement

Cold **asphalt roofing cements**, consisting of modified asphalt and/or coal tar products, are used in the installation of composition shingles, underlayment, roll roofing, and cold asphalt BUR. They are available in liquid non-fibered form or in plastic fibered form. The non-fibered forms are usually used in the lapped installation of underlayment and roll roofing and are spread over large areas with a spreader or mop. The plastic fibered cement is used for spot repairs or cementing nail heads, shingle tabs, valley overlaps, and saddle materials, and as an exposed sealer for gaps or flashing. Asphalt cements are generally available in one-gallon and five-gallon pails.

NAILING APPLICATION	⅜" PLYWOOD OR WAFER BOARD	1" SHEATHING
STRIP OR SINGLE (NEW CONSTRUCTION)	⅞"	1¼"
OVER OLD ASPHALT LAYER	1"	1½"
REROOFING OVER WOOD SHINGLES	–	1¾"

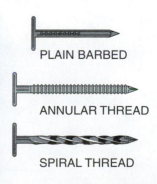

202F17.EPS

Figure 17 ◆ Typical composition shingle nails.

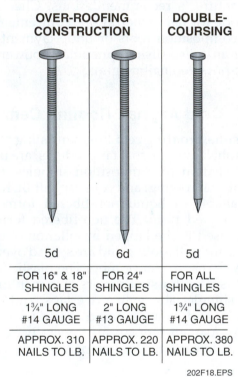

NEW ROOF CONSTRUCTION

3d	3d	4d
FOR 16" AND 18" SHINGLES		FOR 24" SHINGLES
1¼" LONG	1¼" LONG #14½ GAUGE	1½" LONG #14 GAUGE
APPROX. 376 NAILS TO LB.	APPROX. 515 NAILS TO LB.	APPROX. 382 NAILS TO LB.

OVER-ROOFING CONSTRUCTION		DOUBLE-COURSING
5d	6d	5d
FOR 16" & 18" SHINGLES	FOR 24" SHINGLES	FOR ALL SHINGLES
1¾" LONG #14 GAUGE	2" LONG #13 GAUGE	1¾" LONG #14 GAUGE
APPROX. 310 NAILS TO LB.	APPROX. 220 NAILS TO LB.	APPROX. 380 NAILS TO LB.

202F18.EPS

Figure 18 ◆ Typical wood shake or shingle nails.

3.0.0 ◆ TOOLS

Many of the tools used for roofing are common to other trades. Some of these common tools are:

- Backsaw
- Power circular saw
- Crowbar
- Hand saw
- Carpenter's level
- Nail apron
- Sliding T-bevel
- Keyhole saw
- Pop riveter
- Chalkline
- Tape measure
- Power saber saw
- Angle square
- Power drill
- Caulking gun
- Tin snips
- Prybar
- Scribing compass
- Utility knife
- Drill bit sets (regular and masonry)
- Framing square
- Claw hammer
- Pneumatic nailers
- Flat spade or spud bar (for roofing material removal)

Other tools that are specific to the installation of certain types of roofing are also used. Some of these tools are shown in *Figure 19*.

The roofing hammer, also referred to as a shingle hatchet, is used primarily for wood shingle and shake installation. The hatchet end is used to split shingles or shakes and the top edge is marked or equipped with a sliding gauge to set a dimension for the amount of weather exposure for the shingle or shake.

A composition shingle knife is used to trim or cut all types of composition shingles, including architectural shingles, during installation. It is also used to cut underlayment, cap shingles, roll roofing, and membrane roofing.

Slate roofing installation usually requires three specialized tools: a slater's hammer, a nail ripper, and a slate cutter. The slater's hammer is equipped with a sharp edge for cutting slate and a point for poking nail holes through the slate. The nail ripper has sharp-edged barbs on one end that are used to shear off nails under a piece of slate. To shear nails, the ripper is struck on the face of the anvil with a hammer. The slate cutter aids in the trimming of slate and the punching of nail holes.

FATMAX HOOK KNIFE

SHINGLE HATCHETS

SLATE CUTTER

ESCO SLATE CUTTER

HAMMER ANVIL

NAIL RIPPER

NAIL PULLER

SHARP TRIMMING EDGE

SLATER'S HAMMER

HEAVY ROLLER

202F19A.EPS

Figure 19 ◆ Special roofing tools. (1 of 2)

Various types of tile cutters and nibblers are used in the installation of tile roofs for splitting or trimming tiles. A nail ripper, like the one used for slate roofing, can be used to shear off nails under a tile. A wet saw with a diamond wheel can be used on large projects for flat or shaped tile cutting.

Portable brakes are used for the custom bending of flashing material for any type of roof installation. Some roofers use heavy rollers to flatten underlayment to eliminate buckling under the finish roof and for the application of cold-cement,

fully-adhered roll roofs, BURs, or single-ply membrane roofing. These rollers are sold as vinyl flooring rollers in weights ranging from 75 to 150 pounds.

Hand grinders with diamond wheels are used to cut slots in masonry for flashing installation.

In addition to tools, other equipment such as scaffolding, material movement equipment, ladders, and ladder jacks or pump jacks may be required. All roofing installation jobs will require some type of fall protection system.

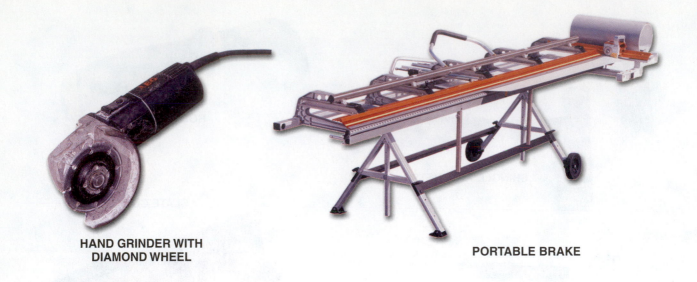

**HAND GRINDER WITH
DIAMOND WHEEL**

PORTABLE BRAKE

PORTABLE WET SAW

**NIBBLING
TOOL**

SCORE-AND-SNAP TILE CUTTER

202F19B.EPS

Figure 19 ◆ Special roofing tools. (2 of 2)

Power Nailers

Make sure that the nailer is equipped with a flush-mount attachment or that the impact pressure of the tool can be regulated at the tool to prevent overdriving the nails and cutting through the roofing material.

PNEUMATIC ROOFING NAILER WITH PLASTIC WASHER ATTACHMENT FOR UNDERLAYMENT APPLICATION

TYPICAL PNEUMATIC ROOFING NAILER

202SA03.EPS

4.0.0 ◆ SAFETY

Worker safety is important on a construction site. Every work site must have a fall protection plan for working on roofs or at certain heights off the ground.

4.1.0 Safety Summary

The following guidelines must be observed to ensure your safety and the safety of others:

- Wear boots or shoes with rubber or crepe soles that are in good condition.
- Always wear fall protection devices, even on shallow-pitch roofs.
- Rain, frost, and snow are all dangerous because they make a roof slippery. If possible, wait until the roof is dry; otherwise, wear special roof shoes with skid-resistant cleats in addition to fall protection.
- Brush or sweep the roof periodically to remove any accumulated dirt or debris.
- Install any required underlayment as soon as possible. Underlayment usually reduces the danger of slipping. On sloped roofs, do not step on underlayment until it is properly fastened.
- On pitched roofs, install necessary roof brackets as soon as possible. They can be removed and repositioned as shingle-type roofing is installed.

- Remove any unused tools, cords, and other loose items from the roof. They can be a serious hazard.
- Check and comply with any federal, local, and state code requirements when working on roofs.
- Be alert to any other potential hazards such as live power lines.
- Use common sense. Taking chances can lead to injury or death.

When working outdoors or in high-heat conditions for extended periods of time, take precautions to avoid heat exhaustion and exposure to the sun's ultraviolet rays. Preventive measures include the following:

- Wear a hard hat.
- Wear light clothing that is made of natural fibers.
- If possible, wear tinted glasses or goggles.
- Use a sun protection factor (SPF) 30 or higher sunblock on exposed skin.
- Drink adequate amounts of water to prevent dehydration, especially in arid parts of the country.

4.2.0 Scaffolding and Staging

Scaffolding and staging have been the causes of many minor and serious accidents due to faulty or incomplete construction or inexperience on the part of the designer or craftsperson constructing them. Therefore, to avoid hazards caused by faulty or incompetent construction, all scaffolding and staging should be designed and constructed by competent, certified persons. Scaffolding and staging must be inspected on a daily basis by a certified, competent person. The inspector must tag the scaffolding/staging for safety every morning or at every change of shift.

Even though you have no part in the design or construction of scaffolding, for safety's sake you should be familiar with safety rules and regulations that govern its construction. If you are going to use it, you should know how it is built. The following safety factors should be thoroughly understood and adhered to by everyone on the job site:

- Any type of scaffold used should have a minimum safety factor ratio of four to one; that is, it should be constructed so that it will carry at least four times the load for which it is intended. Roofing material weight adds up quickly when placed in one location.
- All staging or platform planks must have end bearings on scaffold edges with adequate support throughout their lengths to ensure the minimum safety factor ratio of four to one.
- Scaffolding timbers, if used, must be carefully selected and maximum nailing used for added strength.
- Because scaffolds are built for work that cannot be done safely from the ground, makeshift scaffolds using unstable objects for support such as boxes, barrels, or piles of bricks are prohibited.

When the scaffolding is placed on a solid, firm base and erected correctly, the roofing applicator should be able to work in confidence.

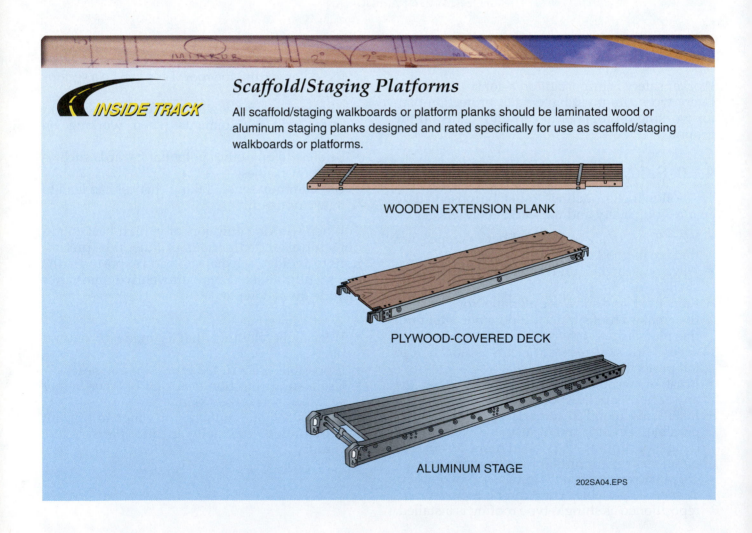

INSIDE TRACK

Scaffold/Staging Platforms

All scaffold/staging walkboards or platform planks should be laminated wood or aluminum staging planks designed and rated specifically for use as scaffold/staging walkboards or platforms.

WOODEN EXTENSION PLANK

PLYWOOD-COVERED DECK

ALUMINUM STAGE

202SA04.EPS

Any scaffolding assembled for use should be tagged. Three tag colors are used:

- *Green* – A green tag identifies a scaffold that is safe for use. It meets all OSHA standards.
- *Yellow* – A yellow tag means the scaffolding does not meet all applicable standards. An example is a scaffold where a railing cannot be installed because of equipment interference. A yellow-tagged scaffold may be used; however, a safety harness and lanyard are mandatory. Other precautions may also apply.
- *Red* – A red tag means a scaffold is being erected or taken down. You should never use a red-tagged scaffold.

Other more common types of scaffolding include ladder jacks (*Figure 20*) and pump jacks (*Figure 21*). Pump jacks and, to a lesser extent, ladder jacks are useful for applying the starter strip and lower courses of roofing. Ladder jacks can usually support a 2'-wide adjustable-length platform up to 10' or 18' long. Ladder jacks, which must not be used for heights over 20', can be attached to either side of a ladder and must be separated by not more than 8' intervals along the length of the platform. Pump jacks using aluminum posts for heights under 50' are movable platform supports that are raised or lowered vertically. They are operated by a foot lever and can raise a person plus a rated load.

4.3.0 Ladders

Ladders are useful and necessary pieces of equipment for roofing application and do not cause accidents if properly used and maintained. Ladder accidents are caused by:

- Improper use of ladders
- Ladders not secured properly
- Structural failure of ladders
- Improper handling of objects while on a ladder

Figure 22 shows a ladder erected correctly in relation to the roof eaves. If the ladder is to be left standing for a long period of time, it should be securely fastened at both the top and bottom.

202F20.EPS

Figure 20 ◆ Ladder jack and adjustable platform.

INSIDE TRACK

Extended Platforms

Aluminum-pole pump jack systems may be extended across the length of a wall using appropriately placed poles supporting properly rated platforms.

202SA05.EPS

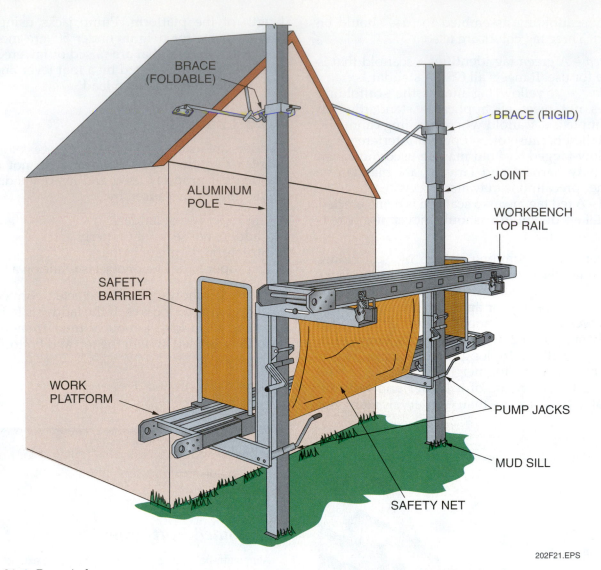

Figure 21 ◆ Pump jacks.

202F21.EPS

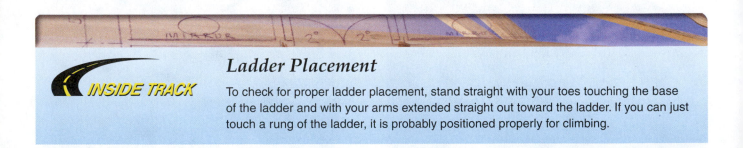

Ladder Placement

To check for proper ladder placement, stand straight with your toes touching the base of the ladder and with your arms extended straight out toward the ladder. If you can just touch a rung of the ladder, it is probably positioned properly for climbing.

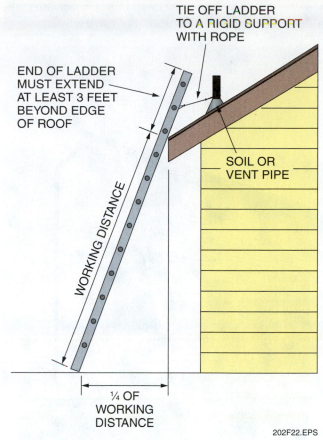

TIE OFF LADDER
TO A RIGID SUPPORT
WITH ROPE

END OF LADDER
MUST EXTEND
AT LEAST 3 FEET
BEYOND EDGE
OF ROOF

SOIL OR
VENT PIPE

WORKING DISTANCE

¼ OF
WORKING
DISTANCE

202F22.EPS

Figure 22 ◆ Safe ladder placement.

The normal purpose of a ladder used in a roofing application is solely to gain access to the roof itself. For ladder safety, follow these precautions:

- Always use a Type 1 (250 lbs/rung) OSHA-rated ladder that is 15" or more in width with rungs that are 12" apart.
- Fiberglass ladders should always be used to reduce the possibility of accidental electrocution resulting from contact with power lines. Avoid using aluminum ladders. Aluminum ladders, if used, should never be raised or placed in situations where they can fall or accidentally come into contact with power lines.
- For longer ladders, two people should carry, position, and erect the ladder.
- Always face a ladder and grasp the side rails or rungs with both hands when going up or down.
- Take one step at a time.

- Remember that an ordinary straight ladder is built to support only one person at a time.
- Before using a ladder, be sure there is no oil, grease, or sand on the soles of your shoes. Due to the tread composition, some shoe types easily attract foreign objects that can cause you to slip on a ladder.
- Never carry tools or materials up or down a ladder. A rope or other device should be used to raise or lower everything so that you can always have both hands free when climbing the ladder.
- Make sure that the base of the ladder is level and has adequate support. Shim the legs or use levelers, if necessary.
- Make sure the ladder is at the correct angle. Always tie off the ladder prior to use (see *Figure 22*). The ladder needs to be secured at its top to a rigid support.
- Never overreach.
- Fixed ladders must be provided with cages, wells, ladder safety devices, or self-retracting lifelines where the length of climb is less than 24', but the top of the ladder is at a distance greater than 24' above lower levels.

4.4.0 Material Movement

Many mechanical devices are used to move materials on a job site. These devices include conveyor belts, power ladder conveyers (attached to a ladder), forklift trucks, and truck-mounted hydraulic lifts. Make sure all safety devices are in place before starting.

Exercise extreme caution to ensure that the lift does not make contact with the roof surface or the person unloading the bundles. Immediately distribute the bundles around the roof area. Do not pile them in any one spot. Roof structures have been designed to carry a specific dead load. Placing unnecessary strain on the roof structure by concentrating the shingles in one place on the roof can jeopardize the safety of the workers. The end result might prove to be disastrous with the collapse of the roof itself. An immediate dispersal of the bundles will prevent any problems from occurring and also make the installation more efficient because you only want to move and place the load once.

Ladder Conveyors

For residential or light commercial work, a ladder conveyor system greatly reduces manual lifting of roofing materials to a roof. Note the safety railings and danger signs.

202SA06.EPS

4.5.0 Roofing Brackets

Roofing brackets provide firm footing and material storage points on steep-slope roofs. If any type of roofing bracket is to be used, the decision will be based on the slope of the roof. Most roofers feel comfortable on a roof with a 4 in 12 or a 5 in 12 slope. When the slope increases to 6 in 12, more strain is placed on the feet and the body. Therefore, the roofer has to be very conscious of the height off the ground and be careful with each and every movement. *Figure 23* shows two types of roofing brackets that can be used with a 10' to 14', defect-free, 2" thick plank.

Both types of brackets can be nailed firmly to the roof, but the adjustable bracket, which is installed to correspond to the slope of the roof, makes standing and moving around more comfortable. Never get overconfident due to bracket usage. Be aware of the height at which you are working and be cautious.

When installing roof brackets, make sure that they are nailed to the rafters, not just to the **roof sheathing**.

Roof brackets and toe boards alone are not sufficient to meet OSHA fall standards. A proper rail or safety harness is required above six feet.

4.6.0 Power Nailers

There are many types of power nailers. In the application of asphalt or fiberglass shingles, the use of one of these tools can cut the labor time in half. Portable units are used by carpenters on the construction site. Portable units are electric or pneumatic (air or carbon dioxide gas powered).

Some of the listed safety practices for power nailers are as follows:

- Always wear safety glasses. They must be an OSHA-approved type and are usually recommended by the manufacturer of the unit being used.
- Because operating principles vary, study the manufacturer's operating manual.
- Be certain to use the type of fastener required by the manufacturer.

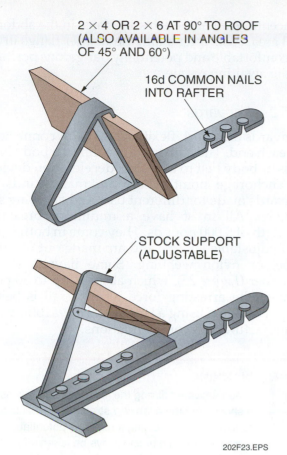

2 × 4 OR 2 × 6 AT 90° TO ROOF
(ALSO AVAILABLE IN ANGLES
OF 45° AND 60°)

16d COMMON NAILS
INTO RAFTER

STOCK SUPPORT
(ADJUSTABLE)

202F23.EPS

Figure 23 ◆ Roofing brackets.

- If pneumatic, make sure that air pressure can be adjusted at the nailer.
- Treat the machine as you would a gun. Do not point it at yourself or others.
- Always keep the unit tight against the surface to drive the fastener correctly.
- When not in use, disconnect the unit from the power source to prevent accidental release of fasteners.
- Keep air lines untangled on the roof to prevent tripping.

4.7.0 Fall Protection Equipment

Roofers spend a major part of their time working on sloped roofs. Most construction injuries and deaths are caused by falls. Falls from high places can cause serious injury or death when the wrong type of fall protection equipment is used, or when the right equipment is used improperly. A fall protection plan must be prepared for any project where workers will be more than 6' off the ground.

There are three common types of fall protection equipment: guardrails, personal fall arrest systems, and safety nets. This section will focus on personal fall arrest systems. These devices and their use are governed by *OSHA Safety and Health Standards for the Construction Industry, Part 1926, Subpart M*. The rules covering guardrails on scaffolds are contained in Subpart L. Basically, OSHA requires that all workers use guardrail systems, safety net systems, or personal fall arrest systems to protect themselves from falling more than 6' (1.8 meters) and hitting the ground or a lower work level.

When describing personal fall arrest systems, the following terms must be understood:

- *Free-fall distance* – The vertical distance a worker moves after a fall before a deceleration device is activated.
- *Deceleration device* – A device such as a shock-absorbing lanyard, rope grab, or self-retracting lifeline that brings a falling person to a stop without injury.
- *Deceleration distance* – The distance it takes before a person comes to a stop. The required deceleration distance for a fall arrest system is a maximum of 3½'.
- *Arresting force* – The force needed to stop a person from falling. The greater the free-fall distance, the more force is needed to stop, or arrest, the fall.

The following sections discuss equipment used in personal fall arrest systems:

- Body harnesses and belts
- Lanyards
- Deceleration devices
- Lifelines
- Anchoring devices and equipment connectors

4.7.1 Body Harnesses

Full body harnesses (*Figure 24*) with sliding back D-rings are used in personal fall arrest systems. They are made of straps that are designed to be worn securely around the user's body. This allows the arresting force to be distributed via the harness straps throughout the body, including the shoulders, legs, torso, and buttocks. This distribution decreases the chance of injury. When a fall occurs, the sliding D-ring moves to the nape of the neck, keeping the worker in an upright position and helping to distribute the arresting force. This keeps the worker in a relatively comfortable position while awaiting rescue.

Selecting the right full body harness depends on a combination of job requirements and personal

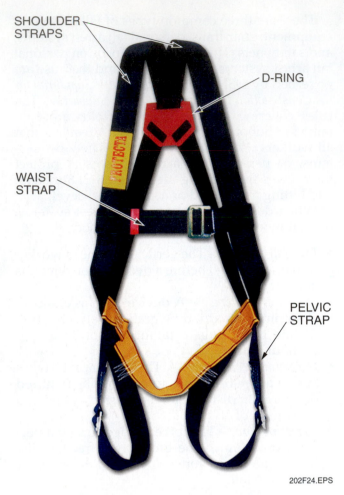

SHOULDER STRAPS

D-RING

WAIST STRAP

PELVIC STRAP

202F24.EPS

Figure 24 ◆ Full body harness.

concentrate all of the arresting force in the abdominal area. Also, after a fall, the worker hangs in an uncomfortable and potentially dangerous position while awaiting rescue.

4.7.2 Lanyards

Lanyards are short, flexible lines with connectors on each end. They are used to connect a body harness or body belt to a lifeline, deceleration device, or anchorage point. There are many kinds of lanyards made for different uses and climbing situations. All must have a minimum breaking strength of 5,000 pounds. They come in both fixed and adjustable lengths and are made out of steel, rope, or nylon webbing. Some have a shock absorber (*Figure 25*), which absorbs up to 80 percent of the arresting force when a fall is being stopped. When using a lanyard, always follow the manufacturer's recommendations.

> **WARNING!**
>
> When activated during the fall arresting process, a shock-absorbing lanyard stretches as it acts to reduce the fall-arresting force. This potential increase in length must always be taken into consideration when determining the total free-fall distance from an anchor point.

4.7.3 Deceleration Devices

Deceleration devices limit the arresting force that a worker is subjected to when the fall is stopped suddenly. Rope grabs with shock-absorbing lanyards and self-retracting lifelines are two common deceleration devices. A rope grab (*Figure 26*) connects to a shock-absorbing lanyard and attaches to a lifeline. In the event of a fall, the rope grab is pulled down by the attached lanyard, causing it to grip the lifeline and lock in place. Some rope grabs have a mechanism that allows the worker to unlock the device and slowly descend the lifeline to the ground or surface below.

preference. Harness manufacturers normally provide selection guidelines in their product literature. Other types of full body harnesses can be equipped with front chest D-rings, side D-rings, or shoulder D-rings. Harnesses with front chest D-rings are typically used in ladder climbing and personal positioning systems. Those with side D-rings are also used in personal positioning systems. Personal positioning systems are systems that allow workers to hold themselves in place, keeping their hands free to accomplish a task. Per OSHA regulations, a personal positioning system should not allow a worker to free-fall more than 2', and the anchorage to which it is attached should be able to support at least twice the impact load of a worker's fall or 3,000 pounds, whichever is greater. Harnesses equipped with shoulder D-rings are typically used with a spreader bar or rope yoke for entry into and retrieval from confined spaces.

Note that in the past, body belts were frequently used instead of a full body harness as part of a fall arrest system. As of January 1, 1998, OSHA banned them from such use. This is because body belts

SHOCK ABSORBER

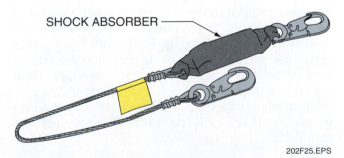

202F25.EPS

Figure 25 ◆ Typical shock-absorbing lanyard.

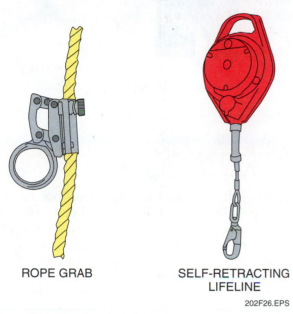

ROPE GRAB   SELF-RETRACTING LIFELINE

202F26.EPS

Figure 26 ◆ Deceleration devices.

Self-retracting lifelines (*Figure 26*) allow unrestricted movement and fall protection while climbing and descending ladders or when working on multiple levels. Typically, they have a 25' to 100' galvanized steel cable that automatically takes up the slack in the attached lanyard, keeping the lanyard out of the worker's way. In the event of a fall, a centrifugal braking mechanism engages to limit the worker's fall. Per OSHA requirements, self-retracting lifelines and lanyards that limit the free-fall distance to 2' or less must be able to support a minimum tensile load of 3,000 pounds. Those that do not limit the free-fall distance to 2' or less must be able to hold a tensile load of at least 5,000 pounds.

4.7.4 Lifelines

Lifelines are ropes or flexible steel cables that are attached to an anchorage. They provide a means for tying off personal fall protection equipment.

Vertical lifelines are suspended vertically from a fixed anchorage at the upper end to which a fall arrest device such as a rope grab is attached. Vertical lifelines must have a minimum breaking strength of 5,000 pounds. Each worker must use his or her own line. This is because if one worker falls, the movement of the lifeline during the fall arrest may also cause the other workers to fall. Vertical lifelines must be terminated in a way that will keep the worker from moving past its end, or they must extend to the ground or the next lower working level.

Horizontal lifelines are connected horizontally between two fixed anchorage points to which a fall arrest device is attached. Horizontal lifelines must be designed, installed, and used under the supervision of a qualified and competent person. The required strength of a horizontal line and its anchors increases substantially for each worker attached to it.

4.7.5 Anchoring Devices and Equipment Connectors

Anchoring devices, commonly called tie-off points, support the entire weight of the fall arrest system. The anchorage must be capable of supporting 5,000 pounds for each worker attached. Eye bolts, overhead beams, and integral parts of building structures are all types of anchorage points.

The D-rings, buckles, and snaphooks that fasten and/or connect the parts of a personal fall arrest system are called connectors. OSHA regulations specify how they are to be made, and require D-rings and snaphooks to have a minimum tensile strength of 5,000 pounds. All such components should be designed for use with the attached hardware. As of January 1, 1998, only locking-type snaphooks are permitted for use in personal fall arrest systems.

4.8.0 Procedures for Safely Using Personal Fall Arrest Equipment

Before using fall protection equipment on the job, your employer should provide you with training in the basics of fall protection and the proper use of the equipment. In addition, a job-specific fall protection plan must be available for the project. All equipment supplied by your employer must meet OSHA standards for strength. Before each use, always read the instructions and warnings on any fall protection equipment. Inspect the equipment using the following guidelines:

- Examine harnesses and lanyards for mildew, wear, damage, and deterioration.
- Make sure no straps are cut, broken, torn, or scraped.
- Check for damage due to fire, chemicals, or corrosives.
- Check that hardware is free of cracks, sharp edges, or burrs.
- Check that snaphooks close and lock tightly and that buckles work properly.
- Check ropes for wear, broken fibers, pulled stitches, and discoloration.
- Make sure lifeline anchors and mountings are not loose or damaged.

4.8.1 Wearing a Full Body Harness

The general procedure for using a sliding back D-ring, full body harness is as follows:

Step 1 Hold the harness by the back D-ring, then shake the harness, allowing all the straps to fall into place.

Step 2 Unbuckle and release the waist and/or leg straps.

Step 3 Slip the straps over your shoulders so that the D-ring is located in the middle of your back (*Figure 27*).

Step 4 Fasten the waist strap. It should be tight, but not binding.

Step 5 Pull the straps between each leg and buckle the straps.

Step 6 After all the straps have been buckled, tighten all friction buckles so that the harness fits snugly but allows a full range of movement.

Step 7 Pull the chest strap around the shoulder straps and fasten it in the mid-chest area. Tighten it enough to pull the shoulder straps taut.

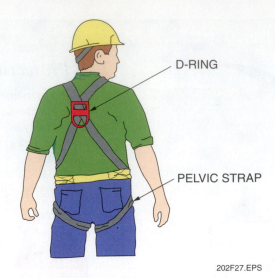

202F27.EPS

Figure 27 ◆ Position D-ring properly.

4.8.2 Selecting an Anchor Point and Tying Off

Once the full body harness has been put on, the next step is to connect it either directly or indirectly to a secure anchorage point by the use of a lanyard or lifeline. This is called tying off. Tying off is always done before you get into a position from which you can fall. Follow the manufacturer's instructions on the best tie-off methods for your equipment. When tying off, ensure that your anchorage point has the following characteristics:

- Directly above you
- Easily accessible
- Damage-free and capable of supporting 5,000 pounds per worker
- Never on the same point as a workbasket tie-off

Be sure to check the manufacturer's equipment labels and allow for any equipment stretch and deceleration distance.

When tying off, consider the following:

- Tie-offs that use knots are weaker than other methods of attachment. Knots can reduce the lifeline or lanyard strength by 50 percent or more. A stronger lifeline or lanyard should be used to compensate for this effect.
- To protect equipment from cuts, do not tie off around rough or sharp surfaces. Tying off around H-beams or I-beams can weaken the line because of the cutting action of the beam's edge. This can be prevented by using a webbing-type lanyard or wire-core lifeline.
- Never tie off in a way that would allow you to fall more than 6'.

A shorter fall can reduce your chances of falling into obstacles, being injured by the arresting force, and damaging your equipment. To limit your fall, a shorter lanyard can be used between the lifeline and your harness. Also, the amount of slack in your lanyard can be reduced by raising your tie-off point on the lifeline. The tie-off point to the lifeline or anchor must always be higher than the connection to your harness.

4.8.3 *Rescue after a Fall*

Every elevated job site should have a rescue and retrieval plan in case it is necessary to rescue a fallen worker. Planning is especially important in remote areas that are not readily accessible to a telephone. Before there is a risk of a fall, make sure that you know what your employer's rescue plan calls for you to do. Find out what rescue equipment is available and where it is located. Learn how to use the equipment for self-rescue and the rescue of others.

If a fall occurs, any employee hanging from the fall arrest system must be rescued safely and quickly. Your employer should have previously determined the method of rescue for fall victims, which may include equipment that lets the victim rescue himself or herself, a system of rescue by co-workers, or a way to alert a trained rescue squad. If a fall rescue depends on calling for outside help, such as the fire department or rescue squad, all the needed phone numbers must be posted in plain view at the work site. In the event a co-worker falls, follow your employer's rescue plan. Call any special rescue service needed. Communicate with the victim and monitor him or her constantly during the rescue.

4.9.0 Testing Fall Protection Systems and Equipment

The testing of fall arrest equipment should be performed regularly to make sure it complies with OSHA requirements. Depending on the prevailing state and local laws, the tests may be either voluntary or mandatory. Guidelines for testing personal fall arrest equipment and systems are given in *OSHA Safety and Health Standards for the Construction Industry, Part 1926, Appendices C and D to Subpart M*. A good practice is to tag or label all items of fall protection equipment with the date when the equipment was last tested and the date it is due for the next test.

Safety nets should be drop-tested at the job site after the initial installation, whenever relocated, after a repair, and at least every six months if left in one place. The drop test consists of dropping a 400-pound bag of sand into the net from at least 42" above the highest walking/working surface at which workers are exposed to fall hazards.

5.0.0 ◆ PREPARATION FOR ROOFING APPLICATION

Regardless of the type of roofing to be installed, the amount of material required must first be estimated. After the amount of material has been determined and obtained, the roof deck must be prepared before the finish roofing material is raised to the roof deck and installed.

5.1.0 Estimating Roofing Materials

If the building plans for the structure are available, the roof dimensions can be determined directly from the plans. If the plans are not available, the length and width of each section of the roof can be measured. Then, the area of each section of the roof is calculated and a percentage is added for waste. The result is converted into the number of squares (100 sq ft) of material required.

Step 1 Measure the length and width of each triangular and rectangular roof section of the structure, including any **overhangs** (*Figure 28*).

Step 2 Calculate the area for one half of each roof section and add the areas together. Then, multiply by two to obtain the total roof area and subtract any triangular areas covered by roof intersections.

To calculate the total area of the roof shown in *Figure 28*, proceed as follows:

30' × 35'	=	1,050.0 sq ft (Area A)
60' × 35'	= +	2,100.0 sq ft (Area B)
15' × 35' × ½	= +	262.5 sq ft (Area C)
		3,412.5 sq ft (½ roof area)
3,412.5 × 2	=	6,825.0 sq ft
35' × 20' × ½	= −	350.0 sq ft (Area D*)
	=	6,475.0 sq ft (roof area)

*Roof projection area

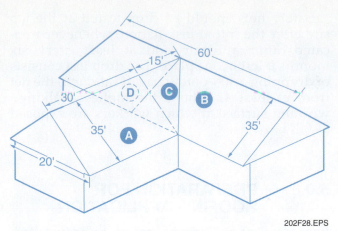

Figure 28 ♦ Roof example (including overhangs).

202F28.EPS

Step 3 Add an average of 10 percent for hips, valleys, and waste. For a complicated roof with a number of valleys or hips, more than 10 percent may be required. For a plain, straight gable roof, less is usually required. In addition, for wood or slate roofs, an additional 100 square feet is usually required for each 100 lineal feet of hips and valleys. Assuming that our example is a simple roof with standard three-tab composition shingles, we will use only the extra 10 percent figure.

$$\text{Total material} = (10\% \times 6{,}475) + 6{,}475$$
$$= 648 + 6{,}475$$
$$= 7{,}123 \text{ sq ft}$$

Step 4 Convert the total square feet of material required to squares by dividing by 100 sq ft (if using wood roofing, round up or down to the nearest bundle or square). For standard three-tab composition shingles, the rounding would be to the nearest ⅓ or ⅔ of a square (one or two bundles) or whole square.

$$\text{Squares} = 7{,}123 \div 100 = 71.23 = 71\tfrac{1}{3}$$

Step 5 The number of rolls of underlayment required is determined from the same total material requirement of 7,123 square feet. Starter strips, eave flashing, valley flashing, and ridge shingles must be added to complete the estimate. All of these are determined with linear measurements.

5.2.0 Roof Deck Preparation

A typical roof installation is shown in *Figure 29*. Before the finish roofing is applied, the roof deck must be flashed with a drip edge along the eaves and any valleys must be flashed. Then, an underlayment and/or a waterproofing membrane is usually installed and capped at the rake edges with metal drip edge flashing. On bare wood roof decks, the underlayment/membrane must be applied on dry wood as soon as possible. If the wood is moist due to rain or morning dew, allow it to dry before applying the underlayment/membrane. If the roof deck is damp, the membrane may not adhere to the roof deck, or the underlayment will buckle and cause the final roof to appear wavy. The underlayment/membrane prevents the finish roof materials from having direct contact with any damaging resinous or corrosive areas of the roof deck and helps resist or eliminate any water penetration into the roof deck. *Figure 30* and *31* show the recommended underlayment/waterproof membrane placement and drip edge installation.

Normally, the drip edge is installed along the length of the eaves first, followed by any valley flashing. The drip edge should be held against the fascia and nailed to the roof deck every 8" to 10". When installing valley flashing, it should overhang the valley at the upper and lower ends. The flashing is nailed every 6" to 8" on both sides, ½" from the edges.

After the flashing is secured, both ends are carefully trimmed flush with the roof deck. After the eave drip edge is in place, the exposed nail heads are covered with asphalt. Starting at the bottom of the roof, the underlayment and/or a waterproof membrane is rolled out and flattened with a roof roller before being tacked to the roof. In valleys, the waterproof membrane should extend over the flashing nails. The membrane will adhere and seal to the valley flashing; however, the underlayment should be trimmed to cover the flashing nails and should be cemented to the valley. In some cases on lower sloped roofs, the underlayment is half-lapped and cemented with asphalt to provide more of a water barrier. After the underlayment/waterproof membrane is in place, the rake edges of the roof are capped with a drip edge that is nailed every 8" to 10" to the roof deck. The bottom end of the rake drip edge overlaps the eave drip edge, and the fascia flange is cut to interlock behind the fascia flange of the eave drip edge. The nail heads should be covered with asphalt cement.

After the roof preparation is complete, the finish roof materials can be lifted and distributed equally over the roof deck.

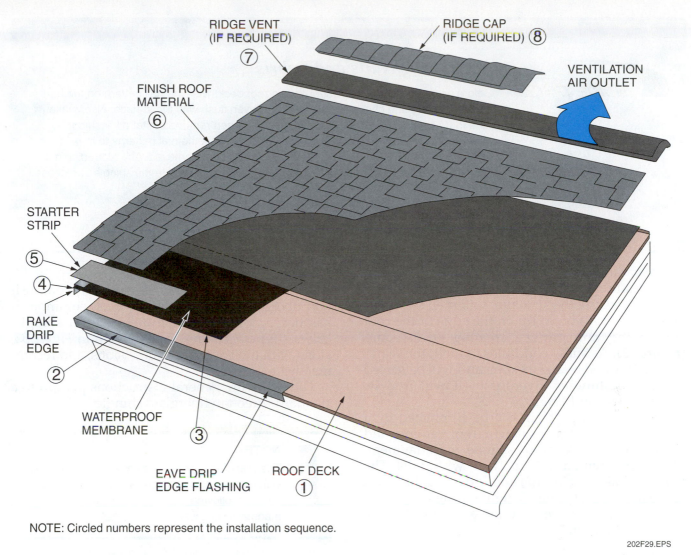

NOTE: Circled numbers represent the installation sequence.

202F29.EPS

Figure 29 ◆ Typical roof installation.

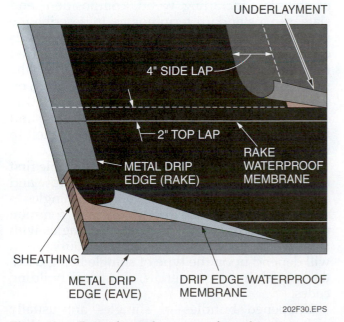

202F30.EPS

Figure 30 ◆ Drip edge and waterproof membrane placement.

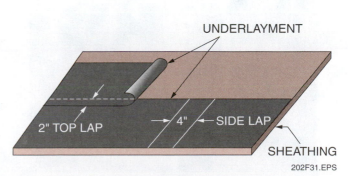

202F31.EPS

Figure 31 ◆ Underlayment or waterproof membrane placement over roof deck.

Protruding Nails and Debris

INSIDE TRACK

Before any roofing material is applied to a roof deck, walk the nail pattern on the bare deck sheathing and check that all nails are driven flush with the surface. Also, make sure that all debris including small pebbles has been removed. After all flashing, underlayment, and drip edges are installed, again walk the nail patterns to make sure all nails are driven flush. Check that all debris has been removed. Any protruding nails or debris under a composition shingle may eventually penetrate the shingle and cause leaks.

5.3.0 Protection Against Ice Dams

In areas subject to heavy snow, the snow will accumulate on the roof. Heat rising through the roof from inside the structure will melt the snow and cause ice to build up on the edge of the roof and in the rain gutters, creating an ice dam (*Figure 32*). As snow continues to melt, the water will be trapped by the ice dam and will be forced under the shingles. Eventually, it will find its way into the building.

This ice dam problem can be eliminated by a combination of attic insulation, roof venting, and the use of a waterproof shingle underlayment. This underlayment comes in 36" wide rolls. The material has a sticky side and is designed to stick to the roof deck, forming a tight seal against water penetration. It will seal around any nails that are driven through it.

6.0.0 ◆ COMPOSITION SHINGLE INSTALLATION

At one time, the three-tab, square-butt composition fiberglass shingle was the most common. It has been replaced by the architectural shingle. Various types of shingles are shown in *Figure 33*, along with their weights, dimensions, and recommended exposures.

The following general instructions pertain to a standard three-tab fiberglass shingle.

> **NOTE**
>
> The manufacturer provides a set of instructions with each bundle of shingles. These instructions must be followed. Failure to do so may void the manufacturer's warranty.

The instructions for the installation of all types of shingles, including wood, composition, and slate, use a standard terminology to describe the placing of the shingles. This terminology is explained in *Figure 34*. Roof shingles can be placed from the left side of the roof to the right or the right side of the roof to the left, depending on the preference of the roofer. On wide roofs, the courses are sometimes started in the middle and laid toward both ends. In this module, the left to right convention is used.

All strip shingles are started with a double first row, which may be made up of a starter row and a row of shingles or a double course of shingles in which two joints have been offset. A common practice is to place a starter row of shingles with the tabs cut off. The type of starter course used will depend upon the type of shingle being used, the availability of materials, and local building codes.

Unopened bundles of shingles are usually placed at various points on the roof for the roofing applicator.

202F32.EPS

Figure 32 ◆ Ice dam.

	PRODUCT*	CONFIGURATION	PER SQUARE			SIZE		
			APPROXIMATE SHIPPING WEIGHT	SHINGLES	BUNDLES	WIDTH, INCHES	LENGTH, INCHES (NOMINAL)	EXPOSURE, INCHES
ARCHITECTURAL	WOOD APPEARANCE STRIP SHINGLE; MORE THAN ONE THICKNESS PER STRIP 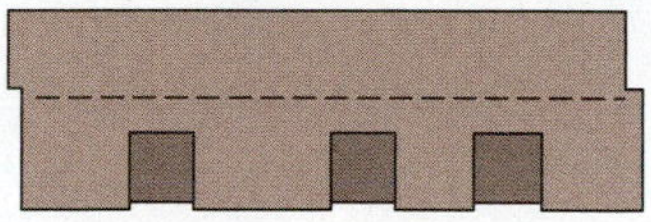 PRELAMINATED OR JOB-APPLIED LAYERS	VARIOUS EDGE SURFACE TEXTURE AND APPLICATION TREATMENTS	285# TO 390#	67 TO 90	4 OR 5	11½ TO 15	36 OR 40	4 TO 6
ARCHITECTURAL	WOOD APPEARANCE STRIP SHINGLE; SINGLE THICKNESS PER STRIP 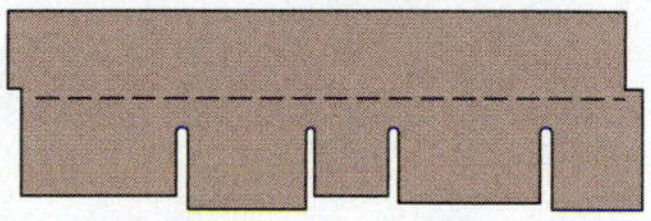	VARIOUS EDGE SURFACE TEXTURE AND APPLICATION TREATMENTS	VARIOUS, 250# TO 350#	78 TO 90	3 OR 4	12 OR 12¼	36 OR 40	4 TO 5⅛
STANDARD	3-TAB SELF-SEALING STRIP SHINGLE 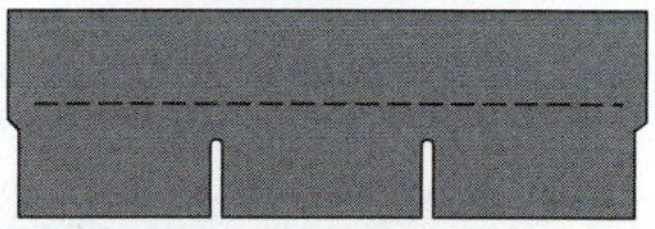	CONVENTIONAL 3-TAB	205# TO 240#	78 OR 80	3	12 OR 12¼	36	5 OR 5⅛
STANDARD	2-TAB (OR 4-TAB) VERSION 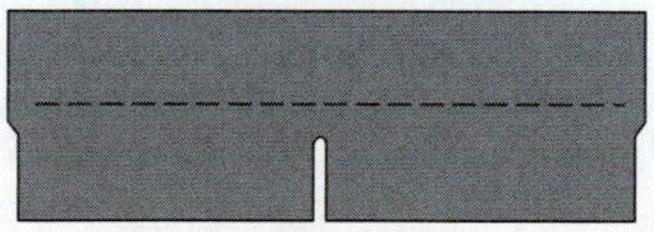	2- OR 4-TAB	VARIOUS, 215# TO 325#	78 OR 80	3 OR 4	12 OR 12¼	36	5 OR 5⅛
STANDARD	SELF-SEALING STRIP SHINGLE 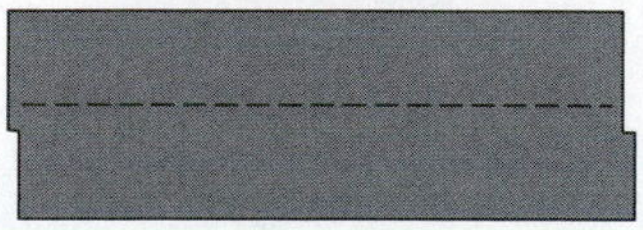 NO CUTOUT	VARIOUS EDGE AND TEXTURE TREATMENTS	VARIOUS, 215# TO 290#	78 TO 81	3 OR 4	12 OR 12¼	36 OR 40	5

*Other types available from some manufacturers in certain areas of the country. Consult your regional asphalt roofing manufacturers' association.

202F33.EPS

Figure 33 ◆ Typical composition shingle characteristics.

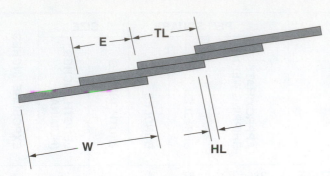

W = Width: The total width of strip shingles or the length of an individual shingle.

E = Exposure: The distance between the exposed edges of overlapping shingles.

TL = Top Lap: The distance that a shingle overlaps the shingle in the course below.

HL = Head Lap: The distance from the lower edge of an overlapping shingle to the upper edge of the shingle in the second course below.

202F34.EPS

Figure 34 ◆ Roofing terminology used in instructions.

CAUTION

Failure to scatter unopened shingle bundles can result in a broken rafter or collapsed roof, due to the weight of the bundles.

NOTE

If you are going to stand on a scaffold, it should be erected in such a manner that you will be working approximately waist high with the eave line. This places you in a safe, comfortable position to install the critical double course of shingles along the eaves.

Under normal circumstances, the shingle can be fastened with four aluminum or galvanized roofing nails positioned at a nailing line from the bottom of the shingle, one at each end and the other two above and adjacent to each cutout. See *Figure 35*. Depending upon the area of the country in which you live, this cutout may be referred to as a notch or gusset.

In areas with very high winds, the number of nails above the cutout can be increased to two, positioned about 3" apart and forming a triangle with the top of the cutout as a low point. In high wind areas, staples are not normally used.

When laying a shingle, butt the shingle to the previous shingle in the course and align the shingle. Then fasten the butted end to the roof. Keep the shingle aligned and fasten across the shingle to the other end. Alignment can be done by using the guide built into a roofing hammer or by chalking lines.

6.1.0 Gable Roofs

This section describes the installation of long and short runs of standard shingles on gable roofs. It does not include instructions for metric or architectural shingles.

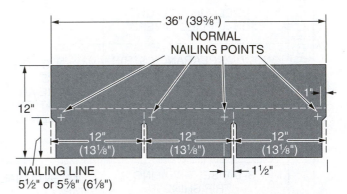

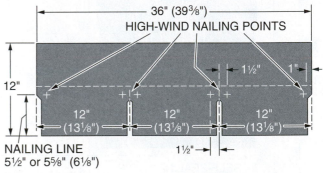

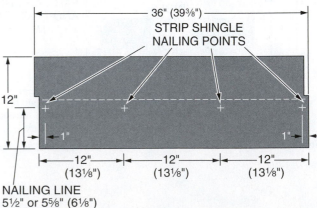

NOTE: Numbers in parentheses represent metric size shingle dimensions in English units.

202F35.EPS

Figure 35 ◆ Nailing points.

6.1.1 Gable Roofs—Long Runs

On large roofs, start applying the shingles at the center of the long run. The following procedure explains how to lay out and mark the roof. By beginning in the center, there is less chance of misalignment as you proceed in both directions. Shingle manufacturers suggest that you run shingles horizontally first rather than stacking them one row above another. The color of the shingles will blend better that way.

To install a long run on a gable roof, proceed as follows:

Step 1 Measure along the length of the roof and find the halfway point. Do this along the ridge and along the eaves. Mark both of these places. Snap a chalkline vertically up the roof at these two marks.

Step 2 Measure 6" to the right and left of this center line at the ridge and at the eaves. Snap a line vertically for these marks as well. This will provide three lines to start the rows on.

Step 3 Starting at the eaves at both ends of the roof, measure 6" up from the eave drip edge and place a mark. Then go up to 11" and make a mark there. Proceed up the tape and mark every 5" interval. Once both sides are done, snap lines horizontally across the roof to align the shingles.

Step 4 For the starter shingles, cut the tabs off a tabbed shingle. Place the remaining part of the shingle so that the tar strip is nearest the edge of the roof. With standard shingles, you should have a starter row that is 7" wide by 36" long. If the starter strip is not done this way, then the first full row of shingles will not be sealed down. The starter strip can be started on one of the vertical lines in the center of the roof. Make sure that the starter shingle stays on the first line snapped horizontally across the roof. If drip edges have been installed, position the starter strip with a ½" overhang on the eave and rake drip edges; oth-

erwise, position the strip with a ¾" overhang. Place and fasten the starter strip both ways from center.

Step 5 Take a full shingle and place it directly over the starter strip and 6" to the right or left of the vertical line that was used for the starter strip. This will ensure that the cutouts will be spaced 6" away from the joints of the starter strip. This is considered the first full row of shingles.

Step 6 The second row is in the center, 6" to the right or left of the starting place where the first row was started. Proceed right and left from that point.

NOTE

Starting 6" over helps to minimize waste. Most of the time the scrap you have left over on the right end will work on the left and vice versa.

6.1.2 Gable Roofs—Short Runs

To install a short run on a gable roof, proceed as follows:

Step 1 Cut the tabs off of the number of strip shingles it will take to go across the roof. Mark a spot 6" up on both eaves. Snap a line across the roof at this location. This line will keep the starter strip running straight across the roof. With the first left starter strip shortened by 6", place the starter strip shingles so that the tar strip is nearest the eave. If drip edges have been installed, position the starter strip with a ½" overhang on the eave and rake drip edge; otherwise, position the strip with a ¾" overhang. Nail the starter close to the top in four locations. After the starter strip is laid as far to the right as possible, return to the rake on your left and start to double up this first course by placing a full shingle directly over the top of the first upside down shingle. See *Figure 36*. Continue this

Starter Strips

Pre-cut starter strips are available from many manufacturers. In many areas of the country, they are available in two sizes: 5" wide for roofing over existing shingles and 7" wide for new and tear-off installations.

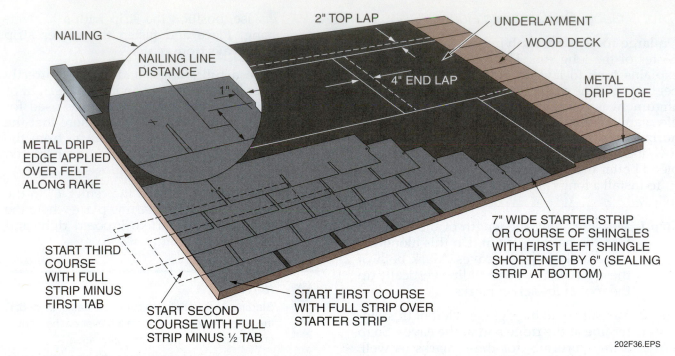

Figure 36 ◆ Shingle layout – 6" pattern.

202F36.EPS

process with full shingles as you move to the right and end before reaching the last starter strip shingle. You will observe that all cutouts and joints are covered with the full 12" tab. Make sure to nail the shingles correctly, as shown in *Figure 37*.

Step 2 Start the second course with a full strip minus 6". This layout is called a 6", half-tab pattern, or a 6-up/6-off layout, which means that half a tab is deliberately cut off and produces vertical aligned cutouts (refer to *Figure 36*). Overhang the cut edge at the rake by the same ¼" margin as the first course. Nail this shingle in place and proceed to the right with full shingles. The gusset openings can be used as a checking procedure to obtain the proper 5" exposure. End the second course before reaching the end of the first course. This is called stair stepping and allows the maximum number of shingles to be placed before moving ladders or scaffolding.

Step 3 Start the third course with a full strip minus a full tab (12"). Follow the same procedure with the same rake overhang. Again, nail the shingles in place moving to the right and gauging your exposure by the 5" gusset.

Step 4 Returning now to the fourth course, start this row with half a strip (for example, with 18" removed). This will show a 6" tab. Nail this in place with the same overhang margin on the rake. Proceed with full shingles, nailing to the right and ending short of the previous course while constantly checking the alignment.

Step 5 Start the fifth course. This row starts with a full tab only (12"). Use the same overhang margin on the rake and full shingles as you move to the right. End before reaching the last shingle on the previous course.

Step 6 The sixth course starts with a 6" tab, the same overhang margin at the rake, and then full shingles. It continues to the right as the nailing proceeds. Again, end before reaching the last shingle on the previous course.

Step 7 As the process is repeated, the seventh course starts with a full shingle. Each successive course of shingles is shortened by an additional 6". This continues as previously described until the twelfth course.

Depending upon the individual or working team, two or more courses may be carried or nailed at a time as the shingling proceeds across

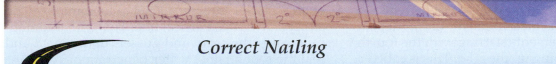

CORRECT
(STRAIGHT
AND FLUSH)

INCORRECT
(CROOKED)

INCORRECT
(INADEQUATE
PENETRATION)

INCORRECT
(TOO DEEP;
CUTS SHINGLE)

202F37.EPS

Figure 37 ◆ Correct and incorrect nailing.

Correct Nailing

Improper setting of nails and crooked nails can prevent the shingles from tabbing correctly and allow the wind to lift or tear the tabs of the shingles. Practice your fastening procedures so you drive the nail straight. The head should be flush with the surface of the shingle. Since your goal is to make the roof watertight, no pinholes or breaks are acceptable. If an accident should happen, a dab of asphalt cement spread with a putty knife will remedy the problem.

Other Shingle Alignment Methods

Several other alignment methods can be used for exposure. One popular method begins by snapping a chalkline along the top edge of the shingle of the first course, or 12" above and parallel with the eave line. Snap several other chalklines parallel with this first line. If you make the lines 10" apart, they can be used to check every other course by aligning the top of the shingle.

High Wind Areas

In high wind areas, the fifth and sixth courses of the 6" pattern are usually eliminated because of the possibility of the small starter tabs being torn off the roof. The fifth course tab (12" long) should be saved for use as a ridge cap. The sixth course tab (6" long) would be discarded. The shingles would also be secured using double fastening at each cutout.

the roof. When two people are working together, they usually work out their own system for speedy, accurate installation.

The procedure described above uses the left-hand rake of the roof as a starting point. Keep in mind that the entire application can be reversed by starting from the right-hand rake (this applies to gable roof construction). On small roofs, strip shingles may be laid starting at either end with a successful result since the roof measurement is usually symmetrical.

To obtain different variations of roof patterns using tabbed shingles, only a change of starting measurement is required. *Figure 38* shows one possibility using a course with a full shingle (36") followed by a second course using a reduced size shingle (32"). The third course would be reduced again to a shorter measurement (28"). Repeat these three measurements starting with the fourth course and continuing up the rake. This is called a 4" pattern and produces a diagonal cutout pattern.

Ribbon courses (*Figure 39*) are a way to add interest to a standard 6" pattern. After six courses have been applied, cut a 4"-wide strip lengthwise off the upper section of a full course of shingles. Fasten the strips as the seventh course correctly aligned with the cutouts of the sixth course. Then reverse the 8"-wide leftover pieces of shingle and align them directly over the 4" strip. Fasten the 8" pieces to the roof deck at the top of the tabs. Cover both with a full-width seventh course of shingles. This creates a three-ply edge known as the ribbon. Repeat the pattern every seventh course.

Due to its simplicity, the first pattern mentioned (half tab = 6") is the most commonly used in the field. The full strip asphalt shingle eliminates all pattern problems and alignment concerns on the vertical plane because it contains no gussets or cutouts.

6.2.0 Hip Roofs

When you encounter a hip roof, the basic nailing procedures remain the same, but the shingle layout starting point has to be at the center of the roof, as described for long gable roofs.

To begin, the starter strip is applied as previously described for long gable roofs. Return to the vertical line and use a full shingle for the doubling of this starter course. Offset this shingle 6" on either side of this vertical line. This will automatically cover the seam and gussets underneath and seal the roof against water penetration. You have the option of continuing this dual shingle starting course in either direction until it terminates at the hip rafter. See *Figure 40*.

At this point, the shingles should be cut to match the angle of the hip rafter and covered with a hip cap (the same as a ridge cap), completing the installation of the shingles. The hip cap is centered on the hip rafter and usually consists of a 12" tab showing a 5" exposure.

Exposed nails in the last caps should be covered with roof sealant. The ridge cap should cover the hip cap to prevent leaks.

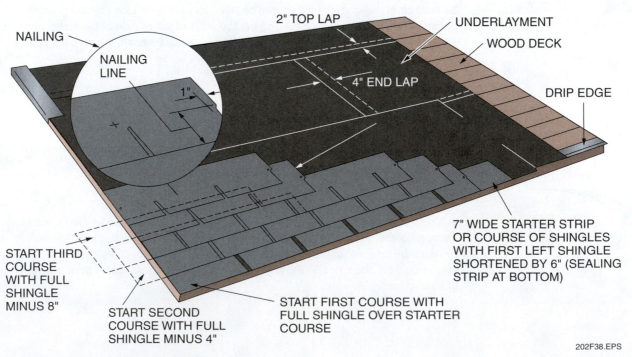

Figure 38 ◆ Shingle layout—4" pattern.

202F38.EPS

3-PLY RIBBON COURSES

202F39.EPS

Figure 39 ◆ Ribbon courses.

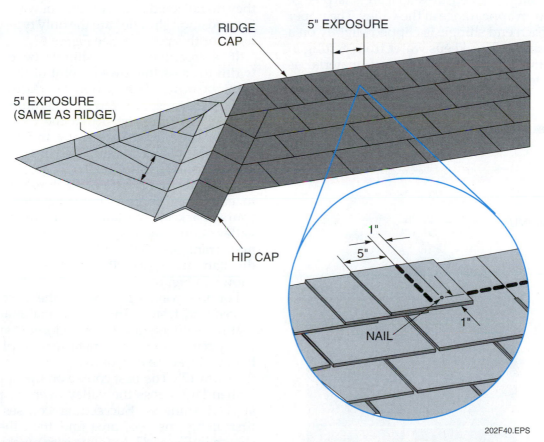

RIDGE CAP

5" EXPOSURE

5" EXPOSURE (SAME AS RIDGE)

HIP CAP

1"

5"

NAIL

1"

202F40.EPS

Figure 40 ◆ Hip and ridge layout.

6.3.0 Valleys

If the building you are constructing is not a perfect rectangle, you may encounter an L or T shape, which calls for another variation of shingling procedures. Where two sloping roofs meet, this intersecting valley has to be able to carry a high concentration of water drainage. Shingling becomes very critical, and the application must be done with extreme care.

6.3.1 Open Valley

With the valley flashing installed as previously described, snap two chalklines the full length of the valley. They should be 6" apart at the ridge or uppermost point. This means they should measure 3" apart when measured from the center of the valley. The marks diverge at the rate of ⅛" per foot as they approach the eaves. For example, a valley 8' in length will be 7" wide at the eaves; one

Unequally Pitched Roof Intersections

If the roof valley is formed by an intersection of two unequally pitched roofs, the woven valley will creep up one side, making it nearly impossible to maintain the correct overlap of shingles. The open valley or the closed-cut valley should be used with unequally pitched roof intersections.

16' long will be 8" wide at the eaves. The enlarged spacing provides adequate flow as the amount of water increases, passing down the valley. See *Figure 41*.

The chalkline you have snapped serves as a guide in trimming and cutting the last shingle to fit in the valley. This ensures a clean, sharp edge and a uniform appearance in the valley. The upper corner of each end shingle is clipped slightly on a 45-degree angle. This keeps water from getting in between the courses. The roofing material is cemented to the valley lining and to itself where an overlap occurs. Use plastic asphalt cement and spread a 6" to 8" bed. Do not overdo it, and clean up all excess cement so no tar shows.

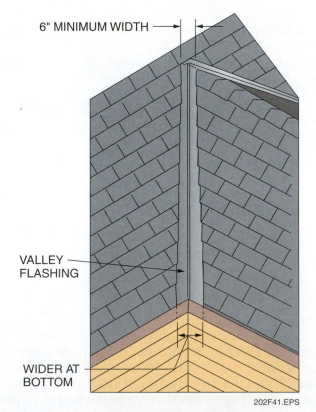

Figure 41 ◆ Open valley flashing (steep pitch).

202F41.EPS

6.3.2 *Closed-Woven Valley*

Some applicators of asphalt shingles prefer to use a closed-woven valley design, sometimes called a full-weave or laced valley. It is faster to install, and some feel it gives a tighter bond. Others believe closed valleys are inferior to open valleys because they do not shed high volumes of water very well. Composition shingles are the only type that can be used for this pattern. See *Figure 42*.

It is essential that a shingle be of sufficient width to cross the lowest point of the valley and continue upward on each roof surface a minimum of 12". Because of the skill required for this process, it is suggested that the two converging roofs be completed to a point 4' to 5' from the center of the valley. Then the weaving process can be accomplished carefully.

To create the 12" extension, it may be necessary to cut some of the preceding shingles in the course back to two tabs. To ensure a watertight valley, either a strip of 36"-wide, waterproof membrane, or 50-lb or heavier roll roofing over the standard 15-lb felt is placed in the valley, as shown in *Figure 42*.

For the weaving process, the first course is placed and fastened in the normal manner. Note that no fasteners are located closer than 6" to the valley center line. An extra fastener is placed at the high point at the end of the strip where it extends the extra 12". The first course on the opposite side is then laid across the valley over the previously applied shingles. Succeeding courses alternate, first along one roof area and then the other, as shown in *Figure 42*. Extreme care must be taken to maintain the proper exposure and alignment. As the shingles are woven over each other, they must be pressed tightly into the valley to provide a smooth surface where the roof surfaces join.

6.3.3 *Closed-Cut Valley*

In a closed-cut valley, sometimes called a half-weave or half-laced valley, the underlayment and valley flashing materials are the same as for the woven application.

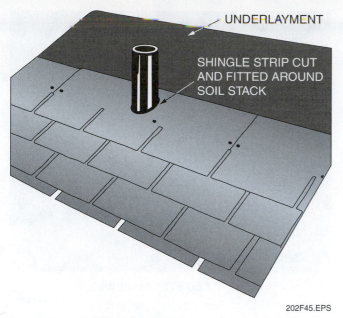

UNDERLAYMENT

SHINGLE STRIP CUT AND FITTED AROUND SOIL STACK

202F45.EPS

Figure 45 ◆ Layout around stack.

SHINGLE COURSES LAID OVER UPPER PORTION OF FLANGE

202F47.EPS

Figure 47 ◆ Covering flashing.

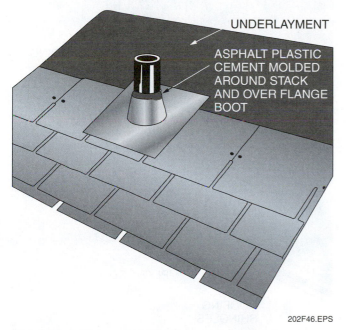

UNDERLAYMENT

ASPHALT PLASTIC CEMENT MOLDED AROUND STACK AND OVER FLANGE BOOT

202F46.EPS

Figure 46 ◆ Placement of flashing.

This turn-up bend has to be done very carefully to maintain the seal and overlap of material without creasing or tearing the felt or membrane. Regular shingling procedures are used as each course is brought close to the vertical wall. **Wall flashing** (step flashing) is used when the rake of the roof abuts the vertical wall.

Metal step flashing shingles are applied over the end of each course of shingles and covered by the next succeeding course. The flashing shingles are usually rectangular. They are approximately 6" to 7" long and from 5" to 6" wide. When used with shingles laid 5" to the weather, they are bent so half of the flashing piece is over the roof deck with the remaining half turned up on the wall. The 7" length enables one to completely seal under the 5" exposure with asphalt cement and provide a 2" overlap up the entire length of the rake.

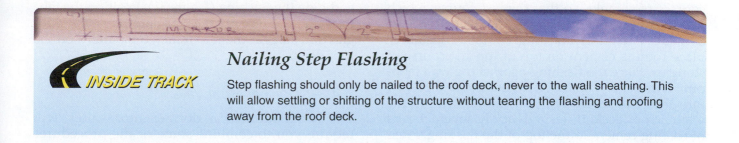

Nailing Step Flashing

INSIDE TRACK

Step flashing should only be nailed to the roof deck, never to the wall sheathing. This will allow settling or shifting of the structure without tearing the flashing and roofing away from the roof deck.

NAILING STEP FLASHING TO ROOF DECK

SINGLE NAIL AT UPPER OUTER EDGE

WATERPROOF MEMBRANE TURNED UP SIDE WALL

INSTALLED STEP FLASHING

202F48.EPS

Figure 48 ◆ Wall (step) flashing.

A careful study of *Figure 48* shows that each flashing shingle is placed just up the roof from the exposed edge of the shingle that overlaps it. It is secured to the deck sheathing with one nail in the top corner.

When the finished siding or clapboards are brought down over this flashing, they serve as a **cap flashing**, sometimes referred to as counter flashing. Usually, a 1" reveal margin is used and the ends of the boards are fully painted or stained to exclude dampness and prevent rot. With proper application of flashing and shingles, the joint between a sloping roof and a vertical wall should be watertight.

On a horizontal abutment (*Figure 49*), continuous flashing must be applied horizontally across the entire top of the abutting roof and against the vertical wall under the siding.

Continuous flashing can be formed with a metal brake or by hand, as shown in *Figure 50*. The flashing should be at least 9" wide and bent to match the angle of the joint to be flashed. Position the bend so that there will be at least 4" of flashing on the roof and 5" on the wall.

Before applying the flashing, adjust the last two courses of shingles so that the last course, which will be trimmed to butt against the wall, is at least 8" wide. After this abutting course is installed, place roofing cement on top of the last course of shingles. Place the flashing against the wall (slipping it under any siding, if necessary) and press it into the cement. Do not nail the flashing to the wall or roof deck. If desired, apply several beads of roofing cement to the top of the flashing. Press the tabs cut from shingles into the cement to cover

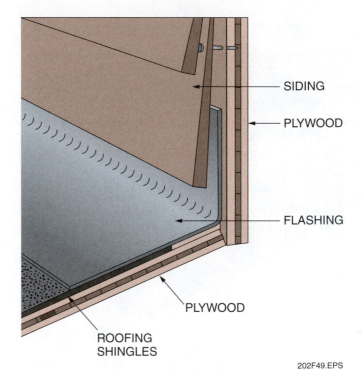

SIDING

PLYWOOD

FLASHING

PLYWOOD

ROOFING SHINGLES

202F49.EPS

Figure 49 ◆ Continuous flashing.

and hide the flashing (*Figure 51*). Position the tabs the same distance apart as the cutouts on the shingles and stagger them to match the pattern on the roof deck.

6.4.3 *Dormer Roof Valley*

The installation of a dormer roof valley will require you to combine some of the procedures previously covered. *Figure 52* shows an open

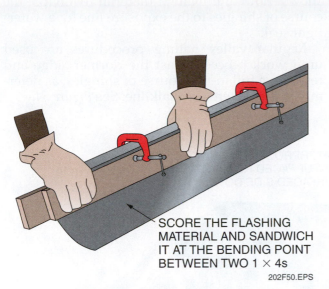

SCORE THE FLASHING
MATERIAL AND SANDWICH
IT AT THE BENDING POINT
BETWEEN TWO 1 × 4s

202F50.EPS

Figure 50 ◆ Bending continuous flashing.

FLASHING   SHINGLE TABS CUT AND
             CEMENTED TO FLASHING

202F51.EPS

Figure 51 ◆ Covering flashing.

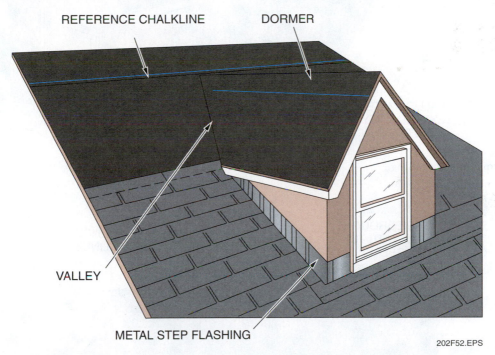

REFERENCE CHALKLINE   DORMER

VALLEY

METAL STEP FLASHING

202F52.EPS

Figure 52 ◆ Dormer flashing.

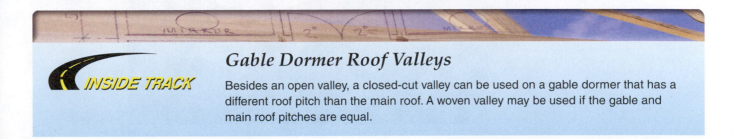

INSIDE TRACK

Gable Dormer Roof Valleys

Besides an open valley, a closed-cut valley can be used on a gable dormer that has a different roof pitch than the main roof. A woven valley may be used if the gable and main roof pitches are equal.

valley for a gable dormer roof. Note that the shingles have been laid on the main roof up to the lower end of the valley.

Extreme care must be used during the installation of the last course against the vertical wall to ensure a tight, dry fit. *Figure 53* displays the standard valley procedures for dormer flashing and shows how the valley material overlaps the course of shingles to the exposure line for a watertight seal.

Regular valley nailing procedures are used until work proceeds past the dormer ridge and resumes a full in-line course of shingles, as determined by a reference chalkline. See *Figure 54*.

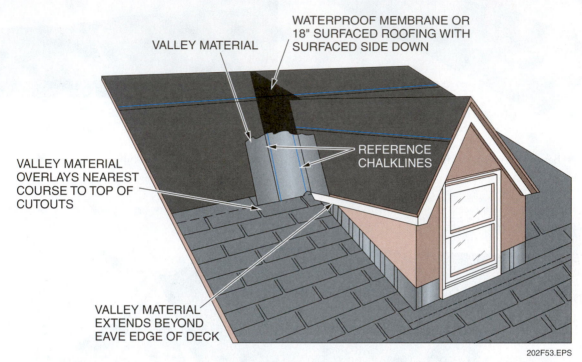

Figure 53 ◆ Dormer valley flashing.

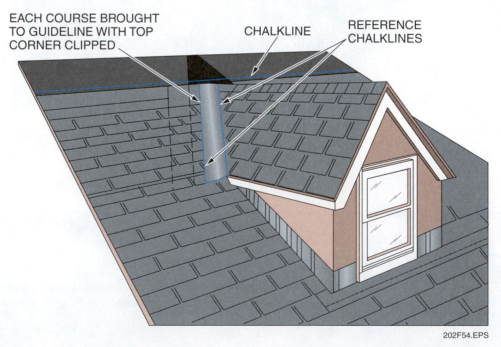

Figure 54 ◆ Dormer valley coverings.

6.4.4 Chimneys

Chimneys are subject to varying loads and certain opposing structural movements due to winds, temperature changes, and settling. Therefore, roofing materials and **base flashing** should not be attached or cemented to both the chimney and roof deck. The process of shingling around chimneys must be approached with extreme care. Due to the size of the opening in the roof and of the chimney itself, additional work must be done on the roof deck around chimneys prior to shingling. A cricket, also called a saddle, must be made. See *Figure 55*.

A cricket placed behind the chimney keeps rainwater or melting snow/ice from building up in back of the chimney. It steers flowing water around the chimney. The cricket is usually supported by a horizontal ridge piece and a vertical piece at the back of the chimney, as shown in *Figure 55*. The height of the cricket is typically half the width of the chimney, although these requirements will vary. Check local codes. The ridge,

which is level, extends back to the roof slope. On wide chimneys, it may be necessary to frame the cricket, as shown in *Figure 56*. Either type of cricket may be covered with two triangular pieces of ¾" exterior plywood cut to fit from the ridge to the edge of the chimney and the roof slope. Heavy-gauge metal can also be used to form the cricket. The covering is then nailed to the support and roof deck.

Flashing at the point where the chimney comes through the roof requires something that will allow movement without damage to the water seal. It is necessary to use base flashing. The counter or cap flashing is secured to the masonry. Metal is used for base flashing and cap flashing.

To apply the chimney flashing, proceed as follows:

Step 1 Apply shingles over the roofing felt up to the front face of the chimney (*Figure 57*). Cut the base flashing for the front cut according to the pattern shown in *Figure 58*.

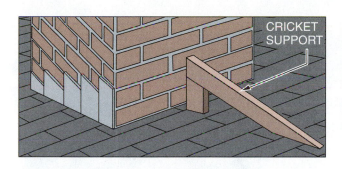

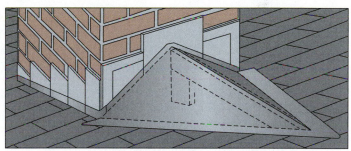

202F55.EPS

Figure 55 ◆ Simple chimney cricket.

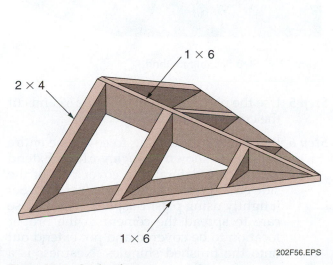

202F56.EPS

Figure 56 ◆ Cricket frame.

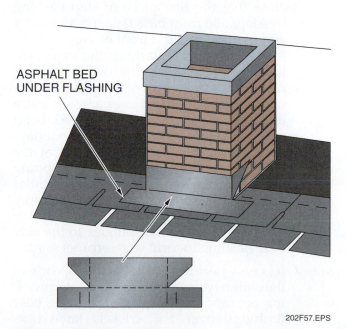

202F57.EPS

Figure 57 ◆ Front base flashing.

Combustible Material Spacing Requirements for a Cricket

Some building codes require that the wood framing and sheathing for a cricket must be spaced up to 1" from the chimney masonry. Always check your local codes for spacing requirements pertaining to combustible materials near chimneys.

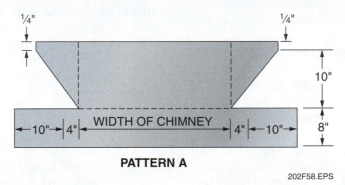

PATTERN A

202F58.EPS

Figure 58 ◆ Pattern for front base flashing.

Step 2 The base flashing for the front cut is applied first. The lower section is laid over the shingles in a bed of plastic asphalt cement. Bend the triangular ends of the upper section around the corners of the chimney.

Step 3 The sides of the chimney are base-flashed next. Either step flashing or continuous flashing can be used. Step flashing is applied as the shingles are applied up to the top side of the chimney (*Figure 59*). Note that the first piece of step flashing overlaps the front base flashing and is cut and bent around the front of the chimney. Like the front base flashing, the step flashing is fastened only to the roof deck with a nail or, if desired, roof cement. The continuous flashing method uses a single piece of metal, cut as shown in *Figure 60*. Bend and cement the flashing to the underlayment along the slope at the sides of the chimney, with the lower end overlapping the front base flashing. Bend the triangular end pieces around the chimney. Apply shingles up the roof to the top side of the chimney. Cement the shingles to the side base flashing to form a waterproof joint.

Step 4 It is now necessary to go to the top side of the chimney and complete the waterproofing operation by cutting and fitting base flashings over the cricket, known as cricket flashing, as shown in *Figure 61*.

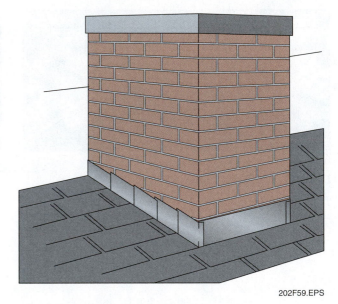

202F59.EPS

Figure 59 ◆ Step flashing method.

Step 5 Use the pattern shown in *Figure 62* and cut the cricket base flashing.

Step 6 Bend the base flashing to cover the entire cricket, as shown in *Figure 61(C)*. Extend the flashing laterally to cover part of the side base flashing previously installed. Set it tightly using plastic asphalt cement. Use care to spread the cement in the proper location to be covered and not extend out onto the finished shingles. Neatness is a must. Bend the ends around the chimney.

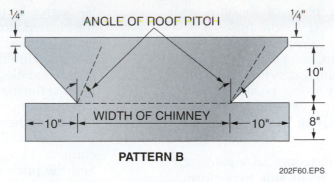

Figure 60 ◆ Pattern for continuous side flashing.

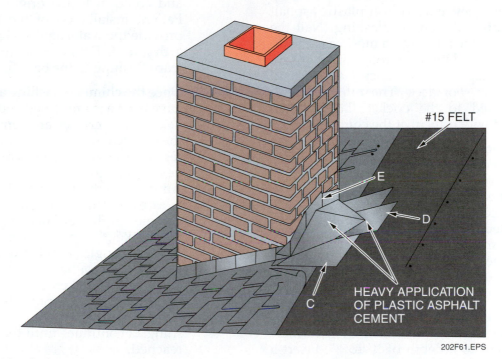

Figure 61 ◆ Cricket flashing.

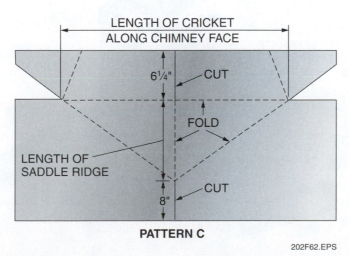

Figure 62 ◆ Cricket base flashing pattern.

Step 7 Cut a rectangular piece of flashing, as shown in *Figure 61(D)*, and make a V-cutout on one side to conform to the rear angle of the cricket. Center it over that part of the cricket flashing extending up to the deck. Set it tightly using plastic asphalt cement. This piece provides added protection where the ridge of the cricket meets the deck.

Step 8 Cut a second small rectangular piece of flashing. Cut a V on one side to conform to the pitch of the cricket, as shown in *Figure 61(E)*. Place it over the cricket ridge and against the flashing that extends up the chimney. Embed it in plastic asphalt cement to the cricket flashing. Nail the edges of the flashing. In most cases, similarly colored pieces of roll roofing are cut to overlap the entire cricket and extend onto the roof deck. The roll roofing is cemented to the cricket flashing and sealed with cement at the chimney edge.

Step 9 For completion, cap flashing (also called counter flashing) is installed. It is usually made of sheet copper 16 ounces or heavier. It can also be made of 24-gauge galvanized steel. If steel is used, it should be painted on both sides. *Figures 63* and *64* show metal cap flashing on the face of the chimney and on the sides. Cap flashing is secured to the brickwork, as shown in *Figure 65*.

Step 10 *Figure 65* shows a good method of securing the cap flashing. Cut a slot in the mortar joint to a depth of ¼" to ½". Insert a 90-degree bent edge of the flashing into the cleared slot between the bricks using an elastomeric sealant or mortar in the slot to secure the flashing to the masonry. When installed, the cap should lie snugly against the masonry. The front unit of the cap flashing should be one continuous piece. On the sides and the rear, the sections are similar in size. They are cut to conform to the locations of mortar joints and the pitch of the roof. If the sides are lapped, they must lap each other by at least 3". The slots are refilled with the brick mortar mix or elastomeric sealant and conform to the original brickwork. Patient installation of the flashing will provide the watertight seal necessary for a dry roof. Do not cement the counter (cap) flashing to the base flashing.

Step 11 Once the chimney flashing and shingling have been accomplished and the shingles next to the cricket are cemented under the edges to make a waterproof joint, the regular shingling process resumes on the next full course above the cricket. Another method of finishing the cricket is to extend the horizontal composition shingles of the roof deck up the pitch of the cricket. Then use step flashing and cement shingles parallel with the cricket ridge to form a half-weave valley at the edges of the cricket. This second method requires cementing cap shingles over the ridge of the cricket. The application of the shingles continues until the roof ridge is reached.

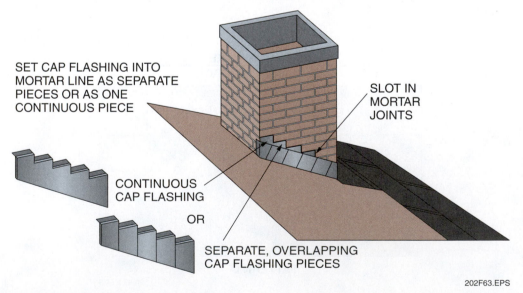

SET CAP FLASHING INTO MORTAR LINE AS SEPARATE PIECES OR AS ONE CONTINUOUS PIECE

SLOT IN MORTAR JOINTS

CONTINUOUS CAP FLASHING

OR

SEPARATE, OVERLAPPING CAP FLASHING PIECES

202F63.EPS

Figure 63 ◆ Cap flashing methods at sides.

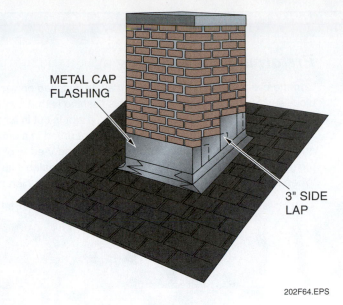

Figure 64 ◆ Flashing cap and lap.

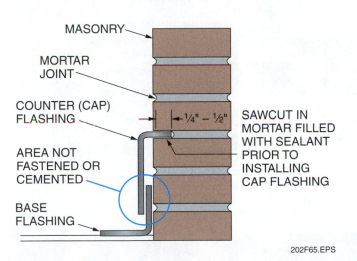

Figure 65 ◆ Counter (cap) flashing installation.

202F65.EPS

202F66.EPS

Figure 66 ◆ Architectural cap shingle.

CUT TABS FROM WHOLE SHINGLES AND
TAPER EACH TAB AS SHOWN

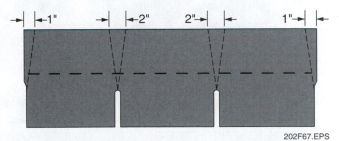

202F67.EPS

Figure 67 ◆ Cutting cap shingles.

6.4.5 Hip or Ridge Row (Cap Row)

Special shingles are required to complete the hip or ridge rows. In some cases, ridge caps and hip caps are premanufactured.

Architectural shingles have a matching cap row shingle (*Figure 66*) that must be used. Cap row shingles cannot be cut from architectural shingles. Most of the time if the roof was shingled with standard three-tab shingles, cap rows are cut from the shingles and the 12×12 tab is used (*Figure 67*). The tab can be reduced to 9×12, but nothing less.

Since the hip or ridge is a potential spot for water leaks, precautions must be taken. If ridge venting will be used, do not apply the ridge caps.

Preformed Cap Flashing

Commercial preformed cap flashing may be installed on vertical brick, concrete, or block surfaces including chimneys. This flashing is made of an aluminum-coated steel. As shown in the sequenced photographs, a slot is cut in all sides of a chimney using a ¼"-thick diamond impregnated steel wheel mounted in a small, high-speed electric grinder. The flashing is trimmed to shape and the V-edge of the flashing is pressed into the groove. The flashing is formed to shape and sealed with an elastomeric sealant. When set, the sealant and flashing may be painted to blend with the roof.

1. PREFORMED CAP FLASHING

2. CUTTING ¼" GROOVE

3. COMPLETED GROOVES

4. TRIMMING FRONT FLASHING IN GROOVE TO SIZE

5. SEATING V-EDGE OF FRONT FLASHING IN GROOVE

6. FITTING SIDE FLASHING

7. SEATING V-EDGE OF SIDE FLASHING IN GROOVE

8. SIDE FLASHING INSTALLED AND FORMED TO FRONT FLASHING

9. SEALANT APPLIED TO GROOVE AND FLASHING

10. FLASHING PAINTED TO MATCH ROOFING

202SA07.EPS

To install a ridge or hip row, proceed as follows:

Step 1 Butt and nail shingle roofing as they come up on either side of a hip or ridge. On a ridge, lay the last course and trim the shingles, as shown in *Figure 68*. On a hip, trim the shingles at an angle on the hip line.

Step 2 After the cap shingles are cut, bend them lengthwise in the center line. In cold weather, warm the shingles before bending to prevent cracks. Begin at the bottom of any hips. Cut the first tab to conform to the dual angle at the eaves.

Step 3 Lap the units to provide a 5" exposure of the granular surface. See *Figure 69*. Secure with one nail on each side, 6" back from the exposed end and 1" from the edge. As each succeeding tab is nailed going up the hip, the nail penetrates and secures two tabs. This tight bond prevents the wind from getting underneath and lifting the tab.

Nailing of the ridge row is similar to that described for the overall hip. Nailing takes place from both ends of the ridge. A final cap piece joins the ridge together in the center, and the exposed nails are covered with roof cement. An exception to this may occur in a very windy area. In that case, the ridge cap would be started at the point on the roof opposite the wind direction. As each ridge shingle is placed, it automatically allows the wind to pass over it, and there is no possibility of shingles blowing off. The junction of the roof ridge and any hip ridges can be capped with a special molded cap or by a fabricated end cap, as shown in *Figure 70*. Bed the final ridge cap or hip/ridge caps in asphalt and secure with nails as shown in the figure. Cover the nail heads with roof cement or sealant.

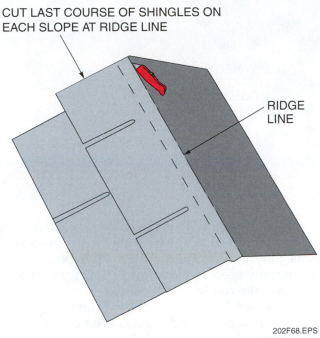

CUT LAST COURSE OF SHINGLES ON EACH SLOPE AT RIDGE LINE

RIDGE LINE

202F68.EPS

Figure 68 ◆ Applying the last course of ridge shingles.

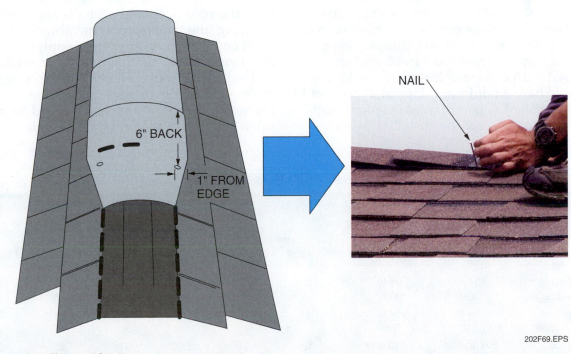

6" BACK

1" FROM EDGE

NAIL

202F69.EPS

Figure 69 ◆ Installing a ridge cap.

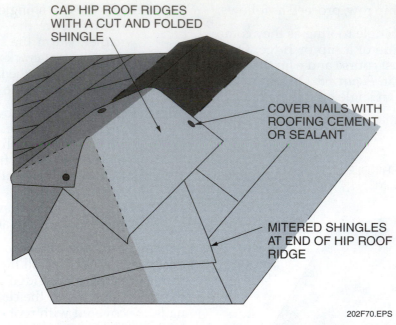

CAP HIP ROOF RIDGES WITH A CUT AND FOLDED SHINGLE

COVER NAILS WITH ROOFING CEMENT OR SEALANT

MITERED SHINGLES AT END OF HIP ROOF RIDGE

202F70.EPS

Figure 70 ◆ Hip and ridge end cap.

7.0.0 ◆ ROLL ROOFING INSTALLATION

Nearly flat roofs can be roofed with a hot-asphalt BUR, a single-ply membrane system, or roll roofing. Roll roofing can be installed on underlayment by itself, as part of a cold asphalt built-up roof, or on a waterproof membrane. The weights, characteristics, side lap, top lap, and recommended exposures for double-coverage, single-coverage, and uncoated roll roofing are shown in *Table 2*.

All flat roofs must have some pitch, either to an edge or roof drains, so that water does not collect. Composition shingles are not used on roofs with a slope of less than 2 in 12; wood shingles are not used with a slope of less than 3 in 12; and shakes are not used with a slope of less than 4 in 12. Flat roofs should have a minimum slope of ¼" per foot. When not installed as part of a cold-asphalt built-up roof, roll roofing can be installed as single-coverage roofing with exposed or concealed nails or as double-coverage roofing with concealed nails. Single-coverage roofing with exposed nails is generally used on slopes of 2 in 12 or more. Single-coverage roofing with concealed nails is used on slopes of 1 in 12 or more. Double-coverage roofing with concealed nails is used on slopes of more than ¼ in 12.

Apply a drip edge and waterproof membrane, or, as a minimum, apply a 15-lb underlayment to the roof deck. Make sure all debris is removed from the roof deck and all nails are flush before applying the waterproof membrane/underlayment and that it is clean before applying roll roofing. Even a very small pebble or protruding nail will eventually poke a hole through single-layer roofing. Flashing of roof projections is accomplished in the same way as described for composition shingles in the previous section.

Table 2 Typical Weights, Characteristics, and Recommended Exposures for Roll Roofing

| Product | Approximate Shipping Weight | | Squares per Package | Length | Width | Side or End Laps | Top Lap | Exposure |
	Per Roll	Per Square						
Mineral surface roll, double-coverage	75# to 90#	75# to 90#	1	36' 38'	36" 38"	6"	2" 4"	34" 32"
Mineral surface roll, single-coverage	55# to 70#	55# to 70#	½	36'	36"	6"	19"	17"
Uncoated roll	50# to 65#	50# to 65#	1	36'	36"	6"	2"	34"

Removing the Curl from Roll Roofing

Before installing roll roofing in cool to cold ambient temperatures, cut the roofing into 12' to 18' sections and stack it for a sufficient length of time to remove any curl. The length of time required will depend on the ambient temperature.

7.1.0 Single-Coverage Roll Roofing Installation

This section describes the installation of single-coverage roll roofing using the exposed and concealed nail methods.

7.1.1 *Exposed Nail Method*

Single-coverage roll roofing with exposed nails is generally applied horizontally over underlayment, as shown in *Figure 71*. To install roll roofing using the exposed nail method, proceed as follows:

Step 1 Protect each valley with 18"-wide metal flashing.

Step 2 For horizontal application (*Figure 72*), snap a chalkline 35½" above the eaves. Apply a 2" band of roof cement to the eaves and rake edges.

Step 3 Using the chalkline as a reference, run the first course so that it overhangs the eaves by ½" and the rake edges by 1". Cement and overlap any vertical seams by 6". Nail all seams and the bottom and rake edges of the first course every 3" with galvanized or aluminum roofing nails. Use a utility knife to trim the roofing edges to the drip edges.

Step 4 Snap another chalkline 3" down from the top edge of the first course and apply a 2" band of roof cement within the band and up the rake edges. Lay the second course to the chalkline and nail every 3" along all seams, the bottom edge, and the rake edges, as shown in *Figure 73*.

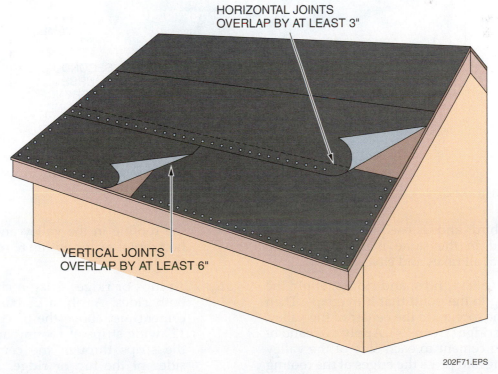

HORIZONTAL JOINTS
OVERLAP BY AT LEAST 3"

VERTICAL JOINTS
OVERLAP BY AT LEAST 6"

202F71.EPS

Figure 71 ◆ Typical roll roofing installation.

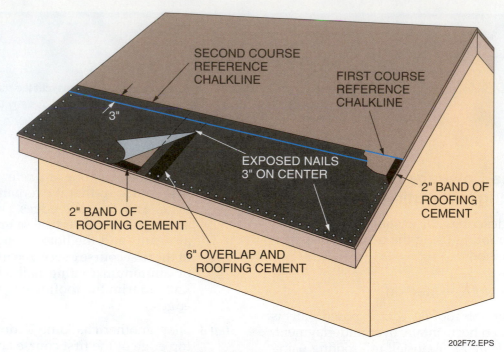

SECOND COURSE REFERENCE CHALKLINE

FIRST COURSE REFERENCE CHALKLINE

3"

EXPOSED NAILS 3" ON CENTER

2" BAND OF ROOFING CEMENT

2" BAND OF ROOFING CEMENT

6" OVERLAP AND ROOFING CEMENT

202F72.EPS

Figure 72 ◆ First course of exposed nail roll roofing.

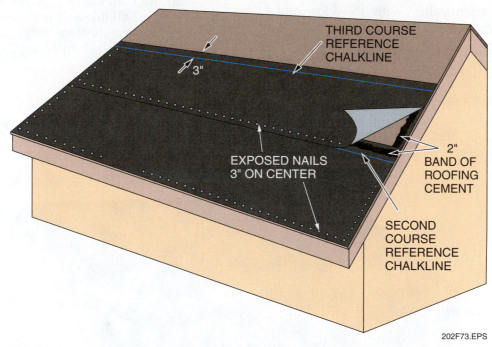

THIRD COURSE REFERENCE CHALKLINE

3"

EXPOSED NAILS 3" ON CENTER

2" BAND OF ROOFING CEMENT

SECOND COURSE REFERENCE CHALKLINE

202F73.EPS

Figure 73 ◆ Second and subsequent courses of exposed nail roll roofing.

Step 5 The third and subsequent courses are applied in the same manner. Trim the edges at all rakes and eaves.

Step 6 At all valleys, hips, and ridges, apply the roofing to the point that it overlaps. Then trim the roofing to the center of the valley, hip, or ridge (*Figure 74*). Apply a 6" band of roofing cement to each side of the valley flashing and press the edges of the roofing into the cement. Do not nail the edges of the roofing in the valley and do not nail horizontal seams within 6" of the center of the valley.

Step 7 At hips or ridges, snap a chalkline 6" on both sides. Apply a 2" band of roofing cement just above the 6" chalklines. Cut 12"-wide strips of roofing and nail down the strips through the cement on both sides of the hip or ridge. As necessary, overlap the strips by 6" and cement.

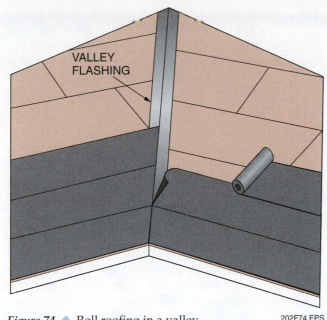

Figure 74 ◆ Roll roofing in a valley.

202F74.EPS

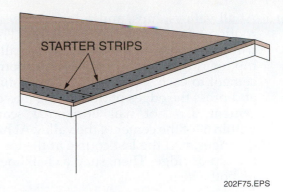

Figure 75 ◆ Roll roofing starter strips.

202F75.EPS

7.1.2 Concealed Nail Method

To install roll roofing using the concealed nail method, use the following procedure.

Step 1 Cut and install 9"-wide roofing material starter strips along the eaves and rakes (*Figure 75*). Nail the strips to the roof deck on both edges with galvanized or aluminum nails spaced ¾" from the edges and 4" apart. These strips provide a surface for adherence of roofing cement. Protect each valley with 18"-wide metal flashing.

Step 2 Snap a chalkline 35½" above the eaves. Using the chalkline as a reference, run the first course so that it overhangs the eaves by ½" and the rake edges by 1". Nail along the top edge, ¾" from the edge, every 4". Then apply a 6" band of roof cement to the eaves and rake edges (*Figure 76*). Press the roofing down into the cement with a roof roller. Cement and overlap any vertical seams by 6". Use a utility knife to trim the roofing edges to the drip edges.

Step 3 Snap another chalkline 6" down from the top edge of the first course. Lay the second course to the chalkline and nail along the top edge, ¾" from the edge, every 4". Apply a 6" band of roof cement under the bottom of the second course and up the rake edges. Press the roofing down into the cement with a roof roller.

Step 4 The third and subsequent courses are applied in the same manner (*Figure 77*). Trim the edges at all rakes and eaves.

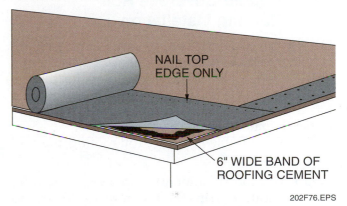

Figure 76 ◆ First course of concealed nail roll roofing.

202F76.EPS

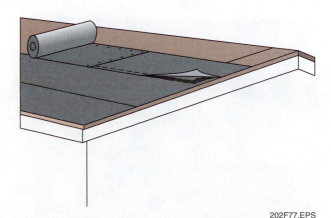

Figure 77 ◆ Third and subsequent courses of concealed nail roll roofing.

202F77.EPS

Step 5 At all valleys, hips, and ridges, apply the roofing to the point that it overlaps. Then trim the roofing to the center of the valley, hip, or ridge. Apply a 6" band of roofing cement to each side of the valley flashing and press the edges of the roofing into the cement. Do not nail horizontal seams within 6" of the center of the valley. At hips or ridges, nail the last course at the top of the hip or ridge. Then, snap a chalkline 6" on both sides.

Step 6 Apply a 6" band of roofing cement just above the 6" chalklines (*Figure 78*). Using 12"-wide strips cut from the roofing material, press the strips down into the cement on both sides of the hip or ridge with a roof roller. As necessary, overlap the strips by 6" and cement.

7.2.0 Double-Coverage Roll Roofing Installation

Double-coverage roll roofing is available for both hot and cold asphalt application. Make sure that only the cold asphalt version is used for the following installation. Double-coverage roll roofing has a 19" overlap (called the selvage) and a 17" mineral-coated exposure.

To apply double-coverage roll roofing, proceed as follows:

Step 1 Cut the 17" mineral-coated exposure from enough strips of the roofing to extend across all eaves. The mineral-coated pieces will be used as starter strips. Save the 19" selvage strips for use when the ridge caps are installed.

Step 2 Snap a chalkline 18½" from the eaves and use it to position the top of the starter strip so that it overlaps the eaves by ½" and the rakes by 1". Nail down the starter strip at both edges with nails spaced 3" at the bottom edge and 12" at the top edge. Nail the center at 12" intervals.

Step 3 Starting from one end, position one strip of the first course over the starter strip and nail it into place using two rows of nails spaced 4½" and 13" from the top of the strip in the selvage area. Space the nails in the rows about 12" apart.

Step 4 Roll back the strip and thickly coat the starter strip underneath with non-fibered liquid roofing cement (*Figure 79*). Roll the strip back onto the cement and press it down using a roof roller.

Step 5 Overlap the next strip by 6" and repeat the nailing and cementing procedure. Vertical seams are cemented, as shown in *Figure 80*, and are not nailed except in the selvage area.

Step 6 Continue applying strips and courses until the roof deck is covered. Trim the edges at all rakes and eaves.

Step 7 At all valleys, hips, and ridges, apply the roofing to the point that it overlaps. Then trim the roofing to the center of the valley, hip, or ridge.

202F78.EPS

Figure 78 ◆ Covering hip or ridge joints.

202F79.EPS

Figure 79 ◆ Coating the starter strip.

Figure 80 ◆ Cementing a vertical seam.

Step 8 Apply a 6" band of roofing cement to each side of the valley and press the edges of the roofing into the cement. Do not nail horizontally to within 6" of the center of the valley.

Step 9 At hips or ridges, nail the last course at the top of the hip or ridge. Snap a chalk-line 6" on both sides.

Step 10 Cut 12"-wide pieces from a roll of double-coverage roofing; include both the selvage and the mineral-coated exposure. These pieces are treated like the shingle tabs used to make a ridge cap, but are applied in double-coverage just like the roof.

Step 11 To begin, cut the selvage from one shingle. Then, starting at one end of the ridge or the bottom of a hip, place the selvage piece over the ridge and nail it down, spacing the nails 1" from the edges and at 4" intervals. Coat the selvage piece with cement.

Step 12 Next, place a full shingle over the cement with the mineral side up and press it into the cement. Nail the selvage of this shingle like the starting piece of selvage. Coat the selvage of that shingle with cement and apply another shingle.

Step 13 Repeat the process until the hip or ridge is completed. The junction of two hips and a ridge can be end capped in the same way as for a shingle roof.

8.0.0 ◆ WOOD SHINGLES AND SHAKES

Wood shingles and shakes for finish roofing are available in Grades 1 through 3.

- *Grade 1* – Grade 1 is the best and is the grade normally used. It is cut from heartwood and is clear (knot-free) and straight-grained. It is generally more rot resistant and also more expensive than the other grades.
- *Grade 2* – Grade 2 has a limited amount of sapwood, which is less rot resistant. It also has some knots and is flat grained. Grade 2 is acceptable for residential roofing.
- *Grade 3* – Grade 3 should only be used on outbuildings.

Shingles (not shakes) are also available as Grade 4 (utility). They are suitable only for use as starter courses or for shim stock because they have big knots.

Wood shingles are not recommended for roofs with a slope of less than 3 in 12, and shakes are not recommended for roofs with a slope of less than 4 in 12. The reason is that voids between the courses are not protected from wind-blown snow or water.

Exposure also must be limited for slight pitches. For example, with a 3 in 12 slope, 16" shingles must have a maximum of 3¾" exposure (5" on a 4 in 12). A 24" shingle can have a 5¾" exposure on a 3 in 12 and 7½" on a 4 in 12.

Table 3 shows the grades, characteristics, sizes, and coverage for 16", 18", and 24" wood shakes at different exposures.

Table 4 shows the grades, characteristics, sizes, and coverage for 16", 18", and 24" wood shingles at different exposures.

Table 3 Summary of Wood Shakes

Grade	Length and Thickness	Courses per Bundle	Bundles per Square	Description
		18' Pack*		
No.1–Hand split and Resawn	16" starter-finish	9/9	5	These shakes have split faces and sawn backs. Cedar logs are first cut into desired lengths. Blanks or boards of the proper thickness are split and then run diagonally through a band saw to produce two tapered shakes from each blank.
	18" × ½" mediums	9/9	5	
	18" × ¾" heavies	9/9	5	
	24" × ⅜" mediums	9/9	5	
	24" × ½" mediums	9/9	5	
	24" × ¾" mediums	9/9	5	
No. 1–Tapersawn	24" × ⅝"	9/9	5	These shakes are sawn on both sides.
	18" × ⅝"	9/9	5	
No. 1–Tapersplit	24" × ½ "	9/9	5	Produced largely by hand using a sharp-bladed steel froe (a cleaving tool) and a wooden mallet. The natural shingle-like taper is achieved by reversing the block, end-for-end, with each split.
		20' Pack*		
No. 1–Straight split	18" × ⅜" True-Edge**	14 straight	5	Produced in the same way as taper split shakes, except that by splitting from the same end of the block, the shakes acquire the same thickness throughout.
	18" × ⅜"	19 straight	5	
	24" × ⅜"	16 straight	5	

*Pack used for majority of shakes.

Table 4 Summary of Wood Shingles

Grade	Length	Thickness at Butt	Courses per Bundle	Bundles/Cartons per Square	Description
No. 1–Blue Label	16"	.40"	20/20	4 bundles	The premium grade of shingles for roofs and side walls. These top-grade shingles are 100% heartwood, 100% clear, and 100% edge grain.
	18"	.45"	18/18	4 bundles	
	24"	.50"	13/14	4 bundles	
No. 2–Red Label	16"	.40"	20/20	4 bundles	A good grade for many applications. Not less than 10" clear on 16" shingles, 11" clear on 18" shingles, and 16" clear on 24" shingles. Flat grain and limited sapwood are permitted in this grade.
	18"	.45"	18/18	4 bundles	
	24"	.50"	13/14	4 bundles	
No. 3–Black Label	16"	.40"	20/20	4 bundles	A utility grade for economy applications and secondary buildings. Not less than 5" clear on 16" and 18" shingles; 10" clear on 24" shingles.
	18"	.45"	18/18	4 bundles	
	24"	.50"	13/14	4 bundles	
No. 4–Undercoursing	16"	.40"	14/14 or	2 bundles	A utility grade for undercoursing on double-coursed side wall applications or for interior accent walls.
	18"	.45"	20/20	2 bundles	
			14/14 or	2 bundles	
			18/18	2 bundles	

In most cases, shingles or shakes are installed on open or spaced roof sheathing (*Figure 81*). This type of decking costs less to install and allows the shingles/shakes to dry out, preventing rot. In high-moisture or snowy areas of the country, a waterproofing membrane is applied up the roof to extend 24" beyond the inside wall. In areas of high humidity, shingles or shakes applied over completely sheathed roofs (closed sheathing) that are covered with underlayment or rigid insulation are usually spaced off the roof with horizontal pressure-treated 1 × 3s nailed to the roof deck and spaced at the exposure of the product. For additional ventilation or when applied over rigid insulation, vertical furring strips are applied and centered over the rafters prior to the horizontal strips. In addition, the roof is ringed with pressure-treated 1 × 3s nailed parallel to the eaves, ridge, and rakes. These ventilation strips, sometimes called skip sheathing, allow the wood shingles or shakes to dry out. Roof projections are flashed in the same way as composition shingle roofs except that flashing projections under and up the wood shingles and shakes must be wider and longer.

8.1.0 Roof Exposure

The exposure represents the area of the shingle or shake that contacts the weather. See *Figure 82*. It depends upon the pitch of the roof. A good shingle or shake installation is never less than three layers thick. There are three lengths of shingles: 16", 18", and 24".

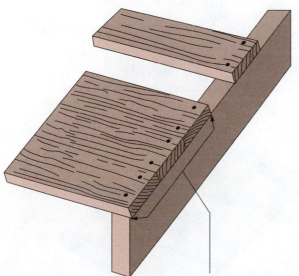

Eave edge solid sheathed for support and, if exposed, for underside appearance. Solid sheathing may extend up the roof as necessary for application of waterproofing membrane.

202F81.EPS

Figure 81 ◆ Open or spaced roof sheathing.

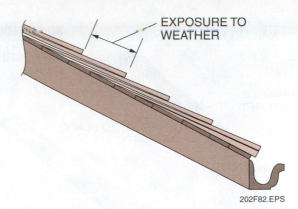

EXPOSURE TO WEATHER

202F82.EPS

Figure 82 ◆ Roof exposure.

If the roof slope is 4 in 12 or steeper (three-ply roof), use the following exposure guidelines:

- For 16" shingles, allow a 5" exposure.
- For 18" shingles, allow a 5½" exposure.
- For 24" shingles, allow a 7½" exposure.
- For 18" shakes, allow a 7½" exposure.
- For 24" shakes, allow a 10" exposure (⅜" thick shake is limited to a 5" exposure).

If the roof slope is less than 4 in 12, but not less than 3 in 12 (four-ply roof), use the following exposure guidelines:

- For 16" shingles, allow a 3¾" exposure.
- For 18" shingles, allow a 4¼" exposure.
- For 24" shingles, allow a 5¾" exposure.

Recommended exposures on low slopes are for No. 1 grade shingles. If applying a different grade, such as No. 3 shingles, make sure you check with the manufacturer. Normally, if the roof slope is less than 3 in 12, shingles are not recommended.

8.2.0 Application of Wood Shingles

The open or spaced sheathing begins at the eave line with a double snug fit or triple nailing of roof boards close together to provide a solid surface for the initial doubling of the first course of wood shingles. A 36"-wide strip of 30 lb felt or waterproof membrane should be applied at the eaves.

When the roof sheathing is completed, the roof application is ready to begin. The tools of the trade for wood shingling are minimal. A straightedge and a shingle hatchet are normally all that are necessary. The shingle hatchet should be lightweight and have a sharp blade and heel and a gauge for checking shingle exposure. See *Figure 83*.

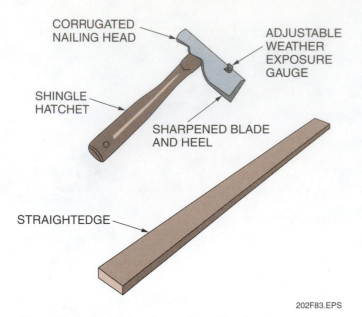

Figure 83 ◆ Shingle hatchet and straightedge.

When using scaffolding, it should be waist-high to the eaves to allow you to do a proper job of doubling up the first course and comfortably reaching successive courses. *Figure 84* shows the recommended shingle spacing on building roofs.

When applying shingles, only rust-resistant nails should be used. Hot-dipped, zinc-coated nails, which have the strength of steel and the corrosive resistance of zinc, are recommended. Sizes of nails for various jobs were discussed earlier in this module.

Most carpenters prefer to use a shingle hatchet to lay wood shingles. The sharpened blade and heel enable them to split and trim the shingles. The adjustable gauge for weather exposure makes spacing easier and quicker.

The first course of shingles at the eaves should be doubled or tripled. The starter course can be a double layer of 15" starter shingles overlayed with a third layer of regular shingles. All shingles should be spaced ¼" apart to provide space for expansion when they become rain-soaked. Let the shingles protrude over the edge to ensure proper spillage into the eave trough or gutter. See *Figure 85*.

Use only two nails to attach each shingle. Use care when nailing wood shingles. Drive the nail just flush with the surface. The corrugated nail head enables you to accomplish this due to its raised surface. The wood in shingles is soft and can be easily crushed and damaged under the nail heads.

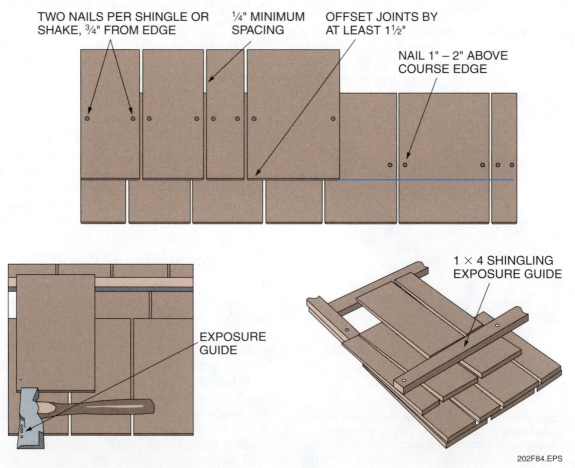

Figure 84 ◆ Shingle/shake nailing and spacing guide.

The Hazards of Working with Shakes and Shingles

Hatchets, roofer's knives, and hammers can be dangerous. Use caution when working with these tools. On some occasions, a small portable trimming saw might be useful. For safety's sake, always work on a dry roof and wear sneakers, crepe soles, or other good-gripping shoes when working on a roof. Roof jacks should always be used on wood roofs so that you or the materials do not slide off the roof. Wood becomes very slippery when wet or damp.

Dust from western red cedar can become an allergen over time and cause respiratory ailments including asthma and rhinitis, or other disorders including dermatitis, itching, rashes, and conjunctivitis. Take precautions to avoid inhaling the dust or getting it on your skin or in your eyes.

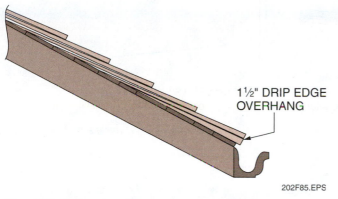

Figure 85 ◆ Eave overhang.

Proper nail placement is very important. The nails should be near the butt line of the shingles in the next course that is to be applied over the course being nailed, but should never be driven below this line so they will be exposed to the weather. Driving the nails 1" to 1½" above the butt line is good practice, with 2" as an allowable maximum. Each nail should be placed not more than ¾" from each side of the edge of the shingle. When nailed in this manner, the shingles will lie flat and provide good service.

The second layer of shingles in the first course should be nailed over the first layer so the joints in each course are offset by not less than 1½". See *Figure 86*.

Beveled Rake

Some roofers use beveled siding placed lengthwise up the rake edges with the butt of the siding even with the rake edge. This deflects water toward the roof and prevents it from dripping from the rake edge. A good shingler will use care in breaking the joints in successive courses so they do not match up in three successive courses. Joints in adjacent courses should be 1½" apart. Unless staggered-length shingles are being installed, it is a good practice to use a board as a straightedge to line up rows of shingles. Tack the board temporarily in place to hold the shingles until they are nailed. This makes the work faster and the results look professional.

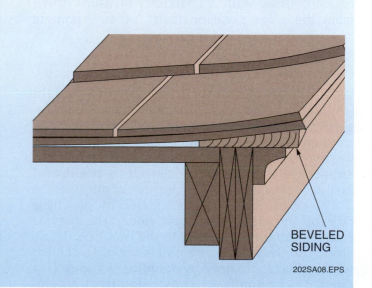

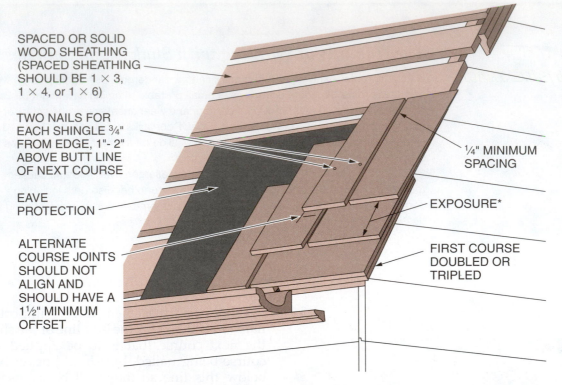

SPACED OR SOLID WOOD SHEATHING (SPACED SHEATHING SHOULD BE 1 × 3, 1 × 4, or 1 × 6)

TWO NAILS FOR EACH SHINGLE ¾" FROM EDGE, 1"- 2" ABOVE BUTT LINE OF NEXT COURSE

EAVE PROTECTION

ALTERNATE COURSE JOINTS SHOULD NOT ALIGN AND SHOULD HAVE A 1½" MINIMUM OFFSET

¼" MINIMUM SPACING

EXPOSURE*

FIRST COURSE DOUBLED OR TRIPLED

*EXPOSURE – For ¼ pitch and steeper roofs, use 5", 5½", or 7½" for 16", 18", or 24" long shingles, respectively. For flatter pitches, use 3½", 4½", or 5½" for the same shingles, respectively.

202F86.EPS

Figure 86 ◆ Application of wood shingles.

Although it is not required, some contractors prefer to separate each course with underlayment strips that extend 4" above the top of a shingle down to and over the nails.

Two people often work together; one distributes and lays the shingles along the straightedge while another nails them in place. As the shingling progresses, check the alignment every five or six courses with a chalkline. Measure down from the ridge occasionally to be sure shingle courses are parallel to the ridge.

8.2.1 Special Effects

Wood shingles are normally applied in a straight single course, but variations are possible. By staggering or building up wood shingles, usually in random patterns, shadow lines and textures can be emphasized. This is a feature sometimes applied to contemporary as well as traditional architecture. *Figure 87* shows three applications.

The ocean wave effect is created by placing a pair of shingles butt-to-butt under the regular course and at right angles to the butt line. These cross shingles should be about 6" wide. The Dutch weave effect is obtained by doubling shingles. This effect can be emphasized by using two shingles

OCEAN WAVE   DUTCH WEAVE   PYRAMID

202F87.EPS

Figure 87 ◆ Shingle effects.

instead of one and is generally referred to as a pyramid pattern. The joints are always broken by at least 1½".

8.2.2 Hips and Valleys

Procedures for nailing wood shingles on hip roofs are similar to the procedures for asphalt shingles. Start at the center and work to the hip rafter. Use the largest shingles to finish off at the hip and carefully miter cut on the center line of the hip rafter.

Staggered Shingles/Shakes

Another special effect that creates a very rustic appearance is the random staggering of the bottom edge of the shingles or shakes. Some of the architectural composition shingles are designed to create this effect. With wood shingles or shakes, it can be achieved by extending the exposure of every other shingle/shake by random distances of between ¾" and 1" instead of using a fixed exposure distance. Another method is to use different lengths of shakes in a random pattern. This method, shown in the illustration, was achieved using synthetic wood shingles.

SIMULATED WOOD SHINGLE ROOF

202SA09.EPS

Valley shingling works in the opposite direction. Start from the valley and work outward. Select wide shingles for this application. Carefully cut the proper angle at the butts, following the guidelines snapped on the flashing. Valley flashing should be heavy galvanized steel or copper W-metal. Valley construction and proper cuts are shown in *Figure 88*.

8.2.3 Ridge and Hip Caps

Tight ridges and hips are required to avoid roof leakage. In good ridge and hip construction, nails are not exposed to the weather. Shingles of approximately the same width as the exposed shingle surface are sorted out to be used as caps. Two lines are then marked on the shingles on the roof the correct distance back from the center line of the ridge or hip. Waterproof membrane or flashing is applied over the ridge or hip.

Figure 89 shows the standard application of hips and ridges. Be sure to use longer nails, which will penetrate into the sheathing. The alternate overlap pattern ensures a dry seal. Factory-assembled ridge and hip units (*Figure 90*) are available.

ON ROOFS FLATTER THAN ½ PITCH, VALLEY SHEETS SHOULD EXTEND AT LEAST 10" FROM VALLEY CENTER

ON ½ PITCH AND STEEPER ROOFS, VALLEY SHEETS SHOULD EXTEND AT LEAST 7" FROM VALLEY CENTER

202F88.EPS

Figure 88 ◆ Valley layout—wood shingles.

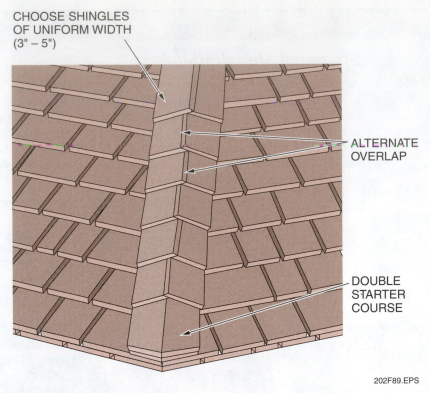

CHOOSE SHINGLES
OF UNIFORM WIDTH
(3" – 5")

ALTERNATE
OVERLAP

DOUBLE
STARTER
COURSE

202F89.EPS

Figure 89 ◆ Hip layout—wood shingles.

8.3.0 Application of Wood Shakes

Shakes are made in several lengths, thicknesses, and widths. Because of their length and thickness, they have greater exposure than shingles. *Figure 91* shows various types of wood shakes.

Shakes must be applied to roofs that have sufficient slope to ensure good drainage. Typical maximum weather exposures are 8½" for 18" shakes, 10" for 24" shakes, and 13" for 43" shakes. Application is similar to that for wood shingles.

Start the application by placing a 36" strip of 30-lb roofing felt or waterproof membrane along the eaves. The beginning or starter course is doubled, just as it is for regular shingles. After each course is applied, an 18" (or wider) strip of 30-lb felt is applied over the top portion of the shakes. The 18" strips are cut from the full 36" roll using a center line. The strips extend onto the sheathing.

The bottom edge of the felt is placed above the butt a distance equal to twice the exposure. For example, if 24" shakes are being laid at a 10" exposure, place the roofing felt 20" above the cuts of the shake. As you can see in *Figures 92* and *93*, the felt will then cover 4" of the top end of the shake and the remaining 14" will rest on the sheathing.

202F90.EPS

Figure 90 ◆ Prefabricated hip and ridge units.

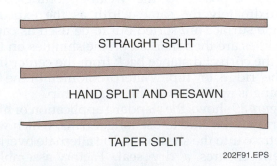

STRAIGHT SPLIT

HAND SPLIT AND RESAWN

TAPER SPLIT

202F91.EPS

Figure 91 ◆ Types of wood shakes.

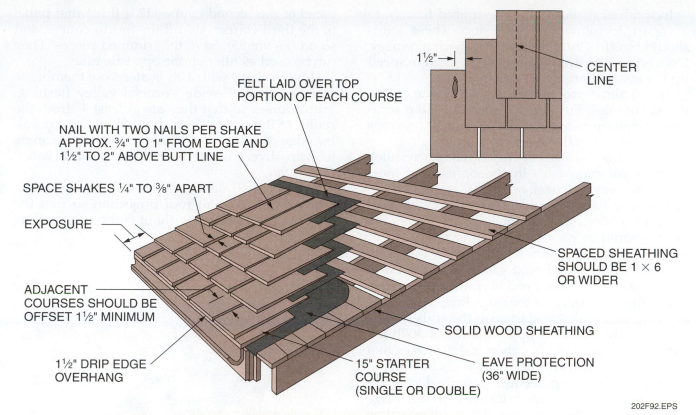

FELT LAID OVER TOP PORTION OF EACH COURSE

NAIL WITH TWO NAILS PER SHAKE APPROX. ¾" TO 1" FROM EDGE AND 1½" TO 2" ABOVE BUTT LINE

SPACE SHAKES ¼" TO ⅜" APART

EXPOSURE

ADJACENT COURSES SHOULD BE OFFSET 1½" MINIMUM

1½" DRIP EDGE OVERHANG

1½"

CENTER LINE

SPACED SHEATHING SHOULD BE 1 × 6 OR WIDER

SOLID WOOD SHEATHING

EAVE PROTECTION (36" WIDE)

15" STARTER COURSE (SINGLE OR DOUBLE)

202F92.EPS

Figure 92 ◆ Shake application.

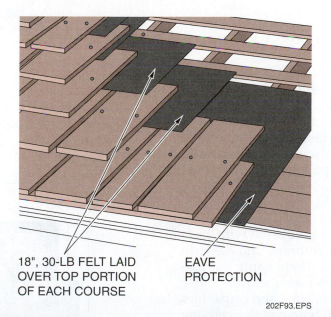

18", 30-LB FELT LAID OVER TOP PORTION OF EACH COURSE

EAVE PROTECTION

202F93.EPS

Figure 93 ◆ Shake felt underlayment.

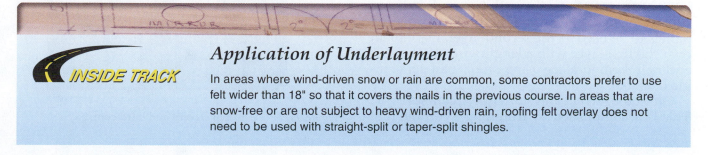

Application of Underlayment

In areas where wind-driven snow or rain are common, some contractors prefer to use felt wider than 18" so that it covers the nails in the previous course. In areas that are snow-free or are not subject to heavy wind-driven rain, roofing felt overlay does not need to be used with straight-split or taper-split shingles.

Individual shakes should be spaced from ¼" to ⅜" apart to allow for expansion. These joints should be offset by at least 1½" in adjacent courses. Once again, the straightedge can be used to speed up the installation.

The nailing procedures are the same as for wood shingles. Emphasis is placed on the accuracy of nailing to make sure the nails in previous courses are correctly covered.

Valleys may be open or closed. The open valley is more common and, in the opinion of most builders, more practical.

The open valley should be underlaid with 30-lb felt and a W-metal sheet at least 20" wide. The metal should be galvanized copper (26 gauge or heavier). The open portion of the valley is usually about 4" wide and should gradually increase in width toward the lower end to control water flow.

For the final course at the ridge line, try to select uniform size shakes and trim off the ends so they meet evenly. Carefully apply a strip of 30-lb felt along all ridges and hips, then install shakes that have a uniform width of about 6". Nail these in place following the procedure described for regular wood shingles. Prefabricated hip and ridge units are available. The use of these will save time and provide uniformity.

8.4.0 Panelized Shingle/Shake Application

Panelized shingles or shakes are much faster to apply than individual shingles or shakes. Each panel has a felt backing and a score line that is used to align courses. One fastener per shingle or shake holds each panel in place. If the felt backing extends to the right of the panel, the panels must be laid left to right. If the felt extends to the left, the panels must be laid right to left.

If the roof will be applied directly over sheathing, make sure that the roof is covered with 15-lb felt and/or a waterproof membrane and has drip edges at the eaves and rakes. Then, apply two layers of panels over each other as a starter course (*Figure 94*). If the roof deck is open and drip edges are not used, make sure that both panels overhang the eaves by 1½" and the rakes by 2". Nail the first layer of the starter course with two nails halfway up both ends. Remove the felt backing from the panels to be used as the second layer. Place them over the first layer with a lengthwise offset of 1½" and nail them in place using one 2" nail per shingle or shake.

Offset the second and subsequent courses by 6". This is accomplished by trimming 6" off the first panel of the second course, 12" off the first panel of the third course, 18" off the fourth course, and so on (*Figure 95*). Save the trimmed pieces. They can be used as filler on the opposite rake.

Cover valleys with a 36" waterproof membrane and install a 24"-wide W-metal valley flashing. Trim courses so that they are at least 4" from the center of the valley. Trim about 2" off the top valley edge of each panel at about a 45-degree angle to help divert water running down the valley (*Figure 96*).

Adjust the depth of the courses when approaching a small roof projection so that the panel can be notched without cutting the nailing bar (*Figure 97*).

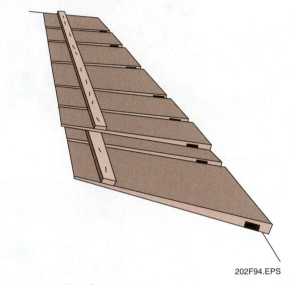

202F94.EPS

Figure 94 ◆ Panel starter course.

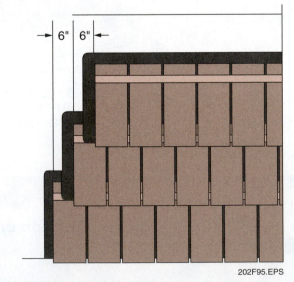

202F95.EPS

Figure 95 ◆ 6" offset of second and third course.

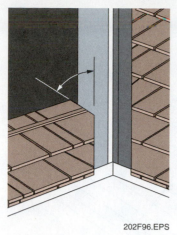

Figure 96 ◆ Trimming shingle/shake panels for an open valley.

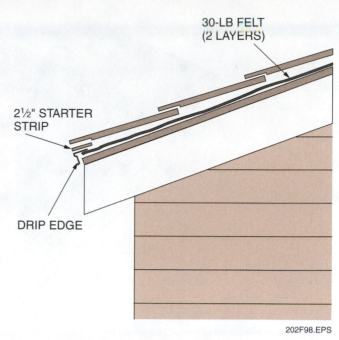

Figure 98 ◆ Starter strip and panel overlap.

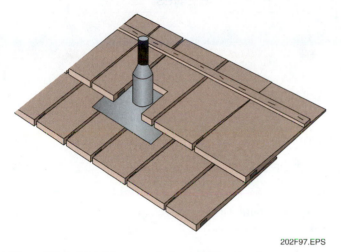

Figure 97 ◆ Notching panels for a roof projection.

8.5.0 Hardboard Shingle/Shake Panel Application

These wood fiber panels, which mimic wood shingles/shakes, are another product that is easy to apply. The panels are 4' long and are laid horizontally only on a closed roof deck. The bottom edge forms a lap joint over the top of the panel below, and the side edges form a shiplap joint with the adjacent panels. They must only be used on roofs with a slope of 4 in 12 or more.

First, make sure that the roof is covered with two layers of 30-lb felt or a waterproof membrane and has drip edges at the eaves and rakes. Cut 2½"-wide starter strips lengthwise from panels and nail them to the eaves so that they overlap the rakes by 2" and the eaves by 1" (*Figure 98*). Apply

the first course of panels starting at the 2" overhang at the left side of the roof (*Figure 99*). Fasten each panel with eight nails or staples along the nailing guideline at the top of the panel. Start fastening at the top center of the shiplap joint on the right end and stop 3" from the left end.

After the first course is laid, shorten the first panel of the second and subsequent courses from the right end of the panel (see *Figure 99*) to stagger the joints of the panels. Save the trimmed left end of the panels for use on the right-hand rake edge. When laying the second and subsequent course panels, make sure that the lap joint at the bottom edge of the panel is fully seated over the panel below before nailing the top edge of the panel.

Cover valleys with a 36" waterproof membrane and install a 24"-wide W-metal valley flashing (*Figure 100*). Trim courses so that they are at least 4" from the center of the valley. Trim about 2" off the top valley edge of each panel at about a 45-degree angle to help divert water running down the valley. Adjust the depth of the courses when approaching a small roof projection so that the panel can be notched less than half the width of the panel.

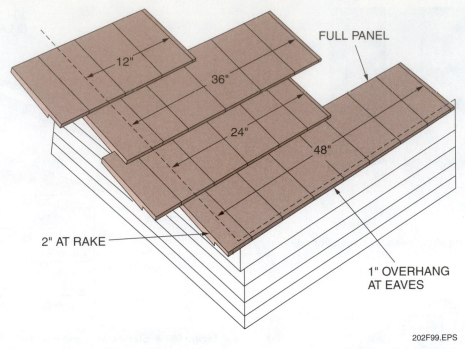

FULL PANEL

12"

36"

24"

48"

2" AT RAKE

1" OVERHANG
AT EAVES

202F99.EPS

Figure 99 ◆ Staggering of hardboard panels.

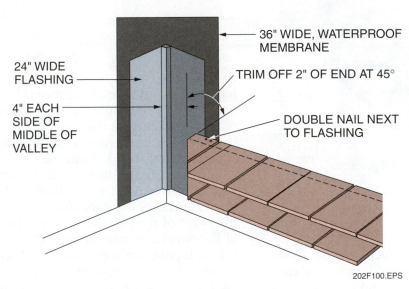

36" WIDE, WATERPROOF
MEMBRANE

24" WIDE
FLASHING

TRIM OFF 2" OF END AT 45°

4" EACH
SIDE OF
MIDDLE OF
VALLEY

DOUBLE NAIL NEXT
TO FLASHING

202F100.EPS

Figure 100 ◆ Trimming hardboard panels for an open valley.

9.0.0 ◆ COMMON METAL ROOFING

This section covers various types of metal roofing systems.

9.1.0 Corrugated Metal Roofing

Another type of roofing material is corrugated metal roofing, or galvanized metal roofing. Only galvanized sheets that are heavily coated with zinc (2.0 oz. per square foot) are recommended for permanent construction.

Galvanized sheets may be laid on slopes as low as a shallow 3" rise to the foot (⅛ pitch). If more than one sheet is required to reach the top of the roof, the ends should overlap by at least 8". When the roof has a pitch of ¼ or more, 4" end laps are usually satisfactory. To make a tight roof, sheets should be overlapped by 1½ corrugations at either side. See *Figure 101(A)*.

When using roofing that is 27½" wide with 2½" corrugations and a corrugation lap of 1½", each sheet covers a net width of 24" on the roof. If 26-gauge galvanized sheets are used, supports may be 24" apart. If 28-gauge galvanized sheets are used, supports should not be more than 12" apart. The heavier gauge has no particular advantage

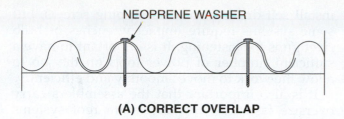

NEOPRENE WASHER

(A) CORRECT OVERLAP

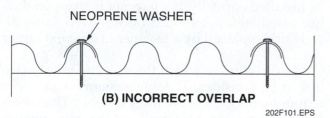

NEOPRENE WASHER

(B) INCORRECT OVERLAP

202F101.EPS

Figure 101 ◆ Corrugated roofing.

except its added strength, because the zinc coating is what gives this type of roofing its durability.

When 27½" roofing is not available, sheets of 26" width may be used. When laying the narrower sheets, every other one should be turned upside down so that each alternate sheet overlaps the two intermediate sheets, as shown in *Figure 101(A)*.

For best results, galvanized sheets should be fastened with neoprene-headed nails, galvanized nails and neoprene washers, or screws with neoprene washers. Fasteners are used only in the tops of the corrugations to prevent leakage. To avoid unnecessary corrosion, use the fasteners specified by the roofing manufacturer.

Corrugated metal roofing panels (*Figure 102*) are used on garages, storage buildings, and farm buildings. They are available in widths up to 4'-0" and lengths up to 24'-0". Normally, these panels are used on roofs with slopes of 4 in 12 or steeper. They can be used on 2 in 12 roofs if a single panel reaches from the ridge to the eave.

The panels are fastened to purlins. A purlin is a structural member running perpendicular to the rafters. See *Figure 102*. Usually, 2 × 4 wood stock is used. The spacing should follow the directions specified by the manufacturer. Filler strips are sold with the panels. They are set at the eave and the ridge. Normally, the panel is cut so it overhangs 2" to 3" at the eave. The installation of metal roofs should be done in accordance with the manufacturer's specifications.

9.2.0 Simulated Standing-Seam Metal Roofing

The character of each standing-seam roof system dictates the amount and type of planning. Each system has different components and slightly different requirements for tools and equipment.

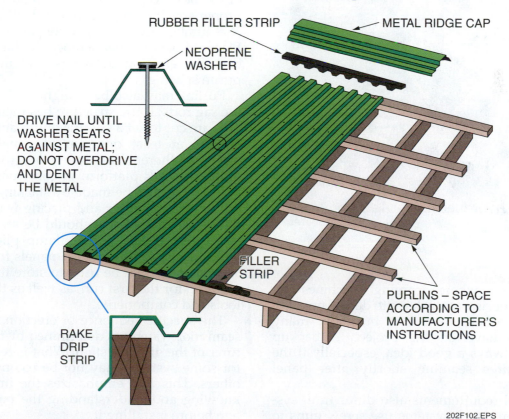

RUBBER FILLER STRIP

METAL RIDGE CAP

NEOPRENE WASHER

DRIVE NAIL UNTIL WASHER SEATS AGAINST METAL; DO NOT OVERDRIVE AND DENT THE METAL

FILLER STRIP

RAKE DRIP STRIP

PURLINS – SPACE ACCORDING TO MANUFACTURER'S INSTRUCTIONS

202F102.EPS

Figure 102 ◆ Corrugated roof layout.

Corrugated Roofing Light Panels

Corrugated translucent fiberglass light panels can be used to allow daylight into the interior of a structure. They can also be used as the entire roof, if desired. These panels are the same size and are installed in the same manner as corrugated metal roofing.

FIBERGLASS PANEL

202SA10.EPS

install self-drilling and self-tapping screws, but some systems require impact wrenches or bulb rivet guns for fastening. It is important to have a sufficient number of power tools on the job to allow the work to move smoothly and efficiently.

It is also important that the assemblers carry oversize fasteners. Standing-seam roof systems minimize through-the-panel fasteners by up to 90 percent. Therefore, it is crucial that the fasteners be installed correctly. If a fastener is stripped during installation, it should be removed immediately and replaced by a fastener of the next larger size.

The direction in which the sheeting takes place has to be considered. Some systems have strict requirements; others are more flexible. The sheeting direction can be found on the construction drawings.

Perhaps the biggest demand each standing-seam roof system makes is to be installed by competent assemblers. Improper installation techniques cause the majority of standing-seam roof failures. This underscores the need for special training in the particular system being used.

Before roofing can begin, the structure must be plumb and level. The purlins must also be straight. Z-purlins, in particular, have a tendency to roll. If there are no purlin braces, use wood blocking. Most manufacturers suggest that wood blocks be driven tightly between purlins to ensure proper spacing and recommend either 2 × 6 or 2 × 8 lumber, depending upon the purlin depth. Place at least one row of blocks in the center of the bay. The erection drawings contain the proper purlin spacing.

Purlins may also be straightened by adjusting the sag rods, if the structure has them. Many sag rods are cut to set a specific width automatically.

Until at least one run of panels has been installed, there is no safe place to work from unless a work platform is constructed. A work platform should be made by stacking two panels on top of each other and placing walk boards in the center. The panels should be attached to the structure with locking C-clamp pliers or some other means to prevent the panels from moving. This platform can be used to store the insulation required for the first run, as well as the necessary tools and components.

The specific sequence of erection of standing-seam roof systems is determined by the manufacturer of the given system. What is recommended for some systems may not be recommended for others. This fact emphasizes the importance of knowing and understanding the particular system before installing it.

In some cases, where seaming machines are required, supervisors may have to decide not only when to lease the machines, but how many machines to have on the project. A backup machine is always a good idea, especially if the system requires seaming shortly after panel installation.

Power tool requirements also differ from system to system. Most systems use screw guns to

INSIDE TRACK

Simulated Standing-Seam Roofing Systems

A number of different methods of joining metal roofing panels are employed for these types of systems. Older systems used galvanized steel or copper standing seams that were soldered together to form a weather seal. Other systems require crimping machines or use snap-type seals.

MACHINE-CRIMPED PANELS

SNAP-TYPE SEAL

202SA11.EPS

Usually, a blanket of insulation is stretched across the structural members prior to sheeting. The blanket width depends on the panel. Keep the stapling edge ahead of the panel, but no more than a foot ahead of it.

As previously mentioned, installing metal roofing requires special tools and experience. The following is an overview of the installation procedure for one type of metal roofing that is designed to be laid over a closed, fully-sheathed roof deck. It is applied in panels that are 12" to 16½" wide and that are precut to run from the eaves to the ridge of the structure. The joints between panels are weatherproofed by means of a C-clip with a neoprene seal.

Cover the roof with a 30-lb felt or waterproof membrane. The membrane or felt must overlap the edges of the eaves and rakes.

The eave trim (*Figure 103*) is then screwed in place before panels are applied. The panels are placed one at a time and secured to the roof deck with a T-clip on one side of the panel.

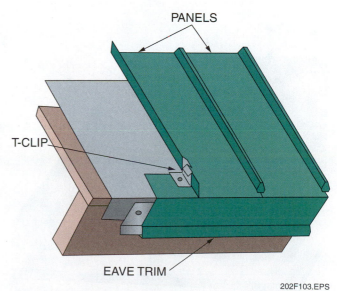

202F103.EPS

Figure 103 ◆ Eave trim.

The next panel is then inserted under the T-clip that is holding the first panel and secured on the opposite side with another T-clip (*Figures 104* and *105*). At the joint of any two panels, the joint is weather-sealed with a C-clip running the full length of the joints, as shown in *Figures 104* and *105*. The T-clips are used every 12" in high-wind areas and every 18" elsewhere. After panels are placed under each side of the T-clips, the T-clip wings are bent down, and the C-clip is forced down on the seam. The neoprene flaps inside the C-clip seal the seam against water penetration, and the bent-down wings of the T-clip keep the C-clip from being dislodged from the seam.

After the first panel is placed at one of the rakes, a rake edge is applied, as shown in *Figure 106*. If necessary at the opposite rake, the panel is trimmed off and Z-strips are sealed and secured to the panel before the rake edge is installed.

At valleys, the panels are trimmed to the angle of the valley. Channel strips running parallel in the valley are sealed and screwed to the valley and hold the edges of the panels (*Figure 107*).

At the ridge, Z-strips are fastened and sealed to the panels between the seams. They are also sealed to the seams. The cover flashing is then sealed and secured to the Z-strips (*Figure 108*).

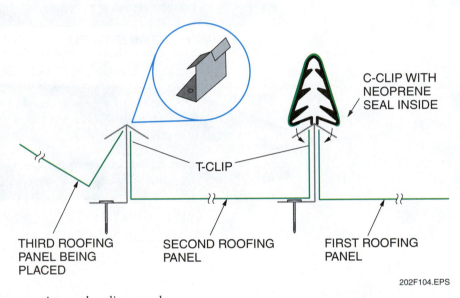

202F104.EPS

Figure 104 ◆ Placing, securing, and sealing panels.

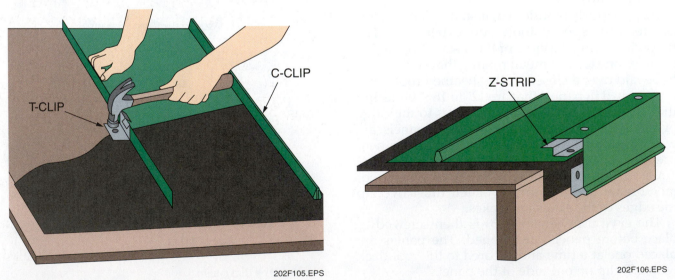

202F105.EPS

Figure 105 ◆ Nailing a T-clip to a roof deck.

202F106.EPS

Figure 106 ◆ Rake edge.

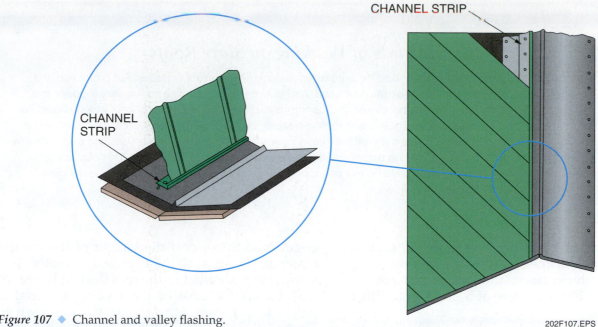

CHANNEL STRIP

Figure 107 ◆ Channel and valley flashing.

202F107.EPS

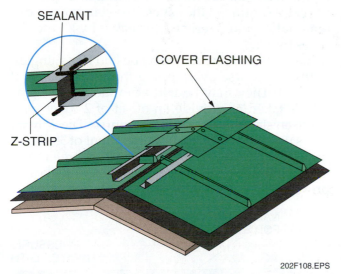

SEALANT

COVER FLASHING

Z-STRIP

202F108.EPS

Figure 108 ◆ Ridge flashing.

9.3.0 Snug-Rib System

Another type of roofing sheet is called the snug-rib system. It utilizes a concealed fastener, which is leak resistant and eliminates through fasteners. This combination of a V-beam industrial sheet and the snug seam joint makes for a greater beam strength and a deeper corrugation than other roofing profiles. Because of the greater strength of this type of roofing panel, the purlin spacing can be increased.

The snug-rib joint is a highly efficient weathertight joining system of a simple nature. The joint is created by engaging the hooked edges of two panels into a Y-shaped extruded spline, previously measured and anchored to the purlins with a self-templating clip. A neoprene gasket is then

rolled into the extrusion between the panel edges, where it holds the panel edges securely in place and creates a watertight seal.

NOTE

No primary fasteners penetrate the weatherproofing membrane.

There is a 19½" covering width. The material has a V-shaped corrugation that has a 4⅞" pitch and a 1¾" depth. Lengths vary from 77" to 163", depending upon the gauge of the material. The end laps must be a minimum of 12", located over roof purlins, and staggered with the end laps in adjacent panels. Install this system according to the manufacturer's specifications.

10.0.0 ◆ SLATE AND TILE ROOFING

This section covers the installation of various types of slate and tile roofing.

10.1.0 Slate or Synthetic Slate Roofing Installation

Roofing slate is hard, tough, and usually has a bright metallic luster when freshly split. Slate has a great variety of colors such as gray, green, dark blue, purple, and red. It is available in any size from 6" to 14" wide, 12" to 24" long, and ⅛" to 2" thick. The most common sizes are 12" × 16" and 14" × 20", and either ³⁄₁₆" or ¼" thick.

Roofs may be constructed of pieces of slate of uniform size, thickness, and color; however, random sizes, thicknesses, and colors can also be used. *Figure 109* is a view of a slate roof with a mitered hip.

An alternative to natural slate is a fiber-cement material that looks and installs the same as slate. This synthetic slate is somewhat lighter and will last 40 to 50 years.

To begin the installation, 30-lb felt and/or a waterproof membrane is laid over a roof deck engineered to support the weight of the roofing material. Then, a ¼"-thick pressure-treated lath starter strip is nailed to the roof deck at the eaves and a starter course of the roofing material is nailed lengthwise along the eaves, as shown in *Figure 110*. The upward tilt or extra pitch caused by the wood starter strip is referred to as providing extra canting at the eaves. The starter course of material should overhang the eaves by ½" and the rakes by 1".

Slate or synthetic slate may be laid like shingles, with each course overlapping the course below it (*Table 5*). They may be laid at random as long as care is taken to provide an offset of 2" over gaps between slates in the previous course. The 3" **head lap** is usually laid on an underlayment of 30-lb saturated felt; however, some contractors prefer individual felt strips under each course (like shakes) to provide extra cushioning and weatherproofing.

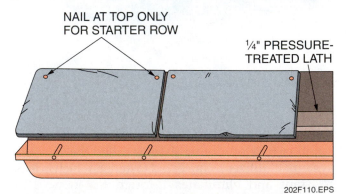

202F110.EPS

Figure 110 ◆ Shimming and applying starter course.

202F109.EPS

Figure 109 ◆ Slate roof with mitered hip.

Table 5	Slate Shingle Head Lap
Roof Slope	**Minimum Lap**
4 in 12 to 8 in 12	4" (102 mm)
8 in 12 to 20 in 12	3" (76 mm)
20 in 12 or more	2" (51 mm)

Slate or synthetic slate shingles are nailed through two prepunched or predrilled holes into wood sheathing or solid plywood decking (*Figure 111*). Drive nails flush with the shingle, but not tight. Copper or slater's nails are preferred for fastening, although redipped galvanized nails and copper-coated nails are often used where permitted by local building codes. Make sure gaps in a covering course are offset by at least 2" from the gaps in the course below. Also, make sure that the gap between shingles is at least ¹⁄₁₆".

Slate is joined at the ridges by a saddle ridge or a combed ridge, both requiring extreme care in installation. Slate is brittle and must be cut carefully to avoid waste.

Typical capping of a ridge using a saddle ridge is shown in *Figure 112*. The slates are all the same size and the flush overlap at the peak is alternated from one side to the other. The slate is positioned horizontally like the starter course and is sometimes referred to as combing slate. The slate is fastened to the roof with two nails at one end. Holes must be drilled with a ¹⁄₈" masonry bit through the underlying slate to accommodate the nails.

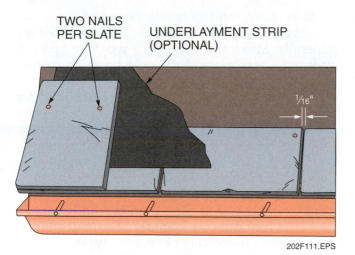

202F111.EPS

Figure 111 ◆ Installing the slates.

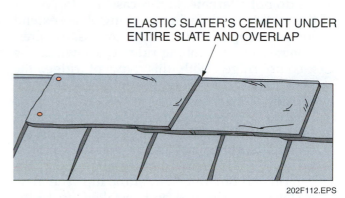

202F112.EPS

Figure 112 ◆ Capping a ridge.

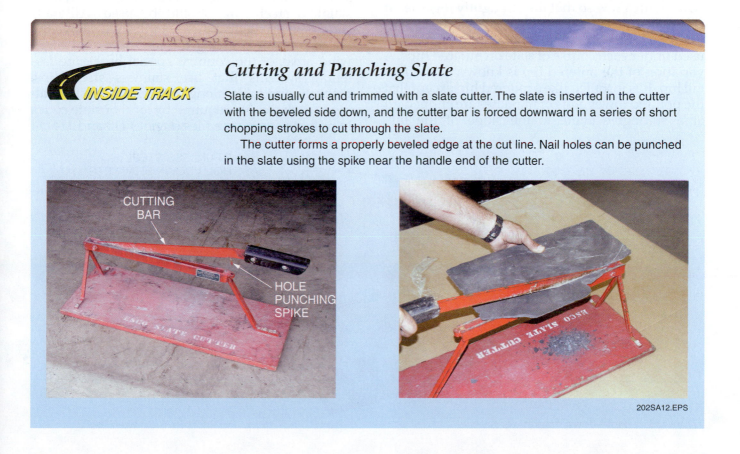

Cutting and Punching Slate

Slate is usually cut and trimmed with a slate cutter. The slate is inserted in the cutter with the beveled side down, and the cutter bar is forced downward in a series of short chopping strokes to cut through the slate.

The cutter forms a properly beveled edge at the cut line. Nail holes can be punched in the slate using the spike near the handle end of the cutter.

202SA12.EPS

Before placing the caps, snap a chalkline the width of the cap shingles on both sides of the ridge. Then, either waterproof membrane or three layers of generously cemented felt must be placed across the open ridge between the chalklines. Elastic slater's cement should be used. After the ridge underlayment is placed, cover with elastic slater's cement and set the slate. Make sure the overlap is also covered with elastic cement.

Other versions of ridge caps include the strip-saddle ridge and combing ridge. A strip-saddle ridge is laid in a similar manner to the saddle ridge except that the slate is butted end-to-end and does not overlap. A combing ridge is laid like a saddle ridge except that the top edges of the slates do not alternate. In this case, the top edges of the east or north facing combing slates extend beyond the opposite slates by 1" or less. An alternate version of a combing ridge is known as the Cox-comb ridge. With this type of ridge, the combing slate extensions project alternately on either side of the ridge.

There are a number of methods of capping hips on slate roofs. Saddle or strip-saddle hips are installed like a ridge cap except that they are fastened to 3"-wide cant strips installed along the length of, and on both sides of the hip. The main roof slates are trimmed and installed up to the cant strips. Another type of hip is the mitered hip. This type of hip cap is formed by mitering the slates with a saw so that they fit tightly together. It is installed on a cant strip like strip-saddle or saddle hips. Sometimes, slip-sheet underlayment is used under each course of slate on mitered hips. A variation of the mitered hip is known as the fantail hip. It is installed like a mitered hip except that the pointed lower ends of the slates are clipped off at a 90-degree angle to the slope of the hip.

Another type of hip cap is known as the Boston hip. This type of cap is like a saddle hip, but it is woven in with the regular slate courses and does not require cant strips.

In dry regions, slate or synthetic slate shingles may be nailed to properly spaced wood strips without a felt underlayment. These roofs should not be used on slopes of less than 4 in 12.

10.2.0 Clay, Ceramic, and Concrete Tile Installation

Clay, ceramic, and concrete (synthetic) tile are available in a variety of shapes. The clay types are generally Mission (barrel or S-type) or Spanish styles; the ceramic and concrete types are generally flat tiles.

Conventional tiles are not generally suited for application in areas where large amounts of ice or snow buildup are expected. In some limited cases, they can be used if a waterproof membrane and other water-sealing methods are used on the roof deck. Depending on local or state code requirements, tiles should not be applied to roofs with a slope of less than 3 in 12 or 4 in 12. In addition, the roof structure must be capable of supporting the load, especially the clay or ceramic types.

Clay, ceramic, and concrete tiles are normally fastened with copper nails if allowed by local or state building codes; otherwise, galvanized or stainless steel nails should be used. All nails should be driven to a nickel's thickness above the tile to prevent cracking the tile. In certain high-wind areas of the country, special hurricane fastening systems that use stainless steel clips, brackets, and nails, along with silicon mastic sealing of each tile, are required by local or state codes in order to secure the tiles (*Figures 113* and *114*).

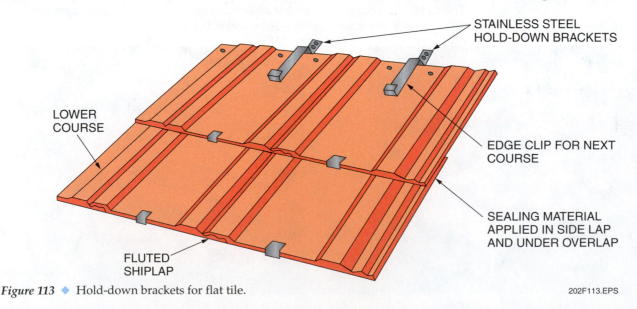

STAINLESS STEEL HOLD-DOWN BRACKETS

LOWER COURSE

EDGE CLIP FOR NEXT COURSE

SEALING MATERIAL APPLIED IN SIDE LAP AND UNDER OVERLAP

FLUTED SHIPLAP

Figure 113 ◆ Hold-down brackets for flat tile.

202F113.EPS

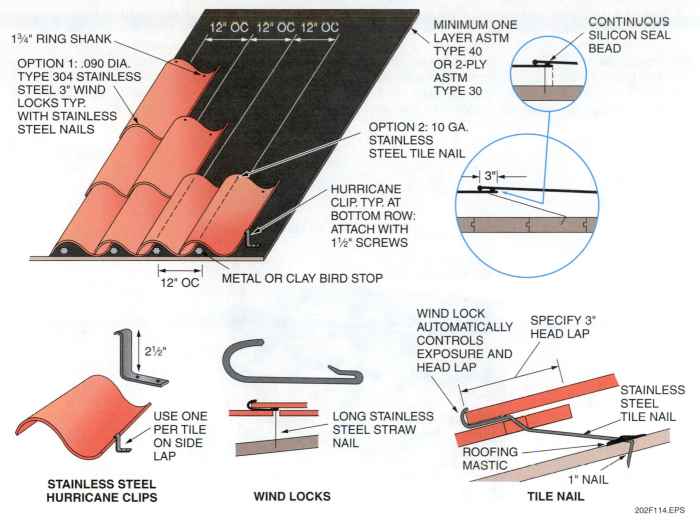

12" OC 12" OC 12" OC

1¾" RING SHANK

OPTION 1: .090 DIA.
TYPE 304 STAINLESS
STEEL 3" WIND
LOCKS TYP.
WITH STAINLESS
STEEL NAILS

MINIMUM ONE
LAYER ASTM
TYPE 40
OR 2-PLY
ASTM
TYPE 30

CONTINUOUS
SILICON SEAL
BEAD

OPTION 2: 10 GA.
STAINLESS
STEEL TILE NAIL

3"

HURRICANE
CLIP. TYP. AT
BOTTOM ROW:
ATTACH WITH
1½" SCREWS

12" OC

METAL OR CLAY BIRD STOP

2½"

USE ONE
PER TILE
ON SIDE
LAP

WIND LOCK
AUTOMATICALLY
CONTROLS
EXPOSURE AND
HEAD LAP

SPECIFY 3"
HEAD LAP

LONG STAINLESS
STEEL STRAW
NAIL

STAINLESS
STEEL
TILE NAIL

ROOFING
MASTIC

1" NAIL

**STAINLESS STEEL
HURRICANE CLIPS**

WIND LOCKS

TILE NAIL

202F114.EPS

Figure 114 ◆ Examples of a high-wind, hold-down system for barrel-style tiles.

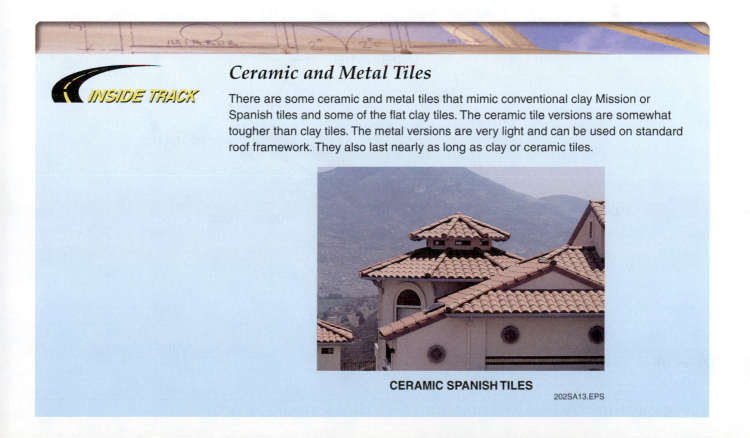

INSIDE TRACK

Ceramic and Metal Tiles

There are some ceramic and metal tiles that mimic conventional clay Mission or Spanish tiles and some of the flat clay tiles. The ceramic tile versions are somewhat tougher than clay tiles. The metal versions are very light and can be used on standard roof framework. They also last nearly as long as clay or ceramic tiles.

CERAMIC SPANISH TILES

202SA13.EPS

The roof deck preparation will vary depending on the tile manufacturer's instructions and the type of tile. Some may not require underlayment, depending on the deck material. However, underlayment or, preferably, waterproof membrane should be used on wood-sheathed roof decks. Some tiles can be fastened directly to the roof deck surface, while others require battens to be laid first to anchor the tiles. Battens are usually pressure-treated pine spaced at intervals that match the tile exposure (*Figures 115* and *116*). They are placed with horizontal gaps of ½" to 1" every 3' to 4' to allow air circulation and drainage of any incidental water down the slope to the eaves.

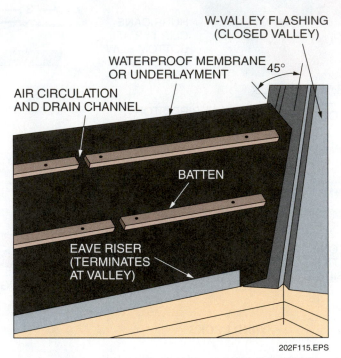

202F115.EPS

Figure 115 ◆ Typical flat tile deck preparation.

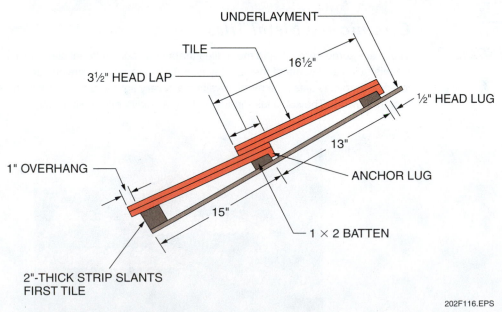

202F116.EPS

Figure 116 ◆ Typical batten spacing for flat tile.

Valleys should be underlaid with 36"-wide waterproof membrane and 24"-wide W-metal flashing made of copper, stainless steel, or galvanized steel. Further preparations may also be required, including 2 × 2s along all ridges and hips, starter strips, and flashing, or 1 × 3s nailed to rake rafters to allow the tiles to extend further sideways. Always check the manufacturer's instructions for specific requirements.

Flat styles of tile have a plain or fluted shiplap along the sides, and sometimes a fluted overlap along the top and bottom as well (*Figure 117*). The fluted shiplaps and overlaps help to prevent water leakage and wind-driven rain penetration between the tiles.

Flat tiles that use battens have a lug on the bottom of the tile that hooks over the batten to help hold the tile in place after it is fastened down (*Figures 116* and *117*). Flat tiles usually require a starter strip and some type of flashing at the eaves, as shown in *Figure 117*.

The roof projection flashing for flat tile roofs is accomplished in essentially the same way as for shingle roofs (*Figure 118*). Some high-profile tiles will require a very soft copper flashing that can be molded and cemented to the surface of the tile.

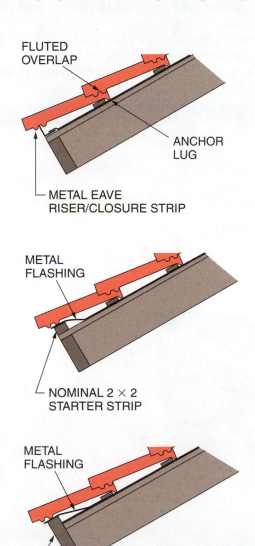

202F117.EPS

Figure 117 ◆ Typical board application.

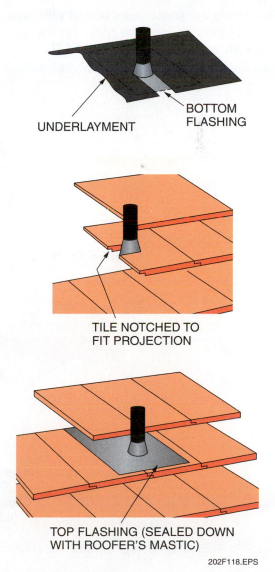

202F118.EPS

Figure 118 ◆ Typical roof projection flashing for flat tiles.

Eliminating Leakage Points

Hips, valleys, and ridges are the most likely areas for leaks in tile roofs. The manufacturer's instructions must be followed exactly to prevent these kinds of leaks.

Figure 119 shows typical rake tiles and a hip or ridge cap treatment used for flat tile roofs.

Barrel-style tiles (Mission or Spanish) are applied as a two-piece pan and cover (*Figure 120*) or as a one-piece pan combined with a cover (*Figure 121*).

These tiles are sometimes sealed with a mastic at the side junctions of the tiles. Typical installations are shown in *Figures 120* and *121*.

Common types of barrel-style flashing are shown in *Figure 122*. These types of flashing may be constructed of copper, galvanized steel, or stainless steel and are used to protect roof projections, valleys, and walls.

Extreme care must be exercised at hips and valleys to ensure perfect angle cuts. A power saw with a masonry blade must be used to obtain these cuts.

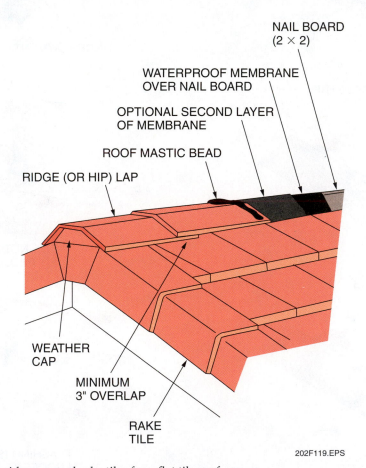

202F119.EPS

Figure 119 ◆ Typical hip or ridge cap and rake tiles for a flat tile roof.

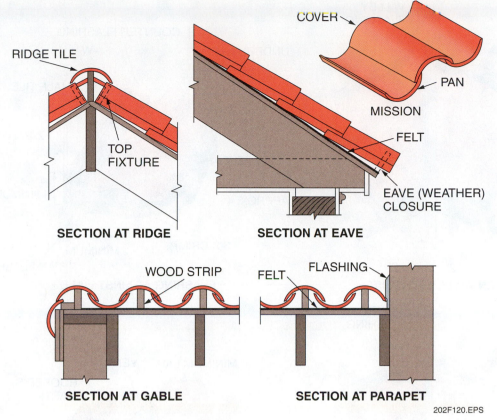

COVER

PAN

MISSION

RIDGE TILE

TOP FIXTURE

SECTION AT RIDGE

FELT

EAVE (WEATHER) CLOSURE

SECTION AT EAVE

WOOD STRIP

FELT

FLASHING

SECTION AT GABLE

SECTION AT PARAPET

202F120.EPS

Figure 120 ◆ Mission barrel-style tile.

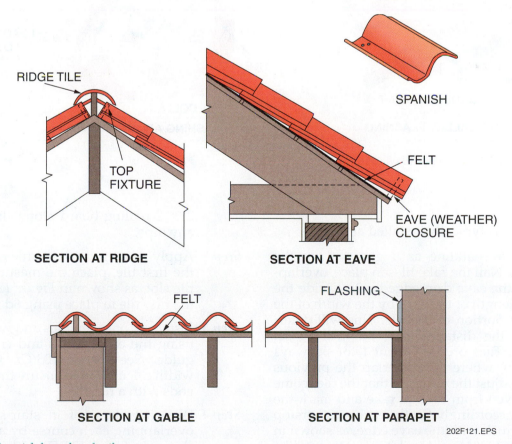

SPANISH

RIDGE TILE

TOP FIXTURE

SECTION AT RIDGE

FELT

EAVE (WEATHER) CLOSURE

SECTION AT EAVE

FELT

FLASHING

SECTION AT GABLE

SECTION AT PARAPET

202F121.EPS

Figure 121 ◆ Spanish barrel-style tile.

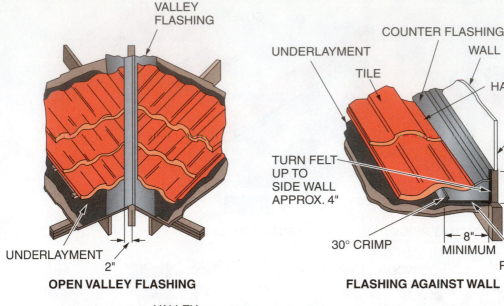

OPEN VALLEY FLASHING

VALLEY FLASHING

UNDERLAYMENT

2"

COUNTER FLASHING
WALL
UNDERLAYMENT
TILE
HALF TILE
Z-BAR
4" MINIMUM
TURN FELT UP TO SIDE WALL APPROX. 4"
30° CRIMP
8" MINIMUM
PAN FLASHING

FLASHING AGAINST WALL

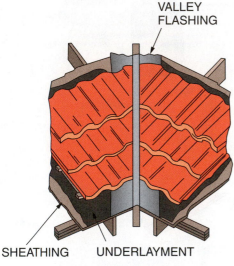

VALLEY FLASHING

SHEATHING   UNDERLAYMENT

CLOSED VALLEY FLASHING

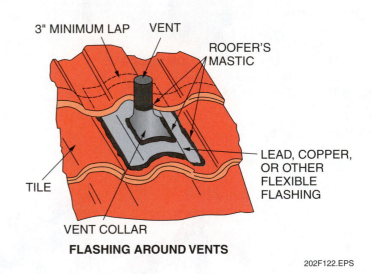

3" MINIMUM LAP   VENT
ROOFER'S MASTIC
TILE
LEAD, COPPER, OR OTHER FLEXIBLE FLASHING
VENT COLLAR

FLASHING AROUND VENTS

202F122.EPS

Figure 122 ◆ Typical barrel-style tile flashing.

Spanish tile is typically installed as follows:

Step 1 Snap a chalkline as a guide along the eaves. Nail the rake tiles in place, overlapping the eave drip edge by 2". Divide the total length of the eave by the width of the cover portion of the tile. This will determine the distance each tile will be set apart. Tiles provide about 1" of sideways leeway where they overlap the previous tile. Adjust the width so that the tiles come out even from rake to rake and mark the eave accordingly. Set the eave weatherstop tiles flush with the eave edge, as shown in *Figure 123(A)*, and nail it into place. Nail a

2 × 2 nailing board along the ridge and any hips.

Step 2 Apply roofing mastic at tile overlaps. For the first tile, place the mastic in the rake tile slot, as shown in *Figure 123(B)*. Secure the first tile in place using 6d nails.

Step 3 Install six or seven tiles and weather-stops using the eave marks and chalkline as a guide. See *Figure 123(C)*. Check your width calculation to ensure that the course ends with a full tile.

Step 4 Work up the roof in stair-step fashion, overlapping each course by at least 3", as shown in *Figure 123(D)*.

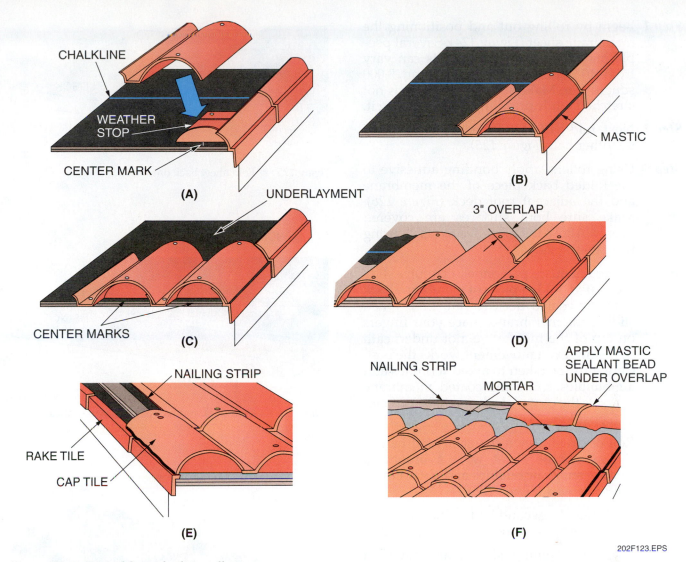

CHALKLINE

WEATHER STOP

CENTER MARK

(A)

UNDERLAYMENT

MASTIC

(B)

CENTER MARKS

(C)

3" OVERLAP

(D)

NAILING STRIP

RAKE TILE

CAP TILE

(E)

NAILING STRIP

MORTAR

APPLY MASTIC SEALANT BEAD UNDER OVERLAP

(F)

202F123.EPS

Figure 123 ◆ Typical Spanish tile installation.

Step 5 When the opposite rake is reached, stop at the next to the last tile and nail the rake tiles in place. Then, nail a 1 × 3 nailing strip from the eaves to the ridge. Position it so that it will line up with the holes at the top of the last tile. Apply mastic to the rake tile slot and place a ridge (cap) tile with mastic on the opposite edge into the rake slot and the pan of the previous tile, as shown in *Figure 123(E)*. Nail the tile into place through the nailing strip. Repeat for each course of tile.

Step 6 When the roof is complete to the ridge 2 × 2 nailing board, lay a bed of tinted mortar and set the ridge starter tiles. Then, lay mortar along both sides of the nailer and over the roof tiles. See *Figure 123(F)*. Position the cap tiles, setting them into the mortar and nailing them to the ridge nailing board. When installing the cap tiles, make sure that they overlap by at least 3".

11.0.0 ◆ SINGLE-PLY ROOFING APPLICATION

Among the membrane-type roofing systems, non-heat bonded single-ply roofing is usually the easiest to install. In most cases, it is accomplished by roofing contractors experienced with the product and its application.

Normally, single-ply membrane roofing is applied over several layers of insulation board on flat roofs. In the following description, the methods of applying only a few sections of fully-adhered EPDM-type membrane are covered without consideration for flashing of roof projections, walls, or eaves and with the assumption that the insulation has been installed.

Most manufacturers of membrane roofing systems market specific accessories such as flashing and vent boots for use with their product. The installation of these accessories varies considerably among manufacturers, so the manufacturer's specific instructions must be followed exactly. Single-ply roofing is installed as follows:

Step 1 Begin by rolling out and positioning the membrane sheet (*Figure 124*). Several people may be needed. The sheets can vary from 10 square feet to more than 5,000 square feet. Make sure the sheeting is not stretched while unrolling or positioning it.

Step 2 Next, fold half of the sheet back on top of the other half (*Figure 125*).

Step 3 Using rollers, apply bonding adhesive to the folded back piece of the membrane and the adjacent roof deck (*Figure 126*). Make sure both surfaces are covered evenly. Allow the adhesive to start curing to a tacky state.

Step 4 Using adequate personnel, carefully slide the coated, folded piece of membrane back onto the coated deck (*Figure 127*). When sliding the membrane, place your fingers on top of the membrane, not underneath it. A slow, even movement works the best. Care must be taken to avoid wrinkles and air bubbles. Once the coated membrane contacts the coated deck, it will be difficult to reposition.

Step 5 Roll or brush down the cemented half of the membrane sheet with a broom to remove small air bubbles and obtain maximum contact (*Figure 128*).

Step 6 Fold back the second half of the sheet and repeat the above process.

Step 7 Position adjoining sheets, as shown in *Figure 129*, allowing for the specified overlap splice. Before splicing, bond the sheet to the deck. Avoid getting bonding adhesive on the splice area of both sheets. Apply splicing cement to the splice area and overlap the sheets. After the splice is complete, protect the splice with lap sealant.

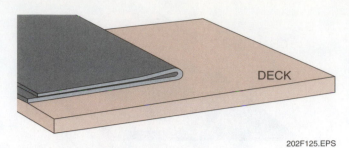

202F125.EPS

Figure 125 ◆ Fold sheet back onto itself.

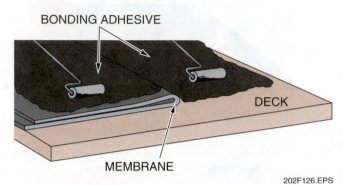

202F126.EPS

Figure 126 ◆ Coat deck and half of sheet with adhesive.

202F127.EPS

Figure 127 ◆ Slide half of sheet onto coated deck.

202F124.EPS

Figure 124 ◆ Place and position membrane sheet.

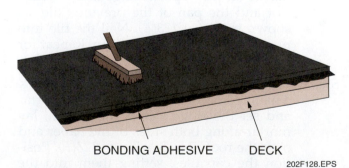

202F128.EPS

Figure 128 ◆ Brush cemented half of sheet to obtain contact.

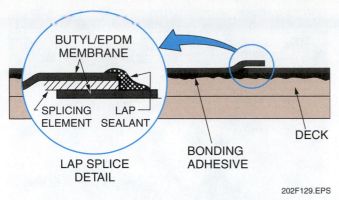

Figure 129 ◆ Apply and splice adjacent sheets.

Step 8 If desired, a fully-adhered EPDM roof can be covered with round ballast stone about ¾" to 1½" in diameter; however, it is not necessary.

12.0.0 ◆ TORCH-DOWN ROOFING APPLICATION

One of the most popular types of membrane roofing material for flat-roof residential applications is an APP product (torch-down roofing). This type of material is installed using the following guidelines:

Step 1 Install a manufacturer-specified underlayment starting at the lowest edge of the roof deck. Overlap the remaining courses of underlayment as recommended by the manufacturer. Fasten the underlayment to the roof deck with washer-headed nails at the manufacturer's recommended spacing (*Figure 130*).

Step 2 After igniting a propane torch (*Figure 131*), start applying the roofing material at the lowest edge of the roof deck (*Figure 132*). Heat the exposed bottom-side asphaltic surface of the roofing roll until it is melted. Be careful not to overheat the material so that the surface begins to run or catches fire. While the surface section is melted, unroll that section on to the underlayment and press it down to bond it to the underlayment. Progressively heat and unroll until the roof edge is covered or the roll is exhausted. If a new roll must be spliced, heat the end of the new roll and overlap the end of the previous roll by about 6" or as recommended by the manufacturer. Once the first course is completed, continue applying roofing material up the roof, overlapping each lower course by about 6" or as recommended by the manufacturer.

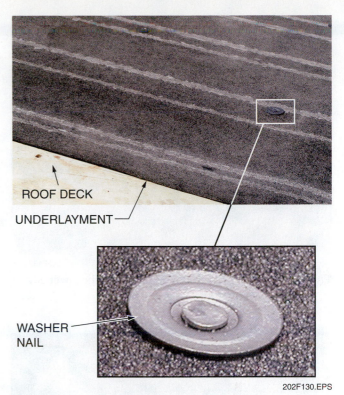

Figure 130 ◆ Underlayment installation.

> **WARNING!**
>
> Be careful when using the torch to avoid setting fire to the roofing material or underlayment. Have a fire extinguisher available in case of fire.

Step 3 Once the roof deck is covered, seal all overlapping seams by melting the surface of the roofing material adjacent to the edge of each seam and troweling the melted material over the seam edge (*Figure 133*).

Figure 131 ◆ Torch-down roofing propane torch.

202F132.EPS

Figure 132 ◆ Applying roofing material to the lower edge of the roof.

DRIP EDGE

202F134.EPS

Figure 134 ◆ Drip edge installation.

202F133.EPS

Figure 133 ◆ Sealing roofing-course seams.

202F135.EPS

Figure 135 ◆ Applying drip-edge strips.

Step 4 Apply manufacturer-recommended drip edge to the roof deck (*Figure 134*).

Step 5 Cut roofing material into 8"- to 9"-wide strips, then heat-bond the strips over the drip edges and main roofing material (*Figure 135*). Once the strips are applied, melt the surface of the strips and adjacent roofing material and seal the edges with a trowel (*Figure 136*).

CAUTION

Torch-down roofing material usually must be covered with a coating to shield it from the UV rays of the sun. Make sure to follow any manufacturer's instructions for applying protective coatings.

202F136.EPS

Figure 136 ◆ Sealing drip-edge strips.

Figure 137 ◆ Applying a protective coating.

302F137.EPS

Step 6 After the roof has cured for a day or so, coat the surface with a manufacturer-recommended roof coating or material such as a fibered aluminum coating (*Figure 137*).

13.0.0 ◆ ROOF VENTILATION AND ICE EDGING

This section covers the installation of various types of roof venting and ice edging systems.

13.1.0 Roof Ventilator Installation

Proper attic ventilation is necessary to allow heat and moisture to escape so that damage to the roofing and roof deck does not occur. In the winter, ventilation keeps the roof deck cold and reduces the buildup of ice on the eaves. This helps prevent water penetration through the roof and subsequent water damage to the structure. It also carries away moisture so that it does not condense on the roof deck, which can cause rotting. In the summer, excess heat can cause overheating of composition shingles, resulting in early failure of the roof.

Residential attics are generally ventilated by convection vents in the form of gable vents or roof-mounted ridge or box vents. Sometimes, electric-powered or wind-powered turbine vents or fans are used (*Figure 138*). The air used for ventilation is provided by soffit vents at the eaves of the roof. The amount of soffit ventilation (in square feet or inches) must be equal to or greater than the amount of roof ventilation.

In residential structures, the proper amount of convection-type ventilation for an unheated attic space is usually defined as 1 sq ft of ventilation for every 300 sq ft of attic area with 50 percent in the roof for exhaust and 50 percent in the eaves for intake. The attic area is calculated using the exterior foundation dimensions of the structure. For example, the amount of ventilation for a residence with an exterior foundation measurement of 40' × 60' would be calculated as follows:

$$40' \times 60' = 2,400 \text{ sq ft (attic area)}$$

$$2,400 \text{ sq ft} \div 300 \text{ sq ft} = 8 \text{ sq ft}$$
(total ventilation required)

$$8 \text{ sq ft} \times 144 \text{ sq in per sq ft} = 1,152 \text{ sq in}$$

$$1,152 \text{ sq in} \div 2 = 576 \text{ sq in for the ridge}$$
and 576 sq in for the soffits

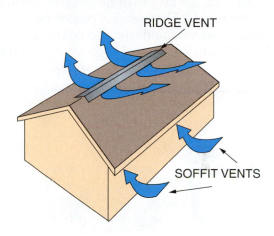

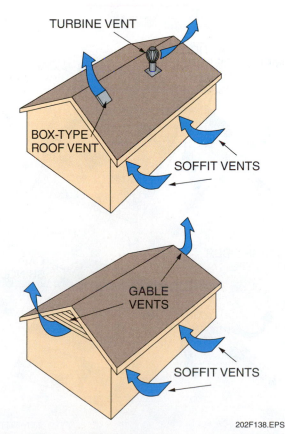

Figure 138 ◆ Residential roof vents.

202F138.EPS

Ventilation Requirements

INSIDE TRACK

Always check local codes for the proper ventilation requirements of a structure. In some areas, the amount of air exchange required may dictate fan-assisted ventilation if the capacity of free-air ventilation devices is not adequate.

Based on the calculated ventilation requirement, appropriately sized roof and soffit convection ventilation devices would be selected and installed.

Ridge vents, box vents, and turbine vents are available in a variety of styles and sizes for residential use. Different types of ridge vents are illustrated in *Figure 139*. The metal or plastic ridge vent is available in several patterns and colors. It is sold in 10' lengths and can be installed over most roofing materials.

The flexible, plastic composition vent is available in 4' lengths. An inert, coarse fiber vent is available in rolls. Flexible plastic and rolled coarse fiber vents can only be used over composition shingles because both are designed to be covered with cap shingles that match the roof. Most residential customers prefer the shingle-covered vents because they tend to blend into the overall roof.

The manufacturer's specifications for a ventilation device must be consulted to determine the amount of free-air ventilation (in square feet or inches) that the device will provide. Ridge ventilators are probably the most efficient and are usually rated in square inches of free-air ventilation per linear foot of the product.

13.1.1 Box Vent Installation

Box vents (*Figure 140*) are easily installed on most roofing materials. On new roofs, the proper size hole is cut in the roof and the hole is surrounded with at least a 24"-wide waterproof membrane. When the roofing courses reach the area of the ventilator, the ventilator is installed with the lower edge of its flashing overlapping the course below it. The flashing is fastened to the roof at the top and sides. Then, roofing courses are applied over and cemented to the flashing at the top and sides of the ventilator.

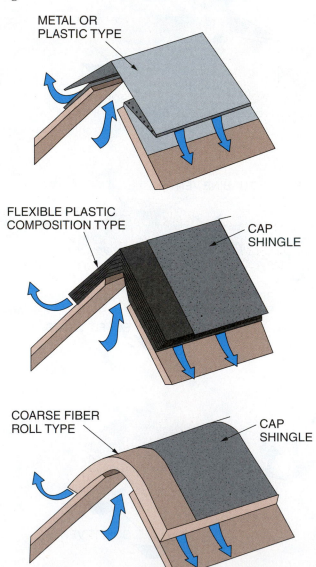

METAL OR PLASTIC TYPE

FLEXIBLE PLASTIC COMPOSITION TYPE

CAP SHINGLE

COARSE FIBER ROLL TYPE

CAP SHINGLE

202F139.EPS

Figure 139 ◆ Typical ridge vents.

202F140.EPS

Figure 140 ◆ Typical box vent.

On existing roofs, the hole is cut, and the upper flashing of the ventilator is slid under the courses above the hole and fastened to the roof. The side and bottom flashing is not fastened; however, the side flashing of the ventilator is completely cemented down to the roofing under the sides. The bottom flashing is usually not cemented.

13.1.2 Ridge Vent Installation

Use the following procedure to install ridge vents:

Step 1　Determine where the roof vent slots will be cut on the ridges and any hips (*Figure 141*). Slots cut along a ridge should start and stop approximately 12" from the gable (terminal) ends, any vertical walls, any higher intersecting roof, any roof projections at the ridge, any valleys, and any hip joints. Slots cut in hips should end 24" above the eaves and should be in 24" sections separated by 12" to maintain maximum roof strength. For appearance, the roof ridges and hips should be covered completely with the roof vent material to maintain a continuous roof line, as shown in *Figure 142*.

Step 2　Next, refer to the manufacturer's specifications and note the width of the slot required for the vent being used. Determine the roof construction (ridge board or no ridge board). If a ridge board is used (*Figure 143*), add 1½" to the required slot width and divide the result by two to find the total slot width to be cut on each side of the peak. If no ridge board is used, only divide the required slot width by two to find the slot width to be cut on each side of the peak.

Step 3　At the peak, measure down the roof on both sides for half of the required slot width as determined above and snap a chalkline along any ridges or hips at both sides of the peak or hip (*Figure 144*). Mark the start and stop point for each slot, as previously specified.

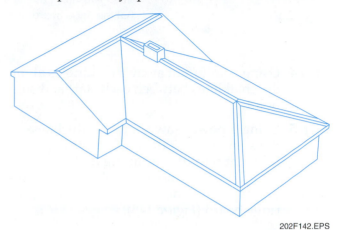

202F142.EPS

Figure 142　◆　Example of a continuous roof line.

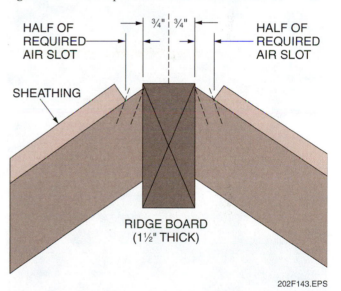

202F143.EPS

Figure 143　◆　Determining total slot width.

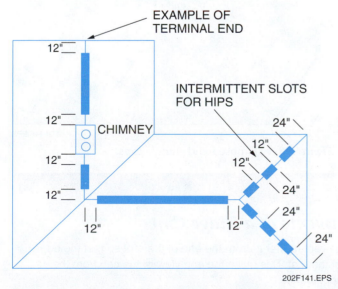

202F141.EPS

Figure 141　◆　Example of slot cutout placement.

202F144.EPS

Figure 144　◆　Snapping a chalkline for slot width.

Precautions When Cutting a Ridge Vent

Use goggles and make sure your footing is solid along the ridges when cutting with a power saw. Make sure the power cord is laid out so that there is no chance of getting entangled with it during the cutting operation. Make sure that the saw depth is set so that the ridge board or rafters are not cut.

Step 4 Using a knife, cut away the shingles along the chalklines between each start and stop point.

Step 5 Using a power saw set to the thickness of the shingles plus the roof sheathing, carefully cut into the sheathing and along the chalklines to remove the sheathing without damaging the rafters or, if present, the ridge board (*Figure 145*).

Step 6 Place a shingle ridge cap at each gable end, before and after each roof projection, at each vertical wall, at each higher intersecting roof joint, and at each eave end for any hips (*Figure 146*). This prevents any water intrusion at or under the exposed ends of the vent from penetrating into or through the sheathing.

Step 7 If packaged in rolls, unroll the vent material and temporarily tack it in place over the ridges and hips or secure the flexible plastic composition sections to the roof over the ridges and hips. Make sure that the vent covers the entire length of all ridges and hips (*Figure 147*). If rigid vent sections (*Figure 148*) are being used, make sure that vents are secured through the prepunched holes.

Step 8 Cut and taper the tabs from a number of three-tab standard or architectural cap shingles for use as cap shingles.

202F145.EPS

Figure 145 ◆ Cutting slots.

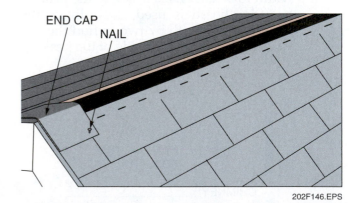

END CAP

NAIL

202F146.EPS

Figure 146 ◆ Exposed-end shingle caps.

Alternate End Treatments of a Ridge Cap

Some contractors cut the ridge vent further in from the end of the roof so that the ridge cap is not flush with the end of the roof. This hides the end view of the cap from the ground for a more attractive appearance.

Plastic or Metal Ridge Vent

If using a rigid plastic or metal vent not intended to be capped with shingles, fasten the vent to the roof through the flanges using a non-corrosive fastener with a neoprene seal washer; then insert an end cap in the exposed ends of the vents, if required.

ROLLED VENT MATERIAL   RIGID VENT MATERIAL

202F147.EPS

Figure 147 ◆ Positioning vent material.

NAILING LINE FOR CAP SHINGLES

PREPUNCHED MOUNTING HOLE

202F148.EPS

Figure 148 ◆ Typical rigid vent section.

Step 9 Starting from the exposed ends of a ridge or hip and using sufficiently long nails, fasten the cap shingles to the roof through the vent material nailing line (*Figure 148*). The cap shingles are applied and mated over the ridges and hips in the same manner as described previously. See *Figure 149*. However, make sure that the shingles are nailed snugly without compressing the vent material. It is advisable to place a bead of roofing cement under the overlapping sections of the shingles to help secure them.

13.2.0 Ice Edging Installation

Many sloped-roof residences in the north have ice damming problems on the roof, usually at the eaves. This is especially true for those residences with finished attics where insulation and venting under the roof deck is limited to the rafter space. In most cases, when a new roof is installed, the application of a waterproof membrane from the eaves up the slope of the roof to a point that is at least 24" beyond the inside wall will prevent water backed up behind any ice dams from penetrating into the structure. However, on a problem residence with an existing roof that lacks a waterproof membrane underlayment, it may be desirable to install an ice edge at the eaves. On most roofing materials, ice dams cannot be easily removed. In addition, ice dams can damage some roofing materials, including slate.

An ice edge is an exposed metal sheeting mounted from the eaves up the slope of the roof from 18" to 36". It provides a shield against water penetration and allows any ice dams that form to be quickly shed from the roof during any brief thaws, thus reducing the chances of a large ice dam buildup. The width of the ice edging used is a factor of how steep the roof is and how thick the ice usually becomes, which determines how much water backs up on the roof. Normally, continuous sheets of plain or tinted/painted aluminum flashing or special, standing-seam aluminum panels are used as the ice edge (*Figure 150*); however, copper flashing can be used where corrosion is a

OVERLAPPING
CAP SHINGLES

202F149.EPS

Figure 149 ◆ Applying cap shingles over a vent.

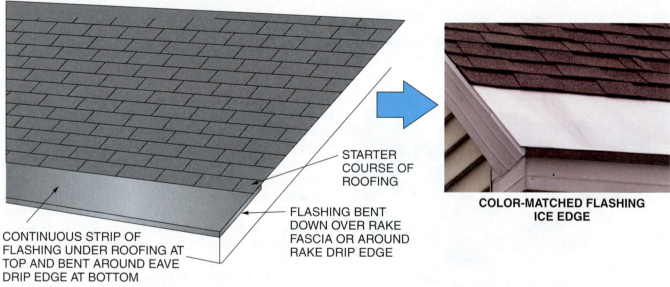

STARTER
COURSE OF
ROOFING

FLASHING BENT
DOWN OVER RAKE
FASCIA OR AROUND
RAKE DRIP EDGE

CONTINUOUS STRIP OF
FLASHING UNDER ROOFING AT
TOP AND BENT AROUND EAVE
DRIP EDGE AT BOTTOM

COLOR-MATCHED FLASHING
ICE EDGE

PREFABRICATED STANDING-SEAM
PANELS UNDER ROOFING AT TOP
AND BENT AROUND EAVE DRIP
EDGE AT BOTTOM

STARTER
COURSE OF
ROOFING

BENT DOWN
OVER RAKE
FASCIA OR
AROUND RAKE
DRIP EDGE

BRIGHT ALUMINUM
STANDING-SEAM PANELS

202F150.EPS

Figure 150 ◆ Types of ice edging.

Residential Secondary Roof Systems

An alternative for roofs that cannot be insulated properly from inside the residence is an insulated, secondary roof system installed over the original roof deck. These systems have spacing between a rigid insulation layer and a second roof deck to allow free airflow over the insulation and under the second roof deck. This greatly reduces the melting of snow and the resulting ice dam on the second roof deck. These secondary roof systems are available from several manufacturers. They can also be used in hot climates to reduce the heat load in a residence with similar insulation problems. The airflow of these systems can also reduce the heat load on the roofing materials used for the second deck.

$7/16$" OSB DECK

VENTILATOR SPACERS

ISOCYANURATE INSULATION BLOCK

BONDED SECOND-DECK ROOFING PANEL

202SA14.EPS

problem. The standing-seam panels are more expensive and are not subject to buckling due to temperature extremes. Their surface appearance is more uniform and does not appear wavy.

There are disadvantages in the use of ice edges. One is that they are generally not considered attractive unless colored to match the roof. The possible exception is plain copper used with slate roofing. Another disadvantage is that for ice edges to be effective in shedding ice, eave troughs cannot be used on the building. This can lead to several severe hazards such as injury or death to people or damage to property or foliage, including damage to the siding of the residence caused by falling ice. In addition, water draining from the roof can collect and penetrate under the foundation and/or into a basement.

Ice Edge Hazards

Before ice edges are considered for use on the eaves of a building, the hazard of falling ice causing injury, death, or property damage under each eave must be carefully evaluated. With ice edging, a long, heavy ice dam (1,000 lbs or more) along an entire roof eave can be released from the eave without warning. In addition, attempts to remove icicles may cause an ice dam along an entire eave to be released.

202SA15.EPS

Summary

Many types of roofing materials and commercial roofing systems are available. They are used to protect a structure and its contents from the elements and vary depending on geographic location. This module covered the most common materials used on residential and small commercial structures, including composition shingles, roll roofing, wood shingles/shakes, slate, tile, metal, and membrane roofing.

While carpenters may not usually be required to install roofing, they can be involved in preparing the roof surface before the installation of roofing materials. Therefore, carpenters should be familiar with the basic preparation and installation of common roofing materials.

Notes

1. Organic asphalt composition shingles have been largely replaced by _____ shingles.
   a. asbestos
   b. wood
   c. fiberglass
   d. synthetic tile

2. Architectural shingles have a typical lifespan of _____ years.
   a. 5 to 12
   b. 15 to 20
   c. 20 to 25
   d. 25 to 40

3. Roll roofing is available in weights of _____ lbs.
   a. 20 to 40
   b. 30 to 70
   c. 50 to 90
   d. 70 to 120

4. The most popular wood shingles and shakes are made of _____ .
   a. yellow pine
   b. red cedar
   c. cypress
   d. redwood

5. Slate roofing weighs about _____ pounds per square foot.
   a. 3 to 7
   b. 5 to 9
   c. 7 to 10
   d. 9 to 12

6. Tile roofing weighs about _____ pounds per square foot.
   a. 3 to 7
   b. 5 to 9
   c. 7 to 10
   d. 9 to 12

7. Terne metal is metal coating comprised of _____ .
   a. 50 percent brass and 50 percent lead
   b. 80 percent lead and 20 percent tin
   c. 60 percent tin and 40 percent copper
   d. 20 percent copper and 80 percent zinc

8. There are _____ general categories of premanufactured membrane roofing systems.
   a. two
   b. three
   c. four
   d. five

9. The most common thermoset single-ply membrane is known as _____ .
   a. APP
   b. SBS
   c. PVC
   d. EPDM

10. Two common thermoplastic single-ply membranes are _____ .
    a. APP and BUR
    b. SBS and PVC
    c. PVC and TPO
    d. EPDM and TPO

11. The standard roofing hammer is also called a _____ .
    a. shingle hatchet
    b. slater's hammer
    c. claw hammer
    d. cricket hammer

12. A slate cutter can be used to _____ .
    a. remove nails
    b. bend metal flashing
    c. punch nail holes
    d. cut tile

13. The edge of a ladder must extend _____ above the edge of a roof.
    a. 1'
    b. 2'
    c. 3'
    d. 4'

14. What is the normal percentage that is added to a roof estimate for waste?
    a. 5 percent
    b. 10 percent
    c. 15 percent
    d. 20 percent

15. The most common composition shingle applied to residential structures is the _____ shingle.
    a. two-tab
    b. three-tab
    c. architectural
    d. strip

16. The abbreviation that defines the distance that a shingle overlaps the shingle below is _____ .
    a. E
    b. TL
    c. HL
    d. W

17. The 6" pattern of laying three-tab shingles means that 6" are _____ .
    a. added to each course
    b. subtracted from each course
    c. subtracted from every other course
    d. subtracted from the second course, 12" from the third course, 18" from the fourth, and so on

18. There are _____ types of closed valleys.
    a. two
    b. three
    c. four
    d. five

19. Step flashing is used on _____ .
    a. valleys
    b. slopes against walls
    c. hips
    d. vent pipes

20. Double-coverage roll roofing with concealed nails can be used on slopes as low as _____ .
    a. ¼ in 12
    b. 1 in 12
    c. 2 in 12
    d. 3 in 12

21. If permitted by local codes, clay, concrete, or ceramic tiles should be installed with _____ .
    a. ungalvanized roofing nails
    b. drywall screws
    c. copper nails
    d. construction adhesive

22. A lug under the top edge of a flat tile is used to _____ .
    a. anchor the tile
    b. space the tile
    c. prevent breakage of the tile
    d. keep the edges of the tile even

23. The flutes in the overlap at the top and bottom of some flat tiles are primarily used to _____ .
    a. hold the tile in place
    b. space the vertical position of the tile
    c. prevent water penetration under the tile
    d. hold mastic for sealing the tile

24. The attic ventilation area is normally divided between exhaust and inlet vents in the proportion of _____ .
    a. 50 percent exhaust, 50 percent inlet
    b. 60 percent exhaust, 40 percent inlet
    c. 70 percent exhaust, 30 percent inlet
    d. 80 percent exhaust, 20 percent inlet

25. One of the hazards of ice edging is _____.
    a. poor appearance
    b. moisture accumulation
    c. falling ice
    d. water penetration

Trade Terms
Introduced in This Module

Asphalt roofing cement: An adhesive that is used to seal down the free tabs of strip shingles. This plastic asphalt cement is mainly used in open valley construction and other flashing areas where necessary for protection against the weather.

Base flashing: The protective sealing material placed next to areas vulnerable to leaks, such as chimneys.

Bundle: A package containing a specified number of shingles or shakes. The number is related to square foot coverage and varies with the product.

Cap flashing: The protective sealing material that overlaps the base and is embedded in the mortar joints of vulnerable areas of a roof, such as a chimney.

Exposure: The distance (in inches) between the exposed edges of overlapping shingles.

Head lap: The distance between the top of the bottom shingle and the bottom edge of the one covering it.

Overhang: The part that extends beyond the building line. The amount of overhang is always given as a projection from the building line on a horizontal plane.

Pitch: The ratio of the rise to the span indicated as a fraction. For example, a roof with a 6' rise and a 24' span will have a ¼ pitch.

Ridge: The horizontal line formed by the two rafters of a sloping roof that have been nailed together. The ridge is the highest point at the top of the roof where the roof slopes meet.

Roof sheathing: Usually 4 × 8 sheets of plywood, but can also be 1 × 8 or 1 × 12 roof boards, or other new products approved by local building codes. Also referred to as decking.

Saddle: An auxiliary roof deck that is built above the chimney to divert water to either side. It is a structure with a ridge sloping in two directions that is placed between the back side of a chimney and the roof sloping toward it. Also referred to as a cricket.

Scrim: A loosely knit fabric.

Selvage: The section of a composition roofing roll or shingle that is not covered with an aggregate.

Side lap: The distance between adjacent shingles that overlap, measured in inches.

Slope: The ratio of rise to run. The rise in inches is indicated for every foot of run.

Square: The amount of shingles needed to cover 100 square feet of roof surface. For example, square means 10' square or 10' × 10'.

Top lap: The distance, measured in inches, between the lower edge of an overlapping shingle and the upper edge of the lapping shingle.

Underlayment: Asphalt-saturated felt protection for sheathing; 15-lb roofer's felt is commonly used. The roll size is 3' × 144' or a little over four squares.

Valley: The internal part of the angle formed by the meeting of two roofs.

Valley flashing: Watertight protection at a roof intersection. Various metals and asphalt products are used; however, materials vary based on local building codes.

Vent stack flashing: Flanges that are used to tightly seal pipe projections through the roof. They are usually prefabricated.

Wall flashing: A form of metal shingle that can be shaped into a protective seal interlacing where the roof line joins an exterior wall. Also referred to as step flashing.

This module is intended to present thorough resources for task training. The following reference works are suggested for further study. These are optional materials for continued education rather than for task training.

Asphalt Manufacturers Association website, *www.asphaltroofing.org*.

National Roofing Contractors Association website, *www.nrca.net*.

Roof Coating Manufacturers Association website, *www.roofcoatings.org*.

NCCER CURRICULA — USER UPDATE

NCCER makes every effort to keep its textbooks up-to-date and free of technical errors. We appreciate your help in this process. If you find an error, a typographical mistake, or an inaccuracy in NCCER's curricula, please fill out this form (or a photocopy), or complete the online form at **www.nccer.org/olf**. Be sure to include the exact module ID number, page number, a detailed description, and your recommended correction. Your input will be brought to the attention of the Authoring Team. Thank you for your assistance.

Instructors – If you have an idea for improving this textbook, or have found that additional materials were necessary to teach this module effectively, please let us know so that we may present your suggestions to the Authoring Team.

NCCER Product Development and Revision
13614 Progress Blvd., Alachua, FL 32615

Email: curriculum@nccer.org
Online: www.nccer.org/olf

❏ Trainee Guide   ❏ AIG   ❏ Exam   ❏ PowerPoints   Other _____

Craft / Level: _____   Copyright Date: _____

Module ID Number / Title: _____

Section Number(s): _____

Description: _____

_____

_____

_____

_____

Recommended Correction: _____

_____

_____

_____

_____

Your Name: _____

Address: _____

_____

Email: _____   Phone: _____

Thermal and Moisture Protection
27203-07

27203-07
Thermal and Moisture Protection

Overview

A properly insulated building will be comfortable to live or work in and will be economical to heat and cool. Without insulation, warm air will escape the building in cold weather, causing the heating system to operate constantly. This results in an increased use of energy. In hot weather, the air conditioning system will have to work harder, with the same results. Proper insulation in walls, floors, and roof decks will minimize this problem. Vapor barriers are also important. They must be used to prevent moisture from penetrating the building. Moisture can cause a variety of serious problems, including wood decay and mold growth. A skilled carpenter will know how to select and install insulating materials and vapor barriers.

Objectives

When you have completed this module, you will be able to do the following:

1. Describe the requirements for insulation.
2. Describe the characteristics of various types of insulation material.
3. Calculate the required amounts of insulation for a structure.
4. Install selected insulation materials.
5. Describe the requirements for moisture control and ventilation.
6. Install selected vapor barriers.
7. Describe various methods of waterproofing.
8. Describe air infiltration control requirements.
9. Install selected building wraps.

Trade Terms

Condensation
Convection
Dew point
Diffusion
Exterior insulation finish system (EIFS)
Perm
Permeability
Permeable
Permeance
Vapor barrier
Vapor diffusion retarder (VDR)
Vapor retarder
Water stop
Water vapor

Required Trainee Materials

1. Pencil and paper
2. Appropriate personal protective equipment

Prerequisites

Before you begin this module, it is recommended that you successfully complete *Core Curriculum*; *Carpentry Fundamentals Level One*; and *Carpentry Framing and Finishing Level Two*, Modules 27201-07 and 27202-07.

This course map shows all of the modules in *Carpentry Framing and Finishing Level Two*. The suggested training order begins at the bottom and proceeds up. Skill levels increase as you advance on the course map. The local Training Program Sponsor may adjust the training order.

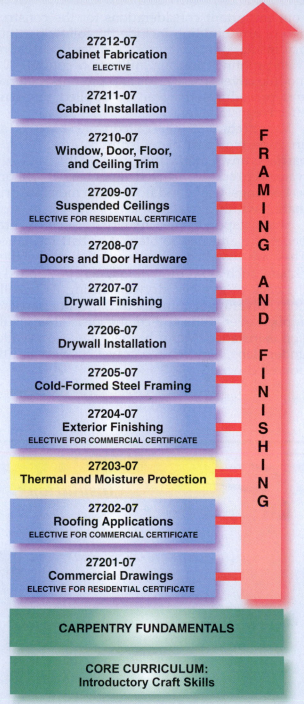

27212-07
Cabinet Fabrication
ELECTIVE

27211-07
Cabinet Installation

27210-07
Window, Door, Floor, and Ceiling Trim

27209-07
Suspended Ceilings
ELECTIVE FOR RESIDENTIAL CERTIFICATE

27208-07
Doors and Door Hardware

27207-07
Drywall Finishing

27206-07
Drywall Installation

27205-07
Cold-Formed Steel Framing

27204-07
Exterior Finishing
ELECTIVE FOR COMMERCIAL CERTIFICATE

27203-07
Thermal and Moisture Protection

27202-07
Roofing Applications
ELECTIVE FOR COMMERCIAL CERTIFICATE

27201-07
Commercial Drawings
ELECTIVE FOR RESIDENTIAL CERTIFICATE

CARPENTRY FUNDAMENTALS

CORE CURRICULUM:
Introductory Craft Skills

FRAMING AND FINISHING

203CMAP.EPS

1.0.0 ◆ INTRODUCTION

Four important considerations for the construction of any building include the following: thermal insulation, moisture control and ventilation, waterproofing, and air infiltration control. This module covers these areas and presents materials and procedures that can be applied to ensure effective installations.

Vapor retarders, also called **vapor barriers,** are an important part of moisture control. A vapor barrier is any material that prevents the passage of water. A properly installed vapor barrier will protect ceilings, walls, and floors from moisture originating within a heated space.

Some vapor barrier materials, such as kraft paper, are attached to blanket or batt insulation. They are installed when insulation is installed. Others, such as aluminum foil, may be applied to the back of gypsum drywall during its installation. Polyethylene film used as a vapor barrier is applied over studs and ceiling joists after insulation is installed. Vapor barriers are also installed under slabs, between the gravel cushion and the poured concrete.

2.0.0 ◆ THERMAL INSULATION

Most materials used in construction have some insulating value. Air is an excellent insulator if it is confined to very small spaces and is kept very still. Manufactured insulation material is based on trapping a large amount of air in a large number of very small spaces to provide resistance to the transfer of heat and sound. Double-pane and triple-pane windows use this method to reduce heat loss.

The amount of insulation in a building directly affects heating and cooling costs. It also affects the value of the building. Some jurisdictions require that a permanent certificate be placed on the building's electrical box stating the insulative properties of material used in the structure. When required, this certificate is completed by the builders or designer.

Insulation

There are many different types of insulation. Each type has specific applications for which it is best suited.

203SA01.EPS

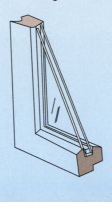 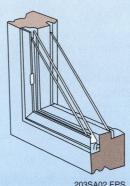

203SA02.EPS

2.1.0 Thermal Resistance Value of Materials

A law of physics is that heat will always flow (or conduct) through any material or gas from a higher temperature area to a lower temperature area.

The term R-value refers to the resistance to conductive heat flow through a material or gas. R-value is expressed as:

$$R = \frac{1}{k} \text{ or } \frac{1}{C}$$

Where

k = amount of heat in British thermal units (Btus) transferred in one hour through 1 sq ft of a material that is 1" thick and has a temperature difference between its surfaces of 1°F; also called the coefficient of thermal conductivity

C = conductance of a material, regardless of its thickness; the amount of heat in Btus that will flow through a material in one hour per sq ft of surface with 1°F of temperature difference

R = thermal resistance; the reciprocal (opposite) of conductivity or conductance

The higher the R-value, the lower the conductive heat transfer. *Table 1* shows the R-values of a number of common building materials including some common insulating materials.

The total heat transmission through a wall, roof, or floor of a structure in Btus per sq ft per hour with a 1° F temperature difference is called the total heat transmission or U-value. It is expressed as follows:

$$U = \frac{1}{R_1 + R_2 + \ldots R_n}$$

Where

$R_1 + R_2 + \ldots R_n$ represents the sum of the individual R-values for the materials that make up the thickness of the wall, roof, or floor

The lower the U-value, the lower the heat transmission.

The R-values of two typical wall structures are shown in *Figure 1*. While the R-values provide a convenient measure to compare heat loss or gain, the total U-value for a structure is used in the calculations for sizing the structure's heating and cooling equipment.

By doubling the R-value of a wall or roof, the conductive heat loss or gain can theoretically be reduced by half. However, it is important to note that as insulation thicknesses are increased, the heat transmission (U-value) is decreased, but not in a direct relationship. Increases of insulation will continue to decrease heat loss, but at lower and lower percentages. At some point, it becomes economically useless to add more insulation. The same is true for double-, triple-, and quadruple-pane windows. It must also be noted that conductive heat loss or gain does not include heat gains or losses due to air leaks or radiation through windows or other openings.

Table 1 R-Values of Common Materials

Material	Thickness	R-Value °F/sq. ft/hr in Btus
Air film and spaces		
Air space bound by ordinary materials	¾"	0.96
Air space bound by ordinary materials	¾" to 4"	0.94
Exterior surface resistance	–	0.17
Interior surface resistance	–	0.68
Masonry units		
Sand and gravel concrete block	4"	0.71
Sand and gravel concrete block	8"	1.11
Sand and gravel concrete block	12"	1.28
Lightweight concrete block	4"	1.50
Lightweight concrete block	8"	2.00
Lightweight concrete block	12"	2.27
Face brick	4"	0.44
Common brick	4"	0.80
Masonry materials		
Concrete, oven-dried sand, and gravel aggregate	1"	0.11
Concrete, undried sand, and gravel aggregate	1"	0.08
Stucco	1"	0.20
General building materials		
Wood sheathing or subfloor	¾"	0.94
Fiberboard sheathing (regular density)	½"	1.32
Fiberboard sheathing (intermediate density)	½"	1.14
Fiberboard sheathing (nailbase)	½"	1.14
Plywood	⅜"	0.47
Plywood	½"	0.62
Plywood	¾"	0.93
Bevel-lapped wood siding	½" × 8"	0.81
Bevel-lapped wood siding	¾" × 10"	1.05
Vertical tongue-and-groove (cedar or redwood)	¾"	1.00
Asbestos-cement	¼"	0.21
Gypsum board	⅜"	0.32
Gypsum board	½"	0.45
Interior plywood paneling	¼"	0.31
Building paper (permeable felt)	–	0.06
Plastic film	–	0.00
Insulating materials		
Fibrous batts (from rock slag or glass)	2" to 2¾"	7.00
Fibrous batts (from rock slag or glass)	3" to 3½"	11.00
Fibrous batts (from rock slag or glass)	5" to 5½"	19.00
SM brand Styrofoam® plastic foam	1"	5.41
TG brand Styrofoam® plastic foam	1"	5.41
IB brand Styrofoam® plastic foam	1"	4.35
Molded polystyrene beadboard	1"	4.17
Polyurethane foam	1"	5.88
Woods		
Fir, pine, and similar softwoods	¾"	0.94
Fir, pine, and similar softwoods	1½"	1.89
Fir, pine, and similar softwoods	2½"	3.12
Fir, pine, and similar softwoods	3½"	4.35
Maple, oak, and similar hardwoods	1"	0.91

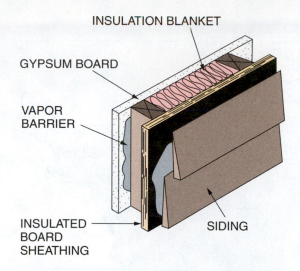

INSULATION BLANKET

GYPSUM BOARD

VAPOR BARRIER

INSULATED BOARD SHEATHING

SIDING

2 × 4 STUD WALL WITH RIGID BOARD:

TYPE	R-VALUE
AIR FILMS*	0.9
¾" WOOD EXTERIOR SIDING	1.0
1" POLYSTYRENE RIGID BOARD	5.0
3½" BATT OR BLANKET INSULATION	11.0
VAPOR BARRIER	0.0
½" GYPSUM BOARD	0.5
	18.4 TOTAL R

2 × 6 INSULATED STUD WALL:

TYPE	R-VALUE
AIR FILMS*	0.9
¾" WOOD EXTERIOR SIDING	1.0
¾" INSULATION BOARD	2.0
5½" INSULATING BLANKET	19.0
VAPOR BARRIER	0.0
½" GYPSUM BOARD	0.5
	23.4 TOTAL R

*Stagnant air film that forms on any surface

203F01.EPS

Figure 1 ◆ R-values of typical wall construction.

2.2.0 Insulation Requirements

Increasing energy costs and mandated government energy conservation have resulted in much higher R-value requirements for new construction. While building code and design standards for insulation have been traditionally based on average low-temperature zones and charts based on the range of low temperatures expected (see *Appendix A*), the requirements are constantly changing. The International Code Council now recommends insulation values be based on climate zones, which are determined by local temperature and humidity levels (see *Figure 2* and *Table 2*).

In warm climates of the country, many codes now require almost the same amount of insulation for air conditioning as the cold climates require for heating. In some cases, codes are using comfort standards similar to those shown in *Table 3*. The all-weather standard requires that the insulation provided must be adequate to maintain a desired interior temperature during periods of extreme high and low outside temperatures. To meet the moderate standard, the insulation provided must be adequate to maintain a desired temperature during periods of average outside temperature extremes.

In general, insulation must be installed where any exterior surface of a structure is exposed to a thermal difference relative to its internal surface. These areas are:

- Roofs
- Above ceilings
- In exterior walls
- Beneath floors over crawl spaces
- Around the perimeter of concrete floors and around foundations

Over-Insulating

INSIDE TRACK

Installing excess insulation wastes money and may cause other problems. If a building is over-insulated and lacks sufficient ventilation and water barrier protection, moisture can collect inside. This promotes the growth of mold and fungus. It is even possible that cancer-causing radon gas could be trapped in the building, accumulating over time to dangerous levels.

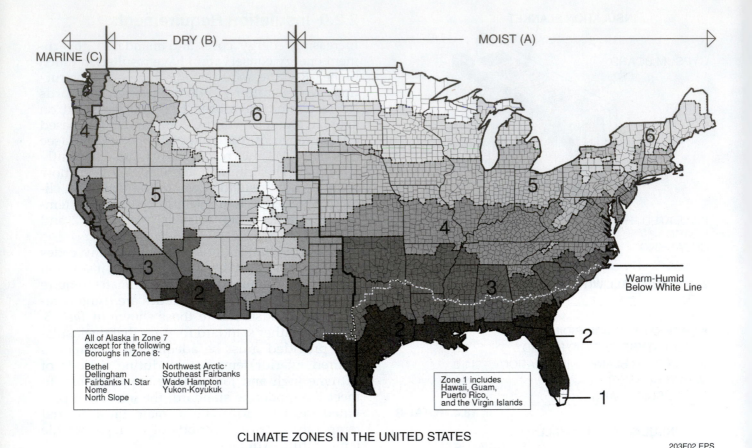

CLIMATE ZONES IN THE UNITED STATES

203F02.EPS

Figure 2 ◆ Climate zones in the United States.

Table 2 Recommended R-Values of Insulation

Climate Zone	Floors	Walls[1]	Ceilings
1	13	13	30
2	13	13	30
3	19	13	30
4 except Marine	19	13	38
5 and Marine 4	30^2	19^2	38
6	30^2	19^2	49
7 and 8	30^2	21	49

[1]Wood frame walls
[2]Alternative options exist

Table 3 Typical Comfort Standards

Comfort Standard	Insulation Location	Insulation R-Value
All-weather	Walls	R-19
	Ceilings	R-30 to R-38
	Floors	R-19
Moderate	Walls	R-13
	Ceilings	R-26
	Floors	R-13
Minimum	Walls	R-11
	Ceilings	R-19
	Floors	R-11

As you study the information in *Tables 1* and *2*, you will notice that the ceiling insulation has the greatest R-value.

2.3.0 Insulation Materials and Types

Insulation materials can be divided into four general classifications, as shown in *Table 4*. These materials are used in the manufacture of five basic categories of insulation: flexible, loose-fill, rigid or semi-rigid, reflective, and miscellaneous.

The R-value of the insulation is marked on the insulation itself or its packaging. See *Figure 3*.

2.3.1 Flexible Insulation

Flexible insulation is usually manufactured from fiberglass in blanket form (*Figure 4*) and fiberglass or mineral wool in batt form (*Figure 5*). In some cases, it is manufactured from wood fiber or cotton and treated for resistance to fire, decay, insects, and rodents. The blankets are available in

Heat Losses

At relatively cold temperatures, a building loses heat in many ways and through different areas. These include heat lost directly through its walls, ceiling, and roof. Heat also escapes through windows, doors, and gaps or cracks in the structure.

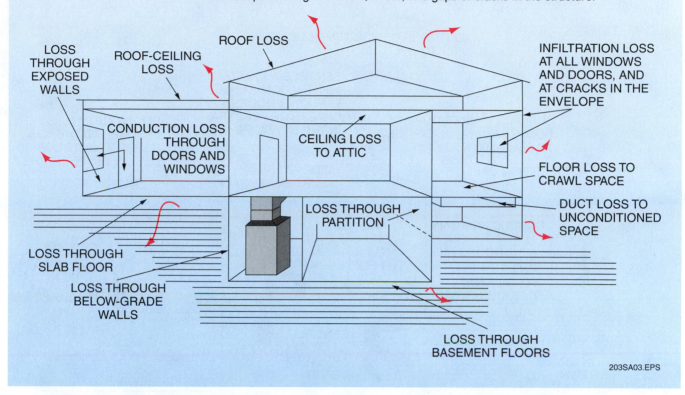

203SA03.EPS

Table 4 Insulation Materials

Classification	Material	Comments
Mineral	Rock Slag Glass Vermiculite Perlite	Rock and slag are used to produce wool by grinding and melting the materials and blowing them into a fine mass.
Natural fiber	Wood Sugar cane Corn stalks Cotton Cork Redwood bark Sawdust or shavings Sheep's wool	Many vegetable products are processed and formed into various shapes, including blankets and rigid boards.
Plastic	Polystyrene Polyurethane Polyisocyanurate Phenolic	
Metal	Foil Tin plate Copper Aluminum	Metallic insulating materials are generally applied to rigid boards or papers and used primarily for their reflective value.

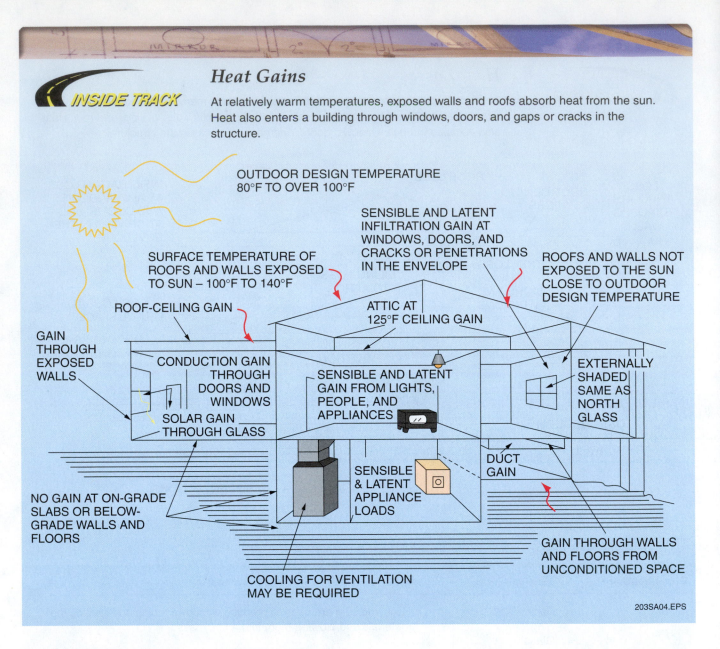

Heat Gains

INSIDE TRACK

At relatively warm temperatures, exposed walls and roofs absorb heat from the sun. Heat also enters a building through windows, doors, and gaps or cracks in the structure.

OUTDOOR DESIGN TEMPERATURE 80°F TO OVER 100°F

SURFACE TEMPERATURE OF ROOFS AND WALLS EXPOSED TO SUN – 100°F TO 140°F

SENSIBLE AND LATENT INFILTRATION GAIN AT WINDOWS, DOORS, AND CRACKS OR PENETRATIONS IN THE ENVELOPE

ROOFS AND WALLS NOT EXPOSED TO THE SUN CLOSE TO OUTDOOR DESIGN TEMPERATURE

ROOF-CEILING GAIN

ATTIC AT 125°F CEILING GAIN

GAIN THROUGH EXPOSED WALLS

CONDUCTION GAIN THROUGH DOORS AND WINDOWS

SENSIBLE AND LATENT GAIN FROM LIGHTS, PEOPLE, AND APPLIANCES

EXTERNALLY SHADED SAME AS NORTH GLASS

SOLAR GAIN THROUGH GLASS

DUCT GAIN

NO GAIN AT ON-GRADE SLABS OR BELOW-GRADE WALLS AND FLOORS

SENSIBLE & LATENT APPLIANCE LOADS

GAIN THROUGH WALLS AND FLOORS FROM UNCONDITIONED SPACE

COOLING FOR VENTILATION MAY BE REQUIRED

203SA04.EPS

Fiberglass Building Insulation

Kraft Faced Roll
6" Nominal Thickness x 23" Width x 39'2" Length 1 Roll 75.07 Sq.Ft.

R-19 Thermal Resistance Insulation Only
COMPLIES WITH FEDERAL SPECIFICATION HI-1-I-521E, Type II
California quality standards

"R" means resistance to heat flow The higher the R-Value the greater the insulation power

Product Code xxxxxxxxxxxx

R-VALUE

203F03.EPS

Figure 3 ◆ Typical R-value identification.

16" or 24" widths and the batts in 15" or 23" widths. Both are furnished in thicknesses ranging from 1" to 12". The batts are packaged in flat bundles in lengths of 24", 48", or 93" and may be unfaced or faced with asphalt-laminated kraft paper or fire-resistant foil scrim (FSK) with or without nailing flanges. The blankets are furnished in rolls that may be encased in an asphalt-laminated kraft paper or plastic film. In most cases, they have a facing with nailing flanges. Some blankets are available with an FSK facing. Blankets and batts with kraft or film casing and/or facings are combustible and must not be left exposed in attics, walls, or floors.

Figure 4 ◆ Flexible blanket insulation.

Figure 5 ◆ Flexible batt insulation.

INSIDE TRACK

Fiberglass Ingredients

The primary ingredients of fiberglass are silica (sand) and recycled (previously melted) glass. The spun glass is held together with a chemical binder. The binder's ingredients include formaldehyde, phenol, and ammonia. The ammonia in the binder sometimes gives fiberglass a strong odor.

While the insulation itself does not burn, the binding material will burn off the glass fibers when the temperature rises high enough (about 350°F). For this reason, fiberglass insulation should not be used in applications that would subject the chemical binder to temperatures approaching its flashpoint.

Fiberglass batt insulation at a 3½" thickness (standard rating or high rating) may be used on exterior walls between the studs. This has been the normal insulation thickness in the past, because of the use of 3½"-wide studs spaced 16" on center (OC). However, in the northern parts of the country, some builders have been using 2 × 6 studs spaced 16" or 24" OC, which has allowed for an increase in the wall insulation to 5½". This thickness is ample insulation for all parts of the United States.

2.3.2 Loose-Fill Insulation

Loose-fill insulation is supplied in bulk form packaged in bags or bales (*Figure 6*). In new construction, it is usually blown or poured and

INSIDE TRACK

Fiberglass Insulation Safety

Flexible fiberglass is probably the first thing that comes to mind when the average person thinks of insulation. Most of us don't realize that this common material must be handled carefully. The tiny strands of glass in fiberglass insulation can irritate skin, injure eyes, and cause a variety of respiratory problems. While insulation installers must wear protective equipment, their responsibilities don't end there. If debris from the installation is not properly removed or if existing insulation is disturbed, fiberglass particles could spread through the building. Fiberglass that enters an HVAC system will be carried to all parts of the building. Always use care when handling fiberglass insulation. This protects you as well as the building's current and future occupants.

203F06.EPS

Figure 6 ◆ Loose-fill insulation.

spread over the ceiling joists in unheated attics. In existing construction that was not insulated when it was built, the material can be blown into the walls as well as the attic.

The materials used in loose-fill insulation include rock or glass wool, wood fiber, shredded redwood bark, cork, wood pulp products such as shredded newspaper (cellulose insulation), and vermiculite. All wood products, including paper, must be treated for resistance to fire, decay, insects, and rodents.

Shredded paper absorbs water easily and loses considerable R-value when damp. In addition to wall surface and/or ceiling vapor barriers, it is essential to install a waterproof membrane along the eaves to prevent water leakage.

The R-value of loose-fill insulation depends on proper application of the product. The manufacturer's instructions must be followed to obtain the correct weight per square foot of material as well as the minimum thickness. Before loose insulation is installed, the area of the space to be insulated is calculated (minus adjustments for framing members). Then, the required number of bags or pounds of insulation is determined from the bag label charts for the desired R-value.

2.3.3 Rigid or Semi-Rigid Insulation

Rigid or semi-rigid insulation is available in sheet or board form and is generally divided into two groups: structural and non-structural. It is available in widths up to 4' and lengths up to 12'.

Structural insulating boards come in densities ranging from 15 to 31 pounds per square foot. They are used as sheathing, roof decking, and wallboard. Their primary purpose is structural, while their secondary purpose is insulation. The structural types are usually made of processed wood, cane, or other fibrous vegetable materials.

Non-structural rigid foam board (*Figure 7*) or semi-rigid fiberglass insulation is usually a light-

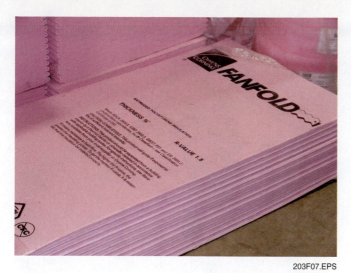

Figure 7 ◆ Rigid foam board.

203F07.EPS

weight sheet or board made of fiberglass or foamed plastic such as polystyrene, polyurethane, polyisocyanurate, and expanded perlite. Most of these products are waterproof and can be used on the exteriors or interiors of foundations, under the perimeters of concrete slabs, over wall sheathing, and on top of roof decks. When used above grade with proper flashing, a protective and decorative coating is sometimes applied directly to the panel. This is known as an **exterior insulation finish system (EIFS)**.

In other cases, it is sealed with an air infiltration film, and normal siding is applied. The foam boards generally range in thickness from 1" to 4" with R-values up to R-30. Because all foam insulation is flammable, it cannot be left exposed. It must be covered with at least ½" of fireproof material.

Some manufacturers also provide rigid foam cores that can be inserted in concrete blocks or used with masonry products to provide additional insulation in concrete block or masonry walls.

The most common rigid and semi-rigid insulation types include the following:

- *Rigid expanded polystyrene* – This material has an R-value of R-4 per inch of thickness. Water significantly reduces this value because rigid expanded polystyrene is not water-resistant. It is not recommended for below-grade insulation. It is the lowest in cost. This material is also called beadboard.
- *Rigid extruded polystyrene* – The R-value of this material is about R-5 per inch of thickness. It is water-resistant and can be used below grade. When used in above-grade applications, it is subject to damage by ultraviolet light and must be coated or covered.
- *Rigid polyurethane and polyisocyanurate* – Initially, these boards have R-values up to R-8 per inch. However, over time the R-value drops to between R-6 and R-7 due to escaping gases. This is referred to as aged R-value. These products are subject to damage by ultraviolet light and must be coated or covered.
- *Semi-rigid fiberglass* – The R-value of this material is about R-4 per inch. Boards of this kind are used on below-grade slabs and walls to provide water drainage as well as insulation. They are also used under membrane roofs as insulation. On below-grade applications, the walls or floors must be waterproofed with a coating or membrane between the wall or floor and the insulation to block water penetration.

2.3.4 Reflective Insulation

This type of insulation (*Figure 8*) usually consists of multiple outer layers of aluminum foil bonded to inner layers of various materials for strength. The number of reflecting surfaces (not the thickness of the material) determines its insulating value. To be effective, the metal foil must face an open air space that is ¾" or more in depth. In some cases, reflective material is bonded to flexible insulation as the inside surface for both insulation and vapor seal purposes.

INSIDE TRACK

Reflective Insulation

Reflective insulation, by itself, can only block radiated heat. At relatively hot temperatures, it helps keep buildings cooler by deflecting heat from the sun. At relatively cold temperatures, it can do little to prevent heat from escaping the building.

Figure 8 ◆ Reflective foil-faced batt insulation.

203F08.EPS

203F09.EPS

Figure 9 ◆ Sprayed-in-place insulation.

2.3.5 Miscellaneous Types of Insulation

There are other types of insulation that do not fit the previous four categories. These types are as follows:

- *Foamed-in-place insulation* – This type of insulation can be applied to new or existing construction using special spray equipment. It can be injected between brick veneer and masonry walls; between open studs or joists; and inside concrete blocks, exterior wall cavities, party walls, and piping cavities. The material must be applied by trained and certified contractors.
- *Sprayed-in-place insulation* – Usually, these types of insulation (*Figure 9*) consist of confetti-like or fibrous inorganic material either mixed with an adhesive or sprayed against a wall with an adhesive coating. They are often left exposed for acoustical as well as insulating properties. Like foamed-in-place products, sprayed-in-place insulation should be applied by trained contractors.
- *Lightweight aggregates* – Insulation material consisting of perlite, vermiculite, blast furnace slag, sintered clay products, or cinders is often added to concrete, concrete blocks, or plaster to improve their insulation quality and reduce heat transmission.

In the 1970s, urea formaldehyde foamed-in-place insulation was injected into many homes. However, due to improper installation, the foam shrank and gave off formaldehyde fumes. As a result, its use was banned in the United States and Canada. Later, it was allowed back on the market in certain areas of the United States. A urethane foam that expands on contact can also be used. It does not have a formaldehyde problem, but it does emit cyanide gas when burned. As a result, it requires fire protection and, like urea formaldehyde, it may also be banned in some areas of the country.

Another foamed-in-place product is a phenol-based synthetic polymer (Tripolymer®–C.P. Chemical Co.) that is fire-resistant and does not drip or create smoke when exposed to high heat. This material does not expand once it leaves the delivery hose of the proprietary application equipment.

3.0.0 ◆ INSULATION INSTALLATION GUIDELINES

Before installation, building plans and codes must be checked to determine the R-values and the types of insulation required or permitted for the

INSIDE TRACK

Sprayed- and Foamed-in-Place Insulation

Sprayed- and foamed-in-place insulation materials are well suited for irregular surfaces. These include walls and ceilings that are curved or that have beams, pipes, or other equipment protruding from them. Foams and sprays can be built up in layers to the desired insulation thickness.

structure being insulated. Then, the required amount of insulation for the structure must be calculated. Any specific instructions provided by the selected manufacturer must be followed when installing the insulation.

3.1.0 Estimating Typical Insulation Requirements

The following is a method of estimating the amount of insulation for the walls, ceilings, and floors of a single-story structure. If no plans are available and the codes specify only a minimum R-value for a structure, refer to the comfort level standards discussed previously and select a desired comfort level based on the occupancy of the structure. Then, perform the following steps to calculate the amount of required insulation material:

Step 1 Determine the square footage of exterior walls to be insulated:

- Determine from the plans or measure the perimeter length of each exterior wall in feet.
- Add the lengths of all exterior walls to find the total perimeter of the structure.
- Multiply the total perimeter by the ceiling height to find the total square footage of the walls:

Exterior perimeter (ft) × ceiling
height (ft) = sq ft of walls

- Determine from the plans or measure the square footage of each opening in the perimeter walls:

Height (in) × width (in) = sq in ÷
144 = sq ft of opening

- Add the square footage of all openings to determine the total square footage.
- Subtract the total square footage of all openings from the total square footage of the walls to find the square footage of insulation required for the walls:

Total wall area (sq ft) − total opening
area (sq ft) = sq ft of wall insulation

Step 2 Calculate the square footage of ceiling/floor to be insulated. The square footage of a floor will be the same as the ceiling. The square footage of either one can be calculated, and the result can be used for both.

- Divide the ceiling or floor plan into rectangular or square areas and determine from the plans or measure the length and width of each area (*Figure 10*).

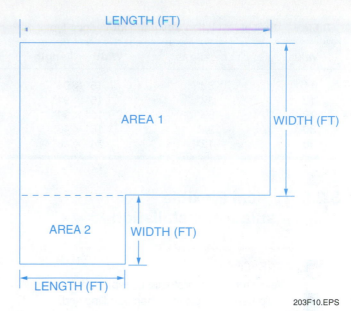

Figure 10 ◆ Dividing a plan view into rectangular or square areas.

- For each area, multiply the length by the width to determine the square footage:

Length (ft) × width (ft) =
sq ft of each area

- Add the square footage of all areas to find the total square footage of insulation required for the ceiling or floor.

Step 3 Add the total insulation square footage required for the walls, the ceiling, and, if required, the floor to determine the total square footage of insulation required for the structure.

Step 4 Divide the square footage of the structure by the coverage per package of insulation for the R-value required. The coverage information will be given in the manufacturer's information for the insulation to be used. See *Table 5* for several examples of package coverage for various R-values and package sizes.

NOTE

The walls, ceiling, and floor of a structure may require different insulation R-values. Check your local code to determine what the requirements are for each installation.

Table 5 Typical Insulation Coverage for Various Types of Packaging and R-Values

R-Value	Thickness	Width × Length	Square Feet per Package	Pieces per Package
11	3½"	15" × 94"	88	9
13	3⅝"	15" × 94"	88	9
19	6¼"	15" × 94"	49	5
30	9½"	16" × 48"	37	7
38	12"	24" × 48"	48	6

3.2.0 Typical Flexible Insulation Installation

WARNING!

Wear proper eye protection, respiratory equipment, and gloves when handling and installing insulation.

Use the following procedure when installing typical flexible insulation:

Step 1 For walls, measure the inside cavity height and add 3". From the wall, lay the distance out on the floor and mark it. Unroll blanket insulation or lay batts on the floor. Use two layers or more. At the cut mark, compress the insulation with a board and cut it with a utility knife. On blanket or faced insulation, remove about 1" of insulation from the ends to provide a stapling flange at the top and bottom.

Step 2 If a separate interior vapor seal will be installed, install blanket or faced insulation so that the stapling flange is fastened to the inside surfaces of the wall studs (*Figure 11*).
  – If the facing of the blanket or batt is the vapor seal, install the stapling flange on the face of the studs and overlap them by at least 1" (*Figure 12*). For faced or blanket insulation, use a power, hand, or hammer stapler to first staple the top flange to the plate.
  – Align and staple down the sides.
  – Staple the bottom flange to the sole plate. Pull the flanges tight and keep them flat when stapling. Space staples about 12" apart if stapling to the face of the studs; on the sides, space them about 6" apart.

  – For unfaced batt insulation, install the batt at the top and bottom first and push it tight against the plates.
  – Evenly push the rest of the batt into the cavity (*Figure 13*). For narrow spaces around windows and doors, stuff the spaces with pieces of insulation and cover it with a plastic or tape vapor seal.

WARNING!

Exercise caution when installing insulation around electrical outlet boxes and other wall openings or devices. Failure to do so may result in electrocution.

Step 3 Faced or blanket insulation for ceilings or floors is usually installed from the bottom in the same manner as the walls. Unfaced batts can be installed from either the top or the bottom.
  – Make sure that ceiling insulation extends over the wall into the soffit area (*Figure 14*). Also make sure soffit baffles (*Figure 15*) are inserted over and cover the ceiling insulation. The baffles should be fastened to the roof deck to hold them in place so that they do not slide down into the soffit and block ventilation.
  – For floors, ensure that the insulation is installed around the perimeter of the floor against the header (*Figure 16*). Floor insulation over a basement is installed with the vapor barrier facing down.
  – Over a crawl space, the vapor barrier faces up. In either case, the insulation can be supported below by a wire mesh (chicken wire), if desired.

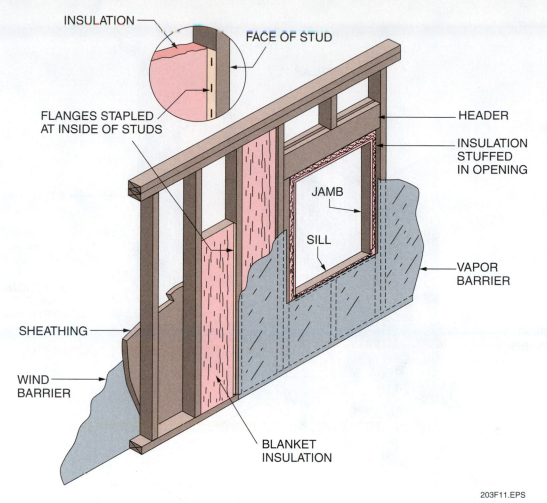

Figure 11 ◆ Blanket installation without integral vapor seal.

203F11.EPS

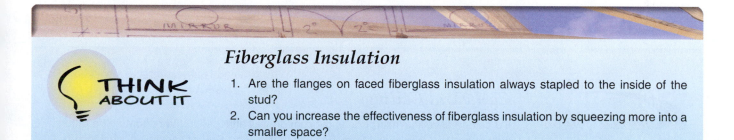

Fiberglass Insulation

1. Are the flanges on faced fiberglass insulation always stapled to the inside of the stud?
2. Can you increase the effectiveness of fiberglass insulation by squeezing more into a smaller space?

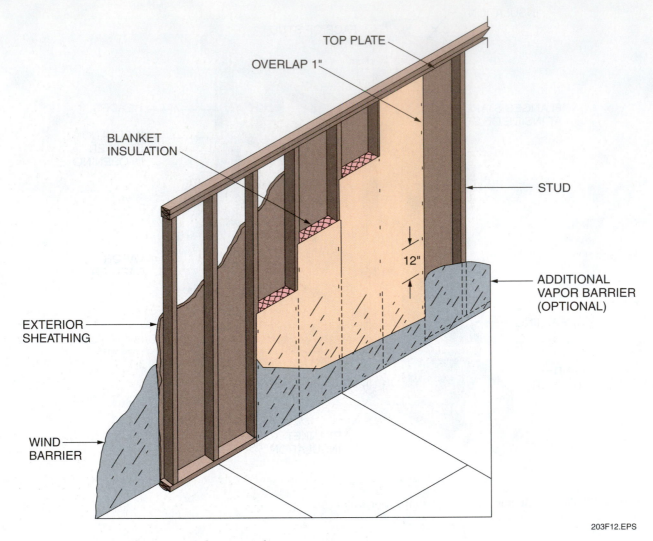

TOP PLATE

OVERLAP 1"

BLANKET
INSULATION

STUD

12"

ADDITIONAL
VAPOR BARRIER
(OPTIONAL)

EXTERIOR
SHEATHING

WIND
BARRIER

203F12.EPS

Figure 12 ◆ Blanket installation integral vapor seal.

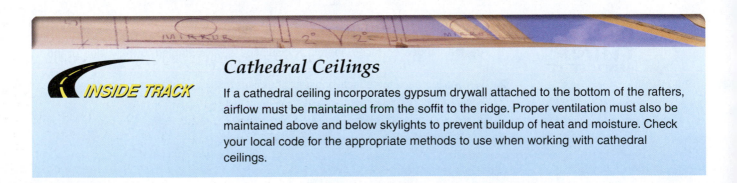

INSIDE TRACK

Cathedral Ceilings

If a cathedral ceiling incorporates gypsum drywall attached to the bottom of the rafters, airflow must be maintained from the soffit to the ridge. Proper ventilation must also be maintained above and below skylights to prevent buildup of heat and moisture. Check your local code for the appropriate methods to use when working with cathedral ceilings.

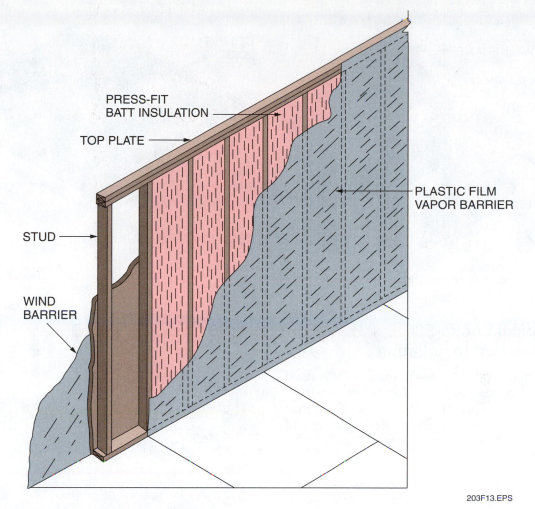

PRESS-FIT
BATT INSULATION

TOP PLATE

STUD

WIND
BARRIER

PLASTIC FILM
VAPOR BARRIER

203F13.EPS

Figure 13 ◆ Batt insulation with separate vapor barrier.

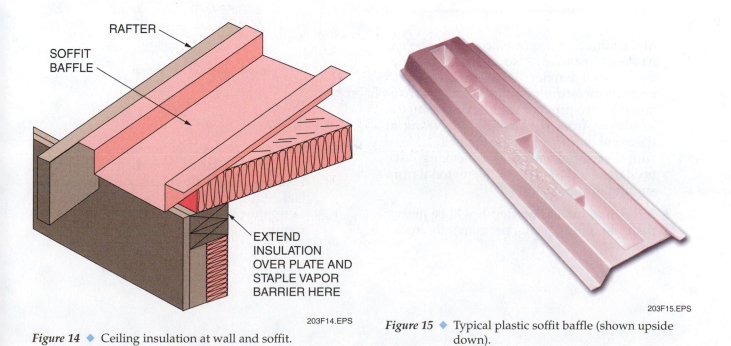

RAFTER

SOFFIT
BAFFLE

EXTEND
INSULATION
OVER PLATE AND
STAPLE VAPOR
BARRIER HERE

203F14.EPS

Figure 14 ◆ Ceiling insulation at wall and soffit.

203F15.EPS

Figure 15 ◆ Typical plastic soffit baffle (shown upside down).

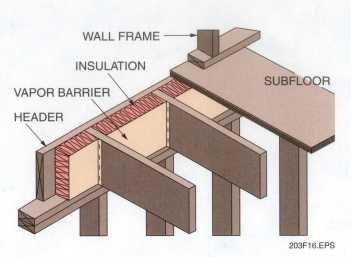

Figure 16 ◆ Perimeter floor insulation.

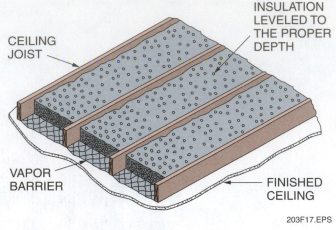

Figure 17 ◆ Loose-fill insulation.

3.3.0 Typical Loose-Fill Insulation Installation

For new construction, loose-fill insulation is used primarily for attic insulation. On older construction, it can also be blown into wall cavities through holes drilled at the center and tops of exterior walls. The following steps only cover attic or ceiling installation.

 WARNING!

Wear proper eye protection, respiratory equipment, and gloves when handling and installing insulation.

Step 1 Make sure that the finished ceiling below has been installed. Also, ensure that a separate vapor barrier has been installed to prevent moisture penetration of the insulation and to prevent the fine dust from the insulation from penetrating the ceiling in the event of future cracks (*Figure 17*). Make sure that soffit baffles and blocking have been installed to prevent the material from spilling into the soffits.

Step 2 If the final insulation depth will be higher than the ceiling joists, permanently install

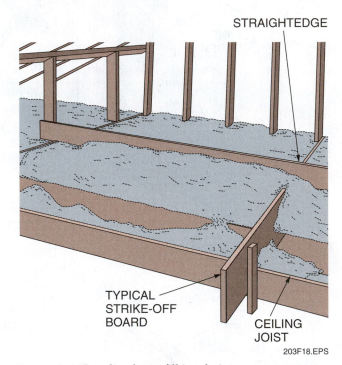

Figure 18 ◆ Leveling loose-fill insulation.

strike-off boards, as shown in *Figure 18*. Pour the insulation from bags or blow the insulation over the ceiling joists using special equipment. Using a straightedge, tamp the insulation and then level it to the required depth for the R-value desired.

Polystyrene Forms

Structural forms made of polystyrene are sometimes used for residential and light commercial construction. Concrete is poured into the forms, which are left in place to provide insulation for the walls. The forms usually provide sufficient insulation by themselves, but check the local code for these requirements.

203SA05.EPS

3.4.0 Typical Rigid Insulation Installation

Rigid insulation panels can be fastened like sheathing over the studs or wood sheathing of a structure. Nails with large heads/washers or screws with washers are used to prevent crushing the insulation.

Rigid insulation panels may be installed on the exterior of a foundation. Typically, the exterior of the foundation is waterproofed first. Then, the panels are applied over special mastic and secured with concrete nails to hold them in place until the mastic sets. For existing construction, the panels may be installed on the interior of the foundation if the walls are adequately waterproofed.

Figure 19 shows typical methods of installing rigid insulation under surface slabs. Usually, the insulation is only applied around the perimeter of the slab, anywhere from 24" to 36" from the edge of the slab and/or down the inside of the slab footings to below the frost line (*Figure 20*). A vapor barrier should be applied under the slab and over any insulation under the slab.

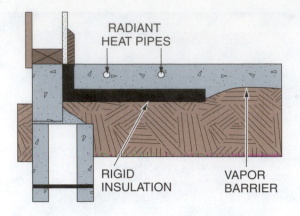

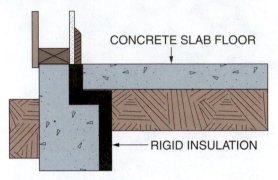

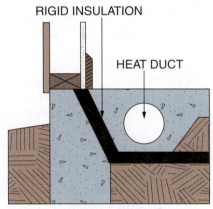

Figure 19 ◆ Rigid insulation installed under a concrete slab.

203F19.EPS

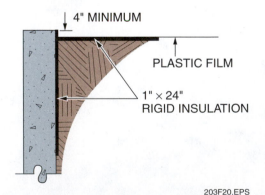

203F20.EPS

Figure 20 ◆ Rigid insulation installed under a slab-and-down footing.

4.0.0 ◆ MOISTURE CONTROL

Water vapor contained in air can readily pass through most building materials used for wall construction. This vapor caused no problem when walls were porous because it could pass from the warm wall to the outside of the building before it could condense into liquid water (*Figure 21*).

When buildings were first constructed with insulation in the walls to cut down on heat loss, moisture in the air passed through the insulation until it reached a point cold enough to cause it to condense. The condensed moisture froze in very cold weather and reduced the efficiency of the insulation. The ice contained within the wall thawed as the weather warmed, and the resulting water in the wall caused studs and sills to decay over time.

For these reasons, it is important to keep cellars, basements, crawl spaces, exterior walls, and attics dry. Moisture in crawl spaces, basements, and attics also encourages woodchewing insects such as termites, as well as the growth of mold. In the case of crawl spaces, moisture often rises from the ground into the crawl space during periods of heavy rain. To prevent the concentration of this damaging moisture, some precautions must be taken in the original design of the structure:

- The earth must slope down and away about 20' from the structure, carrying surface water away.
- The crawl space should be protected from moisture by a vapor barrier on the ground.
- The foundation walls should be penetrated with vents so that moisture will not be trapped in the crawl space.
- A vapor barrier should be installed between the insulation and the subfloor.

Basements usually have the most trouble with **condensation** in summer during humid weather. The earth under the concrete basement floor is comparatively cool, causing the floor of the basement to be a cold surface. The hot air is saturated with moisture and condenses when it comes in contact with the cooler surfaces of the floor and walls. This problem is difficult to control. If the surface of the concrete is rough and porous, the moisture will sink in and not cause a wetness problem. If, however, the floor is dense and smoothly finished, the tightly knit grains of concrete form a vapor barrier of sorts, and the water collects on the slab. This problem can usually be solved by the use of dehumidification devices during the summer months.

Moisture weeping through the concrete floor is a different problem. In new construction, this is

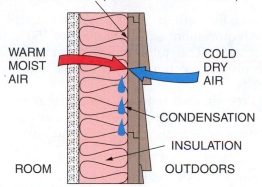

SURFACE WARM (EXCESSIVE HEAT LOSS)

WARM
MOIST
AIR

COLD
DRY
AIR

ROOM   OUTDOORS

**NO INSULATION OR VAPOR BARRIER —
MOIST AIR PASSES THROUGH THE WALL**

SURFACE COLD (REDUCED HEAT LOSS)

WARM
MOIST
AIR

COLD
DRY
AIR

CONDENSATION

INSULATION

ROOM   OUTDOORS

**INSULATION ONLY — MOIST AIR PASSES INTO WALL
AND CONDENSES, CAUSING DAMAGE
TO INSULATION AND BUILDING FRAMEWORK**

SURFACE COLD (REDUCED HEAT LOSS)

WARM
MOIST
AIR

COLD
DRY
AIR

VAPOR
BARRIER

INSULATION

ROOM   OUTDOORS

**INSULATION WITH VAPOR BARRIER —
MOISTURE CANNOT ENTER WALL SPACE**

203F21.EPS

Figure 21 ◆ Effects of insulation and vapor barrier.

controlled by installing perimeter drainage and a vapor barrier under the concrete slab. When installing polyethylene film as an underslab vapor barrier, be careful not to tear, puncture, or damage the film in any way. Any passageways for moisture will defeat the purpose of the vapor barrier. Prior to pouring the concrete slab, make sure the polyethylene film is placed properly and is free of punctures. Keep all construction debris away from the vapor barrier.

To keep moisture from rising up into the basement, 6" of coarse gravel should be placed over the compacted earth to provide drainage to the perimeter drain before the slab is poured. A polyethylene film should be placed on top of the gravel to keep the concrete from penetrating into the gravel and possibly weakening the slab. In very wet areas or areas with a high water table, floor drainage, in addition to a gravel bed, may also be required.

4.1.0 Interior Ventilation

One of the best ways to reduce or eliminate the chances of moisture damage in attics or in the space between the rafters and the finished roof is through proper ventilation. Ventilation provides a stream of outside air to remove trapped moisture before it is allowed to do any damage. In insulated attics, baffles (blocking strips) are used to keep the insulation material from getting into the vented areas. With the increased use of blown-in insulation in attics, baffles are being required by code in some areas.

The amount of ventilation required varies by climate and building codes. Attics and gable and hip roofs may be ventilated with a variety of louvers and vents. Flat roofs are ventilated with a combination of eave vents and roof stacks (*Figure 22*).

INSIDE TRACK

Mold

Moisture accumulating inside a building can damage the structure and promote the growth of mold. While it is not always harmful, this mold may cause allergic reactions or other respiratory problems in some people. Airborne mold spores can also cause infections, primarily in people whose immune systems are compromised.

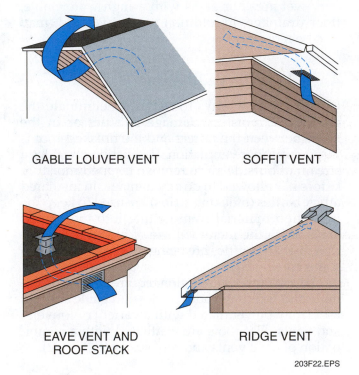

GABLE LOUVER VENT

SOFFIT VENT

EAVE VENT AND
ROOF STACK

RIDGE VENT

203F22.EPS

Figure 22 ◆ Various methods of roof ventilation.

Ice dams can usually be avoided by installing plenty of insulation and providing ample ventilation in the attic.

Properly designed subroof ventilation is the best weapon for preventing water vapor infiltration into a steeply sloped roof, but is less effective on roofs with low slopes because natural **convection** decreases with diminishing roof height. Moisture dissipation occurs through **diffusion** and wind-induced ventilation.

Normally, the ventilation requirement for a gable roof is 1 sq ft of free air ventilation for every 300 sq ft of ceiling area if a vapor barrier exists under the ceiling. If no vapor barrier is present, the requirement is 1 sq ft for every 150 sq ft of ceiling area. The total requirement must be split evenly between the inlet vents and the outlet vents.

Free air ventilation is the rating of the ventilation devices, taking into account any restrictions caused by screening, louvers, and other devices.

4.2.0 Vapor Retarders

Vapor retarders, also known as vapor barriers or **vapor diffusion retarders (VDR)**, are any material or substance that will not permit the passage of water vapor or will do so only at an extremely slow rate. The **permeability** of a substance is a measure of its capacity to allow the passage of liquids or gases. Water vapor permeability is the property of a substance to permit the passage of water vapor and is equal to the **permeance** of a substance that is 1" thick. The measure of water vapor permeability is the **perm**. This equals the number of grains of water vapor passing through a 1 sq ft piece of material per hour, per inch of mercury difference in vapor pressure. All you really have to remember is that any material that has a perm rating of 1.0 or less is considered a vapor retarder and will not allow the passage of any appreciable or harmful amounts of water vapor. Any material with a rating higher than 1.0 is a breathable material that will permit the passage of water vapor in whatever degree its perm rating indicates. The higher the perm number, the greater the amount of water vapor that will pass through the material in a given time; 0.0 is totally impermeable (*Table 6*).

A properly installed vapor barrier will protect ceilings, walls, and floors from moisture originating within a heated space (*Figure 23*).

Table 6	Perm Ratings of Various Vapor Retarder Materials	
Material		**Permeance**
Aluminum foil (1 mil)		0*
Aluminum foil (0.35 mil)		0.05*
Polyethylene (4 mil)		0.08*
Polyethylene (6 mil)		0.06*
Polyester (1 mil)		0.07*
Saturated and coated roll roofing		0.05**
Reinforced kraft and asphalt-laminated paper		0.3**
Asphalt-saturated and coated vapor barrier paper		0.2 to 0.3*
15 lb tarred felt		4.0**
15 lb asphalt felt		1.0**
12.5 lb asphalt		0.5**
22 lb asphalt		0.1**
Built-up membrane (hot-mopped)		0**

*Per *ASTM E96-66, Water Vapor Transmission of Materials in Sheet Form*
**Per *ASTM C355-64, Water Vapor Transmission of Thick Material*

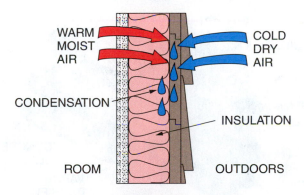

WALL WITH NO VAPOR BARRIER

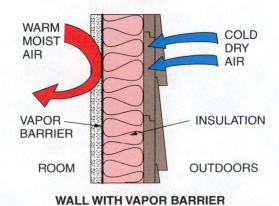

WALL WITH VAPOR BARRIER

203F23.EPS

Figure 23 ◆ Vapor barrier installation.

An insulated wall will divide two temperature gradients. The area on the inside of the structure will normally be warmer than the air on the outside. The vapor barrier is usually located on the warm side to prevent moisture from moving through the insulation to the cool side and condensing.

4.2.1 Materials

Common vapor barrier materials include asphalted kraft paper, aluminum foil, and polyethylene film.

Asphalted kraft paper is usually incorporated with blanket or batt insulation. It serves as a means for attaching the insulation to the building framework and as a reasonably good vapor barrier when installed on the warm side of the wall or ceiling (*Figure 24*).

Aluminum foil may be incorporated with blanket or batt insulation in the same manner as kraft paper. It is also applied to the back of gypsum lath and gypsum wallboard where it works as a relatively effective vapor barrier.

Polyethylene film is applied over the studs and ceiling joists after the insulation is installed. When wallboard with polyethylene film or foil backing is used, the insulation will normally be plain batts or blankets that do not have an integral vapor barrier. As a vapor barrier, polyethylene film is stapled over the studs and also covers the window frames. This helps to keep the window frames and sashes clean during application and finishing of the gypsum wallboard. The film should be overlapped 2" to 4" and sealed with special mastic or tape.

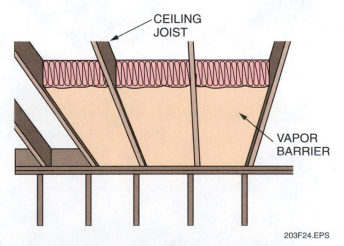

203F24.EPS

Figure 24 ◆ Installing insulation batts between ceiling joists with vapor barrier down.

4.2.2 Installation in Crawl Spaces

The ground under a ventilated crawl space should be covered with a vapor barrier ground cover to protect the underside of the house from condensation (see *Figure 25*). A vapor barrier should also be installed over the subfloor above the crawl space.

Besides installing vapor barriers, crawl spaces should be properly vented to permit the escape of moisture. Usually, this is accomplished by the use of a proper number of screened foundation vents installed in the above-grade foundation surrounding the crawl space. The normal requirement is 1 sq ft of free air ventilation for every 160 sq ft of crawl space area when a vapor barrier ground cover is used.

4.2.3 Installation in Slabs

When allowed to proceed unchecked, moisture will migrate from the ground upward through concrete and into the building, where it can cause moisture problems, damage, and higher energy costs. Even though the water table may be several feet below the slab, moisture vapor will migrate up to and through concrete slabs.

Up to 80 percent of the moisture entering a structure does so by migrating from the ground beneath the structure. Moisture vapor passes through concrete more readily than liquid moisture.

Moisture in a building can cause deterioration of interior finishes, especially floors and equipment. Moisture can also add to energy costs by raising humidity and taxing cooling systems that require dehumidification.

Vapor barriers should be continuous under the slab. Great care must be taken not to tear or puncture the barrier. Keep all construction debris away from the barrier location. Vapor barrier installation must be done by qualified contractors.

When used in thickened-edge slab construction, as shown in *Figure 26*, a vapor barrier is placed between the gravel cushion and the

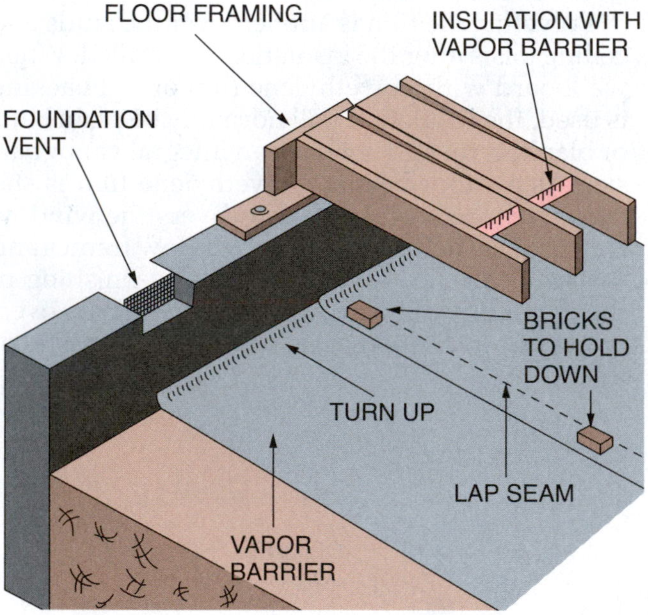

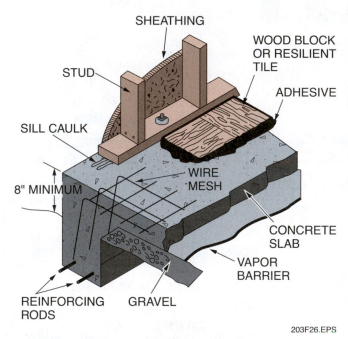

Figure 25 ◆ Vapor barrier installation for crawl spaces.

203F25.EPS

Figure 26 ◆ Thickened-edge slab vapor barrier installation.

203F26.EPS

poured concrete. The same arrangement is used for other types of slab-on-grade construction.

Figure 27 shows a method of constructing a finished floor over a concrete slab, which affords double protection against moisture. The sealer or waterproofer is placed on the slab itself, and a vapor barrier is suspended above the slab.

4.2.4 Installation in Walls

A polyethylene sheet vapor barrier is easy to apply to frame walls where no integral barrier is provided or where a supplementary barrier is preferred. A flap would normally overlap both floor and ceiling barriers to seal the interior off completely. Adjacent sheets of the film are overlapped 2" to 4" and are sealed with a special mastic or tape.

When vapor barriers are applied to walls, particular attention should be paid to fitting the material around electrical outlet boxes, exhaust fans, light fixtures, registers, and plumbing.

Considerable water vapor can escape through the cracks around the equipment, travel from the warm side of the wall to the cold side, and condense on the sheathing or siding. This is especially true if the insulation is poorly fitted at the top and bottom.

4.2.5 Installation in Roofs

A major cause of failure in built-up roofs is condensation of moisture vapor, which rises from inside the building and penetrates the roof deck insulation. When this vapor reaches its **dew point**, which can occur inside the insulation or at the cool outer surface, it condenses. This results in a reduction or total loss of the thermal efficiency of the insulation, as well as dripping and damage. To prevent this, select a vapor barrier that is both easy to apply and resistant to job site abuse. Install it on the warm side of the roof deck insulation.

5.0.0 ◆ WATERPROOFING

The single most critical area for waterproofing construction is the below-grade foundation wall. Rising water tables, hydrostatic heads, structural movement, and ground water all require a special type of protection. A liquid waterproofing system applied by spray methods ensures the high buildup of film thickness needed to cope with these problems (*Figure 28*).

For all below-grade applications of waterproofing, be sure to fill all cracks, crevices, and grooves. Ensure that the coating is continuous and free from breaks and pinholes.

Carry the coating over the exposed tops and outside edges of the footing (*Figure 29*), forming a cove at the junction of the wall and footing.

Spread the coating around all joints, grooves, and slots and into all chases, corners, reveals, and soffits. Bring the coating up to the finished grade.

Do not place backfill for 24 to 48 hours after application. Where possible, backfill should be

VAPOR BARRIER

STRIP FLOORING

NAIL

1 × 4 16" OC

WATERPROOF COATING

CONCRETE SLAB

TREATED 1 × 4 (ANCHOR TO SLAB)

203F27.EPS

Figure 27 ◆ Surface-mounted vapor barrier on a slab.

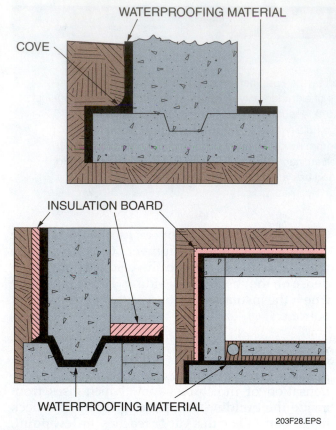

WATERPROOFING MATERIAL

COVE

INSULATION BOARD

WATERPROOFING MATERIAL

203F28.EPS

Figure 28 ◆ Applying waterproofing material.

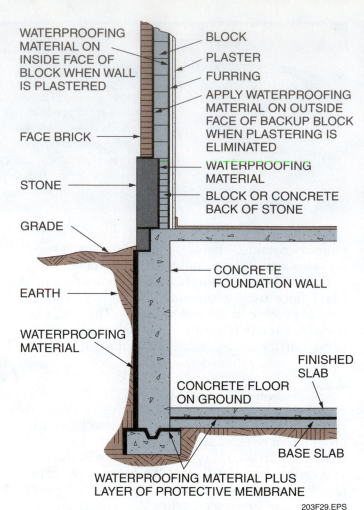

WATERPROOFING MATERIAL ON INSIDE FACE OF BLOCK WHEN WALL IS PLASTERED

BLOCK

PLASTER

FURRING

APPLY WATERPROOFING MATERIAL ON OUTSIDE FACE OF BACKUP BLOCK WHEN PLASTERING IS ELIMINATED

FACE BRICK

WATERPROOFING MATERIAL

STONE

BLOCK OR CONCRETE BACK OF STONE

GRADE

EARTH

CONCRETE FOUNDATION WALL

WATERPROOFING MATERIAL

FINISHED SLAB

CONCRETE FLOOR ON GROUND

BASE SLAB

WATERPROOFING MATERIAL PLUS LAYER OF PROTECTIVE MEMBRANE

203F29.EPS

Figure 29 ◆ Below-grade waterproofing application.

placed within approximately seven days to avoid any unnecessary damage due to construction activities. Take care to place the backfill in a manner that will not rupture or damage the film or cause the coating or membranes to be displaced on the coated surface.

5.1.0 Water Stops

Water stops are thin sheets of rubber, plastic (PVC), or other material inserted in a construction joint to obstruct the seepage of water through the joint. PVC water stops may be used for installations in underpasses, tunnels, tanks, locks, walls, swimming pools, siphons, sewage disposal plants, reservoirs, culverts, sewage treatment plants, channels, drums, filtration plants, foundations, bridges, basements, abutments and decks, mineshafts, aqueducts, retaining walls, and roofs. Refer to *Figure 30* for various applications of water stops.

5.2.0 Joint Treatment

Joints in structures are critical. They must maintain integrity during movement, yet remain permanently waterproof and airtight. Therefore, it is important to select the proper joint treatment system to avoid problems with moisture penetration at the construction joints.

5.3.0 Vapor Barrier for Cold Storage and Low-Temperature Facilities

Cold-storage vapor barriers are designed for use in areas of extremely low temperatures to halt the migration of damaging moisture vapor into and through the insulation. Most cold storage vapor barriers consist of two layers of kraft paper, each extrusion coated with black polyethylene; a layer of aluminum foil; two layers of non-asphaltic adhesive; and two layers of high-tensile-strength reinforcing fibers embedded in the adhesive.

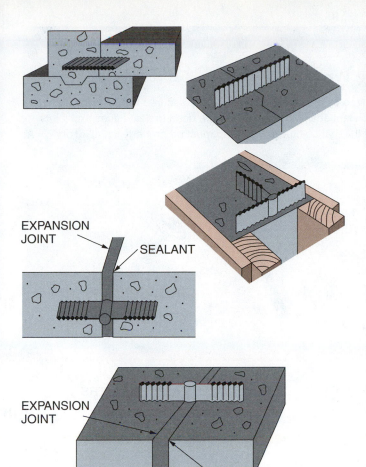

EXPANSION
JOINT

SEALANT

EXPANSION
JOINT

SEALANT

203F30.EPS

Figure 30 ◆ Water stops used in joints.

6.0.0 ◆ AIR INFILTRATION CONTROL

In addition to insulation, the exterior sheathing of a structure should be covered to prevent wind pressure from causing infiltration of outside air into the structure. To achieve maximum energy efficiency in a structure, air infiltration must be strictly controlled.

Traditionally, structures have been covered with water-resistant building paper to help prevent water leakage through the primary barrier (siding) from reaching the structural sheathing or other components of the structure. Additionally, the paper had to be water **permeable** to allow moisture inside the walls of the structure to pass through and evaporate. To some extent, the paper reduced air infiltration of the structure, especially when board sheathing was used.

For a number of years, products called house wraps or building wraps have been used to replace building paper. These products, under brand names such as Tyvek® or ProWrap®, are easier to apply and perform the same functions as building paper (*Figure 31*). When properly applied

RESIDENTIAL APPLICATION

COMMERCIAL APPLICATION

203F31.EPS

Figure 31 ◆ Building wrap.

and sealed, the wraps provide a nearly airtight structure no matter what sheathing material is used. Most versions of these wraps are an excellent secondary barrier under all siding, including stucco and EIFS.

Many of these products are made of spun, high-density, polyethylene fibers randomly bonded into an extremely tough, durable sheet material. They are usually available in several versions and weights for residential and light commercial use. Special versions may be available with vertical water channeling permanently pressed into the material for stucco and EIFS. The material is usually furnished in rolls in various sizes from 18" wide to 10' wide and in lengths from 100' to 200'.

Nails with large heads, nails or screws with plastic washers (*Figure 32*), or 1" wide staples may be used to secure the wrap to wood, plastic, insulating board, or exterior gypsum board. Screws and washers are used for steel construction. Special contractor's tape (*Figure 32*) or sealants compatible with the wrap are used to seal the edges and joints of the wrap.

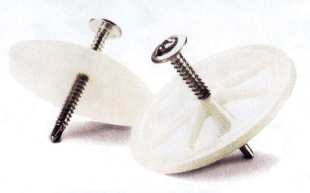

SCREWS WITH PLASTIC WASHERS

 WARNING!

Some building wraps are slippery and should not be used in any application where they can be walked on. Because the surface will be slippery, use pump jacks or scaffolding for exterior work above the lower floor. If ladders must be used, extra precautions must be taken to prevent the ladders from sliding on the wrap.

CONTRACTOR'S TAPE

203F32.EPS

Figure 32 ◆ Building wrap accessories.

Always refer to the manufacturer's instructions for specific installation information. House or building wrap is generally installed as follows:

Step 1 Using two people and beginning at a corner on one side of the structure, leave 6" to 12" of the wrap extended beyond the corner to be used as an overlap on the adjacent side of the structure (*Figure 33*). Align the roll vertically and unroll it for a short distance. Check that the stud marks on the wrap align with the studs of the structure. Also check that the bottom edge of the wrap extends over and runs along the line of the foundation. Secure the wrap to the corner at 12" to 18" intervals.

Step 2 Unroll the wrap two or three more feet and ensure that it overlaps and runs along the line of the foundation. Secure the wrap vertically at 12" to 18" intervals on each stud using the stud marks as a fastening guide. Continue around the structure, covering all openings. If a new roll is started, overlap the end of the previous roll 6" to 12" to align the stud marks of the new roll with the studs of the structure.

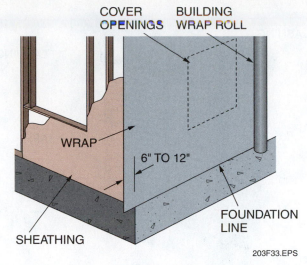

Figure 33 ◆ Starting a roll of building wrap.

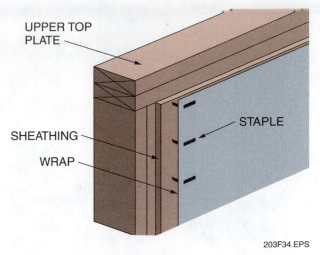

Figure 34 ◆ Top plate detail.

Step 3 If the upper parts of the structure require coverage, repeat Steps 1 and 2, starting above the existing wrap. Make sure that the bottom edge of this layer of wrap aligns along the top edge of the lower wrap and overlaps it by 6" to 12".

Step 4 At the top plate, make sure the wrap covers both the lower and upper (double) top plate (*Figure 34*), but leave the flap loose for the time being.

Step 5 At each opening, use one of the following two methods to cut back the wrap.

NOTE

Always follow the window or door manufacturer's recommendations for flashing windows or doors.

Method 1 – Uninstalled Windows/Doors:
- At the opening, cut the wrap as shown in *Figure 35*. Fold the three flaps around the sides and bottom of the opening and secure every 6". Trim off the excess.
- At the outside, install 6" flashing along the bottom of the opening, then up the sides over the top of the wrap.
- Install head flashing at the top of the opening under the wrap and over the side flashing (*Figure 36*). Tape the flap ends to the head flashing using tape approved by the manufacturer.

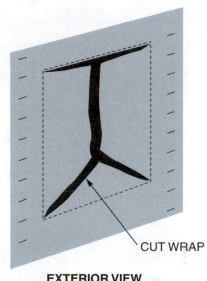

EXTERIOR VIEW

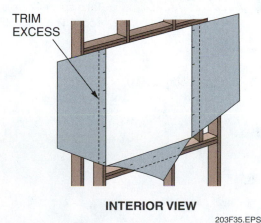

INTERIOR VIEW

Figure 35 ◆ Cutting and folding wrap at an opening.

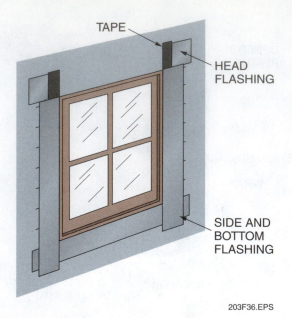

Figure 36 ◆ Installing flashing around an opening.

203F36.EPS

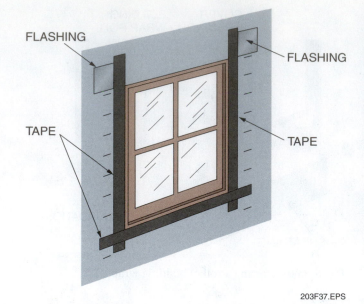

Figure 37 ◆ Installing wrap with a window in place.

203F37.EPS

Method 2 – Windows/Doors with Flanges:
- Create a top flap of the wrap. Insert a head flashing under the flap and over the flange.
- Extend the flashing to the sides about 4" and tape the flap to the head flashing.
- On the remaining sides, trim the wrap to overlap the flange area and tape the edge to the flanges (*Figure 37*).

Step 6 Secure all the bottom edges of the wrap to the foundation with the recommended joint sealer, then fasten the lower edge to the sill. At the top plate, seal the edge to the upper plate with the sealer and fasten the edge to the plate.

Step 7 Seal all vertical and horizontal joints in the wrap with the recommended tape.

Step 8 Before applying the siding, repair any damage or tears in the wrap with tape or sealant.

INSIDE TRACK

Window Flashing System

Take special care to prevent moisture from entering the structure around windows and doors. The Dupont® flashing system is designed for this purpose.

203SA06.EPS

Summary

This module presented the materials and procedures that can be applied to ensure that effective insulation, moisture control, ventilation, waterproofing, and air infiltration control are achieved.

Carpenters will be involved in the application of these materials and must be thoroughly familiar with the materials and the typical installation techniques covered in this module.

Notes

1. Insulation is based on trapping _____ .
   a. large amounts of air in a few small spaces
   b. large amounts of air in a large number of very small spaces
   c. large amounts of air in a few large spaces
   d. small amounts of air in a few large spaces

2. The R-value is a measure of the ability of a material to _____ .
   a. resist the passage of moisture
   b. resist heat conduction
   c. allow cold air to enter a building
   d. convert water vapor into a liquid

3. For thermal transmission control purposes, insulation does not have to be installed _____ .
   a. above ceilings
   b. in exterior walls
   c. in interior walls
   d. beneath floors over crawl spaces

4. The primary purpose of structural insulating boards is insulation.
   a. True
   b. False

5. Strike-off boards are used in the installation of _____ insulation.
   a. loose-fill
   b. rigid
   c. flexible
   d. foamed-in-place

6. Rigid insulation is usually installed around the perimeter of a concrete slab at a distance of _____ from the edge.
   a. 8" to 12"
   b. 12" to 18"
   c. 18" to 24"
   d. 24" to 36"

7. When a vapor barrier is used under the ceiling, proper free air ventilation for a gable roof is defined as 1 sq ft for every _____ sq ft of attic area.
   a. 150
   b. 160
   c. 300
   d. 320

8. Any material with a perm rating of less than _____ is a vapor barrier.
   a. 0.01
   b. 0.05
   c. 0.1
   d. 1.0

9. When waterproofing material is sprayed on exterior below-grade walls, backfilling of the walls should be avoided for _____ days.
   a. one to two
   b. two to three
   c. three to four
   d. five to six

10. The vertical seams of building wrap are usually overlapped _____ at each corner of the building.
    a. 2" to 6"
    b. 3" to 4"
    c. 6" to 12"
    d. 12" to 18"

Trade Terms
Introduced in This Module

Condensation: The process by which a vapor is converted to a liquid, such as the conversion of the moisture in air to water.

Convection: The movement of heat that either occurs naturally due to temperature differences or is forced by a fan or pump.

Dew point: The temperature at which air becomes oversaturated with moisture and the moisture condenses.

Diffusion: The movement, often contrary to gravity, of molecules of gas in all directions, causing them to intermingle.

Exterior insulation finish system (EIFS): A protective and decorative coating applied directly to insulation board.

Perm: The measure of water vapor permeability. It equals the number of grains squared of water vapor passing through a one square foot (sq ft) piece of material per hour, per inch of mercury difference in vapor pressure.

Permeability: The measure of a material's capacity to allow the passage of liquids or gases.

Permeable: Porous; having small openings that permit liquids or gases to seep through.

Permeance: The ratio of water vapor flow to the vapor pressure difference between two surfaces.

Vapor barrier: A material used to retard the flow of vapor and moisture into walls and prevent condensation within them. The vapor barrier must be located on the warm side of the wall.

Vapor diffusion retarder (VDR): See *vapor barrier*.

Vapor retarder: See *vapor barrier*.

Water stop: Thin sheets of rubber, plastic, or other material inserted in a construction joint to obstruct the seepage of water through the joint.

Water vapor: Water in a vapor (gas) form, especially when below the boiling point and diffused in the atmosphere.

Recommended Regional R-Values

Prior to 2006, the International Code Council recommended insulation R-values based on the region's average low temperature (see *Figure A–1* and *Table A–1*). Today, insulation values are based on the region's average temperature and humidity levels. Interestingly, the United States Department of Energy makes more stringent recommendations based on climate zones. It is important for you to remember that the ICC and other agencies, as well as the federal government may recommend insulation requirements for an area, but the local building code is the final authority.

Table A-1 Recommended R-Values of Insulation

Average Low Temperature	Floors	Walls	Ceilings
+10° F to +40° F	0	11	19
0° F to +10° F	19	17	30
0° F and below	19	17	38

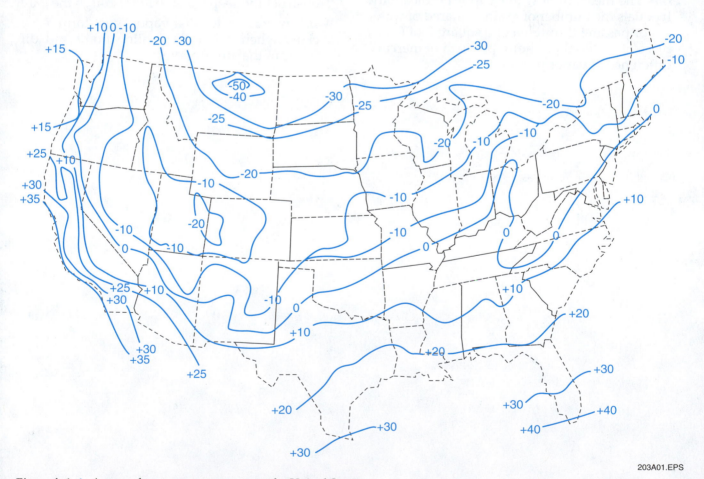

203A01.EPS

Figure A-1 ◆ Average low temperatures across the United States.

Metal and Vinyl Gutters and Downspouts

Occasionally, you may be called upon to install gutters and downspouts after constructing the cornice. Therefore, you should be familiar with available products and have a general knowledge of how they are installed. Always check the local codes. Some localities specify the size and capacity of gutters applied to commercial and/or residential structures.

Gutters are constructed of vinyl or metal, usually aluminum or galvanized steel, but sometimes copper, stainless steel, or baked-on enamel steel. Vinyl and aluminum gutters are prefinished and ready to install. Galvanized metal gutters are usually unfinished and must be painted after they are in place.

Many gutters installed on residential and light commercial buildings are seamless and are installed by companies specializing in the field. The final product is measured, formed, fabricated, and installed in place in the field by the same company.

Some gutters are fabricated on site with a debris guard that requires no cleaning and sheds roof debris, such as twigs and leaves, while directing water into the gutter (*Figure B–1*). Another similar product is available for application to existing gutters. Both types depend on the principle of liquid cohesion and adhesion. Water running off a roof will follow the sharp curve and drop into the gutter because of adhesion to the metal and the surface tension caused by cohesion of the water molecules. Debris, on the other hand, passes by the sharp curve and is shed over the edge of the gutter.

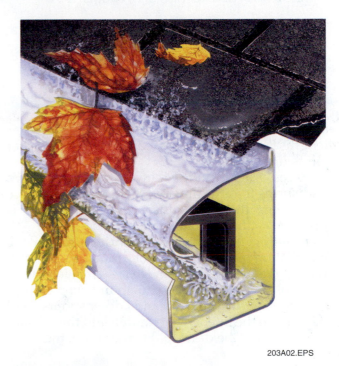

203A02.EPS

Figure B-1 ◆ Debris guard.

PREFABRICATED METAL GUTTERS AND DOWNSPOUTS

Gutters are manufactured in several shapes. The most common shape is the ogee, or K-style, as shown in *Figure B–2*. Another is an older C-shaped (half-round) style.

Downspouts also come in a variety of shapes to match the type of gutter. The most common is the rectangular corrugated type. They are made in

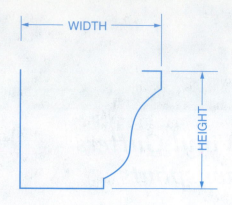

Stock Metal Gutter Section Ogee or Style "K"

Stock Sizes: 2¾" H × 4" W   5¼" H × 7" W
               3¾" H × 5" W   6" H × 8" W
               4¾" H × 6" W

203A03.EPS

Figure B-2 ◆ Typical metal K-style sizes.

standard 8' to 10' lengths. In some cases, longer lengths are available to reduce the number of joints required.

Accessories include elbows of various angles and straps for attachment. One end of each piece is smaller than the other, and each piece is installed with the smaller end down.

There are also many accessories used in the installation of a complete roof drainage system. A typical metal K-style drainage system, along with some of its accessories, is shown in *Figure B–3*.

The size of gutters and downspouts will depend on the intensity of rainfall and the amount of roof area to be drained. Furthermore, the slope

of the gutter and the number of outlets will affect the gutter size. The required sizes will be specified in the construction drawings. The drawings will also indicate where the downspouts are to be located.

The relationship between the gutter and the edge of the finish roofing material is important. A gutter should be centered under the edge of the roof to catch the water off the roof. Gutters should always be placed below the slope of the roof to catch water that runs off, but not to catch snow and ice that slide off, if in areas with regular snowfall. It is necessary that the gutter slope slightly toward the outlet, usually 1" or 2" in 40'. It is a good idea to check on local codes to determine the correct slope.

The exact height varies with the slope of the roof and is measured from a continuation of the slope to the outer lip of the gutter. These dimensions apply to the high point of the gutter, as shown in *Figure B–4*.

When shedding of ice is desired, gutters are sometimes not used on roofs with wide metal ice edging installed on the roof slope over the eaves.

Gutters should be supported on both sides of all outlets, at all ends and joints, and at intermediate points. The maximum spacing of supports is 36" in climates free of snow and ice, while 18" spacing is recommended in areas that have snow.

The tools required to install a typical drainage system are a soldering iron or caulking gun, hacksaw or power saw with a metal-cutting blade, metal shears, hammer, steel tape or folding rule, carpenter's level, putty knife, chalkline, screwdriver, rivet gun, and ladders.

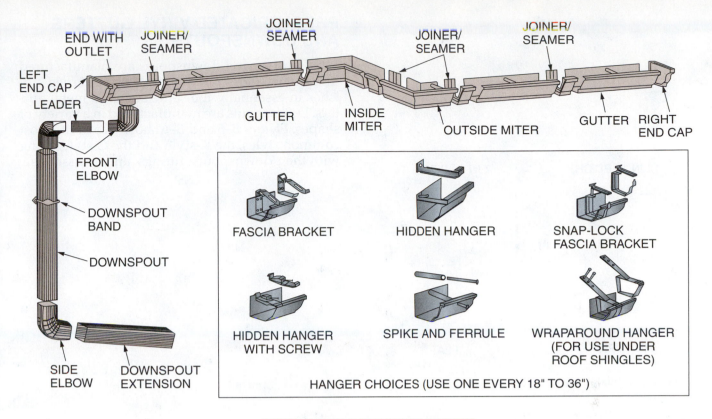

METAL K-STYLE DRAINAGE SYSTEM

ACCESSORIES

203A04.EPS

Figure B-3 ◆ Typical metal K-style drainage system and accessories.

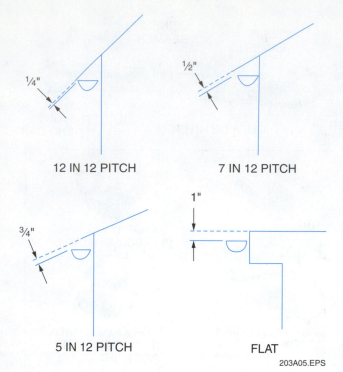

12 IN 12 PITCH   7 IN 12 PITCH

1/4"   1/2"

3/4"   1"

5 IN 12 PITCH   FLAT

203A05.EPS

Figure B-4 ◆ Placement of gutter.

PREFABRICATED VINYL GUTTERS AND DOWNSPOUTS

Vinyl gutters and downspouts are manufactured in various colors with smooth or patterned finishes in essentially the same styles as metal gutters. Downspouts are manufactured in a variety of shapes. *Figures B–5* and *B–6* show two of the most common styles, the K-style and the C-style, along with their downspouts, fittings, and accessories.

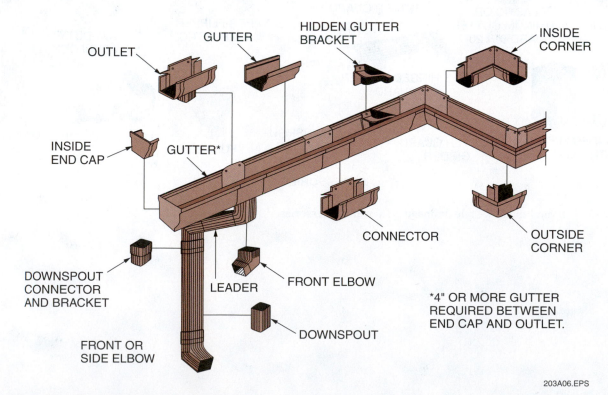

OUTLET   GUTTER   HIDDEN GUTTER BRACKET   INSIDE CORNER

INSIDE END CAP   GUTTER*

CONNECTOR   OUTSIDE CORNER

DOWNSPOUT CONNECTOR AND BRACKET   LEADER   FRONT ELBOW

*4" OR MORE GUTTER REQUIRED BETWEEN END CAP AND OUTLET.

FRONT OR SIDE ELBOW   DOWNSPOUT

203A06.EPS

Figure B-5 ◆ Typical vinyl K-style drainage system.

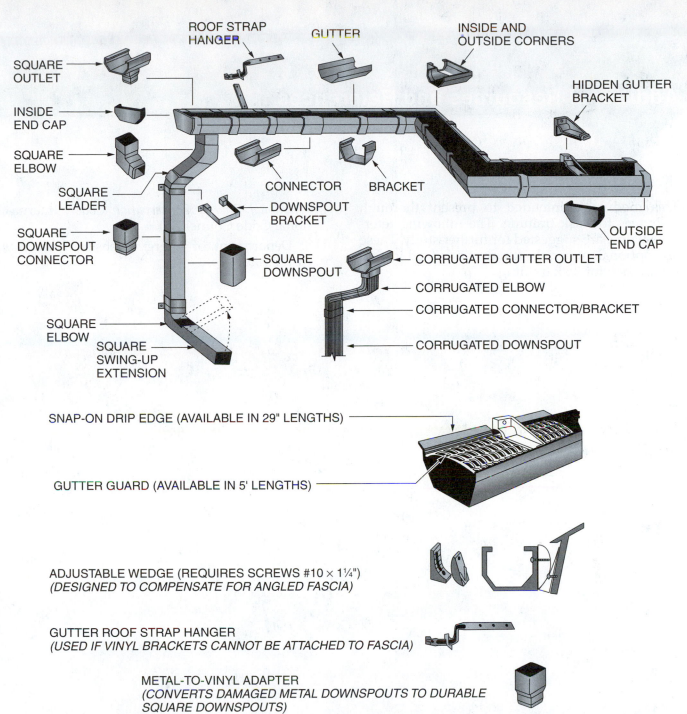

SQUARE OUTLET

ROOF STRAP HANGER

GUTTER

INSIDE AND OUTSIDE CORNERS

INSIDE END CAP

HIDDEN GUTTER BRACKET

SQUARE ELBOW

SQUARE LEADER

CONNECTOR

BRACKET

OUTSIDE END CAP

SQUARE DOWNSPOUT CONNECTOR

DOWNSPOUT BRACKET

SQUARE DOWNSPOUT

CORRUGATED GUTTER OUTLET

CORRUGATED ELBOW

CORRUGATED CONNECTOR/BRACKET

SQUARE ELBOW

SQUARE SWING-UP EXTENSION

CORRUGATED DOWNSPOUT

SNAP-ON DRIP EDGE (AVAILABLE IN 29" LENGTHS)

GUTTER GUARD (AVAILABLE IN 5' LENGTHS)

ADJUSTABLE WEDGE (REQUIRES SCREWS #10 × 1¼") *(DESIGNED TO COMPENSATE FOR ANGLED FASCIA)*

GUTTER ROOF STRAP HANGER *(USED IF VINYL BRACKETS CANNOT BE ATTACHED TO FASCIA)*

METAL-TO-VINYL ADAPTER *(CONVERTS DAMAGED METAL DOWNSPOUTS TO DURABLE SQUARE DOWNSPOUTS)*

203A07.EPS

Figure B-6 ◆ Typical vinyl C-style drainage system.

This module is intended to present thorough resources for task training. The following reference works are suggested for further study. These are optional materials for continued education rather than for task training.

International Energy Conservation Code®, International Code Council, 2006.

U.S. Department of Energy website, *www.eere.energy.gov*.

NCCER CURRICULA — USER UPDATE

NCCER makes every effort to keep its textbooks up-to-date and free of technical errors. We appreciate your help in this process. If you find an error, a typographical mistake, or an inaccuracy in NCCER's curricula, please fill out this form (or a photocopy), or complete the online form at **www.nccer.org/olf**. Be sure to include the exact module ID number, page number, a detailed description, and your recommended correction. Your input will be brought to the attention of the Authoring Team. Thank you for your assistance.

Instructors – If you have an idea for improving this textbook, or have found that additional materials were necessary to teach this module effectively, please let us know so that we may present your suggestions to the Authoring Team.

NCCER Product Development and Revision

13614 Progress Blvd., Alachua, FL 32615

Email: curriculum@nccer.org
Online: www.nccer.org/olf

❏ Trainee Guide   ❏ AIG   ❏ Exam   ❏ PowerPoints   Other _____

Craft / Level: _____   Copyright Date: _____

Module ID Number / Title: _____

Section Number(s): _____

Description: _____

_____

_____

_____

Recommended Correction: _____

_____

_____

_____

Your Name: _____

Address: _____

_____

Email: _____   Phone: _____

Exterior Finishing
27204-07

27204-07
Exterior Finishing

Topics to be presented in this module include:

Overview

Like roofing materials, there is a wide variety of siding materials used to finish the exteriors of homes and some commercial buildings. They include wood, brick, vinyl, metal, and fiber cement board. Each type of siding has its own preparation requirements and installation practices. The siding is not installed just to make the building attractive. It serves to protect the building from the elements. Improperly installed siding can allow water to penetrate the exterior and damage the interior. Siding material that is not installed per the manufacturer's instructions can also result in voiding the warranty. Another element of exterior finish is the closing of roof overhangs with soffit and fascia. This is work that requires skill and attention to detail.

Objectives

When you have completed this module, you will be able to do the following:

1. Describe the purpose of wall insulation and flashing.
2. Install selected common cornices.
3. Demonstrate lap and panel siding estimating methods.
4. Describe the types and applications of common wood siding.
5. Describe fiber-cement siding and its uses.
6. Describe the types and styles of vinyl and metal siding.
7. Describe the types and applications of stucco and masonry veneer finishes.
8. Describe the types and applications of special exterior finish systems.
9. Install three types of siding commonly used in your area.

Trade Terms

Board-and-batten	Lookout
Brown coat	Louver
Building paper	Plancier
Cornice	Rabbet
Course	Rake
Eave	R-value
Fascia	Scratch coat
Felt paper	Shakes
Finish coat	Soffit
Frieze board	Veneer
Ledger	Vent

Required Trainee Materials

1. Pencil and paper
2. Appropriate personal protective equipment

Prerequisites

Before you begin this module, it is recommended that you successfully complete *Core Curriculum*; *Carpentry Fundamentals Level One*; and *Carpentry Framing and Finishing Level Two*, Modules 27201-07 through 27203-07.

This course map shows all of the modules in *Carpentry Framing and Finishing Level Two*. The suggested training order begins at the bottom and proceeds up. Skill levels increase as you advance on the course map. The local Training Program Sponsor may adjust the training order.

27212-07
Cabinet Fabrication
ELECTIVE

27211-07
Cabinet Installation

27210-07
Window, Door, Floor, and Ceiling Trim

27209-07
Suspended Ceilings
ELECTIVE FOR RESIDENTIAL CERTIFICATE

27208-07
Doors and Door Hardware

27207-07
Drywall Finishing

27206-07
Drywall Installation

27205-07
Cold-Formed Steel Framing

27204-07
Exterior Finishing
ELECTIVE FOR COMMERCIAL CERTIFICATE

27203-07
Thermal and Moisture Protection

27202-07
Roofing Applications
ELECTIVE FOR COMMERCIAL CERTIFICATE

27201-07
Commercial Drawings
ELECTIVE FOR RESIDENTIAL CERTIFICATE

CARPENTRY FUNDAMENTALS

CORE CURRICULUM:
Introductory Craft Skills

FRAMING AND FINISHING

204CMAP.EPS

1.0.0 ◆ INTRODUCTION

The primary purpose of any exterior finish is to provide protection from the elements. Some of the most common siding materials are wood, stone, brick, stucco, metal, fiber-cement, and vinyl. Wood, because of its availability and workability, is the most widely used.

Before the installation of any siding, the material to which the siding will be fastened must be made weather resistant. **Building paper**, house wrap, aluminum paper, or insulating boards are usually installed over lumber, plywood, or chipboard exterior sheathing. For a building wrap, a 4" lap at every horizontal and vertical joint is required. At corners, an 8" lap from both sides is usually recommended. All window and door openings must be wrapped with the material to prevent water penetration.

After the sheathing and water barrier are installed, all boxed **cornices**, **rake** sections, windows, and exterior door frames are installed. Exterior window and door trim, if not part of the assembly, should be installed in the same way as interior trim, which is described in another module. For unsheathed structures, jambs and sills of doors and windows should be flashed with either a 6"-wide strip of metal, 3 oz. copper-coated paper, or 6-mil plastic film, as well as other house wrap materials. After any cornices are installed and finished, the siding is applied, and the roof drainage system gutters and downspouts are selected and installed.

Most of the common cornices and wall finishes, along with their installation methods, are covered in this module. Be sure to check the manufacturers' instructions and local building codes, which may require different or more specific construction/installation methods.

2.0.0 ◆ SAFETY

A good carpenter always remembers to put safety first in all situations. Follow all applicable OSHA standards, as well as local and national building codes.

The tools you will be using will most often be of the cutting and hammering type. Power staplers, power hand saws, power or hand sheet metal tin snips, utility knives, and hammers are the most common tools used, and each has its own potential dangers. In most cases, work will be done from the surface of scaffolding or the rung of a ladder, so greater caution is required.

All scaffolding must be placed on a firm footing and leveled. As per OSHA requirements, a ladder must also be placed on a firm footing at the proper

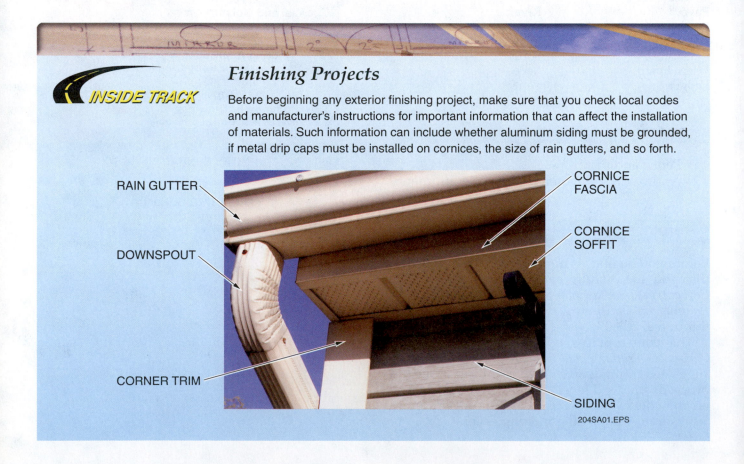

INSIDE TRACK

Finishing Projects

Before beginning any exterior finishing project, make sure that you check local codes and manufacturer's instructions for important information that can affect the installation of materials. Such information can include whether aluminum siding must be grounded, if metal drip caps must be installed on cornices, the size of rain gutters, and so forth.

RAIN GUTTER

DOWNSPOUT

CORNER TRIM

CORNICE FASCIA

CORNICE SOFFIT

SIDING

204SA01.EPS

Building Wrap

One of the more popular building wraps is a spun-bonded olefin material such as ProWrap® or Tyvek®. This material is airtight but breathes to allow water vapor to pass from inside a structure to outside. It is very tough and resists tearing and liquid water penetration. Many local codes do not require building wrap, but it is necessary for a wind and weather-resistant seal. All horizontal and vertical joints must be sealed with a tape approved for that purpose by the manufacturer. Proper installation and taping is especially important to prevent the wrap from being damaged by high winds before the siding is completed.

204SA02.EPS

Scaffolding Systems

Besides sectional, free-standing, manufactured scaffolds that can be assembled in any number of configurations, continuously adjustable aluminum scaffolding systems are available for cornice and siding working heights up to 50'. Typically, an OSHA-recognized system of this type consists of aluminum poles and a standing platform assembly with a safety railing/workbench and safety net. The standing platform can be raised and lowered as an assembly on the supporting poles to obtain the optimum working height. Most systems may be joined both vertically and horizontally for security as well as portability.

204SA03.EPS

angle and be tied off. The usual work clothing may be worn, keeping in mind that loose shirt and trouser cuffs are a hazard. No jewelry should be worn. Work shoes are probably the most important item of clothing. The shoes should be of the safety type, provide good support, and have a thick, non-skid sole for standing on scaffolds and ladders. Always look up for overhead hazards before climbing a ladder. Climbing near power lines must be avoided.

Last, but not least, always consider the weather. The weather conditions are often overlooked, and this can be fatal. When working above the ground, surface wind can be the carpenter's worst enemy. When carrying large pieces of exterior siding, **soffit**, or **fascia** boards, make sure you always point the edge into the wind. If wind strikes the flat surface of these items, it could carry you over the edge of the work area and cause serious injuries. Strong winds can also carry away unsecured ladders and scaffolding, causing severe injury to persons working below. Be aware of the danger of snow, ice, and rain. These conditions create slippery surfaces.

 WARNING!

Health precautions must be observed when cutting certain siding materials, including western red cedar, stucco, masonry coatings, pressure-treated lumber, and fiber-cement siding. The dust resulting from cutting or mixing such products can be hazardous to inhale or may be an allergen. Material safety data sheets (MSDS) furnished by the manufacturers of siding materials must be consulted for any applicable hazards before cutting siding products.

3.0.0 ◆ INSULATION AND FLASHING

This section covers insulation and flashing. Insulation materials and installation procedures are fully covered in the module entitled *Thermal and Moisture Protection*.

3.1.0 Insulation

A conventionally insulated house can suffer a 55 percent heat loss either by losing heat through frame walls or by way of air infiltration. Because of the need for energy conservation, the building industry has been working to devise new products and methods of installation to prevent heat loss in buildings.

Heat loss can be significantly reduced by using insulation sheathing. There are several brands and types of insulation sheathing. These include Dow Styrofoam™, Celotex Tuff-R™ and Celotex Thermax™ sheathing, expanded polystyrene, fiberboard sheathing, Monsanto Fome-Cor™, and thermo-ply foil-faced paper board. When selecting a particular insulating sheathing, the **R-value** and local building codes should be considered.

Installation procedures for insulation sheathing may vary, depending on the type of sheathing selected for use. Insulation sheathing is nonstructural. Adequate corner bracing, such as diagonal 1 × 4 let-in wood bracing, flat or profiled steel bracing, or plywood at corners overlaid with foam sheathing, should be used to comply with local building codes. Dow Styrofoam™ residential sheathing is laminated with a durable plastic film for added resistance to damage and abuse. This sheathing comes with tongue-and-groove edges on ¾" to 1" thicknesses. Celotex foam sheathings (Tuff-R™ and Thermax™) can be easily cut with a utility knife to any shape needed to conform to irregular wall angles or to fit snugly around window or door openings and other projections. Tuff-R™ insulation sheathing is semirigid and can bend around corners to reduce air infiltration.

Siding may be applied directly over the insulating sheathing. Brick, wood, hardboard, aluminum, and vinyl siding are fastened to the wood frame construction by nailing through the sheathing. Care must be taken when driving nails so that the sheathing is not crushed.

Shakes and shingles can also be applied by installing furring strips or a plywood nailer base over the insulating sheathing. A stucco finish can also be applied over an acceptable lath fastened over the sheathing.

Many other factors must be considered when insulating an exterior wall where exterior finish is to be applied. Windows are available with insulating glass. The insulating glass consists of two or three pieces of glass with an air space between them. The edges are sealed and the air space is then turned into a partial vacuum. All materials in a wall system are rated for various insulation and sound values, and different combinations can achieve large differences in insulation and acoustic characteristics.

Aluminum and plastic siding are available with an insulation board backing. This aids in insulating an exterior wall, but is not effective by itself.

3.2.0 Flashing

Before the siding is nailed in place or before the masonry **veneer** is laid up, the flashing must be installed around all openings. Flashing usually consists of galvanized sheet metal, aluminum, or

---

a synthetic material; however, on rare occasions, copper and stainless steel have been used. Normally, aluminum is not used for flashing masonry because of corrosion problems. The primary purpose of flashing is to prevent water that may penetrate the siding from eventually entering the exterior walls and causing rot, water damage of interior surfaces, or mildew. If water does penetrate the siding, flashing is designed to channel it back out again, thus avoiding any water damage. See *Figure 1*.

When brick or stone veneer is used for a frame building, flashing is installed at the base of the sheathing and above door and window openings to channel the water to the outside through weep holes. Frame construction at the water table also requires that flashing be used, as illustrated in *Figure 1*.

Some metal-covered or vinyl-covered windows and doors are manufactured with a flashing flange at the tops and sides and do not require separately installed flashing; however, it is usually a good practice to install flashing as a precaution.

INSIDE TRACK

Box, Closed, and Open Cornices

Of the three types of cornices, box cornices are the most commonly used. A closed cornice is the least desirable type of cornice because it does not allow roof ventilation and provides little protection to the side of the building. Open cornice construction is not commonly used except on porches with directly exposed roof beams. If a ceiling is installed, the roof area above any ceiling cannot be ventilated unless vertical vents through the frieze board are provided. Depending on the rafter tail cut, fascia is sometimes applied to the rafter ends of open cornices to provide better weather protection to the ends of the rafters.

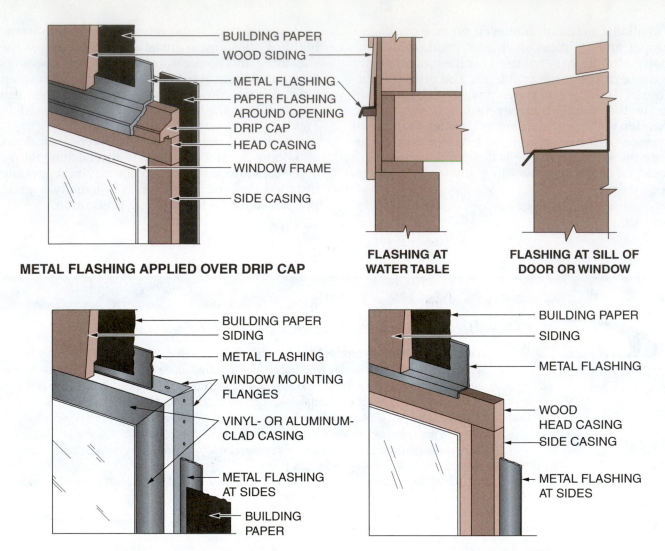

METAL FLASHING APPLIED OVER DRIP CAP

BUILDING PAPER
WOOD SIDING
METAL FLASHING
PAPER FLASHING AROUND OPENING
DRIP CAP
HEAD CASING
WINDOW FRAME
SIDE CASING

FLASHING AT WATER TABLE

FLASHING AT SILL OF DOOR OR WINDOW

METAL FLASHING OVER VINYL- OR ALUMINUM-CLAD WINDOW WITH MOUNTING FLANGE

BUILDING PAPER
SIDING
METAL FLASHING
WINDOW MOUNTING FLANGES
VINYL- OR ALUMINUM-CLAD CASING
METAL FLASHING AT SIDES
BUILDING PAPER

METAL FLASHING OVER WOOD WINDOW CASING

BUILDING PAPER
SIDING
METAL FLASHING
WOOD HEAD CASING
SIDE CASING
METAL FLASHING AT SIDES

204F01.EPS

Figure 1 ◆ Typical flashing installation.

4.0.0 ◆ CORNICES

Cornices are constructed of lumber, as well as aluminum, vinyl, and other man-made products. The type of cornice required for a particular structure is shown on the wall sections of the construction drawings. The three general types of cornices are the closed cornice, open cornice, and box cornice.

A roof with no rafter overhang normally has a closed cornice. See *Figure 2*. This cornice consists of a single strip called a **frieze board**. The frieze board is beveled on its upper edge to fit close under the overhang of the **eaves** and **rabbeted** on its lower edge to overlap the upper edge of the top siding **course**. A strip of wood shingles is used to provide a roof overhang several inches beyond the molding. In this instance, the strip can serve as both a drip edge and starter strip for a wood shingle roof or as a support for the starter strip of an asphalt shingle roof. Some codes may require the

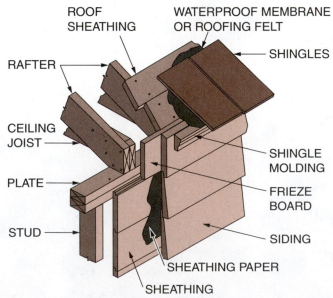

ROOF SHEATHING
WATERPROOF MEMBRANE OR ROOFING FELT
SHINGLES
RAFTER
CEILING JOIST
SHINGLE MOLDING
PLATE
FRIEZE BOARD
STUD
SIDING
SHEATHING PAPER
SHEATHING

204F02.EPS

Figure 2 ◆ Closed cornice.

installation of a metal drip cap on the edge of the strip of shingles. If trim is used, it usually consists of molding installed as shown in *Figure 2*.

A roof with a rafter overhang may have either an open cornice or a box cornice. The simplest type of open cornice consists of only a frieze board cut to fit between the rafters, as shown in *Figure 3*. If trim is used, it usually consists of molding cut to fit between the rafters.

Another type of open cornice consists of a frieze board and a fascia (*Figure 3*). A fascia is a strip nailed to the tail cuts of the rafters. Shingle molding can be attached to the top of the fascia, but it is seldom used. *Figure 4* shows five types of tail rafter cuts.

With a box cornice, the rafter overhang is entirely boxed in by the roof covering, the fascia, and a bottom strip called a **plancier** or soffit. *Figure 5* shows examples of various types of box cornices.

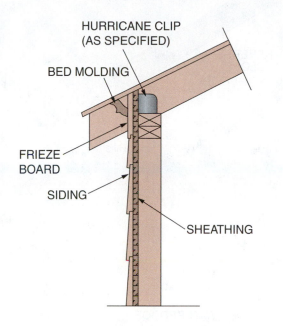

HURRICANE CLIP
(AS SPECIFIED)

BED MOLDING

FRIEZE
BOARD

SIDING

SHEATHING

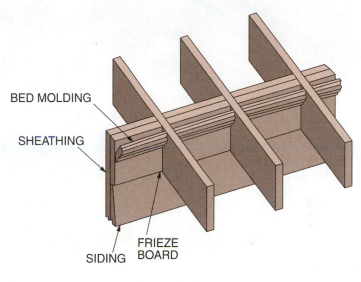

BED MOLDING

SHEATHING

SIDING

FRIEZE
BOARD

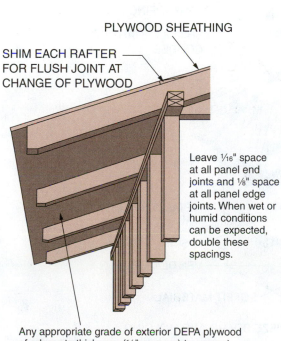

PLYWOOD SHEATHING

SHIM EACH RAFTER
FOR FLUSH JOINT AT
CHANGE OF PLYWOOD

Leave ¹⁄₁₆" space at all panel end joints and ⅛" space at all panel edge joints. When wet or humid conditions can be expected, double these spacings.

Any appropriate grade of exterior DEPA plywood of adequate thickness (½" or more) to prevent protrusion of roofing nails or staples at exposed underside and to carry design roof load.

For aesthetic appearances, ¾" tongue-and-groove 1" × 6" or 1" × 8" boards are sometimes used.

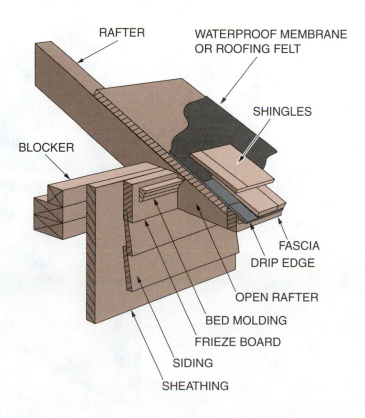

RAFTER

WATERPROOF MEMBRANE
OR ROOFING FELT

SHINGLES

BLOCKER

FASCIA

DRIP EDGE

OPEN RAFTER

BED MOLDING

FRIEZE BOARD

SIDING

SHEATHING

204F03.EPS

Figure 3 ◆ Open cornices.

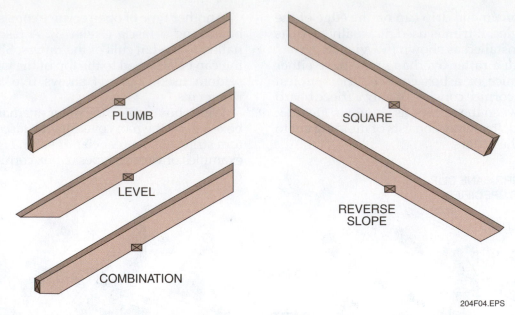

PLUMB

SQUARE

LEVEL

REVERSE
SLOPE

COMBINATION

204F04.EPS

Figure 4 ◆ Types of tail rafter cuts.

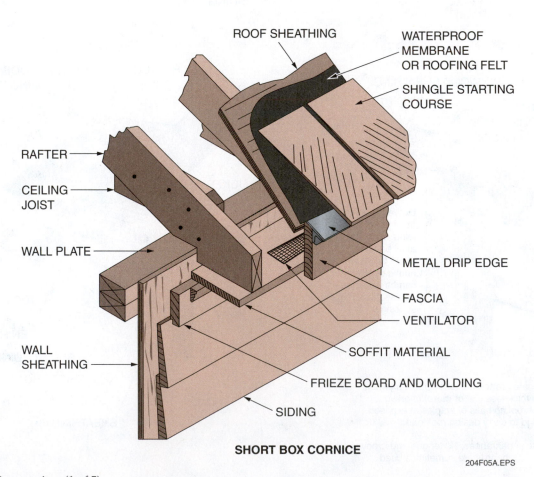

ROOF SHEATHING

WATERPROOF
MEMBRANE
OR ROOFING FELT

SHINGLE STARTING
COURSE

RAFTER

CEILING
JOIST

WALL PLATE

METAL DRIP EDGE

FASCIA

VENTILATOR

SOFFIT MATERIAL

WALL
SHEATHING

FRIEZE BOARD AND MOLDING

SIDING

SHORT BOX CORNICE

204F05A.EPS

Figure 5 ◆ Box cornices (1 of 5).

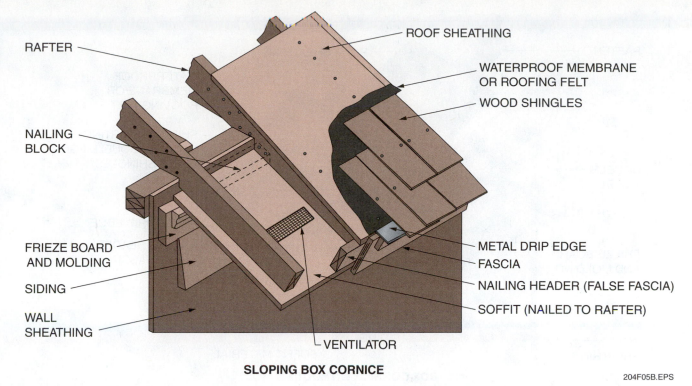

RAFTER

ROOF SHEATHING

WATERPROOF MEMBRANE OR ROOFING FELT

WOOD SHINGLES

NAILING BLOCK

FRIEZE BOARD AND MOLDING

SIDING

WALL SHEATHING

METAL DRIP EDGE

FASCIA

NAILING HEADER (FALSE FASCIA)

SOFFIT (NAILED TO RAFTER)

VENTILATOR

SLOPING BOX CORNICE

204F05B.EPS

Figure 5 ◆ Box cornices (2 of 5).

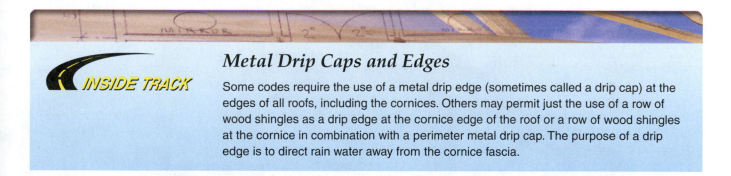

Metal Drip Caps and Edges

Some codes require the use of a metal drip edge (sometimes called a drip cap) at the edges of all roofs, including the cornices. Others may permit just the use of a row of wood shingles as a drip edge at the cornice edge of the roof or a row of wood shingles at the cornice in combination with a perimeter metal drip cap. The purpose of a drip edge is to direct rain water away from the cornice fascia.

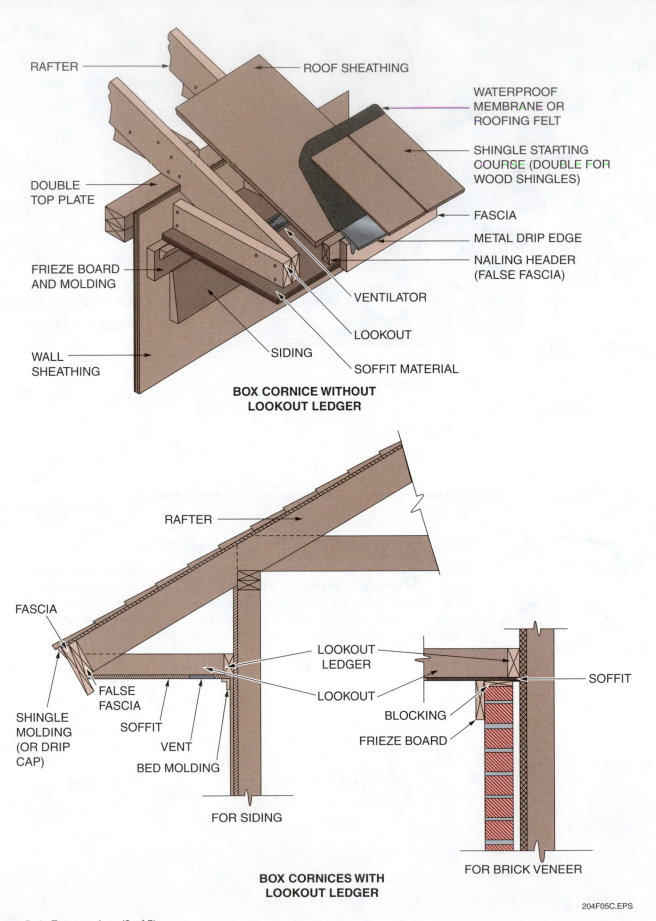

RAFTER

ROOF SHEATHING

WATERPROOF MEMBRANE OR ROOFING FELT

SHINGLE STARTING COURSE (DOUBLE FOR WOOD SHINGLES)

DOUBLE TOP PLATE

FASCIA

METAL DRIP EDGE

NAILING HEADER (FALSE FASCIA)

FRIEZE BOARD AND MOLDING

VENTILATOR

LOOKOUT

WALL SHEATHING

SIDING

SOFFIT MATERIAL

BOX CORNICE WITHOUT LOOKOUT LEDGER

RAFTER

FASCIA

LOOKOUT LEDGER

SOFFIT

LOOKOUT

SHINGLE MOLDING (OR DRIP CAP)

FALSE FASCIA

SOFFIT

VENT

BLOCKING

FRIEZE BOARD

BED MOLDING

FOR SIDING

FOR BRICK VENEER

BOX CORNICES WITH LOOKOUT LEDGER

204F05C.EPS

Figure 5 ◆ Box cornices (3 of 5).

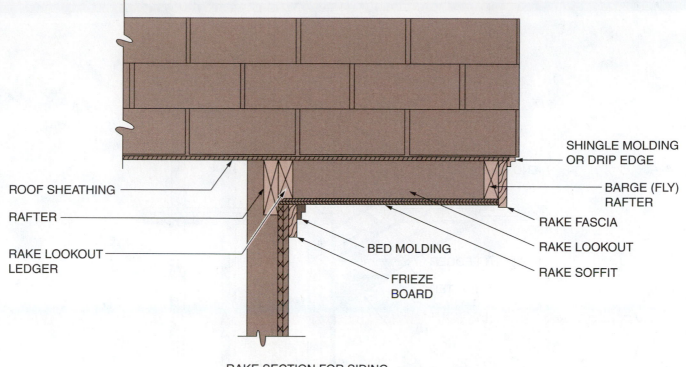

ROOF SHEATHING

RAFTER

RAKE LOOKOUT
LEDGER

SHINGLE MOLDING
OR DRIP EDGE

BARGE (FLY)
RAFTER

RAKE FASCIA

RAKE LOOKOUT

RAKE SOFFIT

BED MOLDING

FRIEZE
BOARD

RAKE SECTION FOR SIDING

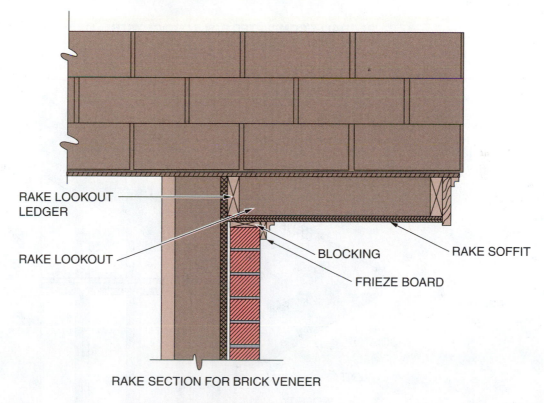

RAKE LOOKOUT
LEDGER

RAKE LOOKOUT

BLOCKING

FRIEZE BOARD

RAKE SOFFIT

RAKE SECTION FOR BRICK VENEER

BOX CORNICE RAKE SECTIONS

204F05D.EPS

Figure 5 ◆ Box cornices (4 of 5).

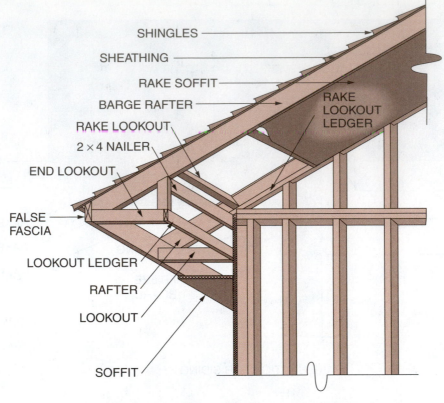

SHINGLES

SHEATHING

RAKE SOFFIT

BARGE RAFTER

RAKE LOOKOUT
LEDGER

RAKE LOOKOUT

2 × 4 NAILER

END LOOKOUT

FALSE
FASCIA

LOOKOUT LEDGER

RAFTER

LOOKOUT

SOFFIT

**PARTIALLY COMPLETED BOX CORNICE WITH
CORNICE RETURN AND RAKE SECTION**

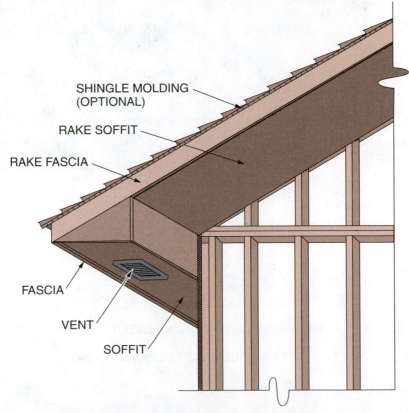

SHINGLE MOLDING
(OPTIONAL)

RAKE SOFFIT

RAKE FASCIA

FASCIA

VENT

SOFFIT

**COMPLETED BOX CORNICE WITH CORNICE
RETURN AND RAKE SECTION**

204F05E.EPS

Figure 5 ◆ Box cornices (5 of 5).

Box cornices use wood, aluminum, vinyl, or exterior gypsum materials for soffits and sometimes for fascia. *Figure 6* shows various types of trim molding used on cornices.

4.1.0 Cornice Installation

The fasteners used in wood cornice work normally consist of various sizes and types of nails. Common steel nails and cement-coated box nails are used in areas where they will be completely enclosed and protected from the weather.

Hot-dipped galvanized nails are commonly used to fasten any cornice member exposed to the weather. These nails should be of sufficient length to hold the material in place and may be any of the types commonly manufactured.

Aluminum and stainless steel nails are available for use on exterior trim to eliminate rust streaks. Stainless steel nails are used on the highest-quality work where the added cost of the nails is incidental.

Figure 7 is a nail spacing chart for plywood used as a closed soffit. Many times, the failure of plywood to maintain a level surface after installation is not due to a product failure, but is caused by improper installation.

The tools used in the installation of the cornice and related finish are the same as those used for rough and finish carpentry.

The following are guidelines for building a box cornice:

Step 1 Start by marking the location of the **ledger** (*Figure 8*), then snap a chalkline to ensure the ledger will be level. Secure the ledger to the wall. The ledger makes it easier to install the **lookouts**.

PLYWOOD THICKNESS	MAXIMUM SPACING SUPPORT	NAIL SIZE	NAIL SPACING EDGE	NAIL SPACING FIELD
⅜"	24"	6d	6"	12"
½"	24"	6d	6"	12"
⅝"	48"	8d	6"	12"

204F07.EPS

Figure 7 ◆ Nail spacing chart.

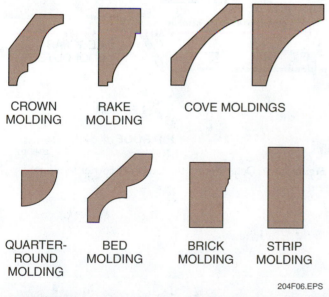

CROWN MOLDING   RAKE MOLDING   COVE MOLDINGS

QUARTER-ROUND MOLDING   BED MOLDING   BRICK MOLDING   STRIP MOLDING

204F06.EPS

Figure 6 ◆ Types of wood cornice trim molding.

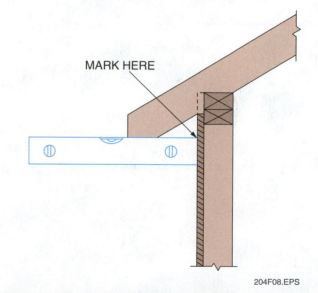

MARK HERE

204F08.EPS

Figure 8 ◆ Determining the ledger position.

INSIDE TRACK

Stainless Steel Nails

While aluminum nails are rustproof, they may be corroded by certain chemicals in industrial or high vehicular traffic environments. Quality stainless steel nails are not affected by any environment.

Step 2 Use a straight board to mark the lookout positions on the ledger (*Figure 9*).

Step 3 Measure the distance from the face of the ledger to the ends of the rafters to determine the length of the lookouts.

Step 4 Trim the corner lookouts to match the slope of the rafters (*Figure 10*).

Step 5 Attach the lookouts to the ledger, then install the assembly, nailing through the sheathing into studs where possible.

Step 6 Once the lookouts are in place, the soffit and fascia boards can be attached (*Figures 11* and *12*).

Step 7 The cornice enclosure must be cut to fit, and the cornice end piece is grooved to fit inside the cornice enclosure (*Figure 13*).

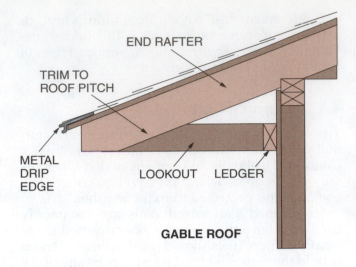

GABLE ROOF

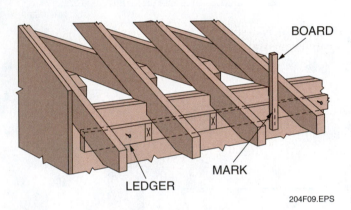

204F09.EPS

Figure 9 ◆ Marking lookout locations on a ledger.

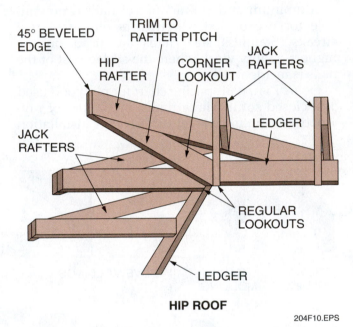

HIP ROOF

204F10.EPS

Figure 10 ◆ Determining the length of corner lookouts.

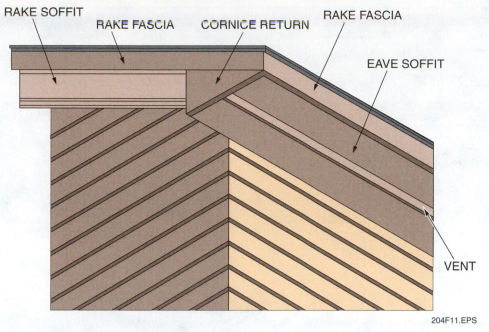

RAKE SOFFIT

RAKE FASCIA

CORNICE RETURN

RAKE FASCIA

EAVE SOFFIT

VENT

204F11.EPS

Figure 11 ◆ Cornice return.

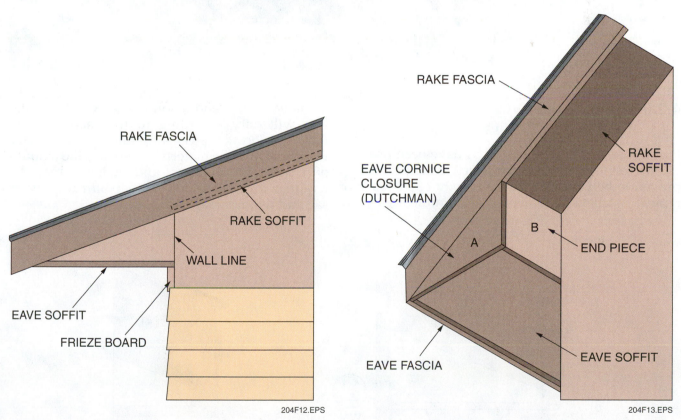

RAKE FASCIA

RAKE SOFFIT

WALL LINE

EAVE SOFFIT

FRIEZE BOARD

204F12.EPS

Figure 12 ◆ Rake soffit cut to wall line.

RAKE FASCIA

RAKE SOFFIT

EAVE CORNICE CLOSURE (DUTCHMAN)

A

B

END PIECE

EAVE SOFFIT

EAVE FASCIA

204F13.EPS

Figure 13 ◆ Boxing in the cornice return.

Cornice Returns

A variety of more complicated cornice returns exists, as shown in these illustrations.

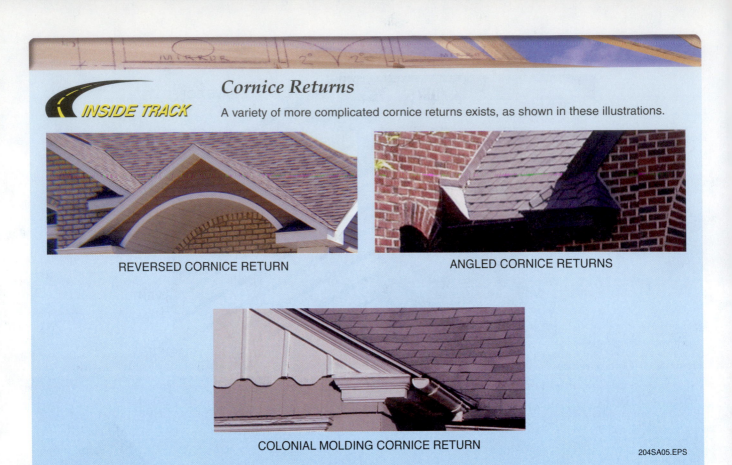

REVERSED CORNICE RETURN

ANGLED CORNICE RETURNS

COLONIAL MOLDING CORNICE RETURN

204SA05.EPS

4.2.0 Aluminum or Vinyl Fascia and Soffits

Prefinished aluminum and vinyl have been used extensively in cornice construction and have proven to be very satisfactory. *Figure 14* shows a typical installation detail of an aluminum or vinyl fascia eave trim and a soffit. The actual installation will vary according to the manufacturer. *Figure 15* shows typical soffit and trim materials. Depending on the materials used and the amount of overhang, lookouts and ledgers may be required for the support of the soffit to prevent sag and wind damage.

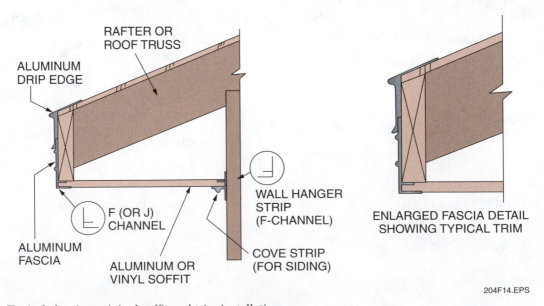

RAFTER OR ROOF TRUSS

ALUMINUM DRIP EDGE

F (OR J) CHANNEL

ALUMINUM FASCIA

ALUMINUM OR VINYL SOFFIT

WALL HANGER STRIP (F-CHANNEL)

COVE STRIP (FOR SIDING)

ENLARGED FASCIA DETAIL SHOWING TYPICAL TRIM

204F14.EPS

Figure 14 ◆ Typical aluminum/vinyl soffit and trim installation.

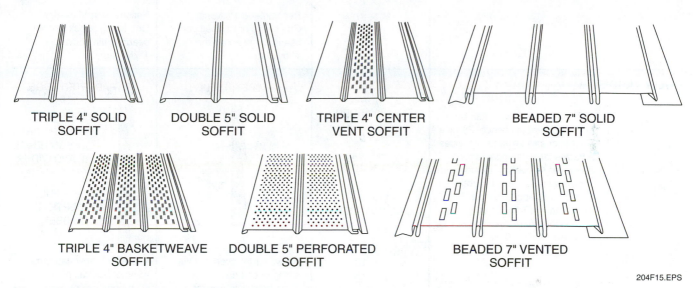

8" – 12"
FASCIA

SOFFIT
COVE
TRIM

F-CHANNEL

J-CHANNEL

H-DIVIDER
BAR

DOUBLE
CHANNEL
LINEAL

EACH SOFFIT PANEL PIECE IS CUT TO THE DESIRED WIDTH AND THEN SLID INTO PLACE FROM THE END
OF THE CORNICE. EACH PIECE INTERLOCKS WITH THE ADJOINING PIECES.

TRIPLE 4" SOLID
SOFFIT

DOUBLE 5" SOLID
SOFFIT

TRIPLE 4" CENTER
VENT SOFFIT

BEADED 7" SOLID
SOFFIT

TRIPLE 4" BASKETWEAVE
SOFFIT

DOUBLE 5" PERFORATED
SOFFIT

BEADED 7" VENTED
SOFFIT

204F15.EPS

Figure 15 ◆ Typical aluminum/vinyl soffit and trim materials.

5.0.0 ◆ WOOD SIDING

The woods used most often for siding are western red cedar (WRC), bald cypress, douglas fir, western hemlock, western larch, ponderosa pine, red pine, southern white pine, sugar pine, and redwood. These woods are shaped into many different siding styles. *Figure 16* is a summary of the most common styles, sizes, and nailing patterns.

Siding, casing, box, finish, ring, or spiral-shanked stainless steel or steel nails with hot-dipped galvanized or noncorrosive coatings (see *Figure 17*) are commonly used to apply wood siding. The size will vary from 6d to 10d. The siding

WARNING!

The dust from western red cedar is an allergen and can cause respiratory ailments including asthma and rhinitis. It can also cause eye irritation and skin disorders, including dermatitis, itching, and rashes. Avoid inhaling the dust or getting it on your skin or in your eyes.

nail is considered the best nail for wood siding except in high-wind areas. Then, a wider-headed face nail is required.

SIDING PATTERNS		SIZES (THICKNESS AND WIDTH)	NAILING	
			6" AND NARROWER	8" AND WIDER

PLAIN BEVELED OR BUNGALOW

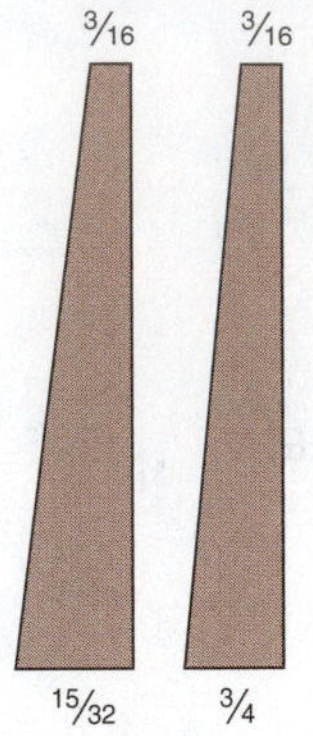

$^3/_{16}$ $^3/_{16}$

$^{15}/_{32}$ $^3/_4$

Bungalow (colonial) is slightly thicker than plain beveled. Either can be used with the smooth or saw-faced surface exposed. Patterns provide a traditional-style appearance. Recommend a 1" overlap. Do not nail through overlapping pieces. Horizontal applications only. Cedar bevel is also available in $^7/_8$" × 10, 12.

Sizes:
$^1/_2$ × 4
$^1/_2$ × 5
$^1/_2$ × 6

$^5/_8$ × 8
$^5/_8$ × 10

$^3/_4$ × 6
$^3/_4$ × 8
$^3/_4$ × 10

6" AND NARROWER — PLAIN
Recommend 1" overlap. One siding or box nail per bearing, just above the 1" overlap.

8" AND WIDER — PLAIN
Recommend 1" overlap. One siding or box nail per bearing, just above the 1" overlap.

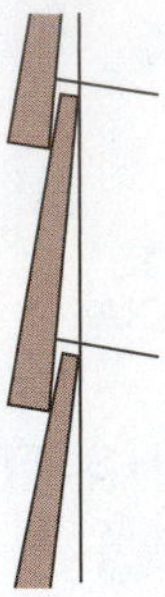

DOLLY VARDEN

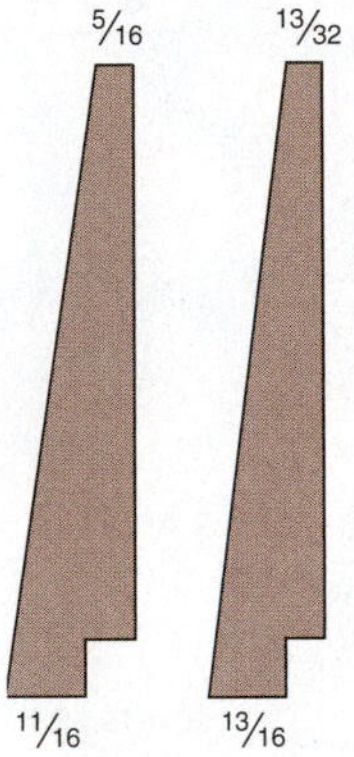

$^5/_{16}$ $^{13}/_{32}$

$^{11}/_{16}$ $^{13}/_{16}$

Dolly Varden is thicker than bevel and has a rabbeted edge. Surface smooth or saw textured. Provides a traditional-style appearance. Allows for $^1/_2$" overlap, including an approximate $^1/_8$" gap. Do not nail through overlapping pieces. Horizontal applications only. Cedar Dolly Varden is also available in $^7/_8$" × 10, 12.

STANDARD DOLLY VARDEN
$^3/_4$ × 6
$^3/_4$ × 8
$^3/_4$ × 10

THICK DOLLY VARDEN
1 × 6
1 × 8
1 × 10
1 × 12

6" AND NARROWER — RABBETED EDGE
Allows for $^1/_2$" overlap. One siding or box nail per bearing, 1" up from bottom edge.

8" AND WIDER — RABBETED EDGE
APPROXIMATE $^1/_8$" GAP FOR DRY MATERIAL 8" AND WIDER
$^1/_2$" = FULL DEPTH OF RABBET
Allows for $^1/_2$" overlap. One siding or box nail per bearing, 1" up from bottom edge.

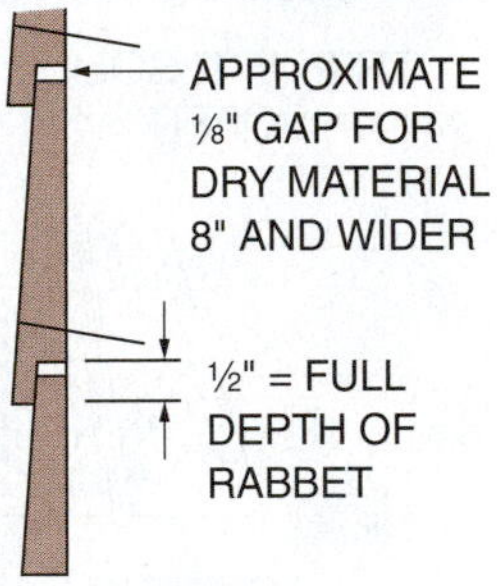

TONGUE-AND-GROOVE (T&G)

T&G siding is available in a variety of patterns. T&G lends itself to different effects aesthetically. Refer to WWPA "standard patterns" (G-16) for pattern profiles. Sizes given are for plain T&G. Do not nail through overlapping pieces. Vertical, diagonal, or horizontal applications.

1 × 4
1 × 6
1 × 8
1 × 10

NOTE:
T&G PATTERNS MAY BE ORDERED WITH $^1/_4$", $^3/_8$", OR $^7/_8$" TONGUES. FOR WIDER WIDTHS, SPECIFY THE LONGER TONGUE AND PATTERN.

6" AND NARROWER — PLAIN
Use one casing nail per bearing to blind nail.

8" AND WIDER — PLAIN
Use two siding or box nails 3" – 4" apart to face nail.

204F16A.EPS

Figure 16 ◆ Common wood siding styles (1 of 2).

SIDING PATTERNS	SIZES (THICKNESS AND WIDTH)	NAILING	
		6" AND NARROWER	8" AND WIDER

DROP

Drop siding is available in 13 patterns, in smooth, rough, and saw-textured surfaces. Some are T&G (as shown), others are shiplapped. Refer to WWPA "standard patterns" (G-16) for pattern profiles with dimensions. A variety of looks can be achieved with different patterns. Do not nail through overlapping pieces. Horizontal or vertical applications.

Sizes: ¾ × 6, ¾ × 8, ¾ × 10

T&G PATTERN / SHIPLAP PATTERN (6" and narrower): Use casing nails to blind nail T&G patterns, one nail per bearing. Use siding or box nails to face nail shiplap patterns 1" up from bottom edge.

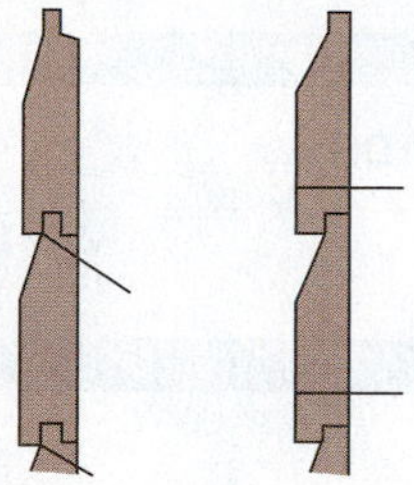

T&G PATTERN / SHIPLAP PATTERN (8" and wider): APPROX. ⅛" GAP FOR DRY MATERIAL 8" AND WIDER. ½" = FULL DEPTH OF RABBET. Use two siding or box nails 3" – 4" apart to face nail starting 1" up from bottom edge.

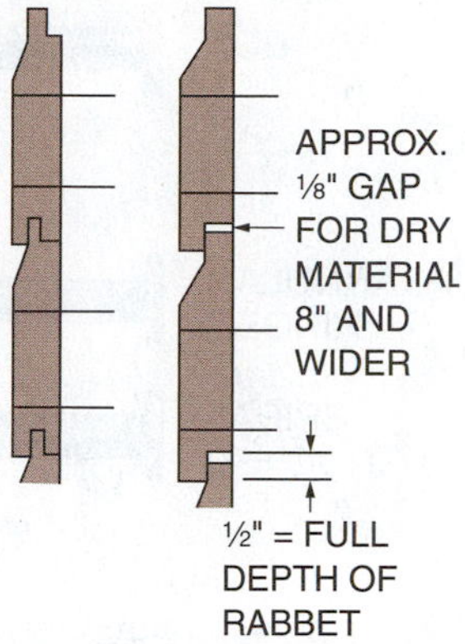

CHANNEL RUSTIC

Channel rustic has a ½" overlap (including an approximate ⅛" gap) and a 1" to 1¼" channel when installed. The profile allows for maximum dimensional change without adversely affecting appearance in climates of highly variable moisture levels between seasons. Available smooth, rough, or saw-textured. Do not nail through overlapping pieces. Vertical, diagonal, or horizontal applications.

Sizes: ¾ × 6, ¾ × 8, ¾ × 10

6" and narrower: Use one siding or box nail to face nail once per bearing, 1" up from bottom edge.

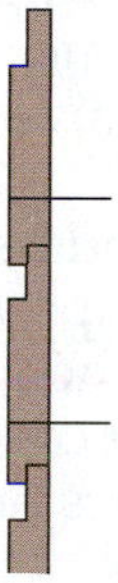

8" and wider: APPROXIMATE ⅛" GAP FOR DRY MATERIAL 8" AND WIDER. ½" = FULL DEPTH OF RABBET. Use two siding or box nails 3" – 4" apart to face nail starting 1" up from bottom edge.

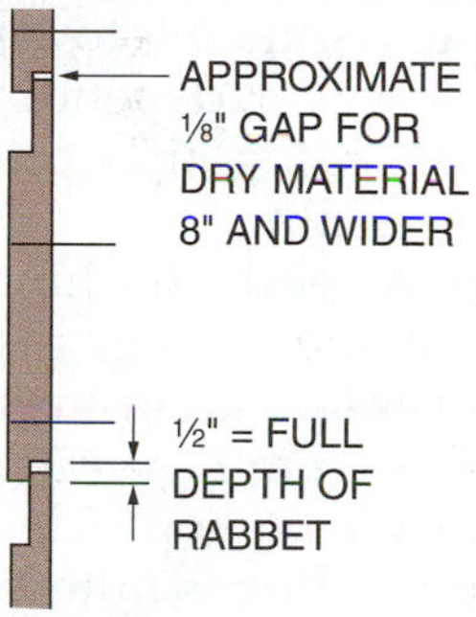

LOG CABIN

Log cabin siding is 1½" thick at the thickest point. Ideally suited to informal buildings in rustic settings. The pattern may be milled from appearance grades (commons) or dimensional grades (2× material). Allows for ½" overlap, including an approximate ⅛" gap. Do not nail through overlapping pieces. Vertical or horizontal applications.

Sizes: 1½ × 6, 1½ × 8, 1½ × 10, 1½ × 12

6" and narrower: Use one siding or box nail to face nail once per bearing, 1½" up from bottom edge.

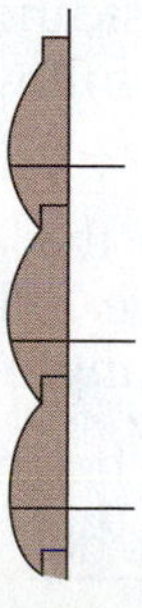

8" and wider: APPROXIMATE ⅛" GAP FOR DRY MATERIAL 8" AND WIDER. ½" = FULL DEPTH OF RABBET. Use two siding or box nails 3" – 4" apart, per bearing to face nail starting 1½" up from bottom edge.

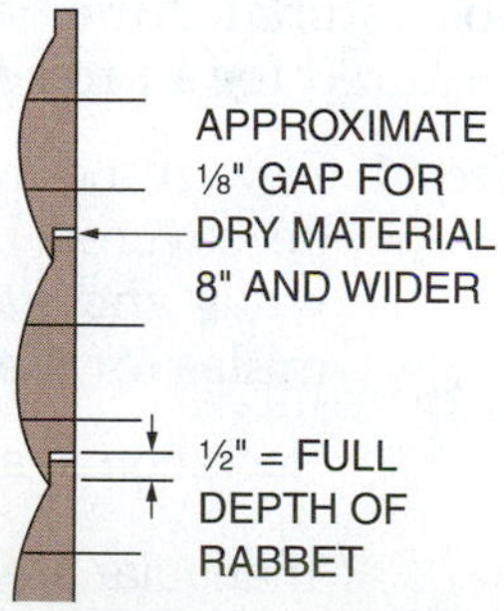

204F16B.EPS

Figure 16 ◆ Common wood siding styles (2 of 2).

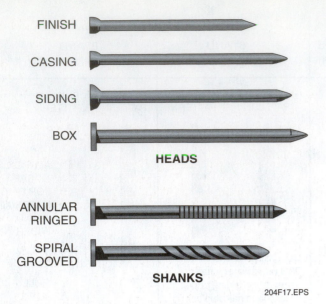

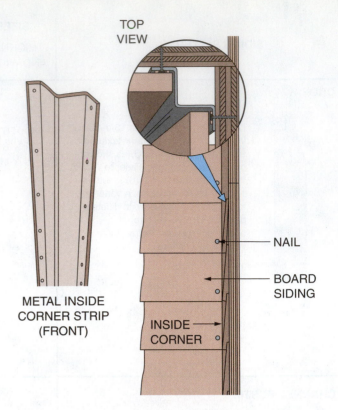

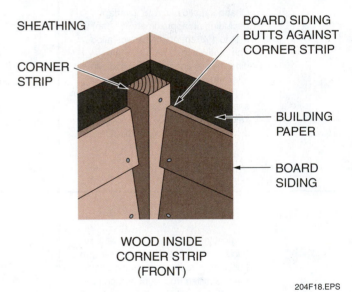

Figure 17 ◆ Commonly used nails.

Pneumatic nailers can be used with a flush-mount attachment or if the pressure of the tool can be regulated to prevent overdriving the nails. However, depending on the type and thickness of the siding, they often cause excessive splitting of the siding.

After the flashing is installed and before applying the siding, inside corner strips are usually installed at all inside corners of the structure, as shown in *Figure 18*. In some more costly projects, inside and outside corner pieces are not used and the siding is mitered to fit.

5.1.0 Estimating Procedure for Panel and Board Siding

To estimate the amount of siding material required for a project, proceed as follows:

Step 1 Determine the total area of the structure to be covered by adding up the areas of all walls and gables using the following formulas for the area of a rectangle or triangle:

Rectangular area = width × height

$$\text{Triangular area} = \frac{\text{width} \times \text{height}}{2}$$

Step 2 Subtract the total area of all openings.

Step 3 For wood board-type siding, add the waste percentages for the size and lap, as listed in *Table 1*. For metal or vinyl siding, add 10 percent.

Figure 18 ◆ Wood or metal inside corner strips.

 NOTE

Except for large openings such as a garage or sliding doors, omit Step 2 for 2 × 8 or 4 × 8 sheet material.

Table 1 Waste Allowances

Siding	Size and Lap	Percent to Add
Beveled	1 × 4 − ¾"	45 percent
	1 × 5 − ⅞"	38 percent
	1 × 6 − 1"	33 percent
	1 × 8 − 1¼"	33 percent
	1 × 10 − 1½"	29 percent
	1 × 12 − 1½"	23 percent
Drop siding and rustic (shiplapped)	1 × 4	28 percent
	1 × 5	21 percent
	1 × 6	19 percent
	1 × 8	16 percent
Drop siding and rustic (dressed and matched)	1 × 4	23 percent
	1 × 5	18 percent
	1 × 6	16 percent
	1 × 8	14 percent
Triangular areas or diagonal installation*		10 percent

*The 10 percent is in addition to other allowances.

Step 4 To determine the number of squares of board-type siding required, divide the total area plus the waste percentage by 100 sq ft. Round up to the next whole square.

$$\text{Number of squares} = \frac{\text{Total area} + \text{waste percentage}}{100}$$

Step 5 For 2 × 8 or 4 × 8 panel siding, divide the total area by the area of a panel to determine the number of panels required. Round up to the next whole panel.

$$\text{Number of 2} \times \text{8 panels} = \frac{\text{total area}}{16}$$

$$\text{Number of 4} \times \text{8 panels} = \frac{\text{total area}}{32}$$

5.2.0 Beveled Siding

Beveled siding is a pattern most often associated with traditional architecture, but it can also be used with success in contemporary structures. Beveled siding comes in plain, bungalow (colonial), and rabbeted (Dolly Varden) styles (see *Figure 19*). Plain beveled and bungalow styles each produce a strong shadow line. Rabbeted

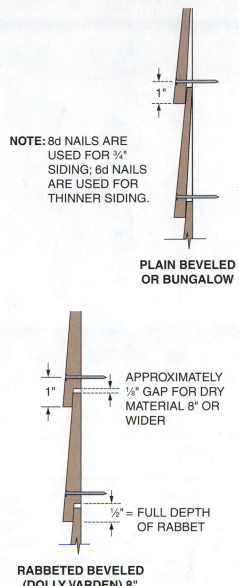

NOTE: 8d NAILS ARE USED FOR ¾" SIDING; 6d NAILS ARE USED FOR THINNER SIDING.

PLAIN BEVELED OR BUNGALOW

APPROXIMATELY ⅛" GAP FOR DRY MATERIAL 8" OR WIDER

½" = FULL DEPTH OF RABBET

RABBETED BEVELED (DOLLY VARDEN) 8" AND WIDER

204F19.EPS

Figure 19 ◆ Beveled siding.

beveled siding provides a somewhat snugger lap and lays up faster, with a greater coverage than beveled siding. The surfaced side is normally used for painted finishes and the rough side for natural finishes and a more informal look.

Beveled siding is face nailed with one siding nail per bearing (8d for ¾" siding and 6d for thinner siding), so that the shank of the nail clears the tip of the under course. Allow ⅛" above the tip for expansion. Lap beveled siding by at least 1" and rabbeted beveled siding by ½". Use aluminum nails, hot-dipped galvanized nails, or stainless steel nails. You will need about 1 lb of 6d nails per 100 square feet of siding or 1½ lbs of 8d nails per 100 square feet of siding. Do not nail through the overlap (*Figure 20*).

Figure 20 ◆ Example of incorrect nailing.

As mentioned previously, plain or bungalow beveled siding must have a minimum lap of at least 1". However, it may be larger to allow for spacing adjustment.

5.2.1 *Beveled Siding Installation*

Before installing beveled siding, make a story pole that is as long as the height of the wall. Draw a line about 1" below the top of the foundation, but at least 6" above grade level. Follow these guidelines to install the siding:

- Use a divider set at the siding exposure to mark the story pole, then transfer the marks to the wall at the corners, windows, and doors. Make sure the story pole is plumb before marking the wall.

◢ INSIDE TRACK

Uses of Story Poles, Transits, and Lasers

It is very important that the horizontal lines of siding on a structure are parallel to the soffit/roof line of the structure even if the structure or soffit/roof line is out of level because of a structural error or settling. This is necessary to prevent the horizontal lines of the siding from being at an angle to the soffit/roof line, which would exaggerate the out-of-level problem. To accomplish this, a siding story pole, measured from the soffits, is typically used to establish vertical location marks for the siding at all corners and at all door and window openings of a structure. The marking is done after application of the external building insulation/wrap and before the application of any siding.

If the soffit/roof lines of the structure are level, a transit can be used as an alternative to set the lower edge of the siding at the corners and intermediate points. In this case, a story pole only needs to be used to place the siding marks up the structure.

On long walls or gable ends of a structure, a transit can be used to set the lower edge of the siding at intermediate points even if the soffit/roof lines are out-of-level. This is accomplished by setting the transit to an out-of-level position (in the plane parallel to the wall) that allows the transit cross hairs to sweep across the same story pole location mark at both outside corners of the wall. However, the transit must be level in the plane perpendicular to the wall.

204SA06.EPS

NOTE

The exposure may be adjusted, as required, so that the spacing to the top comes out even, as long as a minimum overlap of 1" is observed. If possible, also adjust spacing so single pieces of siding will run continuously above and/or below the majority of windows or other wall openings without requiring notching or small slivers of siding above or below the opening.

- It is a good idea to snap a chalkline through the siding location marks on long sides of the building, so that each piece of siding may be nailed to a mark. This will ensure that the siding will have a straight, neat appearance, devoid of waves and sags. If desired, a spacing gauge can be used to place siding, as shown in *Figure 21*. However, a chalkline should be used as a reference.

CAUTION

Before installing siding, make sure that the sheathing and siding are dry. If excessive moisture is present during the application, later drying and shrinking may cause end gaps and stress warping of the siding.

- Start by placing the bottom course of siding around the perimeter of the building.

INSIDE TRACK

Siding Reference Marks

It may be desirable to set nails at the siding reference marks so that a line can be attached to them for siding alignment. They can also be used for a chalkline if that is the method used to mark a reference line for siding alignment.

NOTE

When using plain beveled siding, place a furring strip behind the bottom edge of the starting course. The strip should be approximately the thickness of the top 1" of the siding. This will allow the siding to project the same distance that the next course of siding will project above it (refer to *Figure 21*). Also, it is important that the lowest edge of the siding be at least 6" above the ground level. The high humidity and free water often present at the base of a foundation because of landscaping can cause finish difficulties and structural problems. It is also very important that the end grain of the siding butt joints and the bottom edge of the first course of siding be given a water-repellent treatment.

- When placing the remainder of the siding, do not allow the butt joints to align vertically. Tight-fitting butt joints can be obtained by cutting the siding about ¹⁄₁₆" longer than the measurement. Bow the piece slightly to position the ends, then snap into place.
- If the gable ends will use a different type of finish siding, provide a siding juncture as shown in *Figure 22*.

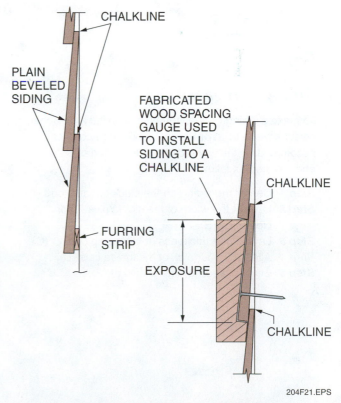

CHALKLINE

PLAIN BEVELED SIDING

FABRICATED WOOD SPACING GAUGE USED TO INSTALL SIDING TO A CHALKLINE

CHALKLINE

FURRING STRIP

EXPOSURE

CHALKLINE

204F21.EPS

Figure 21 ◆ Installation of plain beveled siding.

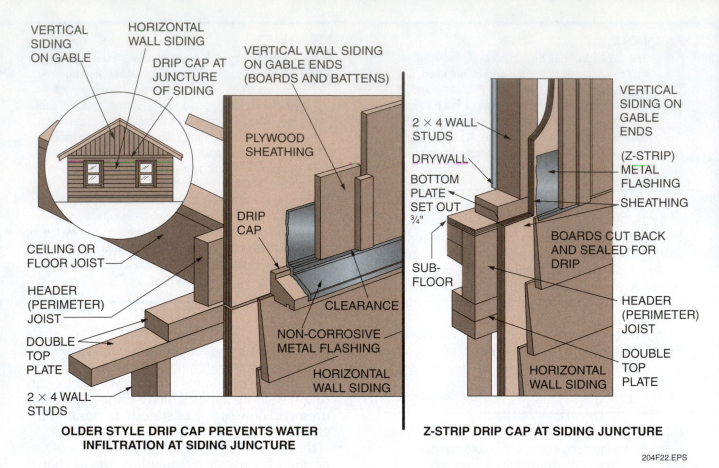

OLDER STYLE DRIP CAP PREVENTS WATER INFILTRATION AT SIDING JUNCTURE

Z-STRIP DRIP CAP AT SIDING JUNCTURE

204F22.EPS

Figure 22 ◆ Siding junctures at gable ends.

- Use a siding gauge when installing siding to fit between two window casings or a door and window casing (*Figure 23*).
- When fitting siding around windows, it may be necessary to notch the siding to fit (*Figure 24*). Flashing must be installed at the top of window and door casings to keep out water.
- The three ways to finish corners for beveled siding are metal corner caps (*Figures 25 and 26*), mitered corners (*Figure 27 and 28*), and corner boards (*Figure 29*).

Mitered corners (*Figure 27*) provide an attractive way to finish corners, but because of the additional labor involved, this method is normally used only on more expensive structures. The angle shown in *Figure 27* is 47 degrees instead of the conventional 45 degrees used for most miter cuts. This prevents gapping at the corners when the siding dries. When applying the siding, force the mitered corners together firmly and nail the mitered ends to the sheathing, not to each other.

Corner boards (*Figure 28*) are thicker than the siding projection and are nailed at all outside corners of the building. The beveled siding simply butts snugly to these boards.

5.3.0 Board-and-Batten Siding

Board-and-batten is an attractive, versatile, squared-edge siding that is widely used and accepted by architects and contractors throughout the building industry (see *Figure 29*).

The amount of nails needed to apply 100 square feet of siding will depend on their size. For 8d nails, 2½ to 3 lbs will be needed.

Board-and-batten is easy to apply and is weathertight. Because it is surfaced on four sides (S4S), it does not require expensive millwork. All of these factors contribute to making it an economical and practical vertical siding.

There are many variations of board-and-batten siding, but the most widely used is the vertical placement of wide boards, with the joints covered by narrow battens. There are a number of different sizes and textures of lumber used.

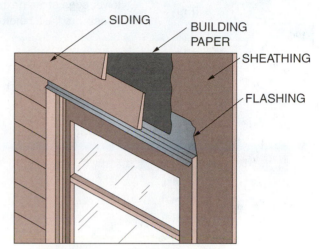

NEW WOOD WINDOW WITHOUT DRIP CAP

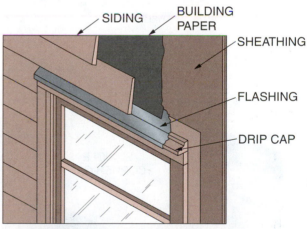

WINDOW WITH OLDER STYLE DRIP CAP

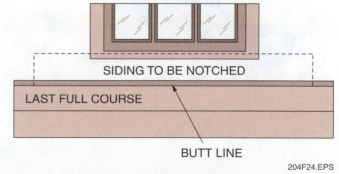

Figure 23 ◆ Using a siding gauge.

204F23.EPS

Figure 24 ◆ Installing siding around windows.

204F24.EPS

Installing Corner Caps

Corner caps are usually used to cover the outside corners of wood siding. They are available in various lengths for different siding widths and are usually installed as each course of siding is applied. Some caps can be installed after all courses are applied. Each course of siding is cut off flush or slightly back from the sheathing on the adjoining wall at each corner and nailed to the sheathing. The lips of the corner cap are tapped up under the course of siding and the cap is nailed to the sheathing above the lower edge of the next course of siding. Always make sure that the cap and siding panels are flush before nailing the cap.

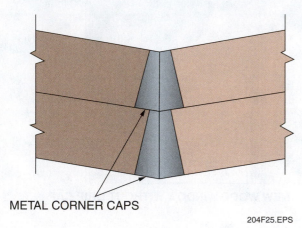

METAL CORNER CAPS

204F25.EPS

Figure 25 ◆ Metal corner caps.

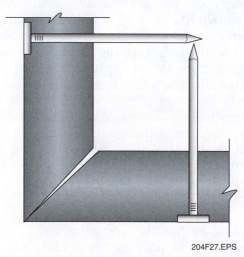

204F27.EPS

Figure 27 ◆ Mitered corner.

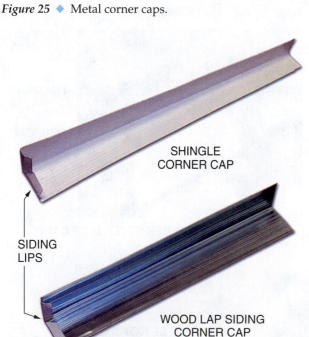

SHINGLE
CORNER CAP

SIDING
LIPS

WOOD LAP SIDING
CORNER CAP

204F26.EPS

Figure 26 ◆ Corner caps.

Corner Flashing

As an added precaution in areas of the country subject to wind-driven rain, vertical flashing is sometimes placed around all outside corners for mitered siding or siding that uses corner boards. The flashing should be wide enough so that it extends 3" to 4" on each side of the corner.

Board-and-Batten Siding as an Architectural Accent

Board-and-batten siding can be used as an architectural accent on a portion of a structure, such as a front-facing gable or part of a wall.

BOARD-AND-BATTEN SIDING

SCALLOPED BOARD-AND-BATTEN ACCENT

204SA08.EPS

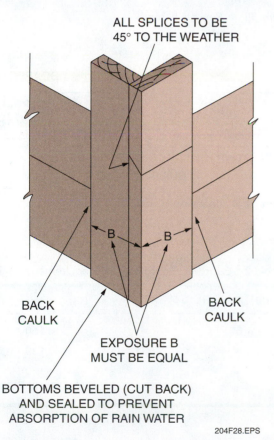

Figure 28 ◆ Outside corner using corner boards.

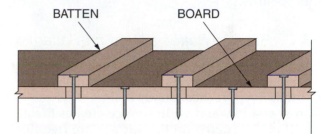

TO AID THE ESTIMATOR, THE LUMBER REQUIREMENTS TO COVER ONE SQUARE (100 SQUARE FEET) ARE INDICATED IN THE FOLLOWING CHART.

WIDTH OF BOARDS	BOARD FEET
6	147
8	139
10	132

204F29.EPS

Figure 29 ◆ Board-and-batten siding.

When the lumber arrives on the job site, it should be placed on blocking off the ground and covered so it will not pick up too much moisture.

When framing an exterior wall that is to receive vertical siding, it is necessary to install horizontal blocking between the studs, from top to bottom at 24" OC.

When applying the siding, space the under-boards ½" apart and drive the nails midway between the edges at each bearing. A major advantage of board-and-batten construction is that with proper nailing, the boards are free to move slightly with changes in moisture content, but are held snugly in place by the battens. To allow for this movement capability, only one nail should be used through the center of a board at each bearing. The nails should penetrate 1½" into

Furring Strips

If Styrofoam™ or wood fiber insulation (⅝" or more) is used on a wall, some plans require the use of horizontal furring strips to provide an adequate surface for nailing the vertical boards. The furring strips can be nailed on 16" or 24" centers.

the studs, the studs and wood sheathing combined, or the blocking. If this depth of penetration is not possible, use annular or spirally grooved nails for their increased holding power. One 8d nail is nailed midway between the edges of the underboard at each bearing.

With the boards in place, fasten the battens using 10d nails. These should overlap each edge of the board underneath by at least 1". The nails should be driven directly through the center of the batten so that the shank passes between the underboards.

5.3.1 Reverse Batten (Board-on-Batten)

Reverse batten (*Figure 30*) is also an attractive vertical siding, giving the building a very sharp, well-defined, deep vertical shadow line. This play of narrow shadow and wide surface creates the illusion that the boards on the surface are free floating. This method can be especially attractive when using rough-sawn boards.

5.3.2 Board-on-Board Siding

Board-on-board is another type of vertical siding. Not only does this method create a vertical shadow line, but it allows the architect or builder to maintain a uniformity in the width of the material used (see *Figure 31*).

For board-on-board siding, apply the underboards first, spacing them to allow a 1½" overlap by the outer boards at both edges. Use standard nailing for underboards, with one 8d nail per bearing. The outer boards must be nailed twice per bearing to ensure proper fastening. Nails having some free length do not hold the outer boards so rigidly as to cause splitting if there is movement from humidity changes. Drive 10d siding nails so that the shanks clear the edges of the underboards by approximately ¼". This provides sufficient bearing for nailing, while allowing clearance for the underboards to expand slightly.

To aid the estimator, the lumber requirements to cover one square (100 square feet) are indicated in *Table 2*.

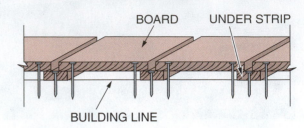

REVERSE BATTEN:
Drive one 8d nail per bearing through the center of the under strip and two 10d nails per bearing through the outer boards.

204F30.EPS

Figure 30 ◆ Reverse batten.

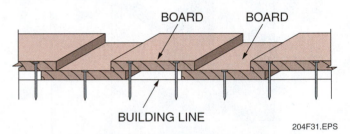

204F31.EPS

Figure 31 ◆ Board-on-board.

Table 2	Board-on-Board Lumber Requirements
Width of Board	**Board Feet/100 ft.²**
6"	150
8"	139
10"	129
12"	123

You will need approximately 25 lbs. of 8d nails per 1,000 board feet (more if using 10d nails).

When applying flat grain boards, orient each board as indicated in *Figure 32*. When the crown surface is exposed to the weather, cupping and grain raising will be prevented.

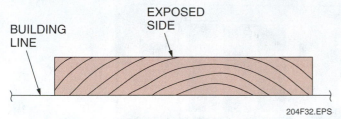

Figure 32 ◆ Applying flat grain board.

5.4.0 Tongue-and-Groove Siding

Tongue-and-groove (T&G) siding (*Figure 33*) can be applied vertically, horizontally, or at an angle and provides a perfectly weathertight wall. It is often installed diagonally. The diagonal application of siding creates an interesting exterior pattern and is pleasing to the eye. As shown, T&G siding is generally available in several styles.

On the exterior elevations of the building plans, the architect will indicate the location and the application angle. The most common slope is 45 degrees, but make sure of this by checking the plans very carefully. In vertically applied siding, do not use boards larger than 1 × 4 or 1 × 6, because using wider boards will cause problems.

Tongue-and-groove drop siding (refer to *Figure 33*) is normally applied only horizontally or vertically. Horizontal T&G drop siding is more water resistant than plain T&G because the top joint is protected by the overhang of the board above, making water penetration of the joint improbable. Like plain T&G siding, T&G drop siding can be blind-nailed.

The advantage that T&G siding has over shiplap siding is that 6" or narrower boards can be blind-nailed, while shiplap siding cannot. Both types of T&G siding are self-aligning, so they take practically no effort to apply after the first piece is set in the correct position.

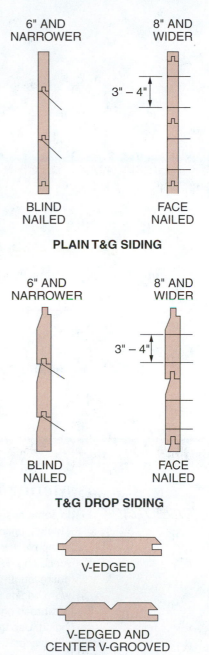

Figure 33 ◆ Tongue-and-groove siding.

Architectural Accents

Vertical or diagonal applications of T&G V-edged or V-grooved and shiplap-style siding are often used as architectural accents on interior or external wall surfaces.

CHANNEL RUSTIC STYLE SHIPLAP SIDING

V-EDGED T&G SIDING

204SA09.EPS

Mating T&G Siding

In some cases, the tongue and groove of T&G siding may have to be forced together to obtain a uniform mating of each course of siding. If necessary, use a hammer and a scrap block of siding on the course being installed to force it onto the tongue of the preceding course. If a board is slightly warped and does not mate evenly along its entire length, secure the board with nails at one or both ends up to the point that the warp begins. Set additional nails in the siding beyond the point of warp. Then, drive a flat, broad chisel into the underlayment-nailing surface with its beveled edge against the siding. Use the chisel as a lever to force the siding into position and then nail the siding into place. Repeat along the length of the board as necessary until the board is seated and secured. Make sure to maintain any inside-groove spacing that may be required.

5.5.0 Shiplap Siding

Plain shiplap siding can be installed vertically, horizontally, or diagonally. In addition to the plain shiplap patterns, it is generally available in four other patterns, with the most common being the V-edged. Plain shiplap lays up with a flush edge. This tends to minimize the direction of the courses and instead accentuates the texture and grain of the wood. On the other hand, V-edged shiplap creates a definite shadow line and indicates the direction of the courses. See *Figure 34*. Other styles include drop, channel rustic, and log cabin. They can be used either horizontally or vertically.

Any style shiplap siding that is 6" or narrower can be face-nailed 1" from the bottom with one nail per bearing. Siding that is wider than 8" should be face-nailed with two nails per bearing. The general rule is that the nails should be long enough to penetrate at least 1½" into the studs, or the studs and wood sheathing combined.

Use 8d nails (25 lbs for 1,000 board feet) for 1" siding and 6d nails (15 lbs per 1,000 board feet) for thinner stock. Nails should be spaced 1½" from the edge of the overlap and 2" from the edge of the underlap for 8" boards. Nail other widths proportionately.

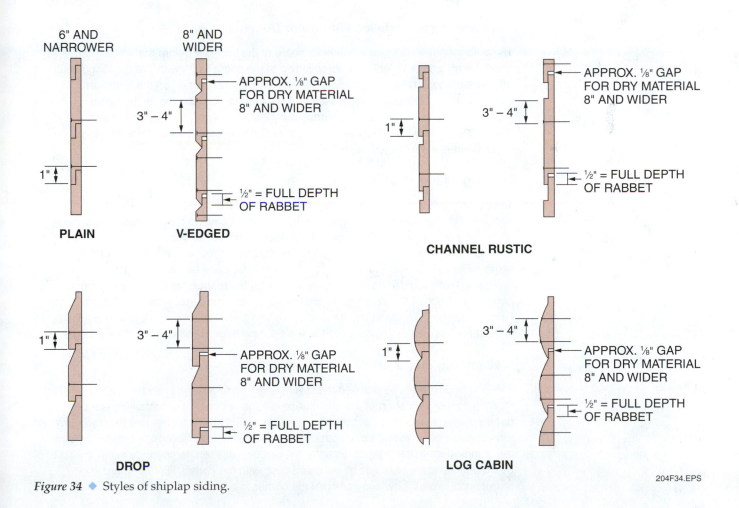

Figure 34 ◆ Styles of shiplap siding.

204F34.EPS

Vertical T&G and Shiplap Application Hints

Here are some application hints for installing 6" to 8" (or less) vertical siding:

Wall Starting Board

Plumb the tongue edge of the board with the grooved edge beyond an outside corner or against an inside corner. Temporarily tack the board in place. Mark the board along the length of the corner. For an outside corner, mark along the backside of the board flush with the corner. For an inside corner, mark along the face of the board using a spacer 6" or 7" long and wide enough to extend beyond the groove depth of the board. Rip off the grooved edge to the mark and slightly back-bevel the edge. Position the board at the corner with ripped edge flush with the corner. Install the board and recheck the plumb of the tongue edge, then face-nail the board at the corner. Blind-nail the tongue edge.

Approaching an Installed Window or Door

Temporarily install and tack a full-width board to the wall just before the window (or door). Using a 6" or 7" length of scrap siding of the same width with just the tongue cut off, mark the top, bottom, and side of the opening, as applicable, on the temporarily installed board. Remove the board and cut out the marked opening. In the same location as the temporary board, permanently install and nail another board with exactly the same width. Then, install and face-nail the cutout board at the opening edge. Blind-nail the tongue edge.

Leaving an Installed Window or Door

Temporarily install and tack two scrap siding pieces (one piece for a door) to the wall as spacers. The piece(s) must be wide enough to extend beyond the opening at positions above and below the opening (above for a door). Temporarily install and tack a board of the same width, cut to full length, against the scrap piece(s) with the tongue of the scrap piece(s) inserted. For a door, plumb the tongue edge before tacking. Using a 6" or 7" length of scrap siding of the same width with the tongue cut off, mark the top, bottom, and side of the opening, as applicable, on the temporarily installed board. Remove the board and cut out the marked opening. Remove the scrap spacer(s). Then, install and face-nail the cutout board at the opening edge. Blind-nail the tongue edge.

Wall Ending Board

The wall ending must be planned to prevent ending with a narrow sliver of siding.

Stop several feet short of the end of the wall and space off the remaining distance to determine the width of the last board. If random widths are available, use them to allow a reasonably wide ending board; otherwise, rip and regroove several boards to achieve the same effect. Install the boards up to the last board. Then, temporarily install the last board and mark the backside of the board flush with the corner. Rip the board and permanently install it by face-nailing at the corner.

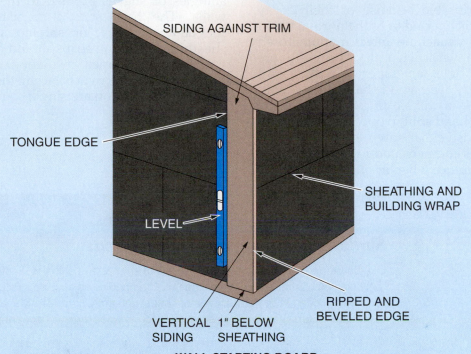

SIDING AGAINST TRIM

TONGUE EDGE

LEVEL

SHEATHING AND
BUILDING WRAP

VERTICAL 1" BELOW
SIDING SHEATHING

RIPPED AND
BEVELED EDGE

WALL STARTING BOARD

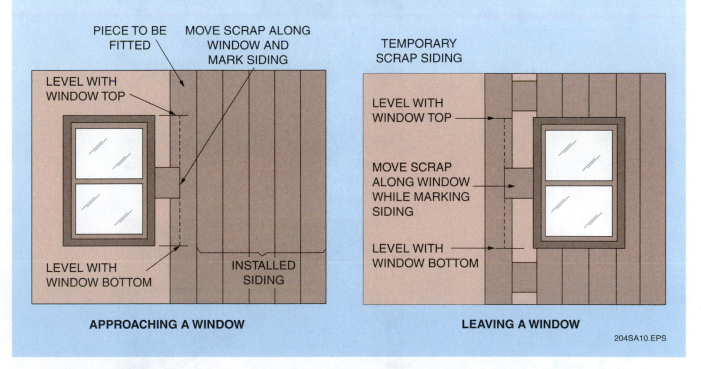

PIECE TO BE
FITTED

MOVE SCRAP ALONG
WINDOW AND
MARK SIDING

LEVEL WITH
WINDOW TOP

LEVEL WITH
WINDOW BOTTOM

INSTALLED
SIDING

APPROACHING A WINDOW

TEMPORARY
SCRAP SIDING

LEVEL WITH
WINDOW TOP

MOVE SCRAP
ALONG WINDOW
WHILE MARKING
SIDING

LEVEL WITH
WINDOW BOTTOM

LEAVING A WINDOW

204SA10.EPS

5.6.0 Shingle Siding or Shakes

The use of shingle siding or shakes as a sidewall finish results in a very attractive, rustic, architecturally interesting siding. Red cedar or cypress is normally used to make the shingles or shakes. They are very durable because of their decay resistance. What is referred to as the normal wood shingle is sawn by machine and is manufactured in 16", 18", and 24" vertical lengths. The widths are random. The shingles are tapered, with a butt thickness of ⅜" to ¾". The wood shake is hand split from a log and is available in taper split form. Because they are hand split, wood shakes are generally more expensive than sawn shingles, but this price difference is compensated for by their beautiful rustic appearance when applied.

Before applying wood shingles or shakes, check the local building codes or with the building inspector in your area and make sure that the fire codes will permit their use.

Solid nailing or stapling is a must for wood shingles or shakes applied to the exterior sidewalls. The base should consist of plywood, tongue-and-groove, or shiplap sheathing. The sheathing or furring strips should first be covered with building paper, insulation board, and/or house wrap. The two basic methods of shingle sidewall application are single course and double course.

INSIDE TRACK

Shake/Shingle Architectural Accents

Like board-and-batten siding, special styles of shakes and shingles, known as fancy-butt shingles or shakes, are sometimes used on gables as accents similar to those used on Victorian-style homes. In other cases, uniformly or randomly spaced staggered-length shakes/shingles are used to enhance a rustic appearance.

STAGGERED-LENGTH SHAKES/SHINGLES

FANCY-BUTT SHINGLE ACCENT

204SA11.EPS

In single-course application, the shingles are applied as in roof construction, but greater weather exposures are permitted. The maximum recommended weather exposures with single-course wall construction are 8½" for 18" lengths and 11½" for 24" lengths. Shingle walls will have two plies of shingles at every point, whereas shingle roofs will have three-ply construction.

The first course of shingles may be doubled at the bottom (*Figure 35*). After the first course is applied, lay out the story pole with all of the courses indicated on it. Make sure you have the courses arranged to line up as closely as possible to the top of all door and window openings. Sometimes it may be necessary to change the exposure slightly on the course at the door and window heads and at the window sills so that they will line up. If such an adjustment is necessary, make sure it is slight so that it is not noticeable when viewing the other exposures.

Using a story pole, transfer those markings to the ends of the building and to all door and window openings. Snap a chalkline at long runs. A furring strip or 1 × 2 straightedge may be temporarily tacked to the building at each course so that the shingle butts may be placed on it for alignment before they are nailed. To form closed joints on outside corners of the sidewalls, shingles in adjoining courses may be alternately overlapped and edge shaved to a close fit.

Another method is to miter the two adjoining shakes in each course, but because of the time consumed in doing this, it is usually too expensive. Inside corners may be woven in alternate overlaps or may be closed by nailing a 1 × 1 square molding or corner board in the corner before the shingles are applied.

The nailing for single coursing is accomplished by using 3d, 1¼" corrosion-resistant nails, such as hot-dipped zinc, aluminum, or stainless steel. Only two nails are used per shingle, and each is placed approximately ¾" from the side edge of the shingle and approximately 1" above the butt line of the next course. Drive the nails flush, but not so hard that the head crushes the wood.

The double-course method of wood shingle sidewall application is much the same as the single-course method, with a few exceptions. Double coursing allows for the application of extended weather exposure shingles over coursing-grade shingles. Double coursing also provides deep, intense, bold shadow lines. When double-coursed, a shingle wall should be tripled at the foundation line by using a double undercourse (see *Figure 36*).

The double-course nailing requires that the outer-course shingle be secured with two 5d (1¾") small head, corrosion-resistant nails, driven 1" to 2" above the butts, approximately ¾" in from each side. Additional nails are driven about 4" apart across the face of the shingle in a straight line.

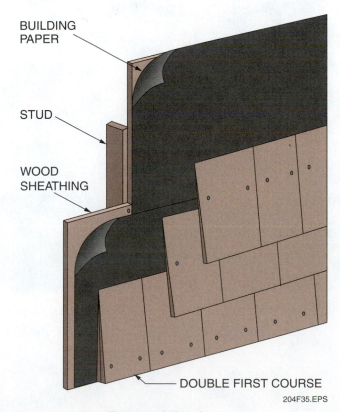

BUILDING PAPER

STUD

WOOD SHEATHING

DOUBLE FIRST COURSE

204F35.EPS

Figure 35 ◆ Installing wood shingles.

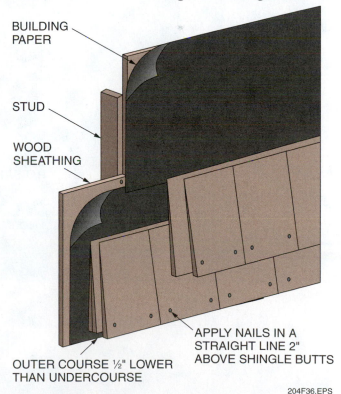

BUILDING PAPER

STUD

WOOD SHEATHING

OUTER COURSE ½" LOWER THAN UNDERCOURSE

APPLY NAILS IN A STRAIGHT LINE 2" ABOVE SHINGLE BUTTS

204F36.EPS

Figure 36 ◆ Double coursing a shingle wall.

Straightedge Shake/Shingle Application

For single coursing, use straightedges tacked at the butt line to rest shingles on for spacing selection. For double coursing, use straightedges with a rabbeted edge so that the outer course is about ¼" below the inner course. Sort the shakes/shingles for proper seam overlap and lay them butt down on the straightedge. Then, nail the shakes/shingles to the wall. For a ribbon-style double coursing, a reversed straightedge with a deeper rabbet can be used to shift the outer shake/shingle up so that about 1" to 1½" of the lower part of the inner shake/shingle is exposed. With any method, use a shingling hatchet to trim and fit the edges if necessary. Butt ends are not trimmed. If rebutted and rejoined shakes/shingles are used, no trimming should be necessary.

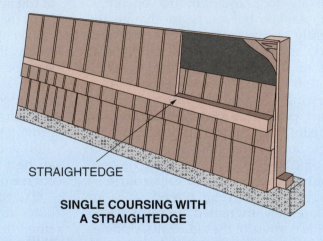

STRAIGHTEDGE

**SINGLE COURSING WITH
A STRAIGHTEDGE**

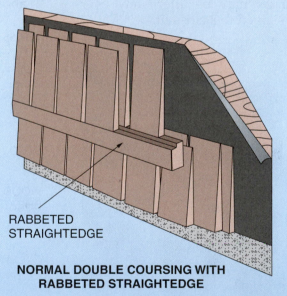

RABBETED
STRAIGHTEDGE

**NORMAL DOUBLE COURSING WITH
RABBETED STRAIGHTEDGE**

204SA12.EPS

The outside corners should be constructed with an alternate overlap of shingles between successive courses. Inside courses may be mitered or woven over a metal flashing, or they may be made by nailing an S4S 1½" or 2" square strip in the corner, after which the shingles of each course are fastened to the strip.

5.6.1 Panelized Shakes/Shingles

In most cases, panelized shakes/shingles are simply shakes glued or stapled to a backer board of plywood. They are available in widths of 4' and 8'. The panels are available in natural wood and are pre-finished in several basic colors of stain. The panels can be applied rapidly, using threaded nails colored to match the panels. The manufacturer's installation instructions should be followed.

5.7.0 Plywood Siding

The use of plywood as an exterior finish siding has been rapidly growing among architects and builders because of the speed of installation and the reliability of the waterproof glues being used. The beauty and diversity of the available surfaces have also added to its growing popularity. Because of its strength, plywood can be nailed directly to the studs, eliminating the need for sheathing. This is another plus in favor of plywood siding, because it saves not only the cost of the plywood sheathing, but also the cost of its installation.

Plywood siding is available in thicknesses of ⅜", ½", ⅝", and ¾"; however, ⅝" is the most commonly used. Some of the common textures and designs of plywood siding are shown in *Figure 37*.

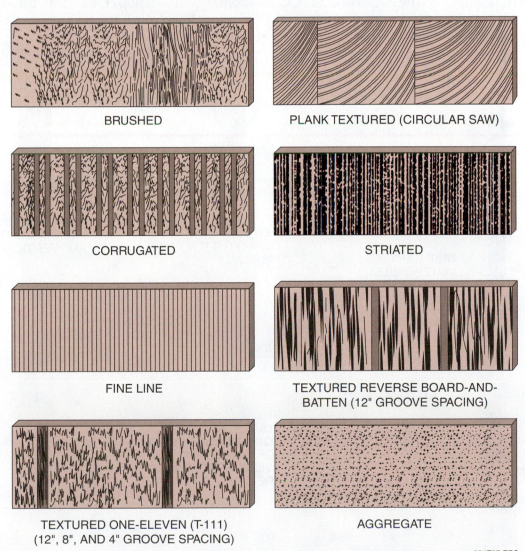

BRUSHED

PLANK TEXTURED (CIRCULAR SAW)

CORRUGATED

STRIATED

FINE LINE

TEXTURED REVERSE BOARD-AND-BATTEN (12" GROOVE SPACING)

TEXTURED ONE-ELEVEN (T-111) (12", 8", AND 4" GROOVE SPACING)

AGGREGATE

204F37.EPS

Figure 37 ◆ Surface textures and designs of common plywood siding.

Manufacturers generally recommend 7d or 9d siding nails or box nails made of hot-dipped galvanized, aluminum, or stainless steel. Ring shank and box nails are the types usually specified. The nailing pattern is 6" OC on the edges and 12" OC in the field. Staples can also be used as long as they are the appropriate size and type.

The spacing of wall studs or nailing supports for the ⅜" plywood is a maximum of 16" OC, but thicker plywood (½", ⅝", and ¾") permits a spacing of 24" OC. Blocking is usually required at all horizontal joints (subject to local building codes). Some of the joint suggestions are indicated in *Figure 38*.

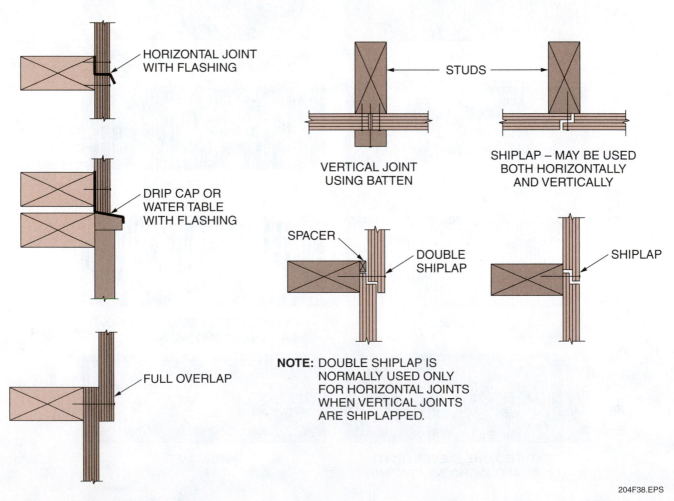

204F38.EPS

Figure 38 ◆ Plywood joint suggestions.

Be sure to refer to the manufacturer's technical specifications prior to installing the siding.

To apply lap plywood siding, follow these suggestions:

- If sheathing is not used, place horizontal blocking at 4'−0" centers.
- Install building paper between the siding and the studs.
- Use a starter strip that is the same thickness as the siding for the first course.
- Coat the edges of the siding with a primer or a water-repellent finish before application.
- Vertical joints should be staggered. These joints must be centered over studs with a tapered wedge at least 1⅝" wide behind the joint.
- Use 8d noncorrosive siding or box nails. Insert one nail at each stud on the bottom edge of the siding.

At all vertical joints, nail 4" OC for siding 12" wide or less. Nail 8" OC for siding 16" wide or more. All nails should be placed ¼" back from the edge of the plywood. Set and putty all casing nails. Box nails are driven flush.

5.8.0 Hardboard and Particleboard Siding

Hardboard siding is fabricated for use as lap siding or as large panels up to 16' in length by 4' in width. It is impregnated with a baked-on tempering compound. This process produces a tough, dense siding that will not split or splinter and is highly resistant to denting. Various pattern grains and profiles are available.

Hardboard siding reinforces wall construction, goes up quickly, and can be easily worked with both power tools and regular woodworking tools. Hardboards take 100 percent acrylic latex paints and stains, as well as gloss or satin oil/alkyd paints. Flat oil/alkyd paints or stains and vinyl acetate or vinyl acrylic copolymer paints and stains are not recommended. This type of siding is available unprimed or factory-primed.

Particleboard can also be used as siding material. It is available in many of the same shapes and surfaces as hardboard or plywood and is applied in the same manner.

Hardboard and particleboard siding is very sensitive to water damage. To achieve satisfactory performance and for warranty purposes, the siding must be installed and finished as specified by the manufacturer. In addition, the siding should never be cleaned using high-pressure or low-pressure washing methods.

The following are guidelines that may apply to hardboard and particleboard sidings, subject to the manufacturer's instructions:

- Hardboard siding may be applied over walls sheathed with wood or insulation and with studs spaced not more than 24" OC or over unsheathed walls with studs spaced not more than 16" OC. The lowest edge of the siding should be at least 6" to 8" above the finished grade level. When cutting, be sure to sand and prime all cut ends.
- In accordance with the manufacturer's instructions, when hardboard siding is applied directly to studs or over wood sheathing, moisture-resistant building paper or **felt paper** (non-vapor barrier) should always be laid directly under the siding. In most cases, a vapor barrier is required on the inside heated wall.
- When applying hardboard siding, use rust-proof siding nails. Nail only at stud locations and on special members around doors and windows. Use 6d nails for nailing directly into studs and 8d nails when nailing over sheathing. Nails must be kept back ½" from the ends and edges of the siding pieces.
- At inside corners, siding should be butted (with approximately 1/16" space) against a 1⅛" × 1⅛" wood, metal, or vinyl corner member. Outside corners may be 1⅛" wood corner boards, or metal/vinyl corners may be used. Caulking should be applied wherever the siding butts against wood corner boards, windows, and door casings. *Figures 39* and *40* show hardboard siding details.

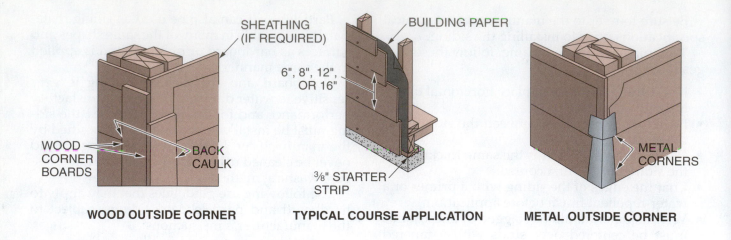

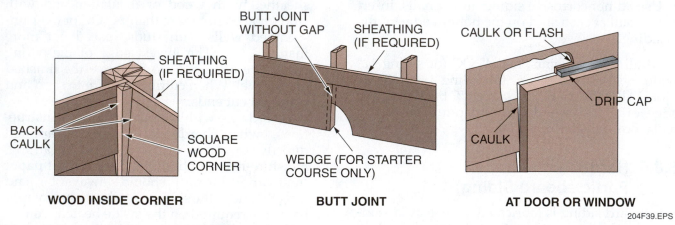

Figure 39 ◆ Typical application of lap siding.

204F39.EPS

5.8.1 *Installing Flat Panels*

Flat panels are installed vertically. All joints and panel edges should fall on the center of the framing members. If it is necessary to make a joint with a panel that has been field cut and the shiplap joint removed, use a butt joint. Butter the edges with caulk and gently attach them. Do not force or spring the panels into place. Leave a slight space where the siding butts against the window or door trim, and apply caulk to the space *(Figure 41)*. Horizontal joints should be spaced ⅛" apart, flashed, and sealed.

5.8.2 *Installing Lap Siding*

Lap siding is installed horizontally. Start the application by fastening a ⅜" × 1⅜" wood starter strip along the bottom edge of the sill. Level and install the first course of siding with the bottom edge at least ⅛" below the starter strip. Fasten the first course by nailing 1½" from the drip edge of the siding and ½" from the butt end.

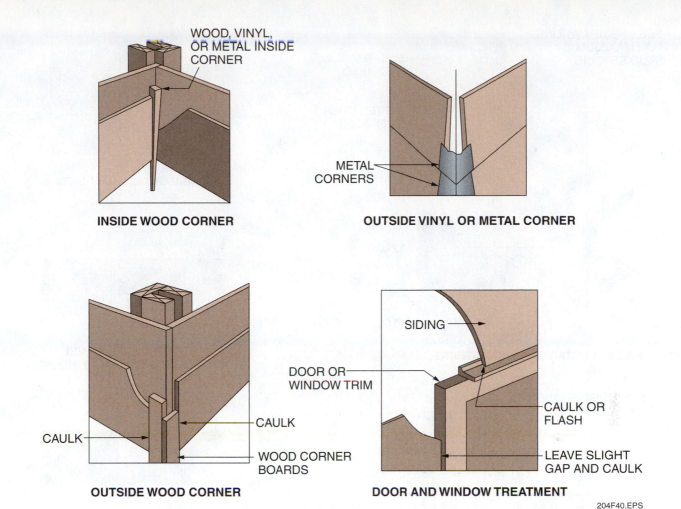

WOOD, VINYL, OR METAL INSIDE CORNER

INSIDE WOOD CORNER

METAL CORNERS

OUTSIDE VINYL OR METAL CORNER

CAULK

CAULK

WOOD CORNER BOARDS

OUTSIDE WOOD CORNER

SIDING

DOOR OR WINDOW TRIM

CAULK OR FLASH

LEAVE SLIGHT GAP AND CAULK

DOOR AND WINDOW TREATMENT

204F40.EPS

Figure 40 ◆ Hardboard siding corner details.

Install subsequent siding courses using a minimum overlap of 1". Butt joints should occur only at stud locations. Metal or vinyl butt joints may be used, if desired (*Figures 42* and *43*). Factory-primed ends should be used for all vertical butt joints that will not be covered. Adjacent siding pieces should just touch at butt joints; if they do not, then a ³⁄₁₆" space may be left and filled with a butyl caulk. Never force or spring the siding into place.

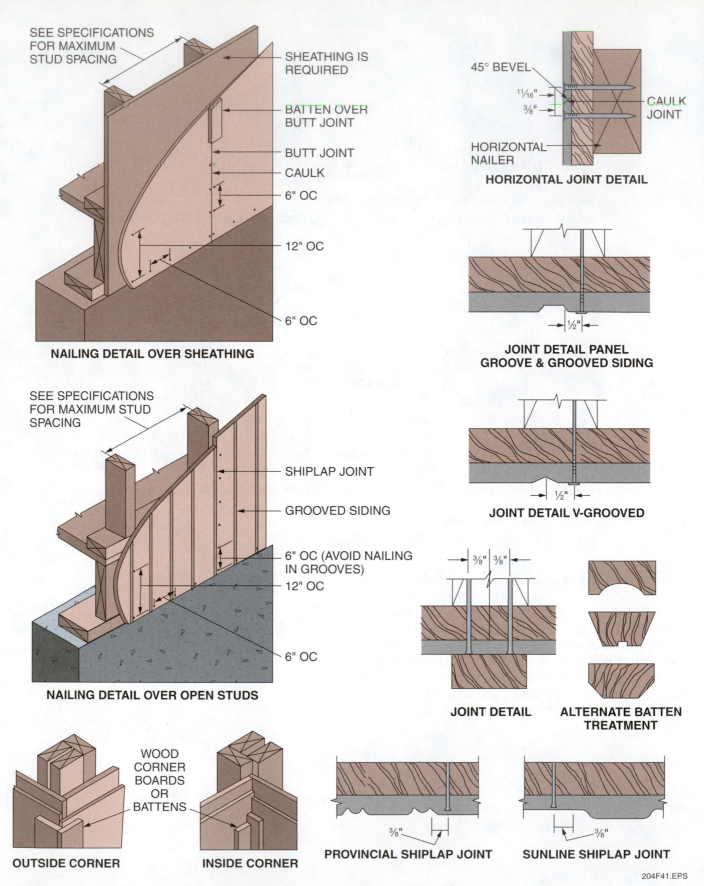

SEE SPECIFICATIONS
FOR MAXIMUM
STUD SPACING

SHEATHING IS
REQUIRED

BATTEN OVER
BUTT JOINT

BUTT JOINT

CAULK

6" OC

12" OC

6" OC

NAILING DETAIL OVER SHEATHING

SEE SPECIFICATIONS
FOR MAXIMUM STUD
SPACING

SHIPLAP JOINT

GROOVED SIDING

6" OC (AVOID NAILING
IN GROOVES)

12" OC

6" OC

NAILING DETAIL OVER OPEN STUDS

WOOD
CORNER
BOARDS
OR
BATTENS

OUTSIDE CORNER

INSIDE CORNER

45° BEVEL

11/16"

3/8"

CAULK
JOINT

HORIZONTAL
NAILER

HORIZONTAL JOINT DETAIL

1/2"

**JOINT DETAIL PANEL
GROOVE & GROOVED SIDING**

1/2"

JOINT DETAIL V-GROOVED

3/8" 3/8"

JOINT DETAIL

**ALTERNATE BATTEN
TREATMENT**

3/8"

PROVINCIAL SHIPLAP JOINT

3/8"

SUNLINE SHIPLAP JOINT

204F41.EPS

Figure 41 ◆ Installing lapped hardboard panels.

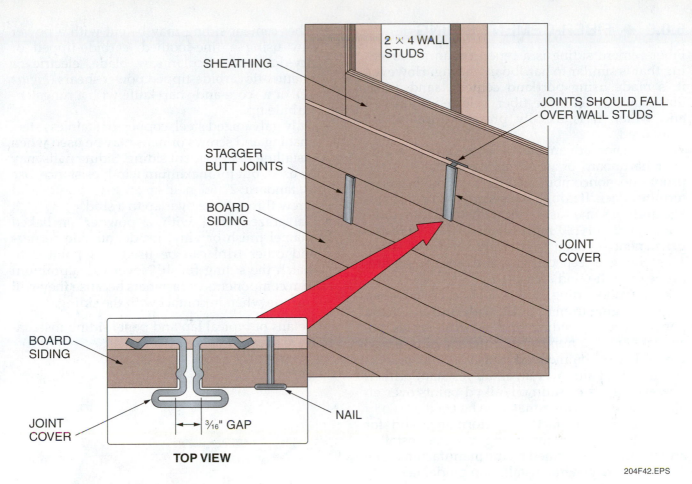

SHEATHING

2 × 4 WALL STUDS

JOINTS SHOULD FALL OVER WALL STUDS

STAGGER BUTT JOINTS

BOARD SIDING

JOINT COVER

BOARD SIDING

JOINT COVER

³⁄₁₆" GAP

NAIL

TOP VIEW

204F42.EPS

Figure 42 ◆ Metal or plastic butt joints.

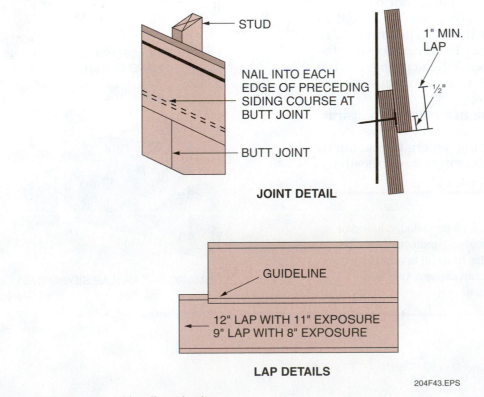

STUD

NAIL INTO EACH EDGE OF PRECEDING SIDING COURSE AT BUTT JOINT

BUTT JOINT

1" MIN. LAP

½"

JOINT DETAIL

GUIDELINE

12" LAP WITH 11" EXPOSURE
9" LAP WITH 8" EXPOSURE

LAP DETAILS

204F43.EPS

Figure 43 ◆ Installing lapped hardboard siding.

6.0.0 ◆ FIBER-CEMENT SIDING

Fiber-cement siding is a type of man-made siding that is similar to hardboard siding. However, it is made using portland cement, sand, fiberglass and/or cellulose fiber, selected additives, and water. It is usually pressure-formed and heatcured.

The major advantage of fiber-cement siding over hardboard or wood siding is that it is rot-proof and noncombustible, and can withstand a termite attack. It is highly resistant to impact damage and, in some cases, can withstand hurricane-force winds of 130 mph or more. It also resists permanent damage from water and salt spray. This siding is especially suited for use in fire-prone or high-wind areas.

Like wood siding, it is available in single-lap siding ranging from 6" to 12" wide and as vertical panels. The lap siding and vertical panels are available with a number of different surface patterns. The recommended finish is 100 percent acrylic latex paint over an alkali-resistant primer; however, gloss or satin oil/alkyd paints over an alkali-resistant primer may also be used.

To achieve satisfactory performance and for warranty purposes, the siding must be installed and finished as specified by the manufacturer. The following are general installation guidelines:

* Fiber-cement siding may be applied over walls sheathed with wood or insulation board up to 1" thick and with studs spaced not more than 24" OC or over unsheathed walls with studs spaced not more than 16" OC. The lowest edge of the siding should not be in contact with the earth or standing water. When cutting, be sure to prime all cut ends with an alkali-resistant primer.
* In accordance with the manufacturer's instructions, moisture-resistant paper or felt may be required under the siding when the siding is applied directly to studs or over wood sheathing.

WARNING!

Because dry material will be drilled, cut, and/or abraded, proper respiratory protection must be used when cutting this material to avoid inhaling toxic silica dust that can cause a fatal lung disease called silicosis.

* Fiber-cement siding may be cut with a power saw using a fine-toothed, carbide-tipped or dry-diamond circular saw blade, electric or pneumatic carbide-tipped power shears (*Figure 44*), or a score-and-snap knife with a tungsten-carbide tip.
* Only galvanized steel, copper, or stainless steel flashing and screws or nails may be used when installing fiber-cement siding. Siding nails may be used, but for maximum wind resistance, use a standard 2" 6d nail or an 8-18 bugle-head screw through the overlap to a stud.
* Galvanized steel with a powder or baked enamel finish or vinyl inside/outside corners and other trim can be used and painted to match the siding finish. Never use aluminum trim components or fasteners because they will corrode when in contact with the siding.

Details of typical lap and panel siding installation are shown in *Figures 45* and *46*.

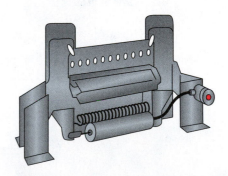

ELECTRIC (OR PNEUMATIC) HAND SHEAR

PNEUMATIC SHEAR (LAP SIDING ONLY)

204F44.EPS

Figure 44 ◆ Power shears.

Fiber-Cement Siding Styles

A number of architectural styles of fiber-cement siding are available and can be used to obtain different effects. Planks are available as smooth or wood grained, and panels are available as plain, stucco, and vertical wood grained.

HARDIPANEL® SMOOTH VERTICAL SIDING

HARDIPANEL® STUCCO VERTICAL SIDING

HARDIPANEL® SIERRA-8 VERTICAL SIDING

HARDIPANEL® SIERRA-4 VERTICAL SIDING

204SA14.EPS

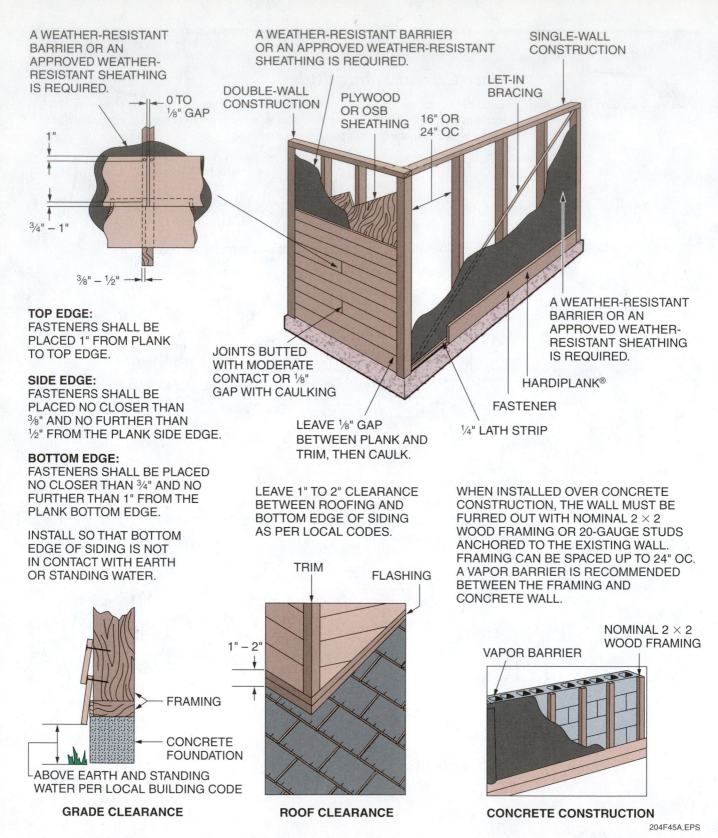

A WEATHER-RESISTANT BARRIER OR AN APPROVED WEATHER-RESISTANT SHEATHING IS REQUIRED.

0 TO ⅛" GAP

1"

¾" – 1"

⅜" – ½"

A WEATHER-RESISTANT BARRIER OR AN APPROVED WEATHER-RESISTANT SHEATHING IS REQUIRED.

DOUBLE-WALL CONSTRUCTION

PLYWOOD OR OSB SHEATHING

16" OR 24" OC

LET-IN BRACING

SINGLE-WALL CONSTRUCTION

A WEATHER-RESISTANT BARRIER OR AN APPROVED WEATHER-RESISTANT SHEATHING IS REQUIRED.

HARDIPLANK®

FASTENER

¼" LATH STRIP

JOINTS BUTTED WITH MODERATE CONTACT OR ⅛" GAP WITH CAULKING

LEAVE ⅛" GAP BETWEEN PLANK AND TRIM, THEN CAULK.

TOP EDGE:
FASTENERS SHALL BE PLACED 1" FROM PLANK TO TOP EDGE.

SIDE EDGE:
FASTENERS SHALL BE PLACED NO CLOSER THAN ⅜" AND NO FURTHER THAN ½" FROM THE PLANK SIDE EDGE.

BOTTOM EDGE:
FASTENERS SHALL BE PLACED NO CLOSER THAN ¾" AND NO FURTHER THAN 1" FROM THE PLANK BOTTOM EDGE.

INSTALL SO THAT BOTTOM EDGE OF SIDING IS NOT IN CONTACT WITH EARTH OR STANDING WATER.

LEAVE 1" TO 2" CLEARANCE BETWEEN ROOFING AND BOTTOM EDGE OF SIDING AS PER LOCAL CODES.

WHEN INSTALLED OVER CONCRETE CONSTRUCTION, THE WALL MUST BE FURRED OUT WITH NOMINAL 2 × 2 WOOD FRAMING OR 20-GAUGE STUDS ANCHORED TO THE EXISTING WALL. FRAMING CAN BE SPACED UP TO 24" OC. A VAPOR BARRIER IS RECOMMENDED BETWEEN THE FRAMING AND CONCRETE WALL.

TRIM

FLASHING

1" – 2"

FRAMING

CONCRETE FOUNDATION

ABOVE EARTH AND STANDING WATER PER LOCAL BUILDING CODE

GRADE CLEARANCE

ROOF CLEARANCE

NOMINAL 2 × 2 WOOD FRAMING

VAPOR BARRIER

CONCRETE CONSTRUCTION

204F45A.EPS

Figure 45 ◆ Typical fiber-cement lap siding installation details (1 of 2).

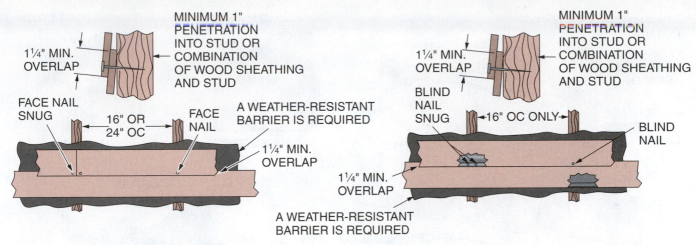

MINIMUM 1"
PENETRATION
INTO STUD OR
COMBINATION
OF WOOD SHEATHING
AND STUD

1¼" MIN. OVERLAP

FACE NAIL SNUG

16" OR 24" OC

FACE NAIL

A WEATHER-RESISTANT BARRIER IS REQUIRED

1¼" MIN. OVERLAP

MINIMUM 1"
PENETRATION
INTO STUD OR
COMBINATION
OF WOOD SHEATHING
AND STUD

1¼" MIN. OVERLAP

BLIND NAIL SNUG

16" OC ONLY

BLIND NAIL

1¼" MIN. OVERLAP

A WEATHER-RESISTANT BARRIER IS REQUIRED

CORROSION RESISTANT NAILS (GALVANIZED‡ OR STAINLESS STEEL)
- 6d (0.118" shank × 0.267" HD × 2" long)
- Siding nail (0.089" shank × 0.221" HD × 2" long)†
- Siding nail (0.091" shank × 0.221" HD × 1½" long)*
- ET & F pin (0.100" shank × 0.25" HD × 1½" long)†

CORROSION RESISTANT SCREWS
- Ribbed bugle-head or equivalent (No. 8-18 × 0.323" HD × 1⅝" long) Screw must penetrate ¼" or 3 threads into metal framing.

CORROSION RESISTANT NAILS (GALVANIZED‡ OR STAINLESS STEEL)
- Siding nail (0.089" shank × 0.221" HD × 2" long)†
- 11 gauge roofing nail (0.121" shank × 0.371" HD × 1¼" long)
- ET & F Panelfast™ (0.100" shank × 0.25" HD × 1½" long)†

CORROSION RESISTANT SCREWS
- Ribbed bugle-head or equivalent (No. 8–18 × 0.375" HD × 1¼" long) Screws must penetrate ¼" or 3 threads into metal framing.

FACE NAILED

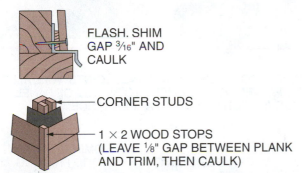

FLASH. SHIM GAP ³⁄₁₆" AND CAULK

CORNER STUDS

1 × 2 WOOD STOPS (LEAVE ⅛" GAP BETWEEN PLANK AND TRIM, THEN CAULK)

BLIND NAILED (NOT APPLICABLE FOR 12" WIDE SIDING)

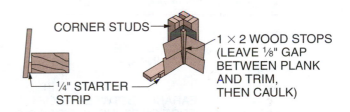

CORNER STUDS

1 × 2 WOOD STOPS (LEAVE ⅛" GAP BETWEEN PLANK AND TRIM, THEN CAULK)

¼" STARTER STRIP

TRIM DETAILS

NOTES:

* For face nail application of 9½" wide or less siding to OSB, fasteners are spaced a maximum of 12" OC.

† The use of a siding nail or roofing nail may not be applicable to all installations where greater windloads or higher exposure categories of wind resistance are required by the Local Building Code. Consult the applicable Building Code Compliance Report.

‡ Hot dipped galvanized nails are recommended.

FASTENING REQUIREMENTS:
- Drive fasteners perpendicular to siding and framing.
- Fastener heads should fit snug against siding (no air space). (Examples 1 & 2)
- Do not underdrive nail heads or drive nails at an angle. (Example 4)
- If nail is countersunk, caulk nail hole and add a nail. (Example 3)

Example 1

SNUG

Example 2

FLUSH

Example 3

COUNTERSUNK, CAULK AND ADD NAIL

Example 4

DO NOT UNDERDRIVE NAILS

Figure 45 ◆ Typical fiber-cement lap siding installation details (2 of 2).

204F45B.EPS

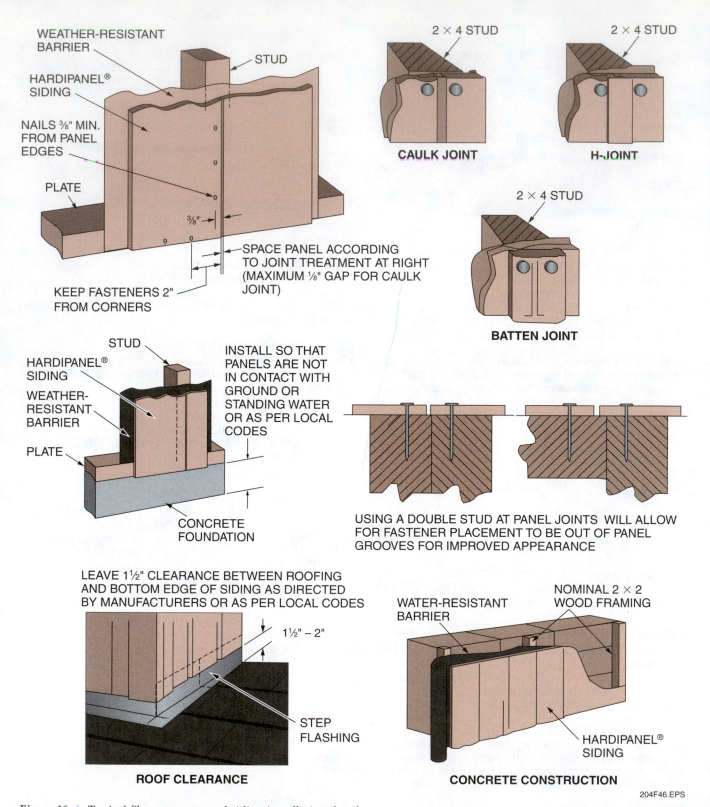

Figure 46 ◆ Typical fiber-cement panel siding installation details.

7.0.0 ◆ VINYL AND METAL SIDING

Vinyl, aluminum, and steel siding are applied in new construction as well as over existing finishes for remodeling work. Both vinyl and metal siding are manufactured to look like beveled siding and are available in many colors and finishes. Inside corner posts, door and window trim, individual corner pieces, starter strips, and butt supports are also available. Metal siding and trim are usually supplied with a baked-on or plastic finish. Many manufacturers offer siding with a rigid insulating backing board. This makes the siding less susceptible to exterior damage and also increases its rigidity.

7.1.0 Vinyl and Metal Siding Materials and Components

Figure 47 shows some of the styles of horizontal siding materials currently available. *Figure 48*

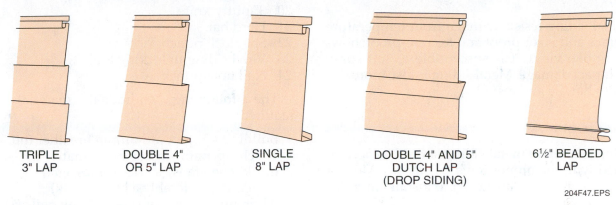

| TRIPLE 3" LAP | DOUBLE 4" OR 5" LAP | SINGLE 8" LAP | DOUBLE 4" AND 5" DUTCH LAP (DROP SIDING) | 6½" BEADED LAP |

204F47.EPS

Figure 47 ◆ Typical vinyl and metal siding materials.

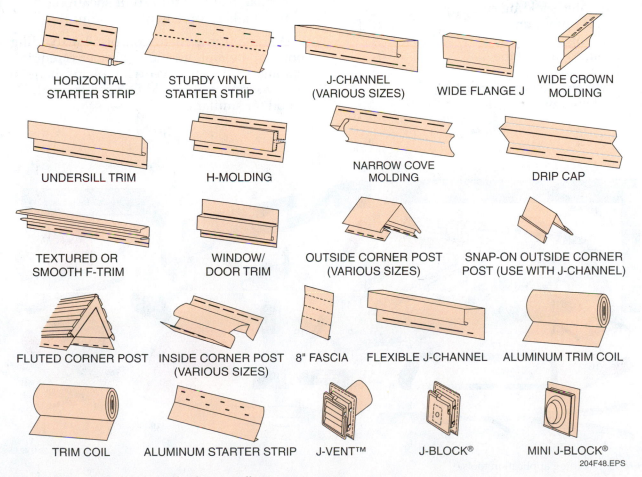

HORIZONTAL STARTER STRIP   STURDY VINYL STARTER STRIP   J-CHANNEL (VARIOUS SIZES)   WIDE FLANGE J   WIDE CROWN MOLDING

UNDERSILL TRIM   H-MOLDING   NARROW COVE MOLDING   DRIP CAP

TEXTURED OR SMOOTH F-TRIM   WINDOW/ DOOR TRIM   OUTSIDE CORNER POST (VARIOUS SIZES)   SNAP-ON OUTSIDE CORNER POST (USE WITH J-CHANNEL)

FLUTED CORNER POST   INSIDE CORNER POST (VARIOUS SIZES)   8" FASCIA   FLEXIBLE J-CHANNEL   ALUMINUM TRIM COIL

TRIM COIL   ALUMINUM STARTER STRIP   J-VENT™   J-BLOCK®   MINI J-BLOCK®

204F48.EPS

Figure 48 ◆ Typical vinyl and metal siding installation components.

shows a variety of installation components used with vinyl or metal siding. Horizontal metal siding is usually limited to single- or double-lap styles.

The major advantages of vinyl siding are its low cost, ease of handling and installation, and resistance to denting. It is also colorfast in sunlight, waterproof, rot-proof, and termite-proof. Its major disadvantage is that its resistance to impact damage is very low in cold temperatures. If installed improperly, it will break during expansion or contraction over wide temperature variations.

Metal siding resists damage from temperature extremes and is rot-proof and termite-proof; however, unlike vinyl, it is susceptible to salt spray and impact damage. Metal siding is also more difficult to handle and install.

7.2.0 Tools

The installation of metal siding should be accomplished with the proper tools *(Figure 49)*. Most of these same tools are also used to install vinyl siding.

1. Carpenter's metal square
2. Carpenter's folding rule
3. 2' level (minimum)
4. Caulking gun
5. Steel measuring tape
6. Fine-tooth file
7. Power circular saw (optional)
8. Claw hammer
9. Chalkline
10. Screwdriver
11. Pliers
12. Tin snips (duckbill type) or power hand shears
13. Aviation shears (double acting)
14. Carpenter's saw (crosscut)
15. Safety goggles
16. Steel awl
17. Fine-tooth hacksaw (24 teeth per inch)
18. Utility knife
19. Line level or water level
20. 3" putty knife
21. Hard hat
22. Snaplock punch
23. Vinyl siding unlocking tool
24. Nail hole punch

The following additional materials are required:

- Building wrap or aluminum breather foil
- Touch-up paint in colors to match the siding (for kitchen fans, service cables, etc.)
- Caulking (preferably a butyl caulk)
- Aluminum, plain, or screw-shank nails (1½" for general use; 2" for re-siding; 2½" or more to nail insulated siding into soffit sheathing; 1" to 1½" trim nails colored to match siding)

A minimum penetration of ¾" (excluding the point of the nail) into solid lumber is required for nailing to be effective with plain shank nails. Screw shank nails could be used through ½" plywood for similar effectiveness.

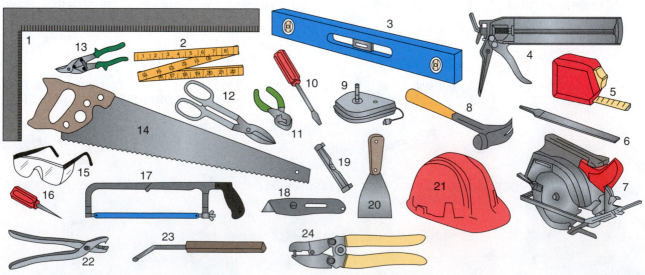

204F49.EPS

Figure 49 ◆ Siding application tools.

7.3.0 Equipment

The following equipment is required when applying metal or vinyl siding:

- *Ladders and scaffolds* – Proper ladders and scaffolds are necessary. The aluminum adjustable scaffolding systems are widely used to provide a working platform. With these systems, the distance from the building facade remains the same from the bottom to the top. Exact specifications on spacing dimensions, planking, permissible heights and loads, and other details are contained in *Occupational Safety and Health Standards for the Construction Industry (CFR 29, Part 1926)*.
- *Portable brake* – For job-site bending of custom trim sections such as fascia trim, window casing, and sill trim, a portable metal-bending machine (brake) is extremely useful (*Figure 50*). Utilizing white or colored coil stock, precision bending, including multiple bends, can be accomplished. These machines are lightweight and can be carried to the job site and set in place. Various sizes and brake styles are available. As shown in *Figure 50*, some are equipped with a lengthwise rolling cutter to allow sizing the trim stock to the desired width.
- *Cutting table* – A cutting table allows a standard portable circular power saw to be mounted in a carrier and held away from the work to avoid damaging the siding. This table can be used for measuring and crosscutting, as well as for making miters and bevels. The table is constructed of lightweight aluminum and can be easily set up on the job site by one worker.

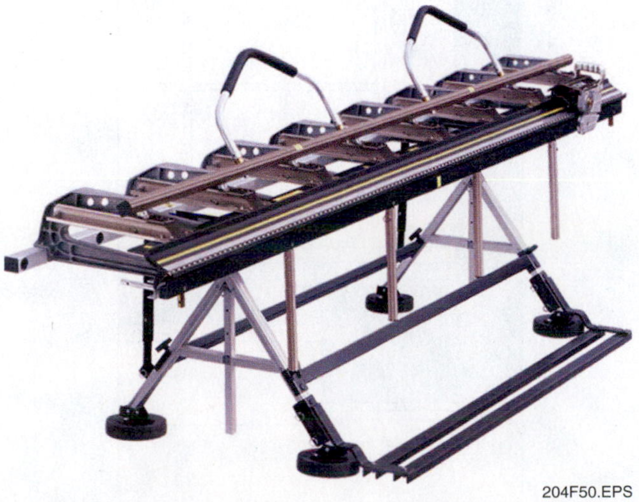

7.4.0 Surface Preparation

The quality of the finished job depends on good preparation of the work surface, especially for remodeling work. Keep the following points in mind when preparing the surface:

- Check for low places in the plane of the wall and build out (shim out) if required.
- Prepare the entire building a few courses at a time. Securely nail all loose boards and loose wood trim. Replace any rotted boards.
- Scrape away old paint buildup, old caulking, and hardened putty, especially around windows and doors where it might interfere with the positioning of new trim. New caulk should be applied to prevent air infiltration.
- Remove downspouts and other items that will interfere with the installation of new siding.
- Tie shrubbery and trees back from the base of the building to avoid damage.
- Window sill extensions may be cut off so that J-trim can be installed flush with the window casing (*Figure 51*). However, if the building owner wishes to maintain the original window design, coil stock can be custom-formed around the sill instead of cutting away the sill extensions.

7.5.0 Furring and Insulation Techniques

Furring strips may be required in order to provide a smooth, even base for nailing on the new siding. Normally, ⅜" thick wood lath strips are used over

![Portable brake](204F50.EPS)

Figure 50 ◆ Portable brake.

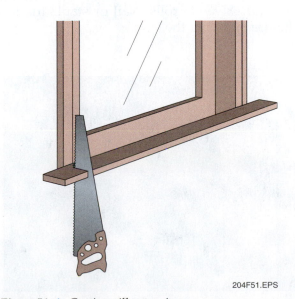

204F51.EPS

Figure 51 ◆ Cutting sill extensions.

wood construction and 1 × 3 strips are used over brick and masonry. Furring is not usually necessary in new construction, but older homes often have uneven walls, and furring out low spots or shimming can help to prevent the siding from appearing wavy. If at all possible, it is preferable to remove any old exterior siding down to the sheathing to avoid furring. If extensive furring is used under vinyl siding, a backer board may be required for reinforcement of the siding.

For horizontal siding, the furring should be installed vertically at 16" OC. The air space at the base of the siding should be closed off with strips applied horizontally. Window, door, gable, and eave trim may have to be built out to match the thickness of the wall furring.

The furring for vertical siding is essentially the same as for horizontal siding, except the wood strips are securely nailed horizontally into structural lumber on 16" to 24" centers. When using 1 × 3 furring, be sure to check what effect, if any, the additional thickness will have on the use of trim (see *Figure 52*).

7.5.1 Aluminum Foil Underlayment

Aluminum reflector foil, if used, is a good insulator and can be used advantageously as an underlayment to siding. It may be stapled directly to the existing wall or over ¾" furring strips to provide an additional air space and better insulation. Reflector foil for remodeling must be of the perforated or breather type to allow for the passage of water vapor.

Foil should be installed with the shiny side facing the air space (outward with no furring; inward if applied over furring). Foil is generally available in 36"- and 48"-wide rolls. Nail or staple the foil just before applying the siding.

When applying foil over furring, be careful not to let the foil collapse into the air space. Place the foil as close as possible to openings and around corners where air leaks are likely to occur. Overlap the side and end joints by 1" to 2".

7.5.2 Window and Door Build-Out

Some trim build-out at windows and doors may be required to maintain the original appearance of the house when using furring strips or underlayment board. This is particularly true when the strips or underlayment board are more than ½" thick. Thicker furring and underlayment generally provide added insulation value and are usually a good investment for the homeowner, particularly if the home is uninsulated.

7.5.3 Undersill Furring

Building out below the window sill is often required in order to maintain the correct slope angle if a siding panel needs to be cut to less than full height. The exact thickness required will be apparent when the siding courses have progressed up the wall and reached this point (see *Figure 53*).

7.5.4 Under-Eave Furring

For the same reason, furring is usually required to maintain the correct slope angle if the last panel needs to be cut to less than full height. The exact thickness required will be apparent when the siding courses have progressed up the wall and reached this point (see *Figure 54*).

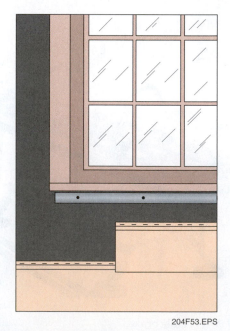

204F53.EPS

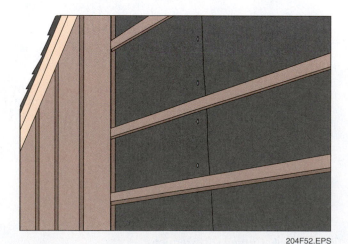

204F52.EPS

Figure 52 ◆ Furring for vertical siding.

Figure 53 ◆ Undersill furring.

Figure 54 ◆ Under-eave furring.

7.6.0 Establishing a Straight Reference Line

The key element in a successful siding installation is establishing a straight reference line upon which to start the first course of siding. The suggested procedure is to measure equal distances downward from the eaves. This ensures that the siding appears parallel with the eaves, soffits, and windows, regardless of any actual settling of the house from true level. See *Figure 55*.

Reference Lines

Like wood lap siding, it is very important that the horizontal lines of vinyl or metal siding are parallel to the soffit/roof line of a structure even if the structure or soffit/roof line is out of level. This is necessary to prevent the horizontal lines of the siding from being at an angle to the soffit/roof line, which would exaggerate the out-of-level structural error. For vinyl or metal siding installation, establishing the initial reference line is even more important than with wood lap siding because the spacing of the courses up to the soffits is not generally adjustable. The position of the line in relation to the soffit/roof lines must be checked and rechecked to make sure that it is equidistant at all points.

On long walls or gable ends of a structure, a transit can be used to set the reference line at intermediate points even if the soffit/roof lines are out of level. This is accomplished by setting the transit to an out-of-level position (in the plane parallel to the wall) that allows the transit cross hairs to sweep across the same reference mark at both outside corners of the wall. However, the transit must be level in the plane perpendicular to the wall.

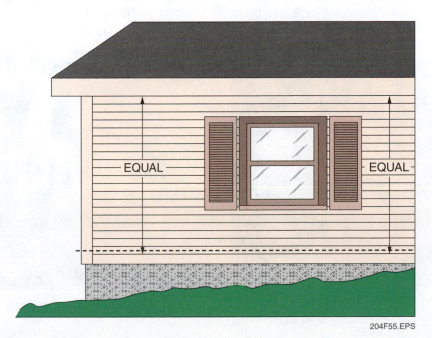

Figure 55 ◆ Straight reference line.

To establish a reference line, find the lowest corner of the house. Partly drive a nail about 10" above the lowest corner, or high enough to clear the height of a full siding panel. Stretch a taut chalkline from this corner to a similar nail installed at another corner. Reset this line based upon measuring down equal distances from points on the eaves. Repeat this procedure on all sides of the house until the chalklines meet at all corners.

Before snapping the chalklines, check for straightness. Be alert to sag in the middle, particularly if a line is more than 20' long. If preferred, lines may be left in place while installing the starter strip, as long as they are checked periodically for excess sag.

If the house is level, an alternative is to use a water level *(Figure 56)* or a transit to set the chalkline approximately 2" (or the width of the starter strip) from the lowest point of the old siding and locate the top of the starter strip at that line. Take the level reading at the corners and centers of the chalkline for best results. The water level can be used for measurements up to 100', and is accurate to ±1⁄16" at 50'.

7.7.0 Inside Corner Posts

The inside corner posts are installed before the siding is hung. Depending on the type of siding (insulated or non-insulated), deeper or narrower posts may be required. The post is set in the corner full length, reaching from 1⁄4" below the bottom of the starter strip up to 1⁄4" from the eave or gable trim. Nail the upper slot at the top of the slot, then nail

approximately 8" to 12" on both flanges with aluminum nails in the center of the slots. Make sure the post is set straight and true. The flange should be nailed securely to the adjoining wall, but do not overdrive the nails so as to cause distortion. If a short section is required, use a hacksaw to cut it. If a long section is required, the posts should be overlapped, with the upper piece outside.

The siding is later butted into the corner and then nailed into place, allowing approximately a 1⁄16" to 1⁄4" space between the post and the siding for expansion purposes (see *Figure 57*).

7.8.0 Outside Corner Posts

If used, the outside corner post produces a trim appearance and will accommodate the greatest variety of siding types. Most outside corner posts are designed to be installed before the siding is hung, in a manner similar to the inside corner post. If desired, old corner posts may sometimes be removed. Set a full-length piece over the existing corner, running from 1⁄4" below the bottom of the starter strip to 1⁄4" from the underside of the eave. If a long corner post is needed, overlap the corner post sections, with the upper piece outside.

Nail the uppermost slot at the top of the slot, then nail approximately every 8" to 12" with aluminum nails on both flanges in the center of the slots. Make sure the flanges are securely nailed *(Figure 58)*, but avoid distortion caused by overdriving nails. Use a hacksaw to cut short sections, if required. If insulated siding is being used, wider corner posts are needed.

ELECTRONIC LEVEL SENSOR UNIT WITH BEEPER PLACED AT DESIRED LEVEL

BEEPER SOUNDS WHEN FREE END IS AT CORRECT LEVEL

204F56.EPS

Figure 56 ◆ Electronic water level.

MOISTURE BARRIER

INSIDE CORNER POST

1⁄16" TO 1⁄4" EXPANSION GAP AT END OF SIDING INSERTED IN SLOT OF INSIDE CORNER POST

204F57.EPS

Figure 57 ◆ Inside corner post.

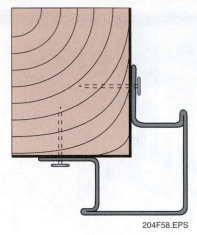

Figure 58 ◆ Correct nailing of flanges.

J-CHANNEL STRIP OR STARTER STRIP AS REQUIRED

1/16" TO 1/4" GAP AT END OF STRIP INSIDE CORNER POST

204F59.EPS

Figure 59 ◆ Installing a starter or J-channel strip.

7.9.0 Starter Strip

Using the chalkline previously established as a guide, take equal distance measurements, as shown in *Figure 59*, and install the starter strip or J-channel strip all the way around the bottom of the building depending on the material at the base of the building. If insulated siding is used, the starter strip should be furred out to a distance equal to the thickness of the backer. It is very important that the starter strip be straight and meet accurately at all corners because it will determine the line of all siding panels installed. Where hollows occur in the old wall surface, shim out behind the starter strip to prevent a wavy appearance in the finished siding.

When using individual corner caps, install the starter strip up to the edge of the house corner. Use aluminum nails spaced not more than 8" apart to fasten the starter strip. Nail the starter strip as low as possible. Be careful not to bend or distort the strip by overdriving the nails. The strip should not be nailed tight. Cutting lengths of starter strip is best accomplished with tin snips. Butt the sections together.

Starter strips may not work in all situations. For example, other accessory items, such as J-channels or all-purpose trim, may work better in starting the siding course over garage doors and porches or above brick. These situations must be handled on an individual basis as they occur.

7.10.0 Window and Door Trim

For a superior job in remodeling work, old window sills and casings can be covered with aluminum coil stock that is bent to fit on the job site. The advantage is freedom from maintenance.

Sometimes, window and door casings need to be built out to retain the original appearance of the house or to improve the appearance. To do this, use appropriate lengths and thicknesses of good-quality lumber and nail them securely to the existing window casings. Remove any storm windows before covering the casings with aluminum coil stock sections custom-formed on the job site.

Forming aluminum sections to fit window casings is done using a portable brake. Door casings are handled in a similar manner.

Figure 60 shows the installation of aluminum window trim.

If there is a step in the wood sill, it can best be covered by bending two separate sill cover pieces with interlocking flanges, as shown in *Figure 61*. By using tin snips and bending flanges on the job, the old sill ends can be boxed in to provide a neat appearance and to prevent water penetration.

J-channel is used around windows and doors to receive the siding. Side J-channel members are cut longer than the height of the window or door and notched at the top. Notch the top J-channel member at a 45-degree angle and bend the tab down to provide flashing over the side members. Caulking should be used behind J-channel members to prevent water infiltration between the window and the channel. See *Figures 62* and *63*.

To provide protection against water infiltration, a flashing piece, which is cut from coil stock or a precut piece of step flashing, is slipped under the base of the side J-channel members. It should be positioned so that it overlaps the top lock of the panel below, as shown in *Figure 64*.

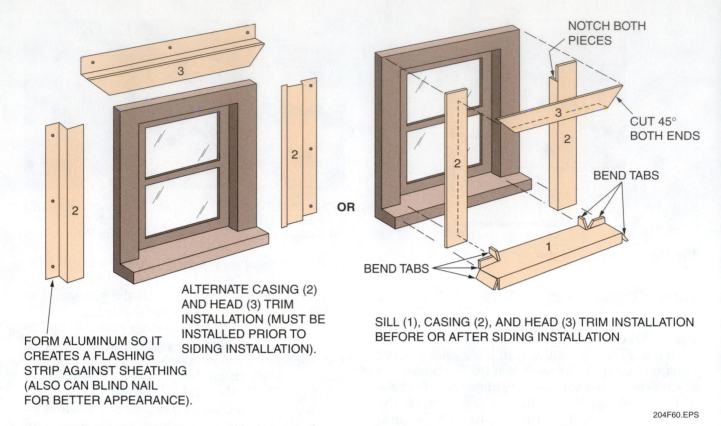

FORM ALUMINUM SO IT
CREATES A FLASHING
STRIP AGAINST SHEATHING
(ALSO CAN BLIND NAIL
FOR BETTER APPEARANCE).

ALTERNATE CASING (2)
AND HEAD (3) TRIM
INSTALLATION (MUST BE
INSTALLED PRIOR TO
SIDING INSTALLATION).

NOTCH BOTH
PIECES

CUT 45°
BOTH ENDS

BEND TABS

BEND TABS

SILL (1), CASING (2), AND HEAD (3) TRIM INSTALLATION
BEFORE OR AFTER SIDING INSTALLATION

204F60.EPS

Figure 60 ◆ Installing aluminum window trim.

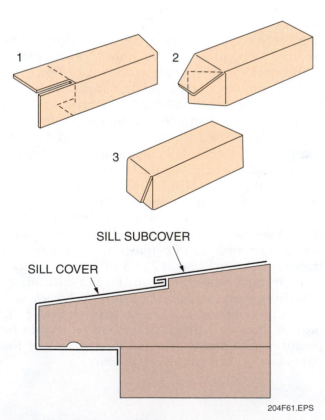

SILL SUBCOVER

SILL COVER

204F61.EPS

Figure 61 ◆ Boxing in sill ends.

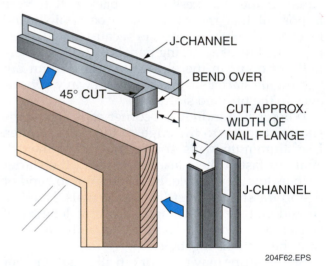

J-CHANNEL

BEND OVER

45° CUT

CUT APPROX.
WIDTH OF
NAIL FLANGE

J-CHANNEL

204F62.EPS

Figure 62 ◆ Cutting J-channel.

7.11.0 Gable End Trim

Before applying the siding, J-channels should be
installed to receive the siding at the gable ends, as
shown in *Figure 65*. Where the left and right sections meet at the gable peak, allow one of the sections to butt into the peak, with the other section
overlapping it. A miter cut is made on the face
flange of this piece to provide a better appearance.
All old paint buildup should be removed before
installing J-channels. Nail the J-channels every 12"
using aluminum nails.

Figure 63 ◆ J-channel.

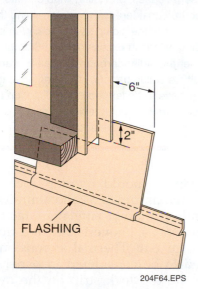

6"

2"

FLASHING

204F64.EPS

Figure 64 ◆ Installing a piece of flashing.

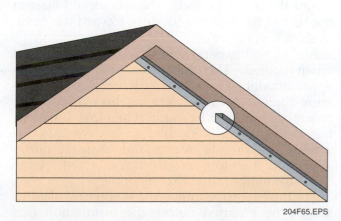

204F65.EPS

Figure 65 ◆ Gable end trim.

7.12.0 Cutting Procedures

For precision cutting, a power saw is the most convenient tool to use. Cutting one panel at a time is recommended. A special jig that will keep the saw base clear of the work is preferred in order to prevent damaging panels. For vinyl, reverse a fine-tooth blade to produce a smoother cut. For aluminum or steel, use a minimum 10-tooth aluminum cutting blade or an abrasive cutting blade. A bar of soap may be rubbed on the blade to produce a smoother cut on the siding panel and prolong blade life. Feed the saw through the work slowly to prevent flutter against the blade. Safety goggles must be worn at all times while operating a power saw.

Individual panels can be cut with tin snips. Start by drawing a line across the panel using a square. Begin cutting at the top lock first (*Figure 66*) and continue toward the bottom of the panel. For metal panels, break the panel across the butt edge and snip through the bottom lock. For metal siding, use a screwdriver to reopen the lock, which may become flattened by the tin snips (*Figure 67*).

Aviation shears are sometimes used to cut the top and bottom locks, and a utility knife is used to score and break the face of the panel and to cut vinyl panels. For straight cuts, the best choice is duckbill snips. For aluminum, a heavy score is made on the panel and the piece is bent back and forth until it snaps cleanly along the score line.

On window cutouts, a utility knife and tin snips may be used for both vinyl and aluminum. Use duckbill tin snips to cut accessories such as all-purpose trim, J-channel, and starter strips. Use a hacksaw to cut accessories such as corner posts.

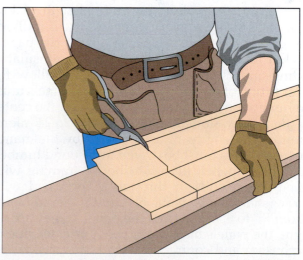

204F66.EPS

Figure 66 ◆ Using tin snips.

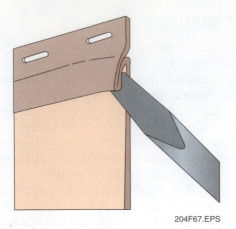

Figure 67 ◆ Reopening a lock after cutting.

204F67.EPS

7.13.0 Siding and Corner Cap Installation

This section covers siding and corner cap installation.

7.13.1 Siding Installation

For metal siding, check with local building codes to see if they require that the first course of siding be grounded to reduce the danger from lightning.

Extra care must be taken when applying the first course of siding because it establishes the base for all other courses. Apply a panel by hooking the bottom lock of the panel into the interlock bead of the starter strip. Make sure the lock is engaged. Do not force it, which might cause distortion of the panel and result in a warped shadow line. Double check for continuous locking along the panel before proceeding further. Also, check carefully for proper alignment at the corners.

At the corner posts, slide the panel into the recess first, then exert upward pressure to lock the panel into place along its entire length. Allow clearance for expansion, as necessary. If individual corner caps are being used, keep the panels back from the corner edges (¾" for non-insulated siding and ¼" for insulated siding) to allow for later fitting of the individual corners. Panels must be hung with aluminum nails through the center of the factory-slotted holes every 16" to 24" along their entire lengths or as specified by the manufacturer. Nails must be driven into sound lumber, such as ¾" penetration into house framing with plain shank nails or through ½" plywood with screw shank nails. Nails or screws should be set about ¹⁄₁₆" to ⅛" away from the panel. If they are set tight, the panels may warp, bend, or crack due to expansion and contraction. On low spots, fasten the panel on both sides of the low spot and allow the panel to float over the low spot.

CAUTION

Do not force the panels up or down when nailing them into position. The panels should not be under vertical tension or compression after being nailed into place.

On the sides of the building, start at the rear corner and work toward the front so that the lapping will be away from the front and less noticeable. On the front of the building, start at the corners and work toward the entrance door for the same reason. For best appearance when lapping, the factory-cut ends of the panels should cover the field-cut ends.

NOTE

In some cases, prevailing wind direction must also be considered when determining the direction of the siding seam laps. With some vinyl siding, winds in excess of 50 to 60 mph entering under the siding through the seams can tear off multiple siding panels. To prevent this, the seams may have to be lapped in the direction of the prevailing wind in high-wind areas. In addition, the maximum nailing distance must not be exceeded.

Metal panels should overlap each other by about ½". A maximum of ⅝" and a minimum of ⅜" is a good rule of thumb. Vinyl manufacturers usually recommend a 1" overlap with a double-size nailing flange cut. Thermal expansion requirements need to be considered in panel overlaps. Cut away the top lock strip on the overlapped panel by twice the amount of the intended overlap (see *Figure 68*).

Avoid short panel lengths of under 24" and make sure that the factory-cut ends are always on top of the field-cut ends. The job should start at the rear of the house and work toward the front, as shown in *Figure 69*.

Siding will expand when heated and contract when cooled. The expansion will amount to approximately ⅛" per 10' length for a 100° temperature change. An allowance for this expansion or contraction should be made when installing siding. If the siding is installed in hot weather, the product is already warm and at least partly expanded; therefore, less room will be required to allow for temperature expansion.

For the best appearance, the staggering of joints should be planned before the installation (see *Figure 70*). Avoid installing siding in a set pattern.

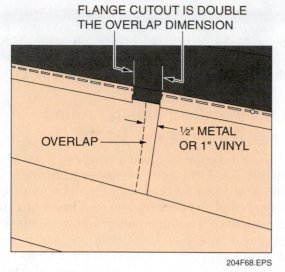

Figure 68 ◆ Overlapping panels.

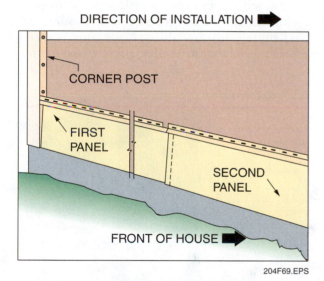

Figure 69 ◆ Sequence of installation.

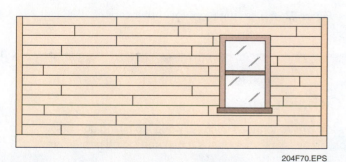

Figure 70 ◆ Proper staggering of joints.

A High-Windload Vinyl Siding

One manufacturer offers a flexible-hem vinyl siding that is designed to be installed under vertical tension and nailed tight. This allows the panel to float over low spots and move to accommodate expansion and contraction. When fastened with staples, its specifications indicate that it will withstand winds of up to 235 mph.

A set pattern may be more labor- and cost-effective, but results in a poor overall appearance. It is best to plan the job so that any two joints in line vertically will be separated by at least two courses. At a bare minimum, separate panel overlaps on the next course by at least two feet. Joints should be avoided on panels directly above and below windows. Shorter pieces that develop as work proceeds can be used for smaller areas around windows and doors.

Backer tabs are used with 8" horizontal noninsulated aluminum siding only. They ensure rigidity, evenness of installation, and tight endlaps. They are used at all panel overlaps and behind panels entering corners. After the panel has been locked into place, slip the backer tab behind the panel with the flat side facing out *(Figure 71)*. The backer tab should be directly behind and even with the edge of the first panel of the overlap. Nail the backer into place.

7.13.2 Corner Cap Installation for Aluminum Siding

Individual corner caps, if used, may be used for 8" horizontal aluminum lap siding instead of outside corner posts. The siding courses on adjoining walls must meet evenly at the corners. To allow room for the cap, install the siding with ¾" clearance from the corner (¼" clearance for insulated siding). Refer to *Figure 72*.

Complete one wall first. On the adjacent wall, install one course of siding, line the course up, and install the corner cap. Each corner cap must be fitted and installed before the next course of siding is installed. A jig can be constructed to facilitate the alignment, or a special tool may be purchased for this purpose.

Install the cap by slipping the bottom flanges of the corner cap under the butt of each siding panel. Use slight, steady pressure to press the cap into place. If necessary, insert a putty knife between the panel locks, prying slightly outward to allow room for the flanges to slip in. Gentle tapping with a rubber mallet and wood block can also be helpful.

When the cap is in position, secure it with 2" or 2½" nails, or nails that are long enough for ¾" penetration into solid wood or sheathing. Nail through at least one of the prepunched nail holes in the top of the corner cap.

7.14.0 Installing Siding Around Windows and Doors

As the siding courses reach a window, a panel will probably need to be cut narrower to fit the space under the window opening. Plan this course of siding so that the panel will extend on both sides of the opening. Hold the panel in place to mark for the vertical cuts. Use a small piece of scrap siding as a template, placing it next to the window and locking it into the panel below *(Figure 73)*. Make a mark on this piece ¼" below the sill height to allow clearance for all-purpose trim. Do the same on the other side of the window, because windows are not always absolutely level.

The vertical cuts are made from the top edge of the panel with duckbills, tin snips, or a power saw. For aluminum, the lengthwise (horizontal) cut is scored once with the utility knife and bent back and forth until the unwanted piece breaks off. For

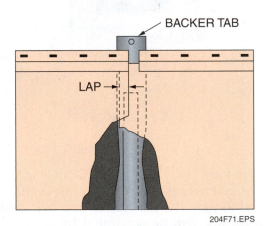

204F71.EPS

Figure 71 ◆ Inserting a backer tab for metal siding.

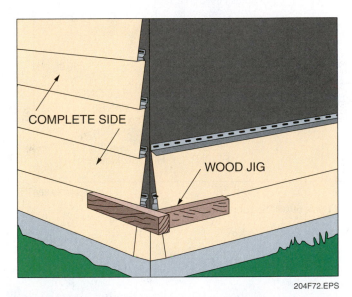

204F72.EPS

Figure 72 ◆ Leaving room for corner caps.

Aluminum Siding Corner Caps

INSIDE TRACK

Aluminum siding corner caps are similar to those used for wood lap siding corners. Before nailing the corner caps, always make sure that the cap and siding panels are flush.

vinyl, the horizontal cut is made with a utility knife. For steel, the horizontal cut is made with tin snips.

The raw edge of the panel should be trimmed with all-purpose trim for the exact width of the sill. First, determine if furring is required behind the cut edge to maintain the slope angle with adjacent panels. Nail the correct thickness of furring under the sill and install all-purpose trim over it with aluminum nails, close up under the sill for a tight fit (see *Figure 74*). On vinyl panels, use a snaplock punch to place raised ears on the raw edge of the panel. Slide the panel upward so as to engage the undersill or J-trim, the J-channels on the window sides, and the lock of the panels below.

Fitting panels over door and window openings is almost the same as making undersill cutouts, except that the clearances for fitting the panel are different. The cut panel on top of the opening needs more room to move down to engage the interlock of the siding panel below on either side of the window. Mark a scrap piece template without allowing clearance and then make saw cuts ¼" to ⅜" deeper than the mark *(Figure 75)*. This will provide the necessary interlock clearance.

Check the need for furring over the top of the window or door in order to maintain the slope angle and install it, if required *(Figure 76)*. Make sure the furring is pressure-treated and is spaced off the bottom of the J-channel.

Cut a piece of all-purpose trim the same width as the raw edge of the cut panel and slip it over this edge of the panel before installing it. Drop the panel into position, engaging the interlocks on the siding panels below. The all-purpose trim can now be pushed downward to close any gap at the juncture with the J-channel. Refer to *Figure 77*.

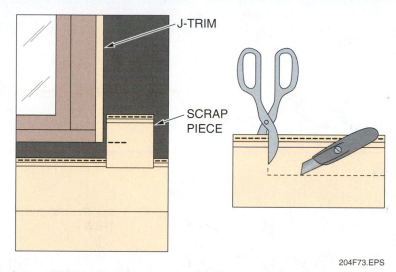

Figure 73 ◆ Cutting a panel around a window.

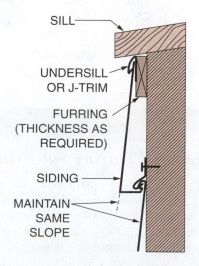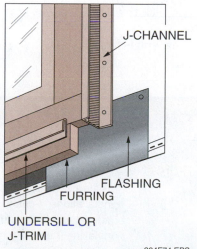

Figure 74 ◆ Fitting panels at door and window sills.

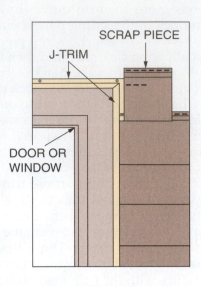

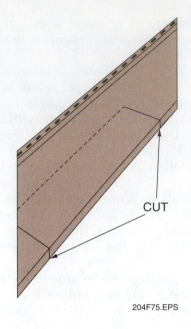

204F75.EPS

Figure 75 ◆ Using a scrap piece to measure clearance.

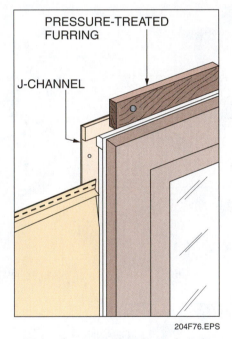

204F76.EPS

Figure 76 ◆ Using furring with J-channel.

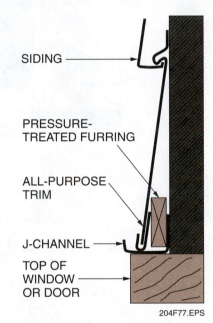

204F77.EPS

Figure 77 ◆ Completing the installation at a window top.

Pressure-Treated Furring and Furring Substitutes

If used, always make sure that any wood furring for vinyl or metal siding is pressure-treated to resist insects and rot. In place of pressure-treated furring in some installations, J-channel is used under the sill. Then undersill trim installed under the siding provides the furring spacing and a panel locking point. At the top of the opening, installing undersill trim under the siding provides the furring spacing and panel locking point.

7.15.0 Installing Siding at Gable Ends

When installing siding on gables, diagonal cuts will have to be made on some of the panels. To make a pattern of cutting panels to fit the gable slope, use two short pieces of siding as templates (*Figure 78*). Interlock one of these pieces into the panel below. Hold the second piece against the J-channel trim on the gable slope. Along the edge of this second piece, scribe a line diagonally across the interlock end panel and cut along this line with tin snips or a power saw.

This cut panel is a pattern that can be used to transfer cutting marks to each successive course along the gable slope. All roof slopes can be handled in the same manner as gable end slopes.

Slip the angled end of the panel into the J-trim previously installed along the gable end. Lock the butt into the interlock of the panel below. Remember to allow for expansion or contraction where required. If necessary, face nail with 1¼" (or longer) painted-head aluminum nails in the apex of the last panel at the gable peak. Touch-up enamel in matching siding colors can also be used for exposed nail heads.

Do not cover existing **louvers** or **vents.** Attic ventilation is necessary in summer to reduce temperatures and in winter to prevent the accumulation of moisture.

7.16.0 Installing Siding Under Eaves

The last panel course under the eaves will almost always have to be cut lengthwise to fit in the remaining space. Usually, furring will be needed under this last panel to maintain the correct slope angle. Determine the proper furring thickness and install it. Nail all-purpose trim or J-channel to the furring with aluminum nails. The trim should be cut long enough to extend the length of the wall (*Figure 79*).

To determine the width of the cut required, measure from the bottom of the top lock to the eave, subtract ¼", and mark the panel for cutting. Take measurements at several points along the eaves to ensure accuracy. Score the panel with the utility knife and bend it until it snaps. For vinyl panels, use a snaplock punch to place raised ears (16" or 24" apart) along the top cut edge so that it will lock into the J-channel.

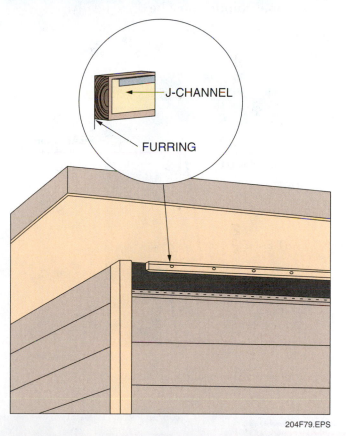

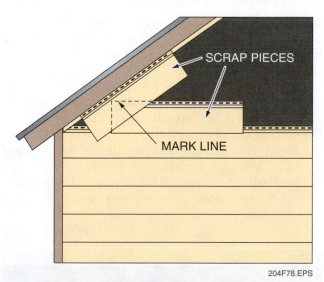

204F78.EPS

Figure 78 ◆ Installing siding at gable ends.

204F79.EPS

Figure 79 ◆ Trim extends the length of the wall.

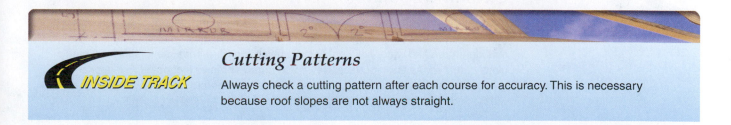

Cutting Patterns

INSIDE TRACK

Always check a cutting pattern after each course for accuracy. This is necessary because roof slopes are not always straight.

For aluminum siding, apply gutter seal to the nail flange of the all-purpose trim. Slide the final panel into the trim. Engage the interlock of the panel below.

On metal panels, the lock may be flattened slightly using a hammer and a 2' or 3' piece of lumber before the final panel is installed so it will grip more securely. Press the panel into the gutter seal adhesive. With this technique, face nails will not be required. Refer to *Figure 80*.

7.17.0 Caulking and Cleanup

In general, caulking is done around doors, windows, and gables where the siding meets wood or metal, except where accessories are used to make caulking unnecessary. Caulking is also needed where siding or siding accessories meet brick or stone around chimneys and walls. Surface caulking required around faucets, meter boxes, and other panel cutouts must be done neatly.

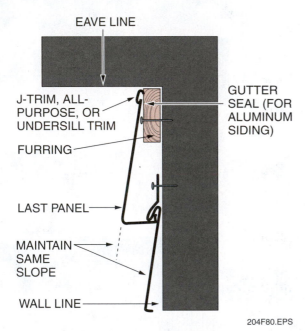

Figure 80 ◆ Installing siding under eaves.

It is important to get a deep caulking bead that is ¼" minimum in depth, not just a wide bead. To achieve this, cut the plastic caulking cartridge tip straight across rather than at an angle. Move the gun evenly and apply steady, even pressure on the trigger. The butyl type of caulking is preferred as it has greater flexibility. Most producers supply caulking in colors to match siding and accessories. Do not depend on caulking to fill gaps more than ⅛" wide, as the expansion or contraction of the siding may cause the caulking to crack.

Reinstall all fixtures, brackets, downspouts, etc. that were removed. Accessories that were not replaced, such as kitchen fan outlets or service cables, may be painted to match the new siding color. Most manufacturers have touch-up paint or matching paint formulas, which can be purchased at a local paint store. All scrap pieces, cartons, nails, and other materials should be removed and the job site left neat and clean each day.

8.0.0 ◆ STUCCO (CEMENT) FINISHES

Stucco is a durable cement-based coating for exterior walls that is normally applied by painters or masons.

When applying stucco over frame walls, use wood sheathing, exterior gypsum, or cement board. Building paper is placed over the sheathing and 2 × 8 or 4 × 8 panels of metal reinforcement mesh, called diamond mesh (1⅜" × ⅜" openings), are placed over that. The diamond mesh is applied with special spacing fasteners that hold it away from the wall slightly so that it will become embedded in the first coat of plaster.

Three coats of cement plaster must be applied *(Figure 81)*. The first coat is called the scratch coat, the second is the brown coat, and the last is called the finish coat. The first two coats can be troweled on with a rough finish. The finish coat can be applied rough or smooth, as desired. In many cases, these coats are applied using texture

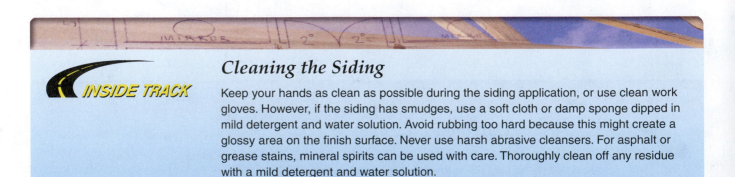

Cleaning the Siding

INSIDE TRACK

Keep your hands as clean as possible during the siding application, or use clean work gloves. However, if the siding has smudges, use a soft cloth or damp sponge dipped in mild detergent and water solution. Avoid rubbing too hard because this might create a glossy area on the finish surface. Never use harsh abrasive cleansers. For asphalt or grease stains, mineral spirits can be used with care. Thoroughly clean off any residue with a mild detergent and water solution.

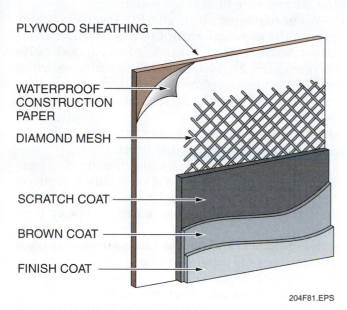

Figure 81 ◆ Stucco section.

PLYWOOD SHEATHING

WATERPROOF
CONSTRUCTION
PAPER

DIAMOND MESH

SCRATCH COAT

BROWN COAT

FINISH COAT

204F81.EPS

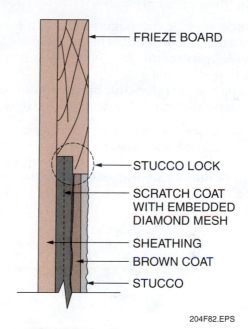

Figure 82 ◆ Stucco lock.

FRIEZE BOARD

STUCCO LOCK

SCRATCH COAT
WITH EMBEDDED
DIAMOND MESH

SHEATHING

BROWN COAT

STUCCO

204F82.EPS

sprayers. The total thickness of these three coats should be approximately 1" (Portland Cement Association). Control joints must be placed in accordance with the manufacturer's specifications.

When applying wood trim next to stucco, such as the frieze board or half timbers used to simulate Tudor architecture, the trim should have a rabbet along the back joining edge, called a stucco lock (*Figure 82*). The stucco lock prevents water penetration around the stucco at the juncture of the stucco and the wood trim.

When applying stucco to concrete block walls, the diamond mesh is not required, but all other requirements apply.

9.0.0 ◆ BRICK AND STONE VENEER

The most durable materials that can be used for exterior wall finishes are brick, synthetic stone, and natural stone masonry. Brick and synthetic stone are man-made products, while stone is a natural material. All three materials are attractive and require little or no maintenance. Brick is made from a clay mixture and is hardened by baking it in a kiln. The nominal size of a brick is $2\frac{1}{4}" \times 3\frac{3}{4}" \times 8"$. This size may vary slightly due to the hardening processes in the kiln. Synthetic stone is made from a cement mixture. Natural stone is either found on the surface or is mined from an open pit quarry.

10.0.0 ◆ DEFS AND EIFS

Direct-applied exterior finish systems (DEFS) and exterior insulation and finish systems (EIFS) are designed as water-managed systems (*Figure 83*). In appearance, they are similar to traditional stucco or masonry finishes, but they employ different types and applications of material.

Water-managed systems are usually defined as wall cladding systems that:

- Provide specific drainage methods for intruding water that penetrates beyond the cladding
- Provide protection for water-sensitive construction elements
- Are applied to water-durable or water-resistant substrates that can tolerate exposure to water

In most cases, the substrate is a fiber-cement or fiberglass-coated and treated gypsum panel that can be used over a wood sheathing with underlayment or non-structural sheathing.

All water-managed systems for framed construction or masonry construction have very specific means and methods for flashing and directing incidental water that enters around or through windows, doors, and other openings. For this reason, and to achieve satisfactory performance as well as warranty protection, the manufacturer's instructions must be rigorously followed when installing the components of these systems. In addition, a vapor barrier is usually required on the interior side of exterior walls.

Water-managed DEFS and EIFS wall claddings are almost identical, except that in an EIFS, insulating boards, usually made of expanded polystyrene (EPS), are fastened to the substrate surface and a mesh reinforcement is bonded to the insulation board under the base coat. The insulation boards can vary from 1" to 4" in thickness, depending on the insulation value required.

In addition to the normal finish for DEFS, several masonry facings can also be used. These include ceramic tile set in a latex-fortified grout on top of a latex-fortified mortar base coat and bond coat or exposed aggregate set in an epoxy base coat.

DEFS and EIFS wall cladding surface finishes are combustion-proof, making them ideal for use in fire-prone areas. With two layers of substrate, as well as fire-rated interior insulation and two layers of fire-rated interior drywall, these systems are rated for up to two hours as a firestop. Like fiber-cement siding, these systems are rot-proof, are termite-proof, and can withstand high winds.

Normally, the base coat and surface finishes for these systems are applied by painters or masons using texture sprayers. Finishes involving thin brick, tile, or aggregate are usually applied by masons. Carpenters normally install the flashing, water barrier, substrate(s), seam tape, insulation, and/or fiber mesh.

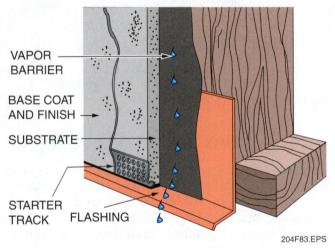

VAPOR BARRIER

BASE COAT AND FINISH

SUBSTRATE

STARTER TRACK   FLASHING

204F83.EPS

Figure 83 ◆ Typical DEFS water management.

The following are typical general application characteristics/guidelines for these systems:

- When applied to standard wood studs or 20-gauge metal studs at 16" OC, the walls can withstand wind loads of 40 pounds per square foot (psf). Greater wind loads are allowable with the use of larger studs or closer stud spacing.
- These systems can be used for ceilings, soffits, curtain walls, bearing walls, panelization elements, and privacy fences.
- The final surface finishes can be tinted any color.
- These systems cannot be used as sill finishes.

- The substrate may not be used as structural sheathing. Racking resistance must be accomplished by separate bracing.
- All windows, doors, and other openings must be properly flashed.
- DEFS on steel framing must be laterally braced. To enhance crack resistance over steel framing, mesh reinforcement is required over the entire substrate to reinforce the base coat.
- Multiple layering of the substrate can be used to achieve various architectural effects such as banding on various levels of a building.
- The drawings and manufacturer's specific instructions must be rigidly followed.

INSIDE TRACK

DEFS/EIFS Codes

Always check state and local building codes before using DEFS or EIFS cladding. Some states and/or localities have prohibited its use on residential structures due to deficiencies in earlier versions of the water-management systems, compounded by faulty installation that resulted in severe damage. Some manufacturers offer improved versions of DEFS and EIFS cladding systems that are designed to alleviate these problems. Make sure that any system used is the latest version offered by the manufacturer. Also, make sure that the manufacturer's instructions for installation are rigorously followed.

Summary

You must be aware of the materials and the general methods of installation for a variety of exterior finishes and roof drainage systems. This module covered the installation methods for flashing and insulation, types of cornices and their fabrication/installation, along with descriptions and installation methods for a variety of wood, metal, and vinyl siding. It also discussed other exterior finishes, including stucco, masonry, and various special exterior finish systems. When installing any type of exterior finish or related accessory, always refer to the manufacturer's installation instructions.

Notes

Review Questions

1. The primary purpose of an exterior finish is to _____.
   a. provide a base for the final finish
   b. prevent entry of water at all openings
   c. provide protection from the elements
   d. allow the use of the most cost-effective interior construction

2. When building wrap is applied to the exterior of a structure, the wrap should be overlapped at the corners by _____.
   a. 2"
   b. 4"
   c. 6"
   d. 8"

3. The cutting of some wood and manufactured siding materials can produce a hazardous dust.
   a. True
   b. False

4. A conventionally insulated house can suffer a heat loss through frame walls or by air infiltration of _____.
   a. 55 percent
   b. 65 percent
   c. 75 percent
   d. 85 percent

5. The most common flashing material is _____.
   a. stainless steel
   b. asphalt building paper
   c. galvanized sheet metal
   d. copper

6. Plancier is another name for a _____.
   a. soffit
   b. ledger board
   c. frieze board
   d. lookout

7. A ledger board is used to simplify the installation of _____.
   a. fascia boards
   b. rafters
   c. frieze boards
   d. lookouts

8. The horizontal framing member that the soffit is attached to is called a _____.
   a. jack rafter
   b. hip rafter
   c. lookout
   d. cornice return

9. To the nearest square, how many squares of 1 × 6 wood siding would be required for a hip roof structure that is 50' long and 35' wide with 8' walls, twenty 3' × 5' windows, two 8' × 7' garage doors, and two 3' × 6½' entrance doors?
   a. 12 squares
   b. 14 squares
   c. 15 squares
   d. 16 squares

10. To the nearest panel, how many 4 × 8 panels of vertical siding would be required for the same structure described in Question 9?
    a. 37 panels
    b. 39 panels
    c. 41 panels
    d. 43 panels

11. Two common styles of beveled siding are _____.
    a. colonial and log cabin
    b. log cabin and rustic
    c. colonial and Dolly Varden
    d. rabbeted and tongue-and-groove

12. A vertical siding where boards overlap boards underneath that are the same size as the overlapping boards is called _____ siding.
    a. board-and-batten
    b. reverse batten
    c. board-on-board
    d. shiplap

13. Of the following, _____ is a style of tongue-and-groove siding.
    a. colonial
    b. Dolly Varden
    c. V-edged
    d. log cabin

14. The type of shiplap siding shown below is known as _____ siding.
    a. channel rustic
    b. drop
    c. V-edged
    d. plain

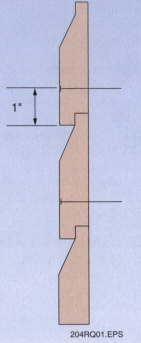

1"

204RQ01.EPS

15. The two fasteners used to apply wood shingles are _____.
    a. nails and staples
    b. nails and drywall screws
    c. staples and construction adhesive
    d. sheet metal screws and roofing nails

16. A plywood siding panel called textured one-eleven (T-111) has a surface pattern that _____.
    a. is striated
    b. is brushed
    c. has widely spaced grooves
    d. has a plank texture

17. When plywood is used as exterior siding, it is _____.
    a. nailed to the sheathing
    b. nailed to the studs
    c. glued to the sheathing
    d. glued to the studs

18. All of the following finishes are recommended for hardboard siding *except* _____.
    a. 100 percent acrylic latex paint
    b. flat alkyd stain
    c. gloss alkyd paint
    d. satin alkyd paint

19. Which of these types of siding should *not* be cleaned using pressure-washing equipment?
    a. Fiber-cement
    b. Vinyl
    c. Hardboard
    d. Aluminum

20. When installing lapped hardboard siding, the minimum overlap is _____.
    a. ½"
    b. ¾"
    c. 1"
    d. 2"

21. The maximum stud spacing for installation of fiber-cement board applied over sheathed walls is _____ OC.
    a. 12"
    b. 16"
    c. 20"
    d. 24"

22. Which of the following metals will corrode when placed in contact with fiber-cement board siding?
    a. Aluminum
    b. Galvanized steel
    c. Copper
    d. Stainless steel

23. When cutting vinyl siding with a power saw, the blade used should be a _____ fine-toothed blade for a smooth finish.
    a. carbide-tipped
    b. reversed
    c. dry diamond
    d. high-speed

24. A stucco lock overlaps the _____ of a stucco finish.
    a. sheathing
    b. scratch coat
    c. brown coat
    d. finish coat

25. In a water-managed system, a substrate is _____.
    a. plywood sheathing
    b. brick veneer
    c. asphalt-impregnated felt
    d. fiber-cement panels

Trade Terms Introduced in This Module

Board-and-batten: A type of vertical siding consisting of wide boards with the joint covered by narrow strips known as battens.

Brown coat: A coat of plaster with a rough face on which a finish coat will be placed.

Building paper: A heavy paper used for construction work. It assists in weatherproofing the walls and prevents wind infiltration. Building paper is made of various materials and is not a vapor barrier.

Cornice: The construction under the eaves where the roof and side walls meet.

Course: One row of brick, block, or siding as it is placed in the wall.

Eave: The lower part of a roof, which projects over the side wall.

Fascia: The exterior finish member of a cornice on which the rain gutter is usually hung.

Felt paper: An asphalt-impregnated paper.

Finish coat: The final coat of plaster or paint.

Frieze board: A horizontal finish member connecting the top of the sidewall, usually abutting the soffit. Its bottom edge usually serves as a termination point for various types of siding materials.

Ledger: A board to which the lookouts are attached and which is placed against the outside wall of the structure. It is also used as a nailing edge for the soffit material.

Lookout: A member used to support the overhanging portion of a roof.

Louver: A slatted opening used for ventilation, usually in a gable end or a soffit.

Plancier: The same as a soffit, but the member is usually fastened to the underside of a rafter rather than the lookout.

Rabbet: A groove cut in the edge of a board so as to receive another board.

Rake: The slope or pitch of the cornice that parallels the roof rafters on the gable end.

R-value: A numerical designation given to a material based on its insulating ability, such as R-19.

Scratch coat: The first coat of cement plaster consisting of a fine aggregate that is applied through a diamond mesh reinforcement or on a masonry surface.

Shakes: Hand- or machine-split wood shingles.

Soffit: The underside of a roof overhang.

Veneer: A brick face applied to the surface of a frame structure.

Vent: A small opening to allow the passage of air.

This module is intended to present thorough resources for task training. The following reference works are suggested for further study. These are optional materials for continued education rather than for task training.

The Vinyl Siding Institute website, *www.vinylsiding.org*.

Cedar Shake & Shingle Bureau website, *www.cedarbureau.org*.

NCCER CURRICULA — USER UPDATE

NCCER makes every effort to keep its textbooks up-to-date and free of technical errors. We appreciate your help in this process. If you find an error, a typographical mistake, or an inaccuracy in NCCER's curricula, please fill out this form (or a photocopy), or complete the online form at **www.nccer.org/olf**. Be sure to include the exact module ID number, page number, a detailed description, and your recommended correction. Your input will be brought to the attention of the Authoring Team. Thank you for your assistance.

Instructors – If you have an idea for improving this textbook, or have found that additional materials were necessary to teach this module effectively, please let us know so that we may present your suggestions to the Authoring Team.

NCCER Product Development and Revision

13614 Progress Blvd., Alachua, FL 32615

Email: curriculum@nccer.org
Online: www.nccer.org/olf

❏ Trainee Guide   ❏ AIG   ❏ Exam   ❏ PowerPoints   Other _____

Craft / Level: _____   Copyright Date: _____

Module ID Number / Title: _____

Section Number(s): _____

Description: _____

_____

_____

_____

_____

Recommended Correction: _____

_____

_____

_____

_____

Your Name: _____

Address: _____

_____

Email: _____   Phone: _____

Cold-Formed Steel Framing

27205-07

27205-07
Cold-Formed Steel Framing

Topics to be presented in this module include:

Overview

This module describes the uses and installation of cold-formed steel framing. In commercial and multi-family residential construction, it is common to use steel framing materials in place of wood studs to frame walls and partitions. It is also becoming more common in single-family residential construction as the price of lumber rises. As the demand for more cost-efficient, more sustainable construction grows, builders and developers are specifying both structural and non-structural applications of cold-formed steel. Recognition and acceptance of cold-formed steel in building codes is leading to a better understanding of its capabilities and benefits. Low- and mid-rise structures using cold-formed steel as the main structural component have been successfully built throughout the United States, creating more durable structures and lowering construction costs. Steel framing, however, requires a carpenter to master tools and joining techniques different from those required in wood-framing construction. In order to work a variety of steel framing projects, from installing loadbearing steel assemblies to non-structural walls, you will need to become familiar with these materials and methods.

Objectives

When you have completed this module, you will be able to do the following:

1. Identify the components of a steel framing system.
2. Identify and select the tools and fasteners used in a steel framing system.
3. Identify applications for steel framing systems.
4. Demonstrate the ability to build back-to-back, box, and L-headers.
5. Lay out and install a steel stud structural wall with openings to include bracing and blocking.
6. Lay out and install a steel stud non-structural wall with openings to include bracing and blocking.

Trade Terms

Blocking
C-shape
Clip angle
Cold-formed steel
Curtain wall
Diaphragm
Header
Knurled
Lateral
Lips

Mil
Panelization
Plenum
Racking
Rim track
Roof rafter
Shear wall
Track
Web

Required Trainee Materials

1. Pencil and paper
2. Appropriate personal protective equipment

Prerequisites

Before you begin this module, it is recommended that you successfully complete *Core Curriculum*; *Carpentry Fundamentals Level One*; and *Carpentry Framing and Finishing Level Two*, Modules 27201-07 through 27204-07.

This course map shows all of the modules in *Carpentry Framing and Finishing Level Two*. The suggested training order begins at the bottom and proceeds up. Skill levels increase as you advance on the course map. The local Training Program Sponsor may adjust the training order.

27212-07
Cabinet Fabrication
ELECTIVE

27211-07
Cabinet Installation

27210-07
Window, Door, Floor, and Ceiling Trim

27209-07
Suspended Ceilings
ELECTIVE FOR RESIDENTIAL CERTIFICATE

27208-07
Doors and Door Hardware

27207-07
Drywall Finishing

27206-07
Drywall Installation

27205-07
Cold-Formed Steel Framing

27204-07
Exterior Finishing
ELECTIVE FOR COMMERCIAL CERTIFICATE

27203-07
Thermal and Moisture Protection

27202-07
Roofing Applications
ELECTIVE FOR COMMERCIAL CERTIFICATE

27201-07
Commercial Drawings
ELECTIVE FOR RESIDENTIAL CERTIFICATE

CARPENTRY FUNDAMENTALS

CORE CURRICULUM:
Introductory Craft Skills

FRAMING AND FINISHING

205CMAP.EPS

1.0.0 ◆ INTRODUCTION

Designers and builders have long recognized steel for its strength, durability, and functionality. They, along with an increasing number of architects, are recognizing steel's important environmental attributes as well. Industry-wide, steel represents an average of 67 percent recycled content, making it the world's most recycled material. In the United States alone, almost 70 million tons of steel were recycled or exported for recycling per year. **Cold-formed steel** framing helps with industry efforts to promote sustainable construction.

By comparison, the wood framing used in a majority of residential construction places heavy demands on timber resources. Typically, one acre of forest is used in the construction of a 2,000-square-foot home.

By virtue of its material characteristics and properties, steel offers the following significant advantages for building construction:

- Steel studs and joists are strong, lightweight, and made from uniform-quality material. Steel has the highest strength-to-weight ratio of any building material.
- Steel framing members are dimensionally stable and will not warp, crack, rot, or split.
- Steel is uniform and provides a flat surface for sheathing or other material attachment.
- Steel framing does not contribute combustible material to feed a fire. The subsequent lower combustion rating results in lower insurance costs for builders and owners.
- Steel can be engineered to meet the strongest wind and seismic ratings specified by building codes.

NOTE

Appendix A contains a list of terms commonly used in conjunction with cold-formed steel framing.

North American Codes and Standards

The use of cold-formed steel (CFS) members in building construction began around 1850. In North America, however, steel members were not widely used until 1946, when the American Iron and Steel Institute (AISI) Specification was first published. This design standard was primarily based on research sponsored by AISI at Cornell University. Subsequent revisions to the document reflected technical developments, and ultimately led to the publishing of the *North American Specification for the Design of Cold-Formed Steel Structural Members 2001 Edition*. AISI, along with the American National Standards Institute (ANSI), American Society of Testing and Materials International (ASTM), and the International Codes Council (ICC) Steel Stud Manufacturers Association (SSMA) govern the design, manufacturing and use of cold-formed steel and framing. All framing members carry a product identification to comply with the minimum sheet steel thickness, coating designation, minimum yield strength, and manufacturer's name.

2.0.0 ◆ TOOLS FOR STEEL FRAMING WORK

Metal stud work requires more use of power tools than wood frame assembly, so be sure to practice power tool safety measures at all times, including grounding of all electrical tools. Some of the common tools that are used in steel framing component assembly are as follows:

• *Screwgun* – The screwgun (*Figure 1*) drives screws to connect steel members and attach sheathing material such as plywood and gypsum board to steel. Also called a power screwdriver, it increases installation speed and efficiency. The preferred screwgun for steel-to-steel connections should have an adjustable clutch and torque setting, with a speed range of 0 to 2,500 revolutions per minute (rpm). A drywall screwgun is recommended for attaching plywood or gypsum board to steel. In addition to standard and cordless screwguns, there is a type powered by compressed air. Feed attachments are also available for screwguns (*Figure 2*). They use collated strips of screws that automatically feed at the end of a bit tip.

CAUTION

If the screwgun is running too fast, the tip of the screw may burn out before it penetrates the steel. Once the screw is properly seated, the screwgun automatically stops spinning the screw, preventing the screw from stripping.

NOTE

A power screwdriver is not just a drill with a screwdriver bit. They normally have an adjustable depth control to prevent overdriving the screws. Many have a clutch mechanism that disengages when the screw has been driven to a preset depth. Some power screwdrivers are designed to perform specific fastening jobs, such as fastening drywall to walls and ceilings.

• *Stud driver* – Two types of stud drivers are used to drive fasteners into concrete slabs, structural steel, and foundations: the powder-actuated fastener gun (*Figure 3*) and the hammer-drill (*Figure 4*). The holding strength of the fastener in concrete depends on the compressive strength of the concrete, shank diameter, and depth of penetration, as well as spacing and edge conditions. Headed or threaded drive pins are available with **knurled** shanks to increase holding power in structural steel material. Fasteners loaded in tension may require washers to prevent the fastener from pulling through the steel track. Carpenters must be properly trained in the use of these tools and must have an operator's certificate before operating a powder-actuated tool. The operator must wear a hard hat, heavy boots, and eye, ear, and face protection. Treat a powder-actuated stud driver as if it were a loaded gun. Before you handle it, determine whether it is loaded with a powder charge, a fastener, or both. Point it away from yourself and away from others at all times. Be sure to recognize when the tool is and is not in locked position.

205F01.EPS

Figure 1 ◆ Standard screwgun.

205F02.EPS

Figure 2 ◆ Screwgun with collated feed.

Figure 3 ◆ Powder-actuated installation tool.

Figure 4 ◆ Hammer drill.

Figure 5 ◆ Locking C-clamps.

WARNING!
Powder-actuated fasteners can be used only by personnel trained and certified in the use of the specific tool. You must carry your certification card with you when using the tool, and the card must cover the tool you are using.

Be sure to select the proper charge for the job in accordance with the manufacturer's instructions. If in doubt, check with your supervisor.

- *Locking C-clamps and bar clamps* – These tools come in a variety of sizes and are used to hold steel members together during fastening. C-clamps (*Figure 5*) prevent separation, also known as screw jacking, when the first layer of steel climbs the threads of the screw. The 9DR locking C-clamp was designed specifically for cold-formed steel framing. Bar clamps are used extensively to hold steel wall members together

until permanent fastenings are applied. They are also used to hold **headers** in place until they are fitted into the top track. All clamps used in steel framing should have regular tips on ends without pads to reach around the steel flanges.

2.1.0 Cutting Tools

Because so many components of steel framing are prefabricated, the amount of steel cutting you will have to do in the field is minimal. However, if field cutting needs to be performed, the following are some common tools and methods.

NOTE
Some contract documents prohibit the use of certain cutting tools. Always check installation specifications before using any one method.

- *Chop saw* – Chop saws are most commonly used for field cutting, using an abrasive blade to cut quickly through the steel (*Figure 6*). While chop saws are very effective for square cuts and for cutting bundled studs, they are very noisy and give off hot flying metal filings. Be aware that the edge produced by a chop saw is very rough, with sharp burrs left on the steel. There are a number of circular saws on the market that can be used with carbide-tipped blades to cut steel studs. Some of them have special guards to catch flying metal chips, which can be a hazard when using a high-speed saw to cut steel.
- *Swivel-head shears* – Shears are manufactured in both electric and battery-operated models that can cut thickness up to 68 **mils**. *Figure 7* shows typical swivel head shears. They are used to cut

Figure 6 ◆ Cutting cold-formed steel with a chop saw.

Figure 7 ◆ Swivel-head shears.

Figure 8 ◆ Aviation snips.

steel studs, runners, sheet metal, and cold-formed steel up to 33 mils. They are portable and make smooth cuts with no abrasive edges. The cutting edge rotates 180 degrees for over-head and side work. Some drawbacks to shears are that they may be difficult to use in cutting tight radii on **C-shapes**, and blades are expensive to replace.

• *Circular saw/dry-cut metal cutting saw* – Newer types of metal cutting saws are becoming available with blades made of titanium and aluminum that produce a smooth edge. One manufacturer produces a dry-cut saw that collects the metal filings right in the tool.

• *Aviation snips* – Aviation snips (*Figure 8*) are hand tools that can cut cold-formed steel up to 43 mils. They are useful when cutting and coping steel, for snipping flanges and for making small cuts. Some brands of snips are color-coded for left, right, and straight. Always wear gloves and safety goggles when using snips.

2.1.1 Tools for Cutting Holes

Most openings in steel studs are pre-punched at the factory. However, it may be necessary to punch additional openings for electrical or telecommunications cabling that may be installed later. Holes in **webs** of studs, joists, and tracks must be in conformance with an approved design or a recognized design standard.

• *Hole punch/steel stud punch* – For small holes, up to approximately 1" to 1½" in diameter, a hole punch can be used for steel members up to 33 mils in thickness. Grommets should be inserted to protect wiring from sharp edges. They may also be used to isolate copper or other dissimilar metals from the steel.

• *Hole saw* – For larger holes up to 6" in diameter, and through material thicknesses greater than 33 mils, hole saws and bits are recommended. The saws are used on a drill motor to cut through the steel. Bits, while more costly, tend to cut through the steel faster.

3.0.0 ◆ FASTENERS FOR STEEL FRAMING WORK

Knowing how to select the right fastener for any steel-framing project will make all the difference in the structural integrity and quality of your work. You must consider the loads to be transmitted through the connection, as well as the thickness, strength, and configuration of the materials to be joined. Fasteners include screws, nails, pins, clinches, and welds, as well as anchor bolts, rivets, powder-actuated fasteners, and expansion bolts. For applications needing increased stability or when working with steel thicker than 43 mils, the components may be joined by shielded metal arc welding to provide additional strength. It is important to follow manufacturer recommendations for fastening systems during all phases of cold-formed steel construction projects.

3.1.0 Screws

In steel framing, the screw is the workhorse of the fasteners. Because attachment holes are not pre-drilled in steel framing, screws used to secure steel framing must have the ability to make their own holes before they engage. They are installed with screwguns and come in a variety of head styles to fit a wide range of structural and cosmetic requirements. Screws have three distinct thread thicknesses: coarse (threads spaced farthest apart), medium, and fine (threads spaced closest together). Screws used in steel framing generally have coarse threads for optimum cutting. Thicker steel, such as 97 mils, requires the use of a fine-threaded screw. Screws are generally finished with zinc or cadmium plating to withstand environmental impacts and resist corrosion. The size and the type of screw needed will depend on the thickness of the sheathing material and steel. Become familiar with the different types of screws to choose from. If the wrong screw is used, the proper connection may not be made, or the screw may break off or not penetrate the steel.

There are two main types of steel framing screws: self-drilling and self-piercing (*Figure 9*). Self-drilling screws are designed to drill through layers of steel before any of the screw threads engage. Self-piercing screws have sharp points that can typically penetrate 18 to 30 mils of material with ease. They are commonly used to attached plywood and gypsum board to these thinner layers of steel. Code and standard requirements state that for all connections, screws must extend through the steel a minimum of three exposed threads. For most steel-to-steel connections, ½" or ¾" screws are acceptable. When applying plywood, gypsum board, or foam insulation, proper screw length is determined by adding together the measured thickness of all materials, with an extra ⅜" for the exposed threads.

Head styles of screws are available in many different forms. The head locks the screw into place and prevents it from sinking through the layers of material. The head also contains the drive type (type of bit tip) to apply the screw. The head type profile is selected to avoid interference with other building components.

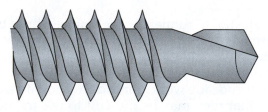

SELF-DRILLING SCREW

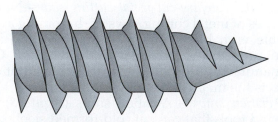

SELF-PIERCING SCREW

205F09.EPS

Figure 9 ◆ Self-drilling and self-piercing screws.

Common self-drilling screws are shown in *Figure 10* and include the following:

- *Gypsum board (bugle-head) screw* – These screws are used to attach gypsum board to steel framing. No. 6, sharp-point bugle-head drywall screws were designed for this application. They should be used with a depth-setting nosepiece to avoid tearing the gypsum board or protective paper. For steel thicker than 33 mils, use a self-drilling bugle-head drywall screw.
- *Flat-head screw* – This screw is used for wood flooring and facings. It is designed to countersink and seat flush without causing splintering or splitting.
- *Wafer-head screw* – This screw is larger than the flat-head screw and is used to secure soft materials to steel studs. The large head provides a greater bearing surface, and seats flush to achieve a clean, finished appearance.
- *Hex washer-head screw* – The most popular head style for steel-to-steel connections with no sheathing or gypsum board is the hex washer. The washer face provides a surface for the driver socket, assuring good stability in driving. If sheathing and gypsum board are applied over the fastener, the head style must have a very thin profile to prevent blow-outs or bumps at the screw locations.

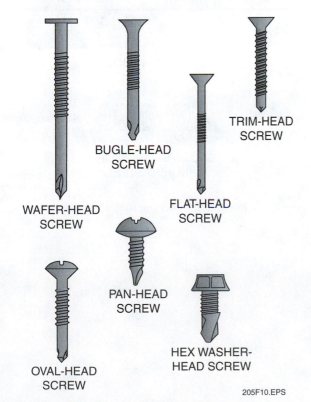

Figure 10 ◆ Typical self-drilling screws.

WAFER-HEAD SCREW

BUGLE-HEAD SCREW

FLAT-HEAD SCREW

TRIM-HEAD SCREW

PAN-HEAD SCREW

OVAL-HEAD SCREW

HEX WASHER-HEAD SCREW

205F10.EPS

- *Trim-head screw* – Small trim-head screws are preferred when installing baseboard and other trim. The small head penetrates the trim, leaving a tiny hole easily filled with putty. Pneumatic and powder actuated trim nailers can also be used.
- *Oval-head screw* – This screw spans clearance holes and has a low profile appearance for attaching accessory items to steel stud walls.
- *Pan-head screw* – This screw is used to fasten studs to runner tracks and to connect steel bridging, strapping, or furring channels to studs or joists.

3.2.0 Pins

Nail guns are commonly used in steel framing to attach plywood, OSB, or siding to walls and roofs. Some contractors use only nails (or pins), while others use screws around the perimeter and pins in the field of the board.

Plywood or OSB sheathing material may be applied using pneumatic pins (*Figure 11*). To be effective, the plywood must be held tightly against the steel before the pin is driven, because firing the pin does not tighten the plywood against the steel. Roof sheathing is more easily installed with pins because the carpenter usually stands on the plywood, keeping it tight to the steel. An alternative method is to tack the plywood to the steel with screws along the perimeter first and then pin the field.

There have been advancements in fastener technology to include pins for steel-to-steel connections as well. Connections are accomplished by driving a pin with a knurled shank (*Figure 12*) into the layers of steel.

3.3.0 Clinching

This method requires no screws or pins, only a pneumatic tool that press-joins pieces of steel together. The clinching tool, in effect, creates a rivet in the metal, as shown in *Figure 13*. Manufacturers of clinching tools provide test reports that verify the strength of the press-joining. Clinching systems work well in **panelization** environments, and are less permanent than welding. If studs are fastened incorrectly with a clinch connection, the stud may be popped out with a screwdriver or drilled out.

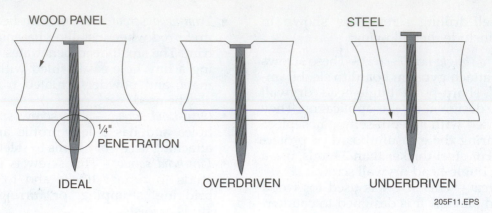

Figure 11 ◆ Proper pin installation.

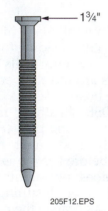

Figure 12 ◆ Knurled shank pin.

3.4.0 Bolts and Anchors

Bolts and anchors are also commonly used to fasten cold-formed steel framing to masonry, concrete, and other steel components. Except for some proprietary anchors, pre-drilling of holes is necessary. Bolts require the installation of a washer, and must meet or exceed the requirements of *ASTM A307*. Expansion anchors are commonly used for connections to concrete or masonry and require information from the manufacturer to determine the capacity and spacing requirements.

CLINCHING TOOL

CROSS-SECTION OF CLINCHED STEEL

205F13.EPS

Figure 13 ◆ Pneumatic clinching tool.

4.0.0 ◆ STEEL FRAMING MATERIALS

Framing components include steel studs, steel joists, and steel roof trusses. Framing systems come with accessories such as the clips, web stiffeners, resilient channels, fastening devices, and anchors required for complete and proper installation of members. Other manufacturers offer specialized products that enable builders to shorten construction times for complicated curves, arches, and other unique architectural features. The vertical and horizontal framing members serve as structural load-carrying components for a large number of low-rise and high-rise structures.

Steel framing materials have several advantages over wood framing, such as lower costs, faster assembly and installation, and design flexibility.

The components are also noncombustible, lightweight, corrosion-resistant, and fabricated to fit together easily. Manufacturing tolerances for cold-formed steel members are governed by ASTM standards. These standards govern length, web depth, flare, crown, bow, and twist. Cold-formed steel components are also coated against corrosion according to ASTM standards. Hot-dipped zinc galvanizing is the most effective coating method. Depending on the thickness of zinc applied to the steel, and the environment in which the steel is placed, zinc coatings can protect the steel for more than 1,000 years.

Steel framing is also compatible with all types of surfacing materials. There are a variety of load-bearing and nonbearing systems, but *Figure 14*

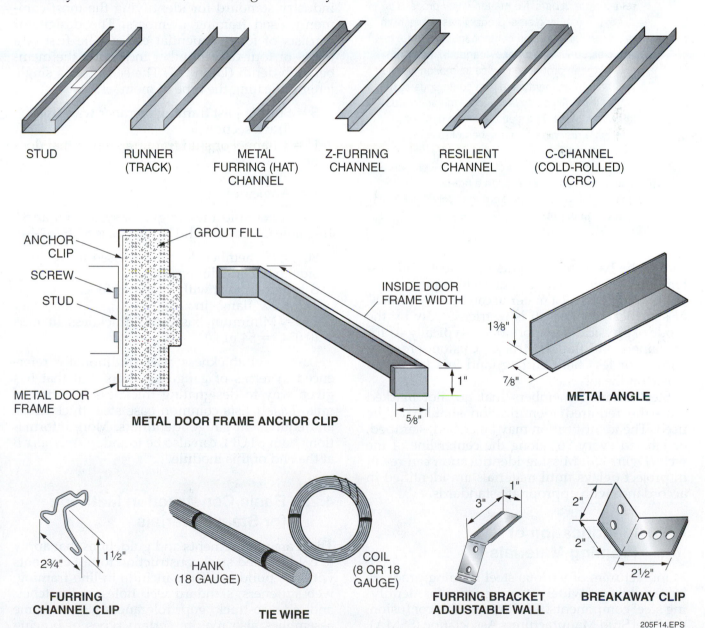

Figure 14 ◆ Components of a metal (steel) stud system.

205F14.EPS

Did You Know?

Modern steel production relies on two technologies: basic oxygen furnace (BOF) or electric arc furnace (EAF). The BOF process uses 25 to 35 percent old steel to make new steel. It produces sheet for products where the main requirement is drawability; that is, products like automotive fenders, steel-framing members, refrigerator encasements, and packaging. The EAF process uses 95 to 100 percent old steel to make new steel. It is primarily used to manufacture products where strength is critical, such as structural beams, steel plates, and reinforcement bars. No matter which process is used, the resulting steel product has a minimum of 25 percent recycled content. Industrywide, the recycled content of steel averages 67 percent.

During the production process, molten steel is poured into an ingot mold or a continuous caster, where it solidifies into large rectangular shapes known as slabs. The slabs are then passed through a machine with a series of rolls that reduce the steel into thin sheets at desired thickness, strength, and other physical properties. The sheets are sent through a hot-dipped galvanizing process, and are then rolled into coils that weigh approximately 13 tons.

Figure 15 ◆ Steel product label.

framing production, developed a universal designator system. The STUFL system has become the industry standard for identifying the most commonly used framing members. The designator consists of four sequential codes. The first is a three- or four-digit number indicating the member web depth (D) in $\frac{1}{100}$". The second is a single letter indicating the type of member:

S = Stud or joist framing member with **lips**
T = Track section
U = Channel or stud framing section that does not have lips
F = Furring channels
L = Angle or L-header

As an example, the designator system for an 8", 16-gauge C-shape with 1⅝" flanges is 800S162-54:

800 = 8" member depth expressed in $\frac{1}{100}$" (outside to outside dimension)
S = Stud or joist with flange stiffeners
162 = 1⅝" flange in $\frac{1}{100}$"
54 = Minimum base steel thickness in mils (0.054" = 54 mils)

Base steel thickness was traditionally referenced in terms of gauge numbers, but that has given way to designating thickness in terms of mils. *Table 1* lists common base steel thicknesses and equivalent gauge identifiers. More information about STUFL can also be found in *Appendix B* at the end of this module.

shows the basic components of a steel stud system. To meet custom material requirements, studs, track, and joist material can be cut within ⅛" of specification. Length is restricted only by the mode of physical transportation—typically 40' for containers and flat bed trucks. Custom ordering allows for less field cutting, and less labor and waste on the job site.

Steel framing members that are not marked with the required identification should not be used. The identification may be etched, stamped, or labeled every 96" along the center line of the web (*Figure 15*). Missing identification can result in project delays until materials are identified in accordance with appropriate standards.

4.1.0 Identification of Framing Materials

Manufacturers of various steel framing products have published widely varying codes for identifying steel components. To eliminate this confusion, the Steel Stud Manufacturers Association (SSMA), representing more than 80 percent of all U.S. steel

4.2.0 Basic Construction Methods for Steel Materials

There are requirements and guidelines that apply to cold-formed steel construction for all elements within a building. They include in-line framing, web stiffeners, standard web holes and patches, and stud-to-track gap tolerances. Steel frame assemblies also require certain types of bracing for stability.

Table 1 Minimum Base Steel Thickness of Cold-Formed Steel Members

Designation (thickness in mils)	Minimum Base Steel Thickness Inches (mm)[1]	Old Reference Gauge Number[2]
18	0.0179 (0.455)	25
27	0.0269 (0.683)	22
30	0.0296 (0.752)	20 – Drywall[3]
33	0.0329 (0.836)	20 – Structural[3]
43	0.0428 (1.09)	18
54	0.0538 (1.37)	16
68	0.0677 (1.72)	14
97	0.0966 (2.45)	12
118	0.1180 (3.00)	10

[1] Design thickness shall be the minimum base steel thickness divided by 0.95.

[2] Gauge thickness is an obsolete method of specifying sheet and strip thickness. Gauge numbers are only a rough approximation of steel thickness and shall not be used to order, design, or specify any sheet or strip steel product.

[3] Historically, 20 gauge material has been furnished in two different thicknesses for structural and drywall (non-structural) applications.

4.2.1 In-Line Framing

In-line framing, or direct alignment, is the preferred and most common framing method for providing a direct load path for transfer from studs to joists, through the framing system, to the ground. In this method, cold-formed steel framing members are aligned vertically so that the center line of the joist web is within ¾" of the center line of the structural stud member below, or the center line of the stud web is within ¾" of the center line of the web joist below (*Figure 16*).

4.2.2 Web Stiffeners

Web stiffeners, or bearing stiffeners (*Figure 17*) in the shape of C-shaped members or track members are used to prevent joists from crippling at the point where the load transfers from a stud into the floor joist under structural walls. The thickness of a stiffener is, at a minimum, the same as the floor joist, and the length of the stiffener is the depth of the joist minus ⅜". Stiffeners are installed across the joist depth of the web and on either side of the web. The stiffener is fastened to the web with either three or four No. 8 screws. Three fasteners are used when the screws are installed in a single row, and four when installed with one fastener in each corner.

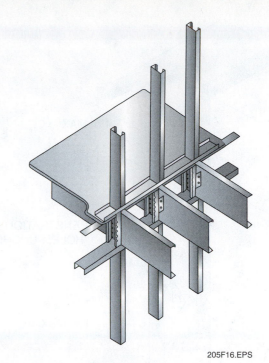
205F16.EPS

Figure 16 ◆ In-line framing detail.

205F17.EPS

Figure 17 ◆ Web stiffener.

4.2.3 Web Holes and Patches

Depending upon the application, framing members may be designed and manufactured with standard punchouts—web holes made during manufacturing (*Figure 18*). Holes in webs of studs, joists, and tracks must conform to an approved design. Standard web hole sizes are typically 1.5" by 4", and are located on the center line of the web of the member at 24". Proprietary products are also produced with much larger holes that have been engineered and approved.

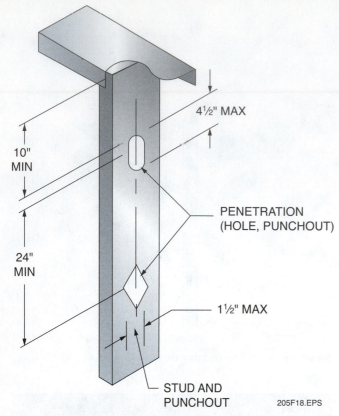

4½" MAX

10" MIN

24" MIN

PENETRATION
(HOLE, PUNCHOUT)

1½" MAX

STUD AND
PUNCHOUT

205F18.EPS

Figure 18 ◆ Standard stud punchouts.

CAUTION

Do not patch or reinforce errors in holes yourself. If a structural member is damaged, a design professional should be consulted for any corrective action.

WARNING!

Field modifications to framing members cannot be attempted without the approval of the design professional. Damaged materials should never be installed, because they may compromise structural integrity.

4.2.4 Built-Up Shapes

The open, C-shaped cross-section commonly used in cold-formed steel framing provides a very low resistance to twisting. Built-up shapes, such as a nested stud and track assembly, provide the increased stiffness of a closed section as well as flat surfaces for attaching finish materials. Built-up shapes are commonly used for door and window jambs, headers, beams, and posts (*Figure 19*). The strength of built-up shapes is determined by

the properties of the individual members, as well as the method of fastening.

4.2.5 Bridging, Bracing, and Blocking

The strength of individual framing members is a function of the bracing provided. This bracing restrains the member from moving laterally or twisting, and may be in the form of any one or a combination of three common methods:

- Cold-rolled channels placed through the web punchouts
- Steel strapping attached to the flanges with periodic solid **blocking**, X-bracing
- Sheathing attached to the flanges

Overall, system bracing anchors the steel members and provides stability to the entire structure.

4.2.6 Thermal Considerations

Products used in construction may require some additional insulation to meet energy codes. *The Thermal Design Guide for Exterior Walls*, produced by the American Iron and Steel Institute, provides designers and contractors with guidance on thermal design of buildings that use cold-formed steel framing members. The thermal performance of a steel-framed structure may also be improved by the batt and other insulating materials within the wall cavity. Some thermal regions will require insulation foam board on the exterior of the frame. Designs should consider the effects of moisture when assessing the application of cavity and continuous insulation.

4.2.7 Pressure-Treated Wood

Be careful when using pressure-treated wood products with cold-formed steel framing, as accelerated corrosion may result. It is preferable not to use pressure-treated wood with steel framing, but if you do, specify a less corrosive pressure treatment, such as sodium borate. Always separate the steel framing material from the pressure-treated wood with a nonabsorbing closed-cell sill seal.

4.2.8 Protecting Piping and Wiring

Copper piping is commonly run through the web holes of wall studs, creating the potential for direct contact with the galvanized steel and producing a galvanic reaction that will compromise the strength and performance of both the steel and copper. This contact of dissimilar metals can and should be avoided by installing non-conductive grommets, plastic bushings, or other materials designed to separate the materials.

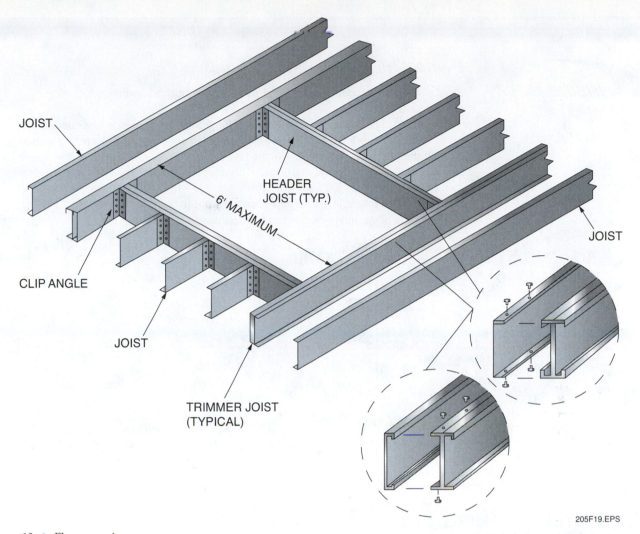

JOIST

HEADER JOIST (TYP.)

6' MAXIMUM

JOIST

CLIP ANGLE

JOIST

TRIMMER JOIST (TYPICAL)

205F19.EPS

Figure 19 ◆ Floor opening.

Plastic piping, on the other hand, does not require protection from contact with steel framing materials. Even so, consideration should be given to installing non-metallic brackets that will isolate the pipe from the hole in order to prevent noise and the potential for an incision in the pipe.

You must ensure that wiring sheathed with a non-metallic coating is separated from the sharp edges typically found in web punchouts of wall studs and joists (*Figure 20*). The *National Electrical Code*® (*NEC*®) states that non-metallic sheathed cable must be protected by bushings or grommets securely fastened in the opening prior to the installation of the cable. Cable that follows the length of the framing member also needs to be secured against movement. This is commonly accomplished by the use of tie-downs such as nylon cable ties and nylon zipper ties. These are attached to the studs at intervals as required by local building codes.

205F20.EPS

Figure 20 ◆ Proper protection of wiring.

Punch Openings in Steel Studs

Steel studs usually come from the factory with pre-punched openings for wiring, conduit, and piping.

A metal stud punch is available for punching additional openings if needed. This tool works with studs up to 33 mil. When electrical or telecommunications cabling is routed through the openings, grommets like the one shown should be inserted in the openings to protect the cabling from sharp edges. Other types of devices are made to eliminate conduit rattle.

Never punch a hole in a structural member without considering the structural requirements of the member. Consult the approved standard.

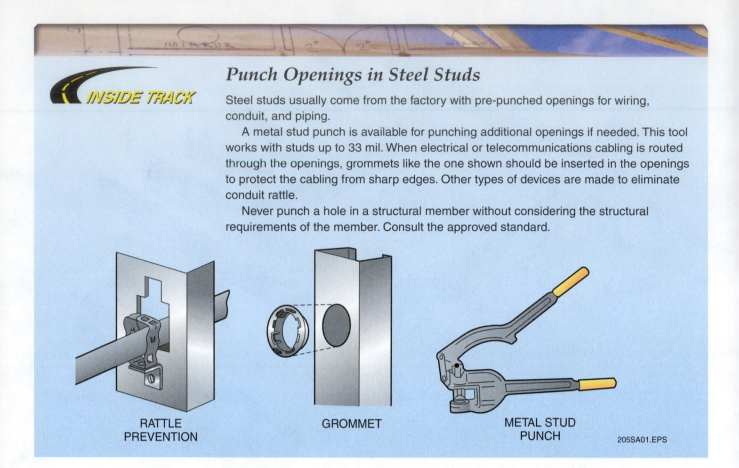

RATTLE
PREVENTION

GROMMET

METAL STUD
PUNCH

205SA01.EPS

5.0.0 ◆ STEEL FRAMING APPLICATIONS

Historically, cold-formed steel framing has been widely accepted for use in non-structural applications or partition walls. Today, it is commonly used for floor, structural wall, and roof assemblies as well. This section covers steel framing as it applies to structural walls used in residential and commercial structures (*Figure 21*).

5.1.0 Framing Walls

Structural walls support the weight of the building and protect occupants from wind loads and other forces of nature. A basic, cold-formed structural steel wall includes structural studs, track, fasteners, bracing, and bridging. Headers for window and door openings may be constructed using structural studs, joists, or proprietary cold-formed shapes. In addition, shear walls are typically framed within the stud wall assembly.

Wall framing members have a material base steel thickness of 33 to 118 mils, with a minimum metallic zinc coating of G60. Typically, structural wall studs are produced in sizes from 2½" to 8", with flanges ranging from 1⅜" to 2½" or more. The return lips depend on the flange size. Structural track is sized to accommodate the web depth of wall studs, and the flanges are typically sized 1¼" to 3". The selection of a particular member will depend on its intended use. A number of proprietary and non-proprietary accessories are also available for structural wall assembly, including cold-formed channel for bridging, as well as clips, angles, and straps.

(A) RESIDENTIAL

(B) COMMERCIAL

205F21.EPS

Figure 21 ◆ Examples of structural wall systems.

5.1.1 Layout

It is important to lay out the wall studs accurately to align with other structural assemblies such as roof and joist framing. Steel framing is typically spaced at 24" on center to take full advantage of the framing member.

Place the top and bottom track members on the straight edge of the panel table. They should be arranged with the webs next to each other, clamped together temporarily. They should fit tightly against the edge of the end stop (the straight edge at the end of the wall). Mark the layout of the wall studs on the flanges of the top and bottom tracks, starting with a wall stud at the end of the wall. Use highly visible ink, such as a black felt-tip marker. Place a line at the web location and an X on the side of the line to indicate the stud flanges.

Mark the next stud location to match the first truss or **roof rafter** location from the end wall. Continue marking every 24" (or 16", depending on the layout) for the full length of the wall. Where the exterior corner walls intersect, the wall that runs to the edge of the foundation has an extra stud. This stud is 3" from the end and acts as a backer to screw to the shorter intersecting wall.

Next, identify the rough openings in the walls. Check the architectural drawings to find door and window locations. If the dimensions are not provided, scale them off the drawing. Mark the location for the center of the openings on the top and bottom tracks. Use a red felt-tip marker to distinguish these marks from the layout marks. Check the door and window sizes on the drawings and verify rough openings with actual window sizes. Add 12" to the width of the window openings to allow for two jack studs on each side of the header with room to spare.

Using a tape measure, center the dimensions over the red marks on the track. Mark each end of the tape measure. This shows the location of the webs of the king studs. Put the X on the side of the mark away from the window. The webs of the king studs will be on the rough opening side. Remember to mark both the top and bottom tracks with 12" added to the rough window openings. This dimension is standardized to simplify **header** ordering and assembly. Two king studs may not be required at every opening, and framers may vary the length of the headers. However, standardizing header lengths helps to simplify cut lists.

All loadbearing studs must be aligned with the trusses, joists, or rafters above or below the wall. Because these walls carry loads, it is important to fit each stud tightly in the track member to allow the stud to properly carry the axial (downward) load from above. The top and bottom wall tracks must be of equal or greater thickness than the studs. Panelized walls may be constructed on a concrete slab, floor deck, or panel table. For all systems of wall construction, make sure that the surface the wall is built on is level. Walls must not be out of plumb more than ⅛" for every 10'.

Walls may be framed in two ways: Full-length walls may be framed up to 40' depending on available manpower at the job site. Be aware that longer walls also tend to twist; they could get damaged if there is not a big enough crew to keep the wall straight. Shorter sections need more plumbing and alignment, but they are easier to assemble and may be built and spliced together. The track will also need to be spliced (*Figure 22*).

Figure 22 ◆ Track splicing.

5.1.2 Wall Assembly

Before beginning assembly of a structural wall, ensure that the foundation or the bearing surface is free of all defects. The bearing surface should be uniform, with a maximum ⅛" gap between it and the track.

After layout is complete, separate the top and bottom tracks, and install a wall stud at each end of the wall between the top and bottom tracks. Clamp the stud flanges to the track flanges with locking C-clamps at each end. Tap the track on one end with a hammer to seat the studs as tightly as possible in the track. Fitting each stud tightly and perpendicular to the track keeps the wall straight. Screw one low-profile No. 8 screw through the flange of the track into the flange of the stud on either side of the stud. If an elevated panel table is used, the framer may be able to install the screw (from underneath) on the other flange as well. If not, once all the studs and headers are in place, and all screws are installed on one side, flip the wall over to install the screws on the other side.

In drywall framing, the framer often only installs screws on one side of the wall. Check local codes. However, for loadbearing walls, screws must be installed on each flange on both sides of the wall to keep the studs from twisting and to provide the proper connection for in-line framing. Continue twisting the studs into the track. Install the studs all the same way, with the open side of the C shape facing the same direction and toward the start of the layout. Align the punchouts in the studs to provide straight runs for the plumbers and electricians. The studs should be aligned so that they all face the same direction on parallel loadbearing walls.

Install the king studs at the rough openings with the hard side of the stud facing the rough opening and punchouts aligned. Do not install studs at the markings between the king studs. These markings will be for the cripple studs when the rough openings are framed. Continue down the length of the wall until all the studs are twisted and screwed in place. Do not remove the wall panel from the table until the headers and rough openings are completed.

During wall assembly, it is important to seat the studs into the track as tightly as possible. Studs must be seated with a gap of no more than ⅛". This ensures that the building loads are transferred through the studs, and not through the fastener (*Figure 23*).

ACCEPTABLE

UNACCEPTABLE

Figure 23 ◆ Stud seating.

Advantages of Metal Framing

Metal framing offers a great deal of flexibility in building design. Fire ratings are easier to achieve because, unlike wood, steel framing is noncombustible. In addition, steel framing can be produced in lengths not readily available with wood.

205SA02.EPS

Structural walls may be constructed using any of the following methods:

- *Stick building* – Walls are framed on the job site, one stud at a time. Walls can be built on a flat surface of the site, such as a concrete slab.
- *Panelization* – This reduces construction time while improving the efficiency and quality of the steel framing assembly.
- *Pre-engineered method* – This assembly typically increases the size and spacing of structural steel members. In some cases, the spacing can be as much as 4' or 6' on center.

Screws are most commonly used to attach track to cold-formed steel floor or roof assemblies, but welding may be used when specified. Powder-actuated fasteners may be required when the track connects to structural steel-framing members or concrete. Expansion bolts are used at jamb locations and corners, while expansion anchors are typically used at shear wall locations.

5.1.3 Wall Installation

If anchor bolts are used in the foundation, measure their locations and place holes in the bottom track of the wall panels so that the walls will fit over the bolts. If strap anchors or other kinds of anchors are used, this will not be necessary. Place temporary bracing material near the foundation in preparation for the wall to be raised. Any stud material at the job site, preferably 12' long, may be used. Caulk the concrete foundation with weatherproof caulking material and use foam closed-cell sill sealer.

Move the wall panel and set the bottom track on the foundation. Position it over the anchor bolt locations and tilt the wall up. Leaving the wall tilting slightly outward, clamp the temporary brace material to the wall studs in two or three locations, depending on the length of the wall. Make sure the braces do not lap past the inside face of the wall. Before removing the clamps, secure a No. 10 hex-head screw through the brace into the stud. Install a brace every 8' to 12' along the wall. Secure the bottom of the brace with a stake driven into the ground or other solid surface. Screw through the stud into the stake to hold it in place. Repeat this process with all wall panels until all loadbearing walls are standing.

WARNING!
Before tilting up a wall assembly, ensure that the braces do not hang past the wall edges. Snagging on the braces is very dangerous and could result in injury.

Some builders choose to frame walls in place, especially for one- or two-person crews. In this case, the track should be cut for the full length of the wall. Mark the top and bottom track for layout, and anchor the bottom track in place, securing the studs in each end of the track. Position the top track at the ends and with intermediate studs. Use a string line and level to position the remaining studs for the wall. Install headers, X-bracing, or plywood with the wall standing.

A balloon frame can be used for some structures (*Figure 24*). This is categorized as a pre-engineered structure. The carpenter and the engineer must work closely together to size the members and develop details for balloon framing.

5.1.4 Shear Walls

Lateral (shear) loads are typically resisted by a system of interconnected shear walls and floor and roof/ceiling **diaphragms** that work together to transfer applied wind and seismic loads to the foundation of the structure. These diaphragms are called shear walls and can be created by attaching gypsum board, wood structural sheathing, steel decking, X-bracing (*Figure 25*) or other materials

to the floor, wall, ceiling or roof framing. In addition to field-fabricated shear wall systems, there are now several proprietary, high-strength pre-engineered systems that may help shorten construction time (*Figure 26*).

The wall design for any steel-framed structure should provide for the placement of shear walls. In multi-story construction, proper alignment (stacking and load path) is also required to ensure there is adequate shear transfer between roof or floor diaphragms and shear walls, or other assemblies such as bar joists, long-span deck, and wood framing. Shear walls are usually connected to the foundation using proprietary or engineered connectors and hold-downs.

205F25.EPS

Figure 25 ◆ X-bracing.

205F24.EPS

Figure 24 ◆ Balloon framing.

205F26.EPS

Figure 26 ◆ Proprietary shear wall.

NOTE

All concealed cavities such as headers in exterior walls must be pre-insulated before they are installed.

5.1.5   Header Assembly

Loadbearing headers are typically boxed, unpunched C-shapes that are capped on the top and bottom with track sections (see *Figure 27*). They must be engineered for bending and shear strength, along with web crippling (crushing) at the locations of the loads from above. The two types of headers that are most commonly built from standard C-shapes are the box and back-to-back headers. A third type is a steel angle in the shape of an L, also called an L-header. Prefabricated headers may also be ordered from manufacturers. Selection of the header type depends on loads and applications. L-headers are being used more frequently in low-rise and multi-family construction because they use fewer fasteners and less material.

To install the header, loosen one of the king studs at the top of the rough opening, and install the header, with the open end up, by inserting it into the top track. This is usually a tight fit. Clamp the header at one end with the bar clamp. Apply pressure, working from one side of the header to the other using the clamp to tightly push the header into the top track. Make sure that the header is fitted tightly into the track before screwing it in place. Reposition the king studs that were loosened, and screw them back into place.

Back-to-back headers are formed by placing two C-shapes with the webs of the members touching each other. They are positioned in the top track of the wall and finished just like a box header.

The L-header (*Figure 28*) consists of one or two angle pieces that fit over the top track. This saves labor because there is no special fabricating, and the number of screws is reduced. The L-shape itself spans the opening for the header.

Once the header is installed in the top track, the remainder of the rough opening may be framed (*Figure 29*).

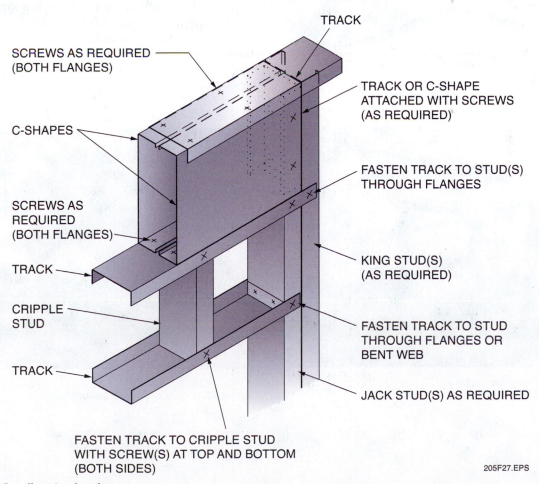

SCREWS AS REQUIRED (BOTH FLANGES)

C-SHAPES

SCREWS AS REQUIRED (BOTH FLANGES)

TRACK

CRIPPLE STUD

TRACK

FASTEN TRACK TO CRIPPLE STUD WITH SCREW(S) AT TOP AND BOTTOM (BOTH SIDES)

TRACK

TRACK OR C-SHAPE ATTACHED WITH SCREWS (AS REQUIRED)

FASTEN TRACK TO STUD(S) THROUGH FLANGES

KING STUD(S) (AS REQUIRED)

FASTEN TRACK TO STUD THROUGH FLANGES OR BENT WEB

JACK STUD(S) AS REQUIRED

205F27.EPS

Figure 27 ◆ Loadbearing header.

Mark the jack (trimmer) studs for the top of the window opening. Another mark should be made on the trimmer studs at the bottom of the window. Cut two track pieces to fit between the jack studs to make up the head and sill pieces. These pieces should be cut 2" longer so the flanges can be clipped 1" on each side.

Clip the flanges and bend the web down toward the flanges. Set the head and sill pieces in

205F28.EPS

Figure 28 ◆ L-header.

205F29.EPS

Figure 29 ◆ Completed rough opening.

the opening, keeping the hard side of the track toward the opening. Screw the tab at the flange of each piece into the jack studs with one No. 8 screw at each tab. The tabs on the webs of the header track should also be screwed into the trimmer studs with two No. 8 screws.

The cripple studs should be cut to fit between the header track and head pieces, as well as between the sill piece and bottom track. The cripples should maintain the spacing layout (16" or 24" on center) for ease in installing gypsum board and sheathing. Screw the cripple studs into place with a No. 8 screw at each track flange on both sides.

Door openings do not require a bottom sill. However, the bottom track should run continuously at the bottom of the door to hold the wall together temporarily. The track can be cut out after the wall is plumb, level, and permanently braced.

5.1.6 Jambs

Loadbearing jambs require a minimum of two studs on each side of the framed opening (one trimmer plus one king), with more if required by the building code or by engineering analysis. The studs must be fastened together to act as one member, either by capping with track and screw fastenings or by stitch welding (*Figure 30*).

205F30.EPS

Figure 30 ◆ Stitch-welded jamb.

Sill members are usually single-track sections that are the same width as the wall stud used to frame an opening. They are clipped and screwed to the jamb studs. In cases where the allowable lateral load is exceeded, sills must be constructed with multiple track sections.

5.2.0 Steel Floor Assemblies

Cold-formed steel floor assemblies typically use standard C-shape floor joists (*Figure 31*), proprietary floor joists, pre-engineered steel floor trusses, **rim track**, web stiffeners, **clip angles**, hold-down anchors, and fasteners. They can be installed on crawl spaces and stem walls, as well as directly to interior structural walls. They are similar to conventional framing and use single or multiple span installation techniques.

5.3.0 Steel Roof Assemblies

Recent years have seen a dramatic rise in the use of cold-formed steel framing members in roof assemblies. They allow for easy and standardized assembly and are durable and non-combustible. In addition to the standard C-shape member, scores of proprietary shapes, fabrication methods, and installation requirements are available from truss manufacturers nationwide. *Figure 32* shows an example of custom-designed steel trusses.

6.0.0 ◆ BRACING STEEL WALLS

Bracing prevents a cold-formed steel-framing member from twisting and buckling. In wall construction, there are several common methods for bracing. They all have different applications and purposes. *Figure 33* through *35* are examples of typical types of bracing.

Intermediate stud bracing – Intermediate stud bracing is used when gypsum board or structural sheathing is not applied to both sides of loadbearing walls, such as garage walls (*Figure 36*).

Shear wall bracing – There are two ways of applying shear wall bracing: structural sheathing

205F31.EPS

Figure 31 ◆ Standard floor joists.

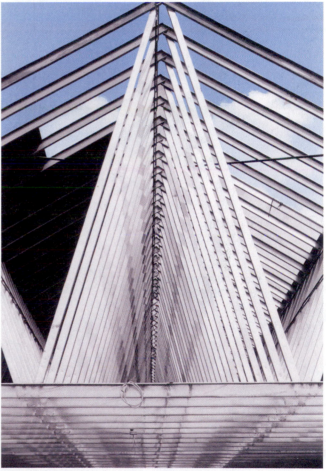

205F32.EPS

Figure 32 ◆ Example of complex framing using steel trusses.

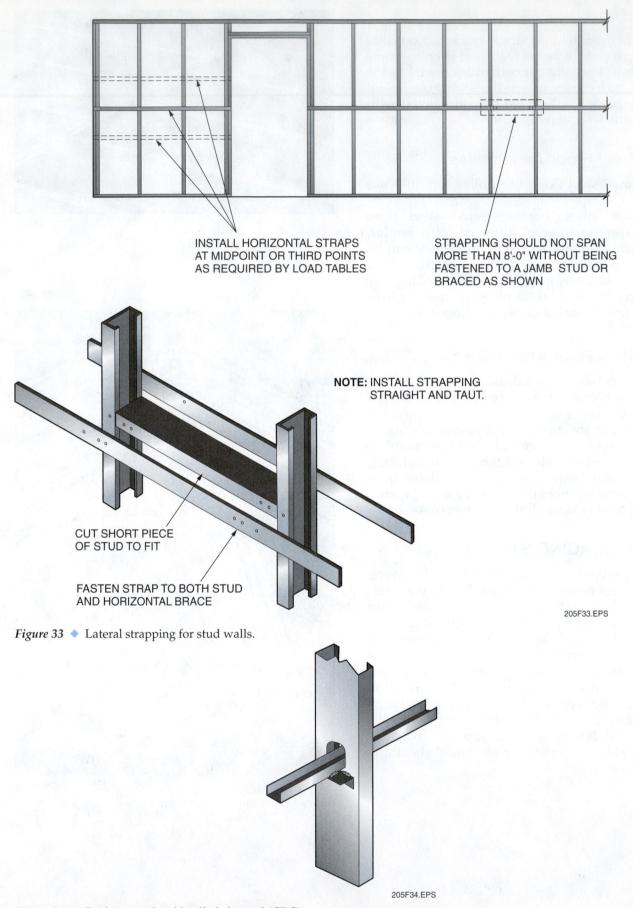

INSTALL HORIZONTAL STRAPS
AT MIDPOINT OR THIRD POINTS
AS REQUIRED BY LOAD TABLES

STRAPPING SHOULD NOT SPAN
MORE THAN 8'-0" WITHOUT BEING
FASTENED TO A JAMB STUD OR
BRACED AS SHOWN

NOTE: INSTALL STRAPPING
STRAIGHT AND TAUT.

CUT SHORT PIECE
OF STUD TO FIT

FASTEN STRAP TO BOTH STUD
AND HORIZONTAL BRACE

205F33.EPS

Figure 33 ◆ Lateral strapping for stud walls.

205F34.EPS

Figure 34 ◆ Bridging with cold-rolled channel (CRC).

NOTES:

• Install strapping as close to 45° as possible.

• Place straps on both sides of stud wall in order to prevent eccentric loading.

• Check for increased axial load applied to wall studs due to tension in straps. Double studs will typically be required at strap ends.

• Install wall straps straight and taut.

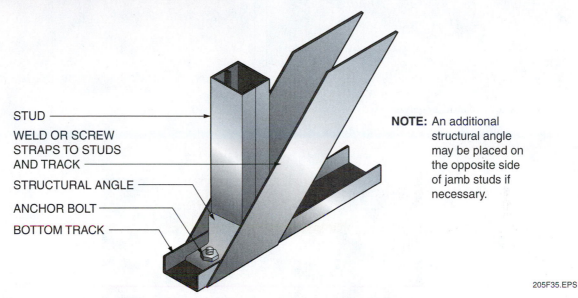

STUD

WELD OR SCREW
STRAPS TO STUDS
AND TRACK

STRUCTURAL ANGLE

ANCHOR BOLT

BOTTOM TRACK

NOTE: An additional structural angle may be placed on the opposite side of jamb studs if necessary.

205F35.EPS

Figure 35 ◆ Diagonal strapping for shear walls.

and X-bracing (*Figures 37* and *38*). Structural sheathing Type II plywood or OSB may be adequate to keep the wall from **racking** depending on the design, as long as there are not excessive openings in the wall. In order for structural sheathing to be effective, it should be installed with the long dimension parallel to the stud framing (vertical orientation). The plywood may be secured to the wall while panelizing, or after the wall is plumb and level. X-bracing is another way to obtain shear strength when structural sheathing is not used. X-braces are diagonal steel straps attached to the walls with screws or welded connections. This bracing must be designed by an engineer, and the straps must be inspected for the correct number of fasteners. Do not tighten the straps until the walls are plumbed and aligned.

Temporary bracing – There are two types of basic temporary bracing: one for panelized walls and one for installed walls. After a straight wall is constructed on a panel table, installing plywood or temporary bracing prevents racking when the wall is removed from the table. Before the wall is taken off the table, check for squareness by diagonally measuring the panel. Adjust if necessary. Lay extra studs or truss material across the wall diagonally, and screw the bracing to the wall studs, especially at door openings where the bottom track is weak. Leave the bracing on the wall until the wall is installed and permanently braced. These precautions will help provide straight walls ready for installation.

Installation of steel framed walls requires adequate temporary bracing in order to resist loads during construction until permanent bracing can be installed (*Figure 39*).

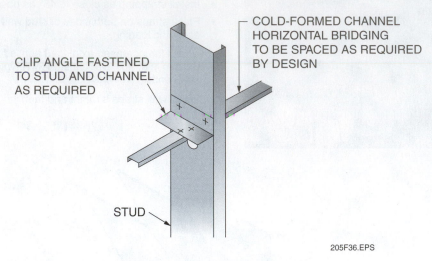

COLD-FORMED CHANNEL
HORIZONTAL BRIDGING
TO BE SPACED AS REQUIRED
BY DESIGN

CLIP ANGLE FASTENED
TO STUD AND CHANNEL
AS REQUIRED

STUD

205F36.EPS

Figure 36 ◆ Example of stud bracing.

205F37.EPS

Figure 37 ◆ Wood structural sheathing.

Figure 38 ◆ Gusset plate X-bracing.

205F38.EPS

Figure 39 ◆ Temporary bracing.

205F39.EPS

7.0.0 ◆ NON-STRUCTURAL (NONBEARING) WALL FRAMING

Framing members used in interior systems may be nonbearing, or may be designed as part of the structural system. This section primarily discusses nonbearing walls that function as space partitions within the exterior walls of a building.

Similar to structural wall systems, non-structural walls are comprised of studs, track, and accessories. The primary differences are the characteristics of the materials and the application of connectors and accessories.

Non-structural framing members typically have a base steel thickness of 18, 27, or 30 mils, compared with a minimum thickness of 33 mils for structural studs. In addition, the minimum stud flange dimension for non-structural framing members is 1.25", and the minimum return (stiffening) lip dimension is ⅛", compared with 1⅝" flange and ½" lip for structural studs. Also, non-structural members typically have a G40 galvanized coating weight, compared with G60 or higher for structural studs. The rules for non-structural framing are different from those for structural walls, and may be found in the gypsum specification, rather than the building codes.

Since non-structural studs are not intended to carry any loads, accessories used in this application are intended for architectural features within the building, including curves and arches. No connection is required between the stud and track for nonbearing walls, except for studs that are adjacent to window and door openings.

7.1.0 Steel Curtain Walls

Since their introduction in the early 1900s, metal and glass **curtain wall** systems have become very popular in the architectural design of modern structures. Unlike interior partitions, wind-bearing curtain walls resist loads from exterior wind pressures that, in some cases, exceed 60 pounds per square foot. Cold-formed steel curtain walls are made up of various components:

- *Angles* – Clip angles and continuous angles are used to connect framing members within the curtain wall system.
- *Clip angle* – A steel angle, generally 3" to 12" long, which makes the transition between a framing member and the component supporting it. These angles are used to connect two framing members.
- *Continuous angle* – A steel angle that makes the transition between a stud curtain wall and the primary frame. The angle is typically of hot-rolled thickness (³⁄₁₆" to ⅜"); however, thinner gauged materials can be used if the span and load requirements are relatively small.

- *Diagonal brace (or kicker)* – A sloping brace used to provide lateral support to a curtain wall assembly. When installed horizontally, this brace is referred to as a strut.
- *Embed* – A hot rolled steel plate or angle, reinforced with shear studs or steel rebar, which is cast into a concrete floor or beam. Embeds allow for the welded attachment of steel supports.
- *Girts* – Horizontal structural members that support wall panels and are primarily subject to bending under horizontal loads, such as wind load.
- *Slide clip* – A connection device that permits deflection of the primary frame to which a stud attaches, while it braces the stud against lateral forces.
- *Slip track* – A track section used in for-fill curtain wall applications (*Figure 40*). Slip tracks accommodate vertical movements of a primary frame (normally ¼" to ¾"), while bracing the wall against lateral forces. Slip tracks may also be specified at the top of interior drywall partitions.

The term curtain wall is used to distinguish this system from loadbearing framing. As described earlier in this section, loadbearing framing requires that the wall members carry the weight of the structure above. With curtain-wall framing, the structure is usually already in place, and the wall framing is filled in between the floor slabs. The stud-to-track gap distance for curtain wall is no more than ¼", unless it is otherwise specified in an approved design. This is different from load-bearing construction, which permits only a ⅛" gap. The only gravity loads that curtain wall framing typically carries are the weight of any cladding or finish materials attached to it.

CAUTION

For curtain wall installation, the use of components formed from steel measuring less than 33 mils should be avoided.

7.1.1 *Construction Methods for Steel-Framed Curtain Walls*

There are four main methods for building assemblies in curtain wall construction:

- *Infill* – This method describes applications where studs are only one story tall, spanning from floor to floor of a structure. Infill framing requires less stud material. Connections are often easier to make, since low-cost powder-actuated pins can be used in many applications. In addition, the spans are often shorter than bypass conditions, so thinner steel or wider spacing can be used. The disadvantage of infill framing is that more track material is needed, and wall sections can be difficult to panelize.
- *Panelization* – This method can be used if field measurements are made after the floor systems are in place, or if a slip connector or telescoping stud system is used. When structural steel is used for main framing, the spandrel beam can get in the way of framing. In that case, framing must be attached to the bottom of the beam. If framing takes place outside the spandrel beam, it can cause insulation support difficulty.
- *Bypass framing (balloon framing)* – This method allows a single stud to be used for framing two or more floors, and it requires less track material. The multiple spans can reduce moment stresses in members, allowing for more widely spaced or thinner framing members. In addition, this method makes it easier to pre-panelize large sections, including multi-story panels. On the other hand, some connections can be more difficult to make, such as bracing and support connections behind columns and spandrel beams. Depending on the condition, double connections may need to be made. This requires clips or slip connectors attached first to the structure, and then again to the stud. This last

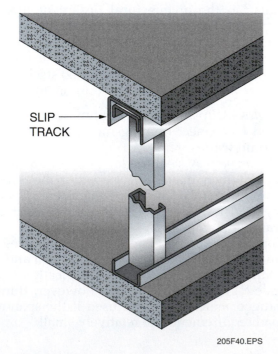

SLIP TRACK

205F40.EPS

Figure 40 ◆ Slip track.

issue can be mitigated somewhat by the use of connectors that friction-fit inside the stud, thus reducing connection time. Finally, if slab or other structural elements extend into the stud cavity, they may need to be chipped away, or stud framing may need to be altered to correctly install the bypass framing.

- *Stacked wall framing* – This method permits bypass framing, while isolating slip connections to one- or two-story segments. As a result, multi-story panels and pre-finished panels can be fabricated and installed, including insulation. However, in order to prevent water infiltration, this approach requires special detailing at slip connections and exterior finishes. Because the entire panel weight goes to fixed connections, usually at every other floor, these fixed connections need to be strengthened to carry the added dead load of the taller panels. Also, some connections are more difficult to reach, such as bracing and support connections behind columns and spandrel beams.

7.1.2 Bracing for Curtain Walls

The most typical cold-formed steel curtain wall shape is the C-stud. Due to its shape and geometric properties, a C-stud tends to rotate under lateral load. Non-braced studs may also move out of plane. This is known as torsional-flexural buckling. Mechanical bridging and/or the sheathing materials can restrain the flanges and prevent this. When using discreet bracing rather than sheathing, decreasing the bridging spacing typically increases the member capacities; increasing the spacing will decrease capacities.

In a typical stud-framed exterior curtain wall, member deflections are the primary serviceability issue. Deflection limits are most often determined by the architectural finishes, and are set forth in the project specifications.

7.1.3 Finishing for Curtain Walls

A wide range of finish systems may be applied to steel stud curtain-wall frames, including the following:

- Brick veneer
- Split-faced block veneer
- Tile or thin-cast brick
- Exterior insulation finish systems (EIFS)
- Glass fiber reinforced concrete (GFRC)
- Metal panel
- Modified portland cement (stucco)
- Fiber-cement board or siding
- Dimensional stone, such as granite or limestone
- Wood siding and many other finish systems

7.2.0 Radius (Curved) Walls

Some partitions that might be difficult to construct with wood framing members can easily be built with steel framing members. A radius wall is an example. Note that the wall forms a quarter circle. Constructing this radius wall with wood framing members would take many hours. Using steel framing members, such a wall could be built in less than an hour. A plywood template may be required for more complex radius walls.

The bend can be achieved using several methods. Curved walls can be framed out of steel using curved track for partitions or exterior loadbearing walls. The track can be bent at the job site by slitting the flanges. Ordering curved track from specialty companies also speeds the construction of a radius wall.

Track may be ordered curved to a specified radius. Some specialty companies use machinery to bend the track without slitting the flanges. Others produce a flexible track that can be ordered and formed on the job site based on the desired effect. This provides a clean, neatly bent track to an exact radius. Wall track is bent around the flanges (*Figure 41*).

7.3.0 Other Non-Structural Wall Assemblies

- *Chase walls* – It is common in commercial and residential construction to build chase walls to conceal plumbing and other utilities. For oversized utilities or acoustical requirements, two separate walls with a void in between may be framed in advance. Cross-bracing between the two walls may be required.

205F41.EPS

Figure 41 ◆ Example of a radius track.

- *Fire-rated assemblies* – Building codes require that certain partitions provide fire-rated separation of interior spaces, with a specific rating as to how long the partition or assembly will last under the design load before being penetrated by a fire. Approved assemblies must be constructed precisely as described in the directories published by the rating agencies.
- *Area separation walls* – Although this term sounds redundant, it refers to a specific group of rated walls designated by code requirements, which are designed to permit structural failure on one side of the wall while still providing fire protection.
- *Head-of-wall conditions* – Some building codes require that wall assemblies accommodate horizontal and vertical movement while still providing a fire-resistive barrier. Proprietary and non-proprietary devices and assemblies have been designed and tested to meet code requirements.
- *Shaftwall systems* – These systems are special types of rated wall systems that may be constructed from one side only. Higher sound and fire ratings may be achieved by attaching additional layers of gypsum board to the outer or inner face. A 1"-core board is specifically made for this purpose. Horizontal shaftwall may also be used for soffits or **plenums** where ratings are required.

7.4.0 Furring

Design requirements may specify the standoff of gypsum board materials from the stud using hat-shaped steel furring members. This provides sound isolation. In this case, the gypsum board is usually furred out using resilient channel (*Figure 42*). Resilient furring channels are used over both metal and wood framing to provide a sound-absorbent spring mounting for gypsum board and should be attached as specified by the manufacturer. They not only improve sound insulation, but they help isolate the gypsum board from structural movement, minimizing the possibility of cracking. Resilient furring channels may also be used for application of gypsum boards over masonry and concrete walls. In wood frame construction, gypsum board can be screwed to resilient metal furring channels to provide a higher degree of sound control. Furring has additional applications:

- Provide additional space for insulation.
- Allow out-of-plane walls or walls of different thickness to match and have a smooth surface.
- Provide additional space to conceal fixtures or structural elements within a wall.

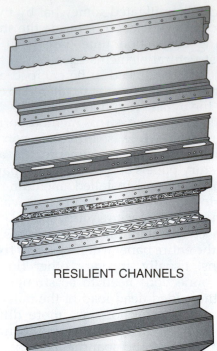

RESILIENT CHANNELS

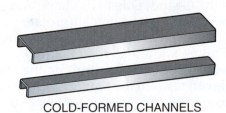

FURRING (HAT) CHANNEL

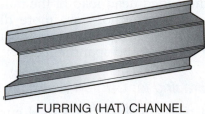

COLD-FORMED CHANNELS

205F42.EPS

Figure 42 ◆ Types of furring channels.

7.4.1 Ceiling Systems

Ceiling systems typically consist of a suspended drywall grid with gypsum board. An alternative method is to use furring and cold-rolled channel to provide a rigid framework for suspended gypsum board and other ceiling assemblies, with the steel members suspended (*Figure 43*). The furring channel is commonly clipped or wire-tied, perpendicular to the underside of the U-channel at appropriate intervals for attaching gypsum board with screws.

This furring channel can also be screw-attached to other structural steel members, and can be attached directly to the bottoms of bar joists. In the latter case, the furring is installed perpendicular to the joists and wire-tied at appropriate intervals. Note that wire ties are required for fire-rated and multi-layer assemblies.

8.0.0 ◆ SLIP CONNECTIONS

Slip connectors (*Figure 44*) are devices that allow for the vertical movement of a structure without imposing additional loads on cold-formed steel framing or other wall components. They are used where a structural system other than steel framing is needed to carry the loads of upper floors and the roof down to the foundation. Loads on the upper portions of a structure cause it to deflect, which in turn may induce loading in walls and wall components that are not designed to carry these loads. Slip connectors are designed to allow for this movement without creating unmanageable stress on components.

Slip connectors are generally located at the top of a wall panel, where it comes to the underside of a structural element, such as a floor slab or beam. Under gravity, seismic, or wind loads, this upper portion of the structure may deflect up or down. The connector is designed to allow this movement, restrain the wall system from out of plane movement, and prevent any additional axial loading on the stud.

Slip connections are also useful at locations where a wall system is continuous, bypassing intermediate floors (*Figure 45*). Where this occurs, a slip connection extends from the side of the structure and supports the wall components laterally. Connections are also installed at roof bypasses, where the wall system extends past a roof structure to form a parapet or high wall. At this location, the slip connection must permit movement of the structure either up due to wind uplift, or down due to gravity loads.

In multi-story construction, movement may occur at the floor below a curtain wall system, causing the entire wall to move down. In this case, the connector at the top of the wall (either a bypass or below structure) must have sufficient capacity to allow the wall system to move down without creating tension on wall components.

205F43.EPS

Figure 43 ◆ Ceiling framework.

205F44.EPS

Figure 44 ◆ Slip connectors.

205F45.EPS

Figure 45 ◆ Slip connector application.

8.1.0 Other Typical Connections

Other than slip connections, the most typical connections used in metal stud curtain wall design include the following:

- *Base connections* – Stud to track, track to concrete or steel deck.
- *Head or sill to jamb connections*
- *Continuous angle connections* – Field-welded connection usually found at spandrel framing.
- *Clip angle connections* – Typically at non-slipped bypass or spandrel framing connections and designed to carry both lateral and self-weight forces.
- *Outrigger clips* – Short lengths of angles designed as an axially-loaded strut to carry stud lateral reactions back to the structure.

- *Wind girt connections* – Usually at taller spandrel framing cases when the use of diagonal braces is impractical. These connections typically are not required to carry any gravity loads.
- *Diagonal braces or kickers (as opposed to X-bracing for shear walls)* – Stud diagonals providing for a bottom spandrel stud reaction back to the structure.
- *Stud-to-stud connections* – Either lapped or track-to-track. On occasion, these connections require movement allowance, which can be accommodated by a slip track or slip-pin type detail.
- *Knee wall base* – This is a moment connection at the bottom of a knee or stub-wall, usually either a freestanding parapet or a long segment of wall under a continuous or ribbon window condition.

Summary

Cold-formed steel framing is much more widespread in the construction industry than it was just a few years ago. Steel will not warp, swell, or split. It is immune to termites and will not add fuel to a fire. It is a perfect fit for non-combustible construction and environmentally conscious building, and is more cost-effective than lumber. In a wide variety of residential and commercial buildings, cold-formed steel is being used not just for partition walls, but as the main structural component. While the layout of cold-formed steel framing members may be very similar to conventional wood framing, there are important differences in components and techniques to remember. The studs, fasteners, and various accessories are designed to best take advantage of the light weight and design flexibility of steel.

Notes

1. One mil is equivalent to _____.
   a. $\frac{1}{10}$"
   b. $\frac{1}{100}$"
   c. $\frac{1}{1,000}$"
   d. $\frac{1}{10,000}$"

2. The maximum speed of screwguns for steel-to-steel screw connections is _____ rpm.
   a. 1,500
   b. 2,000
   c. 2,500
   d. 3,000

205RQ01.EPS

3. The tool shown above is a(n) _____.
   a. channel stud shear
   b. locking C-clamp
   c. aviation snip
   d. stud crimper

4. A steel framing component that has a web, two flanges, and two lips is called a _____.
   a. cripple stud
   b. C-shape
   c. web stiffener
   d. header

5. A cutting tool that can cut steel components of large thickness and leave no abrasive edges is a _____.
   a. chop saw
   b. snip
   c. C-clamp
   d. swivel-head shear

6. When selecting a screw to attach steel framing parts, the screw length should be the thickness of all the material plus _____.
   a. $\frac{1}{8}$"
   b. $\frac{1}{4}$"
   c. $\frac{3}{8}$"
   d. $\frac{1}{2}$"

7. When using bolts to fasten steel framing to concrete, you must install _____.
   a. a washer
   b. a hex washer head screw
   c. caulking
   d. pneumatic pins

8. The organization that created the standard designation system for steel components is the _____.
   a. American Iron and Steel Institute
   b. OSHA
   c. United Steel Alliance
   d. Steel Stud Manufacturers Association

9. In the STUFL system, the letter L represents a(n) _____.
   a. furring channel
   b. track
   c. stud with lips
   d. angle

10. The part of a C-shape framing member that extends from the flange as a stiffening element is called a _____.
    a. lateral
    b. lip
    c. web
    d. knurl

11. The maximum center line to center line tolerance for in-line framing of cold-formed steel is _____.
    a. $\frac{1}{8}$"
    b. $\frac{1}{4}$"
    c. $\frac{1}{2}$"
    d. $\frac{3}{4}$"

12. Structural stud sizes are typically _____.
    a. 1" to 6"
    b. 1⅜" to 6"
    c. 2" to 5"
    d. 2½" to 8"

13. A wall that is designed to resist lateral forces such as those caused by earthquakes or wind is called a _____ wall.
    a. curtain
    b. panel
    c. shear
    d. radius

14. Diagonal steel straps attached to the walls are used to _____.
    a. brace shear walls
    b. temporarily brace panelized walls
    c. provide structural sheathing
    d. support headers

15. All of the following are typical thicknesses for non-structural framing members *except* _____ mils.
    a. 18
    b. 27
    c. 30
    d. 33

Trade Terms
Introduced in This Module

Blocking: C-shaped track, break shape, or flat strap material attached to structural members, flat strap, or sheathing panels to transfer shear forces.

C-shape: A cold-formed steel shape used for structural and non-structural framing members consisting of a web, two flanges, and two lips (edge stiffeners).

Clip angle: An L-shaped piece of steel (normally with a 90-degree bend), typically used for connections.

Cold-formed steel: Sheet steel or strip steel that is manufactured by press braking blanks sheared from sheets or cut lengths of coils or plates, or by continuous roll forming of cold- or hot-rolled coils of sheet steel.

Curtain wall: A light, nonbearing exterior wall attached to the concrete or steel structure of the building.

Diaphragm: A floor, ceiling, or roof assembly designed to resist in-plane forces such as wind or seismic loads.

Header: A horizontal structural framing member used over floor, roof, or wall openings to transfer loads around the opening to supporting structural framing members.

Knurled: A series of small ridges used to provide a better gripping surface on metal and plastic.

Lateral: Running side to side; horizontal.

Lips: That part of a C-shape framing member that extends from the flange as a stiffening element that extends perpendicular to the flange. Also called edge stiffener.

Mil: A unit of measurement equal to $\frac{1}{1,000}$".

Panelization: The process of assembling steel-framed walls, joists, or trusses before they are installed in a structure. Roll-formers normally cut studs to within $\frac{1}{8}$" tolerance. This helps the framer to consistently create straight walls in the panel table that are easy to install in the field.

Plenum: A confined space, such as the space between a suspended ceiling and an overhead deck, which is used as a return for heating, cooling, and ventilation systems.

Racking: Being forced out of plumb by wind or seismic forces.

Rim track: A horizontal structural member that is connected to the end of a floor joist.

Roof rafter: A horizontal or sloped, structural framing member that supports roof loads.

Shear wall: A wall designed to resist lateral forces such as those caused by earthquakes or wind.

Track: A framing member consisting of only a web and two flanges. Track web depth measurements are taken to the inside of the flanges.

Web: That portion of a framing member that connects the flanges.

Common Terms Used in Cold-Formed Steel Framing Work

AISC: American Institute of Steel Construction.

AISI: American Iron and Steel Institute.

Applicable building code: Building code under which the building is designed.

Approved: Approved by a building official or design professional.

Base steel thickness: The thickness of bare steel exclusive of all coatings.

Bracing: Structural elements that are installed to provide restraint or support (or both) to other framing members so that the complete assembly forms a stable structure.

Ceiling joist: A horizontal structural framing member that supports ceiling components and may be subject to attic loads.

Cold-formed sheet steel: Sheet steel or strip steel that is manufactured by press braking blanks sheared from sheets or cut lengths of coils or plates, or by continuous roll forming of cold- or hot-rolled coils of sheet steel. Both forming operations are performed at ambient room temperature; that is, without any addition of heat such as would be required for hot forming.

Cold-formed steel: See *cold-formed sheet steel*.

Cold-formed steel structural framing: The elements of the structural frame, as given in the *Code of Standard Practice, Section B1*.

Component assembly: A fabricated assemblage of cold-formed steel structural members that is manufactured by the component manufacturer, which may also include structural steel framing, sheathing, insulation, or other products.

Component design drawing: The written, graphic, and pictorial definition of an individual component assembly, which includes engineering design data.

Component designer: The individual or organization responsible for the engineering design of component assemblies.

Component manufacturer: The individual or organization responsible for the manufacturing of component assemblies for the project.

Component placement diagram: The illustration supplied by the component manufacturer identifying the location assumed for each of the component assemblies, which references each individually designated component design drawing.

Construction manager: The individual or organization designated by the owner to issue contracts for the construction of the project and to purchase products.

Contract documents: The documents including, but not limited to, plans and specifications, which define the responsibilities of the parties involved in bidding, purchasing, designing, supplying, and installing cold-formed steel framing.

Contractor: The individual or organization that is contracted to assume full responsibility for the construction of the structure.

Cripple stud: A stud that is placed between a header and a window or door head track, a header and wall top track, or a window sill and a bottom track to provide a backing to attach finishing and sheathing material.

Design professional: An individual who is registered or licensed to practice his or her respective design profession as defined by the statutory requirements of the state in which the project is to be constructed.

Design thickness: The steel thickness used in design that is equal to the minimum base steel thickness divided by 0.95.

Drawings: See *plans* and *installation drawings*.

Edge stiffener: That part of a C-shape framing member that extends perpendicular from the flange as a stiffening element.

Erection drawings: See *installation drawings*.

Erector: See *installer*.

Flange: That portion of the C-shape framing member or track that is perpendicular to the web.

Floor joist: A horizontal structural framing member that supports floor loads and superimposed vertical loads.

Framing contractor: See *installer*.

Framing material: Steel products, including but not limited to structural members and prefabricated structural assemblies, ordered expressly for the requirements of the project.

General contractor: See *installer*.

Harsh environments: Coastal areas where additional corrosion protection may be necessary.

In-line framing: Framing method where all vertical and horizontal load-carrying members are aligned.

Installation drawings: Field installation drawings that show the location and installation of the cold-formed steel structural framing.

Installer: Party responsible for the installation of cold-formed steel products.

Jack stud: A stud that does not span the full height of the wall and provides bearing for headers. Also called a *trimmer stud*.

King stud: A stud, adjacent to a jack stud, that spans the full height of the wall and supports vertical and lateral loads.

Lip: See *edge stiffener*.

Material supplier: An individual or entity responsible for furnishing framing materials for the project.

Non-structural member: A member in a steel framed assembly that is limited to a transverse load of not more than 10 lb/ft² (480 Pa), a superimposed axial load, exclusive of sheathing materials, of not more than 100 lb/ft (1460 N/m), or a superimposed axial load of not more than 200 lb (890 N).

Owner: The individual or entity organizing and financing the design and construction of the project.

Plans: Drawings prepared by the design professional for the owner of the project. These drawings include but are not limited to floor plans, framing plans, elevations, sections, details, and schedules as necessary to define the desired construction.

Punchout: A hole made during the manufacturing process in the web of a steel framing member.

Shop drawings: Drawings for the production of individual component assemblies for the project.

Span: The clear horizontal distance between bearing supports.

Specifications: Written instructions, which, with the plans, define the materials, standards, design of the products, and workmanship expected on a construction project.

Standard cold-formed steel structural shapes: Cold-formed steel structural members that meet the requirements of the SSMA Product Technical Information.

Strap: Flat or coil sheet steel material typically used for bracing and blocking that transfers loads by tension and/or shear.

Structural engineer-of-record: The design professional who is responsible for sealing the contract documents, which indicates that he or she has performed or supervised the analysis, design, and document preparation for the structure and has knowledge of the requirements for the load-carrying structural system.

Structural member: A floor joist, rim track, structural stud, wall track in a structural wall, ceiling joist, roof rafter, header, or other member that is designed or intended to carry loads.

Structural stud: A stud in an exterior wall or an interior stud that supports superimposed vertical loads and may transfer lateral loads, including full-height wall studs, king studs, jack studs, and cripple studs.

Stud: A vertical framing member in a wall system or assembly.

Trimmer: See *jack stud*.

Truss: A coplanar system of structural members joined together at their ends, usually to construct a series of triangles that form a stable beam-like framework.

Yield strength: A characteristic of the basic strength of the steel material defined as the highest unit stress that the material can endure before permanent deformation occurs as measured by a tensile test in accordance with *ASTM A 370*.

Interpreting STUFL

The Right STUFL:

Universal Designator System for Cold-Formed Steel Framing Members

The Right STUFL will identify any common cold-formed steel framing members using:

Web Depth (D), expressed in 1/100th inches,
Flange Width (B), expressed in 1/100th inches,
Minimum Base Steel Thickness, expressed in mils (1/1000th inches), and the following designators,

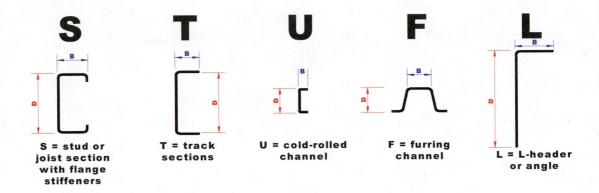

S = stud or joist section with flange stiffeners

T = track sections

U = cold-rolled channel

F = furring channel

L = L-header or angle

Examples

Designation for a 5-1/2" – 16 gauge C-shape with 1-5/8" flanges: 550S162-54

550   S   162   —54

Minimum base steel thickness in mils (.054 in = 54 mils)
1-5/8" flange in 1/100th inches
Stud or joist with flange stiffeners
5-1/2" member depth in 1/100th inches (outside to outside dimension)

Designation for a 3-1/2" – 20 gauge Track with 1-1/4" flanges: 350T125-33

350   T   125   —33

Minimum base steel thickness in mils (.033 in = 33 mils)
1-1/4" flange in 1/100th inches
Track section
3-1/2" member depth in 1/100th inches (inside to inside dimension)

205A01.EPS

The Right STUFL:

Universal Designator System for Cold-Formed Steel Framing Members

The Logic

The dimensions of the webs and flanges in the designators are used because they convey an intuitive picture of the shape, like a "2 x 4" describes a wood stud. The web depth was put first, instead of the flange width (the "2" in "2 x 4") because with steel the strength-to-weight ratio is usually optimized by increasing web depth instead of flange width, all else being equal.

The web depth and flange width are expressed in $1/100^{th}$ or $1/1000^{th}$ inches because fractions are messy, and with such a variety of depths and widths available, rounding to the nearest fractions of an inch is not sufficiently descriptive. Also, the metric system is expected to become universal, in which case the depths and widths would be expressed in millimeters, resulting in the same 3-digit format. Lastly, the logic behind the alpha designators (S for Stud, T for Track, etc.) should be obvious.

Minimum Base Steel Thicknesses

The minimum delivered base steel thickness designators are those used in Prescriptive Standards that are currently in the International Building Code (IBC) and International Residential Code (IRC), and correspond to ICBO's minimum prescribed thicknesses (expressed in mils). These minimum thicknesses are:

Gauge	Inches	Mil
25 gauge	.0179"	18 mil
22 gauge	.0269"	27 mil
20 gauge (Drywall)	.0296"	30 mil
20 gauge (Structural)	.0329"	33 mil
18 gauge	.0428"	43 mil
16 gauge	.0538"	54 mil
14 gauge	.0677"	68 mil
12 gauge	.0966"	97 mil
10 gauge	.1180"	118 mil

Stiffening Lips on C-Sections

The dimension of the stiffening lip by flange width and material thickness are as follows:

Material Thickness	Flange Width	Stiffening Lip
.0179" to .0296"	1-1/4"	3/16"
All thicknesses	1-3/8"	3/8"
All thicknesses	1-5/8"	½"
All thicknesses	2"	5/8"
All thicknesses	2-1/2"	5/8"

The Purpose

With the universal designator system for steel framing members and accepted minimum base steel thicknesses and stiffening lip dimensions, the section properties and load-carrying abilities of any given profile can be calculated and implemented. Regardless of the manufacturer, not only will the designation be identical for a standard steel framing member, but the section properties and load-carrying abilities of that standard member will be uniform throughout the Country. In addition to helping eliminate the confusion in the market stemming from the widely varying properties and loads published by manufacturers making essentially identical shapes, these standards greatly facilitate submittals for plan check, code approvals, prescriptive standards, software development, etc. The intent is to make the products easier to use in existing markets and to accelerate their acceptance in new markets.

205A02.EPS

This module is intended to present thorough resources for task training. The following reference is suggested for further study. This is optional material for continued education rather than for task training.

www.steelframing.org.

NCCER CURRICULA — USER UPDATE

NCCER makes every effort to keep its textbooks up-to-date and free of technical errors. We appreciate your help in this process. If you find an error, a typographical mistake, or an inaccuracy in NCCER's curricula, please fill out this form (or a photocopy), or complete the online form at **www.nccer.org/olf**. Be sure to include the exact module ID number, page number, a detailed description, and your recommended correction. Your input will be brought to the attention of the Authoring Team. Thank you for your assistance.

Instructors – If you have an idea for improving this textbook, or have found that additional materials were necessary to teach this module effectively, please let us know so that we may present your suggestions to the Authoring Team.

NCCER Product Development and Revision
13614 Progress Blvd., Alachua, FL 32615

Email: curriculum@nccer.org
Online: www.nccer.org/olf

❏ Trainee Guide   ❏ AIG   ❏ Exam   ❏ PowerPoints   Other _____

Craft / Level: _____   Copyright Date: _____

Module ID Number / Title: _____

Section Number(s): _____

Description: _____

_____

_____

_____

_____

Recommended Correction: _____

_____

_____

_____

_____

Your Name: _____

Address: _____

_____

Email: _____   Phone: _____

Drywall Installation
27206-07

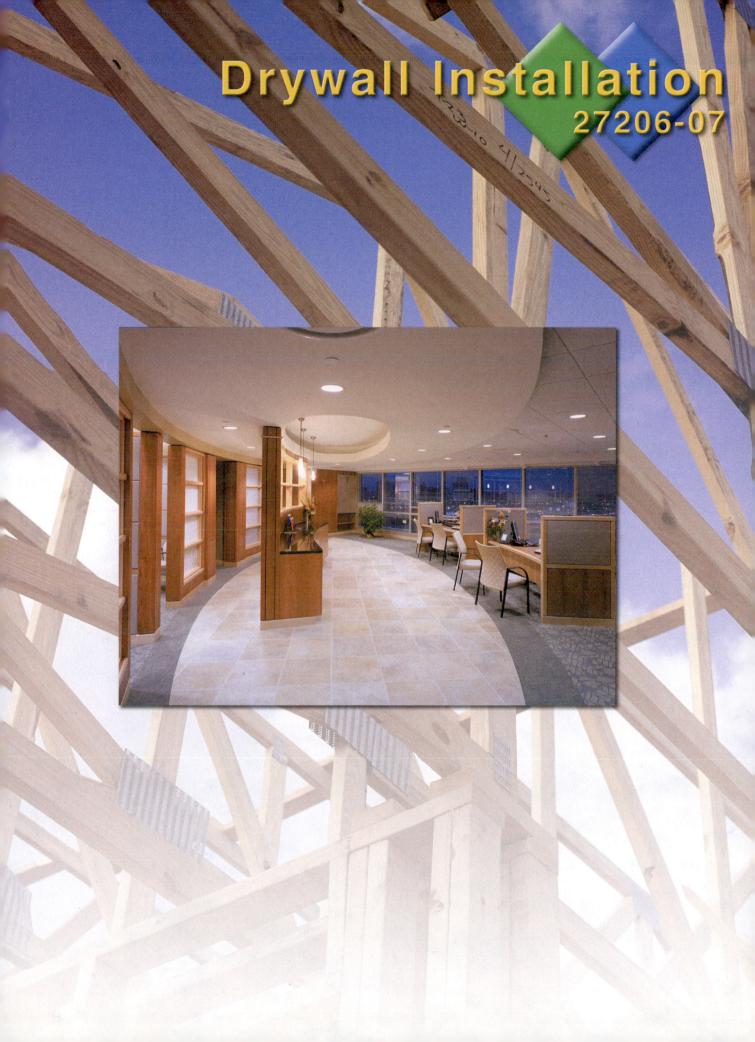

27206-07
Drywall Installation

Topics to be presented in this module include:

Overview

Gypsum drywall is the most common wall finish used in residential and commercial construction. On major projects, drywall installation will probably be done by a drywall contractor. However, this is not always the case. If you are doing remodeling work, you will be responsible for the drywall installation. There are a variety of drywall materials used for different applications, as well as a number of construction methods used to build walls to meet building codes for fire resistance and sound transmission. There are also many different types of fasteners used in drywall installation. The selection of materials, fasteners, and construction methods is controlled by building codes, and therefore must be carefully considered.

Objectives

When you have completed this module, you will be able to do the following:

1. Identify the different types of drywall and their uses.
2. Select the type and thickness of drywall required for specific installations.
3. Select fasteners for drywall installation.
4. Explain the fastener schedules for different types of drywall installations.
5. Perform single-layer and multi-layer drywall installations using different types of fastening systems, including:
   - Nails
   - Drywall screws
   - Adhesives
6. Install gypsum drywall on steel studs.
7. Explain how soundproofing is achieved in drywall installations.
8. Estimate material quantities for a drywall installation.

Trade Terms

Corner bead
Floating interior angle construction
Gypsum board

Joint
Nail pop
Slurry
Substrate

Required Trainee Materials

1. Pencil and paper
2. Appropriate personal protective equipment

Prerequisites

Before you begin this module, it is recommended that you successfully complete *Core Curriculum*; *Carpentry Fundamentals Level One*; and *Carpentry Framing and Finishing Level Two*, Modules 27201-07 through 27205-07.

This course map shows all of the modules in *Carpentry Framing and Finishing Level Two*. The suggested training order begins at the bottom and proceeds up. Skill levels increase as you advance on the course map. The local Training Program Sponsor may adjust the training order.

Course Map — FRAMING AND FINISHING

27212-07 **Cabinet Fabrication** ELECTIVE

27211-07 **Cabinet Installation**

27210-07 **Window, Door, Floor, and Ceiling Trim**

27209-07 **Suspended Ceilings** ELECTIVE FOR RESIDENTIAL CERTIFICATE

27208-07 **Doors and Door Hardware**

27207-07 **Drywall Finishing**

27206-07 **Drywall Installation**

27205-07 **Cold-Formed Steel Framing**

27204-07 **Exterior Finishing** ELECTIVE FOR COMMERCIAL CERTIFICATE

27203-07 **Thermal and Moisture Protection**

27202-07 **Roofing Applications** ELECTIVE FOR COMMERCIAL CERTIFICATE

27201-07 **Commercial Drawings** ELECTIVE FOR RESIDENTIAL CERTIFICATE

CARPENTRY FUNDAMENTALS

CORE CURRICULUM: Introductory Craft Skills

206CMAP.EPS

1.0.0 ◆ INTRODUCTION

Gypsum board, also known as gypsum drywall, is one of the most popular and economical methods of finishing the interior walls and ceilings of wood-framed and metal-framed buildings. When properly installed and finished, gypsum drywall can give a wall or ceiling made from many panels the appearance of being made from one continuous sheet.

The responsibility for drywall installation and finishing varies from job to job and from one locale to another. In some situations, carpenters install the drywall and painters finish it. In others, professional drywall workers do the entire job. The smaller the project, the more likely it is that the carpenter will install and finish the drywall.

At a minimum, it is important for the carpenter to understand the framing techniques that are necessary for the proper installation of drywall.

2.0.0 ◆ GYPSUM BOARD

Gypsum board is a generic name for products consisting of a noncombustible core. This product is made primarily of gypsum with a paper covering on the face, back, and long edges. A typical board application is shown in *Figure 1*.

Gypsum board is often called gypsum drywall, plasterboard, or Sheetrock™, the latter being a trade name of the United States Gypsum Company. Gypsum board differs from products such as plywood, hardboard, and fiberboard because of its noncombustible core. Gypsum is a mineral found in sedimentary rock formations in a crystalline form known as calcium sulphate dihydrate.

One hundred pounds of gypsum rock contains approximately 21 pounds (10 quarts) of chemically combined water. The gypsum rock is mined and then crushed. The crushed rock is heated to about 350°F, driving out or evaporating three-fourths of the chemically combined water in a process called calcining. The calcined gypsum is

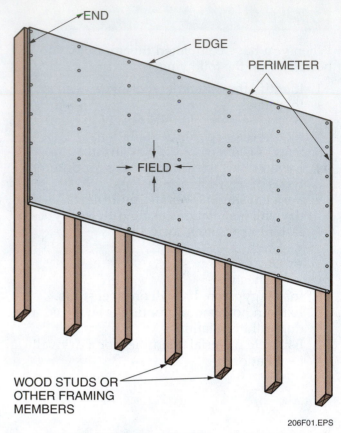

Figure 1 ◆ Typical board application.

then ground into a fine powder used in plaster, wallboard, and other gypsum products.

To produce gypsum board, the calcined gypsum is mixed with water and additives to form a **slurry,** which is fed between continuous layers of paper on a board machine. As the board automatically moves down a conveyor line, the calcium sulphate recrystallizes or rehydrates, reverting to its original rock state.

The paper becomes chemically and mechanically bonded to the core. The board is then cut to length and conveyed through dryers to remove any free moisture.

The Way It Was

Until the 1930s, walls were typically finished by installing thin, narrow strips of wood or metal known as lath between studs, and then coating the lath with wet plaster. Skilled plasterers could produce a very smooth wall finish, but the process was time-consuming and messy. In the early 1930s, paper-bound gypsum board was introduced and soon came into widespread use as a replacement for the tedious lath and plaster process.

2.1.0 Advantages of Gypsum Board Construction

Gypsum board walls and ceilings have a number of outstanding advantages:

- Fire resistance
- Sound insulation
- Durability
- Economy
- Versatility

2.1.1 Fire Resistance

Gypsum board is an excellent fire-resistive material. It is the most commonly used interior finish where fire resistance classifications are required. Its noncombustible core contains chemically combined water, which, under high heat, is slowly released as steam, effectively retarding heat transfer. Even after complete calcination, when all of the water has been released, it continues to act as a heat-insulating barrier.

In addition, tests conducted in accordance with the American Society for Testing Materials International, *ASTM Method E84*, show that it has low flame spread and low fuel and smoke contribution factors. When installed in combination with other materials, it serves to effectively protect building elements from fire for prescribed time periods. Type X board is most often used in fire-rated assemblies. Be sure all local codes and regulations are met.

2.1.2 Sound Isolation

Control of unwanted sound that might be transmitted to adjoining rooms is a key consideration in the design of a building. It has been determined that low-density paneling transmits an annoying amount of noise. Sound-absorbing acoustical surfacing materials, while they reduce the reflection of sound within a room, do not greatly reduce transmission of sound into adjoining rooms. Gypsum board wall and ceiling systems effectively help to control sound transmission.

2.1.3 Durability

Gypsum board makes strong, high-quality walls and ceilings with excellent dimensional stability. Their surfaces are easily decorated and refinished.

2.1.4 Economy

Gypsum board products are easy to apply. They are the least expensive of the wall surfacing materials that offer a fire-resistant interior finish. Both regular and architectural wallboard may be installed at relatively low cost. When architectural board is used, further decorative treatment is unnecessary.

2.1.5 Versatility

Gypsum board products satisfy a wide range of architectural requirements for design. Ease of application, performance, availability, ease of repair, and adaptability to all forms of decoration combine to make gypsum board unmatched by any other surfacing product.

2.2.0 Types of Gypsum Products

Many types of gypsum board are available for a variety of building needs (see *Table 1*). Gypsum board panels are mainly used as the surface layer for interior walls and ceilings; as a base for ceramic, plastic, and metal tile; for exterior soffits; for elevators and other shaft enclosures; and to provide fire protection for architectural elements.

Foil-backed gypsum board reduces radiant heat loss in the cold season and radiant heat gain in the warm season. However, foil-backed gypsum board should not be used as a backing material for tile, as a second face ply on a two-ply system, in conjunction with heating cables, or when laminating directly to masonry, ceiling, and roof assemblies.

Various thicknesses of gypsum board are available in regular, Type X, water-resistant, and architectural boards:

- $\frac{1}{4}$" *gypsum board* – A lightweight, low-cost board used as a base in a multi-layer application for improving sound control, to cover existing walls and ceilings in remodeling work, and for curved walls and barrel ceilings.
- $\frac{5}{16}$" *gypsum board* – A lightweight board developed for use in manufactured construction, primarily mobile homes.
- $\frac{3}{8}$" *gypsum board* – A lightweight board principally applied in a double-layer system over wood framing and as a face layer in repair and remodeling.
- $\frac{1}{2}$" *gypsum board* – A board generally used for single-layer wall and ceiling construction in residential work and in double-layer systems for greater sound and fire ratings. These panels are also available in 54" widths for use with 9' ceilings.
- $\frac{5}{8}$" *gypsum board* – A board used in quality single-layer and double-layer wall systems. The greater thickness provides additional fire resistance, higher rigidity, and better impact resistance. It is also used to separate occupied and unoccupied areas, such as a house from a garage or an office from a warehouse.

Table 1 Types and Uses of Gypsum Wallboard

Type	Thickness	Sizes	Use
Regular, paper faced	¼"	4' × 8' to 10'	Recovering old gypsum walls
	⅜"	4' × 8' to 14'	Double-layer installation
	½", ⅝"	4' × 8' to 16'	Standard single-ply installation
Regular with foil back	½", ⅝"	4' × 8' to 14'	Use as a vapor barrier or radiant heat retarder
Type X, fire-retardant	⅜", ½", ⅝"	4' × 8' to 16'	Use in garages, workshops, and kitchens, as well as around furnaces, fireplaces, and chimney walls; ⅝" is ¾-hour fire rated
Moisture-resistant	½", ⅝"	4' × 6' to 16'	For tile backing in areas not exposed to constant moisture
Architectural panels	5⁄16"	4' × 8'	Any room in the house
Gypsum lath	⅜", ½", ⅝"	16" × 4'	Use as a base for plaster
		2' × 8' to 12'	Use ⅜" for 16" on center (OC) stud spacing; ½" or ⅝" for 24" OC stud spacing
Gypsum coreboard	1"	2' × 8' to 12'	Shaft liner Laminated partitions

- 1" *gypsum board* – A special board also known as coreboard. Either a single 1" board or two ½" factory-laminated boards may be used as a liner or core in shaft walls and in semi-solid or solid gypsum board partitions.

Standard gypsum boards are 4' wide and 8', 10', 12', or 14' long. The width is compatible with the standard framing of studs or joists spaced 16" or 24" on center. Other lengths and widths are available from the manufacturers on special order. The stan-

dard edges are rounded, tapered, beveled, square edge, and tongue-and-groove, as shown in *Figure 2*.

- *Regular gypsum board* – Regular gypsum board is used as a surface layer on walls and ceilings. Type X gypsum board is available in ½" and ⅝" thicknesses and has an improved fire resistance made possible by the use of special core additions. It is also available with a predecorated finish. Type X gypsum board is used in most fire-rated assemblies.

Special-Use Gypsum Board

INSIDE TRACK

Regular ½" and ⅝" gypsum board are the most common types. There are, however, several types of gypsum board designed for special applications. These include:

- Type X gypsum board provides improved fire ratings because its core material is mixed with fire-retardant additives. Type X is often used on walls that separate occupancies. Examples are walls and ceilings between apartments or a wall separating a garage from the living area of a house. Use of Type X is normally specified by local building codes for protection of occupants.
- Flexible ¼" drywall panels have a heavy paper face and are designed to bend around curved surfaces.
- Special high-strength drywall panels are made for ceiling applications. The core of these panels is specially treated to resist sagging.
- A weather-resistant drywall panel is available for installation on soffits, porch ceilings, and carport ceilings.

Gypsum sheathing panels are used in cases where the required fire rating of exterior walls exceeds that available with OSB, plywood, or other types of sheathing. Gypsum sheathing panels have a water-resistant core covered on both sides with water-repellent paper. Gypsum sheathing panels are widely used in commercial construction.

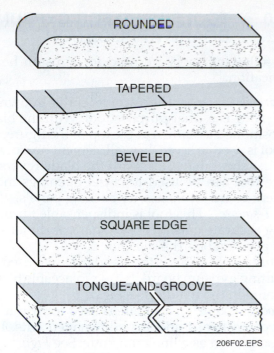

Figure 2 ◆ Standard edges of gypsum board.

Figure 3 ◆ Architectural gypsum board.

- *Architectural gypsum board* – Architectural gypsum board (*Figure 3*) has a decorated surface that does not require further treatment. The surfaces may be coated, printed, or have a vinyl film. Textured patterns are also available. It requires additional trim, divider, and corner pieces. It is also known as predecorated gypsum board.
- *Water-resistant gypsum board* – Water-resistant gypsum board, also known as green board, has a water-resistant gypsum core and water-repellent paper. The facing typically has a light green color. It serves as a base for the application of ceramic or plastic wall tile and plastic finish panels in kitchen and laundry areas. It is available with a regular or Type X core and in ½" or ⅝" thicknesses. Water-resistant gypsum board is not recommended for use in tub and shower enclosures and other areas exposed to water; tile backer is now preferred for these high-moisture applications.
- *Tile backer* – This type of wallboard is replacing water-resistant gypsum board as a backing for tile in damp areas such as baths and shower stalls. One type of this board is known as cement board, which is made from a slurry of portland cement mixed with glass fibers. It is colored light blue for easy recognition. These backer boards, which are available under a variety of trade names, such as Hardibacker,™ Durock,™ and Denshield,™ are very versatile. In addition to their use as a tile backer, they can be used as a floor underlayment, countertop

base, heat shield for stoves, and a base for exterior finishes such as stucco and brick veneer. They are available in 4' × 8' and 3' × 5' panels. Common thicknesses are ¼", ⁷⁄₁₆", and ½".
- *Gypsum backing board* – Gypsum backing board is designed to be used as a base layer or backing material in multi-layer systems. It is available with aluminum foil backing and with regular or Type X cores.
- *Gypsum form board* – Gypsum form board has a fungus-resistant paper and is used as a permanent form and support for poured-in-place reinforced gypsum concrete roof decks.
- *Gypsum coreboard* – Gypsum coreboard is available as a 1" thick solid coreboard or as a factory-laminated board composed of two ½" boards. It is used in shaft walls and laminated gypsum partitions with additional layers of gypsum board applied to the coreboard to complete the wall assembly. It is available in a width of 24" and with a variety of edges, of which square and tongue-and-groove are the most common.
- *Gypsum sheathing* – Gypsum sheathing is used as a protective, fire-resistive membrane under exterior wall surfacing materials such as wood siding, masonry veneer, stucco, and shingles. It also provides protection against the passage of water and wind and adds structural rigidity to the framing system. The noncombustible core is surfaced with firmly bonded water-repellent paper. In addition, a water-repellent material may be incorporated in the core. It is available in 2' and 4' widths and ½" and ⅝" thicknesses. The latter is also available with a Type X core.
- *Gypsum board substrate* – Gypsum board **substrate** for floor or roof assemblies has a Type X

core that is ½" thick. It is available in 24" or 48" widths. It is used under combustible roof coverings to protect the structure from fires originating on the roof. It can also serve as an underlayment when applied to the top surfaces of floor joists and under subflooring. It may also be used as a base for built-up roofing applied over steel decks.

- *Gypsum base for veneer plaster* – Gypsum base for veneer plaster is used as a base for thin coats of hard, high-strength gypsum veneer plaster.
- *Gypsum lath* – Gypsum lath is a board product used as a base to receive hand-applied or machine-applied plaster. It is available in ⅜" or ½" thicknesses and in widths of 16" or 24". Gypsum lath comes in 48" lengths. Other lengths are available on special order.
- *Flexible gypsum board* – Flexible gypsum board is specially designed for radius applications. A ¼" sheet can be laminated on a ¼" sheet of masonite for a tight radius application.
- *Abuse-resistant panels* – Some drywall panels are designed specifically for use in areas where they might be subject to impact or vandalism. These types of panels have heavy-duty face paper and either heavy-duty backing or internal reinforcement.

3.0.0 ◆ TOOLS USED FOR GYPSUM BOARD APPLICATION

The following tools are used for gypsum board application:

- *Carbide cutter* – The carbide cutter shown in *Figure 4* is a dual-purpose tool. With the blade positioned as shown at the top of the figure, the tool is used to score drywall, cement board, and other sheet materials. With the blade reversed, the tool can be used to chisel openings in cement board and masonry backing panels.
- *4' T-square* – This tool is indispensable for making accurate cuts across the narrow dimension of gypsum board. See *Figure 5*.
- *Utility knife* – This is the standard knife used for cutting gypsum board. It has replaceable blades stored in the handle. See *Figure 6*.
- *Hook-bill knife* – This knife is used for trimming gypsum board and for odd-shaped cuts. It is also known as a linoleum knife. See *Figure 7*.
- *Rasp* – The rasp is used to quickly and efficiently smooth rough-cut edges of gypsum board. The tool has both a file for finishing and a rasp for rough shaping. See *Figure 8*.
- *Circle cutter* – The circle cutter has a calibrated steel shaft that allows accurate cuts up to 16" in diameter. The cutter wheel and center pin are heat-treated. See *Figure 9*.

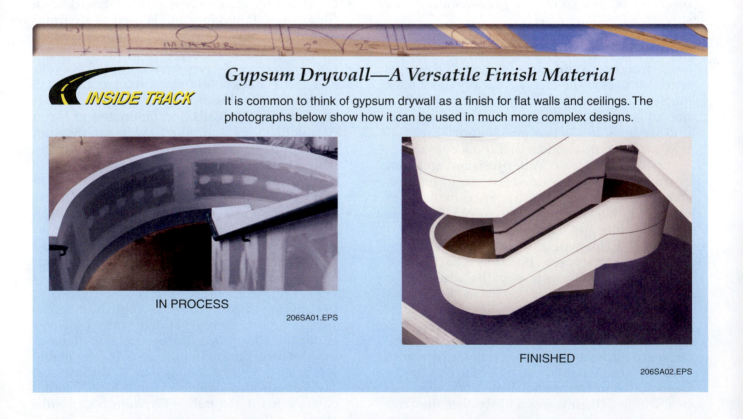

INSIDE TRACK

Gypsum Drywall—A Versatile Finish Material

It is common to think of gypsum drywall as a finish for flat walls and ceilings. The photographs below show how it can be used in much more complex designs.

IN PROCESS

206SA01.EPS

FINISHED

206SA02.EPS

Figure 4 ◆ Carbide cutting tool.

Figure 8 ◆ Drywall rasp.

Figure 5 ◆ T-square.

Figure 9 ◆ Circle cutter.

Figure 6 ◆ Utility knife.

• *Utility saw* – This saw (also known as a keyhole saw) is used for cutting small openings and making odd-shaped cuts. See *Figure 10*. A power cutout tool can be used for the same purpose.
• *Drywall saw* – This saw has a short blade and coarse teeth. It cuts gypsum quickly and easily. The sharp points of the teeth and the stiffness of the blade allow the saw to be punched through the board for starting the cut. See *Figure 11*.
• *Light box cutter* – This tool is used to cut exact hole shapes in gypsum board for various types of electrical boxes. This tool will not damage surrounding gypsum core or paper facings. See *Figure 12*.
• *Gypsum board lifter* – A gypsum board lifter is used to move the board forward as it is being lifted. The lifter can be used for either parallel or perpendicular board applications. See *Figure 13*.

Figure 7 ◆ Hook-bill knife.

Measuring and Marking

INSIDE TRACK

When measuring drywall, use a soft lead pencil to mark the drywall. A ballpoint pen mark may bleed through the joint compound and paint.

POWER CUTOUT TOOL

UTILITY SAW

206F10.EPS

Figure 10 ◆ Power cutout tool and utility saw.

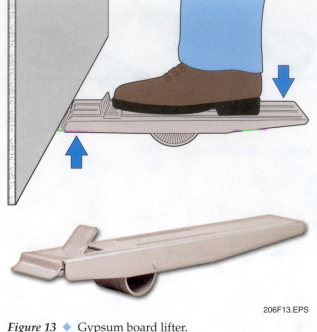

206F13.EPS

Figure 13 ◆ Gypsum board lifter.

206F11.EPS

Figure 11 ◆ Drywall saw.

◢◢ *INSIDE TRACK*

Cutting Small Strips

The drywall stripper shown here is designed to cut narrow strips of wallboard up to 4½". There is a sharp edge on each side of the tool, so it cuts both sides of the panel at once. The drywall stripper makes a cleaner cut than a utility knife when cutting long, narrow strips such as those that might be needed around a window or door.

206SA03.EPS

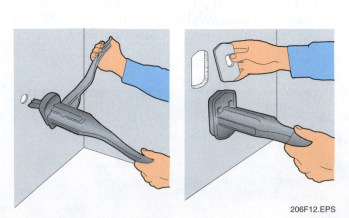

206F12.EPS

Figure 12 ◆ Light box cutter.

Locating Electrical Boxes

The Blind Mark™ system uses magnets to find electrical boxes under drywall. A target magnet is placed in the electrical box before the drywall is installed. After the drywall is up, a locator magnet is used to find the box containing the target.

(1)

(2)

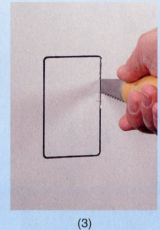

(3)

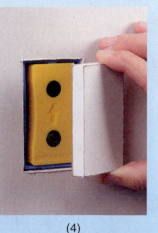

(4)

206SA04.EPS

- *Drywall hammer* – This hammer has a symmetrical convex face designed to compress the gypsum panel face and leave the desired dimple. The blade end is not for cutting. It is dull and is used for wedging and prying. See *Figure 14*.
- *Screw gun* – Electric drywall screw guns (*Figure 15*) are designed to drive steel screws to a precise depth below the gypsum board face; at that point, the drive is disengaged by a clutch mechanism. The depth setting is adjustable. These guns run from 0 to 4,000 rpm. The screws are held in place by a magnetic bit tip.
- *Drywall lift* – This special device is designed to raise and support drywall panels during ceiling or high wall installations. See *Figure 16*.
- *T-brace* – A job-built T-brace is used to hold drywall in place against ceiling joists while fasteners are being installed or while adhesive is setting. See *Figure 17*.

206F14.EPS

Figure 14 ◆ Drywall hammer.

206F15.EPS

Figure 15 ◆ Screw guns.

Figure 16 ◆ Drywall lift.

206F16.EPS

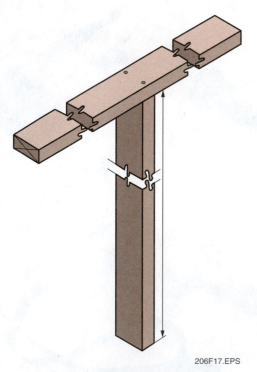

Figure 17 ◆ T-brace.

206F17.EPS

4.0.0 ◆ APPLICATION OF GYPSUM BOARD

Gypsum board panels can be applied over any firm, flat base such as wood or steel framing and furring. Gypsum can also be applied to masonry and concrete surfaces, either directly or to steel or wood furring strips. If the board is applied directly, any irregularities in the masonry or concrete surfaces must be smoothed or filled. Furring is a means to provide a flat surface for standard fastener application. It also provides a separation to overcome dampness in exterior walls.

The most common type of residential interior wall construction is the standard gypsum board system with **joints** between the panels and internal corners reinforced with tape and covered with joint treatment compound to prepare them for decoration.

The term joint is used to describe any point where two drywall panels meet. A butt joint is where two sheets of wallboard with untapered sides meet. A flat joint is the intersection of two bevel-edged wallboards. External corners are normally reinforced with **corner bead**, which, in turn, is covered with joint compound. Exposed edges are covered with metal or plastic trim. The result is a smooth, unbroken surface ready for final decoration. When architectural board is used, no further decoration is necessary, but trim moldings or battens can be used to cover the joints. Special instructions for installing architectural panels will be given later in this module.

4.1.0 Single-Ply and Multi-Ply Construction

In light commercial and residential construction, single-ply gypsum board systems are commonly used (*Figure 18*). Generally, they are adequate to meet fire resistance and sound control requirements. Multi-ply systems, as shown in *Figure 19*, have two or more layers of gypsum board to increase sound isolation and fire-resistive performance. They also provide better surface quality because face layers are often laminated over base layers, thereby reducing the number of fasteners. As a result, the surface joints of the face layer are reinforced by the continuous base layers of gypsum board. **Nail pop** and ridging problems are less frequent, and imperfectly aligned supports have less effect on the finished surface.

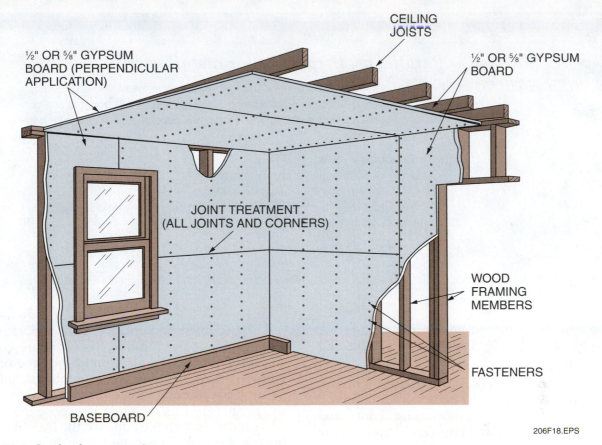

CEILING JOISTS

½" OR ⅝" GYPSUM BOARD (PERPENDICULAR APPLICATION)

½" OR ⅝" GYPSUM BOARD

JOINT TREATMENT (ALL JOINTS AND CORNERS)

WOOD FRAMING MEMBERS

FASTENERS

BASEBOARD

206F18.EPS

Figure 18 ◆ Single-ply construction.

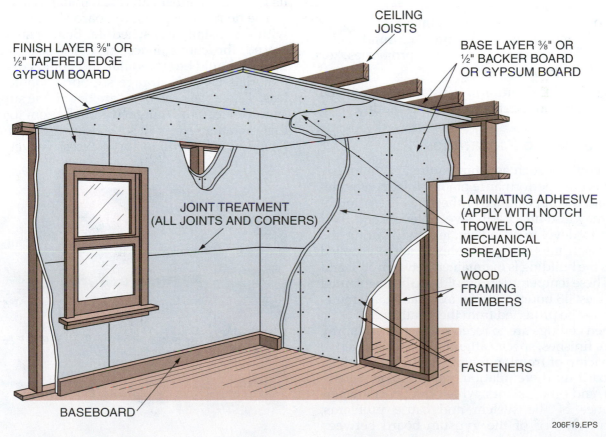

CEILING JOISTS

FINISH LAYER ⅜" OR ½" TAPERED EDGE GYPSUM BOARD

BASE LAYER ⅜" OR ½" BACKER BOARD OR GYPSUM BOARD

JOINT TREATMENT (ALL JOINTS AND CORNERS)

LAMINATING ADHESIVE (APPLY WITH NOTCH TROWEL OR MECHANICAL SPREADER)

WOOD FRAMING MEMBERS

FASTENERS

BASEBOARD

206F19.EPS

Figure 19 ◆ Multi-ply construction.

Satisfactory results can be assured with either single-ply or multi-ply assemblies by requiring the following:

• Proper framing details, consisting of straight, correctly spaced, and properly cured lumber
• Proper job conditions, including controlled temperatures and adequate ventilation during application
• Proper measuring, cutting, aligning, and fastening of the board
• Proper joint and fastener treatment
• Special requirements for proper sound isolation, fire resistance, thermal properties, or moisture resistance

Single-ply and multi-ply installations will be discussed in more detail later in this module.

4.2.0 Job-Site Preparation

Job conditions such as temperature and humidity can affect the performance of joint treatment materials and the appearance of the joint. These conditions may also affect adhesive materials and their ability to develop an adequate bond. During the cold season, interior finishes should not be installed unless the building is maintained between 50°F and 80°F. These temperatures should also be maintained for at least 48 hours after the installation. All materials must be protected from the weather.

When ceilings are to receive water-based spray texture finishes, special attention must be given to the spacing of framing members, the thickness of the board used, ventilation, vapor barriers, insulation, and other factors, which can affect the performance of the system and cause problems, particularly sag of the gypsum board between framing members.

Lumber must be kept dry during storage and installation at the job site. Its moisture content should not exceed 15 percent at the time of gypsum board application. Green lumber should not be used for framing. Since lumber shrinks across the grain as it dries, it tends to expose the shanks of nails driven into the edges of the framing members. If shrinkage is substantial or the nails are too long, separation between the gypsum board and its framing member can result in nail pops.

The delivery of gypsum board should coincide with the installation schedule. Boards should be placed for convenience at the work location. Boards should be stored flat and under cover. The materials used as storage supports should be at least 4" wide. As the units are tiered, the supports should be carefully aligned from bottom to top so that each tier rests on a solid bearing, as shown in *Figure 20*. Be careful to avoid excessive weight.

206F20.EPS

Figure 20 ◆ Gypsum board storage.

Moving Drywall

A drywall dolly like the one shown here is specially designed to transport drywall panels at the job site.

206SA05.EPS

Stacking long lengths on short lengths should be avoided to prevent the longer boards from breaking. Leaning boards against the framing members for a prolonged period of time with the long edges horizontal is not recommended. You should avoid leaning boards during periods of high humidity, as the boards could be subject to warping. All materials should remain stored in their original wrappers or containers until ready to use on the job site. When boards are moved, they should be carried, not dragged, so the edges are not damaged.

4.3.0 Cutting and Fitting Procedures

Any gypsum board installation should be carefully planned. Accurate measuring, cutting, and fitting are very important. In residential buildings with less than 8'-1" ceiling heights, it is preferred that the wallboard be installed at right angles to the supporting members because there are usually fewer joints to finish. On long walls, boards of maximum practical lengths should be used to minimize the number of end joints. Scored, scratched, broken, or otherwise damaged boards should not be used.

Storing and Handling Gypsum Drywall

Figure 20 shows gypsum drywall as it would be stored in a warehouse or building supply store. Just before installation, the drywall panels would be distributed along interior walls and stood on edge.

Drywall panels are sold in pairs. Two sheets of drywall are connected by a strip of paper tape, which can be stripped off to separate the panels. Drywall is heavy. Two ½" panels weigh about 110 pounds, while a pair of ⅝" panels weigh close to 150 pounds. The panels need to be handled carefully so they don't break under their own weight. They should be lifted and carried by the edges rather than the ends. Also, proper lifting procedures must be used by people handling drywall in order to prevent injury.

Measurements should be done accurately at the correct ceiling or wall location for each edge or end of the board. Accurate measuring will usually reveal any irregularities in framing or furring so corrective allowances can be made in cutting. Poorly aligned framing should be corrected before applying gypsum board.

Gypsum board should be cut by first scoring through the paper down to the core with a sharp utility knife, working from the face side. The board is then snapped back away from the cut face.

The back of the paper is broken by cutting it with a utility knife. Gypsum board may also be cut by sawing. All cut edges and ends of the gypsum board should be smoothed to form neat, tight-fitting joints when installed. Ragged cut ends or broken edges can be smoothed with a rasp or sandpaper, or trimmed with a sharp knife. If burrs on the cut ends are not removed, they will form a visible ridge in the finished surface.

The practices listed below should be followed to ensure a sound application:

- Install the ceiling boards first, then the wall panels.
- The panels should fit easily into place without force.
- Always match edges and ends. For example, tapered end to tapered end and square-cut end to square-cut end.
- Plan to span the entire length of ceilings or walls with single boards, if possible, to reduce the number of end joints, which are more difficult to finish.
- Stagger end joints and locate them as far from the center of the ceiling or wall as possible so they will be inconspicuous.
- In a single-ply application, the board ends and edges parallel to the supporting members (framing) should fall on these members to reinforce the joint.
- Mechanical and electrical equipment, such as cover plates, registers, and grilles, should be installed to provide for the final wall thickness when applying the trim.
- Place a shim under the wallboard to keep it from absorbing moisture from the floor.

NOTE

The depth of electrical boxes should not exceed the framing depth, and boxes should not be placed back-to-back on the same stud. Electrical boxes and other devices should not be allowed to penetrate completely through the walls. This is detrimental to sound isolation and fire resistance. Make sure the wires are pushed back into the box.

4.4.0 Drywall Fasteners

Nails and screws are commonly used to attach gypsum board in both single-ply and multi-ply installations. Clips and staples are used only to attach the base layers in multi-ply construction. Special drywall adhesives can be used to secure single-ply gypsum board to framing, furring, masonry, and concrete, or to laminate a face ply to a base layer of gypsum board or other base material. Adhesives must be supplemented with mechanical fasteners.

Where fasteners are used at the board perimeter, they should be placed at least ⅜" from the board edges and ends. Fastening should start in the middle of the board and proceed outward toward the board perimeter. Fasteners must be driven as near to perpendicular as possible while the board is held firmly against the supporting construction.

Also, fasteners must be used with the correct shield, guard, or attachment recommended by the manufacturer. Nails should be driven with a crown-headed hammer, which forms a uniform depression or dimple that is not more than ¹⁄₃₂" deep around the nail head. See *Figure 21*.

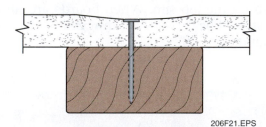

206F21.EPS

Figure 21 ◆ Uniform depression or dimple.

INSIDE TRACK

Special Fasteners

Application of gypsum board requires special fasteners. Ordinary wood or sheet metal screws and common nails are not designed to penetrate the board without damage, to hold it tightly against the framing, or to permit correct countersinking for proper concealment. For high-end installations, sheets can be connected between studs using butt clips.

4.4.1 Nails

Both annular and cupped-head nails are acceptable for gypsum board application (*Figure 22*). Preferably, the nails should have heads that are flat or concave and thin at the rim. The heads should be between ¼" and ⁵⁄₁₆" in diameter to provide adequate holding power without cutting the face paper when the nail is dimpled. Casing and common nails have heads that are too small in relation to the shank; they easily cut into the face paper and should not be used.

Nail heads that are too large are also likely to cut the paper surface if the nail is driven incorrectly at a slight angle. The nails should be long enough to go through the wallboard layers and far enough into the supporting construction to provide adequate holding power. The nail penetration into the framing member should be ⅞" for smooth shank nails and ¾" for annular ring nails, which provide more withdrawal resistance and require less penetration. For fire-rated assemblies, greater penetration is required (generally 1⅛" to 1¼" for one-hour assemblies).

Particular care should be taken not to break the face paper or crush the core by striking it too hard with the hammer.

Gypsum board can be attached by either a single nailing or a double nailing method. Double nailing produces a tighter board-to-stud contact. Whenever fire-resistive construction is required, the nail spacing specified in the fire test should be followed. Always check local codes for nailing requirements.

Single nails should be spaced at a maximum of 7" on center on ceilings and 8" on center on walls along framing members. See *Figure 23*.

Nails are first driven into the center or field of the board and then outward to the edges and ends. In single-ply installations, all ends and edges of gypsum board are placed over framing members or other solid backing, except where treated joints are at right angles to framing members.

In double nailing, the spacing of the first set of nails is 12" on center, with the second nailing 2" to 2½" from the first. See *Figure 24*.

The second set of nails is applied in the same sequence as the first set, but not on the perimeter of the board. The first nails driven should be reseated as necessary following the application of the second set. The general attachment procedure is as follows:

Step 1 Carefully measure and cut the board.

Step 2 Prior to nailing, mark the gypsum board to indicate the location of the framing.

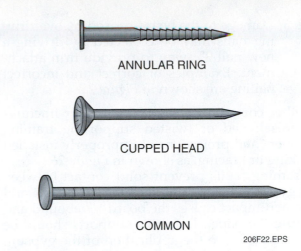

ANNULAR RING

CUPPED HEAD

COMMON

206F22.EPS

Figure 22 ◆ Nails.

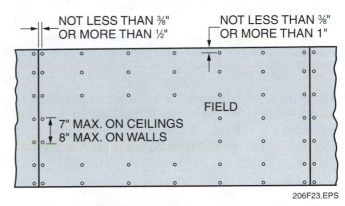

NOT LESS THAN ⅜" OR MORE THAN ½"

NOT LESS THAN ⅜" OR MORE THAN 1"

FIELD

7" MAX. ON CEILINGS
8" MAX. ON WALLS

206F23.EPS

Figure 23 ◆ Single nail spacing.

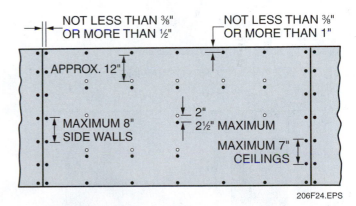

NOT LESS THAN ⅜" OR MORE THAN ½"

NOT LESS THAN ⅜" OR MORE THAN 1"

APPROX. 12"

MAXIMUM 8" SIDE WALLS

2"
2½" MAXIMUM

MAXIMUM 7" CEILINGS

206F24.EPS

Figure 24 ◆ Double nail spacing.

Step 3 To avoid nail pops or protrusions, hold the board firmly against the framing when nailing.

Step 4 Drive the nails straight into the framing member. Nails that miss the framing member should be removed, and the nail hole dimpled and covered with joint compound.

Step 5 Damage to the board caused by overdriving nails may be corrected by driving a new nail 2" away to provide firm attachment. Examples of correct and incorrect nailing are shown in *Figure 25*.

Other common causes of face paper fractures are misaligned or twisted supporting framing members and projections of improperly installed blocking or bracing, as shown in *Figure 26*.

Framing faults prevent solid contact between the gypsum board and framing members and hammer impact causes the board to rebound and rupture the paper. Defective supports should be corrected prior to the application of the gypsum board. Protruding framing members should be trimmed or reinstalled. Shims can be used, if necessary, for receiving framing members. The use of screws, adhesives, or two-ply construction will minimize problems resulting from these defects.

4.4.2 Screws

Drywall screws (*Figure 27*) are used to attach gypsum board to wood or steel framing or to other gypsum board. They have a cupped Phillips head design that is intended to be used with a drywall

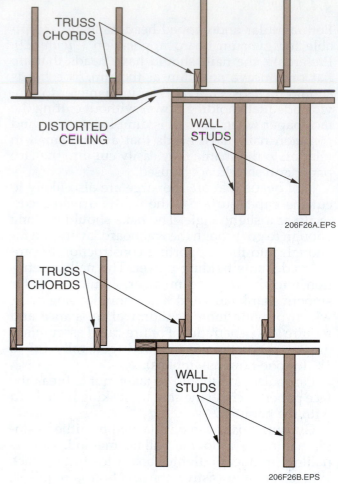

Figure 26 ◆ Incorrect and correct alignment.

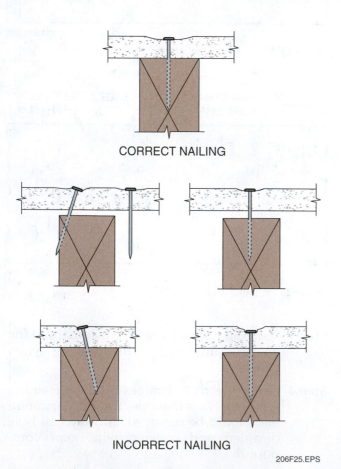

Figure 25 ◆ Correct and incorrect nailing.

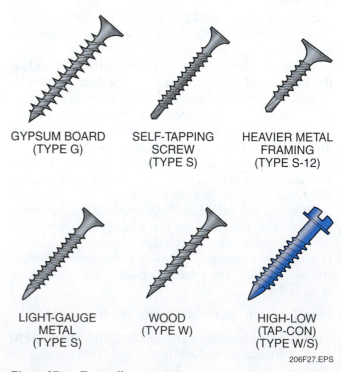

Figure 27 ◆ Drywall screws.

INSIDE TRACK

Selecting Drywall Screws

If you go to a building supply store to get drywall screws, you will find that there are many different types and sizes. It is important to determine exactly what you need before you make the trip.

206SA06.EPS

power screwdriver. These screws pull the board tightly to the supports without damaging the board and minimize fastener and surface defects due to loose boards. The specially contoured head, when properly driven, makes a uniform depression that is free of ragged edges and fuzz.

Type W gypsum drywall screws are designed for fastening gypsum board to wood framing or furring. The Type W screw points are diamond-shaped to provide efficient drilling action through both gypsum and wood, and their specially designed threads provide both quick penetration and increased holding power.

The recommended minimum penetration into supporting construction is ⅝". However, in two-ply construction where the face layer is screw attached, additional holding power is developed in the base ply, which permits reduced penetration into supports down to ½". Type S screws may be substituted for Type W screws in two-ply construction.

Type S gypsum drywall screws are designed for fastening gypsum boards to steel studs or furring. They are self-drilling with a self-tapping thread and generally a mill slot or hardened drill point that is designed to penetrate sheet metal with little pressure. Easy penetration is important because steel studs are often flexible. They tend to bend away from the screws, and the screws tend to strip easily.

INSIDE TRACK

Electric Screwdriver Attachment

This attachment holds a strip of 50 drywall screws. It has a depth control that allows the operator to set and lock in the correct depth for drywall screws.

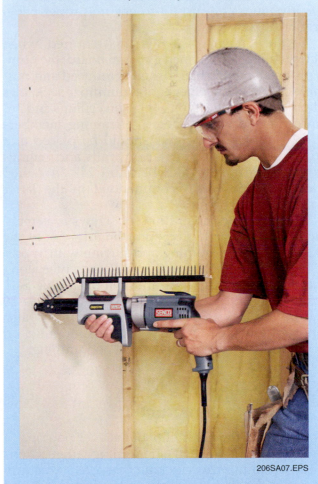

206SA07.EPS

Type G gypsum drywall screws are used for fastening gypsum board panels to gypsum backing boards. They are similar to Type W screws, but have a deeper, special-thread design. They are generally 1½" long, but other lengths are available. Gypsum drywall screws require a penetration of at least ½" of the threaded portion into the supporting board. Allowing approximately ¼" for the point results in the minimum penetration of ¾".

Gypsum drywall screws should not be used to attach wallboard to ⅜" backer board because they do not provide sufficient holding strength. Nails or longer screws should be driven through both the surface layer and the ⅜" backer (base ply) to provide the proper penetration in the supporting wood or metal construction.

For best results, the screw gun should be kept perpendicular to the work surface. Adequate pressure must be exerted to engage the clutch and prevent the screw from slipping (also known as walking). The tool should be triggered continuously until each fastener is seated. A one-piece socket makes driving easier and more efficient than separate socket and extension pieces because it provides a more rigid base and firmer control. Depth gauges are useful to ensure proper penetration.

Because fewer fasteners are required when screws are used to attach gypsum board, the number of fasteners to be finished is reduced and possible application defects are minimized. Screws should be placed 12" on center on ceilings and 16" on center on walls where the framing members are 16" on center. Screws should be placed at a maximum of 12" on center on walls and ceilings where the framing members are 24" on center. Double screws are recommended in the latter case.

The required penetration for screws is as mentioned previously. Gypsum board should be attached to steel framing and furring with Type S screws spaced no more than 12" on center along supports for both walls and ceilings. Type S-12 screws are required for steel framing that is 20 gauge or heavier. A 12" screw spacing is also appropriate when gypsum board is mounted on resilient furring channels over wood framing.

4.5.0 Floating Interior Angle Construction

To minimize the possibility of fastener popping in areas adjacent to wall and ceiling intersections and to minimize cracking due to structural stresses, the floating angle method may be used for either the single-layer or double-layer application of gypsum board to wood framing.

Floating interior angle construction helps to eliminate nail popping and corner cracking by omitting fasteners at the intersections of walls and ceilings. This is applicable for single nailing, double nailing, and screw attachment. *Figure 28* shows a typical single-layer application. The same nail-free clearances at corners should be maintained in double nailing.

In floating interior angle construction where the ceiling framing members are perpendicular to the wall/ceiling intersection, the ceiling fasteners should be located 7" from the intersection for single nailing and 11" to 12" for double nailing or screw applications.

On ceilings where the joists are parallel to the wall intersection, nailing should start at the intersection. Gypsum board should be applied to the ceiling first and then to the walls. See *Figure 29*.

Gypsum board on side walls should be applied to provide a firm, level support for the floating edges of the ceiling board.

Apply the overlapping board firmly against the underlying board to bring the underlying board into firm contact with the face of the framing member behind it. The overlapping board should be nailed or screwed, and the fasteners should be omitted from the underlying board at the vertical intersection.

4.6.0 Adhesives

Adhesives are used to bond single layers of gypsum board directly to the framing, furring, masonry, or concrete. They can be used to laminate gypsum board to base layers of backer boards, sound deadening boards, rigid foam, and other rigid insulating boards. The adhesive must be used in combination with nails or screws, which provide supplemental support.

The adhesives used for applying wallboard finishes are classified as follows:

• Stud adhesives
• Laminating adhesives such as dry powder (including joint tape compound), special drywall laminating adhesives, and drywall contact and modified contact adhesives

4.6.1 Stud Adhesives

Stud adhesives are specially prepared to attach single-ply wallboard to steel or wood studs and are generally used in conjunction with nails. Some permit a significant reduction in the use of mechanical fasteners, but they still require some fastening, at least at the board perimeters. These adhesives should be of caulking consistency so that they bridge framing irregularities. Stud adhesives should meet the requirements of the *Standard Specification for Adhesives for Fastening*

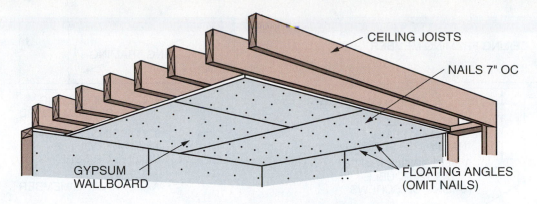

CEILING JOISTS

NAILS 7" OC

GYPSUM WALLBOARD

FLOATING ANGLES (OMIT NAILS)

PERPENDICULAR CEILING APPLICATION (SINGLE NAILING)

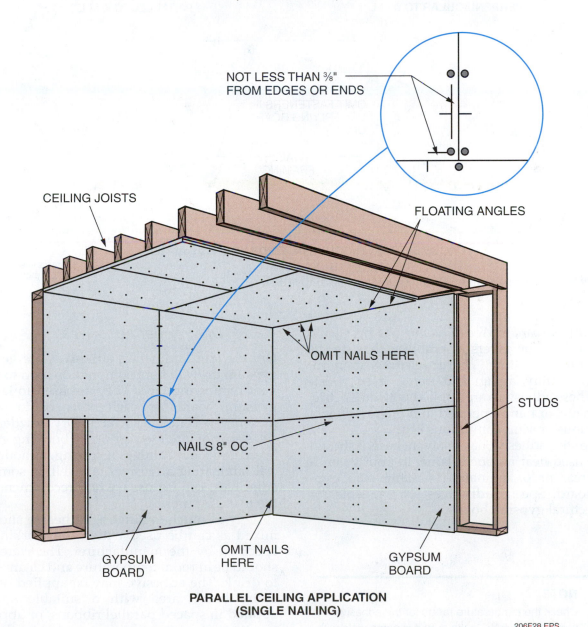

NOT LESS THAN ⅜" FROM EDGES OR ENDS

CEILING JOISTS

FLOATING ANGLES

OMIT NAILS HERE

STUDS

NAILS 8" OC

OMIT NAILS HERE

GYPSUM BOARD

GYPSUM BOARD

PARALLEL CEILING APPLICATION (SINGLE NAILING)

206F28.EPS

Figure 28 ◆ Floating interior angle construction.

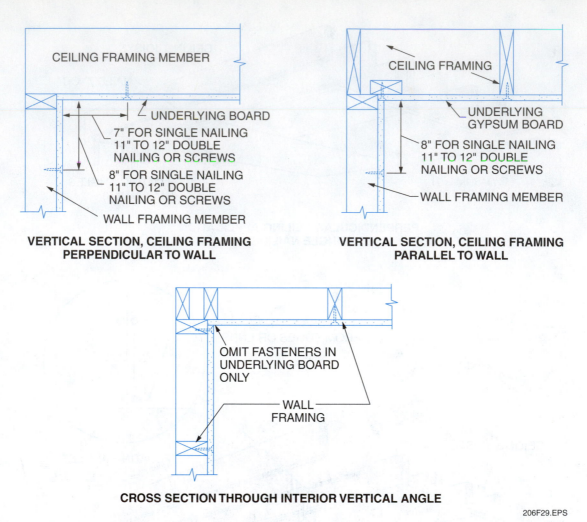

VERTICAL SECTION, CEILING FRAMING PERPENDICULAR TO WALL

CEILING FRAMING MEMBER

UNDERLYING BOARD

7" FOR SINGLE NAILING
11" TO 12" DOUBLE
NAILING OR SCREWS

8" FOR SINGLE NAILING
11" TO 12" DOUBLE
NAILING OR SCREWS

WALL FRAMING MEMBER

VERTICAL SECTION, CEILING FRAMING PARALLEL TO WALL

CEILING FRAMING

UNDERLYING GYPSUM BOARD

8" FOR SINGLE NAILING
11" TO 12" DOUBLE
NAILING OR SCREWS

WALL FRAMING MEMBER

OMIT FASTENERS IN UNDERLYING BOARD ONLY

WALL FRAMING

CROSS SECTION THROUGH INTERIOR VERTICAL ANGLE

206F29.EPS

Figure 29 ◆ Fastener patterns for floating interior angle construction.

Gypsum Wallboard to Wood Framing (*ASTM C557*). This specification covers workability, consistency, open time, wetting characteristics, strength, bridging ability, aging, and freeze/thaw resistance. These adhesives are applied with an electric, pneumatic, or hand-operated gun (*Figure 30*) in a continuous or semi-continuous bead.

If the stud adhesive has a solvent base, it should not be used near an open flame, in poorly ventilated areas, or for lamination of architectural gypsum board. Special adhesives are available for architectural gypsum board.

4.6.2 Dry Powder Adhesives

Dry powder laminating adhesives are generally gypsum drywall joint compounds used to embed joint reinforcing tape. They are used to laminate gypsum boards to each other or to suitable masonry or concrete surfaces. Dry powder adhesives are not intended for use in bonding gypsum board to wood framing or furring, although special laminating adhesives, as well as some stud adhesives, can be used when recommended by the manufacturer.

Only as much laminating adhesive should be mixed as can be used within the working time specified by the manufacturer. The water used should be at room temperature and clean enough to drink. The adhesive may be applied over the entire board area with a suitable spreader, applied in spaced parallel ribbons, or applied in a pattern of spots, as recommended by the manufacturer. All dry powder laminating adhesives require permanent mechanical fasteners at the board perimeters.

NOTE

Check the temperature range for the adhesive you plan to use to make sure it is compatible with the expected operating temperatures of the building.

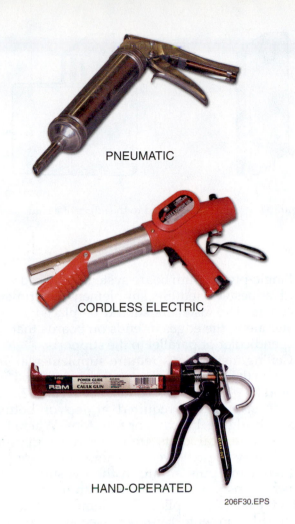

PNEUMATIC

CORDLESS ELECTRIC

HAND-OPERATED

206F30.EPS

Figure 30 ◆ Adhesive applicators.

If the boards are applied vertically on side walls, fasteners are placed at the top and bottom. Face boards may require temporary support or supplemental fasteners until the full bond strength is developed.

4.6.3 Drywall Contact Adhesives

Drywall contact adhesives require permanent mechanical fasteners at least at the perimeters of all boards applied to walls, ceilings, and soffits. If used to apply architectural gypsum board vertically on side walls, permanent fasteners are required only at the top and bottom of the boards, where they can be hidden by base and ceiling moldings or other decorative trim.

Contact adhesives may be used to laminate gypsum boards to each other or to steel studs. The adhesive is applied by roller, spray gun, or brush in a thin, uniform coating to both surfaces to be bonded. For most contact adhesives, some drying time is usually required before surfaces can be joined and the bond can be developed.

To ensure proper adhesion between surfaces, the face board should be impacted over its entire surface with a suitable tool, such as a rubber mallet. No temporary supports are needed while a contact adhesive sets and the bond forms.

One disadvantage of contact adhesives is their inability to fill irregularities between surfaces, which leaves some areas without an adhesive bond. Another disadvantage is that most of these

INSIDE TRACK

Using Adhesives on Drywall

When drywall is applied using adhesive, either drywall adhesive or construction adhesive may be used if it meets the requirements of *ASTM C557*. The adhesive should be allowed to dry for 48 hours before finishing the joints. Using adhesive does not eliminate the need for fasteners; it simply reduces the number of fasteners needed. Check the job specifications or local codes.

Adhesives cannot be used to attach drywall panels to studs if the building has an inside moisture barrier.

When laminating drywall panels for multi-layer installation, either lightweight or standard setting-type joint compound may be used in place of adhesive to laminate the panels.

206SA08.EPS

adhesives do not permit moving of the board once contact has been made. A sheet of polyethylene film or tough building paper can be slipped between the surfaces so gradual bonding of surfaces will occur as the slip sheet is withdrawn. Extra care should be taken when contact adhesives are used. The manufacturer's recommendations should always be followed.

> **WARNING!**
>
> Observe all manufacturer's safety data sheet (MSDS) precautions for adhesives. Extreme caution must be taken when using contact cement as it is highly flammable. It should be used in a well-ventilated area as the fumes can quickly overcome a worker.

4.6.4 Modified Contact Adhesives

Modified contact adhesives provide a longer placement time. They have an open time (up to a half hour) during which the board can be repositioned, if necessary. They combine good long-term strength with a sufficient immediate bond to permit erection with a minimum of temporary fasteners.

In addition, these adhesives have enough bridging ability to cover up minor framing irregularities. Modified contact adhesives are intended for attaching wallboard to all types of supporting construction, such as solid walls, other gypsum boards, and various insulating boards, including rigid foam insulation.

Adhesives are also used for securing drywall materials and paneling to steel studs. The use of adhesives will eliminate some of the fasteners required. Use only the adhesives specified by the manufacturers or suppliers of the particular steel framing and sheathing material being used. Improper adhesives will not only fail to add to the structural integrity, but may be detrimental to system performance.

4.6.5 Application of Adhesives

Stud adhesives should be applied with a caulking gun in accordance with the manufacturer's recommendations. A straight bead, approximately ¼" in diameter, is applied to the face of the studs in the field (center) of the panel. See *Figure 31*. Where two gypsum panels join over a stud, two parallel beads of adhesive should be applied, one near each edge of the stud.

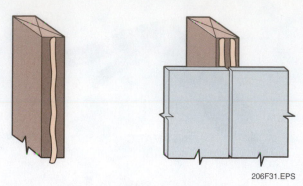

206F31.EPS

Figure 31 ◆ Adhesive applied to the edges of a stud.

Single-ply gypsum board systems attached with stud adhesives require supplemental perimeter fasteners. The fasteners should be placed 16" on center along the edges or ends on boards that are perpendicular or parallel to the supports.

Ceiling installations require supplemental fasteners in the field as well as on the perimeter. They should be placed 24" on center. See *Figure 32*.

Adhesive is not required at top or bottom plates, bridging, bracing, or fire stops. Where fasteners at vertical joints are undesirable, gypsum panels may be prebowed, as shown in *Figure 33*.

Prebowing puts an arc in the gypsum board, which keeps it in tight contact with the adhesive after the board is applied. Supplemental fasteners (placed 16" on center) are then used at the top and bottom plates.

Gypsum board may be prebowed by stacking it, face up, with the ends resting on 2 × 4 lumber or other blocks, and with the center of the boards resting on the floor. Allow it to remain overnight or until the boards have a permanent bow.

Architectural boards can also be installed using adhesive, but care should be taken to avoid adhesive contact with the decorated face. Position the boards within the open time specified for the adhesive and use a rubber mallet to tap the boards along the studs to ensure a continuous bond with the framing. Follow the manufacturer's specifications for architectural gypsum board.

4.6.6 Adhesive Application to Metal Framing

Some stud adhesives, such as those used with steel framing, require fasteners on intermediate supports as well as at the perimeters of gypsum panels. The framing spacing varies both according to the load and the type of board being used. See *Table 2*.

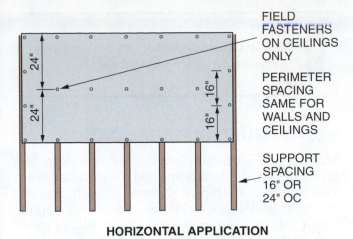

HORIZONTAL APPLICATION

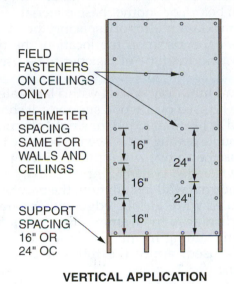

VERTICAL APPLICATION

206F32.EPS

Figure 32 ◆ Supplemental wall and ceiling fasteners.

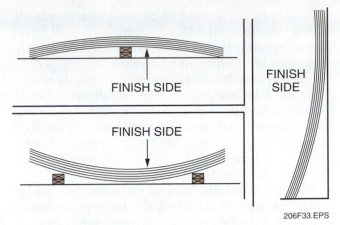

206F33.EPS

Figure 33 ◆ Prebowing of gypsum panels.

4.6.7 *Adhesive Application to Concrete and Masonry*

Gypsum board panels can be laminated directly to above-grade interior masonry and concrete wall surfaces if the surface is dry, smooth, clean, and flat. Gypsum board can be laminated directly to exterior cavity walls if the cavities are properly insulated to prevent condensation and the inside face of the cavity is properly waterproofed.

Prefinished gypsum board with a surface that is highly resistant to water vapor should not be laminated to concrete or masonry, because moisture may become trapped within the gypsum core of the board. The base surface must be made as level as possible. Rough or protruding edges and excess joint mortar should be removed and any depressions filled with mortar to make the wall surface level.

Table 2	Maximum Spacing of Ceiling Framing			
Gypsum Board (Thickness)		**Application to Framing**		**Maximum OC Spacing of Framing**
Base	Face	Base	Face	
⅜"*	⅜"	Perpendicular	Perpendicular or Parallel	16"
½"*	⅜" or ½"	Perpendicular or Parallel	Perpendicular or Parallel	16"
⅝"*	½"	Perpendicular or Parallel	Perpendicular or Parallel	16"
⅝"*	⅝"	Perpendicular or Parallel	Perpendicular or Parallel	24"

Adhesive between plies should be dried or cured prior to any decorative treatment. This is especially important when a spray-applied, water-based texture finish is to be used.

Sidewalls – For two-layer application with adhesive between the plies, ⅜", ½", or ⅝" gypsum board may be applied perpendicularly (horizontally) or parallel (vertically) on framing spaced a maximum of 24" OC.

Base surfaces should be cleaned of all form oil, curing compound, loose particles, dust, and grease in order to ensure an adequate bond. Concrete should be allowed to cure for at least 28 days before gypsum board is laminated directly to it.

Exterior below-grade walls or surfaces should be furred and protected with the installation of a vapor barrier and insulation in order to provide a suitable base for attaching the gypsum board. This is also true for any surface that cannot be prepared readily for direct lamination.

Supplemental mechanical fasteners spaced 16" on center may be used to hold the gypsum board in place while the adhesive is developing a bond.

A variety of clips, runners, and adjustable brackets are available with furring systems to facilitate installation over irregular masonry walls, as shown in *Figure 34*. When special clips are used, the manufacturer's instructions for their use must be followed.

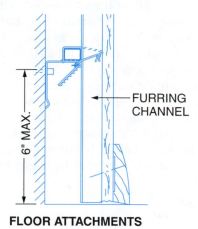

FLOOR ATTACHMENTS

ROUGH OR FINISHED CEILING

- METAL TRIM
- FURRING CHANNEL
- ¼" – MIN.
- 2¼" – MAX.
- ¾" – C.R. CHANNEL
- WIRE TIE
- FURRING BRACKET

6" MAX.

CEILING ATTACHMENTS

206F34.EPS

Figure 34 ◆ Adjustable wall furring.

NOTE

An alternative to anchoring furring to a masonry wall is to build a 1⅝" metal stud wall.

5.0.0 ◆ MULTI-PLY APPLICATION

Multi-ply construction consists of one or more layers of gypsum board applied over a base layer. This results in an improved surface finish, greater strength, and higher fire resistance and sound classifications. The base layer can consist of gypsum backer board (with or without foil), regular gypsum board, or another base material.

The maximum support spacing for multi-ply systems depends upon the location, the type of face ply, and the type of fastener to be used. See *Table 3*.

When a multi-ply system with a laminated face ply is to be used over wood supports, it should be fastened as recommended for single-ply construction. Double nailing is not needed because the fasteners used on a two-ply application will produce a firmly fastened system.

In a normal wall installation, the base ply may be installed with the long edges either parallel or perpendicular to the framing. End joints should occur on framing members. If foil-backed wallboard is used, apply the foil side against the framing.

The face ply is installed perpendicular to the base ply, with the joints offset from the base ply

Table 3	Fastener Spacing			
	Nail Spacing		**Screw Spacing**	
Location	**Laminated Face Ply**	**Nailed Face Ply[1]**	**Laminated Face Ply[2]**	**Screwed Face Ply[1]**
Walls	8" OC	16" OC	16" OC	24" OC
Ceilings	7" OC	16" OC	16" OC	24" OC

Note: Fastener size and spacing for sound-deadening boards vary for different types of fire-rated and sound-rated construction. Follow the manufacturer's recommendations.

[1]Fastener spacing for face ply shall be the same as for single-layer application.

[2]12" OC for both ceilings and walls when supports are spaced 24" OC.

Multi-Ply Wallboard
INSIDE TRACK

The most common reason for using multi-ply wallboard construction is to meet fire-rating requirements.

joints. The exception occurs when the face ply consists of decorated gypsum panels. In that case, the face ply is always installed parallel to the framing, regardless of how the base ply is installed. The wallboard should not be in contact with the floor, as it can wick up moisture that will damage the wallboard.

At inside corners, it is recommended that only the overlapping base ply be nailed or screwed and that the fasteners be omitted from the face ply. The floating corner treatment is better able to resist structural stresses, as shown in *Figure 35*.

A floating angle construction for multi-ply systems has the overlapping side of the base ply nailed only at the interior corners.

5.1.0 Attaching the Base Ply to Metal Framing or Furring

Base ply gypsum board is normally attached to steel framing and furring with screws that are at least ⁵⁄₁₆" longer than the thickness of the board. The board application may be either perpendicular or parallel. In a perpendicular application where no adhesive is used between the plies, the

base ply should be fastened with a single screw in each stud or furring channel at the board edges and with one screw at the middle of the board at each stud or channel.

In parallel applications with no adhesive between the plies, the base ply is fastened with screws located 12" on center along the edges of the board and 24" on center in each stud in the field of the board. When the base ply is to be attached perpendicular or parallel to steel framing that is 16" on center and with adhesive between the plies, the screw spacing should be 12" on center for walls. The maximum screw spacing on steel framing at 24" on center is 12" on center for both walls and ceilings.

5.2.0 Face Ply Attachment

The joints in the face ply should be offset at least 10" from the joints in the base ply. Gypsum board can be applied to framing in either a perpendicular or parallel configuration, whichever results in the least waste of materials. A perpendicular application is preferred on walls, because it usually results in fewer joints. Architectural board is usually installed parallel and does not require joint treatment. Some systems are designed to use decorative battens over the joints. When the face ply is attached with fasteners and with no adhesive between the plies, the maximum spacing and minimum penetration recommended for screws should be the same as for single-ply applications.

5.3.0 Adhesive Attachment

Typical multi-ply construction may employ sheet lamination, strip lamination, or spot lamination to attach the face ply to the base ply. Lamination involves covering the entire back with a laminating adhesive using a notched spreader (*Figure 36*), box spreader, or other suitable tool.

Notched spreaders are used to apply adhesive for sheet lamination. The size and spacing of the notches are determined by the type of adhesive

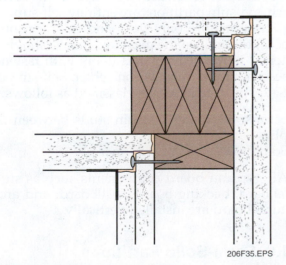

206F35.EPS

Figure 35 ◆ Floating corner treatment.

being used. The gypsum board should be applied using moderate pressure. Any adhesive squeezed out at the joints must be promptly removed.

Strip and spot lamination are preferred to sheet lamination in sound-rated partitions. In strip lamination, the adhesive is applied in ribbons using a special spreader. The ribbons are normally spaced 16" to 24" on center, as shown in *Figure 37*. In spot lamination, spots of adhesive are brushed or daubed on in a regular pattern.

206F36.EPS

Figure 36 ◆ Spreading adhesive using a notched spreader.

206F37.EPS

Figure 37 ◆ Properly spacing adhesive ribbons.

5.4.0 Supplemental Fasteners

In order to ensure a satisfactory adhesive bond, it is necessary to hold the face ply firmly against the base ply with supplemental fasteners, shoring, or bracing while the adhesive is setting up. Generally, the fasteners are applied in the field of each laminated face ply in ceiling applications. On side walls, these fasteners can be placed at the perimeters of the boards where they will be concealed by the joint treatment or trim.

It should be remembered that in fire-rated assemblies, the specific fastener spacing is given for the particular assembly tested and may not be related to whether or not adhesive is used between the plies. Fastener spacing details for fire-rated assemblies are available from the sponsor of the test and, in many cases, from the Gypsum Association.

In sound-rated partitions where fire resistance is not a consideration, an acceptable practice is to attach the face ply vertically over sound insulation board or backing board with permanent fasteners in the back ends. Intermediate fasteners are omitted, and the panels are temporarily braced until the adhesive has developed a sufficient bond.

6.0.0 ◆ SELF-SUPPORTING GYPSUM PARTITIONS

Self-supporting gypsum partitions are nonbearing, noncombustible, multi-ply partitions. They can be fabricated at the job site almost entirely from gypsum board. The inner plies are gypsum studs formed by gypsum board or sheets, with the face plies laminated to them. Unlike framed ceiling and wall assemblies, which depend on wood, metal, or other construction to support the surface finish, gypsum partitions are entirely self-supporting. They require only floor and ceiling runners (tracks) to stabilize the partition. Tracks should be fastened to the ceilings and floors with fasteners spaced not more than 24" on center. Self-supporting gypsum partitions are classified as follows:

- Semi-solid, using gypsum studs between face plies
- Solid, with a coreboard inner ply

All gypsum board components such as studs, coreboard, backing board, wallboard, and architectural board are installed vertically.

6.1.0 Semi-Solid Partitions

Semi-solid partitions provide a vertical interior passage for mechanical and electrical services.

They consist of single-ply or two-ply faces of gypsum board separated by vertical gypsum studs that are 6" wide and preferably not less than 1⅝" thick. See *Figure 38*.

In single-faced partitions, the face ply is attached directly to the gypsum studs. Partitions with two-ply faces consist of gypsum studs, a backing board base ply, and a gypsum board face ply.

Semi-solid walls may be prelaminated or laminated in place. Gypsum studs that have been cut 6" shorter than the wallboard are laminated to the face ply on one side of the partition. The edges of door and window openings should be reinforced with wood or metal. Adhesive is spread on the erected base ply or on the gypsum studs, keeping the adhesive 1" from the stud edges. The stud is then pressed into place. The stud-reinforced panels are secured to wood or metal tracks at the top and bottom using the appropriate drywall screws spaced 16" on center.

Adhesive is then spread on the studs attached to the erected panels, and facing panels are installed over the studs and attached to the runners with screws spaced 12" on center. An adequate bond to the studs is ensured by using 1½",

Type G drywall screws placed along each stud, beginning 12" from the top and bottom and spaced not more than 36" on center. A rapid method for adhering studs to panels is shown in *Figure 39*.

Following assembly, panels with studs attached are erected with adhesive spread on the exposed face of the studs. The erected panel edges are kept in contact with the studs by supplemental fasteners while the adhesive develops a bond.

For partitions with two-ply faces, the studs and base plies are assembled and installed as previously described for two-ply supported partitions using supplemental drywall screws. The screw spacing for fire-rated partitions should be in accordance with the manufacturer's requirements or as recommended by the Gypsum Association.

Separated stud construction to improve sound insulation can be provided by either single- or two-ply partitions. The installation is essentially the same as for regular gypsum stud partitions, but twice as many studs are required. These are spaced 12" on center and staggered on opposite sides of the partition.

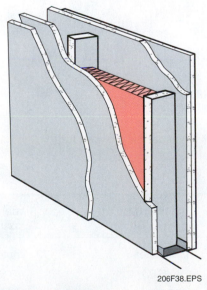

206F38.EPS

Figure 38 ◆ Semi-solid partition.

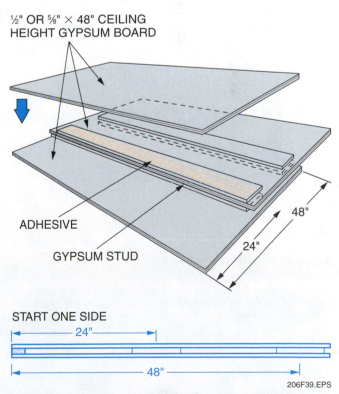

206F39.EPS

Figure 39 ◆ Cross section–application of prelaminated panel assemblies.

6.2.0 Solid Partitions

Solid gypsum partitions have gypsum board faces laminated to each side of the coreboard. Vertical joints in the face ply should be offset 3" from joints in the coreboard. The face ply is typically attached by sheet lamination with supplemental drywall screws. Double- and triple-solid partitions are separated by air spaces, as shown in *Figure 40*.

Double-solid gypsum board partitions have parallel courses attached to two runners separated by an air space. The courses consist of coreboard faced with wallboard.

Triple-solid gypsum board partitions have three parallel courses and provide superior sound control when a mineral wool or glass fiber blanket is added.

The depth of electrical outlet boxes installed in solid partitions should not exceed 1½". The outlet boxes installed in double or triple partitions may exceed this dimension.

7.0.0 ◆ ARCHITECTURAL NONCOMBUSTIBLE PANELING OVER GYPSUM SUBSTRATE

The addition of gypsum board as a substrate when using architectural panels provides increased fire resistance and sound control. In new or existing construction, a ⅜" or ½" gypsum board substrate is recommended before applying noncombustible paneling (*Figure 41*).

Research has shown that increased fire resistance, sound control, and impact resistance may be achieved through the use of a gypsum board substrate under noncombustible paneling. Examples are shown in *Table 4*.

The gypsum board substrate should be attached parallel to the framing using 1⅜" drywall nails, 1" drywall screws, or a drywall stud adhesive. The spacing of fasteners should be the same as mentioned previously. The edges of the board should be centered on framing members. The joints need not be taped and finished. The rigid architectural panels should be applied using a

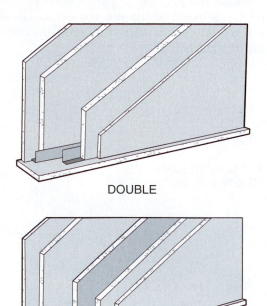

DOUBLE

TRIPLE

206F40.EPS

Figure 40 ◆ Solid partitions.

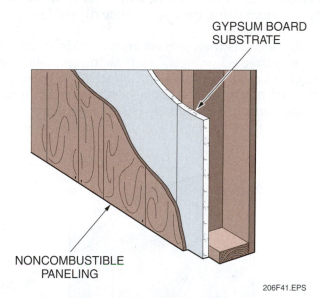

GYPSUM BOARD SUBSTRATE

NONCOMBUSTIBLE PANELING

206F41.EPS

Figure 41 ◆ Gypsum board substrate.

Table 4 Fire Resistance, Sound Control, and Impact Resistance Data			
	¼" Paneling without Gypsum Board	¼" Paneling with ⅜" Gypsum Board	¼" Paneling with ½" Gypsum Board
Burn-through time	8 minutes	42 minutes	73 minutes (plus)
Sound transmission class (STC)	28 STC	40 STC	40 STC
Impact resistance	130 ft-lbs	410 ft-lbs	410 ft-lbs

bead of panel adhesive over each stud and a bead midway between the studs. Joints in the architectural panel should be staggered from the gypsum board. Secure the panel at the top and bottom using 4d finishing nails located 12" on center and one nail at the middle point of each stud.

8.0.0 ◆ RESURFACING EXISTING CONSTRUCTION

Gypsum board may be used to provide a new finish on existing walls and ceilings of wood, plaster, masonry, or wallboard. If the existing surface is structurally sound and provides a sufficiently smooth and solid backing without shimming, ¼" gypsum board can be applied with adhesives, nails, or screws. Drywall nails should penetrate the framing by ⅞". When power-driven screws are to be used, the threaded portion of the screw must penetrate the framing by at least ⅝".

Existing surfaces that are too irregular to receive gypsum board directly should be furred and shimmed to provide a suitable fastening surface. The minimum gypsum board thickness for various support spacing and installation methods should be as previously recommended for new construction over furring. Any surface trim for mechanical and electrical equipment, such as switch plates, outlet covers, and ventilating grilles, should be removed and saved for reinstallation. Electrical boxes should be reset prior to the installation of new gypsum board.

9.0.0 ◆ CORNER BEADS AND CASINGS

Figure 42 shows common beads and casings used around doors, windows, and other openings. They are also used when gypsum board is butted against a different surfacing material. Casing moldings cover the unfinished ends of drywall to protect it, while corner beads protect outside corners. L beads protect the drywall where it abuts dissimilar materials, but they require finishing; J beads serve the same purpose, but do not require finishing.

Figure 43 shows examples of decorative trim used with architectural drywall panels.

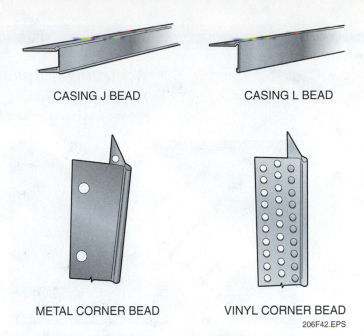

CASING J BEAD   CASING L BEAD

METAL CORNER BEAD   VINYL CORNER BEAD

206F42.EPS

Figure 42 ◆ Examples of protective beads.

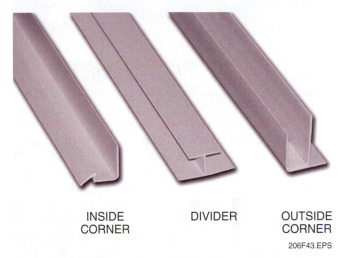

INSIDE CORNER   DIVIDER   OUTSIDE CORNER

206F43.EPS

Figure 43 ◆ Moldings for prefinished drywall.

10.0.0 ◆ FIRE-RATED AND SOUND-RATED WALLS

The construction of walls and partitions is driven by the fire and soundproofing requirements specified in local building codes. In some cases, a frame wall with ½" gypsum drywall on either side

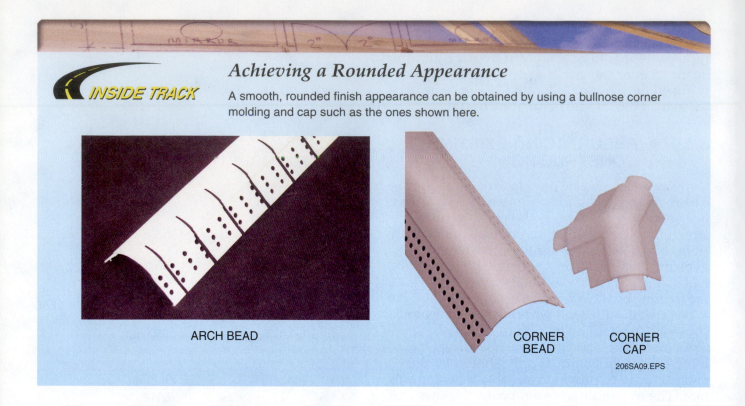
is satisfactory. In extreme cases, such as the separation between the offices and the manufacturing spaces in a factory, it may be necessary to have a concrete block (CMU) wall combined with fire-resistant gypsum board, along with rigid and/or fiberglass insulation, as shown in *Figure 44*. This is especially true if there is any explosion or fire hazard.

While they are only occasionally used in residential construction, steel studs are the standard for framing walls and partitions in commercial construction.

Once the studs are installed, one or more layers of gypsum board and insulation are applied. The type and thickness of the wallboard and insulation depend on the fire rating and soundproofing requirements. Soundproofing needs vary from one use to another and are often based on the amount of privacy required for the intended use. For example, executive offices and medical examination rooms may require more privacy than general offices.

The requirements for sound reduction and fire resistance can significantly affect the thickness of a wall. For example, a steel stud wall with a high sound transmission class (STC) and fire resistance might have a total thickness of nearly 6", while a low-rated wall might have a thickness of only 3" using steel studs and 2¼" using wooden studs.

10.1.0 Fire-Rated Construction

Every wall, floor, and ceiling in a building is rated for its fire resistance, as established by building codes. The fire rating is stated in terms of hours, such as one-hour wall or two-hour wall. The rating denotes the length of time an assembly can withstand fire and provide protection from it, as determined under laboratory conditions (*Figure 45*). The greater the fire rating, the thicker the wall is likely to be.

In multi-family residential construction, such as apartments and townhouses, the walls and ceilings dividing the occupancies must meet special fire and soundproofing requirements. The code requirements will vary from one location to another and may even vary within areas of a jurisdiction. For

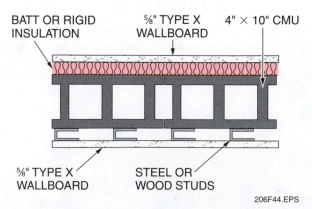

BATT OR RIGID INSULATION

⅝" TYPE X WALLBOARD

4" × 10" CMU

⅝" TYPE X WALLBOARD

STEEL OR WOOD STUDS

206F44.EPS

Figure 44 ◆ High fire/noise resistance partition.

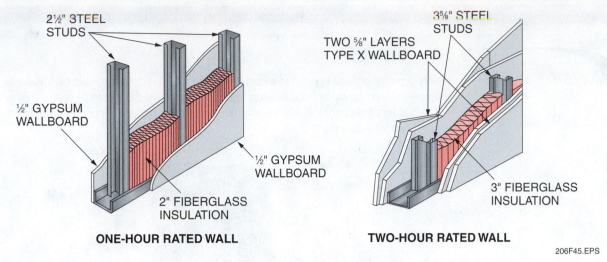

ONE-HOUR RATED WALL

2½" STEEL STUDS

½" GYPSUM WALLBOARD

2" FIBERGLASS INSULATION

½" GYPSUM WALLBOARD

TWO-HOUR RATED WALL

3⅝" STEEL STUDS

TWO ⅝" LAYERS TYPE X WALLBOARD

3" FIBERGLASS INSULATION

206F45.EPS

Figure 45 ◆ Partition wall examples.

example, dwellings in high-risk areas may have stricter standards than those in other areas of the same city or county.

In some cases, the code may require a masonry wall between occupancies. This masonry wall may even be required to penetrate the roof of the building so that if a fire occurs, it is contained within the unit in which it started because it is unable to travel through the walls or across the attic space.

There are many different construction methods for so-called party walls. Each is designed to meet different fire and soundproofing standards. The wall is likely to be more than 3" thick and contain several layers of gypsum board and insulation. A fire-rated wall may abut a non-rated partition or wall. In this case, the rated wall must be carried through to maintain the fire rating. *Figure 46* shows an example of how this can be done.

10.1.1 Firestopping

Firestopping means cutting off the air supply so that fire and smoke cannot readily move from one location to another. You will hear the term firestop used in two different ways.

In frame construction, a firestop is a piece of wood or fire-resistant material inserted into an opening such as the space between studs. This firestop acts as a barrier to block airflow that would allow the space to act as a chimney, carrying fire rapidly to upper floors. It does not put out the fire, but it slows the fire's progress.

In commercial construction and some residential applications, firestopping material is used to close wall penetrations such as those created to run conduit, piping, and air conditioning ducts. If such openings are not sealed, fire will travel through the openings in its search for oxygen.

In order to meet the fire rating standards established by the building and fire codes, the openings must be sealed. The firestopping methods used for this purpose are classified as mechanical and nonmechanical.

Code Compliance

Local codes specify the fire ratings that must be achieved in different occupancies and uses. They may also specify the nailing pattern to be used on drywall panels. Check the local codes before proceeding. Also, keep in mind that electrical and plumbing installations must be inspected in order for the building to receive a certificate of occupancy. Before covering these installations with drywall, make sure the inspection has been performed. An inspection sheet should be at the site.

Fire Ratings

When gypsum drywall is used in a party wall, the architect's plans must be followed precisely. If the inspector finds flaws in the construction, a certificate of occupancy will not be issued. This photo illustrates why building codes place so much emphasis on properly constructed party walls.

206SA10.EPS

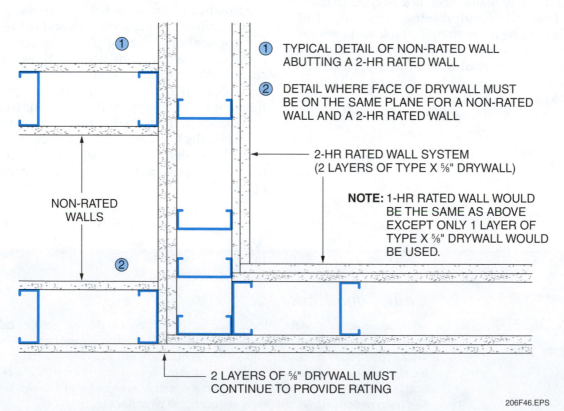

① TYPICAL DETAIL OF NON-RATED WALL ABUTTING A 2-HR RATED WALL

② DETAIL WHERE FACE OF DRYWALL MUST BE ON THE SAME PLANE FOR A NON-RATED WALL AND A 2-HR RATED WALL

2-HR RATED WALL SYSTEM (2 LAYERS OF TYPE X ⅝" DRYWALL)

NOTE: 1-HR RATED WALL WOULD BE THE SAME AS ABOVE EXCEPT ONLY 1 LAYER OF TYPE X ⅝" DRYWALL WOULD BE USED.

NON-RATED WALLS

2 LAYERS OF ⅝" DRYWALL MUST CONTINUE TO PROVIDE RATING

206F46.EPS

Figure 46 ◆ An example of a fire-rated wall abutting a non-rated wall.

Drilling Fire-Rated Walls

There are many wall variations, so you must first establish the type of construction before undertaking any invasive action, such as drilling. If drilling is permitted, you may have to use firestopping materials to seal off the opening.

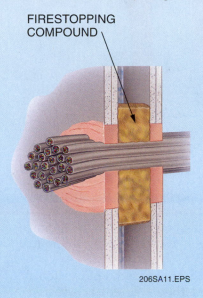

FIRESTOPPING COMPOUND

206SA11.EPS

Mechanical firestops are devices such as the one shown in *Figure 47* that mechanically seal the opening.

Nonmechanical firestops are fire-resistant materials, such as caulks and putties, that are used to fill the space around the conduit or piping. You may be required to install various nonmechanical firestopping materials when working with fire-rated walls and floors. Holes or gaps affect the fire rating of a floor or wall. Properly filling these penetrations with firestopping materials maintains the rating. Firestopping materials are typically applied around all types of piping, electrical conduit, ductwork, electrical and communication cables (*Figure 48*), and similar devices that run through openings in floor slabs, walls, and other fire-rated building partitions and assemblies.

Nonmechanical firestopping materials are classified as intumescent or endothermic. Both are formulated to help control the spread of fire before, during, and after exposure to open flames. When subjected to the extreme heat of a fire, intumescent materials expand (typically up to three times their original size) to form a strong insulating material that seals the opening for three to four hours. Should the insulation on the cables, pipes, etc., passing through the penetration become consumed by the fire, the expansion of the firestopping material also acts to fill the void in

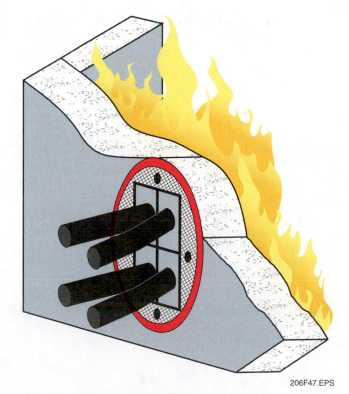

206F47.EPS

Figure 47 ◆ Mechanical firestop device.

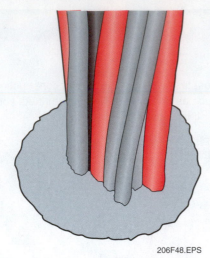

Figure 48 ◆ Fire barrier of moldable putty installed around electrical cables.

the floor or wall in order to help stop the spread of smoke and other toxic products of combustion.

Endothermic materials block heat by releasing chemically bound water, which causes them to absorb heat.

Firestopping materials are formulated in such a way that when activated, they are free of corrosive gases, reducing the risks to building occupants and sensitive equipment.

Firestopping materials are made in a variety of forms including composite sheeting, caulks, silicone sealants, foams, moldable putty, wrap strips, and spray coatings. They come in both one-part and two-part formulations. The installation of these materials must always be done in accordance with the applicable building codes and the manufacturer's instructions for the product being used. Depending on the product, firestopping materials can be applied via spray equipment, conventional caulking guns, pneumatic pumping equipment, or a putty knife.

Any firestopping materials used must meet the criteria of standard *ASTM E814, Fire Test*, as tested under positive pressure. They must also have an hourly rating that is equal to or greater than the hourly rating of the floor or wall being penetrated. Based on *ASTM E814/UL 1479* tests, one of four ratings, measured in time, may be applied to firestopping materials and systems. These ratings are as follows:

- *F rating* – A firestopping system meets the requirements of an F rating if it remains in the opening during the fire test for the rating period without permitting the passage of flames through the opening or the occurrence of flaming on any element of the unexposed side of the assembly.
- *FT rating* – A firestopping system meets the requirements of an FT rating if it remains in the opening during the fire test within the limitations as specified for an F rating. In addition, the transmission of heat through the firestopping system during the rated period shall not have been such as to raise the temperature of any thermocouple on the unexposed surface of the firestopping system by more than 347.8°F (181°C) above its initial temperature.
- *FH rating* – A firestopping system meets the requirements of an FH rating if it remains in the opening during the fire test within the limitations for an F rating. In addition, during a hose stream test, the firestopping system shall not develop any opening that would permit a projection of water from the stream to the unexposed side.
- *FTH rating* – A firestopping system shall be considered as meeting the requirements of an FTH rating if it remains in the opening during the fire test and hose stream test within the limitations as described for FT and FH ratings.

10.2.0 Sound-Isolation Construction

The first step for airborne sound isolation of any assembly is to close off air leaks and flanking paths. Since noise can travel over, under, or around walls, through the windows and doors adjacent to them, through air ducts, and through floors and crawl spaces, these paths must be correctly treated.

Buildings are generally required to meet a sound transmission class (STC) rating. The STC is a numeric rating representing the effectiveness of the construction in isolating airborne sound transmission. The higher the STC rating, the better the sound absorption. Hairline cracks and other openings can have an adverse effect on the ability of a building to achieve its STC rating, particularly in higher-rated construction. Where a very high

STC performance is needed, air conditioning, heating, and ventilating ducts should not be included in the assembly. Failure to observe special construction and design details can destroy the effectiveness of the best assembly. Improved sound isolation is obtained by the following:

- Separate framing for the two sides of a wall
- Resilient channel mounting for the gypsum board
- Using sound-absorbing materials in wall cavities
- Using adhesive-applied gypsum board of varying thicknesses in multi-layer construction
- Caulking the perimeter of gypsum board partitions, openings in walls and ceilings, partition/mullion intersections, and outlet box openings
- Locating recessed wall fixtures in different stud cavities

The entire perimeter of sound-isolating partitions should be caulked around the gypsum board edges to make it airtight, as detailed in *Figure 49*. The caulking should be a non-hardening, non-shrinking, non-bleeding, non-staining, resilient sealant.

Sound-control sealing must be covered in the specifications, understood by all related tradespeople, supervised by the appropriate party, and inspected carefully as the construction progresses.

10.2.1 Separated Partitions

A staggered wood stud gypsum partition placed on separate plates will provide an STC between 40 and 42. The addition of a sound-absorbing material between the studs of one partition side can increase the STC by as much as 8 points. With 5/8" Type X gypsum board on each side, an assembly has a fire resistance classification of one hour. Separated walls without framing can also be constructed by using an all-gypsum, double-solid, or semi-solid partition.

Steel or wood tracks fastened to the floor and ceiling hold the partitions in place. For the attachment of kitchen cabinets, lavatories, ceramic tile, medicine cabinets, and other fixtures, a staggered stud wall rather than a resilient wall is recommended. The added weight and fastenings may short circuit the construction acoustically.

10.2.2 Resilient Mountings

Resilient attachments acting as shock absorbers reduce the passage of sound through the wall or ceiling and increase the STC rating. Further STC increases can result from more complex construction methods incorporating multiple layers of gypsum board and building insulation in the wall cavities.

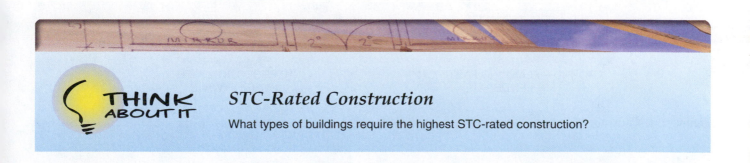

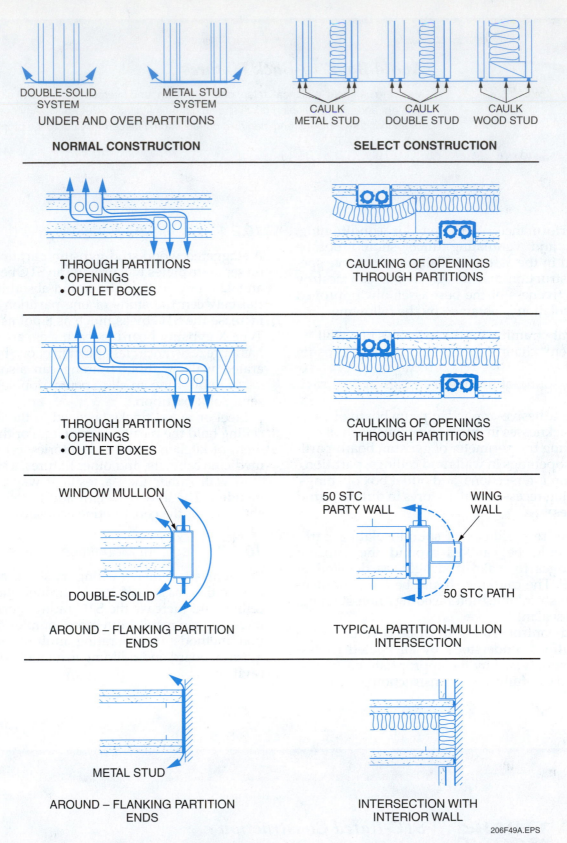

NORMAL CONSTRUCTION

SELECT CONSTRUCTION

DOUBLE-SOLID SYSTEM

METAL STUD SYSTEM

UNDER AND OVER PARTITIONS

CAULK METAL STUD

CAULK DOUBLE STUD

CAULK WOOD STUD

THROUGH PARTITIONS
• OPENINGS
• OUTLET BOXES

CAULKING OF OPENINGS THROUGH PARTITIONS

THROUGH PARTITIONS
• OPENINGS
• OUTLET BOXES

CAULKING OF OPENINGS THROUGH PARTITIONS

WINDOW MULLION

DOUBLE-SOLID

AROUND – FLANKING PARTITION ENDS

50 STC PARTY WALL

WING WALL

50 STC PATH

TYPICAL PARTITION-MULLION INTERSECTION

METAL STUD

AROUND – FLANKING PARTITION ENDS

INTERSECTION WITH INTERIOR WALL

206F49A.EPS

Figure 49 ◆ Caulking of sound-isolation construction (1 of 2).

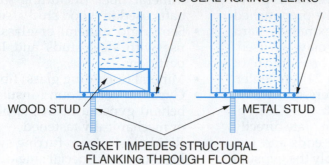

PRE-DESIGN CONSTRUCTION
SIMULATING LABORATORY CONDITIONS

¼" PERIMETER RELIEF AND CAULKING
TO SEAL AGAINST LEAKS

WOOD STUD   METAL STUD

GASKET IMPEDES STRUCTURAL
FLANKING THROUGH FLOOR

TYPICAL FLOOR-CEILING OR ROOF DETAIL

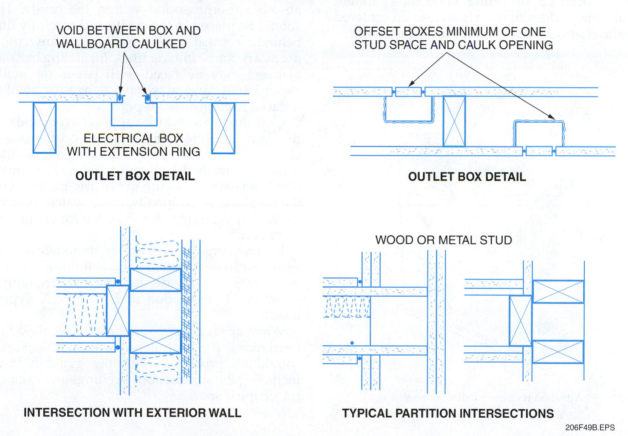

VOID BETWEEN BOX AND
WALLBOARD CAULKED

ELECTRICAL BOX
WITH EXTENSION RING

OUTLET BOX DETAIL

OFFSET BOXES MINIMUM OF ONE
STUD SPACE AND CAULK OPENING

OUTLET BOX DETAIL

INTERSECTION WITH EXTERIOR WALL

WOOD OR METAL STUD

TYPICAL PARTITION INTERSECTIONS

206F49B.EPS

Figure 49 ◆ Caulking of sound-isolation construction (2 of 2).

Resilient furring channels are attached with the nailing flange down and at right angles to the wood stud, as shown in *Figure 50*.

To install furring channels, drive 1¼" Type W screws or 6d coated nails through the prepunched holes in the channel flange. With extremely hard lumber, ⅞" or 1" Type S screws may be used. Locate the channels 24" from the floor, within 6" of the ceiling line, and no more than 24" on center. Extend the channels into all corners and fasten them to the corner framing. Attach ½" × 3" gypsum board filler strips to the bottom plate directly over the studs by overlapping the ends and fastening both flanges to the stud. Apply the gypsum board horizontally with the long dimension parallel to the resilient channels using 1" Type S screws spaced 12" on center along the channels. The abutting edges of boards should be centered over the channel flange and securely fastened.

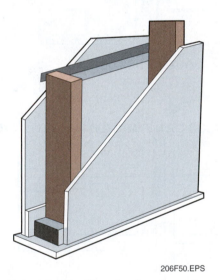

206F50.EPS

Figure 50 ◆ Attached resilient furring channel.

10.2.3 Sound-Isolating Materials

Sound-isolating materials include:

- Mineral fiber (including glass) blankets and batts used in wood stud assemblies
- Semi-rigid mineral or glass fiber blankets for use with steel studs and laminated gypsum partitions
- Mineral (including glass) fiberboard
- Gypsum core sound-insulating board used behind gypsum board, applied with adhesive or mechanically fastened
- Rigid plastic foam furring systems
- Lead or other special shielding materials

Mineral wool or glass fiber insulating batts and blankets may be used in assembly cavities to absorb airborne sound within the cavity. They should be placed in the cavity and carefully fitted behind electrical outlets and around any cutouts necessary for plumbing lines. Insulating batts and blankets may be faced with paper or another vapor barrier and may have flanges or be of the unfaced friction-fitted type.

Gypsum board may be applied over rigid plastic foam insulation. It is applied on the interior side of exterior masonry and concrete walls to provide a finished wall and to protect the insulation from early exposure to fire originating within the building. Additionally, these systems provide the high insulating values needed for energy conservation.

In new construction or for remodeling, these systems can be installed with as little as 1" dimension from the inside face of the framing or masonry (½" insulation and ½" Type X gypsum board).

When applying gypsum board over rigid foam insulation, the entire insulated wall surface should be protected with the gypsum board, including the surface above ceilings and in closed, unoccupied spaces.

Rigid Foam Insulation

Gypsum board applied over rigid plastic foam insulation in the manner described may not necessarily provide the finish ratings required by local building codes. Many building codes require a minimum fire protection for rigid foam on interior surfaces equal to that provided by ½" Type X gypsum board when tested over wood framing. The flammability characteristics of rigid foam insulation products vary widely, and the manufacturer's literature should be reviewed.

Single-ply or double-ply, ½" or ⅝" gypsum board should either be screw-attached to steel wall furring members attached to the masonry or nailed directly into wood framing, as shown in *Figure 51*. Follow the insulation manufacturer's instructions.

Furring members should be designed to minimize thermal transfer through the member and to provide a 1¼" minimum width face or flange for screw application of the gypsum board.

Furring members should be installed vertically and spaced 24" on center. Blocking or other backing as required for attachment and support of fixtures and furnishings should be provided. Furring members should also be attached at floor/wall and wall/ceiling angles (or at the termination of the gypsum board above suspended ceilings) and around door, window, and other openings. Single-ply gypsum board should be applied vertically, with the long edges of the board located over furring members. The installation should be planned carefully to avoid end joints. The fastener spacing should be as required for single-ply application over framing or furring.

In double-ply applications, the base ply should be applied vertically. In horizontal face ply applications, the face ply and end joints should be offset by at least one framing or furring member space from the base ply edge joints.

The fastener spacing should be as required for two-ply application over framing or furring, as discussed previously.

In wallboard applications, mechanical fasteners should be of such a length that they do not penetrate completely to the masonry or concrete. In single-layer applications, all joints between gypsum boards should be reinforced with tape. In addition, gypsum board joints should be finished with joint compound. In two-ply applications, the base layer joints may be concealed or left exposed.

Adhesive should not be used to apply vinyl-faced gypsum board face layers over a wall insulated with rigid foam.

10.3.0 Control Joints

Control joints (*Figure 52*) should be installed in gypsum board systems wherever expansion or control joints occur in the base exterior wall and not more than 30' on center in long wall furring runs. Wall or partition height door or window frames may be considered control joints.

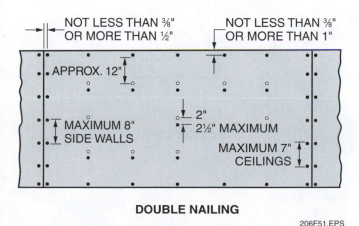

Figure 51 ◆ Nailing patterns for installation over rigid foam insulation.

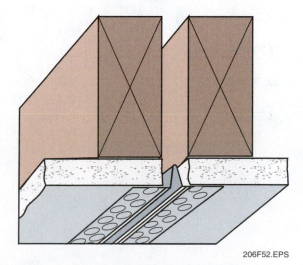

Figure 52 ◆ Typical control joint.

Control joints should also be used over window and door openings. The gypsum board should be isolated from structural elements such as columns, beams, and loadbearing interior walls, and from dissimilar wall or ceiling finishes by control joints, metal trim, or other means. Refer to the specifications and drawings for the locations of control joints.

Movement of the structure can impose severe stresses and cause cracks, either at the joint or in the field of the board. Cracks are more prevalent at an archway or over a door, because this is usually the weakest point in the construction. In new construction, it is wise to wait until at least one heating season has passed before repairing or refinishing.

A source of cracking in non-bearing walls of high-rise or commercial buildings is the modern trend toward less rigid structures. Larger deflections in structural members and greater expansion and contraction of exterior columns can impose unexpected loads on non-bearing walls and lead to cracking. Detail designs for perimeter relief of non-bearing partitions are available to improve this condition. One solution is to use relief runners to attach non-bearing walls to ceiling and column members (*Figure 53*).

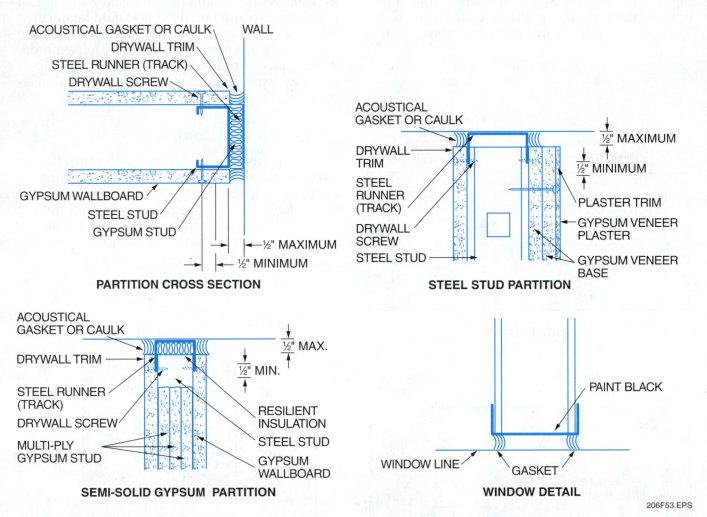

Figure 53 ◆ Designs for perimeter relief.

206F53.EPS

Control Joints

Control (expansion) joints are used in large expanses of wall or ceiling drywall to compensate for the natural expansion and contraction of a building. Control joints help prevent cracking and joint separation. They are common in commercial construction, especially where exterior concrete walls contain expansion joints.

If the control joint is installed in a space where fire rating and/or sound control are important, a seal must be used behind the control joint. The control joint has a ¼" slot that is covered by plastic tape. The tape is removed after the joint is finished, leaving a small recess.

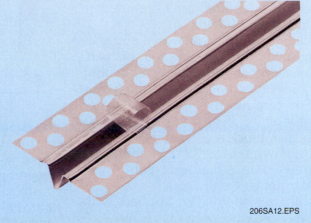

206SA12.EPS

11.0.0 ◆ MOISTURE-RESISTANT CONSTRUCTION

Special consideration must be given when finishing bathrooms, laundries, kitchens, and other areas subject to moisture. Although water-resistant gypsum board can be used in these applications, a waterproof tile backer such as cement board should be used. Unlike water-resistant gypsum board, cement board can also be used in areas of high moisture and humidity such as saunas and gang showers.

As mentioned earlier, gypsum board that will be subjected to moisture should not be foil backed or applied directly over a vapor barrier, as the vapor barrier will trap moisture within the board.

In moisture-resistant construction, the tile backer or gypsum board should be applied horizontally, with the factory-bound edge spaced a minimum of ¼" above the lip of the shower pan or tub. Shower pans or tubs should be installed prior to the installation of the board. Shower pans should have an upstanding lip or flange located at a minimum of 1" higher than the entry wall to the shower. It is recommended that the tub be supported. If necessary, the board should be furred away from the framing members so the upstanding leg of the pan (*Figure 54*) will be on the same plane as the face of the board.

> **NOTE**
>
>
>
> Different types of waterproof boards have different applications and limitations, so it is always necessary to check the manufacturer's product data sheets and installation instructions for the type of board being used.

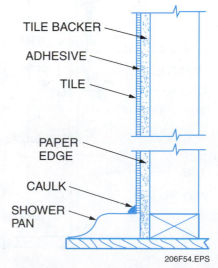

Figure 54 ◆ Pan is on the same plane as the face of the board.

Tile Application

Ceramic wall tile application to gypsum board should meet the *American National Standard Specifications for Installation of Ceramic Tile with Water-Resistant Organic Adhesive.* The adhesives used should meet the *American National Standard for Organic Adhesives for Installation of Ceramic Tile.*

An additional board extending the full height from floor to ceiling is required for a fire-rated or sound-rated construction (*Figure 55*).

Suitable blocking should be provided approximately 1" above the top of the tub or pan. Between-stud blocking should be placed behind the horizontal joint of the board above the tub or shower pan. For ceramic tile applications, use studs that are at least 3½" deep and placed 16" on center. Appropriate blocking, headers, or supports should be provided for tub plumbing fixtures and to receive soap dishes, grab bars, towel racks, and similar items.

Tile backer boards should be applied with nails or screws spaced not more than 8" on center. When ceramic tile more than ⅜" thick is to be applied, the nail or screw spacing should not exceed 4" on center. When it is necessary for joints and nail heads to be treated with joint compound and tape, either use waterproof, non-hardening caulking compound or seal-treat joints and nail heads with a compatible sealer prior to the installation.

NOTE

The caulking compound or sealer must be compatible with the adhesive to be used for the application of the tile. Follow the adhesive manufacturer's instructions.

Interior angles should be reinforced with supports to provide rigid corners. The cut edges and openings around pipes and fixtures should be caulked flush with a waterproof, non-hardening, silicone caulking compound or adhesive complying with the *American National Standard for Organic Adhesives for Installation of Ceramic Tile.* The directions of the manufacturer of the tile, wall panel, or other surfacing material should also be followed.

The surfacing material should be applied down to the top surface or edge of the finished shower floor, return, or tub, and installed to overlap the top lip of the receptor, subpan, or tub.

12.0.0 ◆ ESTIMATING DRYWALL

This section covers the guidelines for estimating wallboard and fasteners. *Table 5* shows rules of thumb for ordering different types of drywall nails and screws.

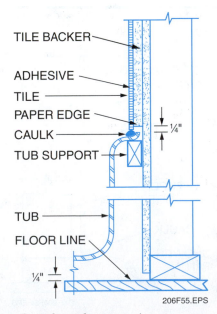

Figure 55 ◆ Sound-rated construction.

Table 5	Fastening Materials Required for 1,000 Square Feet of Drywall
Material	**Amount Required for 1,000 Square Feet of Drywall**
1¼" annular-ring nails	6¼ lb
1⅜" annular-ring nails	6¾ lb
1" drywall screws	3 lb
1¼" drywall screws	4¼ lb
1⅝" drywall screws	5½ lb

12.1.0 Gypsum Board

To estimate how many sheets of wallboard you will need for a job, first determine how many square feet of space the room contains. Multiply the length of each wall by two, add the results, then multiply the total wall space by the wall height to get the total square footage of wall space. If the ceiling is to be covered with wallboard, its square footage would be included as well. Then divide the total square footage by 32, which is the square footage of a 4' × 8' sheet, or 36, which is the square footage of a 4'-6" × 8' panel. For example, a 10' × 12' room with 8' high walls would contain 352 square feet of wall space. Converting this number into sheets (352 ÷ 32) yields 11. Door and window openings are usually figured solid unless there is a large picture window or door. This creates a built-in allowance for waste.

12.2.0 Fasteners

The types and amounts of fasteners will vary from job to job. It is important to consult the job specifications as well as the manufacturer's guidelines for this information. The following will serve as a rule-of-thumb guide to estimating fasteners.

For single-layer application 16" OC, approximately 1,000 screws are needed per 1,000 square feet of wallboard. If the framing is 24" OC, about 850 screws per 1,000 square feet will be needed.

If nails are used, you need to calculate the amount required in pounds. To install a single layer of ⅝" wallboard, you will need about 5 pounds of nails for every 1,000 square feet of wallboard. The size of the nail, and therefore the total weight, depends on the thickness of the wallboard. It can range from 4½ pounds of nails for ¼" wallboard to 7 pounds of nails for 1¼" (two ⅝" sheets laminated) wallboard.

Summary

This module covered gypsum drywall and its installation. Once the drywall is properly installed, the joints must be finished to create a smooth surface for application of the final decorative finish, such as paint or wallcovering.

Gypsum drywall panels installed over wood or steel framing are the most common method of finishing walls. In residential applications, drywall panels are commonly used to finish ceilings as well.

There are different types and sizes of drywall designed for different applications. It is therefore important that the carpenter know the different types, their applications, and their methods of installation.

Notes

1. Which of these is a common use of ¼" gypsum board?
   a. Coreboard in shaft walls
   b. Backing for tile in showers
   c. New single-layer wall system
   d. Base layer in a multi-layer system

2. Which of the following is true regarding the use of water-resistant gypsum board?
   a. It is recommended for use in saunas and steam rooms.
   b. It is not recommended for showers and tubs.
   c. It should have a foil backing.
   d. It should be applied directly over a vapor barrier.

3. The tool commonly used to cut gypsum board panels to size is a _____.
   a. circular saw
   b. cross-cut saw
   c. utility knife
   d. hook-bill knife

4. The point at which the edges of two panels of drywall meet is known as a _____.
   a. bedding seam
   b. joint
   c. gypsum lath
   d. dimple

5. As a general rule, gypsum drywall should only be installed when the building temperature is greater than _____.
   a. 50°F
   b. 60°F
   c. 70°F
   d. 80°F

6. Fasteners should be applied to wallboard working from _____.
   a. top to bottom
   b. edge to edge
   c. center to edge
   d. corner to corner

7. When you are installing fire-rated wallboard, the nails should penetrate the studs by at least _____.
   a. ½"
   b. ¾"
   c. ⅞"
   d. 1⅛"

8. In a double-nailing system, the second set of nails is driven about _____ from the first.
   a. 2"
   b. 4"
   c. 8"
   d. 12"

9. When ceiling drywall is attached with screws, the screws should be _____ on center.
   a. 7"
   b. 8"
   c. 12"
   d. 16"

10. In single-nailing floating interior angle construction where the framing is perpendicular to the wall/ceiling intersection, the ceiling fasteners should be located _____ from the intersection.
    a. 1"
    b. 7"
    c. 10"
    d. 12"

11. When covering existing walls with structurally sound surfaces, _____ gypsum board should be used.
    a. ⅝"
    b. ⅜"
    c. ¼"
    d. ⅛"

12. In semi-solid gypsum partitions, gypsum panels are applied to _____.
    a. metal or wood studs
    b. gypsum studs
    c. plywood
    d. a foam core

13. When noncombustible architectural panels are installed over a ⅜" gypsum substrate, the burn-through time is about _____ times that of a panel applied directly to the framing.
    a. 2
    b. 5
    c. 8
    d. 10

14. Each of the following is a construction method used to control noise except _____.
    a. caulking around outlet box openings
    b. placing air conditioning ducts back-to-back
    c. using separate framing for the two sides of a wall
    d. mounting gypsum board in resilient channels

15. When you are installing cement board, the factory-bound edge should be placed at least _____ above the lip of the shower pan or tub.
    a. ¼"
    b. ½"
    c. ¾"
    d. 1"

Trade Terms
Introduced in This Module

Corner bead: A metal or plastic angle used to protect outside corners where drywall panels meet.

Floating interior angle construction: A drywall installation technique in which no fasteners are used at the edge of the panel in order to allow for structural stresses.

Gypsum board: A generic term for paper-covered gypsum core panels; also known as gypsum drywall.

Joint: A place where two pieces of material meet.

Nail pop: The protrusion of a nail above the wallboard surface that is usually caused by shrinkage of the framing or by incorrect installation. Also applies to screws.

Slurry: A thin mixture of water or other liquid with any of several substances such as cement, plaster, or clay.

Substrate: The underlying material to which a finish is applied.

This module is intended to present thorough resources for task training. The following reference works are suggested for further study. These are optional materials for continued education rather than for task training.

Gypsum Construction Handbook. Chicago, IL: United States Gypsum Company, 2000.

Installing & Finishing Drywall, William Spence. New York, NY: Sterling Publishing Company, Inc., 1998.

NCCER CURRICULA — USER UPDATE

NCCER makes every effort to keep its textbooks up-to-date and free of technical errors. We appreciate your help in this process. If you find an error, a typographical mistake, or an inaccuracy in NCCER's curricula, please fill out this form (or a photocopy), or complete the online form at **www.nccer.org/olf**. Be sure to include the exact module ID number, page number, a detailed description, and your recommended correction. Your input will be brought to the attention of the Authoring Team. Thank you for your assistance.

Instructors – If you have an idea for improving this textbook, or have found that additional materials were necessary to teach this module effectively, please let us know so that we may present your suggestions to the Authoring Team.

NCCER Product Development and Revision

13614 Progress Blvd., Alachua, FL 32615

Email: curriculum@nccer.org
Online: www.nccer.org/olf

❏ Trainee Guide ❏ AIG ❏ Exam ❏ PowerPoints Other _____

Craft / Level: _____ Copyright Date: _____

Module ID Number / Title: _____

Section Number(s): _____

Description: _____

_____

_____

_____

_____

Recommended Correction: _____

_____

_____

_____

_____

Your Name: _____

Address: _____

_____

Email: _____ Phone: _____

Drywall Finishing

27207-07

27207-07
Drywall Finishing

Topics to be presented in this module include:

Overview

When gypsum drywall is first installed, there are visible seams between the drywall sheets and at corners. These seams must be closed in a way that makes them invisible to anyone looking at the finished wall. Look around your home. Although the walls were most likely finished with 4 × 8 sheets of drywall, it should look as if it was done with one solid sheet. The seams are covered with paper or fiberglass tape, which is embedded with a compound commonly called mud. When the mud has dried, the joint is sanded flat. This process is usually done three times for each seam. It sounds simple, but like other areas of construction, it is not. There are many types of compound, a variety of finishing tools, and some specialized skills that can only be learned with practice.

Objectives

When you have completed this module, you will be able to do the following:

1. State the differences between the six levels of finish established by industry standards and distinguish a finish level by observation.
2. Identify the hand tools used in drywall finishing and demonstrate the ability to use these tools.
3. Identify the automatic tools used in drywall finishing.
4. Identify the materials used in drywall finishing and state the purpose and use of each type of material, including:
   - Compounds
   - Joint reinforcing tapes
   - Trim materials
   - Textures and coatings
5. Properly finish drywall using hand tools.
6. Recognize various types of problems that occur in drywall finishes; identify the causes and correct methods for solving each type of problem.
7. Patch damaged drywall.

Trade Terms

All-purpose compound	Ridges
Bullnose	Skim coat
Feathering	Tape
Joint compound	Tapered joint
Lightweight compound	Taping compound
Mud	Topping compound

Required Trainee Materials

1. Pencil and paper
2. Appropriate personal protective equipment

Prerequisites

Before you begin this module, it is recommended that you successfully complete *Core Curriculum*; *Carpentry Fundamentals Level One*; and *Carpentry Framing and Finishing Level Two*, Modules 27201-07 through 27206-07.

This course map shows all of the modules in *Carpentry Framing and Finishing Level Two*. The suggested training order begins at the bottom and proceeds up. Skill levels increase as you advance on the course map. The local Training Program Sponsor may adjust the training order.

207CMAP.EPS

1.0.0 ◆ INTRODUCTION

In some situations, the carpenter may have to install and finish gypsum drywall. A remodeling project is one example. In other cases, it may be up to the carpenter to repair damaged drywall or drywall that was improperly installed by someone else. Therefore, it is essential to be thoroughly familiar with the tools, materials, and procedures used in drywall finishing and repair.

2.0.0 ◆ FINISHING STANDARDS

Drywall requires different levels of finish depending on its location, lighting conditions, and decorative treatment. A hidden surface, such as an attic area, requires far less finishing than one in full view, such as a living room wall. In addition, even minor flaws are apt to be more evident when the surface is exposed to strong lighting conditions or decorated with certain finishes, such as gloss or semi-gloss paints or thin wallcoverings. Generally, the more visible a surface, the more likely its lighting or decoration are to show surface defects, and the more finishing work it requires.

These factors are addressed in *A Recommended Specification for Levels of Gypsum Board Finish*, which was jointly developed by the Painting and Decorating Contractors of America, the Association of the Wall & Ceiling Industries–International, the Gypsum Association, and the Ceilings & Interior Systems Construction Association.

This specification is designed to serve as a standard reference for architects, specification writers, contractors, building owners, and others. It provides them with a specific description of the final appearance of gypsum walls and ceilings finished to different levels before the application of a decorative coating of paint, texture material, or wallcovering.

The specification describes the following six levels of finish and typical applications for each of them:

- *Level 0* – No taping, finishing, or accessories required. This level might be used for temporary construction or where final decoration is undetermined.
- *Level 1* – All joints and interior angles shall have **tape** embedded in **joint compound** (also referred to as **mud** or **taping compound**): Surface shall be free of excess joint compound. Tool marks and **ridges** are acceptable. A Level 1 finish might be specified for attics, areas above ceilings, service corridors, and other areas not generally seen by the public. It provides some degree of smoke and sound control. In some areas, it is called fire-taping.

- *Level 2* – One separate coat of joint compound shall be applied over all joints, angles, fastener heads, and accessories. The surface shall be free of excess joint compound. Tool marks and ridges are acceptable. All joints and interior angles shall have tape embedded in joint compound. Joint compound applied over the body of the tape at the time of tape embedment shall be considered a separate coat of joint compound and shall satisfy the conditions of this level. A Level 2 finish might be recommended for garages, warehouses, and other areas where surface appearance is not critical. It is specified where water-resistant gypsum backing board (*ASTM C 630*) is used as a substrate for tile.
- *Level 3* – All joints and interior angles shall have tape embedded in joint compound and one separate coat of joint compound applied over all joints and interior angles. Fastener heads and accessories shall be covered with two separate coats of joint compound. All joint compound surfaces shall be smooth and free of tool marks and ridges. A Level 3 finish might be specified for surfaces to be finished with a medium or heavy texture before painting or with heavy-grade wallcovering. It is not recommended for light or medium weight wallcoverings or for smooth painted surfaces.
- *Level 4* – All joints and interior angles shall have tape embedded in joint compound and two separate coats of joint compound applied over all flat joints and one separate coat of joint compound applied over interior angles. Three separate coats of joint compound shall be applied over all fastener heads and accessories. All joint compound shall be smooth and free of tool marks and ridges. Light textures or wallcoverings require this level of finish. The specification notes that in critical lighting areas, flat paint over light textures reduces shadowing of finished joints through the surface decoration, but that gloss, semi-gloss, and enamel paints are not recommended for this level of finish. It also notes that the type of wallcovering applied over this level should be chosen carefully to properly conceal joints and fasteners.
- *Level 5* – All joints and interior angles shall have tape embedded in joint compound, two separate coats of joint compound applied over all flat joints, and one separate coat of joint compound applied over interior angles. Three separate coats of joint compound shall be applied over all fastener heads and accessories. A thin **skim coat** of joint compound, or a material manufactured especially for this purpose, shall be applied to the entire surface. The surface shall be smooth and free of tool marks and

ridges. Level 5 is recommended for gloss, semi-gloss, enamel, or non-textured flat paints or severe lighting conditions. This is the highest level of finish and provides the best protection against joints or fasteners being visible through the decorative coating.

The specification, which is available from any of the associations that developed it, notes that for Levels 3, 4, and 5, it is recommended that the prepared surface be coated with a drywall primer prior to the application of finish paint. It also notes that the effects of severe lighting on a surface can be minimized by skim coating the drywall, by decorating it with medium to heavy textures, or by using window coverings to soften shadows.

3.0.0 ◆ DRYWALL FINISHING TOOLS

Proper finishing of walls and ceilings would be nearly impossible without a variety of finishing tools. These hand-operated and mechanical tools not only speed up the finishing process, but also help create walls and ceilings that are smooth and flat. This module will introduce you to many tools that are used specifically in finishing procedures.

3.1.0 Tool Safety

Although they appear simple and harmless, finishing tools can be as dangerous as any other kind of tool when handled improperly. Finishing knives have very sharp edges and corners. Automatic tapers have gears and chains that can catch fingers or hair. Automatic finishers have sharp blades, and some have spring-loaded hatches that can catch fingers, hair, and clothing.

The best way to avoid accidents and tool-related injuries is to treat your tools with respect. *Never* horse around when tools are being used by you or anyone else. Do not handle any tool unless you have been thoroughly trained in its use. If you do not know how to use a tool, ask your instructor, job supervisor, or another carpenter to teach you.

Always keep your finishing tools clean and free of rust or excess compound, both of which can be poisonous if they get into a cut. Do not use knives with chipped blades or broken handles. Not only can they hurt you, but they can also ruin your finishing work.

Treat automatic tools with special care. They must be cleaned and maintained according to the manufacturer's directions. You can hurt yourself

trying to force a mechanical tool to work when the tool is jammed with dried joint compound. Inspect automatic tools before every use; if a tool looks damaged or in bad repair, report the problem to your supervisor.

3.2.0 Hand Tools

The following hand tools are used to cut, hang, and finish drywall:

- 4' straightedge or T-square
- Utility knife with plenty of blades
- Drywall saw
- Circle cutter
- Drywall hammer
- Caulking gun
- Screwdriver
- Broad knife
- Joint trowel
- Corner tool
- Mud pan or hawk
- Sandpaper/drywall screen
- Sanding block, pole sander, or electric sander
- Sponge sander

An easy way to cut drywall is to place the straightedge on the finished side of the panel and score the drywall with a utility knife, cutting through the paper facing and into the gypsum. Then snap the panel apart along the score line by applying pressure from the back of the board. If the backing paper is still intact, use the knife to finish cutting through it.

An alternative method of cutting through drywall is to use a drywall saw. It is good for making straight as well as curved cuts. A saw with a long, thin, pointed blade (*Figure 1*) works well for limited or detail cuts such as when cutting out a hole for an electrical box. The saw can cut in a straight line or in an arc. Use a drill to make a starter hole for the blade.

Finishing knives range in type and size from the 1¼" putty knife (*Figure 2*) to the extremely wide 24" taping knife (*Figure 3*). They are all similar in function, however, because they bed and feather the taping or **topping compound**. Each knife is designed for use in a different situation. For example, the smallest putty knife is used in hard-to-reach spaces, for patching, and for working the compound around windows, cabinets, and doors. The 4", 6", 8", 10", and 12" widths are the most common.

Taping knives generally have blades made of blue steel, which flexes under the pressure of bedding and **feathering** taped joints. They range in size according to blade width.

Figure 1 ◆ Drywall saw with thin, pointed blade for detail cuts.

207F01.EPS

Figure 2 ◆ Putty knife.

207F02.EPS

207F03.EPS

Figure 3 ◆ Finishing knives.

A long-handled broad knife is used to wipe excess compound from freshly taped joints. The broad knife's blade is made of stainless steel or blue steel and may range in width from 7" to 9". Typically, a broad knife's handle is about 10" to 12" long, but handles up to 28" long are available. These long knives are handy tools for cleaning up messy joints and spatters on high walls and ceilings.

Finishing trowels (*Figure 4*) are similar to cement and plaster trowels, but they are available with either a flat or concave finishing surface (also called a blade). The concave blade is very useful for shearing the joint taping compound. These types of trowels are often preferred over taping knives for finishing wallboard joints. Blades range from 10" to 18" in length and are usually 4½" wide.

CURVED BLADE TROWEL

TROWEL IN USE

207F04.EPS

Figure 4 ◆ Finishing trowel.

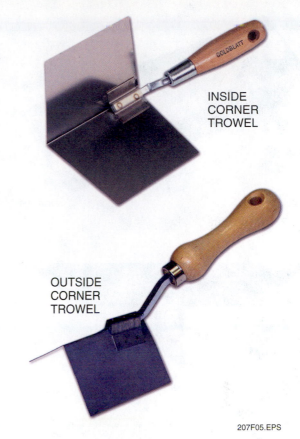

INSIDE CORNER TROWEL

OUTSIDE CORNER TROWEL

207F05.EPS

Figure 5 ◆ Corner trowels.

Blades maintain their bow shape due to the spring steel or flexible stainless steel from which they are made. Flat-blade trowels may be used for applying textured finishes.

Cornering tools enable you to finish both inside and outside corners. They include not only troweling tools, but sanding and bead-attaching devices as well. Corner trowels are made for finishing inside and outside corners. While inside and outside corner trowels are shaped alike, their handles are on opposite sides *(Figure 5)*. These tools are generally not used by professional drywall finishers.

Corner beads can be screwed on or applied with staples. Some corner systems have a paper-faced corner bead that can be coated with joint compound for direct application to a corner.

For attaching corner bead and metal trim to outside corners, the corner clinching tool may also be used. The clincher is struck with a rubber mallet once the corner bead is positioned and the tool is set over it. When struck, the corner clinching tool centers the bead and clinches it to the corner by crimping each side into the wallboard at four points. The entire length of corner bead can be attached by moving the clincher up or down the corner and striking it several times *(Figure 6)*.

A mud pan *(Figure 7)* is simply a long container that holds joint compound (mud) to be knifed or troweled over the wallboard. The pan is commonly made of steel, aluminum, or plastic. It also may have replaceable steel edges mounted on the

207F06.EPS

Figure 6 ◆ Corner clinching tool with a rubber mallet.

long sides for scraping excess compound off the knife or trowel. A typical mud pan looks just like a baker's bread pan.

A hawk *(Figure 8)* is a handheld metal sheet, similar to a mortar board, on which a supply of plaster or compound is placed until used. This

Figure 7 ◆ Mud pan.

207F07.EPS

Figure 8 ◆ Hawk.

207F08.EPS

207F09.EPS

Figure 9 ◆ Pole sander.

207F10.EPS

Figure 10 ◆ Commercial hand sander.

tool was originally used by plasterers, and it is still known by the name given to it by workers in that trade.

Sanding is an important part of the finishing process, and tools such as sanders are available to help you with this task. There are two basic types of sanders: the pole sander *(Figure 9)* and the hand sander *(Figure 10)*. Both use the same principle of fastening down a strip of sandpaper by means of clamps and wing nuts.

With the hand sander, you are better able to sand dried finished joints and fastener heads that are within normal reach. With a pole sander, you can do the same job over your head without relying on scaffolds or stilts.

Hand sanders, also known as sanding blocks, may be purchased or made on the job site. A commercial sanding block consists of a wooden or metal base, around which a sheet of sandpaper is wrapped. A second block is pressed against one side of the base and tightened down with a wing nut to hold the sandpaper in place around the base. The sander allows you to apply even pressure over the sandpaper's entire face while saving your fingers from abrasions and other injuries.

Electric sanders, which will be discussed later, are also available.

3.3.0 Mixing Tools

Both hand tools and attachments to power drills can be used to mix joint compound and other liquified materials at the job site. For hand mixing, you can use a mud masher, which is a lot like a potato masher *(Figure 11)*. The mud is mashed in a pail until it is mixed to a smooth consistency.

If you need to mix a full bucket of compound, it is faster and easier to use a power drill equipped with a long-stemmed mud mixer *(Figure 12)*. There are several types of spinning mud mixers.

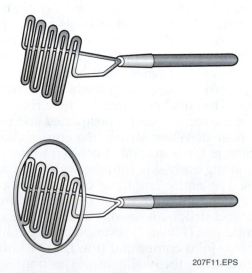

207F11.EPS

Figure 11 ◆ Hand-operated mud mashers.

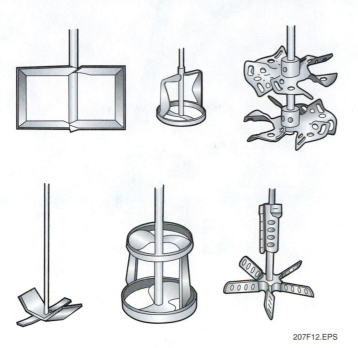

207F12.EPS

Figure 12 ◆ Mud mixers used with a power drill.

The mixer end of the device is placed in the mud once its shaft has been secured in the power drill chuck. The drill is used like a kitchen hand mixer to stir the mud.

3.4.0 Tape Dispensers

A 500' roll of joint tape can be awkward to carry around while trying to apply compound or tape a long joint. Simple tape dispensers and tape holders solve the problem. Many of these lightweight tape-holding devices are designed to hang from belts or shoulder slings, leaving your hands free *(Figure 13)*. Some dispensers even crease the tape for application in corners.

A banjo *(Figure 14)* is a large tape dispenser. It is loaded with a full 500' roll of joint tape and a supply of mud. The banjo applies the mud to the tape as the tape is pulled out. The banjo can be adjusted to change the amount of mud applied to the tape. The banjo may get its name from its similarity in shape to the musical instrument of the same name.

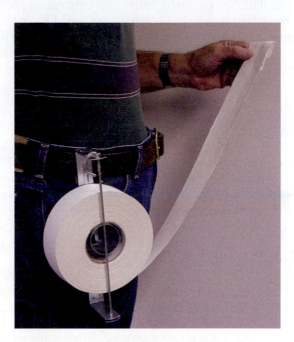

207F13.EPS

Figure 13 ◆ Tape dispenser used with paper or fiberglass mesh tape.

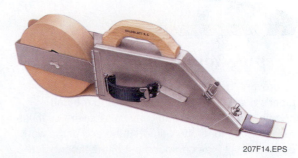

207F14.EPS

Figure 14 ◆ Banjo.

3.5.0 Automatic Finishing Tools

Not only are tools available to help you apply compound and tape and finish the joints, but there are also tools that do many of these operations at the same time—automatically. These tools are operated by hand, but they are referred to as automatic finishing tools or mechanical finishing tools. This is because they use intricate mechanisms to do work that otherwise would have to be done with hand tools and fingers. Although they are operated by hand, the hand operation usually involves simply pushing the tool along a path.

3.5.1 Automatic Taping Tools

One of the most popular mechanical finishing tools is the automatic taping tool (BAZOOKA®), which applies mud and tape to joints *(Figure 15)*.

The automatic taper uses gears, rollers, pulleys, a piston, and levers to quickly guide the tape, coat it with a measured layer of mud, and dispense the tape along the joint. The automatic taper even cuts the tape at the end of a pass and can crease the tape for application in corners.

Using the BAZOOKA®, ceilings up to 10' high can be finished without using a ladder, scaffolding, or stilts. An extension can be mounted to the taper's base to reach even higher ceilings. For closets and other tight spaces, miniature tapers are also available.

You may need a lot of practice to master the use of an automatic taper, but you will probably find it time well spent. Using automatic tools, an experienced carpenter can tape an entire room faster than several carpenters using only hand tools.

Automatic tapers are made primarily of aluminum, plastic, and other rust-resistant materials. This makes it easy to keep the unit clean, which is necessary to guarantee proper operation.

207F15.EPS

Figure 15 ◆ BAZOOKA® automatic taping tool.

The taper is cleaned by filling the empty joint compound chamber with water. Water is then forced out of the unit through the valve that distributes the compound by moving the floating piston up and down inside the compound chamber. The gears and valve openings at the head of the taper should be sprayed clean with a high-pressure water hose.

Automatic taping tools use a small razor blade to cut the joint tape at the end of a pass. This blade must be changed occasionally, as must the cable that operates the unit's piston. The piston moves up and down inside the taper's tube, forcing out compound. You can change the blade or cable in just a few minutes, following the instructions that come with the taping unit. Always follow the manufacturer's instructions for maintaining the taper to avoid breakdowns and lost time on the job.

3.5.2 Nail Spotters

A much simpler automatic finishing tool is the nail spotter *(Figure 16)*. The nail spotter quickly applies compound over nail and screw dimples in fastened wallboard.

INSIDE TRACK

Keeping Your Compound Fresh

To prevent the compound from drying out in the taper's head during use, always stand the unit headfirst in a pail of clean water whenever you must stop working for a short time. At the end of the work day, pump all the compound out of the taper before cleaning it out with water.

The spotter is simply a metal box on a swiveling pole. The pole's swivel allows you to use the spotter at any angle and to push the device along a wall or ceiling while standing still.

The pole is attached to a hinged plate on top of the spotter. As you push the spotter along, the pressure you exert on the pole is transferred to the plate. As the plate sinks into the metal box, it forces compound out through a small opening in the bottom of the box. The compound fills the dimples and excess mud is automatically scraped off by a blade mounted in the trailing edge of the box.

Nail spotters are commonly made in 2" and 3" widths. The pole lets you reach high ceilings and walls without a ladder or stilts. The mechanism is small and light enough to be used in closets and other cramped spaces.

The nail spotter is much easier to use than the automatic taper. Once it is mastered, an entire room can be spotted in just a few minutes. Since the tool automatically scrapes off excess mud and feathers the spot, each dimple must be gone over only once.

The spotter's blade is bowed slightly to leave a small crown of compound over each dimple. The blade is very hard, but can be damaged or broken by an exposed nail or screw head. Broken or worn blades must be replaced. The blade is usually mounted to the unit with a clamp and one or two wing nuts. By removing the nuts, you can back off the clamp and slide out the blade. The new blade is simply slid into place and held down by the wing nuts.

The nail spotter should be cleaned thoroughly after every use by flushing with a high-pressure water hose. The units are usually made of rust-resistant aluminum or stainless steel.

3.5.3 Flat Finishers

The flat finisher is also known as a box. You will generally use more than one box because they come in different sizes *(Figure 17)*. The idea of a flat finisher is to apply topping and finishing coats to taped drywall joints. As each successive layer is applied, you use a wider box. The flat finisher works on the same principle as the nail spotter and dispenses an even, measured strip of compound along any flat joint. Boxes come in 7", 10", and 12" widths. Each applies topping coats of those dimensions. Use the smallest box for the first topping coat after taping; use the 10" or 12" box for the second topping coat if two topping coats are all that are required; or use the 10" for the second coat and the 12" for a third coat.

Like the nail spotter, the flat finisher is a metal box with a hinged lid. When you push the box along the joint, pressure on the handle forces the hinged lid down, squeezing mud out of the box through a small opening at the bottom. The opening can be adjusted to change the amount of mud released. Automatic-feed versions supply compound to the finisher, reducing the amount of effort required.

207F16.EPS

Figure 16 ◆ Nail spotter.

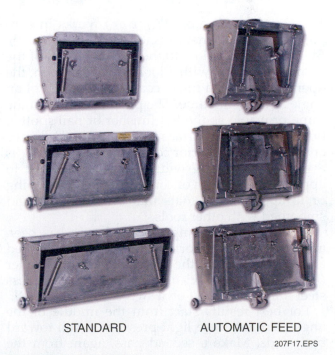

STANDARD   AUTOMATIC FEED

207F17.EPS

Figure 17 ◆ Flat finishers.

Once flat joints have been taped either by hand or with an automatic taper, they can be finished using an automatic finisher. This tool applies a smooth, even coat of mud over the taped joint, automatically feathering the edges and crowning the center.

The flat finisher operates on the same principle as the nail spotter. The device is pushed along, and pressure on the handle forces compound out of a metal box. As you guide the unit along the joint, a metal blade scrapes off excess compound and leaves the desired crown height. The finisher can be adjusted for different crown heights.

Flat finishing tools are available in three widths: 7" for applying a topping coat to a taped joint; 10" for applying a third or finish coat; and 12" for applying a fourth or skim coat.

Flat finishers can be used on wall and ceiling joints running in any direction. The tool is not only easy to operate, but allows you to get the job done in a small fraction of the time it would take to finish joints by hand.

Adjustable handles enable the finisher to reach ceilings up to 12' high. The device is compact enough to use in closets and other cramped spaces. Like most other automatic finishing tools, the flat finishing tool must be kept clean in order to work properly. These applicators must be cleaned thoroughly after each use by squeezing the remaining compound out of the reservoir and flushing the unit with water from a high-pressure hose.

3.5.4 Corner Applicators and Finishers

The corner applicator (*Figure 18*) works in conjunction with a corner finisher, or plow (*Figure 19*). By attaching the plow to the ball/cone end of the corner applicator with a locking retainer clip, the operator can put a finish coat on both sides of an angle at the same time. The plow operates on the same principle as the flat finisher or nail spotter, but its dispensing surface is shaped in a 90-degree angle to fit into interior corners. The applicator is available in 2" or 3" widths and can be used to apply bedding coats of compound prior to taping or to apply finish coats that are feathered and smoothed with other tools.

Once a bed of compound and tape has been applied in interior corners, the tape is smoothed and embedded into the mud using a corner roller (*Figure 20*). This device consists of four stainless steel rollers mounted on a swiveling pole.

For best results start from the middle of the angle joint, and use light pressure to roll toward both ends. Make a second pass, again from the

207F18.EPS

Figure 18 ◆ MudRunner™ corner applicator.

207F19.EPS

Figure 19 ◆ Corner finisher (plow).

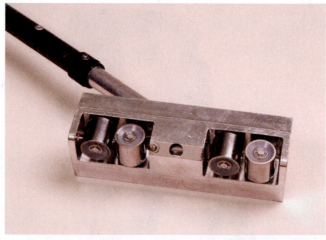

207F20.EPS

Figure 20 ◆ Corner roller.

middle, working toward both ends with firm pressure. This will force excess compound from under the tape. To maintain the corner roller, spray it with a high-pressure hose after use.

3.5.5 Automatic Loading Pumps

Automatic taping and finishing tools are filled with compound through automatic loading pumps *(Figure 21)*. These pumps look a lot like the old-fashioned hand-cranked water pumps used before the invention of the faucet, and they work on the same principle.

The pump's intake nozzle is placed in a five-gallon pail of compound. To stabilize the device, place your foot on the foot plate on the pump's main body, which sticks out of the bucket and sits on the floor.

As you operate the pump's handle, mud is forced out through a J-shaped outlet nozzle. Several attachments can be used to adapt the nozzle to feed various automatic tools. The attachments let you fill the tools quickly and without any over-spill or mess.

Like all automatic tools, the pump must be maintained properly. At the end of the day, remove the device from the pail and force out any compound remaining inside the pump. Flush the pump with water from a high-pressure hose or by placing the intake nozzle in a bucket of fresh water and pumping the water through the compound.

To prevent clogging the pump, make sure the joint compound is properly mixed and free of lumps. Remix the mud periodically if it must stand unused for a while.

The pump's intake nozzle features a small screen that stops lumps or debris from being drawn into the pump. Following the manufacturer's directions, remove and clean the screen after each use. You will probably need to replace the screen occasionally.

INSIDE TRACK

Priming the Pump

A new or fresh compound pump should be primed by pouring ½ cup of water into the outlet.

3.5.6 Vacuum Sanders

The hand-operated vacuum sander works like a hand-operated pole sander. The vacuum sander's pole, however, is actually a rigid hose connected to a powerful vacuum cleaner. A piece of screen-backed sandpaper or a tough sanding mesh is stretched across the hose's flat, hollow head. As the sanding head is pushed back and forth across the dried joint compound, the vacuum sander pulls dust and chunks through the sanding mesh and into the hose. The dust is collected in a large filter bag, as in an ordinary household vacuum cleaner.

There are also power-driven versions of these sanders *(Figure 22)*. They provide a fast, clean way of finishing drywall jobs. The vacuum hose

Figure 22 ◆ Power-driven sander.

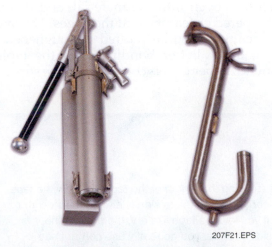

207F21.EPS

Figure 21 ◆ Automatic joint compound loading pump and gooseneck.

is connected to a shop vacuum. The sander uses foam-backed sanding pads that are specially designed for drywall work. The machines are safe, effective, and easy to use. Hoses and attachments enable the finisher to sand normal-height ceilings and walls without stilts or ladders.

4.0.0 ◆ DRYWALL FINISHING MATERIALS

This section covers various types of drywall finishing materials.

4.1.0 Joint Reinforcing Tapes

Three kinds of tape may be used in drywall finishing. They are paper tape, fiberglass mesh tape, and metal edge tape. Each kind may further be divided into those pre-coated with adhesive and those without. The tape most frequently used is paper tape without adhesive.

Standard paper tape is used to cover and reinforce seams, joints, and patchwork. It is a strong paper with feathered edges. There are two types of paper tape available: plain paper and perforated paper.

Paper tape (*Figure 23, left*) can vary in width, but it is generally about 2" wide. The specific width required by most automatic taping tools is

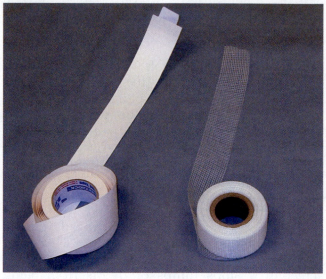

207F23.EPS

Figure 23 ◆ Joint reinforcing tape.

2¹⁄₁₆". The tape usually comes in rolls that range in length from 60' to 500'.

The paper tape may also contain fibers, crisscrossed or woven into the material, and it often comes tapered or feathered at the edges. The overall surface may also be scuffed or roughened in order to allow better bonding with the taping compound. Paper tape is considered to be superior

to fiberglass tape for many applications because paper tape resists stretching and distortion. It also provides a more consistent bond between the face papers of the gypsum boards on each side of most joints.

Perforated paper tape has larger, more visible holes which are designed to allow more compound to ooze through them, producing a better bond once dry. These holes may vary in size, depending upon the manufacturer, but they are usually about ¹⁄₁₆" in diameter.

Fiberglass mesh tape (*Figure 23, right*) is generally self-sticking fiberglass joint tape made of fabric-woven filaments that do not decay. This type of tape is also available without adhesive. Fiberglass mesh makes a strong tape and may be more durable than paper tape under certain conditions, such as in high-moisture areas and when used with moisture-resistant drywall. Fiberglass mesh tape is much more costly than paper tape, but is good for repair work, veneer taping, and other specialized applications. Its use can increase productivity enough to offset the extra cost of the tape.

Metal edge tape (*Figure 24*) is a type of paper tape with the added feature of two galvanized strips of steel down the center, with a small gap between them to allow for the crease. The metal

207F24.EPS

Figure 24 ◆ Metal edge tape.

strips are typically ½" wide. This tape is sometimes referred to as flexible metal corner reinforcing tape. However, the tape is still applied and finished just like regular paper tape.

Metal edge tape is best used for corners with other than 90-degree angles. It is also used for corners formed by radius wall and ceiling intersections, arches, drops, splays, and wherever wallboards need to join in unusual configurations. Metal edge tape makes any outside angled corner straight and sharp, with some reinforcing

Using Fiberglass Tape

Fiberglass mesh tape pre-coated with adhesive is designed to be installed without first applying a bedding coat of taping compound. In other words, adhesive-backed tape is applied directly to the joint and pressed into place with a taping knife or trowel. After the tape is stuck into place, it is covered with layers of compound. Fiberglass mesh tape without adhesive backing is sometimes installed using staples.

Some manufacturers recommend that you do not use fiberglass mesh tape with the usual ready-mixed or powder joint compounds. Instead, special powder compounds, such as quickset compounds, have been developed that work much better with fiberglass tape. Quickset compounds are discussed in more detail later in this module.

An automatic taping tool specifically designed for placing fiberglass tape is shown here.

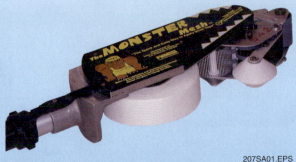

207SA01.EPS

qualities. The edges beyond the metal strips are feathered like those of standard paper tape.

Corner bead is not very flexible, so something else is needed for odd corners made by curved or angled wallboard intersections. The principal idea of metal edge tape is to provide, as much as possible, the same strength and finishing quality that corner bead provides.

Metal edge tape is generally 2⅛" wide and comes in 100' rolls. The gap between the metal strips, allowing for the paper crease, is usually only 1/16". The tape is designed so that the metal side is applied to the wall.

4.2.0 Compounds

Joint compound is more commonly known as mud. It comes in both wet and powder forms. Wet compound has been premixed by the manufacturer, while powder compound *(Figure 25)* is mixed on the job site by adding the proper amount of cold, clean water. Each of these products is designed to accomplish a specific result, depending on the job. Their ingredients are often different, so you should never mix wet compound with dry compound.

Several types of joint compound (mud) are used in finishing drywall. One popular approach is the two-step system in which the tape is embedded in the joint using a taping compound to obtain a strong bond. The joint is then finished with a topping compound, which is much easier to sand. **All-purpose compounds** combine the characteristics of the two-step system into a single compound.

Taping compound is designed for its bonding qualities and strength in bedding and reinforcing taped joints. It is also used as a first coat on metal corner bead or trim, nail or screw heads, and other fasteners. Taping compound is also used as prefill and fill coats, and for repairing surface drywall cracks and cracks in plaster. Taping compound is generally the most likely to shrink. It is also the strongest bonding and most difficult to sand.

Topping compound is used for the second and third coats of the finishing process. This type of compound is softer drying and easier to sand. It also produces less shrinkage. You must never use topping compound for taping because topping compound is not designed to embed and bond a taped joint. A joint bedded with topping compound will crack with the first slight movement or settling of the wall. Topping compound is designed to be molded and sanded flat on top of a joint that has already been fastened together.

All-purpose compound *(Figure 26)* combines the features of both taping and topping compounds, but in so doing, it gives up some of the qualities of each. For example, it loses some of the bonding qualities of taping compound and some of the soft and smooth drying capabilities of topping compound.

However, all-purpose compound is excellent for use in textured finish applications. This type of compound is often used to finish walls with various interesting effects. In that respect, you are almost practicing the art of plastering.

Lightweight compound *(Figure 27)* has the advantages of an all-purpose compound and is 25% to 35% lighter. It also has less shrinkage and sands as easily as topping compound. Lightweight compound can be used to laminate gypsum panels, coat interior concrete ceilings and above-grade columns, and patch cracks in plaster. It can also be used for texturing.

207F25.EPS

Figure 25 ◆ Joint compound.

207F26.EPS

Figure 26 ◆ All-purpose compound.

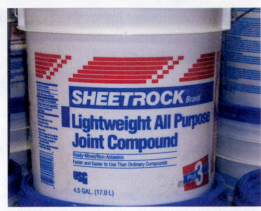

Figure 27 ◆ Lightweight compound.

4.2.1 Powder Compounds

Compounds packaged in dry powder form store better than other forms of compound. Dry powder can be stored at any temperature and in any storage area or warehouse that is kept dry and free from moisture. Since warehouses are rarely heated, powder compounds are best for winter storage. However, it is recommended that powder compounds be moved to a warm mixing room a full day before they are mixed.

All powder compounds must be mixed with clean water in exactly the proportions specified on packaging instructions. Generally, these proportions depend upon the amounts of compound you need for the job. The amounts needed are also determined by the square footage of the joint areas you intend to cover. Once mixed, the compound may be kept in tightly covered containers in storage for many days, as long as the storage area is kept at room temperature. If the powder has been properly mixed to start with, it will not clump up or harden in stored closed containers, but you should always stir or mash it again before use.

Both powder and premixed forms of compound shrink by drying out. As water evaporates from the compound, the compound shrinks to fill in space vacated by the water. This process is different from the chemical hardening process used by special quickset compounds (discussed later), which is more like glue drying.

Taping, topping, and all-purpose compounds are generally available in dry powder form. They may be packaged in bags or cartons. The bags generally contain 25 pounds of powder. Cartons may be measured in gallons or pounds and usually contain more powder than bags.

Current building codes and standards generally forbid any use of asbestos in construction materials. Always make sure that the powder you are going to use is right for the job and complies with local regulations.

Working with Powder Compounds

Here are a few other practical rules you should follow regarding powders:

* Always use clean, cold water for mixing.
* Never vary from the correct proportions of water-to-powder specified for the product you are mixing.
* Always label and date the container you use for mixing. Some carpenters write a description of the contents and the date on a piece of masking tape and attach it to the lid. This is generally better than using ink or magic marker directly on the lid because you can pull the tape off and use the lid again later.
* Never confuse the different mixes. Keep taping compounds separate from topping compounds and so forth. If everything is labeled, you will not have this problem. (Hint: Mix powder taping compound only in empty premix taping compound pails and mix powder topping compound only in empty premix topping compound pails.)
* Be sure to use the oldest dated compound first before mixing a new batch.
* Always avoid mixing one brand of powder with a different brand of powder. Even if they are supposed to be the same kind of powder, different manufacturers make their products differently. They are usually not compatible with similar products made by other companies.
* Follow the dry mix safety instructions given later in this module.

INSIDE TRACK

When Is Five Gallons not Five Gallons?

The traditional five-gallon pail is no longer five gallons. Currently, these full plastic pails contain only 4½ gallons of premixed compound.

4.2.2   Premix Compounds

In general, premix (ready-mix) compounds are formulated, mixed, and packaged by their manufacturers. They are vinyl-based and require little or no mixing. This feature reduces the need for readily available water on the job site. Premix compounds also reduce the waiting or soak times after application, and they offer good crack resistance after drying.

These products will freeze, however, so you need to take precautions in cold weather. If vinyl premix does freeze, thaw it only at room temperature, and do not apply any additional heat. Always use premix products in their packaged state of consistency as much as possible to minimize shrinkage. Follow the specific instructions on the label.

Premix compounds are available in both plastic pails and sealed cartons. They are also quite heavy because, as the name implies, the water is already mixed in. The full pails weigh over 60 pounds; full cartons might be 50 or 60 pounds.

The advantage of premix compounds is that they can be used at job sites where there is no supply of fresh water, which must be available in order to use powders. Of course, powders can be mixed where there is fresh water and then transported in closed pails to sites without water. Generally, however, most contractors simply use the premix in these situations.

The disadvantage of premix is that it must be stored where it will not freeze during the winter. Finishing compounds must be used only in room-temperature environments. In colder climates in winter, you cannot do drywall finishing until the job site interiors are closed up and heat is installed and working.

Even though the premix is already mixed at the factory and ready to use on the job, you will want to mix or mash it again before you use it. Hand mashing is done by forcing the masher down repeatedly in the center and around the sides of the pail. Use smooth, complete downward strokes. On the upward stroke, scrape the masher along the sides of the pail to loosen any compound sticking to the edge.

Only two or three minutes of mixing is usually enough to ensure a smooth, even consistency all through the compound.

You will need to dilute the premix compound slightly for use with automatic finishing tools. Generally, add ½ cup of clean, cold water to 4½ gallons of premix compound. For the automatic taper, mix in two cups of clean, cold water per 4½ gallons of compound. Note that some manufacturers have compounds specifically designed for use with their line of taping and finishing tools.

The full range of compounds are available in premix form: taping, topping, all-purpose, and specialty compounds. Specialty compounds are all-purpose compounds that offer enhanced bonding, shrinking, and sanding capabilities. They often eliminate the need for a third topping coat, and they are good for laminated applications. Specialty compounds are also useful for texturing applications.

4.2.3   Quickset Compounds

Quickset compound, commonly called 20-minute mud, 30-minute mud, or hot mud, is a compound that sets up very quickly in comparison to other compounds because it hardens chemically instead of by water evaporation. Shrinkage is reduced considerably, so quickset makes an excellent filler for metal trim, repairs, and around pipes.

Quickset compound is available only in powder form; it needs to be mixed with water on the job. It is absolutely essential that the water be clean and cold. It is also very important to mix only as much as you can apply in the allotted set-up time.

Quickset compound is good for small jobs, prefill, corner beads, and finishing bathrooms and other high-moisture areas. It is especially good in humid weather. Because it sets so quickly, you need to wipe off excess compound immediately. Sanding the dried compound is also difficult, so take care to apply and wipe it as smooth as possible before it dries. Accelerators are available to make quickset compounds set even faster for special needs.

Quickset compound is packaged and sold by its setup time, which generally ranges from 20 minutes to 360 minutes (6 hours). (Note that the compound shown in *Figure 28* sets up in 90 minutes.) Because it works chemically, once it is hardened it will not shrink, even though it is not dry. This

207F28.EPS

Figure 28 ◆ Quickset compound.

allows a strong bond to form and remain in high-humidity environments. It also enables successive coats to be applied even before previous coats have completely dried, allowing faster finishing and greater cost savings.

Quickset compounds are also very good for laminating applications, especially for laminating wallboard layers together. Quickset compounds can be used for coating concrete walls and ceilings (above ground), for filling in cracks and holes, for skim coating, and even for surface texturing. They are also preferred for finishing exterior ceiling boards and for presetting joints of veneer finish systems.

4.2.4 Dry Mix Safety

Whenever you mix dry compounds, be aware of the dust level and try to keep it to a minimum. You must also use the proper respirator when mixing dry powder compounds. In addition:

- Make sure all mixing containers are clean and free of residue.
- Use only clean, drinkable water for mixing.
- Make sure all tools and mixing blades are clean.
- Mix these compounds only according to the directions on their labels.
- If you use a power mixer or mixers operated by a power drill, use a slow speed such as 300 rpm to 500 rpm.
- Do not try to mix different types of compounds together. Their chemical makeup often differs from manufacturer to manufacturer. Even with the same manufacturer, the different types of compound are not compatible with each other.

4.2.5 Estimating Joint Treatments

The approximate quantities of materials needed for 1,000 square feet of wallboard are as follows:

- *Joint tape* – 370'
- *Joint compound* – 83 pounds of conventional powder or 9.4 gallons of lightweight ready-mix

These figures can be used to calculate the requirements for other square footages by reducing the square footage to a decimal percentage of 1,000 and multiplying the area required by the above amounts. For example, 2,000 square feet would require twice the amount listed above and 1,400 square feet would require 1.4 times the amount listed above.

4.3.0 Trims

Trim comes in a variety of shapes and sizes, each one having a particular function. It can be made of metal or vinyl. Corner trim, or corner bead as it is usually called, is used to protect and reinforce exposed outside corners of wallboard, as well as provide a straight guide for finishing.

4.3.1 Trim Materials

Corner bead serves as a reinforcement for outside corner joints. It consists of edging that strengthens drywall corners.

One type of corner bead is nailed or clinched onto the framing members through the drywall panels. Other types might have a **bullnose** (a metal corner bead with rounded edges) and/or metal mesh flanges, either in regular or expanded widths *(Figure 29)*. Still another type has paper flanges attached to the corner bead, which is applied with joint compound. It usually receives a three-coat finishing process to obtain a smooth surface and to conceal any fasteners. The exposed nose of outside corners or the edge of inside corner bead provides a guide for making the finish a flush surface.

L-bead and J-bead (sometimes called casings) are metal or plastic pieces shaped like Ls or Js *(Figures 30* and *31)*. They provide maximum protection and neat finished edges to wallboard at window and door jambs and other abutments. They are available in sizes to accommodate all thicknesses of gypsum board. They are especially useful and attractive when gypsum board abuts dissimilar objects.

Metal L- and J-bead may be either finished or unfinished, which is to say that they may or may not be treated like regular corner bead and covered with compound and sanded. Finished trims are not covered with compound or texture.

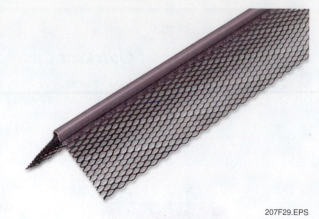

207F29.EPS

Figure 29 ◆ Corner bead with mesh flanges.

207F30.EPS

Figure 30 ◆ L-bead.

207F31.EPS

Figure 31 ◆ J-bead.

Flex tape is another type of drywall trim designed to reinforce edges and corners that serve as decoration. This material might be used as border or edging on arches, splayed angles, door and window frames, or to reinforce in a decorative way various odd intersections of ceiling or wall panels.

Expansion joints are control joints used to relieve stresses caused by expansion and contraction in large ceiling or wall expanses in interior drywall systems. They are made of roll-formed zinc, which makes them corrosion resistant. *Figure 32* shows one type of expansion joint.

Expansion joints are similar to J-bead in that part of the expansion joint, the ¼" open slot, remains visible after finishing. In a room, it simply looks like another panel dividing line, but the joint contracts and expands as the large drywall sections expand and contract. It helps prevent these large expanses of drywall from cracking.

Reveal trim is another type of finishing or decorating material often classified as decorative trim or architectural molding. In general, reveal trim works like decorative spacing between gypsum board panels. It may also function as decorative batten strips or borders or moldings around room perimeters, door frames, windows, archways, and other architectural features.

There are many different types of decorative trims, but all of them generally attempt to eliminate the need for the usual joint finishing procedures. In place of taping and topping, a decorative piece can be installed by drywall finishers. However, some reveal trims need to be installed at the time the wallboard is hung. Such trims usually have a design feature that allows them to stay covered up until the rest of the room or project is finished. Then the protective covering is removed to show off the reveal trim between the wallboard panels.

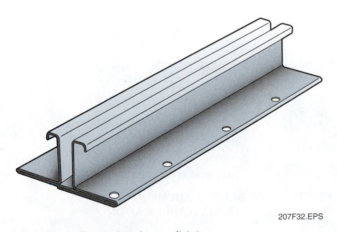

Figure 32 ◆ Expansion (control) joint.

207F32.EPS

4.4.0 Sanding Materials

Sanding operations may be done with sponge sanders, hand sanders, pole sanders, or power sanders, which were described earlier. The sanders require sheets or strips of sandpaper to be fitted to them. Sanding is an important part of the finishing process, and tools are available to help you with this task. With the hand sander, you are better able to sand dried finished joints and nail spots that are within normal reach. With the pole or power sander, you can do the same job over your head without having to bother with scaffolds or stilts.

Sandpaper is rated by coarseness, called grit, which varies by degree of fineness. The lower the grit number, the coarser the paper. Coarse paper is used for first sanding jobs where the surface is very rough and you are trying to smooth it out fairly quickly. Fine paper is used for finish sanding where you are making the surface as smooth as possible, usually in preparation for painting.

Sanding is typically done with 180-grit sandpaper. Sandpaper is also available in grit numbers from 80 to 150 or more. It is packaged in sheets or rolls for hand sanders and in discs for power sanders. The precut sheets are designed to fit specialized tools such as angle sanders and wall sanders (sanding poles). Precut sheets are usually sold in packages of 100.

Sandpaper rolls are designed for you to cut off pieces of whatever length you need to fit onto a hand sander or pole sander. These rolls may vary in total length from 30' to 150' and in width from 3⅓" to 12".

Other sanding materials include sanding cloth, abrasive mesh cloth, and open mesh cloth, which are used chiefly for cutting joint mud. These materials may also be packaged in rolls or as individual sheets.

Mesh cloth, commonly called sanding screen, is preferred by many finishers for use on pole sanders. For first sanding of a joint or seam, this cloth prevents raising the nap (tiny hair-like fibers) on the face paper of the drywall. Do not

Wet Sanding

If only minimal sanding is required, a wet sponge can be used. This method produces no dust and will not scuff the paper. Use a polyethylene sponge, which looks something like carpet padding. Wet the sponge with clean water that is cool to lukewarm. Wring out the sponge enough to prevent dripping, then rub the joints to remove high spots using as few strokes as possible. Clean the sponge frequently during use.

raise this nap at all, if possible, because it makes for better painting if the nap is smooth. An advantage of mesh cloth is that it does not load.

Film-backed drywall abrasives are also effective. A major advantage is that they do not raise the face of recycled paper.

4.5.0 Textures

A texture is any wall or ceiling coating that serves as its own finish. It may also serve to hide taping and other finishing so that the surface need not be sanded smooth or otherwise prepared for painting, wallpapering, or other treatments. *Figure 33* shows examples of texture finishes.

Decorative textures are very popular in both residential and commercial construction. Interesting patterns, simulated acoustical effects, and light or heavy finishes may be applied with rollers, brushes, stencils, sprayers, and other tools and equipment. *Figure 34* shows a roller designed to apply a particular texture finish pattern, along with some of the many roller patterns available. Some of the more common equipment used in texture finishing is shown in *Figure 35*.

1. *Glitter gun* – Used to embed glitter in wet texture ceilings. The hand-crank model shown is most economical, but not as efficient as an air-powered type.

2. *Drywall mud paddle* – Used with an electric drill at less than 400 rpm to mix drywall mud. It is designed to reduce the entrapment of air bubbles in the mixture.
3. *Stucco brush* – Used to create a variety of textures from stipple to swirl. Other variations can be achieved with thicker application and deeper texturing.
4. *Texture brush* – Available in many sizes and styles; tandem-mounted brushes cover a large area to speed a texturing job.
5. *Wipe-down blade* – Has a hardened steel blade and long handle to speed cleaning of walls and floors after application of texture materials. The blade has rounded corners to avoid gouging.
6. *Long-handled roller* – A standard paint roller adapted to the particular type of finish required. Available roller sleeves include short nap, long nap, looped, foam stipple, and carpet types in professional widths.
7. *Texture roller pan* – Used with rollers. Some models can hold up to 25 pounds of mixed texture.
8. *Flat blade knife* – Used to apply texture material and for troweled finishes.
9. *Circular patterned sponge* – Used to achieve patterned swirl finishes.
10. *Sea sponge* – Used to achieve free-form texture finishes.

LIGHT STIPPLE TEXTURE

MEDIUM LIGHT FINISH APPLIED BY SPRAY OF MULTI-PURPOSE TEXTURE FINISH

BOLD SHADOWING WITH ROLLER APPLICATION

MEDIUM STIPPLE TEXTURE

SWIRL FINISH

CROW'S FOOT

207F33.EPS

Figure 33 ◆ Examples of texture patterns.

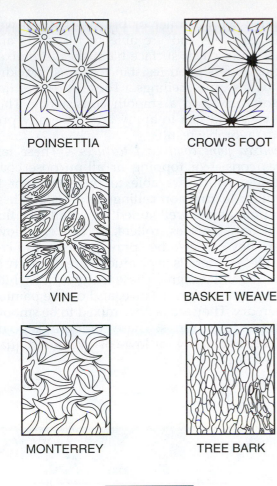

POINSETTIA

CROW'S FOOT

VINE

BASKET WEAVE

MONTERREY

TREE BARK

PALM LEAF

207F34.EPS

Figure 34 ◆ Texture rollers.

11. *Whisk broom* – Similar to a stucco brush but stiffer. It can be used to produce a bolder brush pattern.

12. *Short-handle roller* – Same as the long-handled roller with the same sleeves. A looped texture roller sleeve is shown.

207F35.EPS

Figure 35 ◆ Typical texture finishing equipment.

13. *Trowel* – Used to apply texture materials and for achieving troweled or knockdown finishes.

14. *Texture paddle* – Similar to the drywall mud paddle, but designed for mixing texture materials at 300 rpm to 600 rpm.

Texture materials may be manufactured in both powder and ready-mixed forms. The four types are described as follows:

• *Powder textures* – These textures may be either aggregated or unaggregated, which means there may or may not be other particles mixed into the powder. In aggregated products, particles of such substances as perlite, vermiculite, and polystyrene are used to make textured effects on primed surfaces, particularly ceilings. Aggregated powder products are designed for spray application. They also have a good solution time, only minimum-to-moderate fallout, and good bonding power and crack resistance. When properly sprayed on, these textures hide substrate imperfections very well. Unaggregated powder products may either be sprayed or hand applied to primed walls or ceilings. Several of these products are limited to hand applications only, so that you can produce crow's foot, stipple, or other pattern texture effects. Crow's foot is a design produced by a roller, which makes a kind of random bird-track pattern across the textured surface.

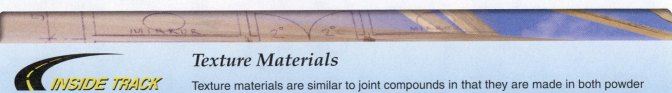

Texture Materials

INSIDE TRACK

Texture materials are similar to joint compounds in that they are made in both powder and premixed forms. In fact, some joint compounds can be easily used for texturing.

- *Powder joint compound textures* – Generally, powder joint compound textures are the same as the topping or all-purpose powder compounds used for normal joint finishing. For texturing with these products, you use a brush, roller, or trowel to produce light and medium hand-formed textures on walls or ceilings. Typically, you use swirling motions to make random patterns. Powder joint compounds are easily mixed in the usual way. They are smooth working and easy to texture by hand. They also hide surface imperfections very well, produce good bonding, and resist cracking. The color is white after drying, which can be left as is or painted another color, depending upon the finish specifications.

- *Premixed textures* – These include thick, heavy-bodied, vinyl-based materials, which are able to produce smooth, very deep textures. They generally dry to hard white finishes that are often left unpainted, especially on ceilings. These textures may be sprayed, troweled, or applied with a roller or brush. They go well over concrete and are able to fill voids and cracks and cover surface blemishes. Premixed texture offers good resistance against cracking on walls and ceilings. These textures are factory-mixed to a smooth consistency. They are easy and fast to apply and generally produce favorable results.

- *Premixed joint compound textures* – These textures consist of topping or all-purpose compounds, which are able to produce light to medium textures on ceilings and walls. These compounds are well suited for small jobs that need only brushes, rollers, or trowels. However, they can also be spray applied. You can use these materials to produce a great variety of patterns and designs. These textures dry white, offer good crack resistance, and can be painted when dry. They are factory mixed to be smooth and free from lumps. These compounds go on quickly and easily for low-cost, yet good quality results.

INSIDE TRACK

Spray Texturing Machines

Spray texturing machines are available for large jobs. They range from a hopper with a pneumatic spray nozzle to a self-contained machine like this one with its own built-in compressor.

207SA02.EPS

5.0.0 ◆ DRYWALL FINISHING PROCEDURES

Drywall finishing involves taping, topping (also known as buttering), and sanding the wallboard seams and joints, whether on walls or ceilings, so these surfaces can be made ready for final decorating. Some jobs use texturing and have a reduced need for taping and sanding. Other jobs require a large amount of butt and seam taping, three or more topping or skim coats, and a lot of sanding to produce the expected results.

5.1.0 Site Conditions

Job site conditions such as temperature and humidity affect the performance of most finishing materials. During winter conditions, drywall finishing should not be attempted unless the building has heat in a somewhat controllable range between 50°F and 80°F. Furthermore, all materials must be protected from the weather at all times. If the humidity is excessive, ventilation must be provided. Windows should be kept open to provide air circulation. In enclosed areas without natural ventilation, fans should be used to move the air. When drying is slow, additional drying time should be allowed between applications of joint compound. During hot, dry weather, drafts should be avoided so the joint compound will not dry too rapidly.

5.1.1 Drywall Inspection

The professional always inspects installed drywall before finishing it. This determines if the drywall is ready for joint treatment. Improperly installed drywall is difficult to finish.

Examine the hung drywall. Nail and screw heads should all be dimpled. This means that each fastener head should have been driven into the wallboard so that a slight depression is made (*Figure 36*). No part of the fastener head should be above the rest of the board's surface. You can check this fairly easily by running your bare hand over the rows of fasteners.

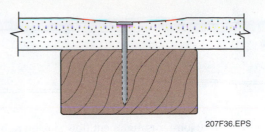

207F36.EPS

Figure 36 ◆ Dimple or uniform depression.

Determine if the fasteners are holding the wallboard panels tightly against the framing members. You can detect loose fasteners by placing a finger over the fastener head. Press the adjoining drywall area in toward the framing with your other hand. When loose, movement will be felt through the fastener head. If you discover this, install an extra fastener above and below the original fastener to make the drywall more secure.

Check the butted joints and outside corners. Remove loose paper and broken board from the drywall edges. Cut the paper back to where it still adheres; do not pull it.

Examine the board fields and the butted joints. Check for torn areas and large gaps. Mark all areas with a lead pencil only; ink or crayon will bleed through. These damaged areas must be repaired.

Examine the inside and outside corners. How well do the wallboard panels butt together? Make sure the panels are properly aligned. When one board sticks out farther than the other, the joint is difficult to tape and finish smooth. Another way of saying this, especially for butt joints, is that there might be a high side and a low side. A high side is produced by a butted panel that sticks out too far from the framing. A low side is then produced in the other abutting panel, which does not stick out as far. This is shown in *Figure 37*. This condition might be caused by a twisted stud. Part of the panel rests against a part of the stud that is not even with the part of the stud to which the other board is attached.

Maintaining the Proper Temperature

If drywall finishing is done during cold weather, the building must be heated to 55°F minimum, and the heat must be maintained during the entire finishing process and until the finish is dry. Avoid sudden changes in temperature, which can cause cracking due to thermal expansion.

Finishing compounds lose strength if they are subjected to freeze-thaw cycles. If a compound has been frozen, it may have to be discarded.

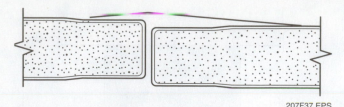

Figure 37 ◆ Butt joint misalignment (low side on right).

207F37.EPS

Since at this stage it is too late to correct the framing, the best you will be able to do is hide the offset. You can hide it by applying a little more compound and finishing the joint a little wider than usual. In fact, any butt joint at all is finished wider than a *tapered joint* because there is no tapered depression to hold the compound. The best you can do with any bad butt joint is to camouflage it.

All these inspections help you determine which procedures to use first in any given situation. If a major repair is required, inform your supervisor. If drywall needs serious correction, such as reframing or rehanging, it is much better to get it done before you begin taping and trying to hide everything with compound. However, if the repair is not that serious, you should be able to fix it.

5.2.0 Overview of the Finishing Process

As a general statement, the ideal drywall finishing process requires five steps. Depending on the drying time, it may take five different trips to the job site:

Step 1 Complete repairs, cut-outs, and prefill. Apply bedding tape at all joints and seams. Complete corner bead and trim installations. Top outside corners. Use an 8" knife on headers and spot fastener heads.

Step 2 Apply first topping coat over taped joints and seams. Use a 10" box. Make double-wide topping coats at all butt joints. Top all corners and angles.

Step 3 Sand lightly, and scuff angles and flats. Apply a second (flash) topping coat to fill in all pits, gaps, depressions, or shrinkage. Apply a straddle coat on the butt joints. Apply another coat on the fastener heads.

Step 4 Perform light sanding and scuffing. Apply a third (skim) topping coat. Use an 18" knife on flat seams and butts. Use a 3" plow for angle topping. Apply a flash coat on headers, seams, and outside corners.

Step 5 Complete pole and hand sanding.

5.3.0 Automatic Taping and Finishing Procedures

Taping and finishing is a multi-step process that varies between three and five different stages of finishing of each board joint or seam. Joints are generally considered to be any place where two edges of wallboard come together. Seams are places where tapered board edges meet each other. Butt joints are places where square (nontapered) board edges meet each other. Flat joints are places where two beveled edges meet.

Generally speaking, joint taping includes prefilling the joint, taping and bedding, topping and skim coats, and sanding. After each step, the compound is allowed to dry, usually overnight.

Joint compound and tape shrink as they dry. This shrinkage results in slight depressions that need to be filled out again by applying topping, skim, or finishing coats of compound. Using an automatic taping tool system, the actual processes at each joint or seam are greatly speeded up. The whole idea is to make every joint or seam as flat as the rest of the wall or ceiling. The idea is also to make these joints and seams undetectable once the decorating is complete.

Large jobs may require drywall finishing equipment. These specialized tools, which were discussed earlier in this module, enable drywall finishers to tape and finish drywall uniformly and efficiently. Basic tool components of such a system are the loading pump, nail spotter, automatic

Drying Time

Atmospheric conditions and other factors always play a part in how fast the taping and topping coats dry. Another factor might even be the wallboard face paper itself. New and recycled paper might well have different drying rates, and different compound materials will vary as to how fast they dry. Therefore, drying times might be longer than just overnight.

Thinning the Compound

Experience has shown that best results come from adding ½ cup of clean, cold water to 4½ gallons of compound for all automatic tools except the automatic taper. For the automatic taper, mix in two full cups of clean, cold water per 4½ gallons of compound. Use the compound full strength for all hand tool applications, nail spotting, prefill, and skim coating.

taper, corner roller, flat finisher, corner finisher, and corner applicator. The finishing process using automatic tools is outlined as follows:

Step 1 Apply the tape using the automatic taper.

Step 2 Press the tape into corners using the corner roller.

Step 3 Wipe down excess compound and embed the tape using a broad knife or taping knife.

Step 4 After the bedding coat has dried, apply a topping coat using a flat finisher for seams and butts. Use a corner applicator/finisher for angles.

Step 5 After the first topping coat has dried, apply a second topping coat using a wider flat finisher for seams and butts, and a corner applicator and finisher for angles.

Step 6 After this topping coat has dried, another skim coat may be applied using taping knives.

Step 7 After all coats have dried, sand the areas treated with compound. Wipe them down with a damp sponge after sanding. This helps the paper fibers to lie down.

207F38.EPS

Figure 38 ◆ A BAZOOKA® being loaded with an automatic loading pump.

5.3.1 Pump Loading Procedures

The loading pump *(Figure 38)* has nozzles of different sizes. Nozzle selection depends upon the equipment and material being used. The pump

Neatness Counts

Always set up your mixing pails and other equipment on a large scrap sheet of wallboard. This helps keep splashes off the floor and will greatly simplify your cleanup efforts. You should also keep at least one full bucket of water handy in this area to soak automatic tool heads when not in use.

has a replaceable screen at the loading pump intake. This screen prevents large particles from passing through the pump. The pail holder is designed for a standard five-gallon pail. A gooseneck attachment mounts on the pump to fill the automatic taper. The loading pump without the gooseneck attachment fills the nail spotter, flat finisher, and corner applicator.

The pump is simple and rugged. It is designed to fill automatic taping and finishing tools with compound. It requires very little training for use and fills the application tools quickly and surely without pumping air.

Be sure to mix all compound thoroughly before using and especially before pumping into any automatic tool. Be sure to mix out any lumps, especially when using powder compounds. Use a power drill with mixer attachment, if available, or a mud masher. With a mud masher, you should plunge it down into the center of the pail and bring it up, scraping against the sides. Rotate around the pail as you mix. This keeps lumpy residues from forming on the sides of the pail. Also, be sure to remix any compound that has been left standing for a period of time.

Keep the pump screen clean and free of lumps or dried compound. It may need to be replaced quite often, along with the O-ring on the gooseneck.

A final recommendation is to have two loading pumps at each job site, one with the gooseneck attachment, and the other with a fill adapter. This will speed up the operations of filling the different types of automatic tools.

5.3.2 Pumping

Set the pump into a full standard-sized premix compound pail, step on the pump's footplate outside the pail, and pump the compound using the pump handle. Before you pump any compound into an automatic tool, however, you may need to add water to improve the consistency of the pre-mixed compound. Taping compound should have a thinner consistency for use with the automatic taper, and a thicker consistency for topping applications. Always follow the recommended mixing instructions on the labels of the products you are using.

Also, before pumping any compound into an automatic tool, be sure to pump the handle a few times to clear out any air. Without the gooseneck attachment, you will simply be pumping compound back into its own pail. With the gooseneck attached, you should use another container to catch these first few pumps. Never attach a taper to the gooseneck until you are sure all the air is out. Then attach the taper upside down to the fitting and pump it full.

5.3.3 Automatic Nail Spotter Procedures

You may use a nail spotter (*Figure 39*) to fill countersunk nail and screw heads with joint compound. A complete row of fastener heads can be filled in one pass with this tool, which is normally available in 2" and 3" widths. The tool allows you to fill rows of fastener head dimples with compound, on both walls and ceilings, while working from the floor.

Use the loading pump to fill the nail spotter with compound. Set the pump into the pail, making sure the compound is well mixed and free of lumps. The pump needs only to be pumped full of compound; no priming is needed. Use the adapter spout (not the gooseneck) to fill the nail spotter at its opening.

Each row of dimples can be filled in one pass. Be sure to make positive contact with the wallboard surface at the beginning of each row. Draw the tool smoothly along the entire row, applying some pressure to force the compound out into the surface. The blade skims off excess compound and leaves a slight crown over each dimple as the tool floats along on the rocking skid.

After you pass over the last dimple, gradually break contact with the surface by using a sweeping motion. This procedure will fill the dimples without leaving excess compound that needs to be removed by hand.

The nail spotter is a simple tool to learn to use. Generally, you will have the procedure down by the end of your first day of using it.

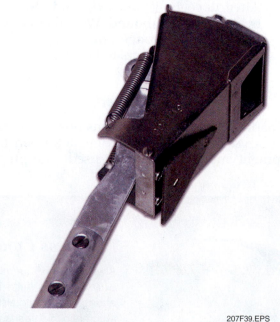

207F39.EPS

Figure 39 ◆ Nail spotter.

5.3.4 Using the BAZOOKA® Automatic Taper on Ceiling Joints

Fill the automatic taper with joint compound using the gooseneck adapter with the loading pump. Close the gate valve and turn the tool upside down to fit its opening over the gooseneck opening. Put one to two fingers of your free hand into the end of the taper. Pump in the compound and stop pumping when your fingers feel the piston. This is to avoid overfilling. If you do happen to overfill, relieve pressure by depressing the filler valve stem with a nail.

 WARNING!
Start slowing down at about six or seven pumps so you will not injure your fingers against the piston if it rises too quickly.

You will likely find that it takes about 9½ pumps to fill an empty taper.

Step 1 After loading, open the gate valve and turn the key counterclockwise until joint compound covers the leading edge of the tape. This is only necessary the first time you tape after each filling.

Step 2 Hold the taper with one hand on the control tube, called the slide, which is similar in operation to a shotgun pump (see *Figure 40*). Put your other hand on the bottom of the mud tube so you can operate the creaser control lever. You might even put several fingers into the end of the mud tube if you find this gives you greater control.

207F40.EPS

Figure 40 ◆ Using an automatic taping tool (BAZOOKA®).

Step 3 Start by taping ceiling butt joints, and then tape the ceiling flat joints. Use both drive wheels for the first 4" to 6" in order to secure the tape to the ceiling. Start at one end of the seam and work to the other end. After taping the first 6", tilt the taper at about a 20- or 30-degree angle away from the plane you are working in so that only one drive wheel is pressed to the board surface. This helps limit the amount of compound and prevents air bubbles. Walk backwards as rapidly as possible, leading with the head of the tool.

Step 4 As you approach the end of the joint, gradually bring the tool back to alignment with your vertical working plane. At about 3" to 4" from the end, stop and pull down sharply on the control tube, which will cut the tape. You will need to slow down and stop completely in order to do this because the tape or blade will jam if you do not stop to make this cut.

 **INSIDE TRACK**

Using the Automatic Taper

Except for beginnings and ends of tape runs, always hold the automatic taper at about a 20° or 30° angle to the wall or ceiling and operate on one wheel only. It is important to have at least one wheel pressing against the board surface at all times while you are taping. These wheels control compound flow onto the tape. If you simply push the tape along without engaging a wheel, you will produce air pockets or bubbles, which are gaps in the tape where compound is missing. You should correct this immediately by going back, tearing out the unbuttered tape, and retaping. If you do not, the topping will dry over the air pocket and will eventually crack apart and crumble or fall off the surface completely.

Step 5 Return the slide to its neutral position and bring back the other drive wheel so both wheels press against the surface once again. Keep both wheels rolling to maintain the continuous buttering of compound needed to press the last bit of tape onto the end of the joint. At this ending sequence (when both wheels are rolling), push the slide forward to eject the end of a new tape. It will be buttered with the correct amount of compound that allows you to start directly into the next run.

Step 6 To finish the ceiling taping procedure, close the gate valve on the automatic taper and put it head first into a pail of clean water. Take a long-handled broad knife and wipe down all the ceiling butt and flat joints. The bedding joint compound should be wiped down while it is still wet, so that it is easily workable and excess moisture does not soak into the wallboard.

Step 7 Use firm pressure and hold your knife blade at about a 45-degree angle to the surface. Wipe along each taped joint, laying the tape flat and forcing out excess compound from underneath. Start in the middle of each seam and work first toward one end, then the other. This helps to avoid wrinkles and bunching up of the tape. Be sure to catch all the excess compound you squeeze out on your knife and scrape it off on your hawk or into a mud pan. The whole process is meant to embed the tape, fill the joint, and leave a generally flat surface.

5.3.5 Using the Automatic Taper on Wall Joints

After the ceiling is wiped down, the next process is to tape the horizontal and vertical wall joints. Again, start with the butt joints. Remember to open the gate valve on the taper.

Step 1 For vertical wall joints, place the automatic taper at the bottom of the joint, about parallel to the floor. Push the control tube forward one or two inches to make a tape leader. Start the vertical taping with the leader overlapping the floor a little. As you proceed upwards, the tape will be drawn up as you go. With a little practice, you will know exactly how much of a leader to use in order to get the tape to end up exactly at the floor line after you have taped that joint. As soon as you can, when moving upwards off the floor, maneuver the taper so you are leading with the head. Also, shift so that you are tracking with just one wheel in contact with the wall.

Step 2 At 3" or 4" from the top of the wall, pull back the control tube to cut the tape and continue rolling to the ceiling intersection on both wheels. To start the next joint, roll the wheels slightly against the surface, starting the flow of compound while ejecting a new leader with the control tube.

Step 3 For taping horizontal joints, push forward on the control tube to produce a 2" or 3" tape leader. Place the leader, again with a slight overlap, at the beginning of the horizontal seam. Except for the start and end of a joint, always hold the taper at an angle to the wall (base of the tool angled downward) so only the bottom drive wheel is rolling against the surface. At the beginning of each seam, however, you need to push both wheels against the wall for about 6".

Step 4 As you come to within 3" or 4" of the end, pull back on the slide tube to cut the tape and continue on both drive wheels to the end of the joint while pushing forward on the slide. This applies the last several inches of tape while feeding out another leader for the next joint.

Teamwork

INSIDE TRACK

On large finishing crews, one carpenter operates the taper, another follows with a broad knife to wipe down excess compound, another spots fastener heads, another comes with the corner roller, and so on. If you are operating the automatic taping tool alone, however, you will need to put down the taper after taping a joint in order to wipe down excess compound.

Step 5 For outside corners (where you are not using corner bead), simply follow the same procedure as explained above for vertical wall joints, but this time only apply tape to one side of the corner. Let the other side remain exposed to the air. When you have completed the vertical run, close the gate valve and fold the tape over the corner using your broad knife.

Step 6 At ceiling angle (or wall intersection) joints, you need to use the creaser wheel, which is extended by pulling on the trigger near the end of the automatic taper. You can also use the creaser wheel to help roll the tape against a flat seam, which is critically important when taping inside corners. Bisect the angle with your tape and make sure both wheels press equally on each side of the angle as you roll. You must be sure to track in a straight line. Avoid twisting the automatic taper as you move. Again, start with a 1" or 2" leader to allow for the tape to be pulled toward the joint end (sometimes called creeping). You may have to push the leader into position in the angle using your fingers before you are able to proceed. Otherwise, taping inside corners and ceiling intersections is the same process that is used for vertical wall joints.

NOTE

Be sure you do not continue using the automatic taper for more than ten minutes if no one is following behind you and bedding the tape.

Step 7 If you are taping alone, stop using the taper after about 10 minutes, close the gate valve, and place the head into the pail of water. Then proceed to wipe down all the tape you just installed to embed it and remove the excess compound.

5.3.6 Using the Corner Roller

After the tape and joint compound have been placed in ceiling and corner angles, use the corner roller tool *(Figure 41)*. This device embeds the tape in the joint compound at inside corners. It forces out excess compound from the tape. At the corners, you will need to remove excess compound with a broad knife. Then, after the angles are rolled, you need to go back over the full length of the angle with a corner finisher called

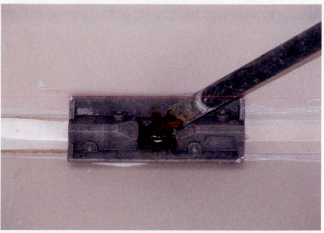

207F41.EPS

Figure 41 ◆ Using a corner roller.

a plow. The general sequence at all corners is as follows:

- Taping
- Rolling
- Plowing
- Corner applicator finishing

Four metal rollers in the head of the corner roller will embed and smooth the tape while forming a sharp corner crease. The tool is easy to use. Work it from the middle of the taped joint out toward the ends of the joint. This will force any overlap of tape due to stretching out to the end of the joint, where it can be trimmed off. This method stretches the tape in place. It also helps to prevent bunching up, which can easily happen if you start rolling at one end and go toward the middle instead of the other way around.

5.3.7 Corner Plow Operation

The corner finisher, or plow, is used to take out excess compound from angles after the corner roller has embedded the tape. This tool is available in widths of 2" and 3". You may use either size to wipe down the excess compound after the corner roller is used. Simply snap one of these tools onto the associated ball on the end of the corner applicator handle and use it like a plow *(Figure 42)*.

With the arrow end leading, work from top to bottom for vertical angles, and from one end to the other for ceiling intersections. Wipe down these angles further on both sides with a 6" taping knife.

The plow is also used for topping, together with the corner applicator *(Figure 43)*. It smooths and finishes both taped and topped corners. The corner finisher feathers the joint compound out

Figure 42 ◆ Using a corner plow.

Figure 43 ◆ Using a corner plow with a corner applicator.

from the corner and onto the drywall. It finishes both sides of the corner at once.

Another nickname for this tool is a butterfly, probably due to its shape. It has four skimming blades and is designed to wipe down and feather both sides of inside corners and other angles at once.

The plow has a spring action that compensates for corners slightly over or under 90 degrees. The blade design produces a smoothly feathered joint.

When you use this tool, be sure the three tips, or arrows, are pointing in the forward direction of travel.

5.3.8 Flat Finisher Procedures

The flat finisher (box) adjusts the amount of topping compound applied to the joint (*Figure 44*). The compound is automatically feathered out from crown to edges. The crown runs down the

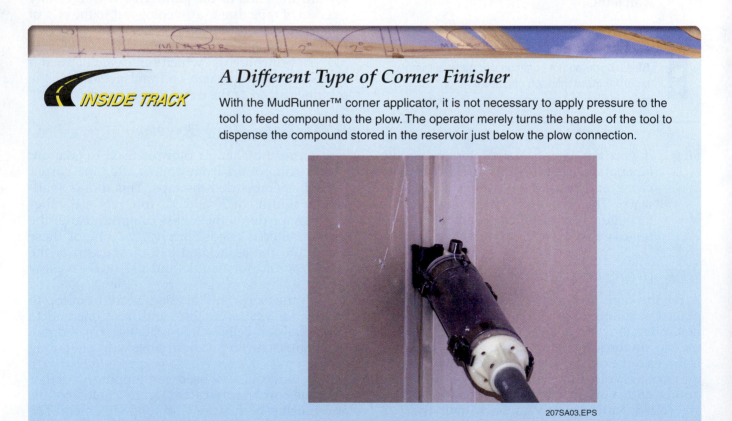

> **INSIDE TRACK**
>
> ## A Different Type of Corner Finisher
>
> With the MudRunner™ corner applicator, it is not necessary to apply pressure to the tool to feed compound to the plow. The operator merely turns the handle of the tool to dispense the compound stored in the reservoir just below the plow connection.
>
> 207SA03.EPS

Figure 44 ◆ Using a flat finisher.

center of the taped joint. This raised area is balanced out by the shrinkage that normally occurs when the compound dries in the joint. You can use a box to apply topping to both tapered and butt joints.

Under the box by the handle connection, you will find a small dial. This dial controls the size of the crown left by the box on the surface. It also controls the amount of topping applied to the surface. As you turn this dial, you will see the blade that controls the crown raise or lower. Establish the setting you will need for your particular application. In general, more crown is needed for earlier coats and less crown for later coats.

The flat finisher comes with handles of various sizes, usually from around 3' to 6'. Longer handles allow you to work at most higher wall and ceiling levels right from the floor without needing stilts or scaffolds.

Always run the box with the wheels leading and the blade trailing. Adjust your grip in relation to your body so that you lead the box with the handle, except at the joint ends. Also note that the end of the handle has a special gripping lever. This grip locks the box in whatever position you desire (in relation to the handle) as you move across the surface.

To finish flat joints, proceed as follows:

Step 1 Load the box through its opening behind the blade using the adapter spout, or nozzle, on the loading pump. The box loads in the same way as the automatic nail spotter. In general, the topping compound should be a little thicker than the taping compound.

Step 2 Apply topping compound on wall or ceiling taped joints by drawing the box steadily along the joint while applying pressure to the back of the box with the handle. This forces out the compound evenly through the opening, depending on the crown you set on the dial. The blade also serves to feather the compound thinly out to the edges, leaving the crown in the center. Always start from one end of the seam and go straight across to the other end without stopping. It should always be a smooth process, and you should always make sure you have enough compound in the box before you begin the run.

Step 3 Ceiling joints should be topped first. The first ceiling joints to receive the topping are the butt joints. Butt joints receive a first topping coat, one coat on each side of the butt, using the 12" box on each side. *Do not* put this coating across the center of the butt; leave the center alone for the first coat. Set the dial on the 12" box to #1 (fullest) crown. Think of this as giving each butt joint a double-wide treatment. The reason for this is that you deliberately finish butt joints much wider than regular tapered (flat) seams to hide the very slight crown already caused by the butted wallboard panels. The finished crown is so gradual and so slight that it will hardly be noticeable.

Step 4 The flat joints receive topping after the butt joints. Again, start with the ceiling. Use the 7" flat finisher set at #3 (medium) crown. Start at one end of the joint and apply even pressure to the middle of the joint. Lock onto the grip as soon as possible after starting the run. Lead with the handle.

Step 5 When you approach the middle of the run, keep the lock on the handle and gradually release the pressure. Then remove the box in a sweeping motion from the surface.

Step 6 Next, reverse hand positions and start at the other end of the run. Repeat the process described above for beginning the run. Lead with the handle toward the middle of the joint where you stopped before. Again, keeping the lock on, slightly overlap the stopping point and remove the box with a sweeping motion.

5.3.9 Flat Finishing for Other Joints

For wall vertical flat joints, start at the bottom and lock onto the handle grip right away; then remove the box in a sweeping motion about 2' to 3' above the floor line. Start at the top of the joint and apply pressure down to your previous stopping point. Again, with the handle locked, slightly overlap that point and then sweep off the surface, which should neatly finish the two topped sections.

When applying topping to joints near doors, windows, and other openings, always work from the corner and move towards the opening. Just before the wheels reach the opening, keep the handle locked and lift the wheels. Then sweep away out into the opening so that topping compound is applied all the way to the edge.

5.3.10 Finishing Coats with the Flat Finisher

Second and third topping coats are applied in the same way as the first topping coat, but in thinner and wider layers. These finishing coats are applied to fill out minor shrinkage and unevenness that could produce shadows and other imperfections after painting. Always allow each coat to dry overnight. In cold or humid conditions, each coat might take even longer to dry.

Before starting each finishing coat, scuff the surfaces. This means lightly sanding the dried compound areas to remove any crumbs, burrs, globs, and so forth. This also prevents any of these things from producing scratches in the surface when you make another pass over them with a box. Be sure to wipe down all sanded areas.

Fill the 10" box, set the dial to #3 (medium), and then apply a straddle coat on all butt joints, starting with the ceilings. Apply this coat of topping compound right down the center (on top of the taped seam) between each of the previous double-wide coats. A correctly finished butt joint should always have a final width of at least 25".

Do the flat seams next. Use the 12" box if only one coat is required, or use the 10" box for the second coat, and the 12" box for the third. Reset the dial to #5 (least) crown and cover the ceiling and wall flat joints in the same way as for the first topping coat.

5.3.11 Corner Applicator Operation

In topping operations, the corner applicator provides a final finish for ceiling and inside corner angles. Use it after the bedding coat has dried. Attach the corner finisher to the corner applicator by snapping it in place. The plow becomes the skimming or troweling blade for the tool. This runs in the topping coat at the corner angles. The corner applicator smooths and feathers the final coat. The corner applicator does for angles and corners what the flat finisher does for flat and butt joints.

To use a corner applicator, proceed as follows:

Step 1 To load the corner applicator, first remove the nozzle end from the chain-slung filler adapter that goes on the loading pump. Inside the nozzle housing is a rubber O-ring that prevents leakage during filling. The filler valve on the corner applicator is inserted into the housing against the O-ring, which seals it. Pressure from the pumped compound pushes the tool's filler valve open.

Step 2 Once the corner applicator is full of topping compound, attach the 3" plow over the round opening. Place the tool at one end of the corner angle. Then with the nose of the plow leading, draw the tool along the angle, applying steady pressure with the handle on the back of the box. This forces the compound out where the plow distributes and feathers it along the run.

Step 3 As you near the end of the run, sweep the tool away from the surface in the same way as with a flat finisher. Reverse hand and body positions and start again at the unfinished end. Draw the tool back to the previous stopping point. Overlap just a little and then sweep away from the surface. This should neatly join both sections of the run. Make sure you apply the plow to neatly bisect the angle. Also, keep the tool at as close to a right angle to the corner as possible.

Step 4 To apply compound to vertical angles, start at the top of the wall and draw the tool downward, sweeping away from the surface at about knee height. Then place the head in the bottom of the angle seam and draw the tool back upward to barely overlap the previous stopping point. Gradually sweep away from the surface, neatly joining both sections of the run.

Step 5 Detail out the corner intersections with a broad knife. Feather any excess compound away from the angle on both sides, also with a broad knife.

When the plow is removed from the corner applicator and a ribbon dispenser attachment is fixed on, it becomes a tool known as a flat applicator. The flat applicator is an alternative tool to the automatic taper. It applies a bedding coat of taping compound and allows you to then attach tape to the surface by hand.

This semi-automatic taping method is useful in places where you physically cannot work with an automatic taper. There is also a mini-taper that might work just as well in confined areas. The flat applicator is convenient for emergencies, for hand operations, or for use by one carpenter when another carpenter is using the only taper on the job.

All these automatic taping and topping tools must be kept clean. Keep the tool heads submerged in a bucket of clean water whenever they are not being used.

5.4.0 Hand Finishing Procedures

Drywall finishing procedures start with gathering all the proper tools, equipment, and materials necessary to do the job. Decisions are then made about sequencing the various tasks: what is done first, second, and so on. These tasks include:

- Inspecting, repairing, and prefilling
- Taping flat joints, corners, and other angles
- Installing bead and trim pieces
- Spotting fastener heads
- Topping, scuffing, and sanding

To finish drywall, proceed as follows:

Step 1 Dimple the nail and screw heads, and cover them with joint compound. You can sometimes create a dimple by banging in the nail with the butt end of your knife handle. Damaged drywall must be patched. Do only a minimum of scuffing or sanding on the paper so as not to roughen it or raise the nap.

Step 2 Once the drywall has been properly installed, carefully inspected, and fixed where necessary, the next step is to prepare for spotting and taping. The usual finishing sequence is to spot the fastener heads; pre-fill gaps, damaged areas, and butt joints; tape the ceiling joints; and then tape the corners, other angles, and finally the flat joints.

Step 3 Prepare the joint compound according to directions and the job site conditions. Put down a suitable sheet of scrap wallboard first. Do all your mixing at this one place on top of the protective scrap board.

Step 4 For hand taping, load the mixed compound into the bladed mud pan or onto the hawk. Obtain the proper compound

consistency. It should generally be the consistency of soft putty. Properly mixed compound does not fall off the hawk when it is tilted for a short time.

Step 5 Apply the compound to the bare joint with a broad knife. (If self-adhesive mesh tape is used, it is applied without a bedding coat.) While the joint compound is still wet, apply the joint reinforcing tape. Press the tape into the compound *(Figure 45)*. Smooth the tape with a broad knife as it is applied. Force the excess compound out from under the tape and remove it with the knife. This bonds the tape to the compound. For perforated tape, force the excess compound up and out through the holes. Again, wipe away the excess with the knife. Make sure there is enough mud under the tape or bubbles will result.

Step 6 Spread a thin coat of joint compound over the top of the tape. Allow these bedding coats of joint compound to dry overnight. Special precautions need to be taken when finishing butt joints. For hand finishing, it is critically important to look at the taped butt before you finish it. Follow these guidelines when finishing butt joints:
  – If the butt joint has a high side and a low side, coat the low side.
  – If the wallboard butt forms a hollow, fill it back to the normal plane by adding much more compound than usual.
  – If the butt joint is regular, it still needs to be finished with the double-wide method *(Figure 46)*. For the first topping coat, do not cover the tape. On day one, make a crown of compound on each

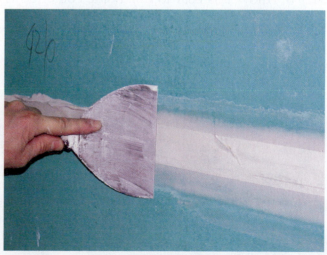

207F45.EPS

Figure 45 ◆ Applying tape.

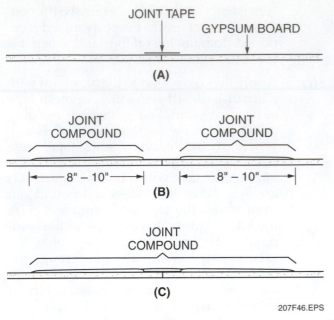

(A)

JOINT TAPE

GYPSUM BOARD

JOINT COMPOUND   JOINT COMPOUND

|← 8" – 10" →| |← 8" – 10" →|

(B)

JOINT COMPOUND

(C)

207F46.EPS

Figure 46 ◆ Finishing a butt joint using the double-wide method.

side of the tape. Then, on day two, apply the straddle coat to cover the tape between the crowns, thereby making one slightly larger crown in the middle of the seam.

Step 7 Apply a thin coat of joint topping compound. Feather the compound's outer edges. Feathering spreads the compound thinly from the center of the taped joint outwards beyond each edge of the tape, causing a slight crown over the center. However, the smaller the crown and the finer the feather, the less sanding will be required.

Step 8 Allow this coat to dry. Some compounds require 24 hours to dry; some take even longer. Drying times also depend upon atmospheric conditions within the structure. If the building is only partially closed off, the finishing work will be affected by the outside weather. If it is too cold or wet,

the compound might not dry out at all until the weather changes. Finishing work depends upon a controlled interior.

Step 9 Once the coat has dried thoroughly, sand it smooth. Remove the sanding dust. Depending on the situation, you may be applying another topping coat, or one or more skim coats. These determinations are almost always made by your supervisor.

5.4.1 Sanding

Always wipe down after any light or heavy sanding to remove sanding dust and tiny particles of compound or other debris. Always check the coated surface to see if it is straight and smooth. Use a straightedge or level to do this.

Always have a hawk or mud pan available filled with topping compound, no matter what finishing procedure you are doing. This way, whenever you see something that needs a little touch-up, like a pit or scratch, you can take care of it immediately to avoid poor-quality finishing.

For general sanding and scuffing procedures, 100-grit, 120-grit, or 150-grit sandpaper is recommended. Sand screen is also used for scuffing as well as for final sanding, because it does not raise the nap on drywall face paper. Sandpaper of less than 100 grit should be avoided, and any sandpaper coarser than 80 grit is unacceptable.

Remember, you are sanding a dried, porous wall joint covering material in order to smooth out tiny bumps and spaces so it will hold a final decoration as well as wallboard face paper. Guard against oversanding, which tends to grind out hollows.

WARNING!

Be sure to wear eye protection and appropriate respiratory equipment when sanding. Check the MSDS for the applicable drywall to learn about the safety hazards associated with that material.

INSIDE TRACK

Correcting Oversanding

Excessive sanding or use of coarse sandpaper can cause the paper fibers of the drywall to stand up. If the problem is not too severe, light sanding with a very fine sandpaper or wiping the panel down with a damp sponge or cloth can correct it. Otherwise, use a light skim coat of topping or all-purpose compound to correct the problem.

5.4.2 Spotting Fastener Heads

Check the drywall nails. Be sure they have been dimpled (set below the substrate with a hammer). Apply the first coat of joint compound on top of the nail heads. Allow it to dry. Sand the dried coat with an abrasive cloth or sand screen. Apply a second coat of joint compound. Allow this coat to dry. Sand it smooth and apply a third coat. The covered area should be smooth and level with the substrate.

Many contractors prefer hand spotting and will not allow the use of a nail spotter tool for this task. Their reasoning is based on several factors:

- Fasteners may not always be below the substrate level, so the nail spotter's blade is frequently damaged. This causes downtime to change blades or tools. When there are no more replacement blades or tools, hand spotting will be the only remaining option.
- Spotting by hand is a reasonably fast method when done by an experienced carpenter.
- Hand spotting is thorough. It forces the compound more effectively onto the head and completely fills the dimple. It also packs compound down into the crisscross of Phillips-head screws better than the automatic tool usually does.
- Hand spotting is something even the newest apprentice can master almost at once. It helps them appreciate the nature of finishing work faster than any other process. It can also build confidence, speed, and an eye for detail.

All fastener heads need to receive three coats of topping so they are undetectable when the surface is finished. Use a 5" knife for the first two coats and a 6" or 8" knife for the third coat. Make sure the compound fills in the crossed indentations in Phillips-head drywall screws.

5.4.3 Outside Corners

To finish outside corners, proceed as follows:

Step 1 Attach the metal corner beads to the outside corner angles. Fasten them with drywall nails or a clinch-on tool or by applying tape with compound. Staples are sometimes used, although many contractors try to avoid them. Dimple the nail heads. Apply joint compound to each flange with a broad knife right after fastening.

Step 2 Spread the compound 7" from either side of the nose (center outside corner). Cover the metal edges with compound. Allow the compound to dry. Sand it lightly, then

apply the next coat. Feather the coat out two inches from the first coat. Let this dry and sand it smooth. Also apply and smooth a third coat of topping at outside corners. Use an 8" knife for all three coats, a 6" knife for the first coat and an 8" knife for the others, or an 8" knife for the first two coats and a 10" knife for the third coat.

The goal with each coat is to fill the corner so it is flat, not concave. Too much compound will make the corner concave. Finish each outside corner so that the corner bead is completely invisible.

5.4.4 Inside Corners

To finish inside corners, proceed as follows:

Step 1 Cut the joint tape to the length of the corner angle. Apply joint compound to each side of the tape angles. Apply small amounts of compound to both sides of the corner angle. This prevents thin cracks from occurring at the angle. Crease the tape length along the center.

Step 2 Use a 4" broad knife to press and embed the tape in the compound. Apply enough pressure to wipe the compound from under the tape. Feather the compound 2" beyond the tape edges. Let it dry and then sand. A corner tool is available as an alternative *(Figure 47)*.

Step 3 Apply and feather the next coat about 2" beyond the first coat. Dry and sand. Then apply a third coat where needed.

207F47.EPS

Figure 47 ◆ Using a corner taping tool.

Step 4 For inside corners, it is better when hand finishing to apply compound first to one side, wait overnight, and then apply compound to the other side. With the automatic corner applicator/finisher tool, you only need apply one coat of topping to inside corners. This is called plowing or glazing the angles.

You might find using trowels more comfortable than taping knives. There are many veteran carpenters in the trade who were trained on trowels. They find taping knives too stiff and awkward. Other carpenters, however, cannot imagine finishing a taped joint with a trowel. Either tool can be used by carpenters to produce equally attractive results. Try both types of tools, if you like, until you decide which is best for you.

5.4.5 Safety and Good Housekeeping

Follow all recommended safety practices for drywall finishing. Maintain your tools and equipment. If you use stilts instead of a ladder, practice safety; always put stilts on and take them off when you are able to lean against a wall, preferably in a corner. Never climb stairs or try to pick up something off the floor while wearing stilts. Ask for help from someone who is not wearing stilts.

Be aware that taping and topping tools, knives, and trowels carry with them a certain degree of hazard. For example, even a dull blade can cut or injure an eye or face. Any tool, if mishandled, can cause an accident or injury. If tools are allowed to clutter up a work area, they can also cause an accident or injury. Keep unused tools in a safe place out of everyone's way.

One of the most common hazards when finishing is slippery conditions, often caused by wet compound carelessly spilled on the floor. This is especially hazardous for someone on stilts. The best rule is if you spill something, clean it up—no matter what it is. If wearing stilts, either remove them and clean up the spill or ask someone who is not on stilts to clean it up for you; do not attempt to bend over on stilts. The same is true if you drop a tool.

Store finishing materials in a cool, dry, protected location. Provide adequate ventilation during dry sanding. Always wear a face mask or

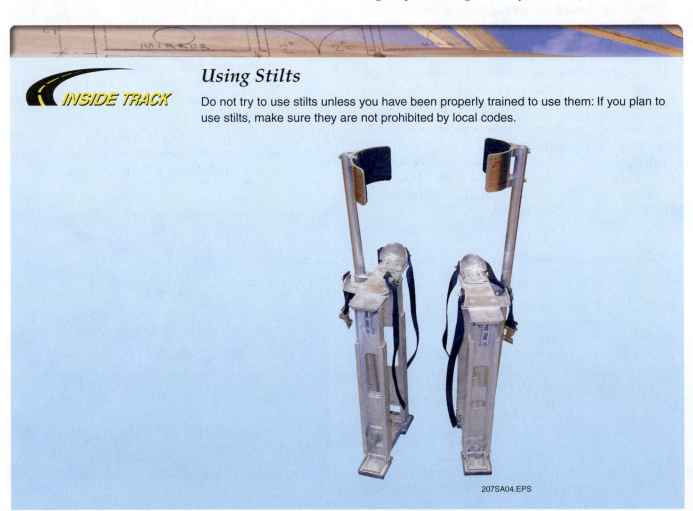

INSIDE TRACK

Using Stilts

Do not try to use stilts unless you have been properly trained to use them: If you plan to use stilts, make sure they are not prohibited by local codes.

207SA04.EPS

respirator to prevent inhalation of sanding dust. Wash hands after applying joint compound as well as after sanding. Proper safety and housekeeping procedures minimize illness and injury.

Final cleanup is always your responsibility. This means a complete sanding and wipedown of all ceilings and floors, and a thorough scraping and sweeping of all floors.

Keep in mind that sanding dust travels. If you are doing remodeling or repair in a finished building, secure the area in which you are working by covering doors and other wall openings with plastic. If you are working in a room that contains furniture, equipment, or other items, cover them.

6.0.0 ◆ PROBLEMS AND REMEDIES

The true test of your finished work will come not from how well you avoid making mistakes, but in how well you repair your mistakes. You may also have to repair those problems left for you by someone else. This section explains the areas where the most common problems are likely to occur and gives you information on their causes and how you can fix them. The four main problem areas are:

- Joints
- Compounds
- Fasteners
- Gypsum board panels

6.1.0 Finished Joint Problems

The common joint problems found in drywall work include:

- Ridging
- Tape photographing
- Joint depressions
- High joints
- Discoloration
- Tape blisters
- Cracks in the joint

6.1.1 Ridging

When a ridge occurs along a joint between two boards, it is often because there has been movement at the joint. Ridging is also sometimes known as beading or picture framing because a visible ridge that surrounds a board resembles a frame surrounding a picture. There are three probable causes of ridging:

- *High humidity, poor heat distribution, or not enough ventilation* – This results in expansion and contraction of the framing and boards.
- *Drywall that is not properly installed* – Improper installation includes misaligned framing and butt joints that do not fit well together. If you force two boards together, joint compound may be squeezed out, forming a ridge. Make sure the joint is not overstressed by too tight a fit.

You may have to cut some of the drywall away at the joint to make a little gap between the boards. Do this by cutting with a knife, making several passes, or use a hand saw or chisel to remove a sliver of one of the boards. If the space left between the boards is too wide, the joint will be weaker than a properly-fitted joint. Ideally, the entire width of the joint tape is bonded to a drywall surface. The tape itself only needs to span a small space between the boards. If the space between the boards is too wide, less of the tape is bonded to the drywall. This can weaken the joint and promote ridging.

You may have to overlap two pieces of tape to strengthen the joint. If the joint is very wide, it may be because one board, or both, is not securely fastened to the framing at the joint. Perhaps neither board is directly over the framing member. You may have to install fasteners at an angle through the boards into the framing in order to make the joint tight against the framing.

This presents a problem; one of the most basic rules is always to drive the fasteners straight so that the edges of the heads will not protrude. In this case, however, they will. Repair the problem by using a hammer to straighten the head (by bending the shank) after the fastener is installed. You may also need to cut and install a thin strip of gypsum board to bridge a space that is too wide.

- *Too much joint compound* – To correct ridging caused by too much mud, first sand the ridge smooth, then apply a finishing coat of joint compound. Hold a light at an angle to the area to make sure you have eliminated the ridge and left a smooth surface.

6.1.2 Photographing

If the joint is still visible even after the wall is finished and painted, this condition is known as photographing. The joint may show through as a slightly different color than the finished wall. Or it may be the same color as the finished wall but have a higher or lower gloss (shine) to it. Photographing can also occur over fasteners if there was insufficient joint or topping compound spotted on the heads.

The usual causes of joint photographing are:

- The installer failed to force excess joint compound out from under the joint.
- High humidity conditions delayed drying of the second and third topping coats of compound.
- The tape absorbed too much moisture from the compound, causing the mud to shrink and conform to the shape of the joint tape. Avoid this by wetting the tape before installation.

To correct photographing, sand the tape edges to feather them into the surface of the wall or ceiling. Then cover the tape with thin coats of joint or topping compound. Use thin coats so as not to rewet the tape too much, or seal the tape with a primer after sanding and before applying the final coats of mud. This keeps the tape from drawing too much moisture out of the mud.

6.1.3 Joint Depressions

A joint depression is a valley that occurs at a seam or joint. It will be most obvious when a light strikes the drywall at an angle. (Hint: Use a 10" or 12" knife to help find high and low spots.) There are two common causes of joint depressions:

- There may not be enough joint compound over the joint. This can happen when the joint compound mixture is too thin or when not enough joint mud is applied to the joint.
- The joint may be sanded too deeply.

The cure for joint depressions is to add more joint or topping compound to the joint. Then smooth it and sand again to get a flat surface at the joint. Make sure the joint is flush with the surface of the wall or ceiling.

6.1.4 High Joints

A high joint is the opposite of a joint depression. It occurs when a wide section of joint is raised above the rest of the drywall surface. Like the joint depression, a high joint is most noticeable when a light strikes it from an angle. High joints are the result of too much mud built up underneath or on top of the tape and/or improper feathering of each coat. The edge of the coat must be feathered into the wallboard surface. When this is not done, high joints will result.

To repair a high joint, sand the area as flush as possible without sanding into the tape. Then apply one or two final skim coats of topping compound. Feather each coat into the board surface. Make each coat wider than the previous coat to conceal the area of the joint *(Figure 48)*.

6.1.5 Discoloration

Joints may discolor or turn lighter or darker than the rest of the finished surface. There are several common reasons for joint discoloration:

- Moisture may be trapped inside the joint. Until a joint is sealed, it can absorb water. If a joint is sealed before it is dry, water is sealed inside the joint. Trapped water will degrade the finish and discolor the surface. Be sure the joint is dry before sealing it.
- The joint was painted in conditions of excess humidity. To prevent this, reduce the room humidity before any painting is done.
- A poor-quality paint was used. Always be sure to use a good-quality paint. Cheap paint often gives uneven coverage and sealing. This increases discoloration.

6.1.6 Tape Blisters

A tape blister is really an air bubble in the surface of a joint. It can be several inches long or as small as a dime. Tape blisters occur when the bond fails between the tape and the first bedding coat of joint compound. One way a bond can fail is when there is no bond to start with. This may happen if care is not taken when using an automatic taper and sections of tape come out without a mud coat underneath.

207F48.EPS

Figure 48 ◆ Coating with topping compound.

Tape blisters can also happen if the joint is too wide, because the tape was not properly embedded in the joint compound, or because the tape draws moisture too quickly from the mud. Another cause of a blister occurs when topping compound is used instead of joint compound to embed the tape.

To repair a tape blister, proceed as follows:

Step 1 Slit the blister with a knife. If the blister is large, cut and remove the entire section of tape that came unbonded.

Step 2 Sand or scrape out enough of the dried mud so you can embed a new section of tape.

Step 3 Work joint compound underneath the tape, smoothing the slit in the old or new section of tape into the joint compound as you go. This embeds the blistered section. This is a hand procedure only. Do not attempt to do this with another run of the automatic taper.

Step 4 Apply a skim coat of mud over the tape. When this coat is dry, apply the required number of topping coats, always allowing enough drying time in between coats. Sand enough to produce a smooth finish that is flush with the surrounding surface.

6.1.7 Cracking

The two common types of joint cracks are those that run along the edges of a joint and those that run along the center of a joint. Each has its own causes.

Edge cracks are cracks along joint edges that occur when the air temperature is high and the humidity is low when the joints are finished. This causes the joint compound to dry too quickly and unevenly, resulting in uneven shrinkage. To slow down the drying rate, run a wet roller over the joint or spray it with a fine water mist from an atomizer.

Edge cracks can also be caused by tape that has a thick edge or by joint compounds applied in coats that are too thick.

The procedure for repairing edge cracks depends on whether the crack is thin or wide. If the crack is small and thin, coat it with a latex emulsion or a thin coat of joint or topping compound. Then sand it as needed.

If the crack is wide, you may have to gouge out some of the joint compound to prepare the surface. Paint the gouged-out crack with a primer, then fill with joint compound and sand smooth.

Center cracks are cracks running along the center of a joint that occur for several reasons:

- If the tape is still intact, the crack is probably the result of applying joint compound too thickly. Also, low humidity may have caused the mud to dry too quickly.
- If the tape under the crack has been torn, it is possible that the structure is settling or another type of movement caused the crack.

To repair cracks along the center of the joint, follow these procedures:

- If the tape is still intact and the crack is narrow, apply latex emulsion to the crack. If the crack is wide, use joint or topping compound to bridge the space.
- If the tape is torn, you may need to remove a section of tape and old joint compound before making repairs. Then retape the joint following the usual finishing procedures.

6.2.0 Problems with Compound

Compound has its own special set of problems. It can debond, grow mold, become pitted, sag, and shrink.

6.2.1 Debonding, Flaking, or Chipping

When the joint or topping compound will not bond to the tape or the board or becomes unbonded from either one, the condition is known as compound debonding, flaking, or chipping. Common causes of compound debonding include:

- A foreign substance was on the gypsum drywall surface or on the surface of the tape when the mud was applied. Examples of foreign substances include dirt, oil, sanding dust, and incompatible paint.
- The mud was mixed improperly, or the wrong ratio of water to dry powder was used to mix the compound.
- Too much water was added during mixing, or incompatible compounds were mixed with each other in order to add to the working supply, combine containers for storage, etc.
- Dirty water was used to mix the compound, or dirty tools were used to mix or apply it.
- Hot or heated water was used to mix the compound. One reason for avoiding hot water is because of possible sediment problems associated with hot water heaters.
- The installer used old or expired compound.

You can avoid most compound debonding problems by following the manufacturer's mixing and usage instructions exactly. Some manufacturers request that you let the compound sit for a while after mixing it. There is a good reason for this, so do not take any shortcuts. They will end up costing excessive time and waste in the long run.

Be sure to always use only clean, cold water to mix any compound. Also use clean mixing and application tools and equipment. Remember that automatic taping tools need to be kept in pails of clean water between uses. Always make sure the gypsum drywall surface, tape, and all mixing pails are clean, too.

Repairing compound debonding is very much like repairing tape blisters, only on a larger scale. First, separate the debonded section of tape from the dried mud. Then remove enough of the old mud to allow you to apply a new layer in which to embed the tape. If the old compound crumbles easily, remove all of it. You will also have to remove whatever mud was used to feather the joint. Apply new compound and tape as you would for a new joint.

6.2.2 Moldy or Contaminated Compound

Using contaminated water, dirty containers or tools, or letting the compound stand too long can result in mold, bacteria, and bad odors in the compound. Hot and humid weather also contributes to the growth of mold and bacteria.

Always be careful to examine every pail before mixing anything in it. You may be surprised at what you might find in a supposedly empty bucket. The best remedy here is simply to look before you mix. If you discover that your batch of compound has become moldy or contaminated, throw it out. Then be sure to soak your tools and containers in a solution of chlorine bleach and clean water at least overnight.

Be sure to clean and wash all tools and equipment components at the end of every working day. The mud intended for use the next day must be stored in covered containers and kept at room temperature overnight. That means warm room temperatures, not freezing cold or scorching hot.

6.2.3 Pitting

Small pits may appear in the finish of the mud after it dries. Pitting has several common causes:

- Air escaped that was trapped in the mud mixture. This can happen if you mix the compound too vigorously or for too long.
- The mud mixture was too thin.
- Not enough pressure was used to apply the mud to the joint; that is, it was not embedded or wiped down properly.
- Mud was not adequately mixed prior to application.

To prevent pitting, mix the compound thoroughly using a slow, steady motion. Set power mixers, if used, on slow speed. You are trying for a smooth mixture that is free of lumps. When you apply the mud, use enough force to establish a good bond to the surface, smooth it out, and feather the edges.

Repair a section of pitted compound by simply skim-coating with another topping layer in order to fill the pits. You may need to sand a little to form a smooth base for applying the new mud. Then apply the new compound as you would apply a topping coat to the joint. Feather it out to conceal the joint area. You may have to feather it wider than the original topping coat in order to completely hide the joint.

6.2.4 Sagging

When compound sags or shows evidence of runs, these conditions are usually present:

- The mud was too thin. When mixed properly, compound is thick and smooth. Be sure to follow the mixing instructions exactly.
- Water added to the mud or to the dry powder compound was too cold to mix completely. Again, cold water is essential, but do not use ice cold water. Remember that finishing is a room-temperature process. Anything colder than what normally comes out of a faucet in a warm room is just too cold.

To repair sags and runs, sand them very smooth after they dry. Then, recoat with layers of joint or topping compound as needed.

6.2.5 Excessive Shrinkage

If the mud shrinks too much when it dries, it is probably the result of one of the following:

- Mud mixed too thin
- Insufficient drying time between coats
- Too much mud applied at any one time

To prevent this problem, use lightweight mud, which tends to shrink less. This problem is similar to the joint depression problem discussed earlier. As in that case, remedy excessive compound shrinkage by applying more mud. However, ensure that each previous coat is thoroughly dry before you begin any repair by adding more compound.

6.2.6   Delayed Shrinkage

Delayed shrinkage is caused when too much time elapses before the correct amount of shrinkage occurs. The mud is not shrinking enough and tends to resist drying out. Delayed shrinkage has several common causes:

- Atmospheric conditions (slow drying capabilities and very high humidity)
- Insufficient drying time between coats of compound (trying to rush the job before it is actually ready for each finishing procedure)
- Excess water added to the mud mixture
- Heavy fills (adding too much mud as a prefill or trying to fill large spaces in the gypsum drywall with compound instead of slivers or strips of wallboard)

One way to prevent delayed shrinkage is to use a faster-drying compound, perhaps a quickset compound, which sets up chemically and does not depend on water evaporation. Quickset compounds were discussed earlier in this module.

A remedy for this condition is to allow extra drying time and then to reapply a full cover coat of a heavy-mixed mud over the tape. Most shrinkage will generally take place on this heavy topping coat. With the right mud, the coat will dry faster and allow you to continue finishing procedures in the usual way.

The best defense against delayed shrinkage is to use a faster-drying compound in the first place. There is very little you can do to mud that needs more drying time, except to give it more time to dry.

6.3.0 Fastener Problems

Two common fastener problems that may be encountered are nail pops and fastener depressions.

6.3.1   Nail Pops

When drywall nail heads work up from under the finished surface after the installation is complete, the job is said to have nail pops. Nail pops are unsightly, protruding fastener heads.

If enough nails pop out, the drywall will loosen and sag. The nails can be driven in again and the hole refinished, but the best remedy is preventing nail pops before they happen. Here are the primary reasons for nail pops:

- Wood framing with relatively high moisture content will shrink as the lumber dries out. As the wood shrinks, the nails lose their tight holding power (Figure 49). When the wallboard is no longer securely attached, a space develops between the board and the stud; the nail shank is exposed at that point. Then almost anything that puts pressure against the wallboard will push it against the stud and the nail—which does not move—will actually pop right out of the panel along with the compound covering it.

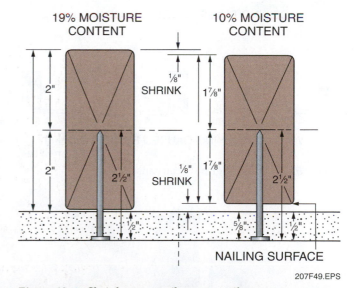

Figure 49 ◆ Shrinkage contributes to nail pops.

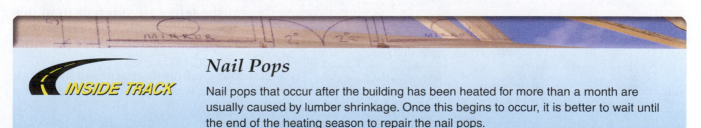

Nail Pops

Nail pops that occur after the building has been heated for more than a month are usually caused by lumber shrinkage. Once this begins to occur, it is better to wait until the end of the heating season to repair the nail pops.

- When drywall is fastened to framing that is out of alignment, stress on the drywall causes fasteners to work up above the surface (*Figure 50*).
- Gravity acting on ceilings and vibrations acting on walls will tend to work the nails loose (*Figure 51*).
- The drywall may not have been installed properly (*Figure 52*).
- If a building has poor ventilation or an inadequate heating system, large temperature fluctuations will cause expansion and contraction of the framing and drywall. If there is too much of that, the fasteners will begin to loosen.

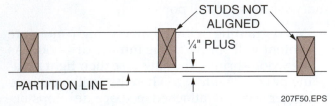

Figure 50 ◆ Non-aligned framing.

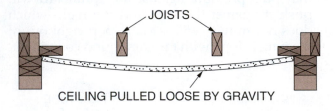

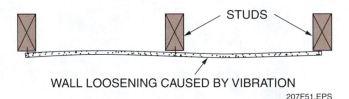

Figure 51 ◆ Force of gravity can cause nail pops.

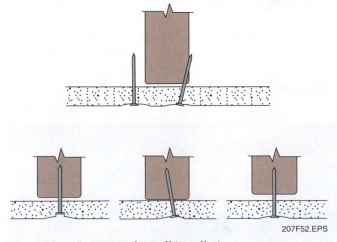

Figure 52 ◆ Improper drywall installation.

6.3.2 *Preventing Nail Pops*

Nail pops may show up days or weeks after installation is complete or gradually over a period of many months. How soon they appear depends on the degree of misalignment, the type of fasteners used, the moisture content of the framing at the time of installation, the amount of vibration present, and the temperature cycles to which the drywall and framing are exposed. Once you know the causes of nail pops, prevention is easier. To prevent nail pops, follow these key rules:

- Make sure your framing lumber is dry before you fasten any drywall to it. The builder should provide enough ventilation to speed up the drying process. When working in cold or humid weather, use portable heaters or blowers to warm or circulate the air. It may take only a few days to reduce the moisture content of lumber to an acceptable level, depending on temperature and humidity. Lumber is too green for hanging drywall if any wet spot appears when the lumber is hit sharply with the head of a hammer. Test several lengths before forming an opinion. The amount of moisture can vary from one piece to the next.
- Make sure the framing members are aligned in the same plane. Sight along the edges of the studs and joists to see that they are in a straight line. You can also check alignment by holding a long straightedge up against the studs and joists.

  If framing is out of alignment, repair it. Applying boards to framing that is out of alignment will prove to be a mistake. The framing will eventually spring back to its original position and the nails will pull out from the framing.
- Use ring-shank drywall nails or screws to fasten the drywall boards to the framing. They have more gripping power than plain-shank nails and offer more protection against nail pops. Also, consider using double rather than single fasteners. Double nailing gives more holding power than single nailing. Use floating corner angles to reduce stress on the drywall. Finally, you can use adhesive in addition to nails or screws to fasten the drywall boards.
- Always nail from the center of the drywall board toward the edges. If you nail from one edge of the board to the other, the board may not rest firmly against the framing for the entire length of the board.

  For example, assume that you are installing the final board in a wall. The other boards along the wall are already in, and so are the boards that cover the wall that forms the other half of the corner. On this final board, if you start installing the fasteners at one edge instead of at the center, the drywall may move slightly toward the opposite edge. This particular board is trapped by a corner and other drywall boards. Any movement after you start nailing will stress the board, eventually causing it to bow and pop the fasteners.

A similar problem can occur if you work from the other edge toward the center of the board. The center may actually spring away from the framing. If you fasten the center of the board first and work toward the edges, the board will not be able to move once the first fasteners are in place. From the moment the first nail is driven home, the board is forced flat against the framing in the center and along the edges. Be sure to hold the drywall board tightly against the framing as you install the fasteners. This ensures that the board stays flat. Do not worry about installing a board with a slight bow. With proper installation, any stress put on the board by flattening its bow will relax in a short time.

Before you cover the nail heads on any drywall finishing job, check to be sure the nails are tight. Re-drive any loose nails. It is also a good idea to drive another nail on each side of a nail that has worked loose. Drive them about 1½" away from the old nail. After you re-drive any loose nails and add extra nails, go back and check all the nail heads again. The vibration caused by hammering may have loosened more nails. The few seconds you spend now can save you a few hours later. Also, if you nailed boards to both sides of a wall, driving the nails on one side may have loosened the nails on the other side. Be sure to check the first side again.

Attention to detail should prevent most nail pops. Take the time to check your framing lumber carefully and install your panels properly.

6.3.3 Repairing Nail Pops

When nail pops occur after you have finished a job, perhaps even after the texturing and painting are done, fixing them takes a little more time. When the nail head works out of the framing, it will show above the surface of the drywall. It may even lift the compound from the depression around the nail head. Repair it as follows:

Step 1 If the nail head has worked loose and become visible, just driving it back may not solve the problem. It is better to pull out the nail and replace it. The best and most permanent replacement is a drywall screw; otherwise, use a longer nail and/or another nail within 1½" of the first.

Step 2 Once the fasteners are in, fill the depression with mud and let it dry. If necessary, use a second coat of compound.

Step 3 You may have to repair the texture, depending on the surface's original texture. Be sure the texture of the patch matches all the surrounding texture.

Step 4 When the new texture has dried, paint it with an oil-based primer. If you skip this step, the paint may soak into the repaired spot when it is later repainted and will make the repaired area very evident.

6.3.4 Fastener Depressions

A fastener depression is a depressed area over the fastener head. This is the opposite of a nail pop. The joint compound over a nail or screw has sunk lower than the surface of the surrounding drywall.

Fastener depressions can be caused by several problems:

- Nails were dimpled too deeply or screw heads driven in too far.
- Not enough mud was applied to the fastener heads to cover them properly.
- The framing lumber was extremely dry. Dry lumber will absorb moisture, squeezing the board between the nail head and the edge of the stud or joist and pulling the fastener head deeper into the drywall.
- The installer used too few fasteners to hold the drywall firmly against the framing, allowing the drywall to flex independently of the framing and forcing the fastener heads deeper into the surface.

Repairing Nail Pops

Another way to repair a nail pop is to drive a Type W screw about 1½" from the popped nail. Then, place a broad knife over the popped nail and hit the knife with a hammer to drive the nail back down.

To prevent fastener depressions, avoid driving fasteners through the facing paper. Install the correct number of fasteners and space them properly. Spot the fastener heads with two coats of compound, sanding lightly between coats, if necessary.

Repairing fastener depressions is a simple matter. First, make sure you have installed enough fasteners. If you need more nails or screws to hold the drywall firmly against the framing, add them. Second, spot the fastener heads with joint compound to bring the surface flush with the surrounding drywall.

6.4.0 Problems with Wallboard

Common problems with gypsum drywall sheets include blisters, damaged edges, water damage, board bowing, board cracks, fractures, and brittleness.

6.4.1 Board Blisters

When the facing paper becomes unbonded from the surface of a piece of gypsum board, it is known as a board blister. It may be caused by a manufacturing defect, or it may be the result of careless handling or improper storage. The gypsum filler tends to break apart inside the wrapped board, causing the facing paper to loosen.

There are two common ways to repair board blisters:

- Inject an aliphatic resin glue, such as yellow or white carpenter's or wood glue, into the blister, and then press the paper flat. This is the best remedy where the blister is small or where the blister is not discovered until after the wall has been textured and/or painted.

WARNING!

Before using any adhesive, check the manufacturer's instructions and applicable MSDS to identify any hazards. Wear protective equipment and apparel as specified by the manufacturer.

- Cut out the entire blistered area and finish it with tape and joint compound. Follow the usual procedure for embedding tape and finishing joints. If one width of tape is not going to be enough to cover the blistered area, add as many other strips as necessary.

6.4.2 Damaged Edges

Improper handling of gypsum drywall sheets is what generally causes damaged surfaces and edges. Such carelessness may cause the facing paper to tear or the gypsum core to crumble.

The only way to repair such damage is to cut off the damaged area back to sound gypsum board prior to installation.

If a board has already been installed and you detect damage along an edge or joint, cut out the damaged area back to sound board and prefill with mud. If this produces too large an area, install a filler strip of good gypsum drywall either laminated to a board layer beneath or attached with screws to the framing. Prefill around the strip and finish the joints as usual.

6.4.3 Water Damage

When gypsum drywall becomes wet, the core becomes soft and is easily deformed. Also, the facing paper may come unbonded (blistered) from the core.

If a board has been exposed to water, let it dry thoroughly before using it. Be very sure it is completely dry before installing it on the framing. If it is so badly warped that even screw attaching will not straighten it, then, after it is dry, put it under a stack of new boards lying flat on the floor.

If a board is already installed and then becomes so wet that it warps away from the framing, drive in some additional screws to hold it. If additional screws do not help, take that board off the framing and replace it.

6.4.4 Board Bowing

Board bowing is similar to the warping problem discussed above. In this case, a board may have been forced into too small a space on the framing, causing the board to bow or warp.

Whenever you discover this problem, the best remedy is to trim the board edges in order to relieve the stress that caused the bowing. You may have to remove the board to do a proper trim job on the edges. Reattach the board when it has been trimmed to fit properly, so that you do not have to force or pry it into place.

6.4.5 Board Cracks and Fractures

A gypsum board can crack along its face, or it may fracture all the way through to the other side. There are various causes and cures for cracks and fractures.

Manufacturing Defects

If you suspect that the drywall panels you are installing are defective, immediately stop work and get instructions from your supervisor. Suppliers and manufacturers will usually replace defective material. There is no sense in putting up defective material only to rip it out later.

Board cracks may occur along the face of any drywall board, but they are most likely to show up over a doorway, where there is a smaller and weaker section of board. If a crack is over ⅛" wide, treat it just as you would a regular joint. Repair it by taping and feathering the joint compound and topping compound until the crack does not show.

This type of cracking is often caused by movement or settling of the building. Many larger buildings, such as skyscrapers, have a built-in flexibility that may contribute to the cracking of interior drywall. In a building with a flexible frame, the best choice is non-bearing interior partitions that have a clearance at the top of every wall. The tracks are fastened to the ceiling to hold the tops of metal studs, which may or may not be actually fastened to the track.

There can be up to ½" clearance between the wallboards and the ceiling. Fill this space with caulk or a specialty gasket or trim such as the type shown in *Figure 53*. A control joint or expansion joint *(Figure 54)* might also be used. Place such metal or plastic trim around the appropriate board edges to give a finished appearance to the room and add protection to the walls.

Any one of three possible reasons can contribute to gypsum board fractures:

- The board was attached across the wide face of the structural framing members, such as the headers. If the framing is wood and the lumber shrinks, the board is compressed and it will crack. If the framing is steel and not very adequate, loads put on it may stress and crack some of the boards attached in this way.
- The wallboard was improperly handled or stored.
- The face paper was scored past the edge of a cutout.

To repair broken or fractured boards, completely cut out the damaged sections and replace them. If the damage was produced by scoring the facing paper beyond the cutout edges, simply repair this score with tape as you would any other joint.

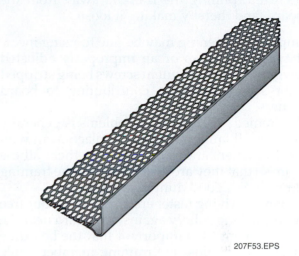

207F53.EPS

Figure 53 ◆ Veneer L-trim casing bead.

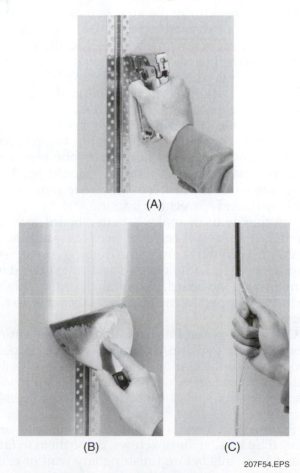

(A)

(B)   (C)

207F54.EPS

Figure 54 ◆ Applying an expansion joint.

6.4.6 Loose Boards

Loose gypsum drywall boards might be caused by any of the following:

- The boards were improperly fastened.
- The framing members were misaligned, uneven, or warped (in the case of wood).
- The screws or nails were not driven in all the way or else (with lumber) some shrinkage has occurred, pulling the framing away from the board, and thereby making it loose.

Improper fastening may be due to using incorrect types of screws or an improperly adjusted screwgun. This can result in screws being stripped or not seated properly, contributing to board looseness.

The remedy for fastener problems is generally to remove all faulty fasteners. Replace them with correct fasteners and properly drive them all the way in, so that they are well fixed into the framing and produce a good dimple in the surface.

When re-driving fasteners, make sure your free hand is pushing solidly against the board near the fastening point. It is important that the board be perfectly flat against the framing member while you are driving the fastener.

Double check to make sure you are using the correct type of drywall screw. Also, readjust the clutch on your screwgun to give you the proper depth into the board. You do not want to tear the face paper, but you need a dimple of about ¹⁄₁₆" to allow proper spotting and finishing. If nails are used, use the double nailing method.

If the cause of loose boards is poor framing, which may be out of alignment, twisted, or warped, your re-driven fasteners alone may not pull the board flush to where it should be. The only way to fix the problem may be to remove all the boards and correct the framing.

Another way to make a better board attachment is to use adhesive as well as additional screws to hold the board to the framing. However, if the framing is badly warped and you succeed in firmly fixing the board, your wall or ceiling might be just as warped as the framing. It is better to fix the framing.

One other possibility is to laminate an entire new layer of gypsum drywall over the warped layer using adhesive or other material. Not only will this provide new drywall laminate, but it will also fill in any spaces caused by the first layer's warping.

Finally and most easily, if loose boards are caused by loose nails or screws, drive them in farther. Check this before finishing any wall or ceiling. Pushing with your hands against the board (even while you are spotting with mud) will indicate if any board is loose. If it is, stop and re-drive the fasteners. You can often drive them by simply using the butt end of your broad knife. Add other fasteners if necessary, then continue spotting and finishing. The time you take to interrupt your finishing and fix the board hanging problem will prevent you or anyone else from having to do the job over again.

As in all repairs, you want the fix to stay fixed. Do not settle for shortcut methods. If you have to rip off the boards and reset the framing, it is better to do it now than to have the problem reported later. If the general contractor or customer discovers the poor framing, it could cost your employer their business and could cost you your job. Fix these problems right from the start.

6.4.7 Patching Drywall

Drywall defects such as holes and dents require patching. For holes two inches or less in diameter, apply joint compound and reinforcing tape over the hole. An additional tape layer may also be needed. Once the bedding coat and tape have dried, apply a topping coat, feathering the edges. Apply a finish coat, if necessary.

Large holes require a different method of repair. Square off and cut out the defective area. Bevel the edges of the squared opening so that the bevels face you. Measure and cut out a patch of new wallboard to fit this opening. Bevel the patch edges to mate with the opening's edges. Use joint compound to cement the patch in place, then tape and finish the edges as you would normal butt joints.

For holes 12" or larger, square off and cut out a whole wallboard section back to the framing members (Figure 55). Cut a fresh patch to fit, cement the patch in place, and use fasteners through the patch into the framing members. Only one drywall screw in each corner of the patch should be necessary. Tape and finish the patch edges like butt joints.

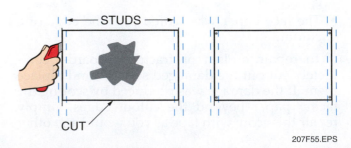

Figure 55 ◆ Patching a large hole.

207F55.EPS

To repair a wallboard dent, first sand the dented section. This raises the nap, but it also permits the joint compound to grip the drywall face paper. Fill the dent with one or more layers of compound. Allow each layer to dry before lightly sanding and then applying the next layer. Finally, sand the filled dent smooth and level with the surrounding wallboard.

Another patching technique is the hot patch or blowout patch, as shown in *Figure 56*.

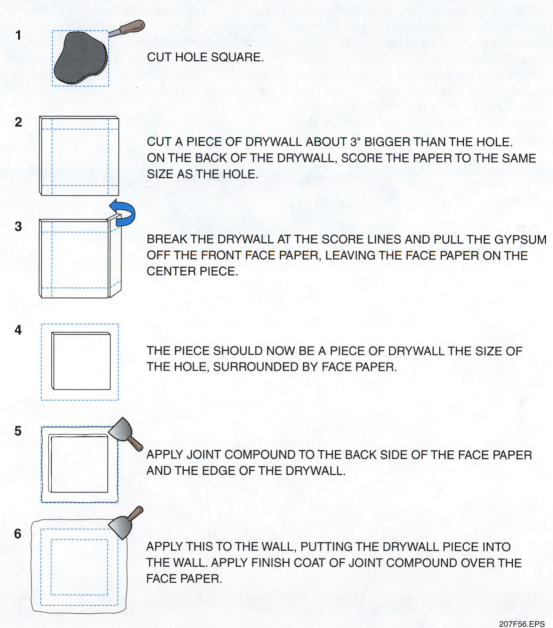

1 CUT HOLE SQUARE.

2 CUT A PIECE OF DRYWALL ABOUT 3" BIGGER THAN THE HOLE. ON THE BACK OF THE DRYWALL, SCORE THE PAPER TO THE SAME SIZE AS THE HOLE.

3 BREAK THE DRYWALL AT THE SCORE LINES AND PULL THE GYPSUM OFF THE FRONT FACE PAPER, LEAVING THE FACE PAPER ON THE CENTER PIECE.

4 THE PIECE SHOULD NOW BE A PIECE OF DRYWALL THE SIZE OF THE HOLE, SURROUNDED BY FACE PAPER.

5 APPLY JOINT COMPOUND TO THE BACK SIDE OF THE FACE PAPER AND THE EDGE OF THE DRYWALL.

6 APPLY THIS TO THE WALL, PUTTING THE DRYWALL PIECE INTO THE WALL. APPLY FINISH COAT OF JOINT COMPOUND OVER THE FACE PAPER.

207F56.EPS

Figure 56 ◆ Patching a hole in drywall using the hot patch (blowout) patching technique.

Summary

Drywall finishing is the difference between craft and mechanics. Once it is finished and painted, the properly finished seam cannot be distinguished from the wall surface itself. Modern finishing equipment, if correctly used, can allow a crew to complete rooms much more quickly than a crew with hand equipment. However, you must use and maintain the equipment correctly in order to get good results. You also need to be able to finish by hand as well as with automatic tapers as there will be places where manual work is the only way to get good results.

The textured surface, more and more popular, eliminates finish sanding, but a poorly taped or fastened joint will still show through. Different kinds of corner beads and tape require different mud mixes and techniques. Drywall that is properly hung is much easier to finish well, preventing nail pops, alignment problems, and gaps. Remember, the walls are out in plain sight; any craftsman or customer will see and recognize shoddy work.

Notes

Review Questions

1. A drywall finishing job with tape embedded in joint compound, one separate coat of compound on joints and interior angles, and two separate coats of compound over fastener heads and accessories meets the _____ requirements.
   a. Level 1
   b. Level 2
   c. Level 3
   d. Level 4

2. A _____ is commonly used to cut drywall.
   a. hacksaw
   b. bandsaw
   c. circular saw
   d. utility knife

3. Which of these tools is used by striking it with a rubber mallet?
   a. Finishing trowel
   b. Corner clinching tool
   c. Pole sander
   d. Mud masher

4. The automatic tool that applies a smooth finish with feathered edges and a center crown to a taped seam is the _____.
   a. flat applicator
   b. banjo
   c. flat finisher
   d. nail spotter

5. Fiberglass mesh tape may be preferred over paper tape in _____ applications.
   a. high-moisture
   b. low-humidity
   c. high-temperature
   d. low-temperature

6. You would be likely to use metal edge tape in each of these locations *except* _____.
   a. outside angled corners
   b. a 90-degree inside corner
   c. the intersection of a ceiling and a radius wall
   d. an arch

7. Each of the following statements regarding topping compound is correct except _____.
   a. topping compound is used for the second and third finishing coats
   b. topping compound is easier to sand because it dries softer than joint compound
   c. topping compound and joint compound are the same thing; you just add more water to joint compound to make topping compound
   d. topping compound shrinks less than joint compound

8. When mixing dry-mix compounds, you are required to wear _____.
   a. gloves
   b. a respirator
   c. a hair net
   d. protective coveralls

9. Which type of texture material has good solution time, minimum-to-moderate fallout, and good bonding power?
   a. Premixed textures
   b. Powder joint compound textures
   c. Aggregated powder textures
   d. Unaggregated powder textures

10. Drywall fasteners should be installed so that _____.
   a. the head penetrates the paper
   b. there is a slight depression, or dimple, in the drywall
   c. the head protrudes ¹⁄₆₄ of an inch from the drywall surface
   d. the head is exactly flush with the drywall surface

11. Which of these is the correct reason for applying multiple coats of compound?
   a. The compound shrinks as it dries, leaving depressions.
   b. A lot of the compound will flake and fall off as it dries.
   c. Extra buildup is needed to compensate for sanding.
   d. Walls look better when the seams are slightly higher than the wall surface.

12. If you find bubbles in tape joints after using the automatic taper, it probably means that _____.
    a. only one wheel was pressing on the drywall surface
    b. neither wheel was pressing on the drywall surface
    c. both wheels were pressing on the drywall surface
    d. the taper was upside down

13. Fasteners normally receive _____ coat(s) of topping.
    a. one
    b. two
    c. three
    d. four

14. When you are able to see a taped joint after the wall has been painted, it is known as _____.
    a. ridging
    b. photographing
    c. high joints
    d. discoloration

15. When mixing joint compound, you should use _____ water in order to avoid bonding problems.
    a. ice
    b. distilled
    c. cold, clean
    d. hot

Trade Terms Introduced in This Module

All-purpose compound: Combines the features of taping and topping compounds. It does not bond as well as taping compound, but finishes better.

Bullnose: A metal corner bead with rounded edges.

Feathering: Tapering joint compound at the edges of a drywall joint to provide a uniform finish.

Joint compound: Patching compound used to finish drywall joints, conceal fasteners, and repair irregularities in the drywall. It dries hard and has a strong bond. Sometimes called mud or taping compound.

Lightweight compound: An all-purpose compound having less weight than standard compounds.

Mud: See *joint compound*.

Ridges: Slight protrusions in the center of a finished drywall joint that are usually caused by insufficient drying time. Also known as beads.

Skim coat: A thin coat of joint or topping compound that is applied over the entire drywall surface. Sometimes required under a high gloss finish.

Tape: A strong paper or fiberglass tape used to cover the joint between two sheets of drywall.

Tapered joint: A joint where tapered edges of drywall meet.

Taping compound: See *joint compound*.

Topping compound: A joint compound used for second and third coats. It dries soft and smooth and is easier to sand than taping compound.

This module is intended to present thorough resources for task training. The following reference works are suggested for further study. These are optional materials for continued education rather than for task training.

Gypsum Construction Guide. Charlotte, NC: National Gypsum Company, 1994.

Gypsum Construction Handbook. Chicago, IL: United States Gypsum Company, 2000.

Installing and Finishing Drywall, William Spence. New York, NY: Sterling Publishing Company, 1998.

Painting and Decorating Craftsman's Manual and Textbook. Fairfax, VA: Painting and Decorating Contractors of America, 1995.

NCCER CURRICULA — USER UPDATE

NCCER makes every effort to keep its textbooks up-to-date and free of technical errors. We appreciate your help in this process. If you find an error, a typographical mistake, or an inaccuracy in NCCER's curricula, please fill out this form (or a photocopy), or complete the online form at **www.nccer.org/olf**. Be sure to include the exact module ID number, page number, a detailed description, and your recommended correction. Your input will be brought to the attention of the Authoring Team. Thank you for your assistance.

Instructors – If you have an idea for improving this textbook, or have found that additional materials were necessary to teach this module effectively, please let us know so that we may present your suggestions to the Authoring Team.

NCCER Product Development and Revision

13614 Progress Blvd., Alachua, FL 32615

Email: curriculum@nccer.org
Online: www.nccer.org/olf

❑ Trainee Guide   ❑ AIG   ❑ Exam   ❑ PowerPoints   Other _____

Craft / Level: _____   Copyright Date: _____

Module ID Number / Title: _____

Section Number(s): _____

Description: _____

_____

_____

_____

Recommended Correction: _____

_____

_____

_____

Your Name: _____

Address: _____

_____

Email: _____   Phone: _____

Doors and Door Hardware

27208-07

201
SOUTH NARCISSUS

© CG Grant 2004

27208-07
Doors and Door Hardware

Topics to be presented in this module include:

Overview

The installation of interior and exterior doors and their companion hardware is an important part of a carpenter's work, whether in residential or commercial construction. Although many doors are prehung in a frame, the installer must be able to install them level and plumb, which is a skill that takes considerable practice. The carpenter must learn to install both wood and metal prehung doors. Not all doors are prehung, however. Sometimes the carpenter has to install the hinges and hang the door in a door opening. This is a specialized skill that requires careful measuring and skillful use of tools.

There are several types of locksets and other hardware used in residential construction, but the number multiplies in commercial construction because security and fire protection come into play.

Objectives

When you have completed this module, you will be able to do the following:

1. Identify various types of door jambs and frames and demonstrate the installation procedures for placing selected door jambs and frames in different types of interior partitions.
2. Identify different types of interior doors.
3. Identify different types of interior door hardware and demonstrate the installation procedures for selected types.
4. Demonstrate the correct and safe use of the hand and power tools described in this module.
5. List and identify specific items included on a typical door schedule.
6. Demonstrate the procedure for placing and hanging a selected door.

Trade Terms

Access	Lockset
Astragal	Molding
Butt	Mortise
Butt gauge	Mortise lock
Casing	Panic hardware
Catches	Plumb
Coordinator	Prefinished
Cylindrical lockset	Prehung door
Deadbolt	Rabbet
Door closer	Rail
Door frame	Rough opening
Door jamb	Scribe
Door stop	Shim
Dustproof strike	Sill
Finish hardware	Smoke gasket
Finishing sawhorse	Sound attenuation
Flush bolt	Sound transmission class
Flush door	(STC)
Hanging stile	Stile
Hardware	Strike
Head	Strike plate
Hinge	Sweep
Jamb	Template
Kerf	Threshold
Knob lockset	Transom
Latch bolt	Weatherstripping
Lock	

Required Trainee Materials

1. Pencil and paper
2. Appropriate personal protective equipment

Prerequisites

Before you begin this module, it is recommended that you successfully complete *Core Curriculum*; *Carpentry Fundamentals Level One*; and *Carpentry Framing and Finishing Level Two*, Modules 27201-07 through 27207-07.

This course map shows all of the modules in *Carpentry Framing and Finishing Level Two*. The suggested training order begins at the bottom and proceeds up. Skill levels increase as you advance on the course map. The local Training Program Sponsor may adjust the training order.

27212-07
Cabinet Fabrication
ELECTIVE

27211-07
Cabinet Installation

27210-07
Window, Door, Floor, and Ceiling Trim

27209-07
Suspended Ceilings
ELECTIVE FOR RESIDENTIAL CERTIFICATE

27208-07
Doors and Door Hardware

27207-07
Drywall Finishing

27206-07
Drywall Installation

27205-07
Cold-Formed Steel Framing

27204-07
Exterior Finishing
ELECTIVE FOR COMMERCIAL CERTIFICATE

27203-07
Thermal and Moisture Protection

27202-07
Roofing Applications
ELECTIVE FOR COMMERCIAL CERTIFICATE

27201-07
Commercial Drawings
ELECTIVE FOR RESIDENTIAL CERTIFICATE

CARPENTRY FUNDAMENTALS

CORE CURRICULUM:
Introductory Craft Skills

FRAMING AND FINISHING

208CMAP.EPS

1.0.0 ◆ INTRODUCTION

This module will identify interior doors used in the construction industry and shows the correct method of installation. Instruction will include placing the door jambs, hanging the doors, and installing the hardware. The cutting, fitting, and placement of door trim is also discussed. The interior finish materials of any building, including doors, must be installed so that they are clean and neat, operate properly, and reflect the skill and pride of the carpenter responsible for the workmanship. The methods of placing a door and its trim may differ from one region of the country to another due to the variations among door manufacturers and local preferences in the sequence of installation.

2.0.0 ◆ SAFETY

A good carpenter is always aware of the safety rules in every area of the construction field. The interior of a building nearing completion sometimes gives the worker a false sense of security. However, the element of danger is always present.

- *Keep work area clean* – Cluttered areas and benches invite injuries.
- *Avoid dangerous equipment* – Do not expose power tools to rain. Do not use power tools in damp or wet locations. Keep the area well lit. Avoid chemical or corrosive environments. Do not use tools in the presence of flammable liquids or gases.
- *Guard against electric shock* – Avoid body contact with grounded surfaces such as pipes, radiators, ranges, or refrigerator enclosures.
- *Keep visitors away* – Do not let visitors come in contact with tools or extension cords.
- *Store idle tools* – When not in use, tools should be stored up.
- *Secure the work* – Use clamps or a vise to hold the work. It is safer than using your hand, and it frees both hands for operating the tool.
- *Use the right tools* – Do not force a smaller tool or attachment to do the job of a heavier tool. Do not use a tool for any purpose for which it was not intended.
- *Dress properly* – Do not wear loose clothing or jewelry. Loose clothing, drawstrings, and jewelry can be caught in moving parts. Rubber gloves and nonskid footwear are recommended. Wear a protective covering to protect long hair.
- *Use safety goggles* – Wear safety glasses or goggles while operating power tools. Wear a face or dust mask if the operation creates dust. All persons in the area where power tools are being operated should also wear safety glasses and face or dust masks.

- *Do not abuse the cord* – Never carry any tool by the cord or yank it to disconnect it from the receptacle. Keep the cord away from heat, oil, and sharp edges. Have damaged or worn power cords and strain relievers replaced immediately.
- *Do not overreach* – Keep proper footing and balance at all times.
- *Maintain tools with care* – Keep tools sharp and clean for safer performance. Follow instructions for lubricating and changing accessories. Inspect tool cords periodically and, if they are damaged, have them repaired by an authorized service facility. Have all worn, broken, or lost parts replaced immediately. Keep handles dry, clean, and free from oil and grease.
- *Disconnect (unplug) tools* – Always disconnect tools when not in use, before servicing, and when changing accessories such as blades, bits, and cutters.
- *Remove keys and adjusting wrenches* – Check to make certain that keys and adjusting wrenches are removed from a tool before turning it on.
- *Avoid unintentional starting* – Do not carry a plugged-in tool with a finger on the switch. Be sure the switch is off when plugging in a tool.
- *Outdoor extension cords* – When a tool is used outdoors, use only extension cords marked as suitable for use with outdoor appliances and store them indoors when not in use.
- *Check damaged parts* – Before using any tool, inspect it to be sure that it will operate properly and perform its intended function. Check for alignment of moving parts, binding of moving parts, breakage of parts, mounting, and any other conditions that may affect its operation. A guard or other part that is damaged should be properly repaired or replaced by an authorized service center unless otherwise indicated in the instruction manual. Have defective switches replaced by an authorized service center.
- *Stay alert* – Watch what you are doing and use common sense. Do not operate a tool when you are tired or while under the influence of medication, alcohol, or drugs.
- *Inspect extension cords* – Inspect extension cords periodically and replace them if damaged.

 WARNING!

Remember that an extension cord with cracked or cut outer insulation or a loose or missing ground pin can kill you. Be careful. Always unplug power tools when changing accessories or making adjustments. Plan ahead. Always put safety first.

3.0.0 ◆ DOOR TYPES AND BASIC CONSTRUCTION

Except for some welded metal **door frames**, interior doors are usually installed after the walls have received their finish covering. Like exterior doors, interior door components consist of the door, jamb, **door stops**, and a **casing** (*Figure 1*), along with the door hardware such as the **hinges** and **locksets**. The jamb is the frame in which the door hangs. The stops halt the door swing when the door is being closed and position the door so that the lockset bolt engages the **strike plate** properly after the door is closed. The casing is used to provide the trim for the door frame.

There are many types of interior doors made of various materials, or combinations of materials, ranging from wood and metal to synthetic materials. However, most doors can be categorized as flush or panel types.

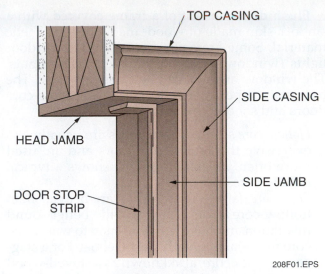

208F01.EPS

Figure 1 ◆ Typical door components.

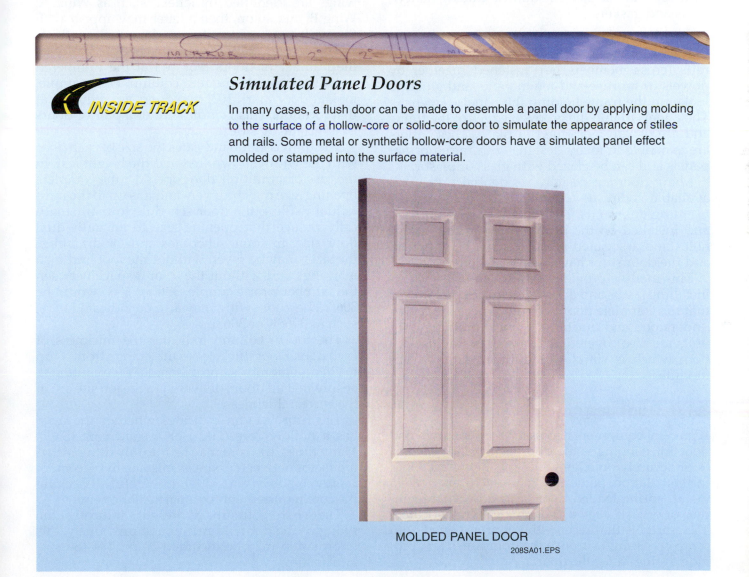

INSIDE TRACK

Simulated Panel Doors

In many cases, a flush door can be made to resemble a panel door by applying molding to the surface of a hollow-core or solid-core door to simulate the appearance of stiles and rails. Some metal or synthetic hollow-core doors have a simulated panel effect molded or stamped into the surface material.

MOLDED PANEL DOOR

208SA01.EPS

Flush doors consist of a frame covered with a smooth skin made of wood, metal, or a synthetic material. Some are equipped (glazed) with door lights (window panes), also called vision lights. The window material can be glass or plastic. The two main types of flush doors are hollow-core doors and solid-core doors.

- *Hollow-core doors* – These doors are usually less expensive than solid-core doors and are used more often. *Figure 2* (top left) shows a typical hollow-core door.
- *Solid-core doors* – These doors are heavier than hollow-core doors. They provide better sound insulation and have less tendency to warp. The core material is usually particleboard or a staggered-block core, also known as a staved-wood core, as shown in *Figure 2*. Special solid-core doors include fire doors that use a mineral core, radiation-blocking doors that use a lead lining, and **sound transmission class (STC)** doors that use special sound-reducing materials and gaskets.

Wood panel doors are also known as stile-and-rail doors because **stiles** (vertical members) and **rails** (cross members) are fastened together by dowels or **mortise**-and-tenon joints and glued to make up the frame holding the panels (*Figure 3*). The panels fit into the grooved edges of the frame and are usually not glued in place. These doors are available in many designs with raised or flat panels and can be glazed with glass or plastic.

Many doors, especially residential doors, are available today as assembled or unassembled **prehung door** units. The hinges are pre-mortised and installed on the jamb and door, the lockset holes and associated mortises have been precut, and the door stops have been installed.

This module provides detailed information on installing jambs and doors that are not prehung so that the complete door hanging procedure can be understood and practiced. Typical prehung door units are also covered, and their general installation, which is simpler, is outlined later in this module.

3.1.0 Door Jack

A piece of equipment necessary to prepare a wood door for hanging is the door jack, which holds a door securely on edge while it is being prepared for installation. Some tool manufacturers make a metal, spring-loaded door jack; however, a simpler door jack (*Figure 4*) can be easily fabricated. The uprights that receive the door are lined with scrap carpeting to protect the door finish and provide a snug fit for the door.

3.2.0 Door Schedules

Each door must be located and identified on the plans or drawings of the building. One way to locate a door is on the door schedule. Almost every set of drawings includes a door schedule. The one shown in *Figure 5* is one example.

NOTE

The door schedule shown is one of many variations used and should not be thought of as standard.

The first column on the left of the schedule indicates the mark of the door by number. The usual method is to number all doors on the first floor in the one hundred series, the second floor in the two hundred series, and so forth. If it is a large building that has wings shown on the floor plan and these wings are identified by letters, such as Wing A, Wing B, and so on, then a letter may appear with the mark number identifying a door, such as 104-A.

The second column on the schedule indicates the number of doors required at the numbered mark indicated in the first column. For example, look at door mark 103; two doors are required at that location. This is indicative of a pair of doors located side by side.

The third column indicates the size of each door with the width shown first and the height last. In the case of a multiple door opening, the complete opening size may be given, and it is up to the individual reading the door schedule to mathematically reduce that opening size to an individual door size. In some schedules and/or drawings, the sizes may be given without the foot and inch markings and without the × or a space between the numbers; for example, 3'-0" × 6'-8" would be 3068. The door swing may also be indicated here, such as 3068R or 3068L.

The fourth column indicates the thickness of each door. Door thicknesses may range from ¾" up to 2½" and more in some instances. Take caution to ensure that all doors delivered to the job site are of the correct thickness.

The fifth column indicates whether the door has a hollow core (H.C.) or a solid core (S.C.). Many times, the core may be the only differentiating factor between doors. A mistake in placement is often made as a result of not carefully checking the core requirement for a particular opening.

The sixth column, if present, indicates the swing of the door. As mentioned previously, this information may be included as part of the size information.

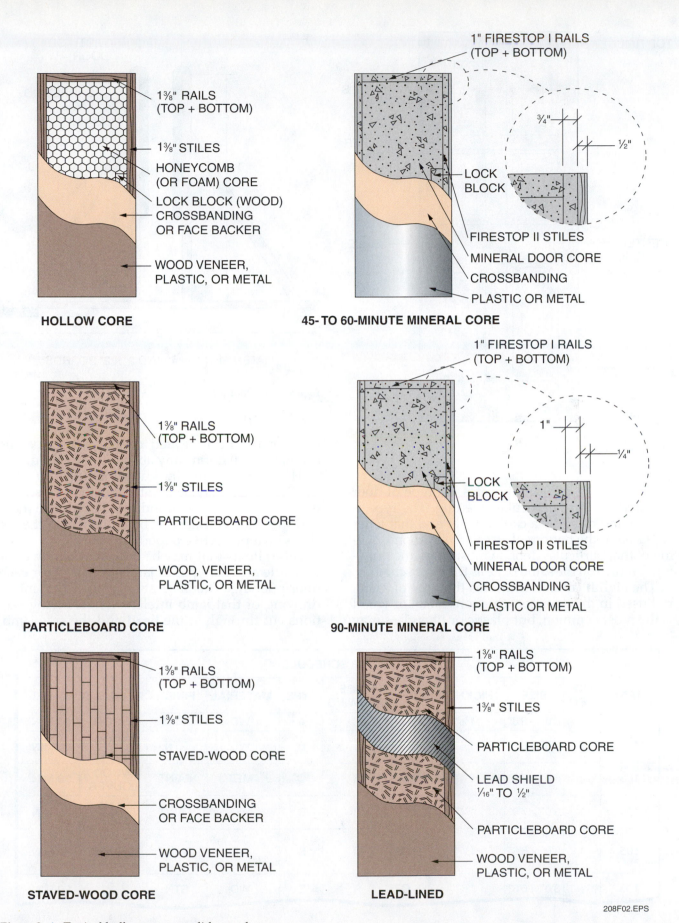

Figure 2 ◆ Typical hollow-core or solid-core doors.

HOLLOW CORE

1⅜" RAILS (TOP + BOTTOM)

1⅜" STILES

HONEYCOMB (OR FOAM) CORE

LOCK BLOCK (WOOD)

CROSSBANDING OR FACE BACKER

WOOD VENEER, PLASTIC, OR METAL

45- TO 60-MINUTE MINERAL CORE

1" FIRESTOP I RAILS (TOP + BOTTOM)

¾"

½"

LOCK BLOCK

FIRESTOP II STILES

MINERAL DOOR CORE

CROSSBANDING

PLASTIC OR METAL

PARTICLEBOARD CORE

1⅜" RAILS (TOP + BOTTOM)

1⅜" STILES

PARTICLEBOARD CORE

WOOD, VENEER, PLASTIC, OR METAL

90-MINUTE MINERAL CORE

1" FIRESTOP I RAILS (TOP + BOTTOM)

1"

¼"

LOCK BLOCK

FIRESTOP II STILES

MINERAL DOOR CORE

CROSSBANDING

PLASTIC OR METAL

STAVED-WOOD CORE

1⅜" RAILS (TOP + BOTTOM)

1⅜" STILES

STAVED-WOOD CORE

CROSSBANDING OR FACE BACKER

WOOD VENEER, PLASTIC, OR METAL

LEAD-LINED

1⅜" RAILS (TOP + BOTTOM)

1⅜" STILES

PARTICLEBOARD CORE

LEAD SHIELD 1/16" TO ½"

PARTICLEBOARD CORE

WOOD VENEER, PLASTIC, OR METAL

208F02.EPS

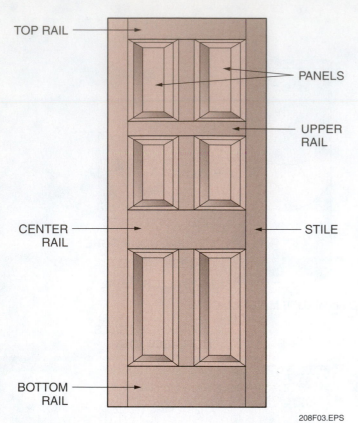

Figure 3 ◆ Panel door.

TOP RAIL

PANELS

UPPER RAIL

CENTER RAIL

STILE

BOTTOM RAIL

208F03.EPS

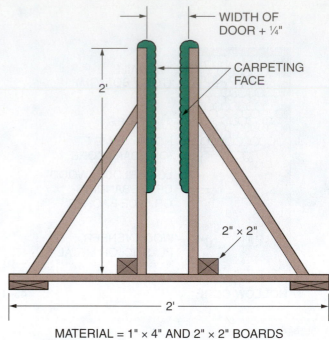

WIDTH OF DOOR + ¼"

CARPETING FACE

2'

2" × 2"

2'

MATERIAL = 1" × 4" AND 2" × 2" BOARDS

208F04.EPS

Figure 4 ◆ Door jack.

The seventh column indicates the type of door required for that particular opening. The type is usually indicated by a capital letter and that letter is placed at the bottom of a scaled elevation drawing of that particular door. In most cases, the drawing is found on the same page as the door schedule.

The eighth column indicates the type of material used in the door construction. Metal or wood is the most common, but plastic or fiberglass may

occasionally be indicated on the schedule. The drawings will define any abbreviations used.

The ninth column indicates the final door finish. The finish may be paint, stain, or varnish. On occasion, a door may be indicated as **prefinished**, and the door will usually be delivered to the job site with a protective paper wrapping.

Other items that may be indicated on the door schedule are the types of jambs required for each door opening. This indication is then tied into a drawing of that jamb in elevation, showing sections cut through at the **head** and the side jamb.

DOOR SCHEDULE										
MARK	NO. REQ	SIZE	THICKNESS	CORE	HINGE OR SWING	TYPE	MATERIAL	FINISH	REMARKS	LABEL
101	1	3'0" × 6'8"	1⅜"	H.C.	L	A	WD.	STN.	WOOD	20 MIN.
102	1	2'8" × 6'8"	1⅜"	H.C.	L	B	WD.	STN.	WOOD LOUVER	20 MIN.
103	2	6'0" × 7'0"	1¾"	H.C.	R	C	MET.	PAINT	WOOD LOUVER	90 MIN.
104	1	3'0" × 7'0"	1¾"	S.C.	R	D	WD.	PAINT	¼" P.P. GLASS	45 MIN.
105	1	3'0" × 6'8"	1¾"	H.C.	R	B	WD.	PAINT	SIGHT PROOF LOUVER	60 MIN.
106	1	3'0" × 6'8"	1¾"	S.C.	R	E	WD.	STN.	¼" P.P. GLASS	20 MIN.

208F05.EPS

Figure 5 ◆ Typical door schedule.

Prehung Door Units

Good prehung door units can save considerable installation time at the job site. On the other hand, the fit-up of inexpensive, low-quality units is often very poor and may cause more expense than installing conventionally hung doors.

208SA02.EPS

The type of **threshold** may also be indicated, and a drawing of each type will be shown corresponding to the schedule. Hardware finish is another item that may occasionally be on a schedule.

The door schedule will also include a Remarks column. This column is reserved for any special information about a door. The schedule might also include a Label column, which lists fire rating information. This will only be included if fire-rated doors are to be used. It shows the number of minutes or the duration of the fire test the door type is capable of enduring before allowing the entry of fire into a protected area.

3.3.0 Door Hand or Swing

The direction in which a door opens is called the hand or swing of the door. Prior to hanging a door, the carpenter must have some knowledge about the hand or swing. The doors that are delivered to the job site must be checked and compared to the door schedule so each door's location may be identified on the floor plan of the drawings. When each door is accounted for and its location is known, the hand or swing must be established according to the plans. A good carpenter should have some knowledge of how the swing is determined. It is very important when ordering assembled or unassembled prehung door units, and it is also important in some instances when ordering hardware to fit a door. Knowing the direction of a particular door swing is critical when giving or receiving instructions on hanging it in the opening. The incorrect determination of the door swing is one of the most common mistakes in construction due to the fact that there is no universal standard for specifying door swing. In some cases, the hand can be defined as either the handle location or the hinge location.

One method of identification dictates that a door swing be described in four ways, as shown in the plan view drawing in *Figure 6*. The door swing is the direction in which a door swings and is determined when facing the door from the outside (public side). This method is used by many contractors and door manufacturers in different

LEFT-HAND SWING:
HINGES AT LEFT.
DOOR OPENS INWARD.
HANDED LOCK = LH

OUTSIDE OF DOOR

LEFT-HAND REVERSE:
HINGES AT LEFT.
DOOR OPENS OUTWARD.
HANDED LOCK = LHR

OUTSIDE OF DOOR

RIGHT-HAND SWING:
HINGES AT RIGHT.
DOOR OPENS INWARD.
HANDED LOCK = RH

OUTSIDE OF DOOR

RIGHT-HAND REVERSE:
HINGES AT RIGHT.
DOOR OPENS OUTWARD.
HANDED LOCK = RHR

OUTSIDE OF DOOR

— **OUTSIDE DOOR DEFINITIONS** —

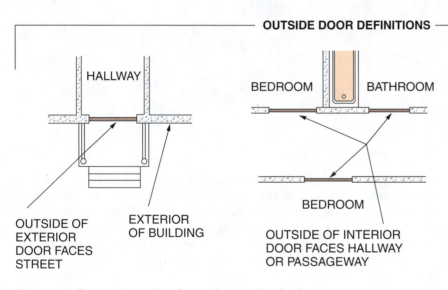

OUTSIDE OF EXTERIOR DOOR FACES STREET

EXTERIOR OF BUILDING

OUTSIDE OF INTERIOR DOOR FACES HALLWAY OR PASSAGEWAY

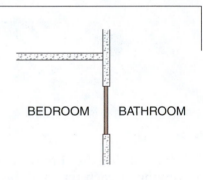

FOR DOORS THAT OPEN BETWEEN ROOMS, THE LOCKABLE SIDE OF THE LOCKSET IS CONSIDERED THE OUTSIDE OF THE DOOR

208F06.EPS

Figure 6 ◆ Door swing when facing the outside of a door.

parts of the country. It is also used by most lockset manufacturers to specify locksets.

In this method, a door can be either left hand or right hand, and either of these types can be a reverse (one that opens out of the room instead of into the room). Reverse doors are prepared for hardware differently than standard doors.

In another method, the hand or swing is determined by mentally placing yourself in the floor plan of the building at the doorway in question so that your back is against the hinge locations. The side of the jamb on which the hinges (**butts**) are located determines the swing of the door, as

shown in the plan view drawing in *Figure 7(A)*. A term used by many carpenters on the job to describe this method is butt-to-butt. Some door manufacturers use other methods and each manufacturer must be checked to determine the specific method used. As a general rule, door swing is based on which side the handle or hinge is on when the door is pulled closed. See *Figure 7(B)*.

To mark the swing of the door for a particular opening, mark the jamb on the side that will have the hinges. Mark the location of the hinges (butts) on the **hanging stile** of the door that corresponds with that opening.

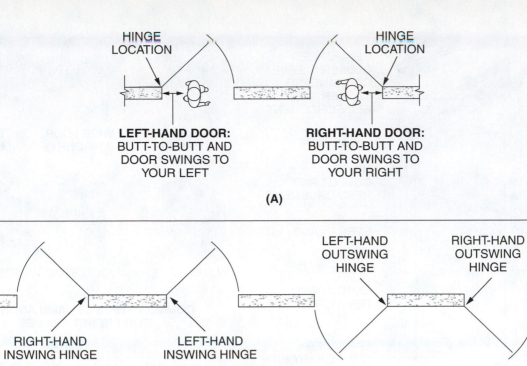

LEFT-HAND DOOR: BUTT-TO-BUTT AND DOOR SWINGS TO YOUR LEFT

RIGHT-HAND DOOR: BUTT-TO-BUTT AND DOOR SWINGS TO YOUR RIGHT

(A)

RIGHT-HAND INSWING HINGE

LEFT-HAND INSWING HINGE

LEFT-HAND OUTSWING HINGE

RIGHT-HAND OUTSWING HINGE

EXTERIOR DOORS

RIGHT-HAND HANDLE OR LEFT-HAND HINGE

LEFT-HAND HANDLE OR RIGHT-HAND HINGE

INTERIOR DOORS

(B)

208F07.EPS

Figure 7 ◆ Hinge location method of determining door swing.

INSIDE TRACK

Prehung Door Unit Swing

When ordering assembled or unassembled prehung door units from the manufacturer, make sure there is an understanding of which method is to be used to determine the swing. This will facilitate delivery of the correct door unit to the job site. Most manufacturers use the convention that when facing the hinge barrel (visible) side of the prehung door unit and the hinge is on the left, the door is a right hand. If the hinge is on the right, it is a left hand.

4.0.0 ◆ METAL DOOR FRAMES

Metal door frames are used extensively in commercial and institutional applications. *Figures 8* and *9* show some of the styles and types of typical metal door frames.

The two basic types of metal door frames are a preassembled, welded unit and an unassembled or knocked down (K.D.) unit. Either type is available in fixed wall widths or adjustable wall widths. The fixed widths typically range from 2½" to 7½" in ⅛" increments. Adjustable units are available to accommodate wall thicknesses ranging from 3¾" to 9¼". Some frames have manufactured and finished snap-on casings made of wood, metal, or synthetic materials. Others can be obtained with door stops that are terminated at an angle above the floor line. These terminated stops are sometimes called hospital or sanitary stops.

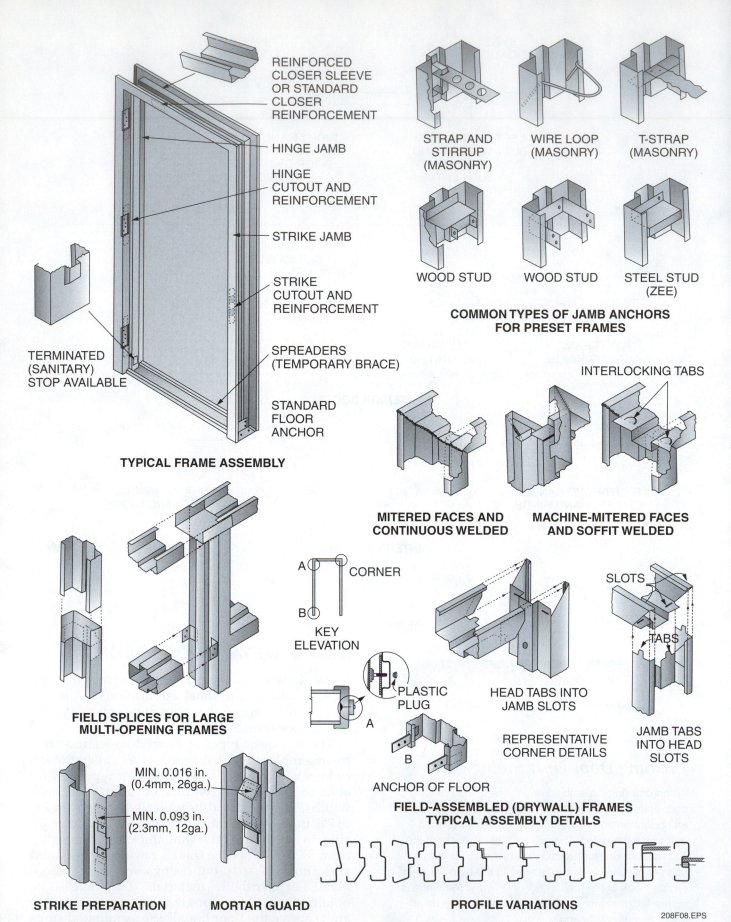

REINFORCED CLOSER SLEEVE OR STANDARD CLOSER REINFORCEMENT

HINGE JAMB

HINGE CUTOUT AND REINFORCEMENT

STRIKE JAMB

STRIKE CUTOUT AND REINFORCEMENT

TERMINATED (SANITARY) STOP AVAILABLE

SPREADERS (TEMPORARY BRACE)

STANDARD FLOOR ANCHOR

TYPICAL FRAME ASSEMBLY

STRAP AND STIRRUP (MASONRY)

WIRE LOOP (MASONRY)

T-STRAP (MASONRY)

WOOD STUD

WOOD STUD

STEEL STUD (ZEE)

COMMON TYPES OF JAMB ANCHORS FOR PRESET FRAMES

INTERLOCKING TABS

MITERED FACES AND CONTINUOUS WELDED

MACHINE-MITERED FACES AND SOFFIT WELDED

FIELD SPLICES FOR LARGE MULTI-OPENING FRAMES

A — CORNER

B

KEY ELEVATION

PLASTIC PLUG

A

B

ANCHOR OF FLOOR

SLOTS

TABS

HEAD TABS INTO JAMB SLOTS

REPRESENTATIVE CORNER DETAILS

JAMB TABS INTO HEAD SLOTS

FIELD-ASSEMBLED (DRYWALL) FRAMES TYPICAL ASSEMBLY DETAILS

MIN. 0.016 in. (0.4mm, 26ga.)

MIN. 0.093 in. (2.3mm, 12ga.)

STRIKE PREPARATION

MORTAR GUARD

PROFILE VARIATIONS

208F08.EPS

Figure 8 ◆ Typical fixed-width, single-piece jamb and casing metal frames.

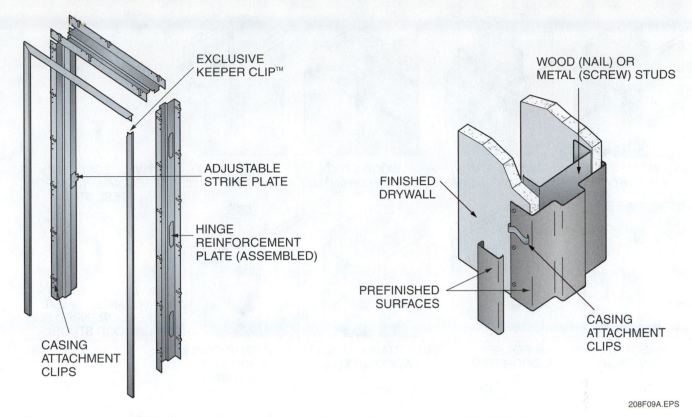

EXCLUSIVE
KEEPER CLIP™

ADJUSTABLE
STRIKE PLATE

HINGE
REINFORCEMENT
PLATE (ASSEMBLED)

CASING
ATTACHMENT
CLIPS

WOOD (NAIL) OR
METAL (SCREW) STUDS

FINISHED
DRYWALL

PREFINISHED
SURFACES

CASING
ATTACHMENT
CLIPS

208F09A.EPS

Figure 9 ◆ Typical fixed-width or adjustable-width metal frame with snap-on casing (1 of 2).

Any type of wood, or a matched metal door, can be attached to the frames, and the frames can be fastened to metal structural studs, wood studs, or masonry. This section will cover the typical installation of one type of assembled and unassembled metal frame in wood or masonry construction.

4.1.0 Typical Welded Metal Door Frame Installation in Wood Stud Construction

Installing a welded metal door frame in wood stud construction is not a difficult task if the following steps are completed. Usually, the metal frame will have wood stud anchors welded to it. Some wood stud anchors will be furnished loose, as shown in *Figure 10*. In some cases, reinforced closer sleeves will be required if the reinforcements have not been welded into the frame.

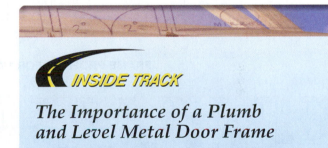

INSIDE TRACK

The Importance of a Plumb and Level Metal Door Frame

Care must be taken to ensure that a metal frame is absolutely plumb and square. An improper frame installation will affect the performance of the door and make fitting the door difficult or impossible. A little extra time spent installing the frame will make door installation much easier.

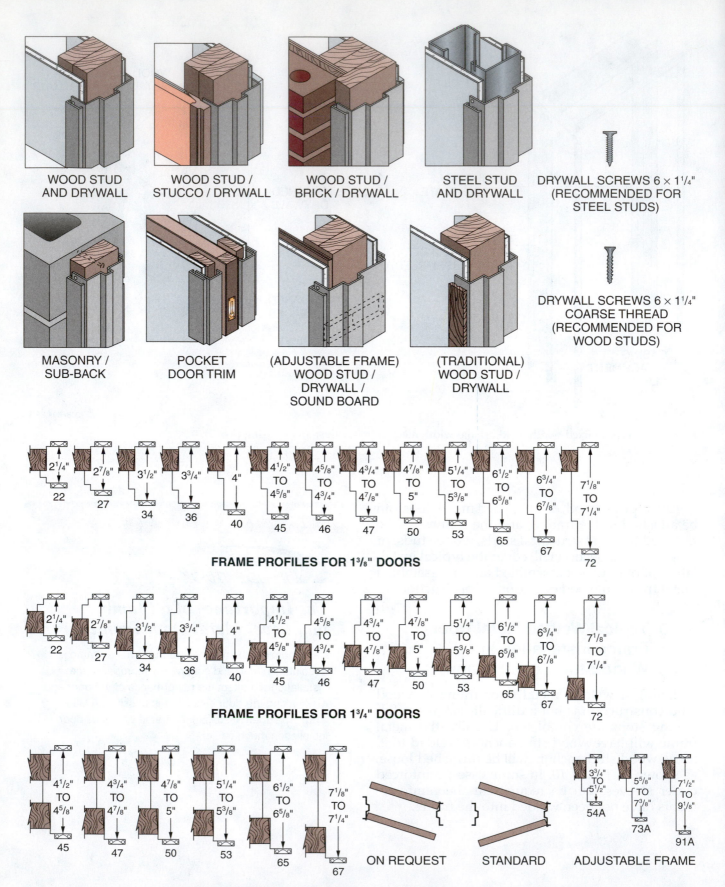

Figure 9 ◆ Typical fixed-width or adjustable-width metal frame with snap-on casing (2 of 2).

208F09B.EPS

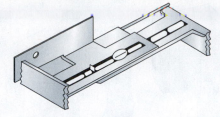

ADJUSTABLE WOOD STUD
ANCHOR FURNISHED LOOSE

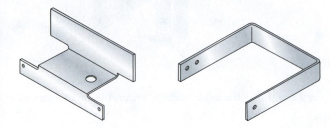

WOOD STUD ANCHORS
WELDED TO FRAME

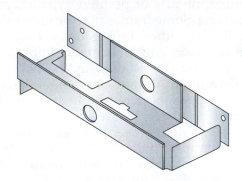

JAMB-FIT TYPE COMBINATION
STUD ANCHOR FURNISHED LOOSE

208F10.EPS

Figure 10 ◆ Wood stud wall anchors.

The loose anchors are installed inside this particular metal frame by forcing or expanding the anchors between the two casing flanges of the frame.

Step 1 Place the anchor into position at an angle and twist it into a horizontal position, as shown in *Figure 11*. If required, place a closer sleeve inside the head jamb frame at the proper location. It should fit tightly to prevent movement.

Step 2 The metal frame may have been shipped using a temporary metal spreader bar that must be removed. Replace the metal spreader bar with a wood spreader at the sill. Cut notches in the ends of the wood spreader for the stops (*Figure 12*). Make sure that the spacing of the jambs at the sill is the same as at the top of the frame.

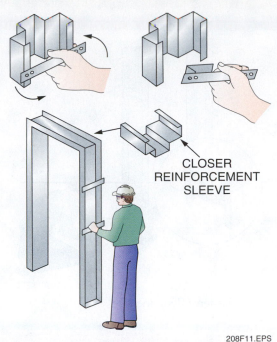

CLOSER
REINFORCEMENT
SLEEVE

208F11.EPS

Figure 11 ◆ Install anchors and closer reinforcement sleeve.

208F12.EPS

Figure 12 ◆ Replace metal spreader bar with a wood spreader.

Step 3 Check the head jamb for level and shim the bottom of the side jambs as required, subject to the manufacturer's recommendations. Check the location of the metal frame again and, using an appropriate fastener, attach the floor anchors (*Figure 13*).

Step 4 The bottom of the door frame must be attached to the floor or the framed opening per the manufacturer's instructions or local code requirements. Add a wood spreader at the center (*Figure 14*).

Figure 13 ◆ Fasten a floor anchor.

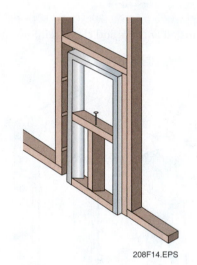

Figure 14 ◆ Install the spreader.

Step 5 Using an accurate level, **plumb** the jambs and make sure all studs are square and true with the metal frame. Bend the jamb anchors around the wood studs, nail them in place, and recheck for plumb and squareness (*Figure 15*). Install a wooden door or matched metal door.

4.2.0 Sound Attenuation and Grouting for Metal Frames

Check the jamb details, specifications, and fire codes to see if grouting is required. Grouting is installed in frames to provide **sound attenuation**

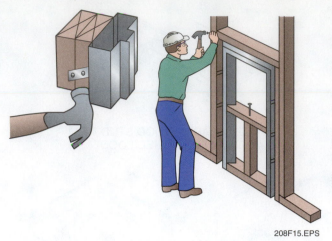

Figure 15 ◆ Bend the jamb anchors around the wood studs.

and/or fireproofing. Grout may be either a sand and cement mixture or perlited gypsum, per the specifications. Some frames may be grouted prior to installation; other frames may have to be grouted in place. Any attachment clips that are necessary must be installed in the door frame prior to grouting. All butt plates must have duct tape placed over them on the inside, and styrofoam must be placed inside at the approximate location of the **door closer** so screw holes can be drilled and tapped in that location. If possible and necessary, install screws for **weatherstripping** so you can back them out and install the weatherstripping later.

4.3.0 Typical Welded Metal Door Frame Installation in Steel Stud Construction

Installing a welded metal door frame in steel stud construction is done in much the same manner as with wood studs. Steel stud anchors are furnished with the metal door frame. Some anchors are welded to the frame and some are loose, as shown in *Figure 16*.

Step 1 The loose anchors are installed inside the metal frame by forcing the anchor between the two casing flanges of the frame. Place the anchor into position at an angle and twist it into a horizontal position, as shown in *Figure 17*. If required, place a closer sleeve inside the head jamb frame at the proper location. It should fit tightly to prevent movement.

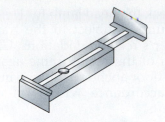

ADJUSTABLE STEEL STUD
ANCHOR FURNISHED LOOSE

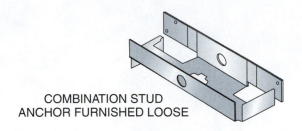

COMBINATION STUD
ANCHOR FURNISHED LOOSE

SNAP OFF FOR 2½" (63.5mm) OR
3½" (88.9mm) STEEL STUD

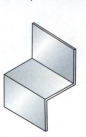

Z CLIP CLOSED STUD ANCHOR
WELDED TO FRAME

HAT CLIP CLOSED STUD
ANCHOR WELDED TO FRAME

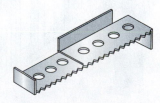

OPEN STUD ANCHOR
WELDED TO FRAME

208F16.EPS

Figure 16 ◆ Anchors.

208F17.EPS

Figure 17 ◆ Place the anchor.

INSIDE TRACK

Metal Door Frame Support

Be sure to use at least three spreaders spaced equally within a metal frame when grouting a frame that has been installed. Make sure that the spreaders are accurately cut to maintain adequate door clearance.

Checking the Plumb and Squareness of a Metal Frame

When plumbing the door jamb, use a 6' level. Check for accuracy before and during the process. One way to check squareness is by measuring diagonally from top to bottom at the corners. Equal measurements indicate that the frame is square.

Step 2 Move the assembled metal door frame into position within the correct location of the metal stud partition, as shown in *Figure 18*. The metal frame may have been shipped using a temporary metal spreader bar. This metal spreader bar must be removed. Replace the metal spreader bar with a wood spreader bar at the sill. Cut out notches for the stops in the wood spreader (*Figure 19*). Ensure that the spacing of the jambs at the sill is the same as at the top of the frame.

Step 3 Check the head jamb for level and shim as required, subject to the frame manufacturer's recommendations. Check the location of the metal frame again and fasten the floor anchor, as shown in *Figure 20* and in accordance with local code requirements.

Step 4 Add a wood spreader and vertical support at the center of the frame (*Figure 21*).

208F18.EPS

Figure 18 ◆ Assembled metal door frame moved into position.

208F20.EPS

Figure 20 ◆ Fasten the floor anchor.

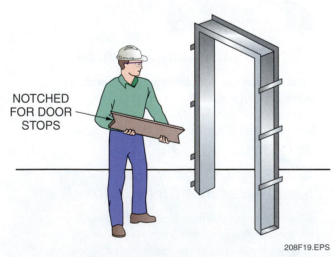

NOTCHED FOR DOOR STOPS

208F19.EPS

Figure 19 ◆ Replace the metal spreader bar.

208F21.EPS

Figure 21 ◆ Spreader installed.

Step 5 Using an accurate level, plumb the jambs and make sure all the studs are square and true with the metal frame. Magnetic levels are available for metal door frames. Care should be taken to ensure that the frame is absolutely plumb and square. An improper frame installation will affect the performance of the door in the opening. Attach the steel studs to the anchors, as shown in *Figure 22*, in which an open stud is shown being wired to an anchor and a closed stud is shown being screwed to an anchor. This will complete the frame installation. A wooden door or a matched metal door is installed after finishing work is complete.

> **NOTE**
>
> Knockdown frames are usually installed after drywall is installed.

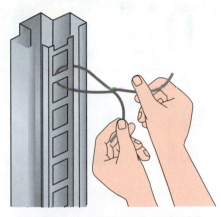

OPEN STUD WIRED TO ANCHOR

CLOSED STUD SCREWED TO ANCHOR

208F22.EPS

Figure 22 ◆ Attach the steel studs to anchors.

4.4.0 Typical Unassembled Metal Door Frame Installation in Drywall Construction

An unassembled metal door frame is shipped to the job site knocked down and must be installed in the opening one piece at a time, so do not assemble the frame. The drywall frame opening must be 1" higher than the metal door and 2" wider, giving it a 1" clearance on the three sides, as shown in *Figure 23*.

Step 1 Attach the sill anchors to the bottom of each jamb, as shown in *Figure 24*. Some frame designs are furnished with a screw hole at the base of each jamb in lieu of the standard base anchor, as shown in *Figure 25*. The screws are usually provided only when the frame is factory-finished so they will match the finish.

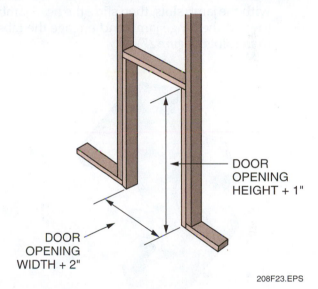

DOOR OPENING HEIGHT + 1"

DOOR OPENING WIDTH + 2"

208F23.EPS

Figure 23 ◆ Door opening height and width.

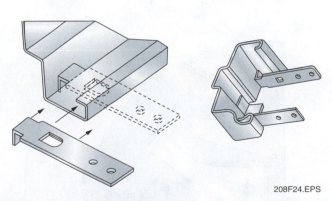

208F24.EPS

Figure 24 ◆ Attach the sill anchors.

Using a Door Buck

A door buck is a cross-braced open frame that is the same size as the finish door. Use a door buck in the frame opening often to check that the door fits during the metal frame installation.

Step 2 Remove the hinge jamb from the knocked down frame parts and slip the jamb into position over the wall, as shown in *Figure 26*. Hold the top in place, then push the bottom in so that it moves toward and over the wall. Check the plans to be sure that the hinge jamb is on the correct side. If a closer will be required and the head jamb frame does not have reinforcement, install a closer sleeve at the proper location. Position the head jamb frame over the wall. Align the head tabs with the jamb slots, then slide the head jamb toward the hinge jamb and engage the tabs in the slots (*Figure 27*).

Step 3 The **strike** jamb is then slipped over the wall, as shown in *Figure 28*. Push the top of the remaining jamb over the wall and mate the jamb slots and head jamb tabs. Push the bottom of this jamb in so it moves toward and over the wall. Level the head jamb.

Step 4 At the base of each side jamb is a shimming angle to be used for vertical shimming (*Figure 29*). If necessary, level the head using these shimming angles.

208F27.EPS

Figure 27 ◆ Slip the head jamb into position.

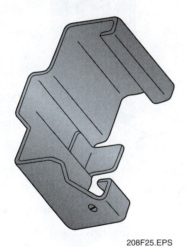

208F25.EPS

Figure 25 ◆ Alternate base anchor.

208F26.EPS

Figure 26 ◆ Slip the hinge jamb into position.

208F28.EPS

Figure 28 ◆ Slip the strike plate jamb into position.

Step 5 Adjust the grip lock anchors in both jambs by turning the anchor screw counter-clockwise until the anchors hit the studs (*Figure 30*). Keep turning the anchors until the metal door frame is rigid and secure.

Step 6 Plumb and square the jambs with an accurate level. When the hinge jamb is plumb, fasten the sill anchor at the hinge jamb with nails or screws, depending on the manufacturer's instructions, as shown in *Figure 31*.

Step 7 Cut an accurate temporary wood spreader the same width as at the top of the frame and install it between the jambs at the base, see *Figure 32*. Adjust the strike jamb to fit firmly against the spreader. Fasten the sill anchor at the strike jamb with nails or screws, depending on the manufacturer's instructions. On some Underwriters Laboratory classified frames, the head may have to be fastened to the side jambs with two screws or 8d 2½" casing nails placed through a hole at the top corners of the head, as shown in *Figure 33*.

208F31.EPS

Figure 31 ◆ Fasten the sill anchor.

SHIMMING ANGLE

208F29.EPS

Figure 29 ◆ Shimming angle.

208F32.EPS

Figure 32 ◆ Wood spreader.

208F30.EPS

Figure 30 ◆ Adjust the grip lock anchor.

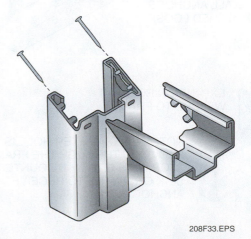

208F33.EPS

Figure 33 ◆ Head jamb fastened to the side jambs.

The metal door frame installation is now complete. Remove the temporary spreader and install a wooden door or a matched metal door.

4.5.0 Typical Welded Metal Door Frame Installation in Existing Masonry Construction

The welded metal door frame will be furnished with an appropriate quantity of one type of existing masonry wall anchor, furnished loose or welded to the frame, as shown in *Figure 34*.

Step 1 The loose anchors are installed in the metal frame by forcing the anchor between the two casing flanges of the frame. Place the anchor into position at an angle and twist it into a horizontal position, making sure the holes through the anchor are lined up with the hole in the frame stop (*Figure 35*). If required, insert a closer reinforcement sleeve inside the head jamb frame at the proper location.

NOTE

See *Figure 36*. The recommended rough opening dimensions in the existing masonry wall should be the frame height plus ¼". The rough opening width should be the frame width plus ½", giving the frame a ¼" clearance on the top and each side.

Step 2 Place the metal door frame in the existing rough masonry opening. The frame may have a spreader bar welded across the sill for shipping purposes. Remove this spreader bar and discard it. Place an accurate wood spreader bar at the sill to match the width of the top of the frame. Two middle spreader bars should be installed only if the frame must be grouted in place. See *Figure 37*.

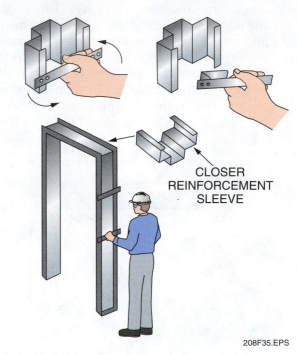

CLOSER REINFORCEMENT SLEEVE

208F35.EPS

Figure 35 ◆ Placing anchors and closer reinforcement sleeve into position.

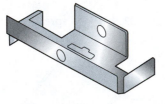

JAMB-FIT COMBINATION EXISTING MASONRY WALL ANCHORS FURNISHED LOOSE

EXPANSION BOLT HAT WELDED TO FRAME

NOTE: A ⅜" DIAMETER HOLE IS PROVIDED THROUGH THE FRAME STOP WITH EITHER A COUNTERSINK OR DIMPLE TO RECEIVE ANCHOR BOLTS

208F34.EPS

Figure 34 ◆ Masonry wall anchors.

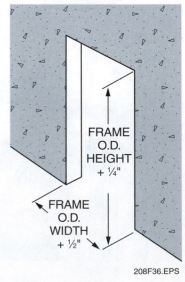

RECOMMENDED ROUGH DIMENSIONS OF EXISTING OPENING IN MASONRY WALL AND FRAME WITH 2" TRIM

FRAME O.D. HEIGHT + ¼"

FRAME O.D. WIDTH + ½"

208F36.EPS

Figure 36 ◆ Rough opening dimensions.

Door Bucks for New Concrete or Masonry Construction

For concrete and masonry walls, a wood or metal door buck is used during wall construction to provide the correct size opening for a metal door frame. Once the walls are completed, the door buck is removed and the metal door frame is installed.

Figure 37 ◆ Wood spreader bar and middle spreader bar.

208F37.EPS

DRILL A ⅜" DIAMETER HOLE NO LESS THAN 1⅜" INTO MASONRY

208F38.EPS

Figure 38 ◆ Drilling into masonry.

Step 3 Plumb the jambs with an accurate level. Level the head and square the corners of the metal frame. Ensure that the frame is absolutely plumb and square. An improper frame installation will affect the performance of the door opening.

Step 4 If the frame is to be installed with expansion shields and machine bolts, the location of the shields must be marked at this time and the frame removed from the wall. Using the marked locations, drill the depth and diameter hole recommended by the shield manufacturer. Install the shields, reinstall the metal frame, and screw it into place. If expansion shields are not to be used, place a ⅜" drill through the ⅜" existing holes in the metal frame and drill no less than 1⅜" into the masonry (*Figure 38*).

Step 5 Using the sleeve anchor shown in *Figure 39*, fasten the metal door frame in the opening. The installation should be complete. Remove the spreaders and install a wooden door or a matched metal door.

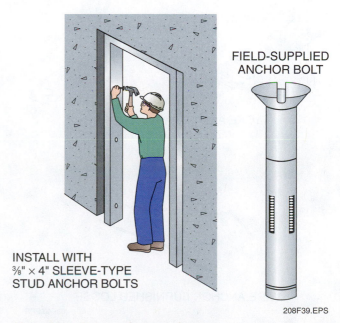

FIELD-SUPPLIED ANCHOR BOLT

INSTALL WITH ⅜" × 4" SLEEVE-TYPE STUD ANCHOR BOLTS

208F39.EPS

Figure 39 ◆ Using a sleeve anchor.

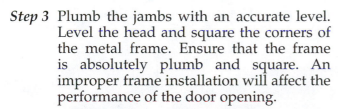

4.6.0 Typical Welded Metal Door Frame Installation in New Masonry Construction

The welded metal door frame will be furnished with an appropriate quantity of one type of masonry wall anchor, furnished loose or welded to the frame, as shown in *Figure 40*. If necessary, install a closer reinforcement sleeve in the head jamb frame.

Step 1 Move the welded metal door frame to the correct location. If the frame has a spreader bar welded across the sill, remove and discard it. Cut an accurate wood spreader bar notched on the ends and place it into position at the sill (*Figure 41*).

Step 2 Check the head jamb for level and shim or adjust as required, subject to the manufacturer's recommendations. Check the location of the metal frame again and fasten the floor anchors, as shown in *Figure 42* and in accordance with code requirements.

Step 3 Place and fasten blocking on the floor. Use telescoping braces to support the frame, as shown in *Figure 43*. Using an accurate level, plumb the jambs and check the level of the head jamb. Square the corners and add two wood spreaders in the center of the frame.

Step 4 Care should be taken to ensure that the frame is absolutely plumb, level, and square. Check it often during placement of the masonry. An improper frame installation will affect the performance of the door opening.

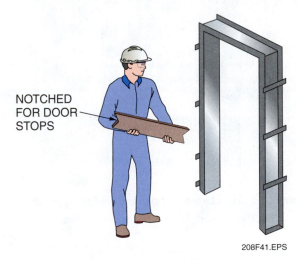

NOTCHED FOR DOOR STOPS

208F41.EPS

Figure 41 ◆ Wood spreader bar installation.

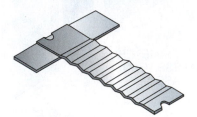

T ANCHOR FURNISHED LOOSE

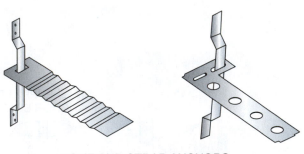

YOKE AND STRAP ANCHORS WELDED TO FRAME

208F42.EPS

Figure 42 ◆ Floor anchor installation.

WIRE ANCHOR FURNISHED LOOSE

208F40.EPS

Figure 40 ◆ Masonry anchors.

INSIDE TRACK

Completing Metal Door Frame Installation in New Masonry

Erection of a block wall around the metal door frame should complete the installation. Prepare the frame for grouting, as described earlier, then install a wooden door or a matched metal door.

208SA03.EPS

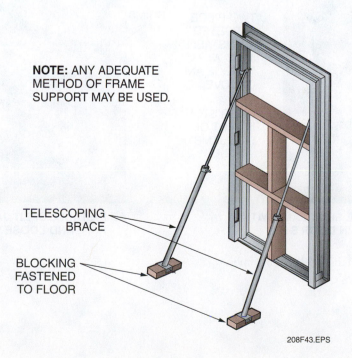

NOTE: ANY ADEQUATE METHOD OF FRAME SUPPORT MAY BE USED.

TELESCOPING BRACE

BLOCKING FASTENED TO FLOOR

208F43.EPS

Figure 43 ◆ Bracing.

5.0.0 ◆ INTERIOR WOOD DOOR JAMBS

The majority of **door jambs** delivered to the job site from the millwork supply house are ¾" thick. In most cases, when a paint finish is required, the jambs will be finger jointed and, when supplied for a natural finish, the jambs will usually be of the same wood as the rest of the interior trim. The width of the door jambs must be the same as the thickness of the wall or partition in which they are to be placed. In some cases, adjustable as well as fixed-width jambs are available. The types of jambs that are most often used on prehung door units are shown in *Figure 44*.

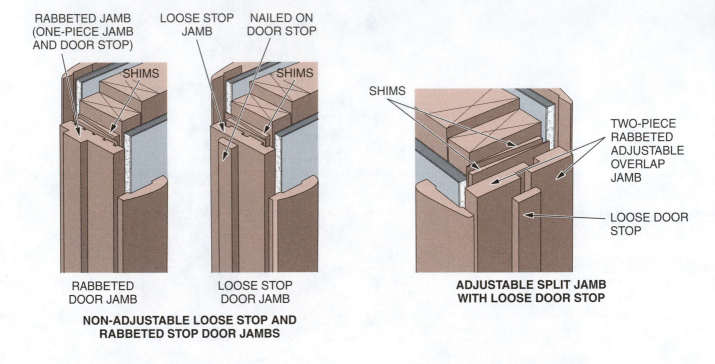

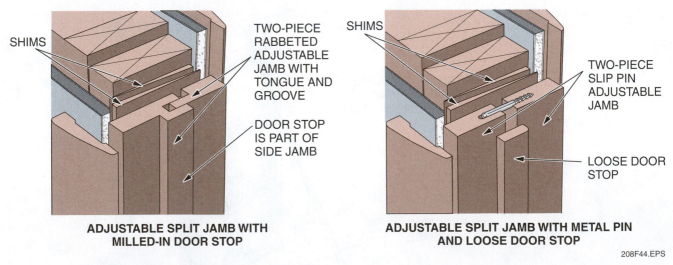

Figure 44 ◆ Fixed- and adjustable-width door jambs.

208F44.EPS

Checking and Correcting the Plumb and Square of Wood Door Framing

Before installing any doors, it is extremely important to check the plumb of all wood door king studs and trimmers in two planes. One plane is parallel to a wall and the other is perpendicular to a wall. If the king studs and trimmers are plumb, the door header must be checked for squareness. If the framing for a door is out of plumb or the king studs and trimmers are badly bowed or crowned, the door cannot be hung properly. Even if it is hung, the door would have to be trimmed at an angle on the top and bottom so that it will close. Other problems are that the door will swing open or closed when it is unlatched or it may strike the floor somewhere in its swing radius.

The best time to check door framing is before the finish walls are installed so that the doorways can be easily reframed if required. Because most doors are installed after the finished walls have been completed, correcting a major framing problem at that point will require removal and replacement of some of the finish wall material in order to reframe the door. However, most minor out-of-plumb conditions can be corrected as illustrated with minor repair to the finish wall material. In addition, a minor crown in the king studs and trimmers can be corrected by cutting both the trimmer and king stud and closing the saw kerf with screws. Several cuts spaced along the length of both studs may be required to achieve an almost straight condition. This type of correction can be accomplished with the finish wall material installed; however, the wall material will require some minor repair.

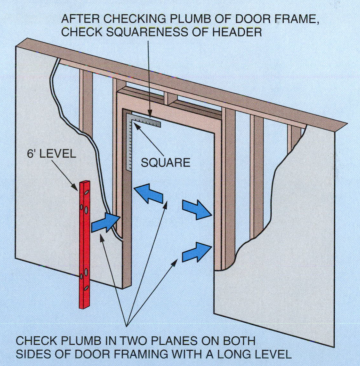

AFTER CHECKING PLUMB OF DOOR FRAME,
CHECK SQUARENESS OF HEADER

6' LEVEL

SQUARE

CHECK PLUMB IN TWO PLANES ON BOTH
SIDES OF DOOR FRAMING WITH A LONG LEVEL

CHECKING PLUMB AND SQUARE OF WOOD DOOR FRAMING

208SA04.EPS

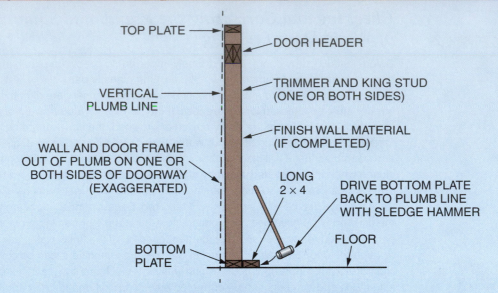

CORRECTING MINOR OUT-OF-PLUMB WALLS

TOP PLATE

DOOR HEADER

VERTICAL PLUMB LINE

TRIMMER AND KING STUD (ONE OR BOTH SIDES)

FINISH WALL MATERIAL (IF COMPLETED)

WALL AND DOOR FRAME OUT OF PLUMB ON ONE OR BOTH SIDES OF DOORWAY (EXAGGERATED)

LONG 2 × 4

DRIVE BOTTOM PLATE BACK TO PLUMB LINE WITH SLEDGE HAMMER

FLOOR

BOTTOM PLATE

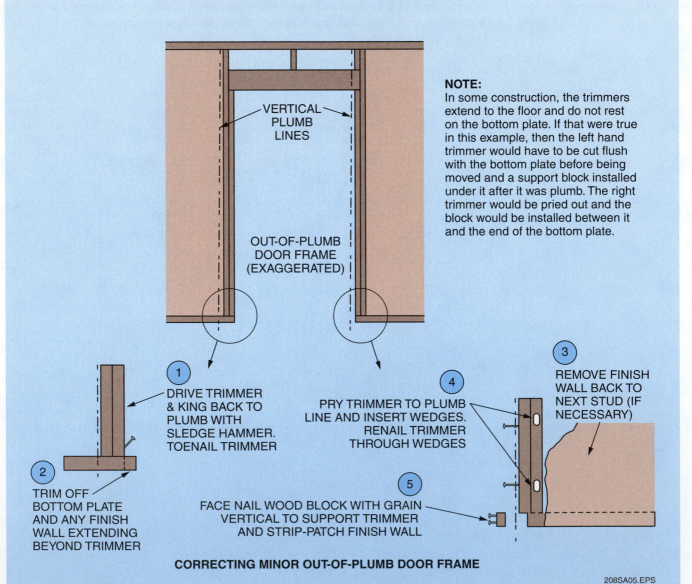

VERTICAL PLUMB LINES

OUT-OF-PLUMB DOOR FRAME (EXAGGERATED)

NOTE:
In some construction, the trimmers extend to the floor and do not rest on the bottom plate. If that were true in this example, then the left hand trimmer would have to be cut flush with the bottom plate before being moved and a support block installed under it after it was plumb. The right trimmer would be pried out and the block would be installed between it and the end of the bottom plate.

① DRIVE TRIMMER & KING BACK TO PLUMB WITH SLEDGE HAMMER. TOENAIL TRIMMER

② TRIM OFF BOTTOM PLATE AND ANY FINISH WALL EXTENDING BEYOND TRIMMER

④ PRY TRIMMER TO PLUMB LINE AND INSERT WEDGES. RENAIL TRIMMER THROUGH WEDGES

⑤ FACE NAIL WOOD BLOCK WITH GRAIN VERTICAL TO SUPPORT TRIMMER AND STRIP-PATCH FINISH WALL

③ REMOVE FINISH WALL BACK TO NEXT STUD (IF NECESSARY)

CORRECTING MINOR OUT-OF-PLUMB DOOR FRAME

208SA05.EPS

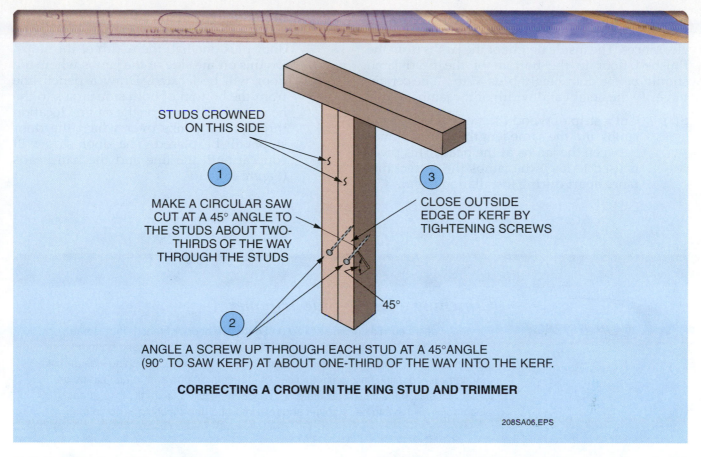

STUDS CROWNED ON THIS SIDE

① MAKE A CIRCULAR SAW CUT AT A 45° ANGLE TO THE STUDS ABOUT TWO-THIRDS OF THE WAY THROUGH THE STUDS

③ CLOSE OUTSIDE EDGE OF KERF BY TIGHTENING SCREWS

45°

② ANGLE A SCREW UP THROUGH EACH STUD AT A 45°ANGLE (90° TO SAW KERF) AT ABOUT ONE-THIRD OF THE WAY INTO THE KERF.

CORRECTING A CROWN IN THE KING STUD AND TRIMMER

208SA06.EPS

5.1.0 Interior Wood Door Jamb Installation

When the jambs are delivered to the job site, they may be assembled or unassembled (knocked down). If they are knocked down, it is a simple matter to place the head jamb into the dado joint on the side jambs and fasten it to the side jambs with 8d box or casing nails. Saw **kerfs** or grooves help prevent warping of jambs. See *Figure 45*.

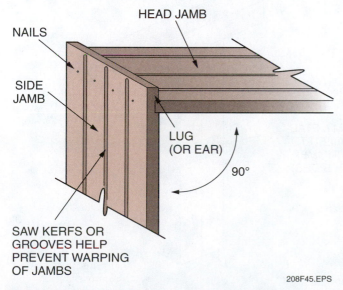

HEAD JAMB

NAILS

SIDE JAMB

LUG (OR EAR)

90°

SAW KERFS OR GROOVES HELP PREVENT WARPING OF JAMBS

208F45.EPS

Figure 45 ◆ Top left corner of interior door jamb.

The opening into which the jamb is to be placed must be checked for height, width, and squareness. For wood jambs, the **rough opening** should normally be 1½" higher and 2" wider than the door. Make sure that the face of the wall is plumb. If the jambs are slightly higher than the rough opening, the lugs or ears can be trimmed off the top so the jamb will fit. If they are trimmed off flush with the head jamb, shimming of the head jamb at the corners during installation is recommended to prevent separation of the head and side jambs. Place the jambs in the rough opening and check the head jamb to see if it is level. If it is not level, cut off the bottom of the highest side jamb by the amount required to make the head jamb level. Then, adjust the jamb height as follows:

- Lift the jamb/casing up off the subfloor at a height equal to the approximate thickness of the finished floor. For floors to be covered with carpeting, it is recommended that the jamb and casing be located approximately ½" off the subfloor.
- Do not place the wood jamb directly on a concrete floor.

The door opening is built to the exact size specified, and the door is cut in the field, if necessary. As shown in *Figure 46*, the inside width of the jambs should be the width of the door specified for the opening plus ³⁄₁₆" for clearance. For example, for a 2'-6" door, the jamb-to-jamb width would be 2'-6" plus ³⁄₁₆". Check the specifications for required clearances. The inside height of the jambs from the finished floor to the bottom of the head jamb should be the door height plus ³⁄₈" to ½". Be certain to verify the height and width of the jambs.

Step 1 Cut a strip of wood the same width as the jambs and the same length as the distance between the jambs at the head. This piece is used to keep the jambs the correct distance apart during installation (*Figure 47*).

Step 2 Measure in from the edge of the jambs by a distance equal to the thickness of the door plus one half the width of the stops. Do this on the side of the jambs where the door will be located. Draw a pencil line from the bottom of both side jambs to the head jamb. This will be the nailing location through the jambs over which the door stop will be placed. The door stop will conceal both the line and the nail heads (*Figure 48*).

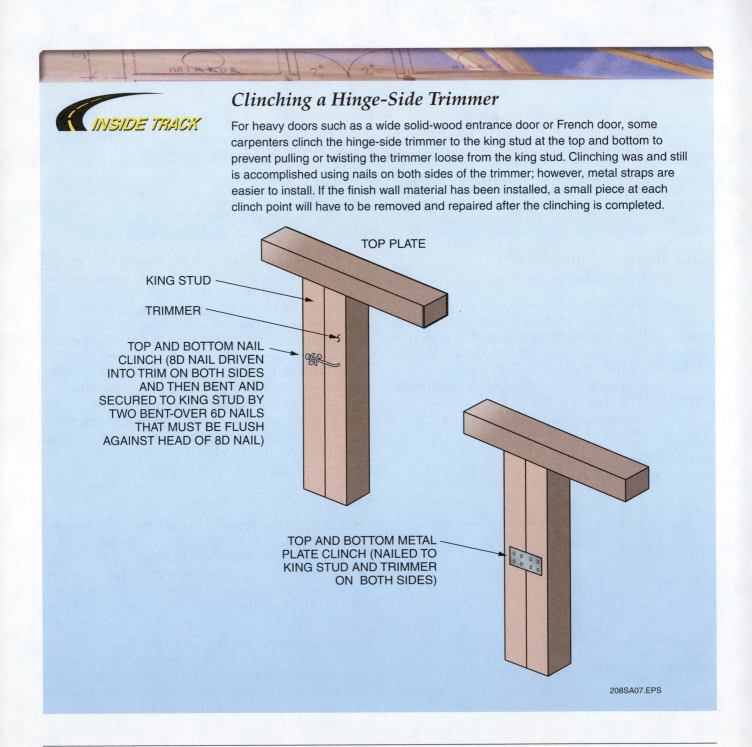

INSIDE TRACK

Clinching a Hinge-Side Trimmer

For heavy doors such as a wide solid-wood entrance door or French door, some carpenters clinch the hinge-side trimmer to the king stud at the top and bottom to prevent pulling or twisting the trimmer loose from the king stud. Clinching was and still is accomplished using nails on both sides of the trimmer; however, metal straps are easier to install. If the finish wall material has been installed, a small piece at each clinch point will have to be removed and repaired after the clinching is completed.

TOP PLATE

KING STUD

TRIMMER

TOP AND BOTTOM NAIL CLINCH (8D NAIL DRIVEN INTO TRIM ON BOTH SIDES AND THEN BENT AND SECURED TO KING STUD BY TWO BENT-OVER 6D NAILS THAT MUST BE FLUSH AGAINST HEAD OF 8D NAIL)

TOP AND BOTTOM METAL PLATE CLINCH (NAILED TO KING STUD AND TRIMMER ON BOTH SIDES)

208SA07.EPS

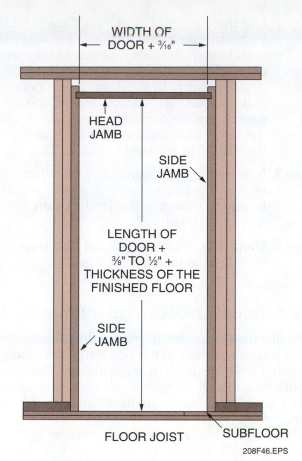

WIDTH OF
DOOR + ³⁄₁₆"

HEAD
JAMB

SIDE
JAMB

LENGTH OF
DOOR +
³⁄₈" TO ½" +
THICKNESS OF THE
FINISHED FLOOR

SIDE
JAMB

FLOOR JOIST   SUBFLOOR

208F46.EPS

Figure 46 ◆ Finding the width and length of the head and side jambs.

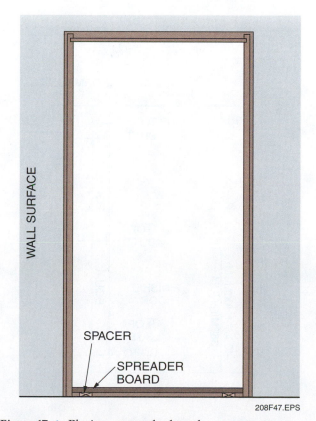

WALL SURFACE

SPACER

SPREADER
BOARD

208F47.EPS

Figure 47 ◆ Placing a spreader board.

Finished Floor Height

If the finished floor has not been installed, determine the type of finished floor and make allowances for the height of the finished floor material. This height may vary considerably, depending on what kind of finished floor the door must clear, such as carpet, ceramic tile, or wood parquet. Review mechanical plans to check whether or not clearance is required at the bottom of the door for air circulation.

Step 3 Using a long level as a straightedge, check again to make sure the faces of the wall on both sides of the opening are plumb. If the wall is not plumb, the door will strike the floor and not open all the way, or the door will swing closed of its own accord, depending on which way the wall is out of plumb. If the wall is not plumb, do not attempt to install the door jambs. If it is a stud wall, it can sometimes be made plumb by bumping the bottom plate with a sledgehammer. Make sure a piece of blocking is used to prevent marring the wall when attempting this.

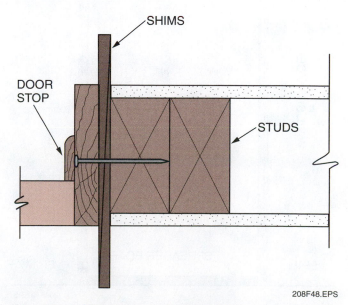

SHIMS

DOOR
STOP

STUDS

208F48.EPS

Figure 48 ◆ Plan section at the side jamb.

Step 4 If the wall faces are plumb, proceed to plumb and level the jambs in the opening. Using wooden shingles as wedges (shims), place two on each side between the jamb and the trimmer stud at the top and bottom. Make sure the edges of the jamb are flush with the finished wall (*Figure 49*).

Step 5 Check the jamb side where the hinges (butts) will be placed using a long level as a straightedge. Make sure the butt jamb is plumb. Adjust the top and bottom shims as required. Fasten the butt jamb with 8d or larger casing nails at the top and bottom using the nailing line to locate the nails. Don't drive the nails flush. Check the **lock** strike side jamb for plumb and nail it in the same manner. Place three pairs of shims behind the butt jamb; one pair goes behind the location of each butt. If only two butts (one pair) are being used, three pairs of shims will still be used. Place shims at the recommended butt locations shown in *Figure 50*. Use 8d or larger casing nails, ½" on each side of the shims along the nailing line to secure the shims. Don't drive the nails flush.

Step 6 Use three pairs of shims on the lock side, with one pair at the lock location and two pairs halfway between the lock location and the top and bottom, as shown in *Figure 51*.

Temporarily fit the door to the opening, making sure the jambs are plumb and square and the correct space exists around the door. Remove the door and complete the nailing of the jambs (*Figure 51*).

Step 7 If necessary, use casing nails on either side of the stop location and ½" from the shim locations to secure the jambs to the trimmer studs (*Figure 52*).

Step 8 Hang the door and install the stops as described in the next sections.

5.2.0 Double Wooden Door Frames

The installation procedure for a double wooden door frame is similar to the procedure for a single wooden door frame. The exception is that for a double wooden door frame, the carpenter uses a cross-story (double-x) method outside of the jamb.

Be certain the frame is plumb, square, and flat with the wall.

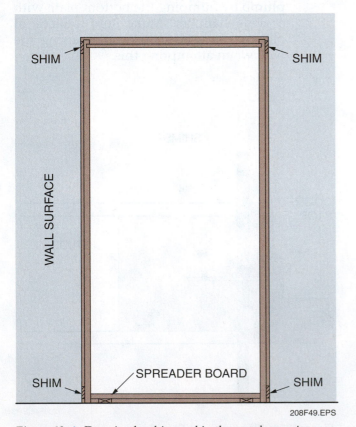

Figure 49 ◆ Door jambs shimmed in the rough opening.

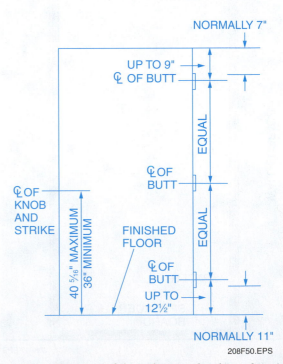

Figure 50 ◆ Elevation of door showing hardware locations.

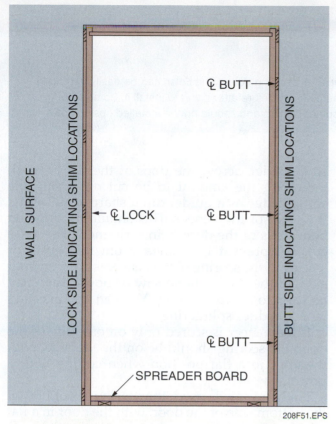

Figure 51 ◆ Elevation of door jambs showing shim locations.

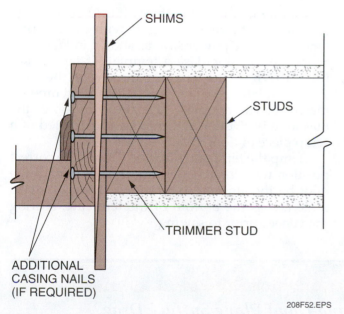

Figure 52 ◆ Additional jamb nailing.

6.0.0 ◆ INTERIOR WOOD DOORS

When doors are delivered to the job site, they should be checked with the plans of the building to make sure they are the correct thickness, width, and height. Make sure the style and type of wood correspond to the specifications.

6.1.0 Handling, Job Finishing, and Installation Instructions

The following is typical of the handling, job finishing, and installation instructions required to comply with one door manufacturer's warranty policy.

- Store the door flat on a level surface in a dry, well-ventilated building. Cover it to keep it clean, but allow for proper air circulation.
- Wear clean gloves when handling doors, and do not drag doors across one another or across other surfaces.
- Deliver doors to the building site only after any plaster or concrete is dry. If doors are stored at the job site for more than one week, the top and bottom edges should be sealed.
- Doors should not be subjected to abnormal heat, dryness, humidity, or sudden changes thereof. They should be conditioned to the average prevailing humidity of the locality before hanging.
- The utility or structural strength of the door must not be impaired when fitting the door, during the application of hardware, or when cutting and altering the door for lights, louvers, panels, or any other special details.
- Use three hinges per door on doors 7'-0" in height or less, and four hinges per door on doors over 7'-0" in height. Hinges should be set flush with the edge surfaces. Be sure that the hinges are set in a straight line to prevent distortion. Allow approximately $\frac{3}{16}$" clearance for swelling of the door or frame during future damp weather periods.
- Immediately after fitting, apply the appropriate weatherstripping and/or threshold and, before hanging any interior or exterior door on the job, the top and bottom edges should receive two coats of paint, varnish, or sealer to prevent undue absorption of moisture.

6.2.0 Fitting Doors in Prepared Openings

After making sure the door is the correct one for the particular opening, measure the opening width and height and compare it to the size of the door to make doubly sure the size matches. Having determined the hinge or butt side of the door, fit it to the hinge jamb of the opening. The butt stile is usually left square and will not interfere with the operation of the door. Some carpenters, however, will slightly bevel the butt stile even though it is not necessary. The use of a power plane to fit the door in the opening will save much

time and energy. If a power plane is not available, it will have to be done with a hand plane. The hand plane usually recommended is the fore plane. The fore plane is 18" in length, and its size is somewhere between the larger jointer plane and the smaller jack plane.

The industry standard is ⅛" for the clearance between the door and the jamb. The clearance at the bottom of the door is determined by the floor finish and, if required, clearance for air circulation. For example, if the floor is a hardwood floor, the clearance over it should be ⅜". If the floor finish is a carpet, the clearance should be the carpet clearance plus ⅜". If there is a threshold under the door, refer to the clearance recommended by the hardware manufacturer. This information is usually included in the hardware manufacturer's installation instructions. Whatever clearance is used between the floor finish and the bottom edge of the door, it is in addition to the 3/32" clearance between the head jamb and the top edge of the door.

If the door must be trimmed at the bottom edge to fit the floor or because of clearance for some type of floor finish, the amount to be removed should be marked lightly with a pencil. Remove the door and carefully place it on carpet-covered **finishing sawhorses**.

The bottom edge of a hollow-core or solid-core door with veneer faces must be cut so no damage is done to the veneer. The damage that can occur is the feathering of the surface veneer away from the base veneers when sawing. To prevent this damage from happening, it is necessary to place a straightedge across the door at the pencil mark indicating the amount to be cut off. Using the straightedge as a guide, run a sharp knife across the grain of the veneer. This should be done on both sides of the door being trimmed. The knife cut will prevent the veneer from feathering or tearing. If the scoring of the veneer is done on both sides of the door, a hand saw or power saw may be used to do the cutting. You can use masking tape to reduce splintering.

If the veneer is scored only on one side of the door, the scoring should be on the top side, with the door in a flat position when cutting with a power saw.

If a hand saw is used, the scoring should be on the bottom side of the door, with the door in a flat position. The saw cut can be smoothed to the knife cut with the use of a power plane, hand plane, or sander.

Another method of cutting off the bottom edge of a door with veneer faces is with the use of a **template** and a power saw, as shown in *Figure 53*. Score the top side. Use a template if tape is not used. This measure will keep the saw off the door. The template is made of ¼" or ⅛" tempered masonite. The template saw guide on top of the masonite is ⅜" thick and may be constructed of a scrap piece of door stop.

Clamp the template onto the door in the correct position with the edge up from the bottom of the door by the amount to be trimmed. Run a sharp knife across the bottom of the template, scoring the veneer prior to sawing.

Sealing the Bottom and Edges of a Door

Always seal the bottom and edges of the door with varnish or paint. Be sure not to get varnish on the door surface. The bottom must be sealed before the door is hung. Otherwise, the door will have to be removed and sealed.

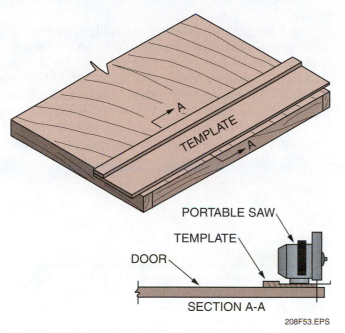

208F53.EPS

Figure 53 ◆ Door sawing template shown in position.

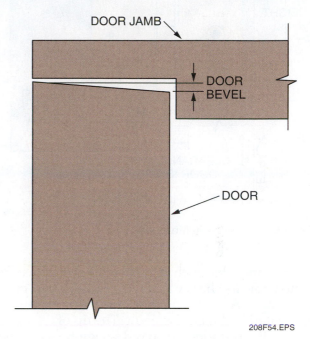

208F54.EPS

Figure 54 ◆ Section of door showing door bevel.

After sawing, place the door in a carpet-covered door jack. Plane the edge of the door to the correct width. Put the door in the opening and, with a shim at the bottom, move the door up against the head jamb.

If the door head is square with the head jamb, nothing more must be done. If the top rail or door head is not square with the head jamb, it must be made square by first scribing it to the head jamb, then sawing and/or planing it to fit. When the door has ³⁄₃₂" clearance at the head and both sides, as well as the proper clearance at the bottom edge and the floor, remove it from the opening and place it in the door jack with the hinge stile down.

Bevel the edge of the lock stile of the door by planing it to an angle of ⅛" in 2", as shown in *Figure 54*.

As shown in *Figure 55*, the door swings on an arc created by a radius that has its center located in the center of the pin of the butt hinge. This center is well forward of the face of the door stop, so the lock edge of the door must be beveled to clear the face edge of the door jamb.

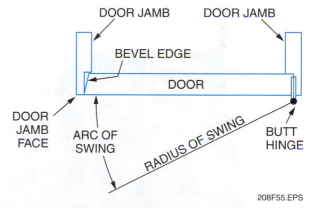

208F55.EPS

Figure 55 ◆ Plan view of door showing swing.

6.3.0 Door Hinges

There is some basic information about the installation of door hinges that you will need to know. The basic dimensions are taken with the hinge in the open position. The two main dimensions are the width and the height, as shown in *Figure 56*.

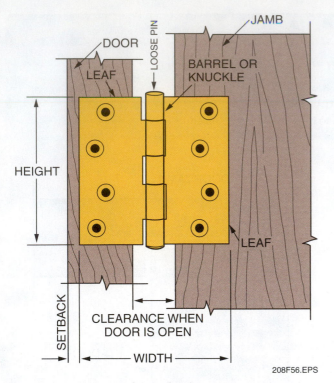

Figure 56 ◆ Loose-pin hinge (butt).

On some jobs, such as a hospital, doors may be quite wide and heavy and will require a ball-bearing hinge. A hinge of this type is permanently lubricated and has a fast joint. The tip of the hinge is usually rounded to avoid snagging clothing, etc., as shown in *Figure 57*.

Prior to installing a door hinge, its size must be determined. The size of a door hinge is always given with the height first and the width last. The height of the hinge is determined by the width and thickness of the door, as shown in *Table 1*.

The width of the hinge is determined by the door thickness and the clearance required for the trim (*Figure 58*). This clearance will allow the door to open parallel to the wall and therefore not mar the door trim.

To determine the width of the hinge, use the door thickness and clearance required, as shown in *Table 2*.

Three hinges per door are normally specified in commercial and institutional work. Extra heavy, high-frequency hinges should be used on doors that have high-frequency usage. This type of door is usually found in commercial buildings such as an entrance to a department store or mall. Examples of institutional doors that receive high-frequency usage are school entrance or school lavatory doors.

Table 1	Hinge Height	
Door Thickness	**Door Width**	**Hinge Height**
1⅜"	To 36"	3½"
1¾"	To 36"	4"
	36" to 41"	4½"
2", 2¼", 2½"	42" to 48"	4½" Extra Heavy
	To 42"	5"
	Over 42"	6"

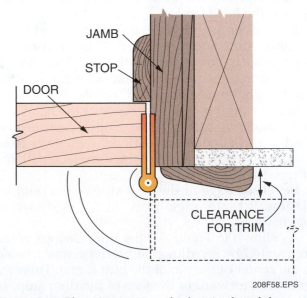

Figure 57 ◆ Ball-bearing hospital hinge with a rounded tip.

Figure 58 ◆ Plan view section of a door jamb and door.

Hinge Nomenclature

Door hinges are sometimes referred to as butts or butt hinges. The flat portions of the door hinge that contain the countersunk holes for the screws are called the leaves. The round portion in the center of the hinge that joins the leaves is called the knuckle. The knuckle is usually divided into five sections, two on one leaf and three on another. A pin running through these five knuckle sections joins the two leaves, creating a five-knuckle butt hinge. In some parts of the United States, the terminology used in the carpentry trade may describe the number of butts used as pairs. For example, a pair and a half of butts would refer to three butts.

In most cases, door hinges used on the interior of a building are made so that the pin can be removed. A door hinge with this type of pin is usually called a loose-pin hinge or butt hinge. A door hung with loose-pin hinges is not very theft-proof. If security is required, hinges with nonremovable pins must be used. A door hinge of this type is called a fast-joint hinge or NRP (nonremovable pin) hinge.

Table 2	Hinge Width	
Door Thickness	**Clearance Required**	**Hinge Width**
1⅜"	1¼"	3½"
	1¾"	4"
1¾"	1"	4"
	1½"	4½"
	2"	5"
	3"	6"
2"	1"	4½"
	1½"	5"
	2½"	6"
2¼"	1"	5"
	2"	6"
2½"	¾"	5"
	1¾"	6"

6.3.1 Door Hinge Installation

The door hinge locations are determined either by the architect or the contractor on the job. Because of the many methods of hinge placement location used throughout the United States, it is difficult to quote a standard hinge location. This hinge placement location also varies from one prehung door manufacturer to another. Many carpenters, however, position the top of the top door hinge 7" down from the top of the door, the bottom of the bottom door hinge 11" above the finished floor, and the third door hinge centered between the other two.

Step 1 Place the door in the opening and, using a wedge at the bottom, move the door upwards until a ³⁄₃₂" gap is at the top. A thin piece of wood, usually a ripping that has been planed to the correct thickness in advance, can be used as a spacer. Next, wedge the door up against the jamb on which the hinges will be located.

Reversible Hinges

Most hinges are reversible, meaning they can be used on either a left-hand or a right-hand operating door. If it is an NRP type of hinge, it can be used with either end mounted in an upright position. There are, however, some hinges that are specifically manufactured for either a right-hand or left-hand door. They cannot be reversed, and caution must be exercised in selecting a hinge for either a right-hand or left-hand opening.

Consistent Hinge Location

All door hinge locations on all doors on the same job should be the same. If there are both prehung door units and doors that must be hung by hand, make sure that the hinge locations on the prehung units are duplicated throughout the job.

Step 2 Measure and then mark the hinge locations on the jamb with a knife and transfer those marks to the edge of the door stile. Be careful not to mark the face of the door or the jamb any deeper than the thickness of one leaf of the butts being used. Remove the door from the opening and place it in the door jack with the stile to receive the hinges in the up position.

NOTE

Hinge mortises may be hand cut and fitted (outlined below) or they may be cut using a template, as described later in this module.

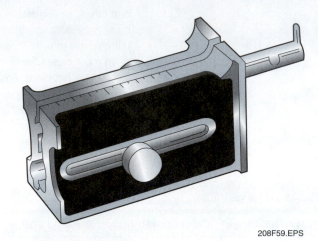

208F59.EPS

Figure 59 ◆ Butt gauge.

There are three measurements to be marked when laying out the mortises:
– Location of the door hinge on the jamb
– Location of the hinge on the door
– Thickness of the hinge leaf on both the door and the jamb

To measure accurately and as rapidly as possible, use a high-quality **butt gauge**, as shown in *Figure 59*, or a butt marker (discussed later).

The door hinge backset should be ¼" for doors up to 2¼" thick and ⅜" for doors over 2¼" thick, as shown in *Figure 60*. Allow ¹⁄₁₆" clearance for the door when nailing the stop in place.

Step 3 Using the butt gauge to square the lines for the door hinge mortises, draw a knife line, as shown in *Figure 61*, for the top and bottom of the hinge. The small machined edges serve like the handle of a tri-square to mark the squared lines. Do this on both the door and jamb for all mortises.

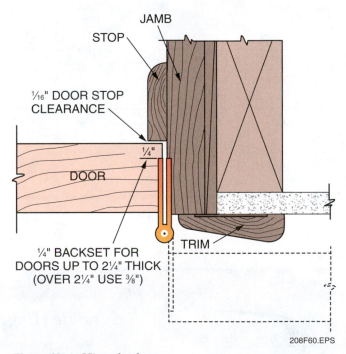

208F60.EPS

Figure 60 ◆ Hinge backset.

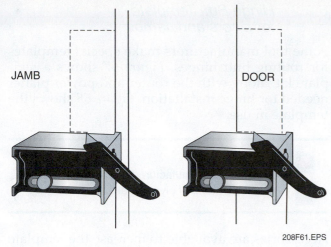

JAMB   DOOR

Figure 61 ◆ Butt gauge (knife line).

208F61.EPS

CAUTION
Verify what you are doing before you mortise for the hinges.

Step 4 Using the butt gauge to mark the width of the mortises, set the spur for the width required (usually the door thickness less the backset), and mark both the door and the jamb within the squared lines, as illustrated in *Figure 62*. The bar with the single spur on the butt gauge is used to mark the depth of the mortise, as shown in *Figure 63*. The depth is the same dimension as the thickness of the door hinge leaf.

Step 5 With the width, height, and depth of the hinge leaf mortise marked on the jambs and the door, remove that portion of the wood with a wood chisel or router. Use a wide, soft-faced hammer or mallet to strike the chisel. Light taps will give sufficient power even on the hardest wood if the chisel is sharp. Using a hard-faced hammer risks damaging your fingers, the wood, or the chisel. Place the cutting edge of the chisel in the marks made with the butt gauge and cut to the depth of the hinge leaf, as indicated by the butt gauge line on the edge.

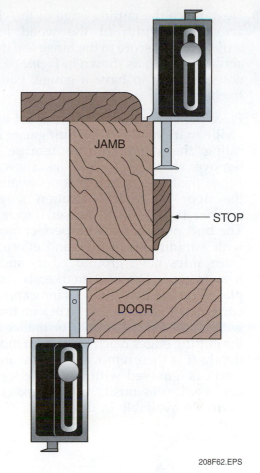

JAMB

STOP

DOOR

Figure 62 ◆ Butt gauge (width).

208F62.EPS

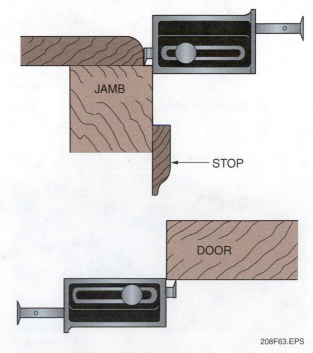

JAMB

STOP

DOOR

Figure 63 ◆ Butt gauge (depth).

208F63.EPS

Step 6 Next, place the cutting edge of the chisel across the width of the wood to be removed, and score to the hinge leaf depth about every ⅛", as shown in *Figure 64*. It is not necessary to have a gauge; you can freehand it.

Step 7 Place the cutting edge of the chisel in the **scribe** mark made by the butt gauge indicating the depth of the mortise and remove all excess wood, as shown in *Figure 65*. If a butt gauge is not available, the door hinge leaf location may be marked using a butt marker (*Figure 66*). The butt marker marks a perfect mortise with a minimum of time and effort. The three sides have ground edges and are heat treated. The butt marker is simply placed on the pencil marks indicating the correct butt location and struck on the top with a soft-faced hammer or mallet until the cutting edges reach the correct mortise depth. It is then removed, and the excess wood is removed with a wood chisel, as described previously. Butt markers are normally available in three sizes: 3", 3½", and 4".

6.3.2 Hinge Butt Template for Doors and Jambs

Some tool manufacturers make special templates for routing butt hinges. *Figure 67* shows a template kit, along with the router and power planer needed for hinge installation. *Figure 68* shows the template in use.

> **NOTE**
>
> In addition to manufactured templates, job-built templates made of plywood are often used.

Accessories are available to increase the template to allow routing for three or four hinges on wood doors and jambs of 7'-0" to 8'-0" or 8'-0" to 9'-0".

Jamb gauges are also available for setting templates to rout wood doors to fit metal jambs or to provide clearance for weatherstripping.

The clearance at the top of the door may be ⅟₁₆" or ⅛". The hinges may be located 5" to 6" from the top of the door and 9" to 10" from the bottom, or 7" to 8" from the top and 10" to 11" from the bottom.

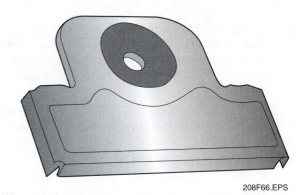

208F66.EPS

Figure 66 ◆ Butt marker.

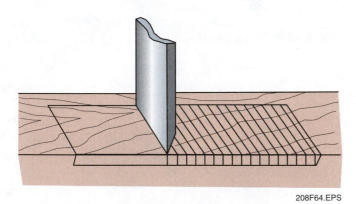

208F64.EPS

Figure 64 ◆ Scoring the wood in the mortise.

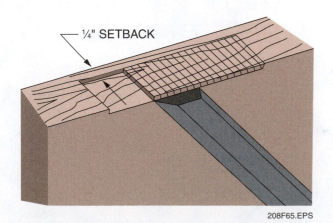

¼" SETBACK

208F65.EPS

Figure 65 ◆ Removing the excess wood from the mortise.

208F67.EPS

Figure 67 ◆ Porter-Cable door hanging equipment.

Figure 68 ◆ Routing the door for hinges.

208F68.EPS

6.3.3 Selecting and Preparing the Router

Determine the size of the corner radius on the hinges to be installed and select a corresponding bit and template guide from those listed in the template manual. To ensure that the correct size mortise is cut, it is important to use the template guide specified for the size of the bit.

If hinges have square corners, the ⅝" diameter bit and corresponding template guide are recommended. A corner chisel, which is used to square the corners of the cut mortise, is available as an accessory.

CAUTION

The following instructions for preparing and using the router are general. Read and follow the instructions in the owner's manual included with your router.

Step 1 Assemble the template guide to the router base with a lock nut and tighten the lock nut securely.

Step 2 Install the selected bit in the router collet and tighten it securely.

Step 3 Set the router on the hinge butt template and adjust the depth of the cut so the bit just touches the door.

Step 4 Set the router depth adjusting ring to the zero position.

Step 5 Lift the router from the template and adjust the depth of the cut so that it is equal to the thickness of the hinge to be installed.

Step 6 Firmly tighten the motor locking device.

6.3.4 Installing Hinges and Hanging the Door

Step 1 Remove the loose pin and place one hinge leaf in the mortise on the door. If the mortise has been properly cut, the hinge leaf may require a few light taps with a soft-faced hammer to seat it to a flush depth. Place the other half of the hinge leaf on the mortised jamb, making sure the hinges on both the door and jamb are in the correct position so the loose pins can be inserted from the top.

Step 2 Place a self-centering screw hole punch or self-centering bit in the countersunk holes of each hinge leaf. If a punch is used, gently tap the pin to mark the location of the pilot hole, as shown in *Figure 69*. The pilot hole will be drilled to receive the hinge leaf screw.

Lubrication of Hinge Screws

If hinge screws are made from a soft metal such as brass, and the door or the jambs are a hardwood such as oak or birch, some of the screw heads are likely to twist off when you install the screws. To prevent this, rub the screw threads across a block of paraffin or beeswax and allow some of the material to collect between the threads. The paraffin or beeswax will act as a lubricant and prevent the screws from binding while being turned in and tightened.

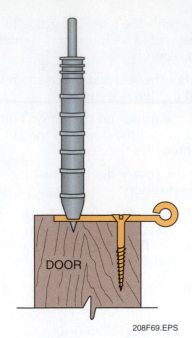

Step 3 Carefully position the door in the opening, engage the door hinge leaves, insert the hinge pins, and check the door for proper operation and clearance. The door should have a clearance of ⅛" to 3⁄32" at the top and 3⁄32" at both sides. If a slight adjustment at the sides is required, the hinge mortises can be cut slightly deeper or the hinge can be adjusted using cardboard strips, as shown in *Figure 70*. All of this adjustment takes time, and time costs money. If care is taken from the start of the door installation, this adjustment will not be necessary. Ensure that all screws are tightened in the door hinges after adjustments are made.

Figure 69 ◆ Self-centering screw hole punch.

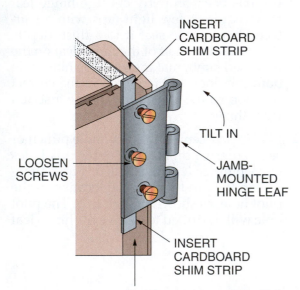

DECREASE HINGE JAMB CLEARANCE BY PLACING THE SHIM STRIP AWAY FROM THE EDGE OF THE JAMB.

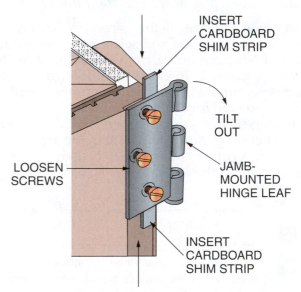

INCREASE HINGE JAMB CLEARANCE BY PLACING THE SHIM STRIP TOWARD THE EDGE OF THE JAMB.

208F70.EPS

Figure 70 ◆ Adjusting the hinge side jamb clearance using thin cardboard shim strips.

7.0.0 ◆ DOOR STOP STRIPS

If the door jambs are not **rabbeted**, loose door stop strips must be installed after the door is hung. The stop on the hinge jamb side of the door must have a 1/16" clearance between it and the door, as shown in *Figure 71*. The stop on the lock jamb side of the door must fit flush so the face of the door is flush with the face edge of the jamb. The stop on the head jamb of the door should line up with both the side jamb stops. Note that by cutting the bottom of the stops on the vertical jambs at a 45° angle, a sanitary (terminated) stop can be created.

To fasten the door stops to the jambs, use 4d finish nails, 1½" in length.

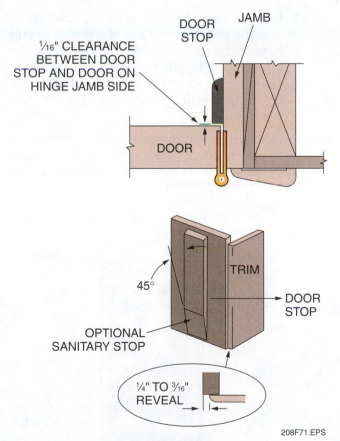

Figure 71 ◆ View of a door jamb and door.

8.0.0 ◆ JOB SITE PREHANGING A DOOR PRIOR TO INSTALLATION

Step 1 The rough door opening must be checked for the correct width and height to be sure the door assigned to that opening will fit. The opening must also be checked for squareness and to see that the wall surface is plumb. If the framed wall surface is slightly out of plumb, it can usually be corrected by bumping the base of it with a sledgehammer. Use a scrap piece of wood against the bottom of the wall to hammer against so the wall will not be marred.

Step 2 The finished thickness of the wall at the rough door opening must now be measured to find out the exact finished width of the jamb. If power tools are available at the job site, the jambs may be cut and assembled there. In most cases, the jambs are pre-cut at a lumber yard or mill and delivered to the job site ready to install.

Step 3 Check the door to be used in the opening to make sure it is the correct thickness, width, and height. Some of the less expensive interior hollow-core doors manufactured today have the lock block only on one side, which will be indicated with a mark on the edge. Some doors have a marked top edge and bottom edge and must be hung in that position. Check all of these conditions prior to fitting the door. Having determined the hinge or butt side of the door, bevel the opposite edge or lock stile of the door by planing it to an angle of 1/8" in 2". The low side of the lock stile bevel should be pointing toward the jamb.

Step 4 Position the door beside the jamb and mark all hinge locations on the jamb. Transfer the marks from the jamb to the edge of the door stile. Mortise the door and the jamb with the hinge butt template and a router. Fasten the hinges to the door jamb and the door.

Step 5 The jambs are now assembled with correct clearances at the top and sides of the door, and the door hung, resulting in a job-built prehung door unit. The installation is identical to that of a factory-built prehung door unit. The trim may be installed on the side opposite the door, and the entire unit is slipped into the rough opening. The unit may also be installed without the trim and the jambs shimmed into position, using the installed door as a guide. The trim may be applied later.

9.0.0 ◆ MANUFACTURED PREHUNG DOOR UNIT INSTALLATION

Manufactured prehung door units may be supplied unassembled, as shown in *Figure 72*, or assembled, as shown in *Figure 73*. They are available in fixed jamb widths or with adjustable width split jambs.

Unassembled prehung door units can be assembled and installed in the same manner as other assembled prehung units, or the jambs can be installed separately and the door hung after the jambs have been secured. In either case, the hinges and lockset mortises and holes have been cut by the manufacturer. Most prehung door units do not have lugs at the top of the side jambs. Consequently, the head jamb is not dadoed into the side jambs, and as a precaution, the head jamb should be shimmed at both top corners when the door is

installed to prevent separation of the head jamb from the side jambs. Before installing an assembled unit, make sure that the lockset side of the door is not nailed in place through the back side of the strike plate jamb. Also, make sure that any shipping braces or tie straps holding the frame together at the bottom of the door will not interfere with the installation and can be removed or cut away when the door is installed and secured in the rough opening.

The installation guidelines for a typical split jamb, milled-stop, prehung door unit are given below. However, always refer to the door manufacturer's instructions for specific information pertaining to any door being installed. Note that the rough opening required for prehung units is typically specified as 2" wider and 1" higher than the door size.

Step 1 Unpack the split jamb door unit and separate the two halves. One of the halves will have the door attached to the jamb by the hinges, and the other will consist of only the assembled top and side jambs (*Figure 74*). Both halves will normally have the trim casing installed.

Step 2 Determine the clearance of the door above the floor and, if necessary, trim the bottom of the jambs and casings to obtain the correct spacing.

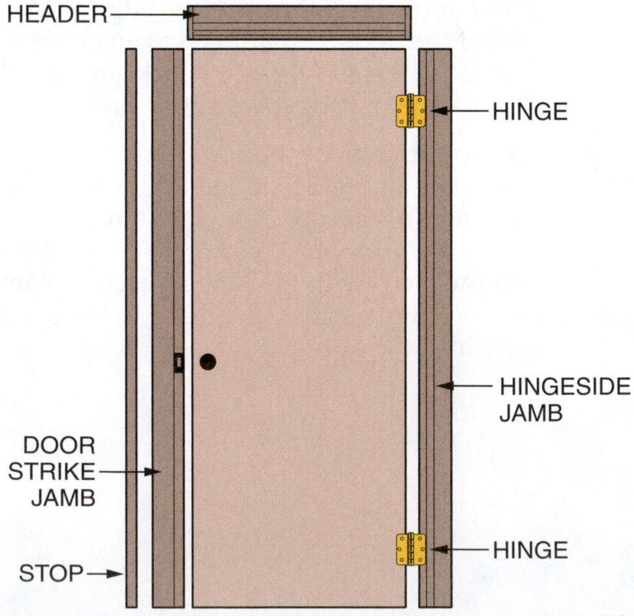

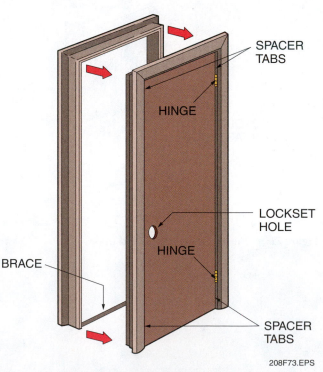

208F73.EPS

Figure 73 ◆ Typical assembled split jamb, prehung door unit.

HEADER

HINGE

HINGESIDE JAMB

DOOR STRIKE JAMB

HINGE

STOP

208F72.EPS

Figure 72 ◆ Typical unassembled prehung door unit.

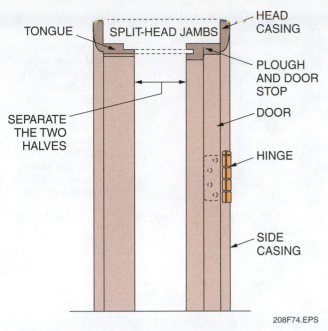

Figure 74 ◆ Unpack and separate split jambs.

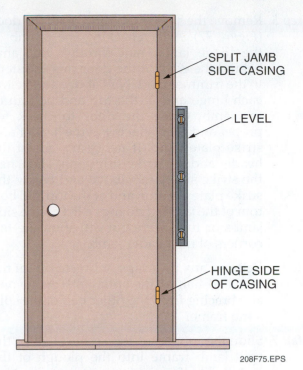

Figure 75 ◆ Position and plumb the split jamb half with the door.

Step 3 Position the split jamb half containing the door into the rough opening, and plumb the jamb by placing a level against the side of the casing on the hinge side of the jamb (*Figure 75*). Check that both jambs rest on the floor. In some cases, the jambs will not rest on the floor. It depends on the type of finished floor. If they do not and a metal door is being hung, insert shims under the jamb and casing that is off the floor. If a wood door is being hung, remove the door unit and trim the bottom of the appropriate jamb and casing so that both jambs will touch the floor (except on concrete). Then, replace the door unit in the opening and replumb the unit. Temporarily nail the casing to the trimmer stud at the top and bottom on the hinge side to hold the door in place. Then, temporarily nail the strike side jamb in place after verifying that the head clearance is correct. Do not drive the nails flush.

Step 4 Move to the other side of the wall and place shims at the top and bottom of both side jambs to hold the unit in place. Shims should span the face of the trimmer studs. Check that the jambs are plumb and renail casings and/or adjust the shims as necessary. Also, place shims behind each hinge location and the strike plate location (*Figure 76*).

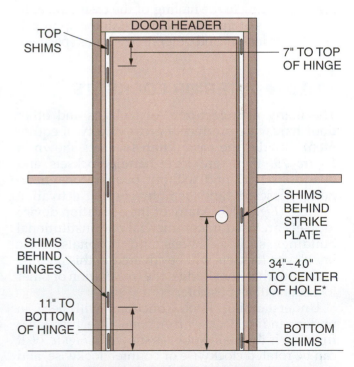

Figure 76 ◆ Shimming the door half of a split jamb frame.

Step 5 Remove the spacer tabs and open the door (*Figure 77*). After checking the plumb of the hinge jamb, nail through the jamb behind the edge of the door (not the stop) to the trimmer stud. Nail above and below each hinge and at the top and bottom of the jamb. Close the door to check for proper and even spacing at the header and strike plate jamb. If necessary, adjust the header and the strike side jamb. Then, nail the strike side jamb above and below the strike plate location and at the top and bottom of the jamb. If no lugs exist on the side jambs or they were cut off, shim the top corners of the header jamb.

Step 6 Remove any tie straps or bracing from the installed half of the split jamb frame and any bracing from the other half of the split jamb frame.

Step 7 Slide the tongue of the second half of the split jamb frame into the plough of the installed half of the frame (*Figure 78*). Nail the casing to the trimmer studs. Then, nail the jamb to the trimmer studs at approximately the same locations as the door half of the jamb. Do not nail through the door stop strip.

Step 8 Complete the nailing of the casing on both sides of the door, then install the appropriate lockset and other hardware as specified.

10.0.0 ◆ INTERIOR LOCKSETS

The many manufacturers of locksets and other door hardware produce a great variety of equipment. Besides the basic **knob locksets** shown in *Figure 79*, there are lever-handle locksets and **deadbolts** with and without pushbuttons, magnetic passcard, and remote control activation devices (*Figure 80*). Many of the activation device locksets are used in commercial and institutional buildings such as hotels and hospitals. Basic knob-type locksets will be covered in this module and, with that knowledge, the installation of other locksets will be possible.

Understanding knob functions will help in understanding lockset functions. The first knob function is the plain knob, as shown in *Figure 79*. It can be rotated clockwise or counterclockwise, and the operation of the **latch bolt** is its only function.

The second knob function is the turn button. The knob function is the same as that of the plain

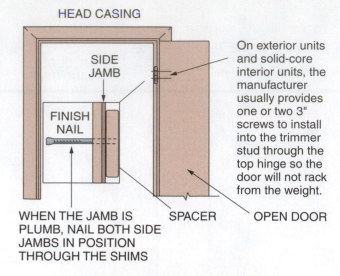

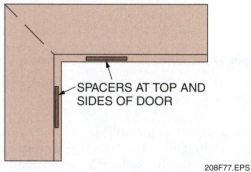

208F77.EPS

Figure 77 ◆ Fastening the jamb to the trimmer stud and removing the spacers.

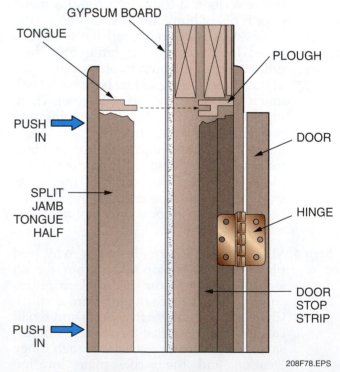

208F78.EPS

Figure 78 ◆ Installing the second half of a split jamb frame.

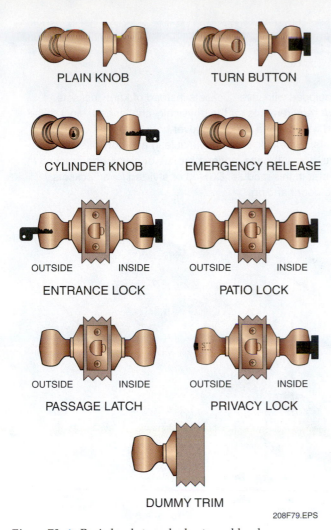

Figure 79 ◆ Basic knob-type locksets and knobs.

PLAIN KNOB

TURN BUTTON

CYLINDER KNOB

EMERGENCY RELEASE

ENTRANCE LOCK
OUTSIDE   INSIDE

PATIO LOCK
OUTSIDE   INSIDE

PASSAGE LATCH
OUTSIDE   INSIDE

PRIVACY LOCK
OUTSIDE   INSIDE

DUMMY TRIM

208F79.EPS

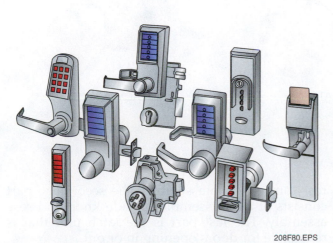

Figure 80 ◆ Various lever-handle locksets with activation devices.

208F80.EPS

knob except that by using the turn button, the opposite knob or both knobs can be locked, depending on the lockset function.

The third knob function is the cylinder knob. The knob function is the same as the plain knob except that by using the correct key, the unit can be locked.

The fourth knob function is the emergency release. This knob function is the same as the plain knob except when an accidental locking occurs (such as a child turning the turn button on the room side of the lock), the latch bolt can be released by inserting a special key or pin in the hole of the outside knob.

10.1.0 Lockset Functions

The entrance lock (*Figure 79*) is used for entrance doors where locking is required. Turning either the outside or inside knob operates the latch bolt. Both knobs can be locked or unlocked by rotating the turn button on the inside knob or by using the key in the outside knob.

The patio lock is used for entrance doors with limited entry. Turning either the outside or inside knob operates the latch bolt. Both knobs can be locked or unlocked by rotating the turn button on the inside knob.

The passage latch is used for doors that do not require locking, such as a closet door. Turning either knob operates the latch bolt at all times.

The privacy lock is used primarily for bedrooms and bathrooms. Turning either knob operates the latch bolt, and rotating the turn button on the inside knob locks the outside knob. The inside knob is always active. An emergency release knob is used as an outside knob in case of accidental locking. Some of these locksets are furnished with different inside and outside finishes.

Dummy trim is a nonoperational knob used for decorative trim, as shown in *Figure 79*. It may be used on a door that operates but is held in a closed position with magnetic catches instead of a latch bolt. It may also be used on the nonoperating half of a pair of doors to balance the hardware.

10.2.0 Lockset Installation

All locksets, regardless of the manufacturer, are packaged with installation instructions and templates to mark drill locations. The installation of most exterior and interior locksets are identical, and most locks require only the drilling of two holes. The entrance lockset, the passage latch, and the privacy lockset are all shown in *Figure 81*.

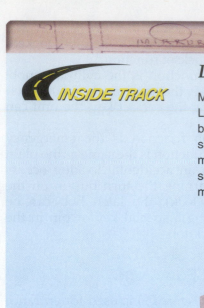

Lever Locksets

Many doors today are being equipped with lever locksets instead of knob locksets. Lever locksets comply with ADA requirements for the physically challenged. Most of the basic locksets can be equipped with either a knob or a lever handle and function in the same manner. Like knob locksets, lever locksets are available in a variety of styles and metal finishes including chrome, brass, bronze, antique bronze or silver, and brushed-satin aluminum, as described later in this module. Curved or stylized lever locksets must be specified based on the door handing.

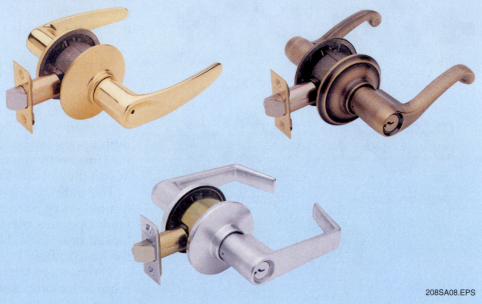

208SA08.EPS

ENTRANCE LOCKSET   PASSAGE LATCH   PRIVACY LOCKSET

208F81.EPS

Figure 81 ◆ Entrance lockset, passage latch, and privacy lockset.

Figure 82 shows an exploded view of a typical residential tubular lockset. Most locksets consist of three interlocking parts that contribute to rapid installation and positive alignment. The primary part consists of a preassembled outside knob, rose, and spindle assembly. The secondary part consists of a preassembled inside knob and rose assembly. The last part is the latch unit that is reversible for doors opening in or out.

Deadbolts

Most residential entrance doors are equipped with single cylinder key-operated deadbolts in addition to an entrance lockset. In some cases, the primary lockset is not key-operated and only the deadbolt can be used to secure the door. Also available are electronically locked residential deadbolts that are remote-controlled from a key-fob type transmitter and manually unlockable from inside the residence. Entrance doors with large door lights or side lights at the strike allow easy access to the turn button release of a single-cylinder deadbolt. Such doors can be equipped with a double cylinder key-operated deadbolt. However, some jurisdictions prohibit their use because they are a safety hazard and do not permit a quick exit from the building. Always check local codes before installing a double-cylinder deadbolt.

INTERCONNECTED
LOCKABLE ENTRANCE
LOCK AND LOCKABLE
DEADBOLT

NON-LOCKABLE
ENTRANCE LOCKSET
WITH LOCKABLE DEADBOLT

208SA09.EPS

Step 1 Remove the lockset from the packing carton and make sure all the parts are there. Check and make sure the lockset is the correct one for the door on which you are working. Read the installation instructions carefully so they are thoroughly understood. Open the door to a position that will allow you to work comfortably on both sides. Place two wedges under the lock side of the door and at right angles to the face of the door, as shown in *Figure 83*. These wedges will hold the door firmly and allow safe work.

Step 2 Measure up from the floor. The standard height to the center of the doorknob is 38", but this may vary. Place a light pencil mark on the edge of the door. Using a square, draw a horizontal light pencil line on the edge and inside surface of the door. Also, place a light pencil mark on the door jamb at the same height of 38" above the floor. With the use of the installation template, mark the center of the cross bore and the center of the door edge, as shown in *Figure 84*. The setback of the cross bore hole is 2⅜" (residential interior) and 2¾"

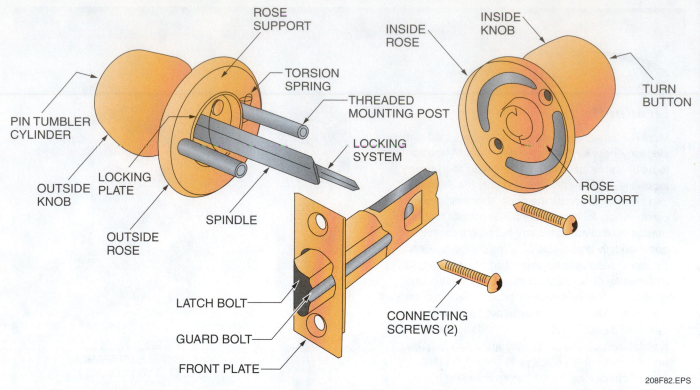

Figure 82 ◆ Exploded view of a typical tubular lockset.

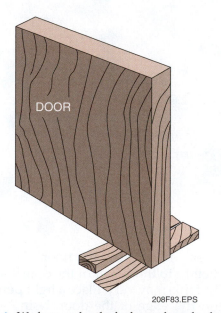

208F83.EPS

Figure 83 ◆ Wedges under the lockset edge of a door.

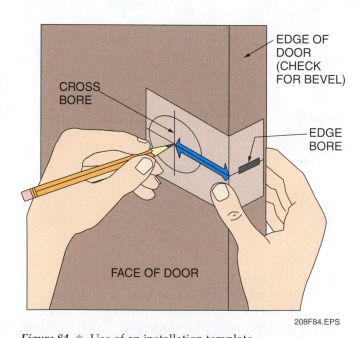

208F84.EPS

Figure 84 ◆ Use of an installation template.

Door Movement During Lockset Installation

If the door moves when the pressure of drilling is applied during lockset installation, it will result in damage to the door and possible injury to the individual operating the drill. Always secure the door to prevent movement.

(commercial and residential exterior) from the edge of the door, as shown. Both may be shown on the template. Keep in mind to adjust the setback if the door edge has a bevel (usually ¹⁄₁₆" difference).

Step 3 To determine the location of the vertical center line of the strike, take one half of the door thickness as measured from the door stop. Using the strike as a template, mark the outline with a sharp knife or pencil. The location of the attaching screws should also be marked at this time, as shown in *Figure 85*.

Step 4 Drill the holes as directed on the installation template. The cross bore hole should be drilled first, and it should be drilled from both sides of the door to avoid splintering the wood. Each side of the hole should have a clean, sharp edge. A hole saw of the correct size will usually give excellent results. Next, using the center mark on the edge of the door, drill the hole for the latch bolt. The hole for the latch bolt should be 3" to 3½" deep from the edge of the door and will extend past the cross bore hole. Insert the latch bolt into the hole in the edge of the door and, using the front trim as a template, mark the outline using a sharp knife or pencil. The location of the attaching screws should also be marked at this time, and pilot holes drilled for them. Chisel out the area marked for the latch bolt to a depth of ⁵⁄₃₂" to allow the front trim to fit flush with the door edge. On the door jamb, drill a 1" diameter hole ½" deep at the center point of the strike. Drill the pilot holes for the attaching screws and chisel out the jamb to a depth of ¹⁄₁₆", so the face of the strike is flush with the face of the jamb.

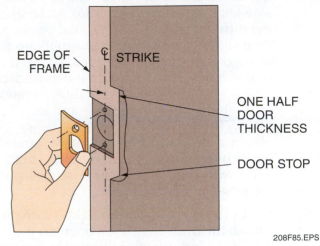

Figure 85 ◆ Center line location of the strike.

208F85.EPS

Cylindrical Lockset Jig

Drilling jigs simplify the installation of cylindrical locksets. These jigs eliminate the use of hand marked templates and ensure accurate, perpendicular boring of lockset holes in the door and jamb.

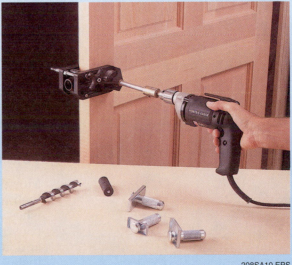

208SA10.EPS

Step 5 Insert the latch bolt unit into the hole in the door edge with the beveled edge of the latch bolt facing toward the door jamb. Fasten the latch bolt to the door with the two attaching screws.

Step 6 Install the outside knob assembly by inserting the spindle and the threaded posts through the holes in the latch bolt unit. Rotate the locking stem so the knob is unlocked.

Step 7 Install the inside knob assembly by aligning it with the tip of the spindle of the outside knob assembly (*Figure 86*). Make sure the turn button is in a horizontal position, then slide the inside knob assembly up to the door surface. Rotate the rose so that the screw holes line up with the threaded posts. Insert the connecting screws through the holes in the rose and into the posts. Tighten the connecting screws to obtain a firm attachment, but be careful not to overtighten. Attach the strike to the jamb with the two attaching screws and the job should be complete.

Step 8 Check the operation of the lockset with the thumb turn, making sure it operates properly. If the lock functions properly, tighten the connecting screws, but do not overtighten as this may cause a bind in the latch mechanism.

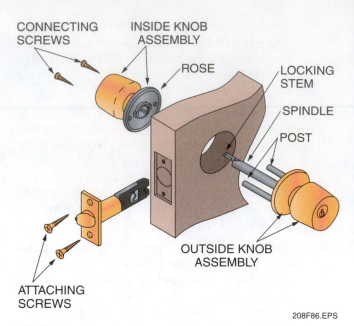

Figure 86 ◆ Disassembled lockset.

Cylindrical locksets (*Figure 87*) are used in industrial, commercial, and institutional construction. They are ruggedly constructed and designed for unlimited use. To simplify architectural specifications and installation, all locks in all functions are usually uniform in size, self-aligning, and simple to install. While both knob and lever locksets are essentially the same mechanically, the lever locksets are compliant with the requirements for disabled persons. Both types of locksets have a pin tumbler cylinder to offer a wide range of keying and control options.

A comparison may be made between the internal mechanism of the tubular lockset, as shown earlier, and the internal mechanism of the cylindrical lockset, as shown in *Figure 88*.

Most manufacturers of heavy-duty locksets also offer a knurled knob, or a knob with an abrasive surface. When called for in the specifications, this type of knob is installed as part of the lockset on doors leading to dangerous areas so these areas can be identifiable by touch for blind persons. A typical knurled knob is shown in *Figure 89*.

Tubular and cylindrical locksets are installed in much the same manner by drilling two holes in the door and mortising for the latch bolt and the strike.

10.3.0 Mortise Locksets

Mortise locksets are very expensive, extremely well made, and available in many styles and finishes. They are usually only used in high-profile construction, such as churches and some offices, and it takes an experienced finish carpenter to install them. An exploded view of a mortise lockset is shown in *Figure 90*.

208F87.EPS

Figure 87 ◆ Heavy-duty cylindrical locksets.

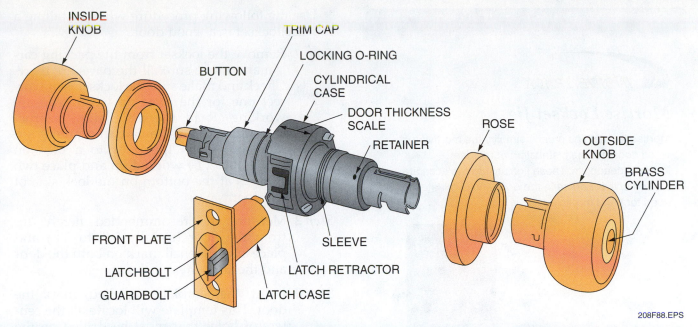

INSIDE KNOB

TRIM CAP

BUTTON

LOCKING O-RING

CYLINDRICAL CASE

DOOR THICKNESS SCALE

RETAINER

ROSE

OUTSIDE KNOB

BRASS CYLINDER

FRONT PLATE

LATCHBOLT

GUARDBOLT

LATCH CASE

SLEEVE

LATCH RETRACTOR

208F88.EPS

Figure 88 ◆ Exploded view of a heavy-duty cylindrical lockset.

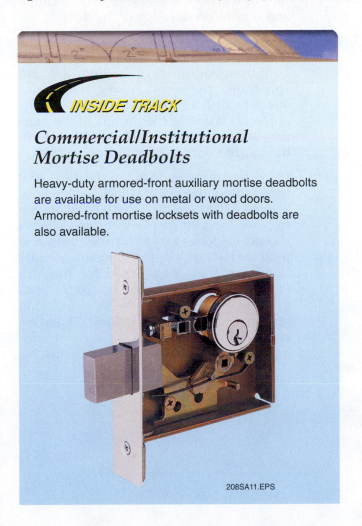
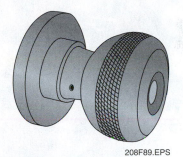

208F89.EPS

Figure 89 ◆ Typical knurled knob.

208F90.EPS

Figure 90 ◆ Exploded view of a typical mortise lockset.

INSIDE TRACK

Mortise Lockset Jig

Mortise lockset jigs greatly simplify the mortising of wood doors for the installation of mortise locksets/deadbolts. These jigs assure the accurate routing of the mortise in the door and eliminate the need for hand mortising.

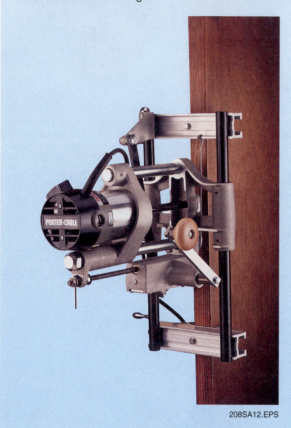

208SA12.EPS

Use the following procedure when installing a mortise lockset in a wood door:

Step 1 Remove the lockset from the packing carton and make sure all the parts are there. Check and make sure the lockset is the correct one for the door on which you are working. Read the installation instructions carefully so they are thoroughly understood. Open the door to a position where it can be worked on and place two wedges at the bottom on the lock side of the door.

Step 2 Measure the recommended height up from the floor (usually 34" to 38") and place a light pencil mark on both the door and the jamb.

Step 3 Using the template supplied, mark the door. This template will locate all the centers of holes that must be drilled on the face of the door and the mortise location and outline on the edge of the door.

Step 4 Using the correct diameter bits, drill the holes through the face of the door first. Drill from both sides of the door to avoid splintering the wood. Using the correct size bits, drill a series of holes in the edge of the door within the outline of the mortise, and clean out the remaining wood in the mortise with a sharp chisel.

Step 5 Place the **mortise lock** in the hole and mark the outline of the lock face on the edge of the door with a sharp knife or pencil. The location of the attaching screws should also be marked at this time and pilot holes drilled for them. Carefully chisel out the area marked for the lock face. Insert the lock in the mortise and fasten it to the door with the two attaching screws. Install the lock cylinder, handles, and strike on the face of the door according to the manufacturer's instructions.

10.4.0 Multiple Lock Keying

The keying of door locks is necessary in large office buildings, hotels, or motels. This is done so one key will open a number of doors in a given location. It eliminates the need for a person to carry more keys than needed.

The following keying terminology should explain most methods used in construction today. Most of the same principles can be applied to magnetic card locks.

10.4.1 Master Keying

There are several types of master key systems, as outlined below:

- *Simple master key systems* – Each lock has its own key, which will not operate any other lock in the system, but all locks in the system can be operated by the master key.
- *Grand master systems* – Each lock has its own individual key, as in the simple master key system. The locks are divided into two or more groups, with each group being operated by a master key, and all groups or the entire system are operated by the grand master key.
- *Great grand master systems* – Each lock has its own individual key, as in the simple master key system. The locks are divided into additional subgroups as needed, including master keys and grand master keys with all groups. All locks in the system can then be operated by a great grand master key.

10.4.2 Construction Keying

Conventional cylinders for all locksets can be provided with construction keying for use by architect and contractor personnel during building construction. A special breakout key is used to permanently void the construction key. This eliminates further use of the construction key.

10.4.3 Hotel Keying

Common hotel keying subgroups are as follows:

- The guest's key operates only the lock or locks of one room or suite.
- The maid's master key operates one group of locks, generally the guest room entrances and linen closets on one floor served by one maid.
- The housekeeper's master key operates a number of maids' groups, generally the entire guest room portion.

- The emergency/shutout key operates all guest room locks even when they are locked from the inside. It is also a shutout key, locking a guest room so it cannot be opened by any other keys in the system except the individual display key.
- The display key locks a guest room door against all other keys except the emergency shutout key. This key is used for guest rooms when the rooms are used as sample rooms or when extra security or privacy is required.
- The grand master key operates all locks in the hotel. The grand master key does not operate as an emergency/shutout key.

10.5.0 Security Hardware

The following sections cover various locking devices and accessories used for entry doors and gates. The electrically operated devices covered are all 12VDC or 24VDC units.

10.5.1 Electric Strikes

Figure 91 shows electric strikes, along with a typical local power module. Electric strikes are very common and provide a remote release of a locked door without requiring the retraction of a latch bolt. They are available as fail-safe or fail-secure. Fail-secure means that on loss of power, the strike remains locked. Fail-safe means that on loss of power, the strike opens. With the possible exception of prisons or mental institutions, most local jurisdictions require a locking device on an exterior exit door to fail-safe upon a fire alarm,

208F91.EPS

Figure 91 ◆ Typical electrical strikes.

sprinkler alarm, or loss of AC power. For interior fire doors or stairway doors, most codes prohibit the use of fail-safe strikes to prevent the spread of fire and smoke. In these applications, electric locks or latches must be used to allow the doors to remain latched when they are electrically unlocked.

Most of these devices can be mounted on either side of the door frame for a right-hand or left-hand opening door. With these devices, a portion of the outer edge of the door frame must be removed to accommodate the back box that allows the strike lip to swing open. This can weaken the door frame, allowing easier penetration.

10.5.2 Electric Bolt Locks

Electric bolt locks (*Figure 92*) are an alternative to electric strikes or magnetic locks because the bolting device that locks a door or gate is mounted on the top and/or sides of the door frame. The door itself has no latch. Multiple electric bolt locks can be used on a door to provide very strong security. The locks are available as fail-secure or fail-safe units. In most cases, neither can be used on

exterior exit doors because of code restrictions. Some of these devices are designed to fit in narrow door frames and do not require removal of a portion of the door frame edge. Others are designed for surface mount or for use as sliding or swinging gate locks.

10.5.3 Electric Locksets (Latches)

The primary use for electric locksets (*Figure 93*), also called electric latches, is in stairway fire doors on each floor of a building. Building codes generally require that stairway fire doors not be locked on the stair side unless they may be remotely unlocked without unlatching. Electric locksets provide the required locking, unlocking, and latching features. While providing controlled access and remote unlocking capability, the doors stay latched even when unlocked, maintaining fire door integrity. Because these locksets are mounted in the door, power transfer hinges like those shown in *Figure 93* are required. The hinges are available with two-, four-, and ten-wire conductors that are protected on the inner face of the hinge.

EXTRA HEAVY-DUTY
COMMERCIAL/INDUSTRIAL
GRADE ELECTRIC BOLT LOCK

SPACESAVER CONCEALED
NARROW ELECTRIC
BOLT LOCK

CONCEALED DIRECT THROW DESIGN
ELECTRIC BOLT LOCK

GATE LOCK

SURFACE MOUNT
DOOR LOCK

208F92.EPS

Figure 92 ◆ Electric bolt locks.

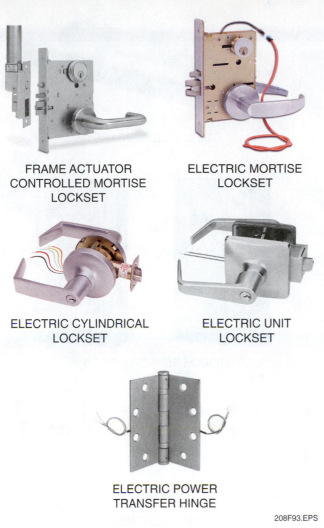

FRAME ACTUATOR
CONTROLLED MORTISE
LOCKSET

ELECTRIC MORTISE
LOCKSET

ELECTRIC CYLINDRICAL
LOCKSET

ELECTRIC UNIT
LOCKSET

ELECTRIC POWER
TRANSFER HINGE

208F93.EPS

Figure 93 ◆ Electric locksets.

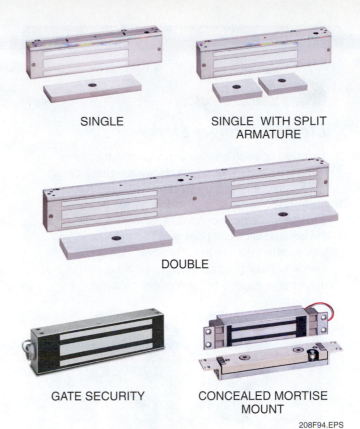

SINGLE

SINGLE WITH SPLIT
ARMATURE

DOUBLE

GATE SECURITY

CONCEALED MORTISE
MOUNT

208F94.EPS

Figure 94 ◆ Electromagnetic locks.

10.5.4 *Electromagnetic Locks*

Electromagnetic locks (*Figure 94*) are fail-safe and can be used on interior doors and exterior exit doors. However, they cannot be used on interior or stairway fire doors. They have no moving parts and are not subject to wear. Electromagnetic locks are available as direct hold and shear hold (concealed) styles. The direct hold styles are graded for use by ANSI as follows:

- *Grade 1* – 1,650 pounds direct holding force, medium security
- *Grade 2* – 1,200 pounds direct holding force, light security
- *Grade 3* – 650 pounds direct holding force, door holding only

There are some electromagnetic locks with 2,000 pounds or more of direct holding force. These locks will stay joined even when the door they secure is destroyed. Shear types have hold-ing forces of 2,700 pounds, but they are rated as only grade 1 because of the 90-degree pulling angle. Most electromagnetic locks have integral door position switches to indicate that the door is locked and secure. The shear locks have relock delay timers activated by the position switch so that the door is at rest before the lock reactivates.

In place of exit switches or readers, touch sense bars and handles (*Figure 95*) and switch bars are commonly used to turn off electromagnetic locks. The touch sense bars and handles are capacitive touch-sensitive switches and have no moving parts. The switch bars have mechanical switches. Some touch sense or switch bars have electronic timers that delay de-energizing an emergency exit for a set amount of time. Armored cables are used to connect the bars or handles to the hinge side of the door frame, where the wiring is routed to a controller or to the electromagnetic lock. These touch bars/handles are part of a group of exit devices that are sometimes called request-to-exit (RTE or REX) devices.

10.5.5 *Delayed-Exit Alert Locks*

Delayed-exit alert locks (*Figure 96*), sometimes called RTE or REX locks, legally delay an exit

Electric Plunger Strike

This type of electric strike is activated in the same manner as electric lip strikes. However, instead of a lip, the device uses a motor-driven plunger to depress the latch bolt and unlock the door. When de-energized, the plunger retracts. When the door closes, the beveled latch bolt rides over the strike plate and falls into the latch pocket. The advantage of this device is that it does not require removing part of the outer frame edge to accommodate the strike.

208SA13.EPS

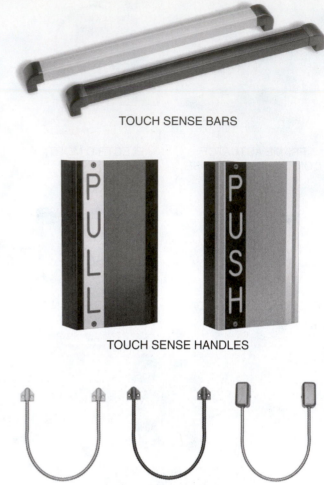

TOUCH SENSE BARS

TOUCH SENSE HANDLES

ARMORED CABLES

208F95.EPS

Figure 95 ◆ Touch sense bars, handles, and armored cables.

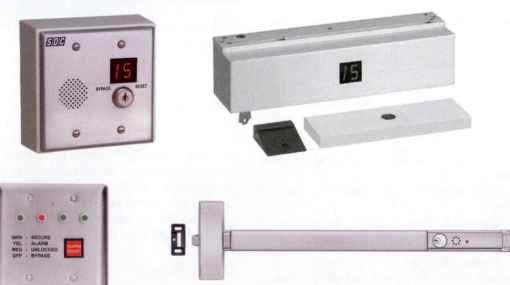

208F96.EPS

Figure 96 ◆ Delayed-exit alert locks.

through exterior exit doors in secure facilities. (By law, exit doors in other public facilities may not be locked or delayed.) When an exit is attempted, an alarm is sounded and a signal is sent to guards for CCTV or physical monitoring purposes. After 15 seconds, the exit door is unlocked, permitting an exit. A signal from a fire alarm system can also release the lock, allowing unrestricted exits during an emergency.

11.0.0 ◆ EXTERNAL DOOR STOPS, HOLDERS, AND DOOR CLOSERS

An external door stop is a rubber-tipped device fastened near the bottom of the door, the wall base, or the floor that prevents the door from striking the wall when the door is fully opened (*Figure 97*). One type of door holder has a plunger device, which is released by foot pressure. The spring holds the rubber tip to the floor. Another type of door holder is screwed into the door. This type wedges the door open when the lever is dropped to the floor.

Many building codes require the use of magnetic door holders such as the one shown in *Figure 97* to control hallway access doors in office buildings and apartment houses. These devices contain an electromagnet that is wired into the building fire alarm system. Under normal conditions, the magnets will hold the door open. If there is a fire alarm, however, the electrical power to the magnet is automatically turned off, and any open doors will close to inhibit the passage of fire and smoke.

Prior to installing a door closer, the hand or swing of the door must be determined. Closer hardware is available for regular arm installation on the pull side of the door only (*Figure 98*). It is also available for regular or parallel arm installation on the push side (**transom** bar or top jamb) of the door only (*Figure 99*). Head frame mounting provides leverage and power to control exceptionally wide doors or doors subject to pressure problems.

Concealed door closers may be used on interior doors where appearance is important. *Figure 100* shows a narrow concealed closer for use in an overhead transom. As shown in *Figure 101*, this concealed closer is mounted inside the head frame, and the track for the arm is in the top of the door. *Figure 102* shows an assembly in which the closer is mounted inside the top of the door and the track is mounted in the head frame.

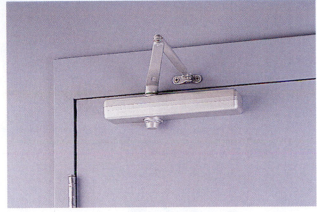

208F98.EPS

Figure 98 ◆ Regular arm installation on the pull side only (door-mounted).

| FLOOR-MOUNTED DOOR STOP | WALL- OR BASE-MOUNTED DOOR STOP | DOOR HOLDER | DOOR HOLDER | MAGNETIC DOOR HOLDER |

208F97.EPS

Figure 97 ◆ Door stop holders.

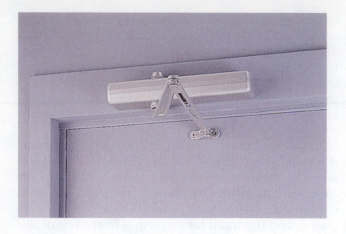

TOP JAMB CLOSER

PARALLEL ARM CLOSER

208F99.EPS

Figure 99 ◆ Regular or parallel arm installation on the push side only (jamb-mounted).

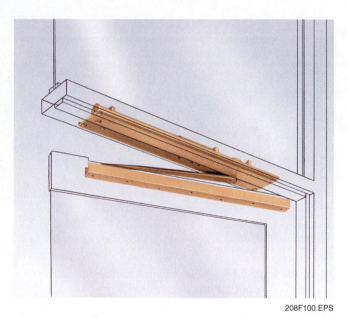

208F100.EPS

Figure 100 ◆ Concealed closer with soffit plate.

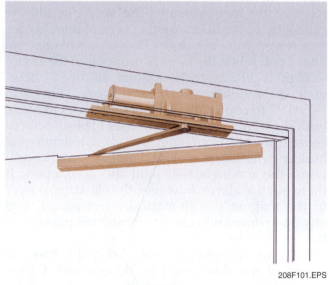

208F101.EPS

Figure 101 ◆ Concealed closer.

Manufacturers offer a wide variety of door closers, some of which are non-handed to permit installation on doors of either hand. Some of the closers can be mounted in different ways. Closers are available in a variety of finishes to complement other door hardware.

New closers offer spring power adjustment, enabling fine tuning of the closer's power. Many manufacturers also offer closers that can be adjusted to independently regulate closing and latch speed.

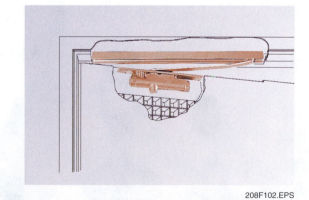

208F102.EPS

Figure 102 ◆ Arm and track fully concealed.

11.1.0 Installation of Door Closers

For installation procedures, the installer will need to refer to the manufacturer's instructions pertaining to the specific unit being installed. Generally, the installation for a door-mounted closer will proceed as follows:

Step 1 Using a supplied template such as the one shown in *Figure 103*, select the angle of the opening desired. Locate and make holes on the door for the closer body and holes on the frame for the arm shoe.

NOTE

The template is shown for illustration purposes only. Use the template that comes with the closer.

Step 2 Install the closer body on the door.

Step 3 Disassemble the secondary arm and shoe assembly from the main arm by removing the elbow screw. Fasten the secondary arm and shoe assembly to the frame face.

Step 4 Place the main arm into the closer pinion shaft, and install and tighten the main arm screw with a ½" wrench.

Step 5 Close the door and adjust the secondary arm assembly so the main arm is perpendicular to the face of the door. Reassemble the secondary arm to the main arm and tighten the screw securely.

Step 6 Adjust the closing tension by using the wrench packed with the door closer on the ratchet. Swing the wrench away from the hinge to wind the spring between three and ten notches, then engage the dog on the ratchet. Increase or decrease the swing power to suit the closing conditions of the door (*Figure 104*).

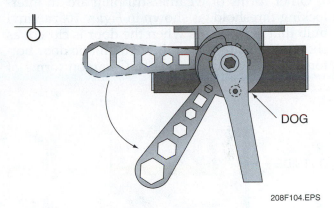

DOG

208F104.EPS

Figure 104 ◆ Adjusting closer tension.

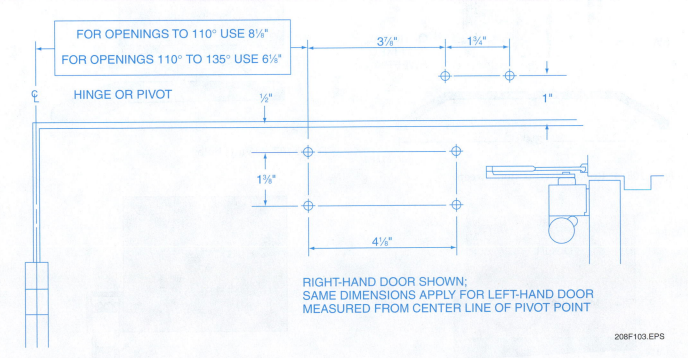

FOR OPENINGS TO 110° USE 8⅛"

FOR OPENINGS 110° TO 135° USE 6⅛"

HINGE OR PIVOT

½"

3⅞"   1¾"

1"

1⅜"

4⅛"

RIGHT-HAND DOOR SHOWN;
SAME DIMENSIONS APPLY FOR LEFT-HAND DOOR
MEASURED FROM CENTER LINE OF PIVOT POINT

208F103.EPS

Figure 103 ◆ Example of a closer mounting template.

12.0.0 ◆ OTHER DOOR HARDWARE

The following section covers other interior finish terms and definitions.

12.1.0 Weatherstripping

Weatherstripping is the application of materials in spaces between the outer edges of doors and windows and the finished frames to keep out air or moisture that would otherwise enter.

There are a variety of methods for achieving this purpose on doors. *Figure 105* shows a rubber fixed-bottom sweep, which is one way to prevent air infiltration.

Other forms of weatherstripping are an interlocking threshold, as shown in *Figure 106*; a vinyl bulb that compresses when the door is closed, as shown in *Figure 107*; and an automatic door bottom, as shown in *Figure 108*. Common forms of weatherstripping might also include a spring metal V-strip, a wood-backed foam rubber strip, or rolled vinyl.

12.2.0 Thresholds

A threshold is a piece of wood, metal, or stone that is set between the door jamb and the bottom of a door opening. An example of an aluminum threshold is shown in *Figure 109*, along with stop strip weatherseals. The threshold must be set in a bead of caulk to keep air from getting under the door.

12.3.0 Touchbar or Crossbar Hardware

Touchbar or crossbar hardware, sometimes called panic hardware, provides secure locking of single

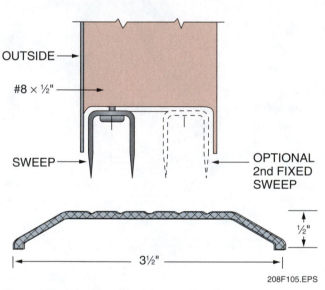

Figure 105 ◆ Rubber fixed-bottom sweep.

Figure 107 ◆ Vinyl bulb.

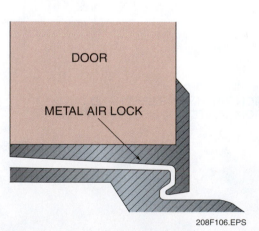

Figure 106 ◆ Interlocking threshold.

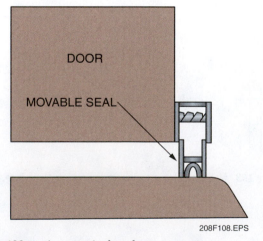

Figure 108 ◆ Automatic door bottom.

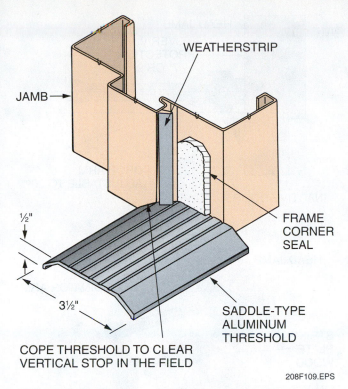

Figure 109 ◆ Aluminum threshold.

JAMB

WEATHERSTRIP

FRAME
CORNER
SEAL

SADDLE-TYPE
ALUMINUM
THRESHOLD

½"

3½"

COPE THRESHOLD TO CLEAR
VERTICAL STOP IN THE FIELD

208F109.EPS

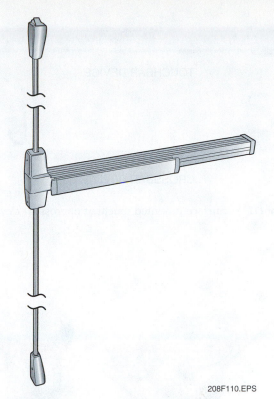

208F110.EPS

Figure 110 ◆ Surface-mounted, double-latch touchbar
device.

or double doors in one direction (from the opposite side of the touchbar), but allows easy passage in the other direction (from the touchbar side). Several types are shown in *Figures 110* and *111*. Touchbar and crossbar hardware is available with or without an opposite side latch and release lock. Touchbars are primarily used on emergency or normal exit doors where local codes prohibit the locking of the doors on the touchbar side at any time. They are also used on internal doors of buildings to control **access** to an area and yet allow rapid egress from the area. Touchbar or crossbar hardware is usually installed in commercial and institutional buildings such as stores, hospitals, schools, and government buildings.

12.4.0 Flush Bolts

A **flush bolt** is a sliding bolt mechanism that is mortised into the door at the top and/or bottom edge. It is used to hold an inactive door in a fixed position on a pair of double doors. For automatic and semi-automatic flush bolts on metal, composite, and wooden doors, a door **coordinator** is required. An automatic flush bolt retracts without manual actuation. A semi-automatic flush bolt engages a door latch when an inactive door closes without any use of a triggering mechanism. The

bolt remains extended until it is retracted by manual release of the bolt actuating lever.

12.5.0 Dustproof Strike

A **dustproof strike** is a metal piece used to receive the latch bolt when the door is closed. The piece has a spring action cover to keep particles out of the strike when the door is open. *Figure 112* shows two examples of dustproof strikes.

12.6.0 Door Coordinator

A door coordinator is a device for use on double doors to hold the active door open until the inactive door is closed. This trip action then gives the active door the function of a single acting door, allowing the normal locking or latching of the lock bolt and the normal overlapping of an **astragal** or rabbeted door, when used (*Figure 113*). Refer to *Figure 114* for a diagram of an astragal set for pairs of doors.

12.7.0 Smoke Gasket

A **smoke gasket** is a rubber cord or strip that goes all the way around the door to keep smoke from penetrating an area in case of fire.

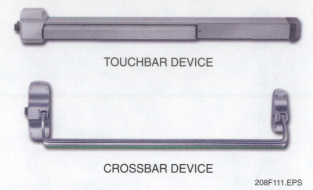

TOUCHBAR DEVICE

CROSSBAR DEVICE

208F111.EPS

Figure 111 ◆ Surface-mounted touchbar or crossbar device.

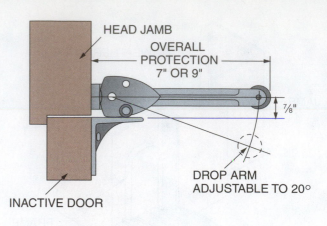

HEAD JAMB

OVERALL
PROTECTION
7" OR 9"

7/8"

DROP ARM
ADJUSTABLE TO 20°

INACTIVE DOOR

FLUSH
FLOOR MOUNT

THREADED
SPRING CARRIER

208F112.EPS

Figure 112 ◆ Dustproof strikes.

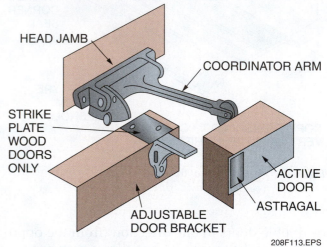

HEAD JAMB

COORDINATOR ARM

STRIKE
PLATE
WOOD
DOORS
ONLY

ACTIVE
DOOR

ASTRAGAL

ADJUSTABLE
DOOR BRACKET

208F113.EPS

Figure 113 ◆ Door coordinator.

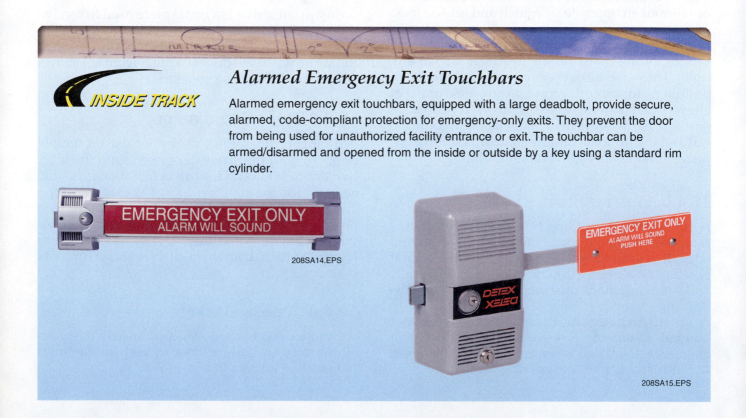

Alarmed Emergency Exit Touchbars

INSIDE TRACK

Alarmed emergency exit touchbars, equipped with a large deadbolt, provide secure, alarmed, code-compliant protection for emergency-only exits. They prevent the door from being used for unauthorized facility entrance or exit. The touchbar can be armed/disarmed and opened from the inside or outside by a key using a standard rim cylinder.

EMERGENCY EXIT ONLY
ALARM WILL SOUND

208SA14.EPS

EMERGENCY EXIT ONLY
ALARM WILL SOUND
PUSH HERE

DETEX

208SA15.EPS

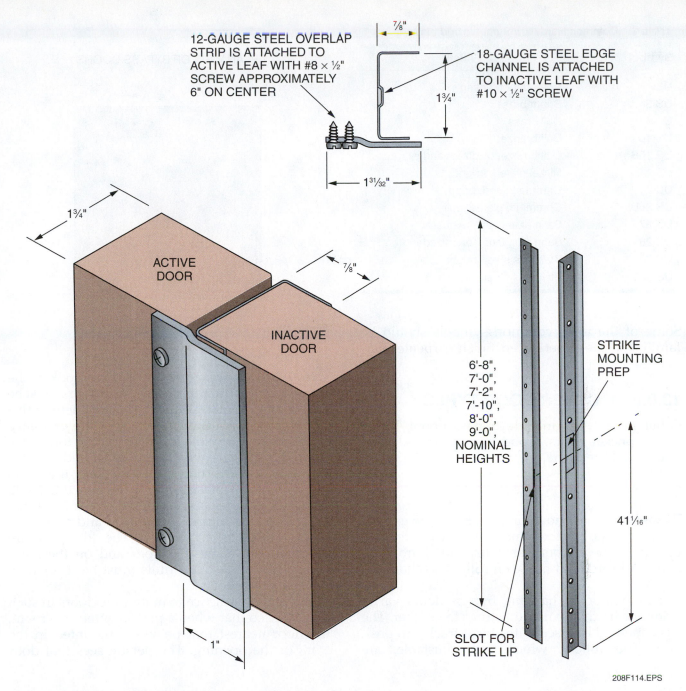

12-GAUGE STEEL OVERLAP STRIP IS ATTACHED TO ACTIVE LEAF WITH #8 × ½" SCREW APPROXIMATELY 6" ON CENTER

18-GAUGE STEEL EDGE CHANNEL IS ATTACHED TO INACTIVE LEAF WITH #10 × ½" SCREW

7/8"

1¾"

1³¹⁄₃₂"

1¾"

ACTIVE DOOR

7/8"

INACTIVE DOOR

1"

6'-8", 7'-0", 7'-2", 7'-10", 8'-0", 9'-0", NOMINAL HEIGHTS

STRIKE MOUNTING PREP

41¹⁄₁₆"

SLOT FOR STRIKE LIP

208F114.EPS

Figure 114 ◆ Diagram of an astragal set for a pair of doors.

12.8.0 Open Back Strike

An open back strike is a strike applied to the inactive leaf of a pair of doors and cut away at the back to permit either leaf to open or close independently. No coordinator or overlapping astragal is necessary.

12.9.0 Hardware Finish

Hardware is divided into two basic classifications: **finish hardware** and rough hardware. Rough hardware includes items such as bolts, screws, nails, or any similar items that are not usually exposed to view in the completed structure. Finish hardware includes hinges, door lock/passage sets, cabinet handles, pulls, and other such items that are exposed to view in the completed structure.

The term hardware finish is used to describe the surface finish on finish hardware. Examples of this include a bright brass finish on door hinges or a satin chromium-plated finish on a kitchen cabinet drawer pull.

The finish symbols have been established into a classification system called US Finish Symbols.

Table 3 Common Hardware Symbols and Finishes

Symbol	Finish
USP	Primed for painting
US 3	Bright brass
US 4	Dull brass
US 10	Dull bronze
US 10B	Dull bronze, oxidized, and oil rubbed
US 14	Nickel plated, bright
US 26	Chrome plated, bright
US 26D	Chromium plated, dull
US 27	Satin aluminum, lacquered
US 28	Satin aluminum, anodized
US 32	Stainless steel, bright
US 32D	Stainless steel, dull

Some of the more common symbols should be familiar to the experienced finish carpenter (see *Table 3*).

13.0.0 ◆ OTHER DOOR TYPES

Other door types include bypass doors, bifold doors, pocket doors, wood folding doors, metal doors, and fire doors.

13.1.0 Bypass Doors

Bypass doors, as shown in *Figure 115*, are usually hung in pairs, but there may be more doors in an opening. These doors are suspended from an overhead track and move on rollers, as shown in *Figure 116*.

The method of hanging bypass doors varies from one hardware manufacturer to another. The basic method is to screw the head track into position. The rollers, which are adjustable, are

PLAN VIEW OF BYPASS DOORS

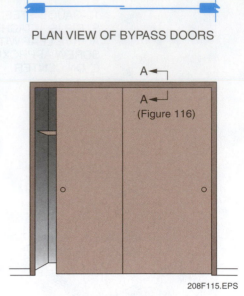

A ←
A ←
(Figure 116)

208F115.EPS

Figure 115 ◆ Elevation of bypass doors.

screwed to the top edge of the door. Look at the installation instructions to see if the doors must be cut off on the bottom edge to shorten them. If this is the case, it is usually a result of the excessive space taken up by the hardware at the top of the door. Preplanning when framing should prevent the necessity of cutting the doors. Hang the doors in position by placing the rollers in the track. Hang the door closest to the rear first. Mark the location of the doors on the floor and screw the nylon guide strip to the floor. The door pulls are mounted 38" above the floor and on the stiles nearest the jambs. The pulls must be flush with the face of the door.

It is a good practice to mount the doors in such a manner so that when a person enters the room, the door nearest that person is mounted in the front of the opening. The person sees that door

BHMA Finish Numbering System

The Builders Hardware Manufacturer's Association (BHMA) has established a sectional classification system for hardware finishes. This standard for materials and finishes contains a description of types of finishes and divides them into categories. A numbering system has been established, which identifies the base material and finish. This system can be more readily used in a computer than the US finish symbols. As an example, the BHMA number 605 is the equivalent of US 3 or bright brass. Some hardware manufacturers are already using the BHMA numbering system. Any finish carpenter who installs a lot of hardware should contact BHMA for a copy of their finish standards.

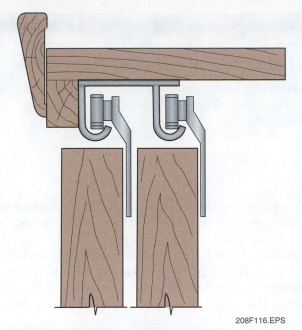

Figure 116 ◆ Head section through bypass doors.

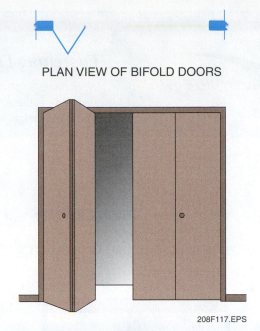

Figure 117 ◆ Plan view of bifold doors.

first, and it covers the edge of the door mounted behind it. Most flush doors are not ordered with matching edge strips, and those strips that are supplied with the door are usually of a lesser grade of wood. A little effort on the part of the carpenter in a situation such as this is appreciated by the customer.

13.2.0 Bifold Doors

Bifold doors (*Figure 117*) allow full access to a closet and, at the same time, permit furniture to be placed almost against the side trim of the closet opening. Usually, this door unit is used with four doors or two pairs, with each pair operating independently of the other. This four-door unit is used for an opening from 4' to 6' in width. For openings of less than 4', only two doors, or one pair, are used.

The door unit is supplied with wood, metal, plastic, or mirrored doors. Hardware such as hinges, pulls, pivot pins, and guide pins are installed on the doors at the factory.

13.2.1 Bifold Installation

Hanging the unit in a prepared opening is simple. A slide guide track is screwed to the bottom of the head jamb according to the manufacturer's instructions. At the base of the side jambs, against the floor on each side, an adjustable jamb bracket is screwed in place.

The pivot pins and guide pins are spring loaded, so it is a simple matter to mount the doors

on the hardware in the opening. A pivot pin and bracket secure the door next to the jamb. The first door and second door are hinged together. If the doors are not plumb or do not meet properly in the center, an adjustment may have to be made at the jamb bracket on the floor. The second door slides along a track on a roller guide. This track is secured to the head jamb.

13.3.0 Pocket Doors

A pocket door unit is delivered to the job site partially knocked down and is rapidly and easily assembled. Pocket door units are designed to fit in a framed opening of 3½" studs. The hardware consists of a track and rollers that are partially concealed in the head pocket of the unit. When the unit is positioned in the rough opening, any 1⅜" thick interior door of the correct size may be hung on the track. The portion of the unit called the pocket or the right-hand side of the unit, as shown in *Figure 118*, may be covered with practically any interior wall finish. The wall finish is applied to the horizontal members of the unit, creating the pocket for the door to slide into.

After the door is positioned in the opening and the rollers are adjusted so that it hangs plumb with the jambs and square with the head, a guide strip is fastened to the bottom edge, as shown in *Figure 119*.

Stops are then nailed on either side of the pocket opening. The thickness of the stops decreases the width of the opening, preventing the door from being pushed out of the opening.

Fastening Finish Wall Material to a Pocket Door Unit

INSIDE TRACK

When fastening the wall finish to the horizontal members of a pocket door unit, make sure the nails, staples, or screws are not so long that they extend into the pocket area where they might scratch the door or even prevent the movement of the door.

PLAN VIEW OF POCKET DOOR UNIT

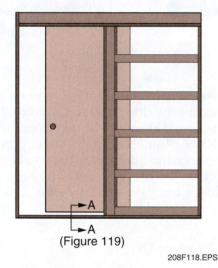

(Figure 119)

208F118.EPS

Figure 118 ◆ Plan view of a pocket door.

The spacing of the stops also acts as a guide for the guide strip on the door bottom. The purpose of the guide strip is to position the door so that the surfaces of the door are not scratched when being slid in and out of the pocket. When the door is completely inside the pocket, the edge of the door facing the opening is flush with the face of the stops.

A special piece of hardware called an edge pull is installed on the edge of the door so that it may be pulled out of the pocket far enough to use the conventional flush pulls mounted on the face of the door. There are many variations of the edge pull, but the simplest is a small metal plate with a finger hole in it that is mortised to the edge of the door 38" above the floor.

13.4.0 Wood Folding Doors

Wood folding doors are assembled of vertical, prefinished, wood door panels operating similar to an accordion (*Figure 120*). The door can be installed to open from either side.

The width of the panels, called the stack width, is 4¼" so it will fit within a typical door frame.

A beveled wood **molding** at the head conceals a surface-mounted head track. The track and molding are drilled for mounting screws supplied with the door. A spring-action catch at the top of the end post (*Figure 121*) engages the nylon head stop for tight stacking of the panels.

The hangers, with free-riding nylon rollers, are attached to alternate panels for quiet, easy operation. The end post hanger has two sets of rollers to keep it aligned with the track (*Figure 122*). The positioning of the rollers allows operation even in openings that are slightly out of plumb.

The panels have special steel alloy springs that are run horizontally through the panels and connect to vertical wood moldings, as shown in *Figure 123*. The spring connectors can be detached to disassemble the door or to replace damaged panels.

Wood folding doors are available in opening widths from 1'-9" to 30'-2¼".

FLOOR SURFACE | DOOR STOP | DOOR GUIDE STRIP | DOOR STOP

DOOR

208F119.EPS

Figure 119 ◆ Door guide strip.

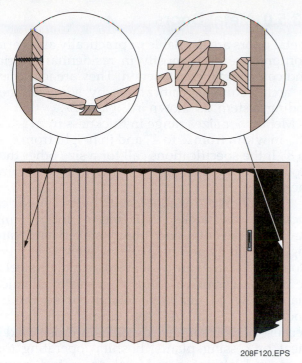

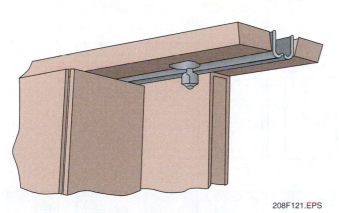

Figure 120 ◆ Elevation of a regular wood folding door.

Wood folding doors can be mounted directly to the ceiling to divide large rooms into smaller rooms as in an office or meeting room area. To do this, the track is simply screwed to supporting members in the ceiling and the beveled wood molding is applied to both sides. A crossover switch track permits the door tracks to cross at right angles, as shown in *Figure 124*.

A curved switch track (*Figure 125*) is available in 90° left or right arcs, and seven radii from 1'-6" to 10'. This type of switch permits the doors to divide a room at several designated points. It may also be used to store stacked doors parallel to a wall.

A simple curved track (*Figure 126*) permits doors to move around structural members and to create curved wall effects. The track is available in standard radii from 1'-6" to 10' and 90° left or right arcs.

Wood folding doors are simple to hang and all hardware is supplied by the manufacturer.

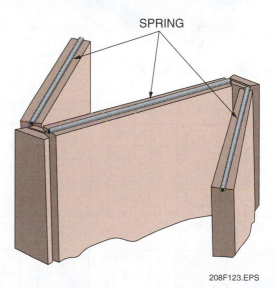

Figure 123 ◆ Wood folding door spring hinging.

Figure 121 ◆ Track molding and head stop of a wood folding door.

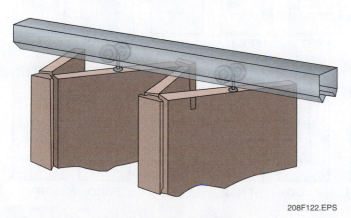

Figure 122 ◆ Wood folding door hangers.

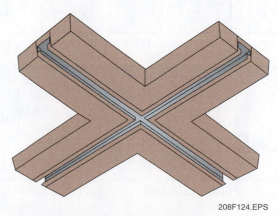

Figure 124 ◆ Crossover switch track.

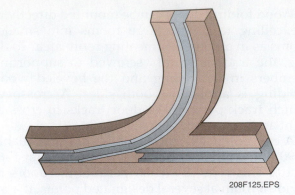

Figure 125 ◆ Curved switch track.

208F125.EPS

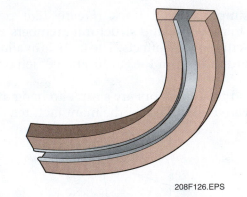

208F126.EPS

Figure 126 ◆ Curved track.

13.5.0 Metal Doors

Metal doors are available in practically any variation and are used mostly in residential exterior and commercial construction. They are identified in architectural specifications by some defined coding system, as shown in *Figure 127*.

Metal door sizes range in thickness from 1⅜" to 1¾", in width from 2' to 4', and in height from 6'-8" to 8'. If the specifications call for a size other than stock, a door can be made to meet those requirements.

Metal doors are available with butts, cylindrical locks, mortise locks, unit (mono) locks, panic devices, flush bolts, and other installed hardware. Light openings are available as well as louvers. Doors and frames are available in stainless steel to be used in corrosive atmospheres as in chemical plants, sewage treatment plants, or swimming pool areas. Stainless steel doors and frames, because of their sanitary qualities, are also used in meat processing plants, hospital operating theaters, and laboratories. Metal doors and frames are manufactured to meet the requirements of practically any architectural opening that can be imagined.

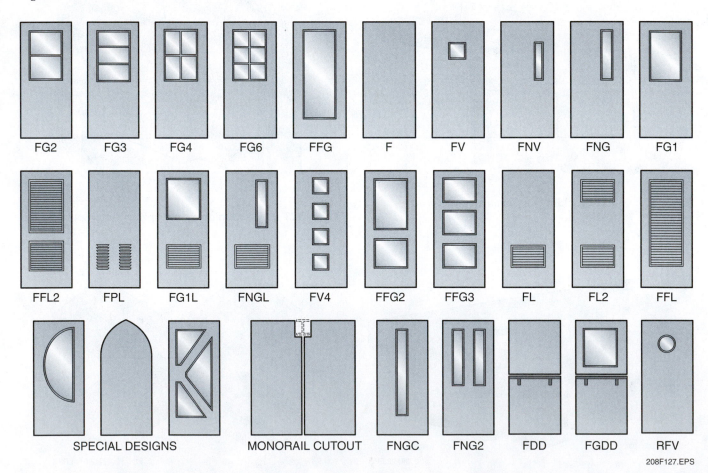

208F127.EPS

Figure 127 ◆ Metal doors.

The internal portion of the door, or core, is also designed to meet most specification requirements, as shown in *Figure 128*. For example, a metal door with a lead core should be used in radiation areas.

13.6.0 Fire Doors and Fire Ratings

Fire doors are manufactured to meet specific conditions and, because of this, the openings are referred to as labeled openings. The letters A, B, C, D, and E are codes used by labeling agencies and refer to the opening locations. Refer to the door schedule for the project. The classification is used to state the opening for which the fire door is considered suitable. The rating time associated with each classification indicates the duration of the fire test the door is capable of enduring. Each location classification requires a specific door rating depending on the fire hazard involved. An example of these classifications and ratings is shown in *Figure 129*.

- *Label A* – Refers to doors located in walls between separate buildings, fire walls, and division walls or curtain walls leading to highly flammable contents such as a paint or chemical storage area.
- *Label B* – Refers to doors located in openings into vertical shafts, such as fire stairs, elevator shafts, incinerator chutes, and walls separating garages from living quarters.
- *Label C* – Refers to doors located in room partition openings and openings in corridors.
- *Label D* – Refers to doors located in openings in exterior walls subject to extreme fire exposure.
- *Label E* – Refers to doors located in openings in exterior walls subject to moderate fire exposure.

NOTE

A labeled door may not be modified and the label must not be painted.

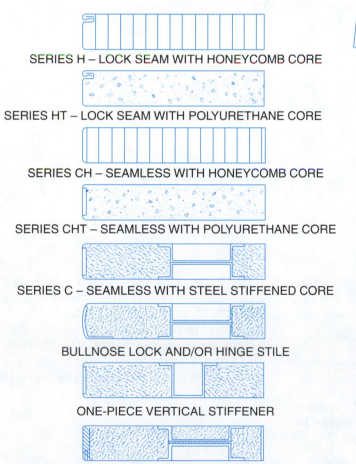

SERIES H – LOCK SEAM WITH HONEYCOMB CORE

SERIES HT – LOCK SEAM WITH POLYURETHANE CORE

SERIES CH – SEAMLESS WITH HONEYCOMB CORE

SERIES CHT – SEAMLESS WITH POLYURETHANE CORE

SERIES C – SEAMLESS WITH STEEL STIFFENED CORE

BULLNOSE LOCK AND/OR HINGE STILE

ONE-PIECE VERTICAL STIFFENER

THERMAL-BREAK DESIGN

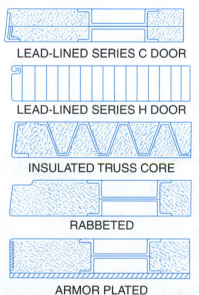

LEAD-LINED SERIES C DOOR

LEAD-LINED SERIES H DOOR

INSULATED TRUSS CORE

RABBETED

ARMOR PLATED

Note: Standard doors are reversible/square edged and will fit both right-hand and left-hand openings. Most designs can be furnished handed and beveled 1/8" in 2" when specified. Some designs must be provided handed and beveled due to construction, hardware, or application considerations.

208F128.EPS

Figure 128 ◆ Standard and engineered door sections.

LOCATION	CLASS	RATING	MAXIMUM GLASS AREA	REGENT	MEDALLION	FUEGO	VERSADOOR
Openings in walls separating buildings or dividing a building into fire areas	A	3 HR.	NONE	NTR	NTR	250° MAXIMUM TEMP. RISE	NA
Openings in enclosures of vertical communication: elevators, stairwells, and in 2-hour rated partitions providing horizontal fire separation	B	1½ HR.	100 SQ. IN. LITE PER DOOR	NTR	NTR	250° MAXIMUM TEMP. RISE	NTR (100 SQ. IN. PER DOOR)
Corridor and room partitions	C	¾ HR.	1296 SQ. IN. PER LITE (EXCEPT VERSADOOR)	NTR	NTR	NTR	NTR (100 SQ. IN. PER DOOR)
Exterior walls subject to severe fire exposure from outside	D	1½ HR.	NONE	NTR	NTR	250° MAXIMUM TEMP. RISE	NTR (100 SQ. IN. PER DOOR)
Exterior walls subject to moderate fire exposure from outside	E	¾ HR.	1296 SQ. IN. PER LITE (EXCEPT VERSADOOR)	NTR	NTR	NTR	NTR (100 SQ. IN. PER DOOR)
Openings where smoke control is a primary consideration ... in partitions between a habitable room and a corridor when the wall is constructed to have a fire resistance rating of more than one hour ... or across corridors where a smoke partition is required		½ HR. ⅓ HR.	1296 SQ. IN. PER LITE (EXCEPT VERSADOOR)	NTR	NTR	NTR	NTR (100 SQ. IN. PER DOOR)

NTR = NO TEMPERATURE RISE

208F129.EPS

Figure 129 ◆ Labeled fire door application chart.

Doors are also available for sound-insulating purposes. X-ray doors, security doors, armor-plated doors, and pressure-resistant doors are also available from most manufacturers.

13.6.1 Hollow Metal Door Construction and Installation

To determine the rough opening, check the manufacturer's instructions for the number of inches to add to the door width and the door height.

Steel doors have prepared hinge mortises and lock holes. Some hollow metal doors are reinforced with a honeycomb pattern core. Others only have steel reinforcement sections for the locks and hinges or other hardware. The holes are drilled and tapped to receive machine screws. *Figure 130* shows an example of metal door construction.

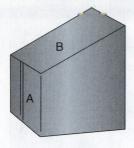

A. EDGE SEAM IS EXPOSED ON BOTH EDGES

G. 18-GAUGE DOOR CLOSER REINFORCEMENT BOX LAMINATED TO INSIDE OF DOOR SKIN

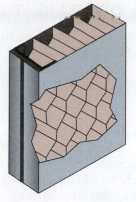

B. FLUSH TOP REINFORCED WITH 16-GAUGE CHANNEL

E. ¾" CELL HONEYCOMB CORE

F. CONTINUOUS 14-GAUGE MORTISE LOCK REINFORCEMENT WITH PROVISIONS FOR GOVERNMENT SERIES 86 MORTISE LOCKSET AND ANSI A115.1 LOCK FRONT (1¼" × 8")

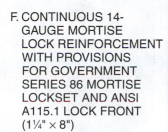

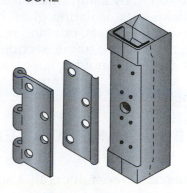

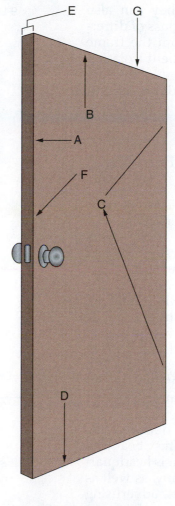

C. CONTINUOUS 11-GAUGE STEEL INTEGRAL HINGE REINFORCEMENT WITH PROVISION FOR 3½" × 3½" FULL MORTISE TEMPLATE-TYPE HINGES

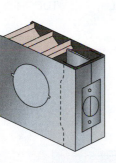

F. CONTINUOUS 14-GAUGE STEEL INTEGRAL CYLINDRICAL LOCK REINFORCEMENT WITH PROVISIONS FOR GOVERNMENT SERIES 160 OR 161 CYLINDRICAL LOCKSETS (2¾" BACKSET) AND ANSI A115.2 LOCK FRONT (1⅛" × 2¼")

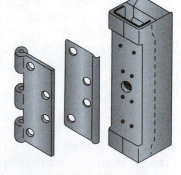

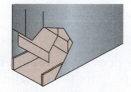

D. BOTTOM REINFORCED WITH 16-GAUGE CHANNEL

AVAILABLE IN 1⅜" OR 1¾" THICKNESSES. EASILY MODIFIED FOR LIGHTS, LOUVERS, AND OTHER OPTIONS

C. CONTINUOUS 11-GAUGE STEEL INTEGRAL HINGE REINFORCEMENT WITH PROVISION FOR 4½" × 3½" FULL MORTISE TEMPLATE-TYPE HINGES

C. STANDARD DOORS ARE NON-HANDED USING HINGE FILLERS

208F130.EPS

Figure 130 ◆ Metal door construction.

14.0.0 ◆ DOOR LIGHTS

Door lights are simply window panes within a door unit. See *Figure 131* for examples of door lights. These are also referred to as vision lights. Vision lights are constructed of different materials, but the primary product used is glass. When a window unit is installed in a door, this is referred to as glazing. Glass is not a good sound retarding material, and double glazing may actually increase the noise level. To reduce the noise level, manufacturers can install sound absorptive material around the perimeter in the space between the two panes. They can also reduce sound by using laminated glass of different thicknesses, isolating glass from the frame with flexible gaskets, or using resilient isolation material between frame sections. Nonparallel panes reduce annoying glare and reflections. Refer to *Figure 132* for an example of snap-in glazing in a steel door panel.

15.0.0 ◆ COMMERCIAL EXTERIOR DOORS

Commercial exterior doors are selected to provide a secure and convenient means of entering and leaving a building. The variety of exterior doors available is enormous. The type and style selected for an exterior door is based on its purpose. Doors used exclusively by employees are usually no-frills doors, while those used on loading docks are strictly utilitarian. The doors used by the public are selected to project a particular image and to make their use convenient. A trendy retail store will use a very different style than a supermarket, and an upscale hotel in New York City will have a different style door than a budget motel located near an interstate highway.

Figure 133 shows a revolving door on the Wrigley Building in Chicago. This beautiful entrance is perfect for a building that is headquarters for the Wm. Wrigley Jr. Company, as well as investment management companies, advertising agencies, marketing firms, a bank, and foreign consulates.

Figure 134 shows a thermally efficient door on a loading dock. This type of door is used when temperature control of the interior of the building is critical.

As you can imagine, the installation of some commercial doors is very complicated. Because of their complexity, commercial exterior doors are often installed by teams that have been specially

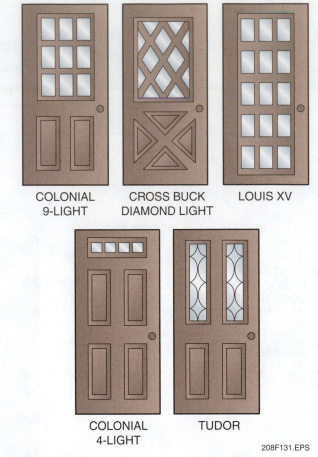

COLONIAL 9-LIGHT   CROSS BUCK DIAMOND LIGHT   LOUIS XV

COLONIAL 4-LIGHT   TUDOR

208F131.EPS

Figure 131 ◆ Door lights.

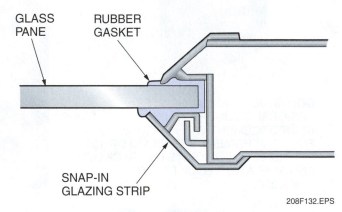

GLASS PANE   RUBBER GASKET

SNAP-IN GLAZING STRIP

208F132.EPS

Figure 132 ◆ Snap-in glazing strip.

trained by the door manufacturer. In these cases, your job is to prepare the installation site according to the building plans and specifications. If you are required to install a commercial exterior door, it is very important that you obtain the door vendor's installation instructions and then carefully follow the instructions.

208F133.EPS

Figure 133 ◆ Revolving door with ornate facade.

208F134.EPS

Figure 134 ◆ Thermally efficient door on a loading dock.

 CAUTION

When you are installing a door, you should perform the work that you are trained and qualified to perform. Other work, such as electrical installations, must be completed by a qualified worker.

You can help to prevent problems during door installation by ensuring that wall construction is done to specification. The wall angle must be plumb and the door opening must be square and level. Although there are a number of fixes the installation team can use to compensate for tolerance variations, it is much better for you to take the time to be sure your work is correct, rather than rely on someone else to fix your errors.

Regardless of whether you will be installing the doors or not, one important task that may fall to you is to accept delivery of the door. It is important that you carefully inspect the door for damage before you accept it. You should note any damage to the door packaging on the delivery invoice. You must not accept delivery of any severely damaged doors, but you should report the damage to your supervisor. Don't remove any packaging unless you suspect the product is damaged. The packaging material will help to protect the door during storage.

Once you have accepted the door, you need to store it in a secure location. The door should be stored in the unopened shipping container to protect it. Store the door in an area that permits good air circulation. Doors should never be stored in an area where they are exposed to the elements. When doors are stored outside, be certain that they are off the ground and protected. If you need to cover the door with plastic, ensure that there is enough of an air gap to prevent humidity build-up. Protect metal doors from corrosives such as lime, and abrasives, such as mortar and cement.

Summary

All carpenters should be familiar with the various steps in a typical door installation, the many kinds of doors used for various openings, and the correct hardware required for each particular situation. Attention to detail and carefully following the manufacturer's instructions are essential to a professional installation.

Notes

1. Methods of placing a door and trim _____.
   a. are standard all over the country
   b. vary from region to region
   c. depend on international code
   d. are governed by fire codes

2. To protect yourself while working, you should do all of the following *except* _____.
   a. lean against a grounded pipe while operating power tools
   b. carefully put away any tool after you have used it
   c. use clamps to hold work so that both of your hands are free
   d. unplug tools before you change blades, cutters, or bits

3. The two main types of flush doors are _____.
   a. hollow-core and solid-core
   b. vision lights and door lights
   c. lead-lined and panel
   d. stile and rail

4. A staved-wood door is a _____ door.
   a. hollow-core
   b. panel
   c. solid-core
   d. mineral-core

5. Solid-core doors have a tendency to _____.
   a. warp less than solid metal doors
   b. be less soundproof than hollow-core doors
   c. warp less than hollow-core doors
   d. be lighter than hollow-core doors

6. A lead-lined door is used for _____.
   a. soundproofing
   b. radiation shielding
   c. fireproofing
   d. waterproofing

7. A door jack is used to _____.
   a. suspend a door
   b. jack up a door
   c. level a door
   d. hold a door on edge

8. In a door schedule, a number such as 3068 for a door would usually indicate _____.
   a. one door, 60" wide by 68" high
   b. two doors, 60" wide by 6'-8" high
   c. one door, 3'-0" wide by 6'-8" high
   d. two doors, each 30" wide by 68" high

9. In a door schedule, a door mark is a _____ designator.
   a. location
   b. core
   c. thickness
   d. finish

10. In a door schedule, the type of door is usually defined by a(n) _____.
    a. letter
    b. number
    c. letter and a number
    d. abbreviation

11. In a door schedule, the core of a door is usually defined by a(n) _____.
    a. letter
    b. number
    c. letter and a number
    d. abbreviation

12. The term swing refers to the _____.
    a. direction that the door opens
    b. side that the hinges are on
    c. side that the door knob is on
    d. type of door hinges used

13. A K.D. unit is a(n _____ metal door frame.
    a. unassembled
    b. assembled
    c. partially assembled
    d. fully welded

14. Adjustable metal door frames usually accommodate wall thicknesses of _____.
    a. 2½" to 7½"
    b. 3" to 8"
    c. 3¾" to 9¼"
    d. 4½" to 10½"

15. A hospital stop is a _____.
    a. hardware device used to hold doors open for wheelchairs
    b. type of one-way door in a hospital
    c. stop strip that is terminated above the floor at an angle
    d. lead-lined door that stops radiation from a hospital X-ray machine

16. Metal doorframes are often _____ for fire-proofing.
    a. painted
    b. locked
    c. grouted
    d. stripped

17. The purpose of the kerfs on the back side of wood jambs is to _____.
    a. allow alignment of the head jamb with the side jambs
    b. space the nails used to fasten the head jamb to the side jambs
    c. mark the cutoff lines to be used when reducing the width of a jamb
    d. help prevent warping of the jambs

18. The rough opening size normally used for conventional job site installed jambs and doors is the door size plus _____.
    a. 1" for height and 2" for width
    b. 1½" for height and 2" for width
    c. 2" for height and 3" for width
    d. 3" for height and 2" for width

19. The clearance used between a wood door and jamb on each side of the door to allow for the final finish is usually _____.
    a. ⅟₃₂"
    b. ⅟₁₆"
    c. ³⁄₃₂"
    d. ⅛"

20. When installing a loose door stop, the stop on the lock jamb side of the door must _____.
    a. fit flush with the door
    b. have a ⅟₁₆" clearance between it and the door
    c. have a ⅛" clearance between it and the door
    d. have a ¼" clearance between it and the door

21. The rough opening size usually specified for a prehung door is the door size plus _____.
    a. 1½" for height and 2½" for width
    b. 1" for height and 2" for width
    c. 2" for height and 3" for width
    d. 3" for height and 2" for width

22. A lockset with a turn button on one knob and an emergency release on the other knob is a(n) _____.
    a. entrance lock
    b. privacy lock
    c. passage latch
    d. patio lock

23. Commercial heavy-duty locksets are known as _____ locksets.
    a. cylindrical
    b. tubular
    c. knob
    d. lever handle

24. Magnetic door holders are often used in hallways of office buildings, so that during a fire alarm the doors automatically _____.
    a. close
    b. lock
    c. open
    d. unlock

25. Touchbar hardware is sometimes called _____ hardware.
    a. entrance
    b. lever handle
    c. panic
    d. passage latch

Trade Terms Introduced in This Module

Access: A passageway or a corridor between rooms.

Astragal: A piece of molding attached to the edge of an inactive door on a pair of double doors. It serves as a stop for the active door.

Butt: Any kind of hinge, except a strap or T hinge.

Butt gauge: A marking gauge normally used to mark the depth and width of mortises for butts.

Casing: The exposed finish material around the edge of a door or window opening.

Catches: Spring bolts used to secure a door when shut.

Coordinator: A device for use with exit features to hold the active door open until the inactive door is closed.

Cylindrical lockset: A lock in which the keyhole and tumbler mechanism are in a cylinder that is separate from the lock case and can be removed to change the keying of the lock.

Deadbolt: A square-head bolt in a door lock that requires a key to move it in either direction.

Door closer: A mechanical device used to check a door and prevent it from slamming when it is being closed. It also ensures the closing of the door.

Door frame: The surrounding case of a door into which a door closes. It consists of two upright pieces called jambs and a horizontal top piece called the head.

Door jamb: See *door frame*.

Door stop: The strip against which a door closes on the inside face of the frame or jamb. It can also be a hardware device used to hold the door open to any desired position or a hardware device placed against the baseboard to keep the door from marring the wall.

Dustproof strike: A metal piece used to receive the latch bolt when a door is closed. The piece has a spring action cover to keep particles out of the strike when the door is open.

Finish hardware: The exposed hardware in a building such as doorknobs, door hinges, door locks, door closers, window hardware, shelf and clothing storage hangers, and bathroom hardware.

Finishing sawhorse: A pair of trestles used to support lumber, molding, doors, stair parts, and other materials while they are being fitted and shaped for installation.

Flush bolt: A sliding bolt mechanism that is mortised into a door at the top and bottom edge. It is used to hold an inactive door in a fixed position on a pair of double doors.

Flush door: A door of any size that has a totally flat surface.

Hanging stile: The door stile to which the hinges (butts) are fastened.

Hardware: Fastenings that allow for the movement and attachment of door components.

Head: The horizontal member at the top of a door or window opening.

Hinge: The hardware fastened to the edge of a door that allows the frame to pivot around a steel pin, permitting the operation of the door.

Jamb: One of the vertical members on either side of a door or window opening.

Kerf: A slot or cut made with a saw.

Knob lockset: A lock for a door, with the locking cylinder located in the center of the knob.

Latch bolt: A spring-loaded bolt in a lock, with a beveled head that is retracted when hitting the strike.

Lock: A mechanical hardware device for securing a door in a closed position. This device is usually designed with a bolt operated by a key, combination, or electrical impulse.

Lockset: The entire lock unit, including locks, strike plate, and trim pieces.

Molding: Long strips of material used for finishing and decorative trim.

Mortise: A measured portion of wood removed to receive a piece of hardware such as a lock or butt.

Mortise lock: A rectangular metal box that houses a lock. It usually has a latch and deadbolt as part of the unit. Options consist of a locking cylinder and/or thumb turn by which it can be secured. It is used on residential entry doors and commercial doors.

Panic hardware: Hardware that provides an emergency escape exit.

Plumb: A true perpendicular/vertical position.

Prefinished: Material such as molding, doors, cabinets, and paneling that has been stained, varnished, or painted at the factory.

Prehung door: A door that is delivered to the job site from the mill already hung in the frame or jamb. In some instances, the trim may be applied on one side.

Rabbet: A rectangular groove cut in the edge of a board.

Rail: A horizontal member of a door or window sash.

Rough opening: Any unfinished door or window opening in a building.

Scribe: To mark wood with a sharp knife or scriber.

Shim: A thin, tapered piece of wood such as a wooden shingle used to fill in gaps and to level or plumb structural components.

Sill: The lowest member at the bottom of a door or window opening. It also refers to the lowest member of wood framing supporting the framing of a building.

Smoke gasket: A rubber strip that goes all the way around the door to keep smoke from penetrating an area in case of fire.

Sound attenuation: The reduction of sound as it passes through a material.

Sound transmission class (STC): The rating by which the sound attenuation of a door is determined. The higher or greater the number, the better the sound reduction.

Stile: The vertical edge of a door.

Strike: A metal piece fastened to a door frame into which the bolt of a lock is inserted so the door can be held securely in place. The strike is mortised into the door jamb.

Strike plate: A metal plate screwed to the jamb of a door so that when the door is closed, the bolt of the lock strikes against it. The bolt is then retracted and slides along the metal plate. When the door is fully closed, the bolt inserts itself into a hole in the plate to hold the door securely in place.

Sweep: A type of weatherstripping. A sweep is a felt or rubber flap mounted in a metal channel to seal door bottoms to prevent air infiltration.

Template: A thin piece of material such as plastic or heavy paper with a shape cut out of it, or the whole of it cut to a shape on its perimeter. A template is used to transfer that shape to another object by tracing it with a pencil or scribe.

Threshold: A piece of wood, metal, or stone that is set between the door jamb and the bottom of a door opening.

Transom: A panel above a door that lets light and/or air into a room or is used to fill the space above a door when the ceiling heights on both sides of the door opening allow.

Weatherstripping: Strips of metal or plastic used to keep out air or moisture that would otherwise enter through the spaces between the outer edges of doors and windows and the finish frames.

This module is intended to present thorough resources for task training. The following reference work is suggested for further study. This is optional material for continued education rather than for task training.

Finish Carpentry. Newton, CT: Taunton Press, 1997.

27209-07
Suspended Ceilings

Topics to be presented in this module include:

Overview

Suspended ceilings are found in most commercial buildings. Unlike fixed ceilings, they provide easy access to wiring, cabling, and air conditioning equipment located in the area between the ceiling and the overhead deck. The ceiling panels are designed to suppress sound transmission. Suspended ceilings are built by first installing a grid that is suspended from the overhead deck. This grid must be both level and straight, as any deviation will make installation difficult and is likely to be very visible to an observer. Ceiling panels are then installed in the grid. There are several types of grids and many types and sizes of ceiling panels. Proper installation of these ceilings is a skill that takes training and practice.

Objectives

When you have completed this module, you will be able to do the following:

1. Establish a level line.
2. Explain the common terms related to sound waves and acoustical ceiling materials.
3. Identify the different types of suspended ceilings.
4. Interpret plans related to ceiling layout.
5. Sketch the ceiling layout for a basic suspended ceiling.
6. Perform a material takeoff for a suspended ceiling.
7. Install selected suspended ceilings.

Trade Terms

Acoustical materials
Acoustics
A-weighted decibel
  (dBA)
Ceiling panels
Ceiling tiles
Decibel (dB)
Diffuser
Dry lines
Fissured
Frequency
Hertz (Hz)
Plenum
Striated

Required Trainee Materials

1. Pencil and paper
2. Appropriate personal protective equipment

Prerequisites

Before you begin this module, it is recommended that you successfully complete *Core Curriculum*; *Carpentry Fundamentals Level One*; and *Carpentry Framing and Finishing Level Two*, Modules 27201-07 through 27208-07.

This course map shows all of the modules in *Carpentry Framing and Finishing Level Two*. The suggested training order begins at the bottom and proceeds up. Skill levels increase as you advance on the course map. The local Training Program Sponsor may adjust the training order.

27212-07
Cabinet Fabrication
ELECTIVE

27211-07
Cabinet Installation

27210-07
Window, Door, Floor, and Ceiling Trim

27209-07
Suspended Ceilings
ELECTIVE FOR RESIDENTIAL CERTIFICATE

27208-07
Doors and Door Hardware

27207-07
Drywall Finishing

27206-07
Drywall Installation

27205-07
Cold-Formed Steel Framing

27204-07
Exterior Finishing
ELECTIVE FOR COMMERCIAL CERTIFICATE

27203-07
Thermal and Moisture Protection

27202-07
Roofing Applications
ELECTIVE FOR COMMERCIAL CERTIFICATE

27201-07
Commercial Drawings
ELECTIVE FOR RESIDENTIAL CERTIFICATE

FRAMING AND FINISHING

CARPENTRY FUNDAMENTALS

CORE CURRICULUM: Introductory Craft Skills

209CMAP.EPS

1.0.0 ◆ INTRODUCTION

Suspended ceilings are widely used in commercial construction and to some extent in residential construction. Modern suspended ceilings serve many purposes. They are designed to help keep outside noise from entering the room and to reduce noise levels occurring within the room itself. In some cases, ceilings are integrated with the electrical and HVAC (heating, ventilation, and air conditioning) functions to provide for correct lighting and temperature control. Use of attractive **ceiling panels** or **ceiling tiles** helps give a warm, relaxed feeling to a room. Complete ceiling systems offer a wide variety of options, both functional and visual. The selection of **acoustical materials**, plans for the acoustical ceiling, and the method of ceiling installation depend on the intended use of the room.

There is a wide variety of suspended ceiling systems, with each system being somewhat different from the others. They use the same basic materials, but their appearances are completely different.

The focus of this module is on the following ceiling systems:

- Exposed grid systems
- Metal pan systems
- Direct-hung concealed grid systems
- Integrated ceiling systems
- Luminous ceiling systems
- Suspended drywall furring ceiling systems
- Special ceiling systems

Also covered in this module is background information relevant to **acoustics** and acoustical ceilings, including information on the propagation of sound waves, acoustical ceiling product terminology, and ceiling-related drawings.

2.0.0 ◆ SOUND WAVE PROPAGATION AND CHARACTERISTICS

Sounds travel through air in a room as a series of pressure waves. These sound pressure waves travel outward in all directions. When sound waves strike a wall or ceiling, some of the sound

Look Up

The next time you are out, take a look at the ceilings in the stores, theaters, malls, and other buildings you enter. You will see that the available styles, designs, materials, and color schemes are more numerous than you ever imagined. The one thing they have in common is that most of them are some form of suspended ceiling using panels set into or attached to a framework.

209SA01.EPS

209SA02.EPS

wave energy is absorbed and some is reflected in wave patterns moving in the opposite direction (*Figure 1*). The result is that you will hear both the original sound and its reflected image. The sound is also transmitted through the air in the wall or ceiling cavity to the opposite surface, causing the surface to vibrate and transmit the sound to any adjoining room(s).

Sound waves have a **frequency**. The frequency, or pitch, measured in **hertz (Hz)**, is the number of vibrations or cycles that occur in the wave in one second. The greater the number of cycles per second (Hz), the higher the frequency and the higher the pitch.

The intensity of sound refers to its degree of loudness or softness. An **A-weighted decibel (dBA)** is a unit of measure used for establishing and comparing the intensity of sound sources. It is used to express the value of all sounds in a range from 0 dBA to 140 dBA and higher.

Table 1 shows some typical examples of noise situations and their relative **decibel (dB)** levels. Note that any sounds greater than 120 dBA can actually produce a physical sensation. Sounds above 130 dBA can cause pain and/or deafness.

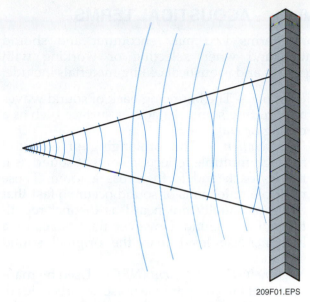

209F01.EPS

Figure 1 ◆ Sound wave reflection.

Table 1 Sound Levels of Some Common Noises

Sound Level (dBA)	Intensity Level	Environment	
		Outdoor	Indoor
140	Deafening	Jet aircraft, artillery fire	Gunshot
130	Threshold of pain	—	Loud rock band
120	Threshold of feeling	Elevated train	Portable stereo headset on high setting
110	Extremely loud	Overhead jet aircraft at 1,000'	Loud nightclub
100	Very loud	Chainsaw, motorcycle at 25', auto horn at 10'	—
90	Loud	Lawn mower, noisy city street	Full symphony band, noisy factory
80	Moderately loud	Diesel truck at 50'	Garbage disposal, dishwasher
70	Average	—	Face-to-face conversation, vacuum cleaner, printers and copiers
60	Moderately quiet	Air conditioning condenser at 15', auto traffic near an interstate highway	Normal conversation, general office
50	Quiet	Large transformer at 50'	—
40	Very quiet	Bird calls	Private office, soft radio music
30	Extremely quiet	Quiet residential neighborhood	Average residence
20	Nearly silent	Rustling leaves	Quiet theater, whisper
10	Just audible	—	Human breathing
0	Threshold of human hearing	—	One's own heartbeat in a silent room

3.0.0 ◆ ACOUSTICAL TERMS

Some terms you may encounter and should understand when selecting or working with acoustics and acoustical ceiling materials include:

- *Reflection* – The bouncing back of sound waves after hitting some obstacle or surface such as a ceiling or wall.
- *Reverberation* – The prolonging of a sound through multiple reflections of that sound as it travels back and forth across a room. These multiple reflections of sound occur so fast that they are usually not heard as distinct repetitions of that sound. However, they can cause a higher noise level than the original sound source.
- *Noise reduction coefficient (NRC)* – Used by manufacturers to compare the noise absorbencies of acoustical products. The higher the number, the better the absorbency. The NRC measures the average percentage of noise a material absorbs at four selected frequencies.
- *Sound transmission loss* – The amount of sound lost as a noise travels through a material. Acoustical ceiling assemblies are rated in terms of sound transmission classification (STC). An STC value of 20 to 25 indicates that even normal speech can be easily understood in an adjoining room. On the other hand, an STC value of 50 to 60 indicates that loud sounds will be heard only faintly or not at all. Acceptable STC ratings range from approximately 39 to 65.
- *Articulation class (AC)* – Rates a ceiling's ability to achieve normal privacy in open office spaces by absorbing noise reflected at an angle off the ceiling into adjacent areas (cubicles). Per the *ASTM E1110* and *E1111* standards, the generally accepted AC ratings for normal privacy in open plan offices is a minimum of 170, with 190 to 210 preferred.
- *Ceiling attenuation class (CAC)* – Rates a ceiling's efficiency as a barrier to airborne sound transmission between adjacent work areas, where sound can penetrate **plenum** spaces and travel to other spaces. CAC is stated as a minimum value. Per *ASTM Standard E1264*, CAC minimum 25 is acceptable in an open plan office, while a rating of CAC minimum 35 to 40 is preferred for closed offices.
- *Absorption* – The energy of sound waves being taken in (entering) and absorbed by a surface of any material rather than being bounced off or reflected.

4.0.0 ◆ READING BLUEPRINTS

In order to avoid mistakes and/or omissions when installing ceilings, an organized and systematic approach should be used when reading the related blueprints. The following is a general procedure for reading blueprints.

Step 1 Check the room schedule on the blueprints.
- Identify the type of material to be used.
- Locate the protrusions into the ceiling.

Step 2 Locate the room on the floor plan.
- Find the room dimensions. If none are found, locate the drawing scale.
- Using the given scale, determine the dimensions of the room.

Step 3 Check to see if there is a reflected ceiling plan. If no reflected ceiling plan is found, check to see if a shop drawing is included. *Appendix A* contains an example of a reflected ceiling plan.

Step 4 Be sure the blueprints are the final revised set.
- Blueprints are often revised several times before the ceiling is ready to be installed. To be sure the blueprints are the final revised set, check the date of issuance against the work order to see if they are the same.
- If work that has already been done is not reflected on the blueprints, chances are the blueprints are not the final revised set. This could have a significant impact on the ceiling installation.

Step 5 Read the specifications.
- Be sure the ceiling to be installed is the same as that listed in the specifications.
- If the job conditions do not agree with what is shown in the specifications and/or plans, call your department superintendent and ask for instructions on how to proceed.

Step 6 Check the mechanical and electrical plans prior to the layout of the ceiling.
- On the mechanical plans, locate the air **diffusers** (HVAC air supply outlets), return grilles, ducts, and sprinkler heads.
- On the electrical plans, locate the light fixtures, fans, and other ceiling protrusions.

Plenum Ceilings

The systems that provide heating and cooling for most commercial buildings are forced-air systems. Blower fans are used to circulate the air. The blower draws air from the space to be conditioned and then forces the air over a heat exchanger, which cools or heats the air. In a cooling system, for example, the air is forced over an evaporator coil that has very cold refrigerant flowing through it. The heat in the air is transferred to the refrigerant, so the air that comes out the other side of the evaporator coil is cold. In homes, the air is delivered to the conditioned space and returned to the air conditioning/heating system through ductwork that is usually made of sheet metal. In commercial buildings with suspended ceilings, the space between the ceiling and the overhead decking is often used as the return air plenum. It is often called an open plenum. (A plenum is a sealed chamber at the inlet or outlet of an air handler.) This approach saves money by eliminating about half the cost of materials and labor associated with ductwork.

One thing to keep in mind is that anything in the plenum space (electrical or telecommunications cable, for example) must be specifically rated for plenum use in order to meet fire ratings. Plastic sheathing used on standard cables gives off toxic fumes when burned. Plenum-rated cable uses non-toxic sheathing.

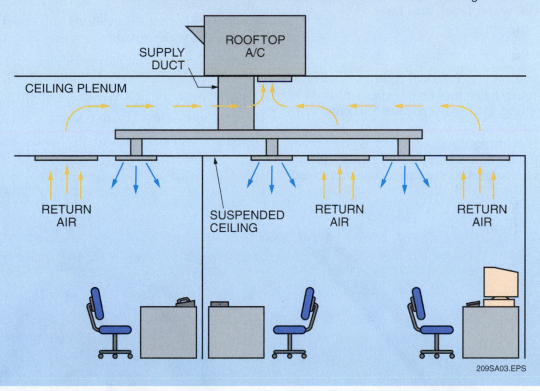

209SA03.EPS

On some large construction jobs, there may be a set of reflected ceiling plans. These show the details of the ceiling as though it were reflected by a mirror on the floor. This view shows features of the ceiling while keeping those features in proper relation to the floor plan. For example, if a pipe runs from floor to ceiling in a room and is drawn in the upper left corner of the floor plan, it is also shown on the upper left corner of the reflected ceiling plan of that same room. Reflected ceiling plans show in detail how the ceiling will be constructed. The plans will indicate the following:

- Layout (direction) of the ceiling panels or tiles
- Location of the center (starting) line
- Size of the borders
- Position of the light fixtures
- Location of the air diffusers

As a rule, reflected ceiling plans also show all items that penetrate the ceiling, including:

- Return grilles
- Diffusers
- Sprinkler heads
- Recessed lights
- Recessed speakers for sound systems

Other information is also used when constructing a ceiling. Your employer may prepare shop drawings that show in detail just how the ceiling should be installed and also indicate the finished appearance. These drawings provide insurance against errors in the details of installation.

5.0.0 ◆ CEILING LEVELING EQUIPMENT

In the installation of suspended ceilings, various types of leveling devices are used to find the level plane of a ceiling. These devices include the carpenter's level, water level, and laser.

5.1.0 Level

In the application of acoustical ceilings, a level, as shown in *Figure 2*, is a tool that comes in handy to check the ceiling grid main runner and cross runner installation.

The most commonly used level is a 6' level. When using a level to install a suspended ceiling, place it at right angles to the runners as you install

them. If the tool is perfectly level, the bubble will appear centered between the cross hatches. The leveling should be checked every 6'.

5.2.0 Water Level

By using the water level shown in *Figure 3*, a level ceiling can be installed even if the floor is not level. Bench marks or elevation points can be established to be used as a starting point for the ceiling.

A water level is a long, clear hose, filled with water and stopped or clamped at both ends. Before using the water level, unplug both ends and tie one end to the rung of the ladder or scaffold; hold the other end at the same height. Pick up the hose approximately 3' from the U or bottom and allow the air bubbles to escape at either end. Once this is complete, it is a true water level. Always do this prior to using the water level. If this is not done, a true and accurate reading will not be obtained.

The following suggestions will assist you in the use of the water level:

Step 1 Remove the stoppers (clamps) from each end.

Step 2 Place a bench mark on a wall near a corner at eye level (5' to 6').

Step 3 Hold one end of the water level against this mark.

209F02.EPS

Figure 2 ◆ Level.

209F03.EPS

Figure 3 ◆ Water level.

Step 4 Have someone take the other end of the water level to the far end of the same wall and hold it at approximately the same height.

Step 5 Watch the bench mark and the water level as it is being raised or lowered at the other end.

Step 6 When the water level at the far end of the wall is at the same height as the bench mark, a level point has been found. Proceed to locate the other level points on the other walls.

5.3.0 Laser Leveling Devices

Many types of laser leveling devices are available that can be used to aid in the installation of suspended ceilings. To use a laser to level a ceiling, follow the manufacturer's instructions for the laser being used. Generally, the procedure involves mounting the laser either on a wall/ceiling mount or on a tall tripod. The laser beam is rotated either at the finished ceiling height or at a reference point. A special target is snapped to a grid, then the grid is moved up or down until the laser beam crosses the target's offset mark. The grid is then secured in place.

A procedure for leveling a ceiling that uses a laser attached to a wall/ceiling mount is given in this section. A laser mounted on a tall tripod can also be used.

WARNING!
OSHA has published *Standard 1926.54* for the safe use of lasers. Some guidelines are:

- Avoid direct eye exposure to the laser beam.
- Only qualified and trained personnel should operate laser equipment.
- Place a standard laser warning sign conspicuously at major approaches to the instrument use area.
- Always turn off the laser when transmission of the beam is not required.

Step 1 Attach the laser to the ceiling mount (*Figure 4*). Align the laser so that the rotating head is at the same end as the securing screws. Secure the laser firmly on the ceiling mount using the securing wheel.

Step 2 Determine the height of the ceiling to be installed. Attach a 15" or longer scrap piece of wall angle about 3" above the determined ceiling height.

Step 3 Slip the slot in the ceiling mount over the wall angle flange and secure the mount by tightening the securing screws (*Figure 5*).

Step 4 Attach the power cable to the laser.

MOUNTING BRACKET

LASER LEVEL

209F04.EPS

Figure 4 ◆ Laser instrument wall/ceiling mount.

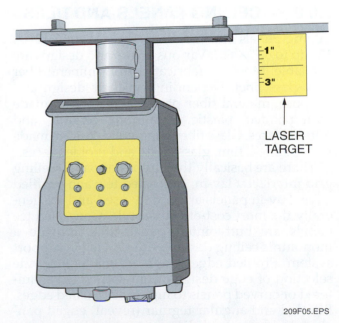

1"

3"

LASER TARGET

209F05.EPS

Figure 5 ◆ Wall/ceiling mount attached to wall angle flange.

Step 5 Plug the battery eliminator into a 110VAC outlet. Attach the alligator clips on the power cable to the battery eliminator (red on red and black on black). Switch the battery eliminator on.

Step 6 Turn the mode switch to bypass and the speed adjustment to 0.

Step 7 Attach a laser target to the wall angle (*Figure 5*).

Step 8 Loosen the securing knob for height adjustment and center the laser beam on the white center line on the target.

Step 9 Turn the mode switch to auto and adjust the rotating head speed for best visibility.

Step 10 Using the center line on the laser target as your height reference, lay out the walls for the wall angle. As the laser revolves, the height of the laser target line can be penciled onto the other walls and a chalkline used to establish the wall angle heights on all four walls. Always put the chalkline at the position of the top of the molding.

Step 11 Once the border lines (wall angle locations) have been established, the laser can be removed and used at another location on the job site. However, the laser can also remain in position and be used instead of the string (dry line) while installing the ceiling grid system. The laser is more accurate than the string method once you become familiar with the instrument.

6.0.0 ◆ CEILING PANELS AND TILES

Ceiling panels and tiles range in size from 12" × 12" up to 60" × 60". Various colors and designs are available. Most are fabricated from mineral fiber and glass fiber. Depending on their design and purpose, mineral fiber panels and tiles are made with painted, plastic, aluminum, ceramic, and mineral faces. Glass fiber panels and tiles are made with painted, film, glass cloth, and molded faces.

There are basically three types of panel/ceiling grid interfaces: lay-in, concealed tee, and profiled edge. Lay-in panels are widely used and are generally the most cost-effective style. Concealed tee panels are butt-jointed panels that provide a monolithic ceiling design with no visible support system. Profiled edge panels are made in a wide selection of edge designs from soft-edged chamfered or curved panels to highly articulated edges. Beveled and angular tegular (reveal) edged panels provide a three-dimensional look in a suspended ceiling.

Many interior spaces have specific requirements for ceiling panel materials and characteristics. These can include sound control, fire resistance, thermal insulation, light reflectance, moisture resistance, maintenance, appearance, and cost considerations. High-performance acoustical panels are used to help prevent noise in open plan and closed types of offices. Three factors contribute to noise distractions in a workplace:

• *General office noise* – The ability of a ceiling material to absorb general office noise is measured using a value known as the noise reduction coefficient (NRC), which is used by panel manufacturers to compare the noise absorbency of their ceiling panel products. The higher the number, the better the absorbency. The NRC measures the average percentage of noise a panel absorbs at various frequencies.

• *Reflected conversational noise that angles off ceilings into adjacent cubicles* – The ability of a ceiling material to absorb reflected conversational noise is measured using a value known as the articulation class (AC), which rates a ceiling's ability to achieve normal privacy in open office spaces by absorbing noise reflected at an angle off the ceiling into adjacent areas (cubicles). Per *ASTM E1110* and *E1111*, the generally accepted AC ratings for normal privacy in open plan offices is a minimum of 170, with 190 to 210 preferred.

• *Sound transmission through cubicles, partitions, walls, and ceilings* – The ability of a ceiling material to absorb sound transmission is measured using a value known as the ceiling attenuation class (CAC), which rates a ceiling panel's efficiency as a barrier to airborne sound transmission between adjacent work areas, where sound can penetrate plenum spaces and carry to other spaces. The CAC is stated as a minimum value. Per *ASTM E1264*, CAC minimum 25 is acceptable in an open plan office, while a rating of CAC minimum 35 to 40 is preferred for closed offices.

Local building codes require fire-safe construction for many building applications. Fire-resistant ceiling panels and support systems are specially made of materials that provide increased resistance against flame spread, smoke generation, and/or structural failure in the event of a fire. Two ratings based on ASTM, ANSI, and NFPA standards are used to evaluate fire-resistant panels: the flame spread rating of the material (*ASTM E84*) and the fire resistance rating of a ceiling assembly (*ANSI/UL 263*, *ASTM E119*, and *NFPA 251*).

Fire-Rated Applications

Suspended ceiling systems can be used in fire-rated applications. Factors such as resistance to fire and flame spread for the panels, as well as the suspension system are taken into account. However, the ceiling materials alone do not determine the fire rating. Rather, the materials and construction methods used in the entire system, including the floor/ceiling or ceiling/roof assembly, all enter into the fire rating determination. If a suspended ceiling is to be part of a fire-rated system, you must consult the manufacturer's product literature to determine the ceiling panel and grid that must be used to meet the rating.

Basically, the flame-spread rating of a ceiling material is the relative rate at which a flame will spread over the surface of the material. This rate is compared against a rating of 0 (highest rating) for asbestos-cement board and a rating of 100 for red oak. Class A ceilings have flame spread ratings of 25 or less, the required standard for most commercial applications. The fire-resistance rating of a ceiling assembly represents the degree to which the entire assembly, not the individual components, withstands fire and high temperatures (measured in hours). Specifically, it is an assembly's ability to prevent the spread of fire between spaces while retaining structural integrity.

High humidity resistance panels are panels that have superior resistance to sagging caused by high humidity conditions. Many also inhibit the growth of mold or mildew that may appear in high humidity conditions. Sagging not only diminishes the attractiveness of a ceiling, but it also causes ceilings to chip and soil more easily and reduces the high light reflectivity of the panels. High humidity resistance panels are typically installed in high humidity climates or areas such as kitchens, locker rooms, shower areas, and indoor pools; buildings where the HVAC systems may be shut down for extended periods; or where the ceiling might be installed early in the construction process before the building is fully enclosed.

Ceiling panels with a high light reflectance (LR) value of 0.83 or greater per *ASTM E1477* help increase effective lighting levels and reduce light fixture costs and energy consumption, especially with indirect lighting systems. Their use also helps to reduce eyestrain. These panels typically have soil-resistant surfaces that stay cleaner longer than standard ceilings, resulting in a much lower loss in light reflectance over time.

Detailed information regarding the various panel characteristics described above can normally be found in ceiling manufacturers' product catalogs and literature. When selecting panels for a particular application, it is best to follow the manufacturer's recommendations.

7.0.0 ◆ EXPOSED GRID SYSTEMS

The following sections describe the components, tools and materials, and a general procedure for installing an exposed grid ceiling system, also called a direct-hung system.

High-Durability Ceilings

Highly durable ceiling panels provide for long life and easy maintenance wherever ceilings are subjected to improper use, vandalism, or frequent removal for plenum access. They are used in applications where resistance to impact, scratches, and soil are major considerations. Ceilings in areas such as school corridors or gymnasiums need to withstand abuse, including surface impact. In any areas where lay-in ceiling panels frequently need to be removed for plenum access, scratch-resistant panels are highly desirable. Otherwise, the panel surface can be scratched, scuffed, or chipped as it is slid across the metal suspension system components. Ceilings installed in laboratories, clean rooms, and food preparation areas are normally required to meet special standards to ensure that they can withstand repeated cleaning.

"Seeing" Through the Ceiling

INSIDE TRACK

It is a good idea to keep a package of colored thumbtacks in your toolbox when installing a suspended ceiling. When you know that other trades may have to get back in to complete their work, you can insert different colored thumbtacks in ceiling panels to mark the locations of electrical and mechanical services. The marking will allow the other trades to locate their components without having to raise a lot of panels, and will therefore help minimize finger smudges and damage to the panels.

> **NOTE**
>
> The procedures given in this module are general in nature and are provided as examples only. Because of the differences in components made by different manufacturers, it is important to always follow the manufacturer's installation instructions for the specific system being used.

7.1.0 Tools Required

The following tools are required when installing an exposed grid ceiling system:

- Aviation snips (tin snips)
- Clamping pliers or vise grips with plastic or rubber corners
- Chalkline
- Dry line
- 50' or 100' tape
- Goggles
- Hammer
- Helmet
- Awl
- Keyhole saw
- Lath nippers
- Level
- Magnetic punch
- Scribes or compass
- Plumb bob
- 6' folding rule
- Straightedge for cutting
- Board
- Tile knife
- Ladders
- Laser leveling unit
- Pop-rivet gun
- Powder-actuated fastening tool
- Scaffold
- Special dies for cutting suspension members
- Water level
- Whitney punch

7.2.0 Materials Required

For an exposed grid suspended ceiling, a light metal grid is hung by wires attached to the original ceiling. Panels that are usually 2' × 2' or 2' × 4' are then placed in the frames of the metal grid. Exposed grid systems are constructed using the components and materials described as follows and shown in *Figure 6*.

- *Main runners* – Primary support members of the grid system for all types of suspended ceiling systems. They are 12' in length and are usually constructed in the form of an inverted T. When it is necessary to lengthen the main runners, they are usually spliced together using extension inserts; however, the method of splicing may vary with the type of system being used.
- *Cross runners (cross ties or cross tees)* – Inserted into the main runners at right angles and spaced an equal distance from each other, forming a complete grid system. They are held in place by either clips or automatic locking devices. Typically, they are either 4' or 2' in length and are usually constructed in the form of an inverted T. Note that 2' cross runners are only required for use when using 2' × 2' ceiling panels.
- *Wall angles* – Installed on the walls to support the exposed grid system at the outer edges.
- *Ceiling panels* – Panels that are laid in place between the main runners and cross ties to provide an acoustical treatment. The acoustical panels used in suspended ceilings stop sound reflection and reverberation by absorbing sound waves. These panels are typically designed with numerous tiny sound traps consisting of drilled or punched holes or fissures, or a combination of both. When sound strikes the panel, it is trapped in the holes or fissures. A wide variety of ceiling panel designs, patterns, colors, facings, and sizes are available, allowing most environmental and appearance demands to be met. Panels are typically made of glass or mineral fiber. Generally, glass fiber

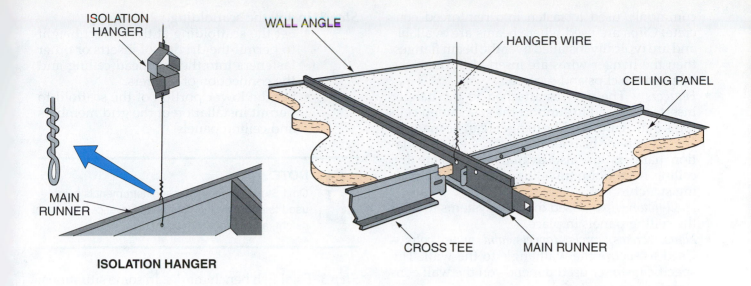

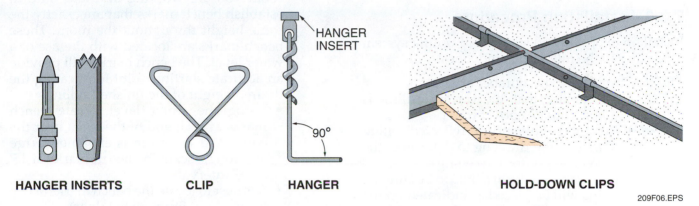

ISOLATION HANGER

HANGER INSERTS **CLIP** **HANGER**

HOLD-DOWN CLIPS

209F06.EPS

Figure 6 ◆ Exposed grid system components.

panels have a higher sound absorbency than mineral fiber panels. Panel facings are typically embossed vinyl in a choice of patterns such as fissured, pebbled, or striated. The specific ceiling panels used must be compatible with the ceiling suspension system due to variations in manufacturers' standards.

- *Hanger inserts and clips* – Many types of fastening devices are used to attach the grid system hangers or wires to the building ceiling located above the suspended ceiling. Screw eyes and star anchors are commonly used and require an electric hammer for installation. Powder-actuated fastening tool eye pin fasteners are also

What's in a Name?

The terms ceiling panel and ceiling tile have specific meanings in the trade. Ceiling panels are typically any lay-in acoustical board that is designed for use with an exposed mounting system. They do not have finished edges or precise dimensional tolerances because the exposed grid system support members provide the trimout. Ceiling tiles are acoustical ceiling boards, usually 12" × 12" or 12" × 24", which are nailed, cemented, or suspended by a concealed grid system. The edges are often kerfed and cut back.

commonly used to fasten into reinforced concrete. Clips are used where beams are available and are typically installed over the beam flanges, then the hanger wires are inserted through the loops in the clips and secured.

- *Hangers* – The devices attached to the hanger inserts and used to support the suspended ceiling's main runners. The hangers can be made of No. 12 wire or heavier rod stock. Ceiling isolation hangers are also available to isolate the ceiling from noise traveling through the building structure.
- *Hold-down clips* – Used in some systems to hold the ceiling panels in place.
- *Nails, screws, rawls, toppets, and molly bolts* – Used to secure the wall angle to the wall. The specific fastener used depends on the wall construction and material.

7.3.0 Installation Procedure

Install an exposed grid suspended ceiling system according to the following guidelines:

Step 1 Check the room number and location.
- Make sure that the correct ceiling is going into the correct room.
- Refer to the blueprints, reflected ceiling plan, or shop drawing to determine the correct height for the ceiling.
- Check the electrical prints to be sure the lights will be placed as indicated on the reflected ceiling plan or shop drawing.
- Check the specification sheets or shop orders for special instructions to ensure that you have the proper hangers and fasteners.

WARNING!

Scaffolds must be used and assembled in accordance with all local, state, and federal/OSHA regulations. OSHA requires the building, moving, or dismantling of all scaffolding to be supervised by a competent person who has the training, knowledge, and experience to identify hazards on the job site and the authority to eliminate them. Mobile scaffolds should only be used on level, smooth surfaces that are free of obstructions and openings. OSHA regulations also require that mobile scaffold casters have positive locking devices to hold the scaffold in place. When moving a mobile scaffold, apply the moving force as close to the scaffold base as possible to avoid tipping it over. Never move a scaffold when someone is on it.

Step 2 Install the scaffolding.
- Set the scaffolding at the correct height to permit the driving of inserts or other fasteners into the overhead ceiling and the connection of hangers.
- Set the lower portion of the scaffold to permit installation of the grid members and ceiling panels.

NOTE

Omit Steps 3 and 4 if a laser instrument is being used to establish the correct and level ceiling height, as previously described.

Step 3 Establish bench marks. In some situations, the floor may not be level in all areas of the room you are working in. Make sure to establish bench marks that are exactly the same height throughout the room. These bench marks are located with the use of a water level. The bench marks will provide an accurate starting point for locating the desired height of the finished ceiling.
- Using the water level, locate bench marks at each end of the room near the corners. If the room is extremely large and long, locate the bench marks at 15' to 20' intervals.
- It is best to locate the bench marks at eye level (about 5' above the floor).
- Always put bench marks on each wall and also on protruding walls. It is better to have too many than not enough.

Step 4 Establish the height for the top of the wall angle.
- Once all bench marks are located and marked, establish a measurement to the bottom of the wall angle.
- To that measurement, add the height of the wall angle.
- Measure from each bench mark and establish a mark at the top of the wall angle. Place a mark at various intervals on the wall.
- Snap a chalkline on those marks. This will establish the top of the wall angle. Be sure to snap the chalkline at the top of the wall angle so that the chalkline will not be visible when the wall angle is installed.

7.3.1 Installation of the Wall Angle

When installing the wall angle:

- Be sure that the top of the wall angle is even with the chalkline at all points.
- Nail or screw the wall angle to the wall. The wall angle should fit securely to prevent sound leaks.
- Miter the wall angle to fit at the corners or use corner caps to cover the joints.

NOTE

The following procedure assumes that the center lines have been laid out on the floor. The procedure would be done in a similar manner if the center lines were laid out on the ceiling. The only difference is that the center lines would be transferred down from the ceiling, instead of up from the floor, to the level of the suspended ceiling grid.

Install the wall angle according to the following procedure:

Step 1 Square the room by first establishing length and width center lines on the floor or ceiling, as appropriate. One procedure for doing this is given in *Appendix B*.

Step 2 Transfer the positions of the center lines from the floor/ceiling to the level required for the suspended ceiling grid as follows:

- At either one of the long walls, use a plumb bob to locate the center of the wall just above the wall angle. Do this by moving the plumb bob until it is directly over the center line marked on the floor (*Figure 7*). Then, mark the wall just above the wall angle flange and insert a nail behind the wall angle at this point. Repeat the procedure at the opposite long wall. Run and secure a taut dry line between the nails in the two long walls.

- Repeat the procedures above for the two short walls and run a second dry line perpendicular to the first.
- Make a final check using the plumb bob at all four points.

When completed, you will have two **dry lines** intersecting at right angles at the center of the ceiling. Use the 3-4-5 method to ensure that the intersection of the dry lines is square. This is important because these intersecting dry lines will be used as the basis for all subsequent measurements used to lay out the ceiling grid system.

7.3.2 Room Layout

In some cases, reflected ceiling plans or shop drawings may indicate a center line and specify border width. In these instances, follow the suggested layout. If you have neither to follow, it will be necessary for you to establish the center line and plan the layout of the ceiling so that the border panels adjacent to the facing wall are the same width and are not less than one half the width of a full panel.

One method for determining the width of the border panels is to convert the wall measurement from feet to inches, then divide this amount by the width of the ceiling panels. For example, assume the room measurements are 42'-6" × 30'-6". What will be the width of the border panels running parallel to the two 42'-6" walls if 2' × 2' panels are being used? To find the answer, proceed as follows:

Step 1 Take the measurement of the short wall (30'-6") and convert feet to inches.

$$30' \times 12" = 360" + 6" = 366"$$

Step 2 Divide that amount by 24" (the width of a tile).

$$366" \div 24" = 15 \text{ tiles with a remainder of } 6"$$

Indoor Swimming Pools

INSIDE TRACK

All-aluminum grid systems are not recommended for use above indoor swimming pools because chlorine gases cause aluminum to corrode.

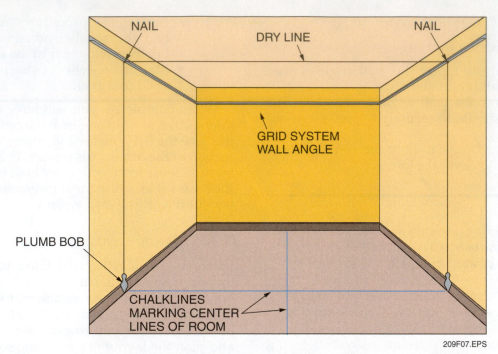

Figure 7 ◆ Transferring room center lines to the height of the suspended ceiling grid.

Step 3 If the division does not result in a whole number, add the width (in inches) of a full board to the remainder (in this case, it is 6").

$$6" + 24" = 30"$$

Step 4 Divide this by 2; the result is 15. This would be the width of a border panel on each side. When adding the width of a panel to the remainder in Step 3, you must delete one full panel from the total. This means that there would be 14 full panels plus a 15" border on each end (*Figure 8*).

$$14 \times 2' + 30" = 30'\text{-}6"$$

Step 5 Determine the width of the border panels for the other two remaining walls in the same manner.

7.3.3 Hanger Inserts

Once all border unit measurements have been established, it is necessary to install eye pins to secure the hangers. Prior to installing the eye pins, you must locate the position of the main runners and cross runners. Line A in *Figure 9* represents the main runner. Line B represents the first cross runner line.

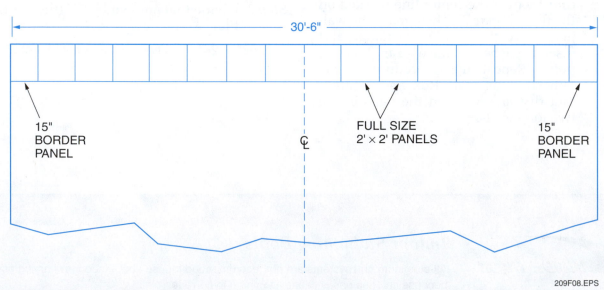

Figure 8 ◆ Fitting the panels to the room.

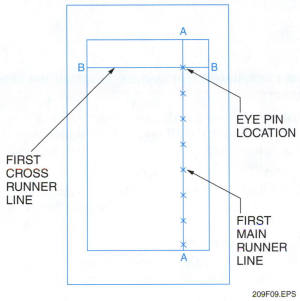

A

B ——— B

EYE PIN
LOCATION

FIRST
CROSS
RUNNER
LINE

FIRST
MAIN
RUNNER
LINE

A

209F09.EPS

Figure 9 ◆ Locating the positions of the first main runner and cross runner.

Step 1 Locate the first dry line (A) as shown in *Figure 9*. The dry line should be installed approximately 1⅛" above the flange of the wall angle. The dry line will also be used to indicate the bend in the hanger wire.

NOTE

Most main runners measure 1⅛" from the bottom of the runner to the hole for the hanger wire.

Step 2 Fasten a second dry line (B) so that it intersects line A at right angles with the proper measurement for the border tiles.

Step 3 The intersection point will be the location of the first hanger wire. Set the first eye pin directly over the intersection of dry lines A and B. Install the other eye pins every 4' on center. Be sure to follow dry line A as a guide for the locations.

WARNING!

Powder-actuated fastening tools are to be used only by trained operators in accordance with the operator's manual. Operators must take precautions to protect both themselves and others when using powder-actuated tools:

- Operate the tool as directed by the manufacturer's instructions and use it only for the fastening jobs for which it was designed.
- To prevent injury or death, make sure that the drive pin cannot penetrate completely through the material into which it is being driven.
- To prevent a ricochet hazard, make sure the recommended shield is in place on the nose of the tool.

As you install each eye pin, suspend the hanger wire from the eye pin. Allow enough wire to ensure proper tying to the eye pin and the main runner. These two ties will require from 10" to 24" of wire. More hanger wire may be required if the ceiling is hung in part from structural members along with the eye pins. All hanger wire is precut and can be obtained in various lengths and gauges. Refer to the specifications for the gauge of hanger wire to use. There are cases in which there may be cast-in-place inserts or hanger wires. In such cases, the insertion of eye pins and installation of wire will not be necessary.

7.3.4 Marking and Bending Hanger Wire

The eye pins and hanger wires should be located on 4' centers in both directions. In most cases, it is best to hang both at the same time, as it requires less movement of the scaffold. To mark and bend the wires, proceed as follows:

Step 1 Mark the hangers where they touch the dry line.

Step 2 Twist the wire using side cutters and bend the other end to the mark (*Figure 10*).

NOTE

At least three turns should be made when twisting the wire. The wire should also be tight to the runner.

7.3.5 Installing the Main Runners

Install the main runners as follows:

Step 1 Measure and cut the main runner so that the cross runner will be at the proper distance for the border panel.

Step 2 Suspend the first length of main runner from the first row of hangers. It is important that the first few hanger wires be straight (perpendicular to the main runner). If they are not, use a Whitney punch and make new holes.

NOTE

As indicated previously, in most exposed grid systems, the main runners come prepunched with holes. The upper edges of the holes generally measure 1⅛" from the bottom of the runner; the dry line will have to be adjusted prior to bending the hanger wire.

Step 3 Continue to install the balance of the first row of main runners; splices will be needed. Note that various types of splices are used with various grid systems and are supplied by the manufacturer of the grid system.

Step 4 After installing the last full length of main runner, measure, cut, and install an end piece from a full length of main runner to complete the first run.

Step 5 After each end of the main runner is resting on the wall angle, insert all hanger wires and twist the wires to secure them in place (*Figure 11*).

7.3.6 Installing the Cross Runners (Cross Tees)

Use the following procedure to install the cross runners:

Step 1 Measure the distance from the main runner to the wall angle. This width was determined earlier when completing the room layout.

Step 2 Using the snips, cut the cross runner to the correct border width. Be sure to cut and save the correct end or it will not slide into the main runner.

Step 3 Insert the factory end into the main runner and let the other end rest on the wall angle (*Figure 12*).

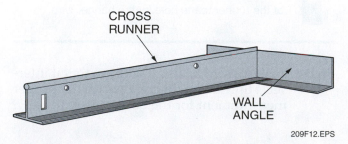

209F11.EPS

Figure 11 ◆ Inserting hanger wires in the main runner.

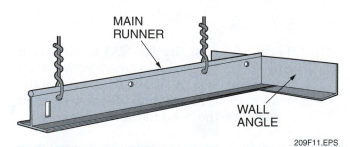

209F12.EPS

Figure 12 ◆ End of cross runner resting on wall angle.

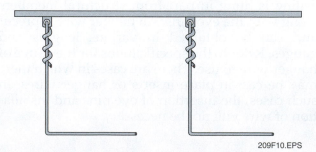

209F10.EPS

Figure 10 ◆ Bending hanger wire.

Hold-Down Clips

Some ceiling panels require hold-down clips to secure the ceiling panels to the grid. For example, clips are used with lightweight panels to prevent them from reacting to drafts. One manufacturer specifies that clips be used if the panels weigh less than one pound. Hold-down clips are not necessarily required for ceilings used in fire-rated applications. Check the manufacturer's instructions.

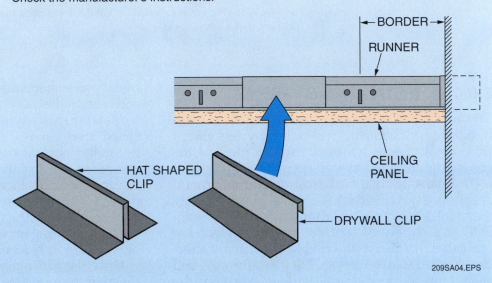

209SA04.EPS

Step 4 When the main runner and the cross runner are perpendicular to each other, lock the cross runner in place. Install the remaining cross runners between the main runners.

Step 5 In order to stabilize the grid system, install pop rivets at every other cross runner.

> **NOTE**
>
> Some grid systems use automatic locking devices to secure the cross runners into the main runners. Other grid systems use clips. Once all cross runners have been installed into the main runners, the grid system is ready to receive the ceiling panel or tile.

7.3.7 *Cutting and Placing Ceiling Panels*

Prior to installing the ceiling panels, it is necessary to wash your hands in order to keep the panels clean. It is a good practice to wear white gloves when handling ceiling panels. Start by cutting the border panels and inserting them into position. After the border panels are installed, proceed to install the full panels. As the full pan-

els are being placed, install the hold-down clips, if required. General installation guidelines include the following:

- Install the ceiling panels in place according to the job progression. Do not jump or scatter panels in the grid system.
- Exercise care in installing panels so as not to damage or mar the surface.
- Always handle panels at the edges and keep your thumbs from touching the finished side of the panels. If gloves are not worn, use cornstarch, powder, or white chalk on your hands.
- If the ceiling panel has a deeply textured pattern, insert it into the grid system so that the directional pattern will flow in the same direction. Such matching of ceiling panels adds beauty to the finished suspended ceiling.

Upon completion of the ceiling, clean up the work area as follows:

Step 1 Dismantle the scaffold.

Step 2 Pick up all tools and equipment.

Step 3 Secure all equipment in a safe place.

Step 4 Pick up all trash.

Step 5 Sweep up any dust or debris.

8.0.0 ◆ METAL PAN SYSTEMS

The metal pan system is similar to the conventional suspended acoustical ceiling system except that metal tiles or pans are used in place of the conventional sound absorbing panel (*Figure 13*).

The pans are made of steel or aluminum and are generally painted white; however, other colors are available by special order. Pans are also available in a variety of surface patterns. Tests have indicated that metal pan ceiling systems are effective for sound absorption. They are durable and easily cleaned and disinfected. In addition, the finished ceiling has little or no tendency to have sagging joint lines or drooping corners. The metal pans are die-stamped and have crimped edges, which snap into the spring-locking main runner and provide a flush ceiling.

The tools, room layout, and installation of hanger inserts, hangers, and wall angle for the metal pan system are basically the same as for the conventional exposed grid suspended ceiling.

Metal pan systems use 1½" U-shaped, cold-rolled steel furring channel members. They are normally installed 4' on center. However, it may be necessary to install them at lesser distances depending on the location of the light fixtures. The furring channels are installed by looping the hanger wires around them. Twist and secure using saddle wire. It is important that the furring channels are properly leveled to ensure a level ceiling.

Once the furring channels have been installed, the main runners (tee bars) must be placed. These bars are manufactured, cold-rolled or zinc-coated steel or aluminum. They are fastened to the furring channels using special clips. The tee bars run at right angles to the furring channels (*Figure 14*).

The tee bar has a spring-locking feature, which grips the metal pan. This feature allows the pan to be removed for access to the area above the pan. If

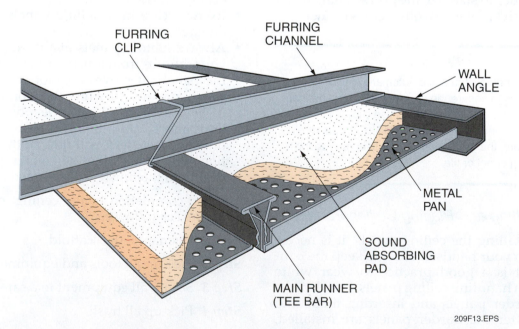

209F13.EPS

Figure 13 ◆ Metal pan ceiling components.

the room plan dimensions are longer than the length of the tee bar, use a tee bar splice to couple the tee bars together, extending them to the required length (*Figure 15*).

Follow the procedure below for installing the tee bars:

Step 1 Install the first tee bar at right angles to the 1½" furring channel. Be sure it is correctly aligned.

Step 2 When in position, place tee bar clips over the furring channel and insert the ends under the flange of the tee bar. Hang additional clips in the same manner along the length of the bar.

Step 3 Install the second tee bar parallel to the first one. The spacing should be from 1' to 4', depending on the size of the metal pans.

After installing all the tee bars in one section of the room (depending on the working area of your scaffold), begin installing the metal pans. Take care in handling the pans. Use white gloves or rub

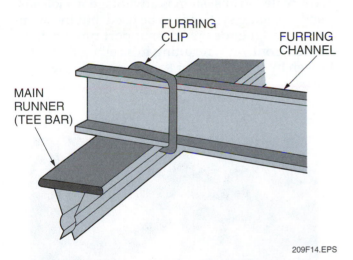

Figure 14 ◆ Main runner (tee bar) clipped to a furring channel.

FURRING CLIP

FURRING CHANNEL

MAIN RUNNER (TEE BAR)

209F14.EPS

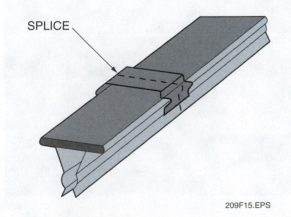

SPLICE

209F15.EPS

Figure 15 ◆ Tee bar splice.

your hands with cornstarch to prevent any perspiration or grease marks from marring the surface of the pans. If care is not taken, fingerprints will be plainly visible when the units are installed. Use the following procedure to remove the pans from their container:

Step 1 Place the pans finished surface face down on some type of raised platform, such as a table, which has been covered with a pad to protect the surface of the pan.

Step 2 Insert the wire grid into the back side of the metal pan. The wire grid is installed between the metal pan and backing pad to provide an air cushion between the two surfaces.

Step 3 Over the grid, place the paper- or vinyl-wrapped mineral wool or fiberglass batt or pad.

NOTE

In some cases, the metal pans come with the wire grid and pad already assembled.

When installing the metal pans, begin at the perimeter of the ceiling, next to the walls, since the units must fit into the channel wall angle. Follow the procedure below:

Step 1 From the room layout or reflected ceiling plan, obtain the width of the border units. Measure the pan to this width.

Step 2 Using a band saw, cut the pan ⅛" short of the desired width along the edge to be inserted into the wall angle.

Step 3 Slide the pan into the wall angle. Do not force the pan all the way into the molding. Leave ⅛" for expansion.

Step 4 Insert two crimped edges of the metal pan into the spring-locking tee bars.

Step 5 When the pan is in position, insert the spacer clip into the channel molding cavity and over the cut edge of the pan. This will prevent the pan from buckling along this cut edge.

Step 6 Cut the wire grid and backing pad to fit the border unit. Place the unit into position.

Step 7 At the corners of the room, install the pans in the order shown in *Figure 16*.

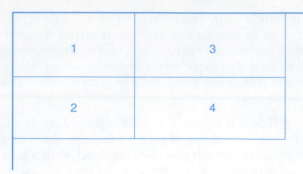

Figure 16 ◆ Panel layout.

Figure 18 ◆ Metal pan ceiling.

Once the perimeter pans are installed in each row, the full size pans can be put into place as follows:

Step 1 Grasping the pan at its edges, force its crimped edges into the tee bar slots. Use the palms of your hands to seat the pan.

Step 2 After installing several of the pans as noted above, slide them along the tee bars into their final position. Use the side of your closed fist to bump the pan into level position if it does not seat readily.

Step 3 If metal pan hoods are required, slip them into position over the pans as they are installed. The purpose of the hood is to reduce the travel of sound through the ceiling into the room.

If a metal pan must be removed, a pan pulling device is available (*Figure 17*). To pull out a pan, insert the free ends of the device into two of the perforations at one corner of the pan and pull down sharply. Repeat this at each corner of the pan. By following this removal procedure, there is no danger of bending the pan out of shape. *Figure 18* shows a finished metal pan ceiling.

9.0.0 ◆ DIRECT-HUNG CONCEALED GRID SYSTEMS

A concealed grid system is advantageous if a suspended ceiling system is to be used, but the architect or designer desires the support runners to be hidden from view, resulting in a ceiling that is not broken by the pattern of the runners (*Figure 19*).

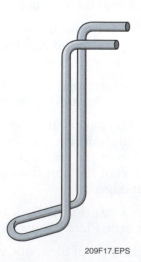

Figure 17 ◆ Pan removal tool.

Figure 19 ◆ Concealed grid system.

The tiles used for this system are similar in composition to conventional tile, but are manufactured with a kerf on all four edges. Kerfed and rabbeted 12" × 12" and 12" × 24" tiles are used with this system. Tiles of various colors and finishes are available. Refer to *Figure 20* for a diagram of the components in a typical concealed grid system.

Step 1 The installation of the concealed grid system begins in the same way as the conventional exposed grid system previously discussed. Once chalklines have been established on the walls at the required height above the floor, the next step is to install the wall angle. This molding, as for the other ceiling systems, provides support for the grid and tile at the wall. The molding should be fastened with nails, screws, or masonry anchors. At corners, miter the inside corner and use a cap to cover the outside corner.

Step 2 Lay out the grid and install the hanger inserts and hangers according to the reflected ceiling plan or lighting layout. Once this has been completed, install the concealed main runners as done for the exposed grid system. The main runners are the primary support members. The method of coupling them to attain a specific length will vary with the manufacturer of the system, but they can all be spliced to the desired length.

Step 3 Install the cross stabilizer bars and concealed cross tees at right angles to the main runners. They rest on the flange of the main tee runners.

Step 4 Place the ceiling tile into position. Cut in the first row of tile at a line perpendicular to the main runners, as previously established. Attach the tile to the wall using spring clips.

Flat anti-breather splines (*Figure 21*) are metal or fiber units that are inserted by the installer into the unfilled tile kerfs between the concealed main runners. Splines are used to prevent dust from seeping through the kerfs in adjacent tiles.

If access is needed to the area above the ceiling, special systems are available, which can be incorporated into the ceiling (*Figure 22*).

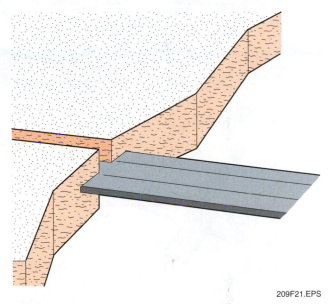

209F21.EPS

Figure 21 ◆ Flat anti-breather spline.

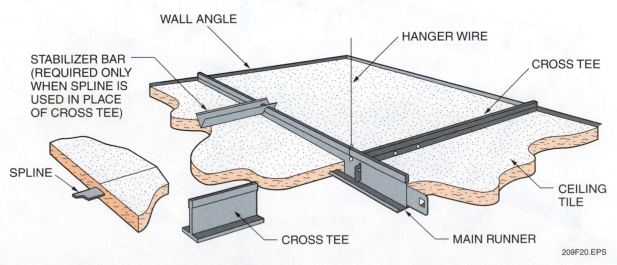

209F20.EPS

Figure 20 ◆ Direct-hung concealed grid system components.

Facts About Ceiling Panels

- Ceiling panels are often called pads.
- Certain lighting fixtures are manufactured in the same sizes as panels to drop directly into suspended ceilings.
- Some panels are treated to inhibit the growth of mold, mildew, fungi, and certain bacteria.
- Special panels are designed for use below sprinkler systems. These panels will shrink and fall out of the grid as the room temperature rises, allowing the water to reach the fire.
- Most ceiling panels can be recycled.

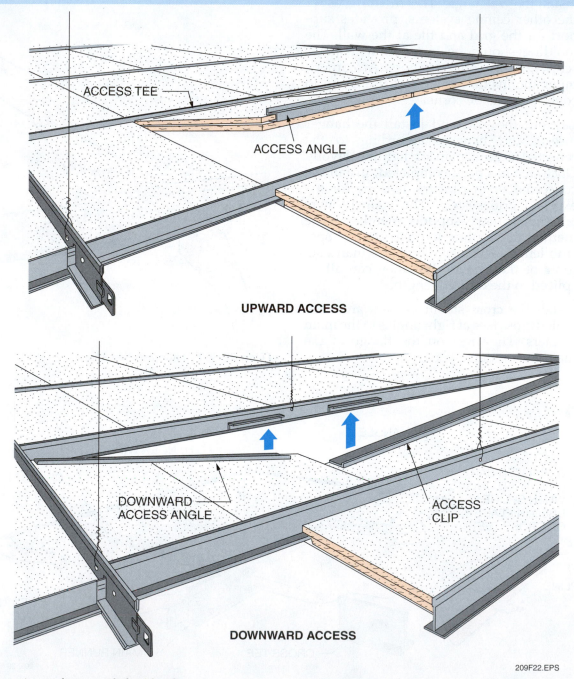

209F22.EPS

Figure 22 ◆ Access for concealed grid ceilings.

10.0.0 ◆ INTEGRATED CEILING SYSTEMS

As noted by its name, the integrated ceiling system incorporates the lighting and/or air supply diffusers as part of the overall ceiling system, as shown in *Figures 23* and *24*.

Some manufacturers of ceiling materials offer specialty ceiling designs. Many integrated ceilings can be used to create distinctive architectural appearances and interior artistic designs. Complete ceiling systems offer dozens of options, both functional and visual.

The functional aspect allows for enhanced lighting and lighting effects. Mechanically, it can incorporate the air supply system through spaced supply air diffusers and return air grilles, all of which have been designed to go beyond function to enhance the artistic appearance of the ceiling.

Integrated ceiling systems are available in units called modules. The common sizes are 30" × 60" and 60" × 60". The dimensions refer to the spacing of the main runners and cross tees.

11.0.0 ◆ LUMINOUS CEILING SYSTEMS

Luminous ceiling systems (*Figure 25*) are available in many styles, such as exposed grid systems with drop-in plastic light diffusers or an aluminum or wood framework with translucent acrylic light diffusers.

Fluorescent fixtures are generally installed above the translucent diffusers. Standard modules of 2' × 2' up to sizes of 5' × 5' are available. It is also possible to purchase custom sizes for special fit conditions. There are two types of luminous ceilings: standard and non-standard. Standard systems are, as their name indicates, those that are available in a series of standard sizes and patterns. Non-standard systems differ in that they deviate from the normal spacing of main supports and may include unusual panel sizes, shapes, and configurations.

All surfaces in the luminous space including pipes, ductwork, ceilings, and walls are painted with a 75 to 90 percent reflectance matte white finish.

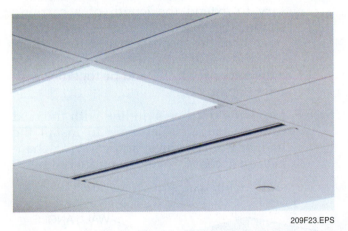

209F23.EPS

Figure 23 ◆ Integrated grid system.

209F25.EPS

Figure 25 ◆ Luminous ceiling system.

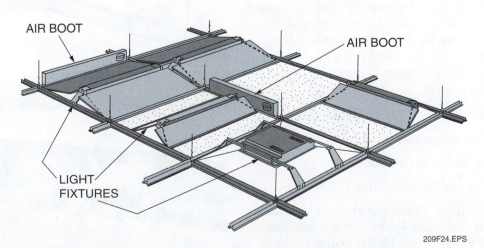

209F24.EPS

Figure 24 ◆ Integrated ceiling schematic.

Any surfaces in this area which might tend to flake, such as fire proofing and insulation, should receive an approved hard surface coating prior to painting to prevent flaking onto the ceiling below.

11.1.0 Installation of Standard Systems

The procedure for installing a standard luminous system is the same as for the exposed grid suspended system with the exception of the border cuts. Luminous ceilings are placed into the grid members in full modules. Any remaining modules are filled in with acoustical material cut to size.

With a 2' × 2' or 2' × 4' standard exposed grid system, use luminous panels, which provide the light diffusing element in the system. These panels are laid in between the runners. There are many sizes and shapes of panels available.

11.2.0 Installation of Non-Standard Systems

The procedure for installing a non-standard luminous ceiling is the same as for the standard system with reference to room layout, hanger insert, installation of hanger wire, main supports, secondary supports, and attachment of some type of wall angle.

Since the non-standard ceiling can deviate so much in terms of size, intricacies of the system, and the exactness of the installation, shop drawings are required. More information on non-standard ceilings can be obtained from ceiling manufacturers.

12.0.0 ◆ SUSPENDED DRYWALL FURRING CEILING SYSTEM

The suspended drywall furring system is used when it is desirable or specified to use a drywall finish or drywall backing for an acoustical tile ceiling.

A general procedure for installing a suspended drywall furring ceiling system is given below.

12.1.0 Installing Carrying Channels

To install carrying channels, proceed as follows:

Step 1 Snap chalklines on all walls.

Step 2 Install the wall angle, if used.

Step 3 Hang dry lines. From the ends of two facing walls, measure 4' along the chalklines. Secure the dry lines to the walls at these two points.

Step 4 Fasten the hanger inserts into the ceiling structure above. Using the dry line as a guide, install a row of hanger inserts 4' on center, or as specified, into the ceiling over the dry line.

Step 5 Install the hangers. Use No. 8 gauge wire or as specified. Hang the wires from the hanger inserts, then twist and secure. Allow enough wire to tie the hanger to the insert and the carrying channel. Generally, 24" to 28" of wire is adequate.

Step 6 Mark where the dry line touches the hanger wires. Mark each hanger wire where it is to be bent. Bend the wire to a 90° angle.

Step 7 When all the hanger wires in one row have been marked and bent, fasten the carrying channel to the hanger wires (*Figure 26*). Loop the wires around the carrying channel at the point where the wires have been bent. Twist and fasten the wires using a saddle tie. It is important that the carrying channels are installed level. If it is necessary to extend the length of a carrying channel because of the room size, face the open U of the second channel toward the U of the one already installed. Overlap for a distance of 8" to 12" and tie together with wire (*Figure 27*).

Step 8 Move the dry line 4' in line with the next row of hangers. Continue to install the hanger inserts, wires, and carrying channels on 4' centers over the entire ceiling.

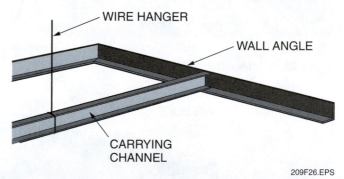

209F26.EPS

Figure 26 ◆ Carrying channel supported by hanger wire.

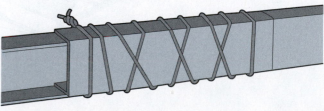

209F27.EPS

Figure 27 ◆ Splicing carrying channels.

12.2.0 Installing Furring Members and Drywall (Gypsum Board)

After the carrying channel support members are installed, proceed with the installation of the cross furring members. There are different kinds of furring members that can be used at right angles to the carrying channels at 16" on center (maximum). For example, a type called a hat furring channel (*Figure 28*) is used when it is desirable to screw the drywall board to the furring channels with self-tapping screws. Install the furring channels as follows:

Step 1 Install the furring channels perpendicular to the carrying channels. Place the first furring channel in line with the first perpendicular dry line. Attach the furring channel to the carrying channel using tie wire.

Step 2 Continue to install the furring channels 16" on center. If it is necessary to extend the furring channels beyond their normal length, lap one end of the channel over the next piece about 9" and wrap the splice well with tie wire (*Figure 29*). Avoid installing furring channels over positions where any fixtures are to penetrate the ceiling.

Step 3 Screw the drywall to the furring channels (*Figure 30*). Be sure to use screws of sufficient length. The screw length should be long enough to go through the board and extend through the furring channel about ¼" to ⅜".

12.3.0 Installing Acoustical Tile

If acoustical tile is to be used as the finish over the drywall, proceed as follows:

Step 1 Lay out the room. Measure the length of one of the two shorter walls. Divide this in half. Mark this halfway point on the board next to this point on the wall. Next, measure the length of the facing wall at the opposite end of the room. Divide this wall length in half and mark this point on the board.

Step 2 Stretch a chalkline across the room and line it up with the two marks on the board. Snap the line.

Step 3 Divide the chalkline in half and set a second half line perpendicular to the first one at this midpoint.

Step 4 Install the ceiling tile per the manufacturer's instructions beginning at the junction of the two center lines.

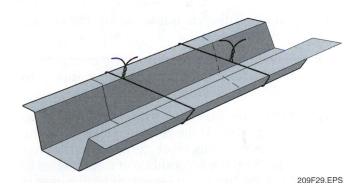

209F29.EPS

Figure 29 ◆ Splicing (lapping) furring channels.

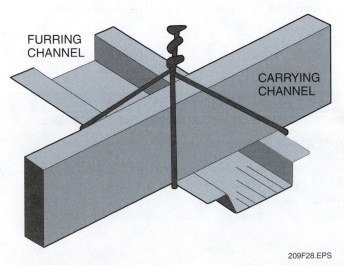

209F28.EPS

Figure 28 ◆ Hat furring channel.

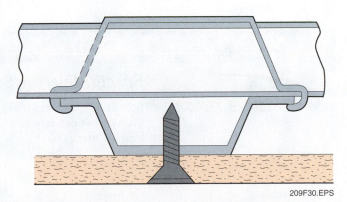

209F30.EPS

Figure 30 ◆ Drywall screwed into a furring channel.

12.4.0 Installing Furring Channels Directly to Structural Members

Sometimes furring channels are installed directly to structural members such as open-web steel beams or wood joists, instead of suspended carrying channels. In these instances, install the furring channels as follows:

Step 1 Mark the walls to indicate the height to be used to level the ceiling. Set the ceiling height control marks to indicate the position of the bottom edge of the furring channels. Check the level of the structural joists. If they are not level, use the low point as the common level for installing the furring channels.

Step 2 Measure up from the bench marks to the height of the lowest point on the joists. Using the distance of the low joist point minus the height of the furring channel, mark it on all walls to indicate a uniform level for the lower edge of all furring channels. Be sure to measure up from the bench marks to establish this common level on all walls.

Step 3 Establish the chalklines using the ceiling control marks.

Step 4 Install the wall moldings (if used). New chalklines may have to be snapped to guide the installation process.

Step 5 Hang the dry lines by securing them to facing walls at the chalkline level. Run the first line at this level right under the first joist. Run the second line at right angles to the first one at the same height. Position it to indicate where to install the first furring channel.

Step 6 Install the furring channels in the same manner as previously described, with the following exceptions:
- When attaching furring channels to steel joists, tie them together with the appropriate gauge tie wire. Wrap the wire around the two so that it bridges the joist and supports the furring channel on either side of the joists. Tie the wire ends together at the side of the union.
- Shimming may be needed between the joists and the furring channel. Be sure to check the level of the furring channel along its entire length. Shim where needed to correct any deviation from level.

12.5.0 Drywall Suspension Systems

Drywall suspension systems are used to install conventional 2 × 2 and 2 × 4 flat panels, as well as curved and domed ceiling systems. These systems support the use of a variety of panel types, including gypsum drywall, which provides the fire resistance needed in some applications. A variety of suspension hangers are available; examples are shown in *Figure 31*.

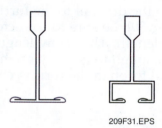

209F31.EPS

Figure 31 ◆ Examples of drywall suspension system hangers.

Special Furring Systems

Some manufacturers make furring system cross tees in 14", 26", and 50" lengths. These sizes can reduce the time it takes to install an F-type lighting fixture from as long as 30 minutes to less than one minute.

13.0.0 ◆ SPECIAL CEILING SYSTEMS

Numerous special ceiling systems differ from those covered in this module. Some of these are special wood systems, special metallic systems, planar systems, and reflective (mirrored) systems (*Figures 32* through 35).

Transparent panels produce a subtly refracted light that extends over all areas of a room's interior (*Figure 36*).

209F34.EPS

Figure 34 ◆ Planar system.

209F32.EPS

Figure 32 ◆ Metallic system.

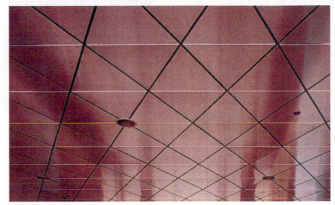

209F35.EPS

Figure 35 ◆ Reflective ceiling.

209F33.EPS

Figure 33 ◆ Pan system.

209F36.EPS

Figure 36 ◆ Transparent panel system.

14.0.0 ◆ LAYING OUT AND ESTIMATING MATERIALS FOR A SUSPENDED CEILING

The estimate of materials for a suspended ceiling should be based on the ceiling plan provided with the architect's drawing set or on a scaled sketch of the ceiling layout. These drawings should show the direction and location of the main runners, cross tees, light panels, and border panels. In a typical suspended ceiling, the main runners are spaced 4' apart and are usually run parallel with the long dimension of the room. For a standard 2' × 4' pattern, 4' cross tees are spaced 2' apart between the main runners. If a 2' × 2' pattern is used, 2' cross tees are installed between the midpoints of the 4' cross tees. Main runners and cross tees should be located in such a way that the border panels on both sides of the room are equal and as large as possible.

If no ceiling plan or sketch is available, you will need to make one in order to determine the required quantity of materials. Sketch a ceiling plan to scale on graph paper. Use a convenient scale; for example, one square equals one square foot. Measure along the ceiling level, including irregular areas such as bays, alcoves, and beams, noting each dimension on the drawing. Then proceed as follows:

Step 1 Determine and sketch the locations of the main runners.
- Convert the dimension for the room width to inches.
- Divide the room width in inches by 48" and add 48" to any remainder.
- Divide the sum by 2 to find the distance from the wall to install the first main runner. Note that this distance is also the length of the border panels or tiles. For example, assume the room has a width of 12'-8" and a length of 20'-8", as shown in *Figure 37*. Changing the room width to inches equals 152". Dividing 152" by 48" yields 3, with a remainder of 8". Adding 48" to 8" equals 56". Dividing 56" by 2 equals 28", the distance from the wall to the first runner. If there is no remainder, the distance from the wall to the first runner is 48".
- Draw a main runner the calculated distance from, and parallel to, the long dimension of the ceiling. Draw the remaining main runners parallel to the first runner at 4' on center. The distance between the last main runner and the

wall should be the same as the distance between the first main runner and the wall.

Step 2 Determine and sketch the locations of the 4' cross tees.
- Convert the long dimension of the room to inches.
- Divide the room length in inches by 24" and add 24" to any remainder.
- Divide the sum by 2 to find the distance in inches from the wall to install the first row of cross tees. Note that this dimension is also the width of the border panels or tiles. For our example room, the length of the room is 20'-8". Changing this dimension to inches equals 248". Dividing 248" by 24" yields 10, with a remainder of 8". Adding 24" to 8" equals 32". Dividing 32" by 2 equals 16", the distance from the wall to the first row of cross tees.
- Draw the first row of cross tees the calculated distance from, and parallel to, the short wall. Draw the remaining rows of cross tees parallel to each other at 2' on center. The distance from the last row of cross tees to the wall should be the same as the distance from the first row of cross tees to the opposite wall.

NOTE

Additional 2' cross tees are required if 2' × 2' panels are used. You will also need one hanger device and wire for about every 4' of main runner.

Step 3 Determine the quantity of ceiling materials. From the ceiling plan or sketch, determine the number of pieces required for the wall angle, main runners, cross tees, and ceiling panels. Normally, main runners come in 12' lengths, cross tees in 4' lengths, and wall angle in 10' lengths.
- Using the ceiling plan or sketch, find the number of main runner sections needed. For our example room (*Figure 37*), six main runner sections are required. This is because main runners are made 12' in length and no more than two pieces can be cut from any one 12' runner. Three main runners, each 20'-8" (20.667') in length, total 62.001'. Therefore, 62.001' ÷ 12' = 5.167 lengths or six lengths when rounded off.

Reflective Ceilings

Reflective ceiling panels can reduce lighting costs because they allow the use of lower-wattage light bulbs.

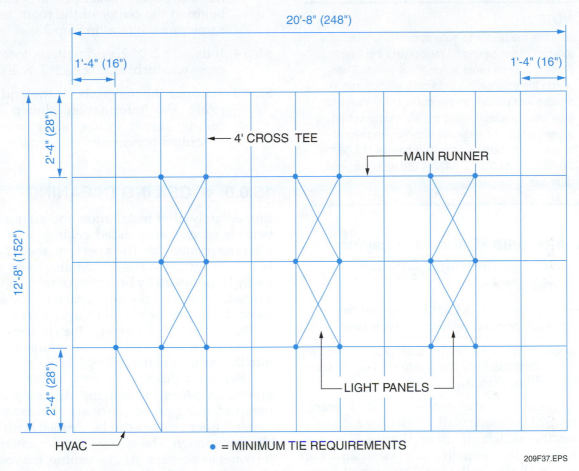

Figure 37 ◆ Completed sketch of a suspended ceiling layout.

- Find the number of 4' cross tees. For our example room, 40 cross tees (10 per row × 4 rows) are required. Note that the border cross tees must be cut from full length cross tees.
- If using 2' × 2' panels or tiles, find the number of 2' cross tees. A 2' × 2' grid is made by installing 2' cross tees between the midpoints of the 4' cross tees. The number of 2' cross tees required for our example room is 44 (11 per row × 4 rows).
- Find the number of sections of wall angle needed. Divide the perimeter of the room [perimeter = (2 × length) + (2 × width)] by 10'. For our example room,

the perimeter is 66'-8" (66.667'). Therefore, seven sections are needed (66.667' ÷ 10' = 6.667 or 7 when rounded off).

- Find the number of ceiling panels or tiles. One method is to count the total number of ceiling panels shown on the ceiling plan or sketch. Note that each border panel requires a full size ceiling panel. Also, subtract one panel for each lighting fixture installed in the ceiling. Assuming the use of 2' × 4' ceiling panels and six light fixtures, our example ceiling would require 38 panels.
- Find the approximate number of hanger wires and hangers needed. Assume one hanger device and wire for about every

4' of main runner. For our example ceiling, approximately 16 hangers are required (62.001' ÷ 4' = 15.5, or 16 when rounded off). Multiply the number of hangers needed by the required length of each hanger wire to find the total linear feet of hanger wire needed.

NOTE

Another method is to determine the total square footage of the ceiling by multiplying the length times the width (area = length × width). Then, divide the total ceiling area by the coverage in square feet printed on the carton of the panels intended for use. If using panels that cover 64 sq ft per carton, our example room ceiling would require 4.09 cartons of ceiling panels (12.667' × 20.667' = 261.789' ÷ 64 = 4.09, or four cartons plus one extra panel).

14.1.0 Alternate Method for Laying Out a Suspended Ceiling Grid System

Another common method for laying out the grid system for a suspended ceiling is given here.

Step 1 Locate the room center line parallel to the long dimension of the room and draw it on the ceiling sketch.

Step 2 Beginning at the center line and going toward each side wall, mark off 4' intervals on the sketch. If more than a 2' space remains between the last mark and the side wall, locate the main runners at these marks. If less than a 2' space remains between the last mark and the side wall, locate the main runners at 4' intervals beginning 2' on either side of the center line. This procedure provides for symmetrical border panels of the largest possible size. Remember to consider the locations of light fixtures and air diffusers in the room.

Step 3 Locate the 4' cross tees by drawing lines 2' on center at right angles to the main runners. To obtain border panels of equal size, begin at the center of the room using the same procedure as in Step 2.

Step 4 If using a 2' × 2' grid pattern, locate the 2' cross tees by bisecting each 2' × 4' module.

Step 5 Estimate the materials for the grid system using the information shown on the ceiling sketch in the same manner as described previously.

15.0.0 ◆ CEILING CLEANING

Immediately after installation and after extended periods of use, suspended ceilings may require cleaning to maintain their performance characteristics and attractiveness. Dust and loose dirt can easily be removed by brushing or using a vacuum cleaner. Vacuum cleaner attachments such as those designed for cleaning upholstery or walls do the best job. Make sure to clean in one direction only as this will prevent rubbing of the dust or dirt into the surface of the ceiling.

After loose dirt has been removed, pencil marks, smudges, or clinging dirt may often be removed using an ordinary art gum eraser. Most mineral fiber ceilings can be cleaned with a moist cloth or sponge. The sponge should contain as little water as possible. After washing, the soapy film should be wiped off with a cloth or sponge slightly dampened in clean water. Vinyl-faced fiberglass ceilings and mylar-faced ceilings can be cleaned with mild detergents or germicidal cleaners.

Cleaning Ceiling Panels

The materials and methods used for cleaning ceiling panels vary widely, depending on the finish and texture of the panels. One manufacturer prescribes eight different cleaning methods for their line of ceiling panels. It is very important to check the manufacturer's instructions before proceeding. When cleaning grids, the ceiling panels should be removed to prevent cleaning solution or dirt from getting on the panels.

16.0.0 ◆ INSTALLATION GUIDELINES

Observing the following guidelines will help you to achieve the desired level of professionalism when installing ceilings:

- All workmanship should meet the highest standards in accordance with the Ceiling and Interior Systems Contractors' Association policy of upgrading quality.
- Ceiling panels should be so arranged that units less than one-half width do not occur unless otherwise directed by the reflected ceiling plans or job conditions.
- All panels, tile joints, and exposed suspension systems must be straight and in alignment.
- All acoustical ceiling systems must be level to ⅛" in 12'.

- Tile must be neatly scribed against butting surfaces and to all penetrations or protrusions where moldings are not required.
- Tile surrounding recessed lights and similar openings must be installed with a positive method to prevent movement or displacement of the tiles.
- Tiles must be installed in a uniform manner with neat hairline-fitted joints between adjoining tiles.
- Wall moldings must be firmly secured, the corners neatly mitered, or corner caps used, if preferred.
- The completed ceiling must be clean and in undamaged condition.

Summary

Although other trades may be responsible for ceiling installations on some jobs, the current trend is for carpenters to do the work. This module covered the major types of acoustical suspended ceiling systems. It is important that you not only understand how the various systems differ from each other, but also how they are alike. In addition, the basic procedures for installing suspended ceiling systems should be understood.

Notes

Review Questions

1. Sound waves travel _____ from their source.
   a. on a line-of-sight
   b. in all directions
   c. vertically
   d. at right angles

2. The intensity of a sound wave refers to _____.
   a. the number of wave cycles it completes in a second
   b. its loudness
   c. its frequency
   d. both its loudness and softness

3. A sound level of 100 dBA is considered _____.
   a. very loud
   b. moderately loud
   c. loud
   d. quiet

4. The term used by manufacturers to compare the noise absorbency of their materials is the _____.
   a. ceiling attenuation class (CAC)
   b. articulation class (AC)
   c. noise reduction coefficient (NRC)
   d. sound transmission classification (STC)

5. On most large construction jobs, there is often _____ ceiling plan supplied in the blueprint set.
   a. a projected
   b. a reflected
   c. a reversed
   d. no

6. Each of the following is a rule for the safe use of lasers when leveling a ceiling *except* _____.
   a. only qualified and trained persons should operate laser equipment
   b. always turn off the laser beam when its use is not required
   c. maintain direct eye contact with the laser beam
   d. place a warning sign at approaches to the area where a laser is being used

7. In an exposed grid ceiling system, hanger wire is used to support the _____.
   a. 4' cross tees
   b. 2' cross tees
   c. wall angle
   d. main runners

8. An exposed grid ceiling system uses the _____ as the primary support members of the grid system.
   a. main runners
   b. 4' cross tees
   c. wall angle
   d. 2' cross tees

9. Acoustical panels and tiles used in suspended ceilings stop the transmission of unwanted sounds by _____ the sound waves.
   a. reflecting
   b. reverberating
   c. absorbing
   d. refracting

10. When laying out eye pins for hanger inserts, two dry lines are used to determine the _____.
    a. exact center of the room
    b. positions of all cross ties
    c. center line of the first ceiling panel
    d. location of the first pin

11. When installing the metal pans in a metal pan ceiling system, begin by installing them _____.
    a. in the middle of the ceiling
    b. at the perimeter of the ceiling next to the walls
    c. at the corners of the room
    d. at the most convenient location

12. Splines installed into the unfilled tile kerfs in a direct-hung concealed grid ceiling system _____.
    a. prevent buckling of the ceiling
    b. are used in place of stabilizer bars
    c. prevent shifting of the tiles
    d. prevent dust from seeping through the kerfs in adjacent tiles

13. What size panel (module) is commonly used in an integrated ceiling system?
    a. 12" × 12"
    b. 12" × 24"
    c. 24" × 24"
    d. 30" × 60"

14. All ceilings, walls, pipe, and ductwork in the space above a luminous ceiling will normally be painted with a _____ paint.
    a. light refracting
    b. 75 to 90 percent reflective matte finish
    c. 90 to 100 percent reflective matte finish
    d. luminous matte finish

15. When installing furring channel members for a suspended drywall ceiling system directly to steel beams or wood joists above the ceiling, _____ can be used against the beams or joists as needed to correct any deviations from level.
    a. shims
    b. spacers
    c. main runners
    d. leveling rods

16. With a suspended ceiling that is 10' × 16', at what distance from the wall will the first main runner be installed if 2' × 4' ceiling panels are to be used with the main runners parallel with the ceiling's long dimension and the 4' length of the panels parallel with the ceiling's short dimension?
    a. 18"
    b. 24"
    c. 36"
    d. 48"

Questions 17 through 20 refer to the ceiling described in Question 16.

17. How many 12' main runners are required for the ceiling?
    a. 2
    b. 3
    c. 4
    d. 5

18. How many 4' cross tees are required for the ceiling?
    a. 19
    b. 20
    c. 21
    d. 22

19. Assuming two light fixtures are used for the ceiling, how many 2' × 4' ceiling panels are required?
    a. 18
    b. 20
    c. 22
    d. 24

20. What are the dimensions (width and length) of the border ceiling panels to be installed along the long dimension of the ceiling?
    a. 18" × 24"
    b. 24" × 30"
    c. 24" × 36"
    d. 36" × 48"

Trade Terms
Introduced in This Module

Acoustical materials: Types of ceiling tile, plaster, and other materials that have high absorption characteristics for sound waves.

Acoustics: A science involving the production, transmission, reception, and effects of sound. In a room or other location, it refers to those characteristics that control reflections of sound waves and thus the sound reception in the area.

A-weighted decibel (dBA): A single number measurement based on the decibel but weighted to approximate the response of the human ear with respect to frequencies.

Ceiling panels: Any lay-in acoustical board that is designed for use with an exposed grid mounting system. Ceiling panels normally do not have finished edges or precise dimensional tolerances because the exposed grid mounting system provides the trimout.

Ceiling tiles: Acoustical ceiling boards that are nailed, cemented, or suspended by a concealed grid mounting system. The edges are often kerfed and cut back.

Decibel (dB): A unit that is used to express differences in sound power. In acoustics, it is equal to ten times the logarithm of the ratio of one sound and a lower-intensity reference sound. One decibel indicates a difference of about 26% and is about the smallest change the ear can detect.

Diffuser: An attachment for duct openings in air distribution systems that distributes the air in wide flow patterns. In lighting systems, it is an attachment used to redirect or scatter the light from a light source.

Dry lines: A string line suspended from two points and used as a guideline when installing a suspended ceiling.

Fissured: A ceiling panel or ceiling tile surface design that has the appearance of splits or cracks.

Frequency: Cycles per unit of time, usually expressed in hertz (Hz).

Hertz (Hz): A unit of frequency equal to one cycle per second.

Plenum: A chamber or container for moving air under a slight pressure. In commercial construction, the area between the suspended ceiling and the floor or roof above is often used as the HVAC return air plenum.

Striated: A ceiling panel or ceiling tile surface design that has the appearance of fine parallel grooves.

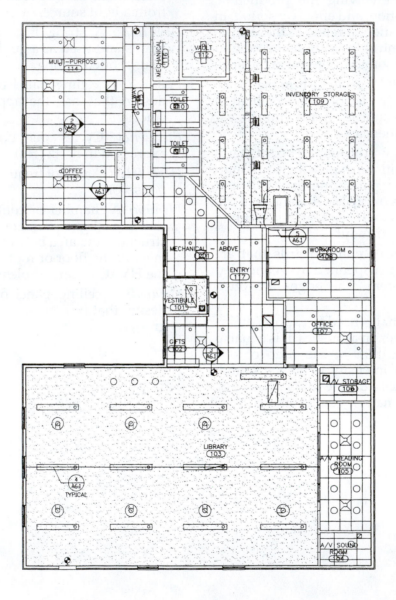

REFLECTED CEILING PLAN

SCALE: 1/8" = 1'-0"

MAIN FLOOR AREA: 5890 SF
MECHANICAL ROOM: 324 SF

209A01.EPS

Establishing Room Center Lines

In order for a grid system to be installed square within a room, it is necessary to lay out two center lines (north-south and east-west) for the room. When correctly laid out, these center lines will intersect each other at right angles at the exact center of the room. If the ceiling of the room located above the proposed suspended ceiling is solid and flat, such as with a plaster or drywall ceiling, then the center lines can be laid out on the ceiling. If the ceiling is not flat, such as an open ceiling with I-beams or joists, the center lines can be laid out on the floor. In either case, the center lines that are laid out on the ceiling or floor can be transferred down from the ceiling, or up from the floor, to the level of the suspended ceiling by use of a plumb bob and dry lines. The procedure given here describes one common method for establishing the center lines in a rectangular room.

Step 1 Measure and mark the exact center of one of the short walls in the room. Repeat the procedure at the other short wall.

Step 2 Snap a chalkline on the ceiling (or floor) between these two marks (see *Figure B-1*, chalkline A-B). This is the first center line.

Step 3 Measure the length of the room along the chalkline. Find its center, then place a mark on the chalkline at this point (point C).

Step 4 From point C, measure a minimum of 3′ in both directions along the chalkline, then place a mark on the chalkline at these points (points D and E).

Step 5 Drive a nail at point D and attach a string to the nail. Extend the string to the side wall so that it is perpendicular to the chalkline, then attach a pencil to the line.

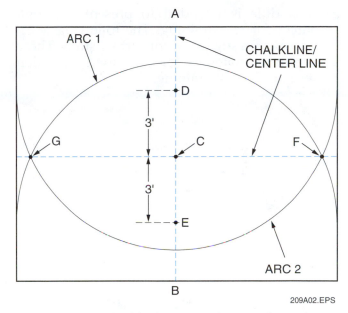

Figure B-1 ◆ Method for laying out the center lines of a room.

Step 6 Making sure to keep the string taut, draw an arc on the ceiling (floor) from the wall, across the chalkline, to the opposite wall (arc 1).

Step 7 Repeat Steps 5 and 6, starting at point E on the chalkline and draw another arc (arc 2).

Step 8 Mark the intersecting points of arcs 1 and 2 on both sides of the chalkline (points F and G).

Step 9 Snap a chalkline from wall to wall on the ceiling (or floor) that passes through points F and G. If done correctly, you should now have two center lines perpendicular to each other that cross in the exact center of the ceiling (floor).

This module is intended to present thorough resources for task training. The following reference works are suggested for further study. These are optional materials for continued education rather than for task training.

CISCA Ceiling Systems Handbook. St. Charles, IL: The Ceiling and Interior Systems Contractors' Association, 1999.

The Gypsum Construction Handbook. Chicago, IL: USG Corporation.

NCCER CURRICULA — USER UPDATE

NCCER makes every effort to keep its textbooks up-to-date and free of technical errors. We appreciate your help in this process. If you find an error, a typographical mistake, or an inaccuracy in NCCER's curricula, please fill out this form (or a photocopy), or complete the online form at **www.nccer.org/olf**. Be sure to include the exact module ID number, page number, a detailed description, and your recommended correction. Your input will be brought to the attention of the Authoring Team. Thank you for your assistance.

Instructors – If you have an idea for improving this textbook, or have found that additional materials were necessary to teach this module effectively, please let us know so that we may present your suggestions to the Authoring Team.

NCCER Product Development and Revision

13614 Progress Blvd., Alachua, FL 32615

Email: curriculum@nccer.org

Online: www.nccer.org/olf

❏ Trainee Guide   ❏ AIG   ❏ Exam   ❏ PowerPoints   Other _____

Craft / Level: _____ Copyright Date: _____

Module ID Number / Title: _____

Section Number(s): _____

Description: _____

_____

_____

_____

_____

Recommended Correction: _____

_____

_____

_____

_____

Your Name: _____

Address: _____

_____

Email: _____ Phone: _____

Windows, Door, Floor, and Ceiling Trim

27210-07

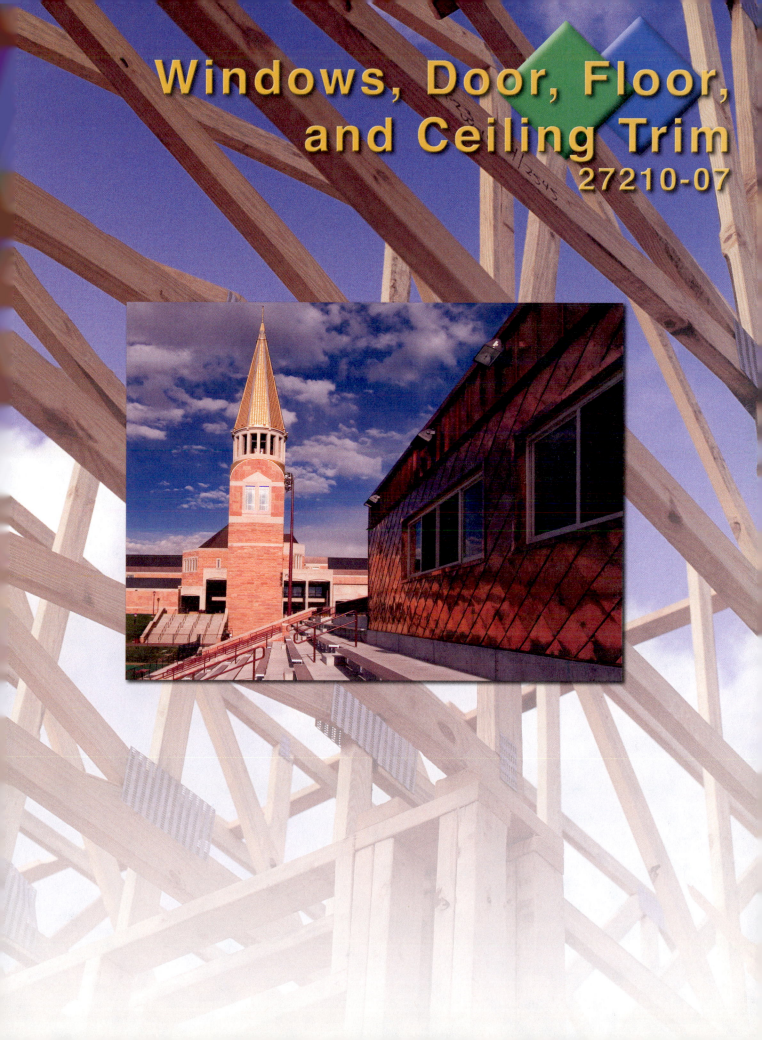

27210-07
Window, Door, Floor, and Ceiling Trim

Topics to be presented in this module include:

Overview

Skilled trim carpenters are always in demand to install floor and ceiling moldings, as well as the finish trim around doors and windows. Quality trim work requires the ability to measure accurately, calculate angles, and make precise, clean cuts. Even a tiny error will be visible in the finished product, so an accomplished trim carpenter is someone who has developed a reputation for careful, accurate work.

Objectives

When you have completed this module, you will be able to do the following:

1. Identify the different types of standard moldings and describe their uses.
2. Make square and miter cuts using a miter box or power miter saw.
3. Make coped joint cuts using a coping saw.
4. Select and properly use fasteners to install trim.
5. Install interior trim, including:
   - Door trim
   - Window trim
   - Base trim
   - Ceiling trim
6. Estimate the quantities of different trim materials required for selected rooms.

Trade Terms

Apron
Compound cut
Coped joint
Finger-jointed stock
Moldings
Reveal

Scarf joints
Square cuts
Stool
Trim
Wainscoting

Required Trainee Materials

1. Pencil and paper
2. Appropriate personal protective equipment

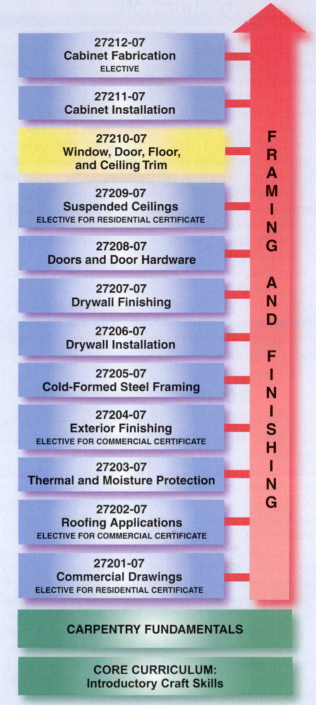

27212-07
Cabinet Fabrication
ELECTIVE

27211-07
Cabinet Installation

27210-07
Window, Door, Floor, and Ceiling Trim

27209-07
Suspended Ceilings
ELECTIVE FOR RESIDENTIAL CERTIFICATE

27208-07
Doors and Door Hardware

27207-07
Drywall Finishing

27206-07
Drywall Installation

27205-07
Cold-Formed Steel Framing

27204-07
Exterior Finishing
ELECTIVE FOR COMMERCIAL CERTIFICATE

27203-07
Thermal and Moisture Protection

27202-07
Roofing Applications
ELECTIVE FOR COMMERCIAL CERTIFICATE

27201-07
Commercial Drawings
ELECTIVE FOR RESIDENTIAL CERTIFICATE

FRAMING AND FINISHING

CARPENTRY FUNDAMENTALS

CORE CURRICULUM:
Introductory Craft Skills

210CMAP.EPS

Prerequisites

Before you begin this module, it is recommended that you successfully complete *Core Curriculum*; *Carpentry Fundamentals Level One*; and *Carpentry Framing and Finishing Level Two*, Modules 27201-07 through 27209-07.

This course map shows all of the modules in *Carpentry Framing and Finishing Level Two*. The suggested training order begins at the bottom and proceeds up. Skill levels increase as you advance on the course map. The local Training Program Sponsor may adjust the training order.

1.0.0 ◆ INTRODUCTION

This module covers the materials, equipment, and procedures used to install trim. Trim is installed to create an attractive finished look and to hide joints at the intersection of walls, floors, ceilings, and different wall coverings. Trim also protects vulnerable edges, holds window sashes in place, and stops the swing of a door in its frame. The specific trim used should harmonize with the overall design of the building or room. For modern residential and commercial buildings, the interior trim typically involves the use of moldings of simple contemporary designs. On the other hand, buildings of traditional design tend to use moldings with more complex shapes and in much greater quantities.

When considering the construction sequence of a building, trim installation is one of the finishing touches. Everybody is going to see the trim; therefore, it has to be installed carefully with tight-fitting joints in order to present a suitable appearance.

2.0.0 ◆ MOLDINGS

Wood moldings are sometimes referred to as woodwork, trim, finish, or millwork. To manufacture a given wood molding profile, lumber of a preselected thickness is sawn or ripped lengthwise into strips of various widths for maximum use of the board. The rips are then sorted by width. Appropriate widths are run through a band saw or resaw where they are converted into shapes called blanks that are ideally suited for specific molding patterns with a minimum of waste. These blanks are sent through a molder, which has four or more cutter heads to shape the profile. The cutter head holds a series of identical knives, each shaped to conform to the exact molding pattern profile. As the blanks are fed into the molder, the cutter heads rotate at high speed, carving or planing the blanks into finished molded shapes. At the outfeed end of the molder, moldings are inspected to ensure conformance with industry quality standards. The finished moldings are then bundled by length, ready for shipment.

Most wood moldings are made of pine because pine can easily be milled into the desired profiles. Premium pine is clear and knot-free and milled for staining. Paint-grade pine moldings, including finger-jointed stock, represent the next step down. Finger-jointed moldings (*Figure 1*) are assembled from shorter lengths. In finger jointing, two additional steps take place between the ripping and resawing operations. First, unacceptable characteristics or defects are trimmed out of lumber rips by crosscutting at the cutoff line. Then, after sorting, the remaining strips of clear wood are passed through a finger jointer, which cuts small fingers in the ends of each piece. These short, fingered pieces are then glued and joined into long lengths of clear finger-jointed lumber, ready to be resawn and molded. Finger-jointed moldings are only used for paint or opaque finishes because the joints show through a stained or natural finish.

Because moldings made of pine can be nicked, dented, or otherwise damaged relatively easily, moldings are also manufactured out of harder woods for use in high-wear areas. Poplar is a clear, slightly more expensive wood that holds an edge better than pine and is a little more resistant to abuse. Moldings are also made out of hardwoods such as oak, cherry, and maple. These moldings are much more expensive than those made of pine. They are also harder to work with than pine because their installation usually requires that all

210F01.EPS

Figure 1 ◆ Finger-jointed molding.

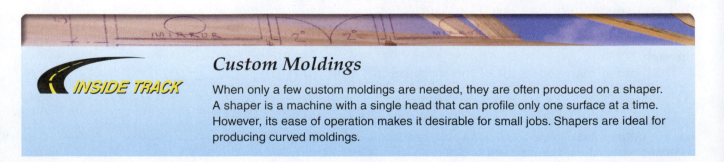

Custom Moldings

When only a few custom moldings are needed, they are often produced on a shaper. A shaper is a machine with a single head that can profile only one surface at a time. However, its ease of operation makes it desirable for small jobs. Shapers are ideal for producing curved moldings.

INSIDE TRACK

Typical Mill Trim Versus Custom Trim

Typical lumberyard or warehouse wood trim must be carefully selected. Much of it is mass-produced by specialty mills that churn out lineal miles of relatively mediocre-grade material. Some inexpensive trim is sanded with shaped abrasive wheels that produce a rounded-edge profile that lacks crispness, as shown in the photo below. Most lumberyard or warehouse trim is softwood trim used only for paint finish and is called paint-grade trim. For natural or stained wood finish, custom millwork shops or yards that carry custom windows and doors should carry the better quality hardwood trim known as stain-grade trim.

CRISP EDGE
PROFILES

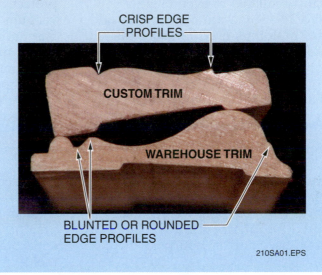

CUSTOM TRIM

WAREHOUSE TRIM

BLUNTED OR ROUNDED
EDGE PROFILES

210SA01.EPS

Polymer Trim

Ornate pre-finished polymer moldings made of polystyrene or other synthetic materials are also available. These cut and nail as easily as pine. Many are glued in place. They will not rot, swell, shrink, or splinter and do not require sanding, priming, or painting.

When cutting polymer trim, place tape over the cut area and cut from the finished side to avoid splintering.

210SA02.EPS

the nail holes be predrilled. Normally, moldings made of hardwoods are used when it is desired to have a natural or stained finish that shows the grain and other characteristics of the hardwood.

Prefinished moldings are often made of medium-density fiberboard (MDF), a stable product that is made of wood fiber. MDF is less likely to split, warp, and cup. For this reason, it is sometimes used in areas where there are big fluctuations in humidity that would cause moldings to gap.

Standard moldings come in a variety of widths and are normally available in even lengths of 8', 10', 12', 14', and 16'. Some are available in odd lengths, such as moldings used for door casing, which are made in 7' lengths to reduce waste. Many types of manufactured molding profiles are available.

Moldings that are to be joined must be carefully selected. Each mill makes moldings of different shapes (profiles), and slight variances occur even in the same profile from the same mill. Make sure that each piece is consistent from end to end and that the profiles of all like pieces to be used in the same room or project are matched. *Figure 2* shows the various places within a room where moldings are typically used for interior trim. The types of moldings widely used for this purpose include the following:

- Base, base cap, and base shoe moldings
- Casing and casing stop moldings
- Crown, bed, and cove moldings
- Quarter round, corner guard, chair rail, and wainscot cap moldings
- Window **stools** and **aprons**

2.1.0 Base, Base Cap, and Base Shoe Moldings

Base moldings (*Figure 3*) are used at the floor level to hide the joint between the wall and floor. They protect the lower wall surfaces from damage by vacuum cleaners, brooms, furniture, and feet. Base moldings are always installed after all door casing moldings are nailed in place. Base moldings come in a variety of sizes and a multitude of contemporary and traditional profiles.

In general, base moldings have an unmolded lower edge, allowing you to add a base shoe molding to conceal any unevenness between the base and floor or to hide edges of carpeting and other flooring. The top profile of the base molding is often decorative, making the use of an additional base cap molding on top unnecessary. A square-edged molding (S4S molding) can be used alone as a simple base molding, or it can be combined with a decorative base cap molding and base shoe molding to create a more traditional look.

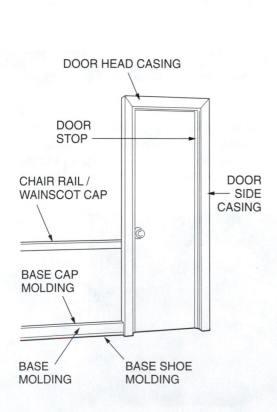

210F02.EPS

Figure 2 ◆ Moldings used to trim the interior of a room.

Figure 3 ◆ Typical baseboard moldings.

5¼"
COLONIAL
BASE

3¼"
COLONIAL
BASE

3"
RANCH
BASE

BASE CAP

S4S

QUARTER ROUND
OR SHOE MOLDING

210F03.EPS

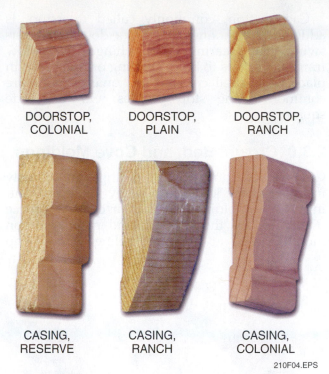

DOORSTOP,
COLONIAL

DOORSTOP,
PLAIN

DOORSTOP,
RANCH

CASING,
RESERVE

CASING,
RANCH

CASING,
COLONIAL

210F04.EPS

Figure 4 ◆ Typical casing and casing stop moldings.

NOTE

Wider base moldings tend to make a room look smaller, so it is best to avoid their use unless the room is spacious. However, a short base is out of proportion in a room with high ceilings, so a wider base should be used in those instances.

2.2.0 Casing and Casing Stop Moldings

Casing moldings (*Figure 4*) come in several widths and are used to trim around windows and doors to cover gaps between the frames (jambs) and the wall. Common profiles include colonial, clam shell, and wedge. Casing profiles often match base moldings, except casing is rounded on both front edges, while the lower edge of a base molding is square to keep dust out of the joint between the molding and the floor.

INSIDE TRACK

Trim Relief

The backside of most baseboard, casing, and chair rail trim molding is relieved to allow the trim to carry over any minor drywall/plaster irregularities. This allows the edges of the trim to lie flat against most surfaces. If major irregularities exist, they must be sanded down to allow the trim to lie as flush as possible against the surfaces. For paint-grade trim, any minor gaps can be filled with a flexible painting caulk.

TRIM RELIEF

210SA03.EPS

Casing stop, commonly called doorstop, is added to door frames to stop the door from over-swinging and tearing out of its hinges. In window frames, it serves to hold a sliding or hung sash in place. Clam shell, colonial, and quarter round are common casing stop profiles with one side square.

2.3.0 Crown, Bed, and Cove Moldings

Crown, bed, and cove moldings are usually installed at the joints formed by walls and ceilings. Cove moldings, with their concave profile (*Figure 5*), are the simplest and most common molding used at the wall/ceiling. A base cap molding is sometimes used beneath a cove molding at the ceiling to create a more intricate built-up profile.

Crown and bed moldings have decorative profiles that provide a more traditional appearance. Crown moldings come in several profiles with the top third portion having a concave profile. The lower side of a crown molding fits against the wall. It has a wider surface than the top side, which rests against the ceiling. Because of their greater widths, these moldings can hide large gaps at the wall and ceiling joint.

Bed moldings have basically the same profiles as crown moldings, except they are usually smaller with a top profile that is rounded.

COVE MOLDINGS

CROWN MOLDINGS

BED MOLDING

FANCY CROWN MOLDING

210F05.EPS

Figure 5 ◆ Cove, crown, and bed moldings.

Hardwood Moldings

Hardwood moldings are available in a variety of species, but their length, width, and thicknesses are limited. The following is a list of maximum lengths and cross sections for different hardwood species as established by the American Woodworking Institute (AWI).

	Length	Cross Section
Cherry	7'-10"	1¹⁄₁₆" × 4¼"
Mahogany	11'-10"	1¹⁄₁₆" × 6"
Maple	11'-10"	1¹⁄₁₆" × 6"
Poplar	11'-10"	1¹⁄₁₆" × 6"
Plain sawn red and white oak	11'-10"	1¹⁄₁₆" × 6"
Rift sawn red and white oak	8'-10"	1¹⁄₁₆" × 4¼"
Walnut	7'-10"	1¹⁄₁₆" × 4¼"

2.4.0 Quarter Round, Corner Guard, Chair Rail, and Wainscot Cap Moldings

Quarter round moldings (*Figure 6*) are sometimes used in place of base shoe moldings to hide the unevenness between the base molding and the floor. They are also used as inside corner trim to hide the seams in corners or under wall cabinets as trim pieces.

All quarter round moldings have a pie-shaped quarter circle profile. They vary only in size, from ¼" to 1⅛". The most common sizes are ⅜", ½", and ¾".

Corner guard moldings are used to protect outside corners, especially in high-traffic areas. They are usually plain wood corner pieces, but some highly decorative styles are made.

Wainscot caps cover exposed end grain in **wainscoting**. Wainscoting is a wall finish consisting of panels applied partway up the wall from the floor. Wainscot cap may include a lipped profile that covers the top of the wainscoting. Back band moldings are similar to wainscot caps.

Chair rail molding is used at chair-back height, 30" to 35" above the floor, to protect walls from chair backs. It may also be used as a cap for wainscoting or as a horizontal dividing line between two surfaces such as wallcovering and paint. Chair rail molding comes in various styles that typically have a bulb-like feature in the top third of the molding. The back side is always flat for installation against walls.

2.5.0 Stools

Stools are a type of window trim. The bottom side of a standard stool (*Figure 7*) is rabbeted at an angle, typically 10 or 14 degrees, to fit on the angled sill of the window frame so that its top side will be level.

3.0.0 ◆ BASIC PROCEDURES AND GUIDELINES FOR INSTALLING TRIM

This section provides some general installation guidelines for cutting and installing trim moldings.

3.1.0 Cutting Trim

Trim carpentry involves making different types of cuts to moldings using special saws and a miter box or a power miter saw. The most common saws used for cutting moldings are the backsaw, dovetail saw, and coping saw (*Figure 8*). A backsaw is normally used with a miter box (*Figure 9*) to make very fine right-angle cuts called **square cuts** and angular cuts called miter cuts. The backsaw is a

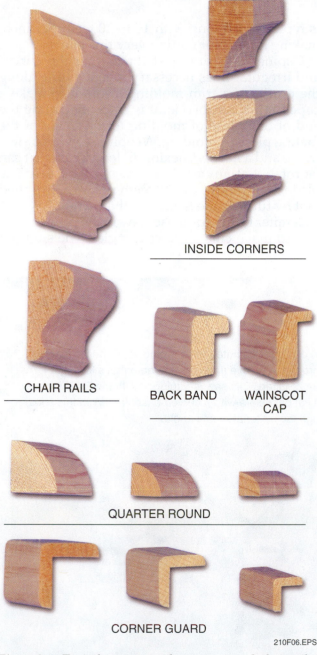

INSIDE CORNERS

CHAIR RAILS

BACK BAND   WAINSCOT CAP

QUARTER ROUND

CORNER GUARD

210F06.EPS

Figure 6 ◆ Typical corner guard, quarter round, chair rail, and wainscot cap moldings.

210F07.EPS

Figure 7 ◆ Rabbeted stool.

DOVETAIL
SAW

COPING
SAW

BACKSAW

210F08.EPS

Figure 8 ◆ Special trim saws.

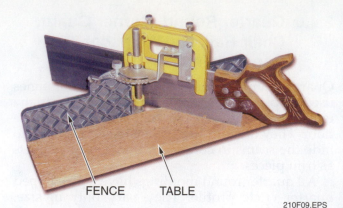

FENCE   TABLE

210F09.EPS

Figure 9 ◆ Metal miter box and backsaw.

crosscutting saw with a squared end and stiffened rib along its back to keep the blade rigid when cutting. Standard blades are 8" to 14" long with 11 to 14 teeth per inch. A dovetail saw is similar to a backsaw, but is smaller with a straight handle and fine teeth set to cut a very narrow kerf. The blade

is typically 10" long with 16 to 20 teeth per inch, making it ideal for cutting very fine joints.

Coping saws are used for cutting the curves and irregular lines necessary when cutting along the profile of a trim molding in order to make a **coped joint**. A coped joint is made by cutting the end of one piece of molding to the shape of the mating piece of molding. A coping saw has a steel frame and a narrow, flexible 6" long blade that can be rotated at any angle to cut small curves. The blade, which has 12 to 18 teeth per inch, is pulled taut by turning the handle of the frame.

A miter box holds a backsaw in the proper position to make cuts at precise angles. The simplest

Bayonet Saws

A bayonet saw (portable jigsaw) can be substituted for a coping saw. Special, narrow scrolling blades that are taper ground and have fine, straight teeth that allow tight turns and produce smooth finishes are required. This saw is equipped with the Collins coping foot, which replaces the normal flat baseplate with a curved baseplate. With this coping foot, the saw can then cut through the material at any angle.

210SA04.EPS

miter box is merely a U-shaped wooden structure with slots through its sides at 90- and 45-degree angles. A more precise metal miter box like the one shown in *Figure 9* typically comes with its own saw and with clamps that hold work of various sizes and shapes. An adjustable mechanism allows the saw to be positioned at the desired angle.

Power miter saws and compound miter saws (*Figure 10*), sometimes called power miter boxes, are also used for cutting trim moldings. This saw combines a miter box or table with a circular saw, allowing it to be used to make square and miter cuts. The saw blade pivots horizontally from the rear of the table and locks into position to cut angles from 0 to 45 degrees right and left. Stops are set for common angles. The difference between the power miter saw and a compound miter saw is that the blade on the compound miter saw can be tilted vertically, allowing the saw to be used to make a **compound cut** (combined bevel and miter cut).

3.1.1 *Making Square and Miter Cuts*

Square cuts are made to square uneven ends of moldings and to make butt joints with walls, door casings, and similar places (*Figure 11*).

Miter cuts are made whenever an angular cut is needed. In basic trim work, 45-degree miter cuts are commonly made at the tops of doors and windows at the joints where the side and head casing meets, at outside corners such as those required when installing baseboards, and when **scarf joints** are needed. A scarf joint is a joint made by overlapping two pieces of molding, with one piece having an open miter cut and the other piece

having a closed miter cut of either 22.5 or 45 degrees. A scarf joint is used whenever it is necessary to join two sections of molding on a wall or ceiling in order to achieve the required length.

Cutting molding using a saw and miter box begins with the proper placement of the molding in the miter box. All moldings should be cut with their face side or edges up or toward you so that the saw splinters out the back side, not the face side.

Flat miters, such as those made in casing for doors and windows, are cut by holding the molding with its face side up and its thicker edge against the back side (fence) of the miter box (*Figure 12*). Some moldings, such as the base, base cap and shoe, or chair rail, are held right side up. Their bottom edge should be against the bottom of the miter box (table) and their back against the back fence of the miter box. Once the molding is properly positioned in the miter box, the required square cut or miter cut can be made with the saw.

When working with crown, bed, and cove moldings, some carpenters position the molding upside down in the miter box (or power miter saw) and at the same 45-degree angle at which it will be installed (*Figure 13*). The lower side of the molding fits against the wall. It has a wider surface than the top side, which rests against the ceiling. When placing the molding in the miter box, position it as if the fence of the miter box were the wall and as if the table of the miter box were the ceiling. To aid in keeping the molding positioned correctly in the miter box while cutting, carpenters use a jig to support the back side of the molding.

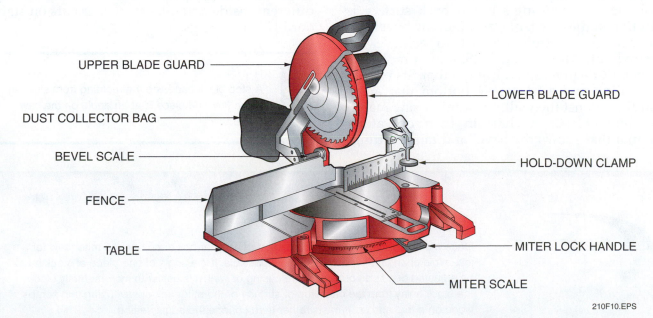

UPPER BLADE GUARD

DUST COLLECTOR BAG

BEVEL SCALE

FENCE

TABLE

LOWER BLADE GUARD

HOLD-DOWN CLAMP

MITER LOCK HANDLE

MITER SCALE

210F10.EPS

Figure 10 ◆ Compound miter saw.

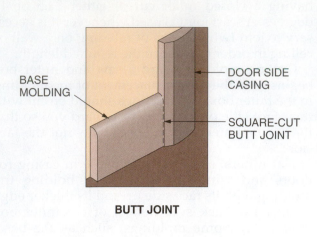

BASE MOLDING

DOOR SIDE CASING

SQUARE-CUT BUTT JOINT

BUTT JOINT

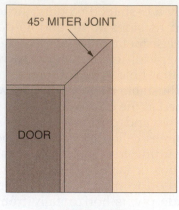

45° MITER JOINT

DOOR

MITER JOINT

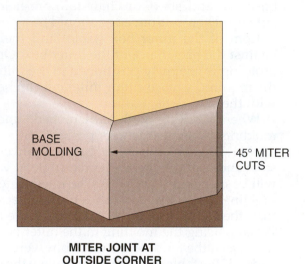

BASE MOLDING

45° MITER CUTS

MITER JOINT AT OUTSIDE CORNER

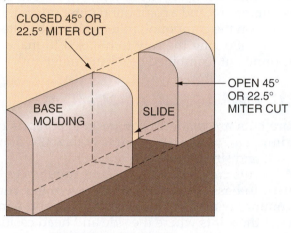

CLOSED 45° OR 22.5° MITER CUT

BASE MOLDING

SLIDE

OPEN 45° OR 22.5° MITER CUT

MITER CUTS MADE TO FORM SCARF JOINT

210F11.EPS

Figure 11 ◆ Typical butt and miter joints.

If using a compound miter saw to cut crown molding, the molding's broad back surface is laid flat on the saw table, and the saw bevel and miter angle controls are set to achieve the desired cut. All standard U.S. crown moldings have a 52-degree top rear angle that fits next to the ceiling and a 38-degree bottom rear angle that fits against the wall. Most miter saw manufacturers provide a chart in their operator's manual that shows the bevel and miter settings to use with a compound miter saw for cutting different inside and outside corner cuts on standard U.S. moldings.

NOTE

A stop block can keep the molding from slipping when the workpiece is at an angle on the saw blade.

Adjusting a Mitered Angle

Once a correct miter angle is established on a miter box or compound miter saw for a job, avoid changing the angle setting if possible. If the angle of one piece of stock must be adjusted due to an out-of-square casing or wall, it is easier to incrementally block the stock away from the backstop or surface of the miter box or saw using thin scraps of wood on a trial and error basis rather than change the angle setting.

BACK SIDE DOWN ⌐   BACK SIDE
                                   AGAINST FENCE

210F12.EPS

Figure 12 ◆ Positioning molding in a miter box for cutting.

A typical chart of this type is shown in *Table 1*. Before making compound miter cuts to crown molding, always make test cuts on scrap material to make sure that the bevel and miter settings being used will produce the desired results.

⌐ CROWN MOLDING   ⌐ STOP BLOCK

210F13.EPS

Figure 13 ◆ Position of crown molding.

3.1.2 *Making a Coped Joint*

When installing baseboard and ceiling moldings, the use of a coped joint on inside corners is recommended rather than a miter joint because mitered joints tend to open at the inside corners when being nailed. Also, eventual shrinkage of the wood may cause the joint to open. Another reason for using a coped joint is that it makes it easier to adjust the angle if the corner is not square.

A coped joint is made by joining a piece of molding with a square cut that is butted against a wall, as shown in *Figure 14(A)*, with a second piece of molding cut to match the profile of the first piece, as shown in *Figure 14(B)*. To cut the second piece to the profile of the first piece, first cut a 45-degree open miter at the mating end of the piece. Then cut the piece at 90 degrees or less with a coping saw, following the edge profile of the first cut, as shown in *Figure 14(C)*. If the corner is greater than 90 degrees, back-cut the cope. After the coped cut is made, use a rat-tail file to finish contouring the edge.

Table 1	Compound Miter Saw Miter and Bevel Settings for U.S. Standard Crown Molding		
Type of Cut	**Bevel Setting**	**Miter Setting**	**Remarks**
Left side–inside corner	33.8°	Right, 31.6°	Position top of molding against miter box fence. Left side of molding is finished piece.
Right side–inside corner	33.8°	Left, 31.6°	Position bottom of molding against miter box fence. Left side of molding is finished piece.
Left side–outside corner	33.8°	Left, 31.6°	Position bottom of molding against miter box fence. Right side of molding is finished piece.
Right side–outside corner	33.8°	Right, 31.6°	Position top of molding against miter box fence. Right side of molding is finished piece.

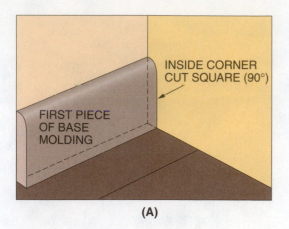

(A)

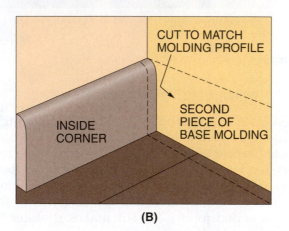

(B)

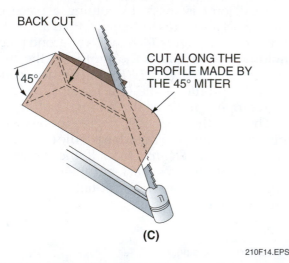

(C)

210F14.EPS

Figure 14 ◆ Making a coped joint.

3.2.0 Fastening Trim

Traditionally, interior trim was typically fastened in place using a lightweight (10 oz.) claw hammer and a nail set to drive finishing nails. Today, trim is commonly installed using pneumatic finish nailers (*Figure 15*). Finish nailers and pneumatically driven finish nails come in three diameters: 15 gauge, 16 gauge, and 18 gauge. They are available in lengths ranging from ¾" to 2¾". The 15-gauge nails are larger in diameter than 18-gauge nails. Typically, 15-gauge nails are used for fastening thicker trim materials, and 16-gauge or 18-gauge nails are used for thinner materials. The thinner the nail, the less likely it is to split the wood. Therefore, a 16-gauge or 18-gauge nailer is commonly used for installing door and window casing, window aprons, and similar trim. Stools, hardwood trim, and similar heavier materials would be installed using a 15-gauge nailer.

The depth to which the nail is driven can be set using a flush stop on some nailers or by adjusting the air pressure. If the nails are set flush by the nailer, then the final set can be accomplished by hand with a nail set.

3.3.0 General Trimming Guidelines

The following guidelines will help to produce professional results when installing trim:

- Make sure all joints are tight-fitting. Measure, mark, and cut carefully. Do not leave a poor fit. Do it over, if necessary.
- If the trim is to be stained, make sure all traces of glue are removed from exposed surfaces using a damp cloth. If allowed to dry, glue will seal the surface and not allow the stain to penetrate, resulting in a blotchy finish.

210F15.EPS

Figure 15 ◆ Typical pneumatic finish nailers.

INSIDE TRACK

Trim Nails

Trim nails should be set to a depth that is at least the width of the head. The head should be primed before the nail hole is filled.

- Make sure any pencil marks left along the edge of a cutline are removed before fastening the trim. Pencil marks in interior trim corners are difficult to remove after the pieces are fastened in position. Pencil marks show through a stained or clear finish and make the joint appear open. When marking interior trim, make light, fine pencil marks.
- All pieces of interior trim should be sanded smooth after they are cut and fitted, and before they are nailed in place. Always sand with the grain, never against the grain. Sanding of interior trim provides a smooth base for the application of stains, paints, and clear coatings.
- All sharp, exposed corners of trim should be rounded slightly. Use a block plane to make a slight chamfer, then round it over with sandpaper.
- Be careful not to leave finish nailer or hammer marks when fastening the trim material.
- Set finish nails below the surface of the trim, and fill the nail holes with putty that matches the paint or stain finish. In the case of prefinished molding, nails of matching color are used, and they are not set below the surface.

4.0.0 ◆ WINDOW TRIM INSTALLATION TECHNIQUES AND GUIDELINES

When trimming out a room, start with the window and door casing. There are two basic methods used to trim windows: conventional and picture frame (*Figure 16*). As shown, the conventional method uses casing at the top and both sides. The bottom of the side casing rests on the stool. Installed below the stool on the face of the wall is a horizontal member called the apron. In the picture frame method, the top, bottom, and sides of the window are all trimmed with casing.

> **NOTE**
>
> The procedures in this section assume that the windows are installed and are plumb and square with the wall.

The first thing to check before trimming a window is that the edge of the jamb is flush with the wall surface. If the jamb projects beyond the surface, you will have to plane or sand it flush. If the jamb is recessed, wood strips called jamb extensions must be nailed over the jamb edge to bring it flush with the inside wall surface (*Figure 17*). If the wall projects slightly beyond the jamb, the wall may require sanding if it does not allow the casing to lie flush against the jamb and wall surfaces.

Jamb extensions may be supplied by the manufacturer with the window unit, they can be ordered from the mill, or they can be made on site. Jamb extensions are not normally installed when the window is installed, but later when the window is trimmed. If supplied with the window, the extensions should be stored in a safe place until it is time to install them. Manufactured jamb extensions are usually precut to length and need only to be cut to the correct width. This is done by ripping the extension to the required width with a slight back-bevel on the inside edge.

The casing material used to trim a window should have the same profile as the door trim. As with most carpentry procedures, installing window and other types of interior trim can be accomplished in more than one way. The conventional method of trimming a window follows. Trimming a window using the picture frame method is done in basically the same way as the conventional method, except instead of using a stool and apron, a piece of casing is installed across the bottom of the window.

To cut and install the stool, proceed as follows:

Step 1 Determine the length of the stool. Typically, the length of the stool is equal to the distance between the inside edges of the window jambs plus the amount of **reveal** on both sides (typically ¼" on each side), plus twice the casing width, plus twice the casing thickness (*Figure 18*). The reveal is the distance that the edge of a casing is set back from the edge of a jamb. For example, if the jamb-to-jamb measurement of the window is 36" and you are using 2¼" × ¾" casing, the length of the stool would be 42½" (36" + ½" + 4½" + 1½").

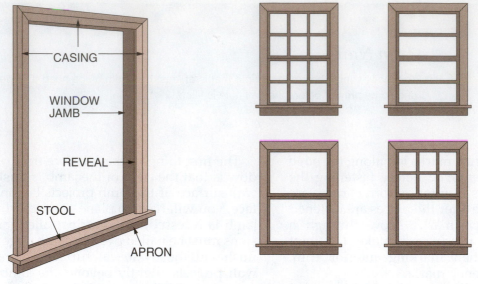

CONVENTIONAL TRIM METHOD

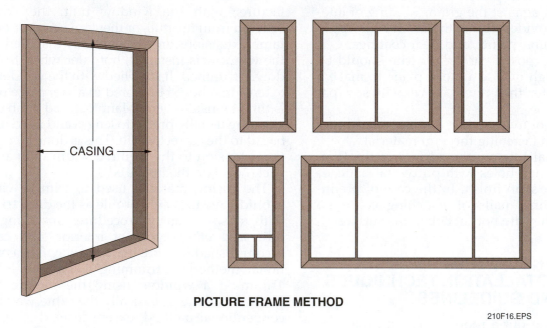

PICTURE FRAME METHOD

210F16.EPS

Figure 16 ◆ Window trimming methods.

Step 2 If it is desired that the wood grain at the ends of the stool be covered with returns to provide a finished appearance, lay out and mark the length of the stool, but do not cut the stool. If return ends are not being used, the stool can be cut to length at this time.

Step 3 Center the stool on the window, hold it level with the sill, and mark the location of the inside edges of the side jambs on the top of the stool using a square and a pencil. Also, mark a line on the face where the stool will fit against the wall surface. For a double-hung window, this line usually falls directly above the square edge of the stool rabbet.

Step 4 At both ends of the stool, carefully cut out that portion of the stool that will butt flush against the wall (*Figure 19*). Then open the lower sash and slide the stool in position.

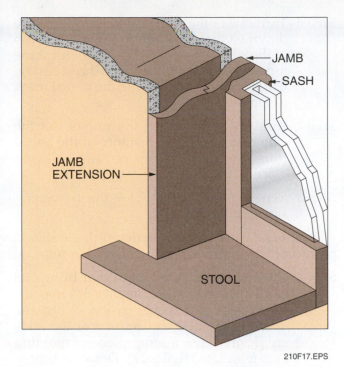

Figure 17 ◆ Installation of jamb extensions.

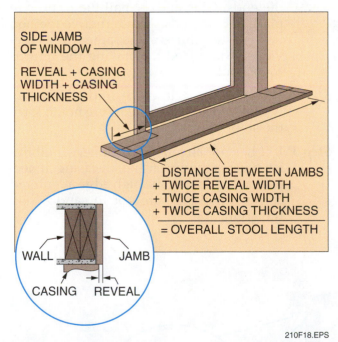

Figure 18 ◆ Method of determining the length of the stool.

Step 5 Lower the sash carefully on top of the stool, and draw or scribe the cutoff line so that the sash will clear the stool. Cut off the excess material. When finished, the stool should fit flush against the wall when the back edge of the stool has ¹⁄₁₆" clearance between it and the bottom rail of the window sash. This clearance is necessary to prevent binding of the sash when raising or lowering it.

Step 6 If using returns on the ends of the stool to provide a finished appearance, mark and cut a return at each end of the stool. The returns are formed by making 45-degree miter cuts at each end of the stool, as shown in *Figure 20*.

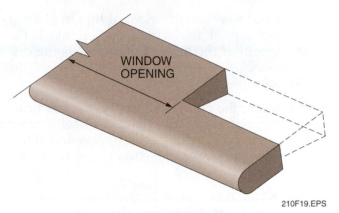

Figure 19 ◆ Window stool showing cutout for one side of the window opening.

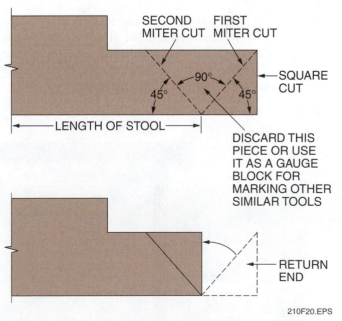

Figure 20 ◆ How to mark and cut a return.

Step 7 Sand all sawed edges, then nail the stool in place, making sure that it is level and perpendicular. Note that some specifications require that the underside of the stool be embedded in caulking compound before being nailed in place. If returns are being used, nail and glue the return pieces to the ends of the stool.

To cut and install the side and top (head) casing, proceed as follows:

Step 1 On the inside face of the side and top jambs, use a marking gauge or measure and mark the distance from the jamb edge to the desired reveal setback. For our example, a ¼" setback for the reveal was used.

Step 2 Hold a section of case molding in place at the top of the window, then mark the length of the head casing. This is the distance between the reveal marks on the window side jambs (*Figure 21*).

Step 3 Using the points marked on the head casing as a guide for where the heel of your miter cuts begin, cut left-hand and right-hand miters on the ends of the head casing.

Step 4 Set the head casing back the distance of the reveal from the edge of the jamb and nail it in place. Near miters, drill holes to prevent splitting the casing unless a nail gun is used.

Step 5 Cut two pieces of case molding slightly longer than needed for the two side case moldings. Cut a left-hand miter on one side case molding and a right-hand miter on the other side case molding.

Step 6 Turn one of the side case moldings upside down so that its mitered end rests on top of the stool, then mark the other end at the top edge (miter cut toe) of the head casing. Cut it off square, allowing the pencil mark to remain.

Step 7 Apply wood glue on the miters, set the side casing back the distance of the reveal from the edge of the jamb, and nail it in place.

Step 8 Repeat Steps 6 and 7 for the remaining side casing.

To cut and install the apron under the stool and perform other finishing tasks, proceed as follows:

Step 1 Cut the apron to a length that brings its ends into vertical alignment with the outside edges of the side casing. When the apron is molded and returns at the ends are desired, use a scrap piece of molding as a template (*Figure 22*). Draw its profile flush with the ends of the apron, then cut out the profile with a coping saw and sand the ends. Glue and/or nail the returns in place at both ends. An alternative method is to cut the returns in the same way as previously described for making the returns for a stool.

Step 2 Nail the apron in place under the stool with 8d finish nails. Make sure not to force the stool upward.

Step 3 If it has not already been done, set all finish nails in the apron and all casing. If necessary, sandpaper the high edge at mitered corners, making sure not to sand across the grain.

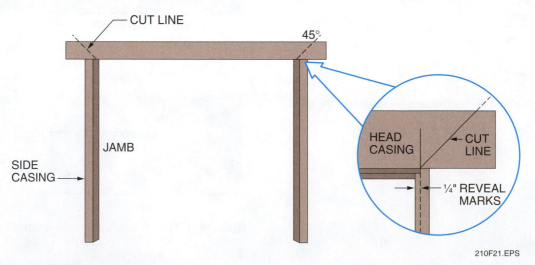

Figure 21 ◆ Marking the head casing for miter cuts.

Marking Gauge

A commercial marking gauge can be set and used to scribe a mark for the setback on the edges of window or door jambs for the reveal. If a commercial gauge is not available, a job site gauge can be made by rabbeting the four edges of a square block of wood for the desired amount(s) of reveal. The block is placed inside the jamb with the correct reveal width overlapping the edge of the jamb. Use a pencil to mark the reveal while sliding the block around the side jambs and head jamb.

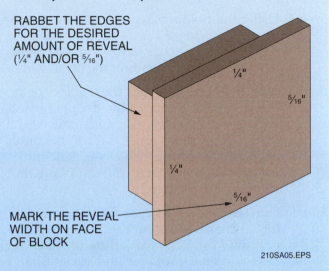

FIXED POINTS

RABBET THE EDGES FOR THE DESIRED AMOUNT OF REVEAL (¼" AND/OR ⁵⁄₁₆")

¼"

⁵⁄₁₆"

¼"

⁵⁄₁₆"

MOVABLE POINT

MOVABLE BLOCK

MARK THE REVEAL WIDTH ON FACE OF BLOCK

210SA05.EPS

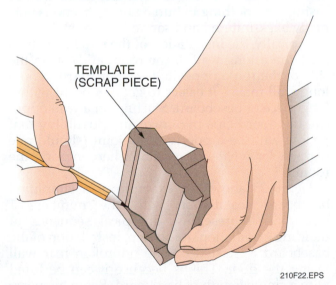

TEMPLATE (SCRAP PIECE)

210F22.EPS

Figure 22 ◆ Mark apron return profile with a scrap piece of molding.

5.0.0 ◆ DOOR TRIM INSTALLATION TECHNIQUES AND GUIDELINES

Interior door trim should always be applied after the finished floor is laid, but before the baseboard trim is installed. Also, the casing material used to trim a door should be the same as that used to trim the windows in the same room.

NOTE

The procedures in this section assume that the doors are installed and are plumb and square with the wall.

Nail Sizes

Nails should be sized so that two-thirds of the length of the nail penetrates the supporting material. For example, if you are attaching a ½" molding to ½" drywall that is attached to the framing, the nails need to be at least 3" long.

Step 1 On the inside face of the side and top door jambs, use a marking gauge or measure and mark the distance from the jamb edge to the desired reveal setback. This is typically ¼".

Step 2 Hold a section of case molding in place at the top of the door, then mark the length of the head casing. This is the distance between the reveal marks on the door side jambs.

Step 3 Using the points marked on the head casing as a guide for where the heels of your miter cuts begin, cut left-hand and right-hand miters on the ends of the head casing.

Step 4 Set the head casing back the distance of the reveal from the edge of the door jamb and nail it in place using 4d finish nails at the jamb edge and 8d finish nails at the outer edge of the casing. Unless a nail gun is used, drill holes near miters to prevent splitting the casing.

Step 5 Cut two pieces of case molding slightly longer than needed for the door's two side case moldings. Cut a left-hand miter on one side case molding and a right-hand miter on the other side case molding.

Step 6 Turn one of the side case moldings upside down so that its mitered end rests on the floor, then mark the other end at the top edge (miter toe) of the head casing. Cut it off square, allowing the pencil mark to remain. If the finished floor is to be carpeted, locate the casing above the subfloor to allow for the carpet and pad.

Step 7 Apply wood glue on the miters, set the side casing back the distance of the reveal from the edge of the jamb, and nail it in place.

Step 8 Repeat Steps 6 and 7 for the remaining side casing.

Step 9 If it has not already been done, set all finish nails. If necessary, sandpaper the high edge at mitered corners, making sure not to sand across the grain.

6.0.0 ◆ BASEBOARD TRIM INSTALLATION TECHNIQUES AND GUIDELINES

Baseboard trim runs continuously around the room. It is one of the last items of interior trim to be installed since it must be installed around door casings, supply and return grilles, cabinet work, and so forth. Baseboard joints made at inside corners should be coped instead of mitered, because mitered inside corners tend to open when being nailed. Joints made at outside corners are mitered.

Before installing moldings, locate the studs within the room and mark their locations on the wall just above the baseboard height or on the subfloor if the walls are finished. Then select and place lengths of baseboard molding around the sides of the room. Sort the pieces so that there will be the least amount of cutting and waste. Try to select lengths that will allow you to make complete runs without joints; if you cannot do so, add two feet to any length that will be joined so that you can cut the joint over a stud.

As mentioned earlier, most carpentry procedures can be accomplished in more than one way. One typical procedure for installing baseboard and related shoe molding is described here. If baseboard shoe molding is being installed, it is installed in the same basic manner as the main baseboard trim, except that it is nailed into the floor instead of the baseboard. This prevents the joint under the shoe molding from opening should shrinkage occur in the baseboard molding. Also, since shoe molding is a small molding and has a solid backing, both inside and outside corners are usually mitered. Where it meets a door casing with nothing to butt against, its end is normally back-mitered and sanded smooth.

Typically, the installation of the baseboard trim starts on a long wall of the room and/or on the side of the room opposite the door. Starting on a long wall makes it easier to get a good fit with a long piece of baseboard. Starting on a wall opposite the door helps minimize any possibility that people will see a poorly fitted joint (should one happen for whatever reason) when entering the room.

Figure 23 shows an example of a simplified baseboard installation with the numbers 1 through 7 representing a typical sequence of molding installation. As shown, installation of the baseboard is started on the unbroken rear wall facing the door. This is because it can be fitted with a single length of baseboard. Begin by measuring the wall, making sure to take your measurements at floor level. Following this, pick a straight length of baseboard molding and make a square cut on one end. Measure the required length, then make a square cut on the other end. Make the length slightly longer than required so that the cut will allow a spring fit. Test the fit on the wall and trim, if necessary, to allow a spring fit. Nail the piece in place at the stud locations along the wall. If carpeting is to be used for the finished floor, fasten the base trim above the subfloor to allow for

Irregular or Wavy Baseboard Mounting Surfaces

If baseboard trim will be installed against a final finished floor without a small shoe molding, it is important that any noticeable gap between the bottom of the baseboard trim and the floor is eliminated. Place the final cut length of baseboard for each side of the room on the floor against the wall and check that it is flush with the floor over its length. If the floor is warped or sags and noticeable gaps exist along the bottom of any portion of a baseboard for each wall, use a pair of dividers set to the widest gap width for any baseboard, and scribe a mark along the bottom edge over the length of each wall baseboard. Trim the bottom of the baseboards to the line with a jig saw or band saw, or sand to the line with a belt sander. As pointed out earlier, if the walls behind baseboards are uneven and the baseboard is paint grade, high spots on the wall can be sanded down and gaps may be caulked after baseboard installation if they are not more than ⅛". If the gaps are larger or the baseboard is stain grade, use a two-piece baseboard with a small cap and force-nail the cap into the gaps; otherwise float the wall using drywall compound to level the gaps, and then nail the cap and/or the baseboard.

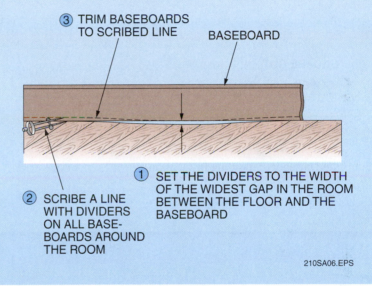

③ TRIM BASEBOARDS TO SCRIBED LINE

BASEBOARD

② SCRIBE A LINE WITH DIVIDERS ON ALL BASE-BOARDS AROUND THE ROOM

① SET THE DIVIDERS TO THE WIDTH OF THE WIDEST GAP IN THE ROOM BETWEEN THE FLOOR AND THE BASEBOARD

210SA06.EPS

the pad and carpet, unless the molding is to be installed after the carpet.

Next, install the second piece of baseboard on the adjoining (left-hand) wall. This piece of baseboard is joined to the first with a coped cut at the inside corner. If necessary, file the contour to get a fit. Once the coped cut is made, measure from the bottom of the first piece of base to the next corner, then square cut the second piece at the other end so that it butts tightly into the intersecting wall surface. Nail the second piece of trim in place. Keep working your way around the room in this manner until all the baseboard molding is cut and nailed in place.

As shown in *Figure 23*, all the inside corners have coped joints and the outside corner has a mitered joint. Where the baseboard molding meets the door casing, square-cut butt joints are used. Be careful when fitting baseboard to the door casing. Do not make the joint so tight that

you push the casing out of position when you spring the baseboard into place.

When making the outside mitered joint for our example, make the coped cut for the inside corner on one piece first, then hold the baseboard in position and mark the back edge for the miter cut, as shown in *Figure 24(A)*. Make a 45-degree miter cut and fasten the piece to the wall, as shown in *Figure 24(B)*. Repeat the procedure for the second piece, as shown in *Figure 24(C)* and *Figure 24(D)*.

NOTE

Few outside corners are exactly 90 degrees, so it is best to use a T-bevel to find the exact angle made by the outside corner, then divide this angle by two and cut a closed miter at each piece of joining base to that angle.

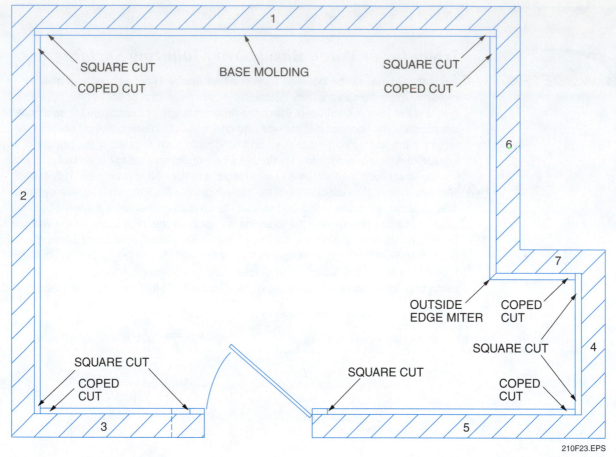

Figure 23 ◆ Simplified baseboard molding installation.

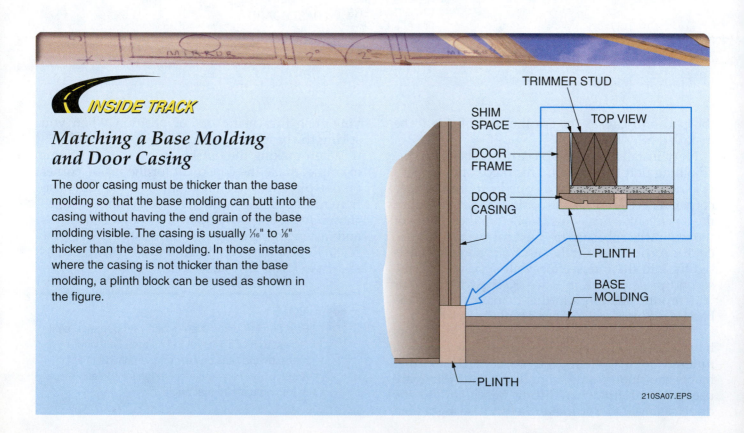

In the event that the job requires two lengths of baseboard be spliced to span a long wall, make the splice using a scarf joint. Set the first piece of molding in place and mark the center point of the stud nearest to the end of the piece. Subtract half the thickness of the molding, then cut a closed 22.5- or 45-degree miter at its end. Install the first piece, but do not nail into the last stud where the piece is mitered. Cut a 22.5- or 45-degree open miter at the end of the second piece. Measure from the face of the miter on the first piece to the corner of the wall, then cut the second piece to this length. Set it in place with the closed miter on the first piece overlapping the open miter on the second piece (*Figure 25*). Apply glue to the joint and nail through both pieces into the stud, then continue nailing the second piece to the corner. It should be pointed out that many carpenters use a biscuit joint cut with a biscuit (plate) joiner to splice long or wide moldings.

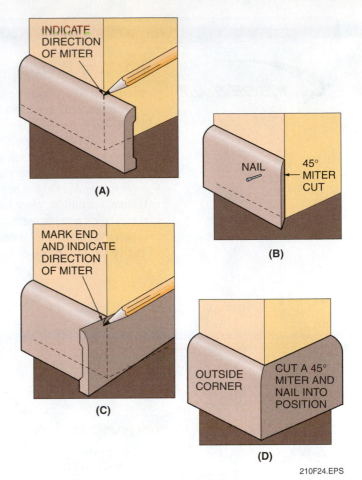

(A)

(B)

(C)

(D)

210F24.EPS

Figure 24 ◆ Procedure for fitting and cutting an outside miter joint.

INSIDE TRACK

Sagging or Uneven Ceilings

Most ceilings are not completely even. When crown moldings or cornices are installed in a room, gaps may show up along their length. In the case of sagging ceilings, the gaps occur at both ends of the ceiling treatment. If the gaps are ⅛" or less and the ceiling treatment is paint grade, the gaps may be caulked after installation of the treatment. If the sag or a long gap is less than ¼" and the ceiling trim is a molding, the molding may simply be bent around the sag or force-nailed tight to the gap. Another solution is to scribe the gap or sag onto the top edge of the molding with a pair of dividers and trim the molding to fit the ceiling. The simplest and best solution for fitting molding or cornices is to install the ceiling treatment straight and level and then float the gaps out using drywall compound.

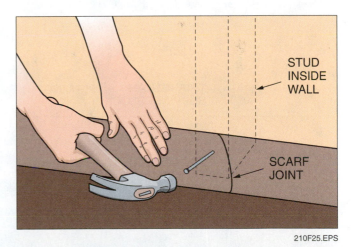

210F25.EPS

Figure 25 ◆ Scarf joint cut and nailed directly over a wall stud.

Backing Boards

If wide wood crown molding (4" or more) is to be installed, wide backing boards (nailers) should have been installed above the edges of ceilings that are parallel with the ceiling joists (A). These boards must be fastened to the top of the top plate in order to supply a nailing support for the upper crown molding edge. If wide backing boards haven't been installed, triangular wood support blocks can be installed behind the crown molding and secured to the wall studs (B). In this case, the crown molding will have to be nailed at the bottom edge and about two-thirds of the way up from the bottom. If a narrow nailer exists, a solution is to fasten two pieces of baseboard trim at right angles and secure the assembly to the wall studs and narrow nailer (C). The crown molding is then installed and nailed to the baseboard trim pieces. Other solutions involve formal cornice construction using both crown and bed molding (D).

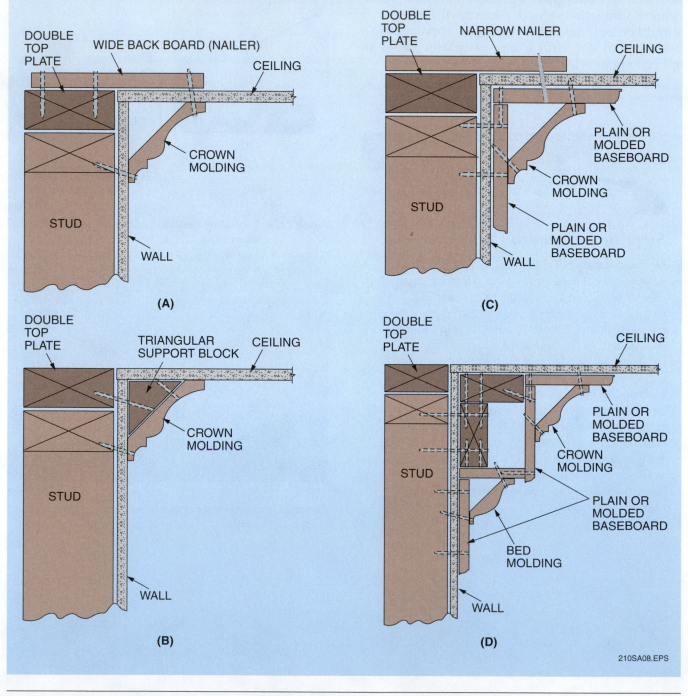

210SA08.EPS

7.0.0 ◆ CEILING TRIM INSTALLATION TECHNIQUES AND GUIDELINES

The general approach, sequence, and procedure for installing ceiling trim is basically the same as that for installing baseboard trim. However, when using crown/bed moldings, the moldings are not applied flat against the wall. Instead, they cover the wall/ceiling joint at a 45-degree angle. If the ceiling has only inside corners, install each piece of trim with a square cut at one end and a coped cut of matching profile at the other end, if possible. As with baseboard trim in rooms where a long wall is opposite the door, run square-cut crown molding from corner to corner and cope the molding of the adjoining walls. This will help to make the coped joints inconspicuous if they open up later due to wood shrinkage. Unfortunately, this may result in a side that requires close fitting and coping of both ends of the molding. Like baseboard trim, always cut crown molding slightly longer than required and spring it into place when installing and nailing. If outside corners are required, compound miter cuts are involved. General procedures for placement and cutting of crown, bed, and cove moldings using a miter box were covered earlier in this module.

When installing crown or cove molding, you must find the distance from the top edge (ceiling) to the bottom edge (wall) where the molding should be installed. One way to do this is to hold a scrap piece of molding against the ceiling and wall to determine the distance from the ceiling to its bottom edge. Once this distance is known, all walls can be marked at both ends and lines snapped between the marks that represent the installation point for the bottom edge of the molding around the room. Another way to find this distance is to measure the molding using a framing square. With the tongue and blade of the framing square representing the wall and ceiling, respectively, align a piece of crown molding in the corner of the square and measure the tongue distance.

8.0.0 ◆ ESTIMATING TRIM MATERIALS

To do a material take-off of the trim materials required for a job, begin by checking the building drawings and specifications for the specific types of trim material to be used for each application, including doors, windows, baseboards, and ceilings. To find the total trim material needed for a typical job, it is necessary to determine how many linear feet of each type of trim will be required.

Trim is generally divided into two categories:

- *Running trim* – This type of trim is installed in random lengths for such applications as baseboards, crown molding, and chair rails.
- *Standing trim* – This trim is specified in particular lengths for purposes such as door and window casings or other special applications where splicing is not acceptable.

Estimating running trim is relatively easy:

- *Baseboard* – This trim is the sum of the perimeter measurements of all rooms, minus the width of the doorways, plus 10 percent for waste. Order the shortest standard mill lengths that will minimize splicing on all but the longest walls. If a very long wall is encountered, order the maximum mill lengths for the wall to minimize splicing, but be aware that they will cost more per foot than standard lengths.
- *Chair rail* – This trim is the sum of the perimeter measurements of all rooms using the chair rail, minus the width of the doorways and any windows that will interrupt the rail, plus 10 percent for waste. Order the shortest standard mill lengths that will minimize splicing on all but the longest walls.
- *Crown molding and/or other cornice materials* – This trim is the sum of perimeter measurements of all rooms, plus 10 percent waste. Order the shortest standard mill lengths that will minimize splicing on all but the longest walls. If a very long wall is encountered, order the maximum mill lengths for the wall to minimize splicing, but be aware that they will cost more per foot than standard lengths.

Estimating standing trim is more difficult:

- *Door casing* – A typical 6'-8" door that is 2'-10" wide requires four 7' pieces of side casing and two 42" head casings, all of which include an extra 4" for each of the ends that require mitering. Therefore, five 7' pieces of casing are required for each door. Since standard lengths are usually 6', 8', 10', 12', 14', and 16', the length to use that yields the most with a minimum of scrap is the 14' length. Three 14' lengths cut in half will case one door with a 7' length left over for the next door. For every two doors then, only five 14' lengths are required. If all the doors are the same, divide the number of doors by two and multiply the result by five to obtain the total number of 14' lengths to order. For wider or narrower doors, a similar analysis must be performed to obtain the optimum ordering lengths. For waste purposes, add one extra of the widest door to the count. Store any leftover casing in the attic for future client use.

- *Window casing* – Windows are analyzed the same as doors except that only one side receives interior trim casing. If the bulk of the windows are the same size, measure the side jambs and the header jamb and add 4" for each end that is mitered. If the windows are picture-framed, add another casing length equivalent to the header jamb plus 8". Determine which standard lengths yield the least amount of scrap when cut to fit one or two of the windows. Then, order the appropriate number of standard lengths to case the windows. Odd-size windows will require a separate analysis.
- *Stool and apron* – The length of material required for the stool and apron is the width of each window plus the width of two side casings, plus 4" for ears and returns. Adjacent double-hung windows (*Figure 26*) are treated as one window for total stool and apron lengths. Determine which standard lengths yield the least amount of scrap for two or three windows and order the required quantity for all the remaining windows.
- *Mullion* – For adjacent double-hung windows, the length of mullion for the center of the windows must be measured and multiplied by the number of sets of adjacent windows. Determine which standard lengths yield the least amount of scrap and order the required quantity.

When ordering the calculated amount of trim, make sure that you specify whether paint-grade or stain-grade trim is required. Remember, paint-grade trim may be finger-jointed from smaller pieces and will be unsuitable for a stained or natural finish.

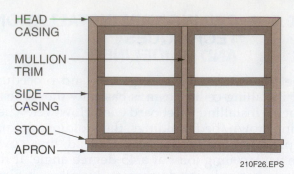

HEAD CASING
MULLION TRIM
SIDE CASING
STOOL
APRON

210F26.EPS

Figure 26 ◆ Mullion trim used with adjacent double-hung windows.

INSIDE TRACK

Mill-Calculated and Prepackaged Trim

If standard sizes of doors are used, single-sided sets of precut trim are available, and it is only necessary to order the desired number of sets. If the building plans are provided, some millwork shops will perform a material take-off and supply precut trim kits for each door and window, regardless of size.

Summary

Interior trim involves the installation of molding around windows, doors, walls, floors, ceilings, and other inside surfaces. Moldings are strips of material shaped in numerous patterns for use in specific locations. Wood is used for most moldings, but some are made of plastic and other materials.

When considering the construction sequence of a building, the trim installation is one of the finishing touches. Everybody is going to see the trim; therefore, it has to be installed carefully with tight-fitting joints in order to present a suitable appearance.

Notes

1. Moldings made of _____ can be nicked, dented, or otherwise damaged relatively easily.
   a. poplar
   b. maple
   c. pine
   d. oak

2. Moldings made of finger-jointed wood are normally _____.
   a. stained
   b. painted
   c. clear finished
   d. stained or clear finished

3. A type of molding used to trim the bottom edge of a wall where the wall meets the floor is base _____ molding.
   a. casing
   b. bed
   c. cap
   d. shoe

4. A type of molding used to trim around windows and doors is _____ molding.
   a. casing
   b. bed
   c. quarter round
   d. base

5. A type of molding sometimes used to cap wainscoting is _____ molding.
   a. quarter round
   b. chair rail
   c. cove
   d. corner guard

For Questions 6 through 13, match the type of molding to its corresponding letter in the figure below.

6. _____ Base cap

7. _____ Colonial base

8. _____ Ranch base

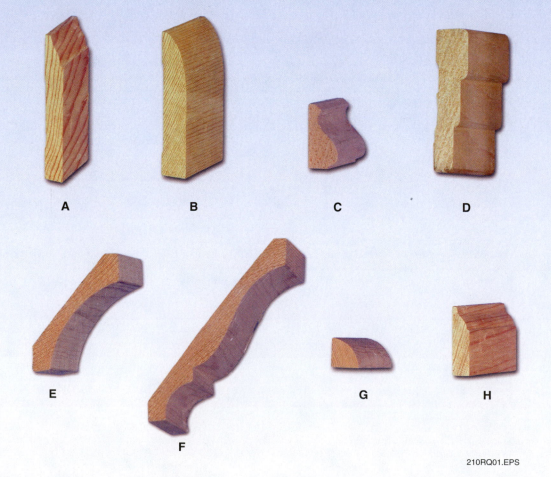

210RQ01.EPS

9. _____ Reserve casing

10. _____ Colonial doorstop

11. _____ Crown molding

12. _____ Cove molding

13. _____ Quarter round

14. A typical height for installation of chair rail moldings is _____ above the floor.
    a. 20" to 25"
    b. 26" to 29"
    c. 30" to 35"
    d. 36" to 40"

15. A type of saw used to shape the end piece of molding so that it matches the profile of the mating molding member is a _____ saw.
    a. dovetail
    b. backsaw
    c. coping
    d. miter

16. A scarf joint is made by overlapping two pieces of molding.
    a. True
    b. False

17. Flat miter joints are commonly made when trimming _____.
    a. chair rail
    b. base cap
    c. crown molding
    d. door and window casing

18. If using both 15-gauge and 16-gauge pneumatic finish nailers to fasten trim, the 16-gauge nailer typically is used to fasten _____.
    a. thinner trim materials
    b. thicker trim materials
    c. hardwood trim
    d. window stools

19. The space between a door or window casing and the edge of the door or window jamb is called the _____.
    a. setback
    b. backset
    c. interval
    d. reveal

20. The joint between two pieces of baseboard molding in inside corners is usually _____.
    a. mitered
    b. butted
    c. coped
    d. bisected

Trade Terms
Introduced in This Module

Apron: A piece of window trim that is used under the stool of a finished window frame. It is sometimes referred to as undersill trim.

Compound cut: A combined bevel and miter cut.

Coped joint: A joint made by cutting the end of a piece of molding to the shape it will fit against.

Finger-jointed stock: Paint-grade moldings made in a mill from shorter lengths of wood joined together.

Moldings: Decorative strips of wood or other material used for finishing purposes.

Reveal: The amount of setback of the casing from the face side of window and door jambs or similar pieces.

Scarf joints: End joints made by overlapping two pieces of molding with 22.5- or 45-degree angle cuts.

Square cuts: Cuts made at a right (90-degree) angle.

Stool: The bottom horizontal trim piece of a window.

Trim: Finish materials such as molding placed around doors and windows and at the top and bottom of a wall.

Wainscoting: A wall finish, usually made of wood, stone, or ceramic tile, that is applied partway up the wall from the floor.

This module is intended to present thorough resources for task training. The following reference work is suggested for further study. This is optional material for continued education rather than for task training.

Finish Carpentry. Newtown, CT: Taunton Press, 1997.

NCCER CURRICULA — USER UPDATE

NCCER makes every effort to keep its textbooks up-to-date and free of technical errors. We appreciate your help in this process. If you find an error, a typographical mistake, or an inaccuracy in NCCER's curricula, please fill out this form (or a photocopy), or complete the online form at **www.nccer.org/olf**. Be sure to include the exact module ID number, page number, a detailed description, and your recommended correction. Your input will be brought to the attention of the Authoring Team. Thank you for your assistance.

Instructors – If you have an idea for improving this textbook, or have found that additional materials were necessary to teach this module effectively, please let us know so that we may present your suggestions to the Authoring Team.

NCCER Product Development and Revision

13614 Progress Blvd., Alachua, FL 32615

Email: curriculum@nccer.org
Online: www.nccer.org/olf

❏ Trainee Guide   ❏ AIG   ❏ Exam   ❏ PowerPoints   Other _____

Craft / Level: _____   Copyright Date: _____

Module ID Number / Title: _____

Section Number(s): _____

Description: _____

_____

_____

_____

_____

Recommended Correction: _____

_____

_____

_____

_____

Your Name: _____

Address: _____

_____

Email: _____   Phone: _____

Cabinet Installation
27211-07

27211-07
Cabinet Installation

Topics to be presented in this module include:

Overview

In residential work, the majority of cabinet installation work involves kitchen cabinets and bathroom vanities. As you have seen in the kitchen of your home, there are base cabinets and wall cabinets. Vanities are simply small base cabinets. Wall cabinets present the most difficult challenge. They must be perfectly level and they must be carefully and securely attached to the wall framing because they are likely to hold a great deal of weight. Cabinet installation requires knowledge of hardware such as door hinges and latches, as well as the ability to select fasteners appropriate to the job. A wider variety of cabinetry is used in commercial work as display cases, reception desks, and storage cabinets. Each of these represents its own installation challenge.

Objectives

When you have completed this module, you will be able to do the following:

1. State the classes and sizes of typical base and wall kitchen cabinets.
2. Identify cabinet components and hardware and describe their purposes.
3. Lay out factory-made cabinets, countertops, and backsplashes.
4. Explain the installation of an island base.

Trade Terms

Backsplash
Escutcheon plate

Furr down
Veneer

Required Trainee Materials

1. Pencil and paper
2. Appropriate personal protective equipment

Prerequisites

Before you begin this module, it is recommended that you successfully complete *Core Curriculum*; *Carpentry Fundamentals Level One*; and *Carpentry Framing and Finishing Level Two*, Modules 27201-07 through 27210-07.

This course map shows all of the modules in *Carpentry Framing and Finishing Level Two*. The suggested training order begins at the bottom and proceeds up. Skill levels increase as you advance on the course map. The local Training Program Sponsor may adjust the training order.

27212-07
Cabinet Fabrication
ELECTIVE

27211-07
Cabinet Installation

27210-07
Window, Door, Floor, and Ceiling Trim

27209-07
Suspended Ceilings
ELECTIVE FOR RESIDENTIAL CERTIFICATE

27208-07
Doors and Door Hardware

27207-07
Drywall Finishing

27206-07
Drywall Installation

27205-07
Cold-Formed Steel Framing

27204-07
Exterior Finishing
ELECTIVE FOR COMMERCIAL CERTIFICATE

27203-07
Thermal and Moisture Protection

27202-07
Roofing Applications
ELECTIVE FOR COMMERCIAL CERTIFICATE

27201-07
Commercial Drawings
ELECTIVE FOR RESIDENTIAL CERTIFICATE

FRAMING AND FINISHING

CARPENTRY FUNDAMENTALS

CORE CURRICULUM:
Introductory Craft Skills

211CMAP.EPS

1.0.0 ◆ INTRODUCTION

The construction of kitchen and other cabinetry falls into three major grades: stock, semi-custom, and custom units.

Stock cabinets, also called modular cabinets, are the least expensive kind of cabinet. They are mass produced in specialized plants and are typically sold off-the-shelf at home centers and builder supply centers. They provide somewhat limited design options and can sometimes be of marginal quality. Stock cabinets are made in three forms: disassembled, assembled but not finished, and assembled and finished. Disassembled cabinets are known as RTA, or ready to assemble.

Semi-custom cabinets are the next grade up. Semi-custom cabinets may be produced in a mill or a cabinet shop. Semi-custom cabinets are normally available in several styles and in a variety of standard sizes. They are offered with many choices in materials, size, and finish. Semi-custom cabinets may be assembled in either unfinished or finished form. They are usually sold by home centers, independent distributors, and mid-range kitchen and bath dealerships.

Custom cabinets are tailor-made units built in a cabinet shop, millwork plant, or on the job site to satisfy a specific application. With these units, a carpenter or cabinetmaker comes to the job site, measures the cabinet area, then builds each cabinet unit to the desired specifications and to fit the exact wall space available.

The first part of this module introduces you to the materials, components, and basic construction methods used by cabinet manufacturers when they build wooden cabinets. This background information is important for you to know so that you can knowledgeably discuss cabinetry with customers, cabinet dealers, manufacturers, and others in the trade. The remainder of this module focuses on the basic procedures for installing factory-built kitchen cabinets at the job site. It also covers the materials and methods involved in the construction and installation of countertops used with kitchen and bathroom cabinets.

INSIDE TRACK

Professionalism

A professional cabinet installation like the one shown here allows a carpenter to take pride in his or her work.

211SA01.EPS

2.0.0 ◆ PARTS OF MANUFACTURED CABINETS

All cabinets, whether designed for commercial or residential use, consist of a case fitted with shelves, doors, and/or drawers. Two types of cabinet construction are commonly used: frameless and face frame (*Figure 1*).

Frameless cabinets, also called European or Eurostyle cabinets, have no support framework or face frame. Support for the shelves and work

surface is provided by the cabinet panels, which are heavy enough to carry the weight of the assembly and materials that will be stored in the cabinet. This style is widely used in applications where a contemporary look is desired. One advantage of frameless cabinets is that they tend to provide more storage capacity because they are built without a space-consuming framework. For example, a frameless four-drawer base cabinet contains about one cubic foot more usable space than a similar size face frame model.

Face frame cabinets do not have framework. They are similar in construction to frameless, but are usually made of lighter stock because the face frame supports the doors. The edges of the cabinet front are covered with a solid lumber face that has openings for doors and drawers. Face frame cabinets are widely used in applications where a traditional look is desired.

Figure 2 shows the components of a basic kitchen cabinet arrangement. The lower unit, called the base cabinet, is formed by fastening a series of individual base cabinet units together to achieve the desired configuration. Most base cabinets are manufactured 34½" high and 24" deep. As shown, a countertop is installed on top of the base cabinet. By adding the usual countertop thickness of 1½", the work surface of the base cabinet is at a

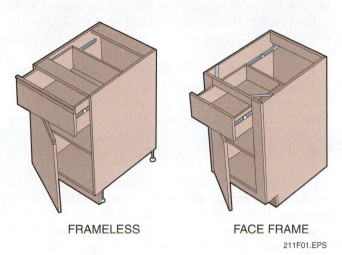

FRAMELESS   FACE FRAME

211F01.EPS

Figure 1 ◆ Cabinet construction methods.

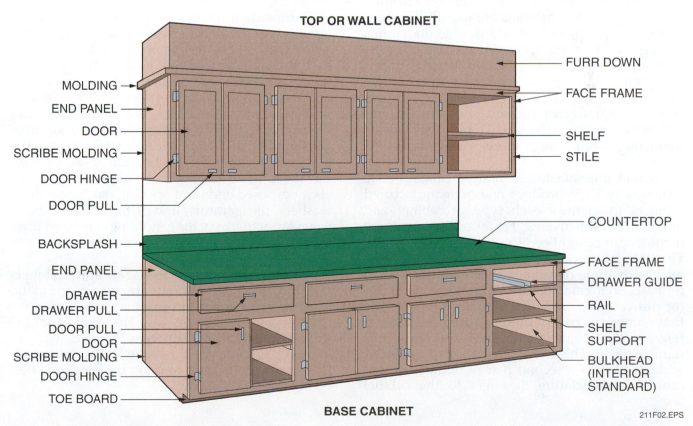

Figure 2 ◆ Components of typical kitchen cabinets.

211F02.EPS

standard height of 36" from the floor. The countertop is normally 25½" deep if a preformed unit or 25" deep if job built.

Standard base cabinets come in several widths that vary in 3" increments. Single-door cabinets typically range from 9" to 24" in width. Most consist of one door, one drawer, and an adjustable shelf. However, some base units have no drawers, while others contain nothing but drawers. Double-door base units range from 24" to 48" in width. They provide access from both sides. Corner units are also available with round rotating shelves (tiered lazy Susans) that make corner storage more accessible. A variety of floor-mounted tall cabinets are also made for use as oven, utility, and pantry units.

Similar to the base cabinet, the top or wall cabinet is formed by fastening a series of individual wall cabinet units together to achieve the desired configuration. Wall cabinets are usually 12" deep and come in single-door and double-door units having the same widths as similar base cabinets. They are made in several standard heights, with a 30" height being typical. Shorter cabinets that are 24", 18", 15", and 12" in height are made for use above sinks, refrigerators, and stoves. In kitchens without a **furr down**, 36" and 42" high cabinets are frequently used when more storage space is desired.

Cabinet manufacturers assign catalog code numbers to their product line of manufactured cabinets to identify each type of cabinet and describe its dimensions. For example, a cabinet manufacturer might code a cabinet as a W361824. The letter(s) preceding the cabinet nomenclature refer to the type of cabinet, such as W for wall, B for base, SB for sink base, CW for corner wall, DB for drawer base, and PB for peninsula base. The first number or pair of numbers after the letter(s) refers to the cabinet width. In our example, the number is 36. This means that the cabinet is 36" wide. If there is a second pair of numbers in the cabinet nomenclature, it refers to the cabinet

height. In our example, the cabinet height is 18". If a third pair of numbers is included, it indicates the depth of the cabinet. In our example, the cabinet is 24" deep.

If the cabinet nomenclature has letters following the sets of numbers, they generally indicate a special feature. For example, a B24SS is a 24" wide base cabinet with sliding shelves. The cabinet nomenclature may also have an L or R after it, indicating left or right hinging of the cabinet doors.

The vertical distance between the countertop on the base cabinet and the bottom of the wall cabinets typically ranges from 15" to 18". However, building codes normally require that there be at least 24" to 30" between the cooking unit and any overhead wall cabinet and a minimum of 22" between the sink and an overhead wall cabinet.

To provide a finished appearance, accessories must be used with the basic cabinet units. Filler strips that match the material and finish of the base and wall cabinets are used to fill any gaps in width between the end base or wall cabinet units and the room walls, or between adjacent cabinet units when no combination of standard sizes can fill an existing space. These filler pieces are usually supplied in 3" or 6" widths and then cut to the required width on the job. Other accessories include finished cabinet end panels and face panels for exposed ends and openings such as at dishwashers, refrigerators, and at the end of cabinet runs. Attractive matching moldings are often used to trim stock or semi-custom cabinets in order to make them more distinctive.

With the exception of the size, vanity cabinets (*Figure 3*) used in bathrooms are constructed in the same way as kitchen base cabinet units.

Most vanity cabinets are 31½" high and 21" deep. Usual widths range from 18" to 36" in increments of 3", then 42", 48", and 60". They are available in a wide variety of door, drawer, and shelf configurations.

Figure 3 ◆ Typical vanity cabinet.

211F03.EPS

3.0.0 ◆ WOODS AND MATERIALS USED IN CABINET CONSTRUCTION

A wide range of materials is used in manufactured cabinets. Low-priced cabinets are usually made from panels of particleboard with a vinyl film or melamine applied to exposed surfaces. The surface is printed with either a solid color or with a wood grain pattern to give the appearance of real wood.

High-quality cabinets are made from **veneers** and solid hardwoods, typically oak, cherry, birch, ash, or hickory. Hardboard, particleboard, plywood, or medium-density fiberboard (MDF) is commonly used for interior panels, drawer bottoms, and as the base for plastic laminate countertops. Some points to look for in higher-grade cabinets include:

- Face frame cabinets should have a ½" to ¾" hardwood face frame, ⅜" to ½" plywood or particleboard side frames, and a ¼" back panel.
- Frameless cabinets should use ⅝" to ¾" particleboard or plywood for the entire cabinet. Wood veneer or high-pressure laminates are preferable to melamine for the exterior finish.

- Drawers should slide smoothly with little play. They should close quietly and solidly. It is typical for metal ball bearing slides to be used. They can be regular or full extension drawers based on the specification. Look for a ½" to ¾" solid wood or 9-ply solid birch plywood drawer box with integral wood dovetail, shoulder, or dado joints. The box is typically attached to the drawer front with screws.
- Door hinges and catch mechanisms should be well made and work without binding.
- Rabbeted joints should be used where the top, bottom, back, and side pieces join.

3.1.0 Solid Woods

A wide variety of softwoods and hardwoods are used in the construction of visible exterior cabinet surfaces such as frames, doors, and drawers. Among the woods commonly used in cabinet construction are ash, beech, birch, cherry, oak, maple, and pine.

The American species of beech wood is not very stable and tends to warp easily. In Europe, they have resolved this problem by steaming it. However, European steamed red beech is very expensive.

Cherry was a favorite furniture wood of early American cabinetmakers and was often used as a substitute for mahogany.

White or red oak is identified more easily by the wood's pore and its radians than by its color. The radians are the lines that run perpendicular to the growth rings when viewing the end of a board. On the surface, they are visible as the distinctive ticking lines between the grain. When comparing white oak to red oak, the radians for white oak are noticeably longer and the pores in the growth rings, thus on the surface, are much smaller and tighter.

Pine is a favorite furniture wood because of its machining qualities. It is used extensively in the production of moldings and other millwork items. The heavier pine woods are made into timbers, and the knotty species are used for paneling and interior finish carpentry.

3.2.0 Plywood

Plywood is an excellent building material and one of the most important modern cabinetmaking products. It has several advantages over lumber, including availability in very large sheets, exceptional strength, and high resistance to warping. It

Sanding Plywood

The face veneers on both softwood and hardwood plywood can be damaged easily because the veneer is so thin. Sanding a light scratch or marred surface in the veneer can be tricky. If done incorrectly, you can ruin an expensive piece of plywood. Before attempting to sand a scratch or mar in the plywood veneer, first moisten the wood and allow it to dry. This will raise the veneer grain, allowing it to be sanded with less chance of damage.

is lightweight, yet very strong in comparison to solid wood. It is also great for cabinets. Expensive woods can be used in plywood because it covers only the top ply; also, plywood does not split as easily as solid wood and is more stable across the grain. Thin wood layers called veneers are glued together to form plywood sheets.

The most commonly used plywood in general construction is softwood plywood, usually Douglas fir. Other softwood species include Western hemlock, redwood, cedar, Southern yellow pine, Western larch pine, white fir, and oak.

All plywood comes in standard 4' × 8' sheets, although larger sizes are available. Nominal thicknesses of softwood plywood are ¼", ⅜", ½", ⅝", and ¾". Softwood plywood is graded according to the appearance of the front and back faces.

The main types of domestic hardwood veneers on plywood sheets are ash, birch, black walnut, cherry, gum, maple, oak, and poplar. Several imported woods such as lauan, mahogany, and teak are also available in specialized lumber yards or through mail order.

Plywood is the best material for most job-built units because of its working ease and availability. In order to take advantage of these special characteristics, different techniques are needed when working with plywood. Most importantly, because plywood comes in large sheets, it is necessary to lay out the project on the plywood, or on paper, before cutting. You must also remember to cut the pieces so the surface grain runs in the proper direction. For example, the grain of both the doors and the sides of cabinets should run in the same direction.

A table saw is the best tool for cutting plywood straight and a portable table saw is a must for a job site workshop. To avoid splintering the edge, work with the good edge up when cutting on the table saw and the good edge down when using a circular saw. Use only a sharp plywood blade or a fine-toothed power saw blade. The saw blade should have very little set (a term that refers to the curve in the blade's teeth) and should project above (or below) the plywood surface just at the height of the teeth. Another method is to tape the plywood with masking tape, draw your line, and cut through the tape. The tape prevents the edges from splintering.

Make sure the panel is supported properly to keep it from sagging while cutting and to prevent the blade from binding and jumping out of the cut. When cutting an angle with a circular saw, it is best to use a guide clamped to the piece so that a straight cut is achieved. Use extreme care when making the initial cuts so the project will fit together properly. The best working combination at a job site is a table saw, two sawhorses at the same height, and a full plywood sheet for a bench top.

3.3.0 Particleboard

Particleboard is manufactured using leftover materials from wood mills and from trees that are not suitable for other purposes. Small chips of wood, sawdust, and even bent and twisted trees can be used. This creates a cheaper panel, making it a very important product in the cabinetmaking industry. There are two kinds of particleboard, mat-formed (flat) and extruded (thin, flat, or shaped). Most particleboard used in the cabinet industry is mat-formed. In either case, the basic raw materials are wood particles left over from planing mills, veneer plants, and pulp mills.

The particleboard used for most cabinets and built-ins will be standard medium density or cabinet grade (45-pound density) and is available in thicknesses from ⅜" to 1". Generally, ¾" particleboard is recommended for most cabinet projects. Extruded particleboard is commonly used by manufacturers to make countertops with pre-formed splash and drop edges. An advantage of this method is that plastic laminate can be bonded to the top as it is being made.

Machining can be carried out with ordinary saws, drills, and routers if they are sharp. Carbide is recommended. The panels are heavy, about 60 pounds for a ¾" × 4' × 8' panel, so be sure to provide adequate support when cutting them.

Particleboard has no grain, so there is less chance of warping. It is perfect for covering with plastic laminates. Although particleboard can be fastened with nails, it does not hold them as well as plywood or solid wood. Cement-coated nails are best, but screws with pilot holes also provide good holding power. Almost any joint can be used with plywood or solid wood. However, the best holding power can be achieved with dado or rabbet joints.

Particleboard can be bought with a liner (usually white) on both sides. This saves time when building the main cabinet box frames. The inside of the cabinet can also be sprayed with sanding sealer to protect it from moisture. Particleboard will raise up and swell if allowed to get wet. Old material should be thoroughly inspected before use. If a carpenter has a job to be covered with a plastic laminate, the best choice of material would be ¾" water-resistant medium-density fiberboard (MDF) because it will not swell if it gets wet.

4.0.0 ◆ CABINET DOORS

Cabinet doors may be merely functional, or they can provide the major decoration of the cabinet unit. Both swinging and sliding doors can be used with cabinets. However, swinging doors are used extensively for kitchen and vanity units. Doors can be made of solid plywood or of panel and frame construction. Panel and frame doors consist of a solid wood frame mounted around a panel of plywood or solid wood. The frame is joined by mortise-and-tenon, mitered spline, or stile joints. Some cabinet doors have a frame covered on each side with thin plywood similar to a hollow-core interior door. Many different trim designs are used with cabinet swinging doors. *Figure 4* shows some widely used styles.

Three main types of hinged swinging doors are used with cabinets: lipped, overlay, and flush (*Figure 5*). These terms refer to the way the door is mounted on the cabinet. As shown, a lipped door is rabbeted along all edges so that when it is mounted on the cabinet, the edges overlap and conceal the cabinet opening. Lipped doors are relatively easy to install because they do not require exact fitting in the opening and the rabbeted edges stop against the face frame of the cabinet.

Overlay doors, also called surface doors, are mounted on the outside frame and overlay the cabinet opening so that the opening is concealed. The overlay door is also easy to install because it does not require exact fitting in the opening and the face frame of the cabinet acts as a stop for the door. For frameless (European) style cabinets, this

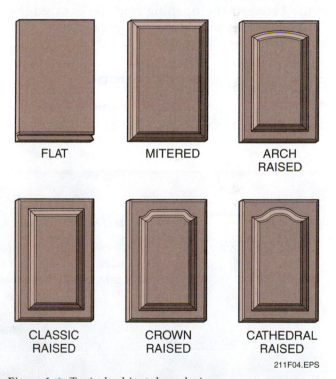

Figure 4 ◆ Typical cabinet door designs.

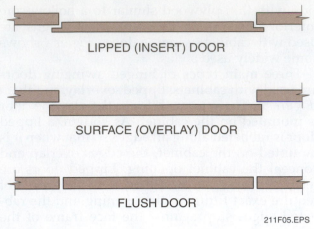

LIPPED (INSERT) DOOR

SURFACE (OVERLAY) DOOR

FLUSH DOOR

211F05.EPS

Figure 5 ◆ Types of cabinet door installations.

type of door completely overlays the front edges of the cabinet.

Flush doors fit into and flush with the face of the cabinet opening. Flush doors are the most difficult to install because they require fitting in the cabinet opening. A clearance of about ⅟₁₆" must be made between the opening and the door edges. Stops must be provided in the cabinet against which the door will close.

5.0.0 ◆ CABINET DRAWERS

Drawer fronts are generally made of the same material as the cabinet doors. The sides and backs are typically made of ½"-thick solid lumber, plywood, or particleboard. The bottom is usually made of ¼" plywood or hardboard. The front and sides can be joined using dovetail, lock shouldered, or rabbeted joints (*Figure 6*). The dovetail joint is the strongest.

Like cabinet doors, cabinet drawers are also classified as overlay, lipped, and flush, depending on the way the drawer front fits the cabinet drawer opening. The overlay drawer (*Figure 7*) is formed by a self-contained drawer unit consisting of the bottom, two sides, a back, and a false front. An overlay front, which conceals the cabinet drawer opening, is fastened to the false front with screws from the inside. The lipped drawer is made in basically the same way as an overlay drawer, except that it does not have a false front. Both ends of the drawer front are rabbeted to receive the drawer sides. Also, the rabbets on each end are made large enough so that they overlap and conceal the drawer opening in the cabinet. The flush door is constructed similarly to the lipped drawer, except the front is cut to fit the overall height and width of the drawer opening. For this reason, a drawer stop is provided at the back of the drawer so that the drawer is flush with the cabinet frame when closed.

Several types of metal drawer guides can be used to support the drawer and limit lateral movement. They also serve to ensure that the drawer does not tilt downward when it is opened. *Figure 8* shows a

Full-Extension Drawer Slides

INSIDE TRACK

Full-extension drawer slides or guides are a little more expensive than ordinary drawer guides, but they are much more convenient. A full extension drawer slide permits access to the entire drawer.

211SA02.EPS

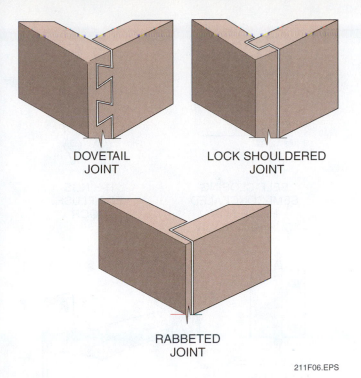

DOVETAIL JOINT

LOCK SHOULDERED JOINT

RABBETED JOINT

211F06.EPS

Figure 6 ◆ Joints used to join drawer pieces.

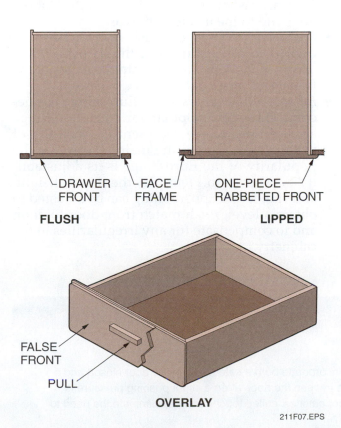

DRAWER FRONT — FACE FRAME

ONE-PIECE RABBETED FRONT

FLUSH

LIPPED

FALSE FRONT

PULL

OVERLAY

211F07.EPS

Figure 7 ◆ Drawer construction.

211F08.EPS

Figure 8 ◆ Typical drawer guide mechanism.

typical drawer guide mechanism. Drawer guides are designed to handle loads of various weights. Generally, side-mounted drawer guides with ball bearing rollers carry more weight than bottom-mounted, single-rail drawer guides. Wood drawer guides are seldom used on manufactured kitchen and vanity cabinets.

6.0.0 ◆ CABINET DOOR AND DRAWER HARDWARE

There are literally thousands of different kinds, shapes, sizes, and finishes of door and drawer hardware. Using the proper hardware for a cabinet is important. The style of a unit can be changed by merely switching the door and drawer hardware. For example, there is hardware designed for French Provincial, Early American, Traditional, Mediterranean, Dutch, Italian, and Spanish styles, just to name a few. In addition to using the correct style, the proper size hardware is also important. If the hardware is too large for a cabinet, it will overpower it and make the cabinet appear smaller. If small hardware is used on a large piece, it will look out of scale.

6.1.0 Hinges

Numerous types of decorative hinges are available for use with swinging cabinet doors. Many types are self-closing, which means that they hold the door closed and eliminate the need for a door catch. *Figure 9* shows a small sample of some

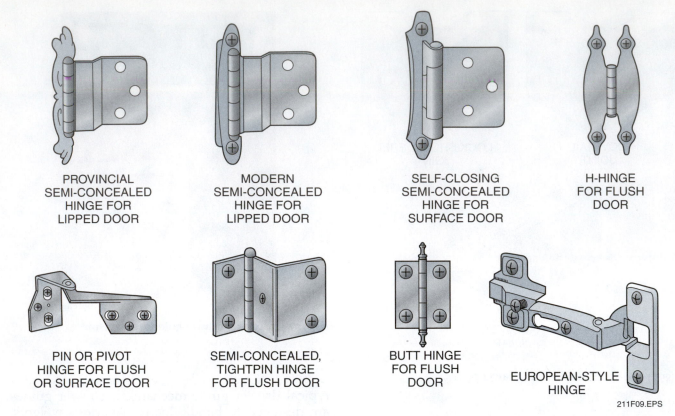

PROVINCIAL
SEMI-CONCEALED
HINGE FOR
LIPPED DOOR

MODERN
SEMI-CONCEALED
HINGE FOR
LIPPED DOOR

SELF-CLOSING
SEMI-CONCEALED
HINGE FOR
SURFACE DOOR

H-HINGE
FOR FLUSH
DOOR

PIN OR PIVOT
HINGE FOR FLUSH
OR SURFACE DOOR

SEMI-CONCEALED,
TIGHTPIN HINGE
FOR FLUSH DOOR

BUTT HINGE
FOR FLUSH
DOOR

EUROPEAN-STYLE
HINGE

211F09.EPS

Figure 9 ◆ Common cabinet door hinges.

common hinges. Cabinet door hinges fall into several categories, including:

- *Surface hinges* – Fastened to the exterior surface of the door and frame. The back side of hinge leaves can be straight for use with flush doors or offset for use with lipped doors.
- *Offset hinges* – Used with lipped doors. A semi-concealed offset hinge has one leaf bent to a ⅜" offset that is screwed to the back of the door. A concealed offset type is one in which only the pin is exposed when the door is closed.
- *Overlay hinges* – Available in semi-concealed and concealed configurations. For semi-concealed types, the amount of overlay is variable. Concealed types are generally made with a specific amount of overlay, such as ¼", ⁵⁄₁₆", ⅜", and ½".

- *Pivot hinges* – Normally used on overlay doors. They are fastened to the top and bottom of the door and to the inside of the case.
- *Butt hinges* – Used on flush doors when it is desired to conceal most of the hardware. The leaves of the hinge are set into the edges of the frame and the door.
- *European-style hinges* – The Euro-hinge has become extremely popular for concealed-hinge applications, which represent the majority of installations. One of the main reasons for the popularity of the Euro-hinge is its adjustability. As shown in *Figure 10*, the depth, height, and side-to-side positioning can be adjusted to obtain a level, flush match from door to door, and to compensate for any irregularities in the cabinetry.

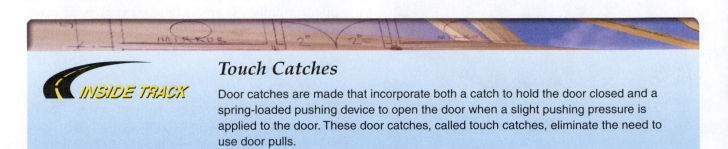

Touch Catches

Door catches are made that incorporate both a catch to hold the door closed and a spring-loaded pushing device to open the door when a slight pushing pressure is applied to the door. These door catches, called touch catches, eliminate the need to use door pulls.

INSIDE TRACK

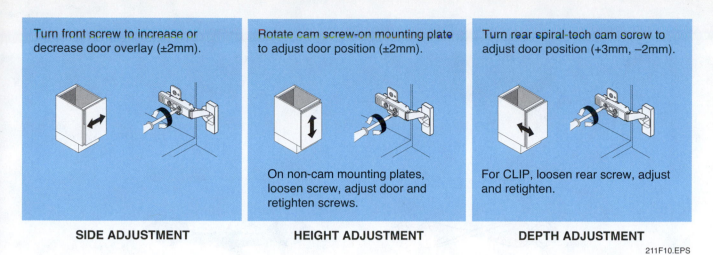

SIDE ADJUSTMENT	HEIGHT ADJUSTMENT	DEPTH ADJUSTMENT
Turn front screw to increase or decrease door overlay (±2mm).	Rotate cam screw-on mounting plate to adjust door position (±2mm). On non-cam mounting plates, loosen screw, adjust door and retighten screws.	Turn rear spiral-tech cam screw to adjust door position (+3mm, –2mm). For CLIP, loosen rear screw, adjust and retighten.

211F10.EPS

Figure 10 ◆ Euro-style hinge.

Self-closing hinges (*Figure 11*) are used in some applications. These hinges contain a spring mechanism that will automatically pull a door shut.

6.2.0 Door Catches

Swinging doors without self-closing hinges require the use of door catches to hold the door closed. Catches are normally installed where they are not in the way. Magnetic, roller, flex, and bullet catches (*Figure 12*) are common. Magnetic catches are made with single or double magnets of varying holding power. The adjustable magnet plate is fastened to the inside of the case, and the metal plate is fastened to the door at the desired location. Roller and flex catches are installed in a similar way as magnetic catches. The adjustable section is fastened to the case and the other section to the door. Bullet catches are spring-loaded catches that fit into the edge of the door. When the door is closed, the catch fits into a recessed plate mounted on the frame.

6.3.0 Knobs and Pulls

Many cabinet doors and drawers require that pulls or knobs (*Figure 13*) be installed in order to open and close them. They come in many styles and designs, and can be made of decorative metal, plastic, wood, porcelain, or other material. Knobs and/or pulls used on the cabinet doors should be the same design as those used for the cabinet drawers.

211F11.EPS

Figure 11 ◆ Self-closing hinge.

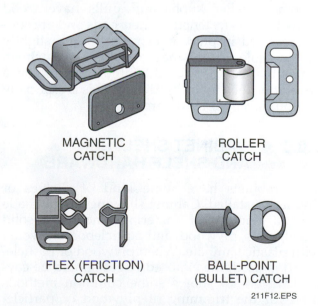

MAGNETIC CATCH

ROLLER CATCH

FLEX (FRICTION) CATCH

BALL-POINT (BULLET) CATCH

211F12.EPS

Figure 12 ◆ Cabinet door catches.

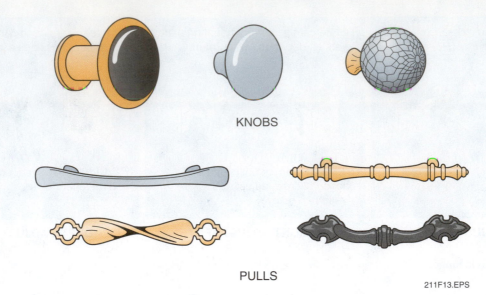

KNOBS

PULLS

211F13.EPS

Figure 13 ◆ Examples of cabinet door and drawer pulls and knobs.

Knobs and pulls are usually installed by boring a hole through the door or drawer front and fastening them using a threaded bolt or screw that is normally supplied with the hardware. If the bolt or screw supplied is not long enough, the hole must be countersunk, or longer bolts provided. When installing knobs or pulls in a number of doors or drawers of the same size, it is best to make a template (*Figure 14*), then use it to accurately locate the holes to be drilled.

When drilling a cabinet drawer for the installation of drawer pulls, drill from the drawer front. Clamp a backing block to the inside of the drawer to prevent blowout (*Figure 15*), or use a brad-point bit. It is also a good idea to reduce the speed of the drill as you approach the pierce point.

Some smaller knobs and pulls have wood screw threads fastened to their ends and are simply screwed in place after predrilling a slightly smaller hole. Knobs may also have **escutcheon plates** that can be placed behind them to add a decorative look to the hardware. This is especially true on the larger hardware.

7.0.0 ◆ CABINET SHELVES AND SHELF HARDWARE

Most cabinets have some kind of shelves or dividers installed. Cabinet shelving can be made from a number of materials, including solid wood, glass, plywood, and particleboard covered with plastic laminate. When plywood or particleboard is used, the exposed edges should be covered. *Figure 16* shows some common methods used for the trimming of plywood or particleboard edges.

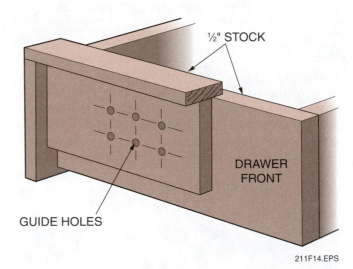

½" STOCK

GUIDE HOLES

DRAWER FRONT

211F14.EPS

Figure 14 ◆ Example of a template for drawer pull mounting holes.

211F15.EPS

Figure 15 ◆ Using a backing block.

Figure 16 ◆ Plywood edge treatment.

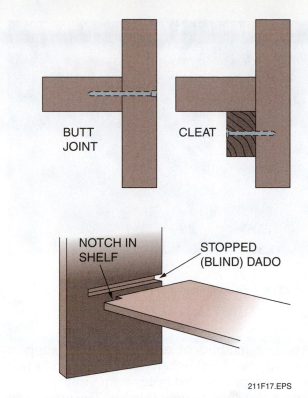

Figure 17 ◆ Methods of installing fixed shelves.

Most wooden shelves should be made of ¾" stock and be no longer than 3' without some sort of bracing. Shelves are joined to cabinets in several ways. One common method used for installing permanently mounted (fixed) shelves is the butt joint where the shelf is butted against the cabinet side, then glued and nailed in place (*Figure 17*). To add more strength to this type of joint, a cleat is placed under the edges of the shelf and fastened to the cabinet sides. The shelf is then tacked to this cleat for proper strength. If the shelf is long, a cleat is normally run along the back edge of the cabinet as well.

A better method used for installing fixed shelving is to fit the shelves into dadoes cut into the cabinet sides. Either a stopped dado is used or facers are installed to cover the dado in the front of the cabinet.

Many shelves used in cabinets are adjustable. A common method of installing adjustable shelves is the use of slotted metal shelf rails (standards) and clips (*Figure 18*). These come in a number of finishes. Four standards are used, two installed on each side of the cabinet interior. The standards are cut to the required length and can be either surface mounted on the sides of the cabinet or recessed in dadoes. If mounted on the sides of the cabinet, the shelves are cut short enough that they can be slipped down between the rails. Two clips fitted into slots on the standards at identical heights are used to support each end of a shelf.

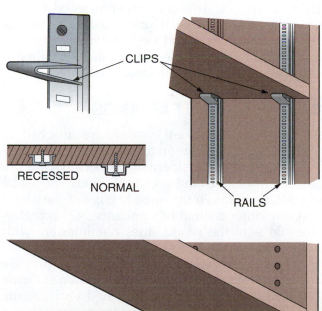

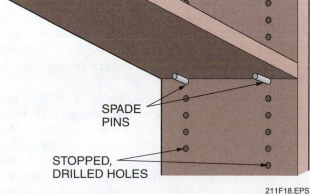

Figure 18 ◆ Rails, clips, and pins used with adjustable cabinet shelves.

(A)

(B)

211F19.EPS

Figure 19 ◆ Free-standing cabinet installation.

Another method of adjustable shelf support commonly used in kitchen cabinets involves two rows of stopped, drilled holes at the front and rear of each side of the cabinet interior. The number of such holes drilled in each row depends on whether the cabinet uses multiple shelves or a single shelf. Typically, a series of holes spaced about 1½" to 2" apart is drilled to cover an area of about 6" in the vicinity of each shelf's location. Two L-shaped plastic or metal pegs or spade pins are inserted into these holes at identical heights to support each end of the related shelf.

8.0.0 ◆ CABINET ISLANDS

Although most cabinet systems are attached to walls, it is not uncommon to see free-standing cabinets, especially in residential kitchens (*Figure 19*). Placement of an island is critical. There should be at least 36" between the outside edge of the island and any other cabinet or appliance; 42" is preferable. Be sure the island does not interfere with door swing or access to large appliances.

In some installations, a frame is built to provide an attaching point for cabinets, as shown in *Figure 19A*. The cabinets are then attached to the frame and the outside and ends of the frame are finished with gypsum drywall (*Figure 19B*).

In other applications, the center island is a prefabricated drop-in. Layout carpenters will mark the location on the slab or floor (*Figure 20*). Piping and electrical wiring may also be provided. Depending on the installation requirements and the manufacturer's instructions, cabinet installers may attach a 2 × 4 frame to the floor. They will then place the island over the frame and attach it to the frame with screws. A frame is not always used, however. If ceramic tile or hardwood flooring is to be installed against the island, the flooring material, combined with the weight of the island, is usually enough to secure the island in place.

211F20.EPS

Figure 20 ◆ Island layout.

9.0.0 ◆ COMMERCIAL CABINETS

Commercial cabinets are likely to be exposed to more abuse and use than home and kitchen cabinets. They are likely to be provided with kickplates covering those sections to which the public may have access. Access is usually one-sided, and doors are frequently large to allow the fullest possible access to storage. Laminates are very popular because of ease of cleaning and relative economy. Relatively high-end materials, such as Dupont's Corian® and marble, are fairly common in countertops, both for appeal to the customers and for ease of cleaning. *Figure 21* shows examples of commercial cabinet installation.

Material used for framing is frequently stronger in commercial work; that is, 2 × 4 top frames instead of 2 × 2. Keep in mind that if you are building for a store or business application, someone may use it to stand or sit on.

Obviously, the customer decides what his or her business needs. The carpenter needs to find a sensible approach to the construction. The use of modular units is one form of economy and may be necessary because of shop space considerations. In the case of retail food and restaurant work, wiring and pipe chases need to be built in, allowing reconfiguration as the customer's needs change. Wiring is an issue that the carpenter needs to consider in the planning stage, as there may be no way to get power to an intended destination without going through the cabinets. Some carpenters install conduit in the same position in all the cabinets to allow the electrician to pull cable as needed. Remember to plan with the electricians and plumbers, so that everybody knows what to do to help each other get the job right.

Exhibit cases are a particular aspect of construction for retail applications, where customers often want lighting installed. Another common commercial structure places spotlights or track lights on the ceiling. In that case, the front edge of the glass exhibit window is minimally framed using sheet metal angle or unistrut to preserve the edge of the glass.

Most commercial contracts, especially the larger ones, are entirely specification driven and are obtained by a detailed bid. Materials will be specified, with several acceptable alternatives, and everything down to the fasteners will be listed by type or brand name "or equal." Those jobs are awarded on competitive pricing, and very few changes will be permitted. They are frequently jobs worked in cooperation between a large cabinet shop and an installation crew, or by the customer's employees. They are also the jobs where deadlines are enforced by liquidated damages, and every day past the deadline invokes cost penalties. Such designs are developed by interior decorators and architects.

Floor installation is fairly common. Some are modules fastened to each other at the ends, some are free-standing display units, and some are on casters. Wall displays can be mounted to the structure, and usually are, because of the possible legal consequences if a customer pulls the display over and is hurt. Remember that construction in retail stores and public access buildings is exposed to any person who comes in; don't assume that everyone who interacts with your work will do so intelligently.

211F21.EPS

Figure 21 ◆ Commercial cabinet installations.

10.0.0 ◆ COUNTERTOPS

Numerous types of countertops can be used with cabinets. These include countertops covered with plastic laminate materials or those made from or covered with solid-surface materials.

10.1.0 Plastic Laminate–Covered Countertops

Plastic laminate is one of the most widely used materials to cover countertops. Made of layers of resin-impregnated kraft paper with a top layer of colored melamine, plastic laminates provide a durable, affordable, and easy to clean surface. Plastic laminate material is available in a wide range of colors and patterns. Sheets come in widths of 30", 36", 48", and 60", and lengths of 6', 8', 10', and 12'. Plastic laminate is referred to as HPL, or high-pressure laminate. HPL for cabinetry comes in three grades:

- *Horizontal (0.048" thick)* – Used for flat countertops.
- *Vertical (0.028" thick)* – Used for walls and doors. It is a lighter weight and not as durable.
- *Postform (0.039" thick)* – Used on factory-built countertops. It is not completely thermoset and can be heated and bent around radii. These preformed countertops are made by bonding flexible postform HPL to a core.

A plastic laminate countertop is usually 1½" thick and overhangs the front and sides of the cabinet(s). Counter overhangs vary, but 1" is common for standard kitchen cabinets that are 24" wide. Tops for peninsulas or islands can be much wider. Bathroom vanity tops range from 19" to 22" wide.

Some plastic laminate countertops are factory built. These preformed countertops are made of plastic laminate bonded to a core material, generally MDF, plywood, or particleboard. The front edge of the countertop is rounded, and the rear edge often curves up to form a **backsplash**. The countertops come in standard widths and are cut to length by the supplier or by the carpenter at the job site. Note that some L-shaped countertops are manufactured in two sections with a mitered corner. These sections are joined with hardware supplied with the countertop. Also supplied are preglued strips of laminate called end caps. These are applied to the ends of the countertop after it has been cut to the proper length.

A typical base for a job-built plastic laminate–covered countertop is built up in two layers, with a total thickness of 1½". The top layer (core) is a sheet of solid ¾" MDF, particleboard, or plywood (*Figure 22*). The second layer is made up of ¾" × 3" strips of plywood or particleboard attached to the front, back, and side edges. These strips reinforce and thicken the top. Cross strips installed at 2' intervals and at sink and appliance locations give the top added strength. Any seams in the core material are reinforced underneath with large squares of plywood or particleboard.

Plastic Laminates

INSIDE TRACK

Some common names of plastic laminates are Formica®, Wilsonart®, and Nevamar®. Each laminate manufacturer produces laminates in a multitude of colors and patterns. To help customers compare and select among their different laminate products, laminate manufacturers readily provide samples for each of their products like the ones shown here.

211SA03.EPS

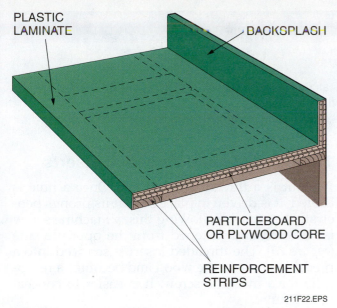

PLASTIC LAMINATE

BACKSPLASH

PARTICLEBOARD OR PLYWOOD CORE

REINFORCEMENT STRIPS

211F22.EPS

Figure 22 ◆ Typical construction of a plastic laminate–covered countertop.

After the base is built, oversized strips of plastic laminate are cut and cemented with contact cement to the exposed edges. They are trimmed flush with a laminate trimmer or router, allowing an oversized top piece of laminate to be cemented in place. After trimming the top piece flush with the laminate trimmer, the sink cutout(s) can be made.

A backsplash that is attached to the back of the countertop is made as a separate piece. Plastic laminate is glued onto the ¾" × 4" backsplash core in exactly the same way as for the countertop core. It is fastened to the back edge of the countertop with screws. When the countertop is installed, the backsplash is fastened to the countertop and adhered to the wall with construction adhesive. The joint between the countertop and backsplash is sealed with matching caulk or a waterproof sealer. For a proper fit, the backsplash may need to be scribed to the wall.

Job-built countertops are usually built in specialty shops and then delivered to the installation contractor. They can be identified by mitered rather than formed (bent) edges.

10.2.0 Solid-Surface Countertops

Countertops made of solid-surface materials are popular because they offer numerous benefits, including an attractive appearance and durability. Recent studies show that about 40 percent of professionally designed kitchens and baths use solid-surface countertops.

Solid-surface countertops come in many colors, have the solid look and feel of stone, resist stains, are easy to renew, and are available in matte, satin, semi-gloss, and high-gloss finishes. A matte or satin finish is best for kitchen countertops because it hides scratches better than shiny surfaces. Light colors and speckled patterns also tend to hide scratches better than dark colors.

The fabrication of countertops using solid-surface materials is labor intensive. Every countertop is usually custom made. All edges, inlays, feature strips, and backsplashes must be created by hand. Often, the same person who fabricates a solid-surface countertop will also install it.

For installation of solid-surface countertops, some manufacturers require that silicone adhesive be applied to the top of the base cabinet frames, then the countertop placed on top of it. Others require that 1" wood strips be attached to the cabinets with the countertop glued to the

Solid-Surfacing Materials

Brand names of prominent solid-surfacing materials currently in use include Corian®, Avonite®, Swanstone®, Fountainhead®, Gibralter®, Surrel®, and Transolid®. While all manufacturers have their own formulations, most solid-surface products are made of resins and fillers, along with pigments, ultraviolet (UV) inhibitors, and other additives. The most common filler is alumina trihydrate (ATH), which can account for up to 70 percent of the product. It helps countertops resist impact as well as water, stains, and chemicals.

Because of the costs involved, rather than taking measurements, templates are made after the cabinets are installed. These templates are used to cut the stone. Consider the lead time involved when setting up the project. Also, stone and concrete tops are very heavy and may require additional bracing in some cabinets.

INSIDE TRACK

Natural Countertops

Countertops may be made of natural materials, including marble, granite, slate, soapstone, ceramic, wood, and concrete.

strips. Some guidelines for a quality installation of a solid-surface countertop include:

- The counter should be leveled with shims rather than simply hiding any gaps with silicone.
- Any seams made at the job site should be inconspicuous and directly supported underneath.
- Inside corners should be cut on a radius and sanded to form a perfect curve, rather than a right angle. This is because corners are stress points, and a radius withstands stress better than sharp angles.
- The finish should be consistent over the entire surface. Check for consistency by examining the surface from several angles.

Many manufacturers of solid-surface countertops require that the installer be trained in the installation of their product. If an uncertified person does the installation, the warranty is voided.

NOTE

One thing to look for when installing solid-surface countertops is proper color matching. When two sections of countertop material are placed side-by-side, it will be immediately obvious if they are mismatched in color or pattern.

Solid-surface countertops should be maintained in accordance with the manufacturer's instructions. Most stains and spills wipe up with soap and water. Stubborn stains can be removed using a mild cleanser and a nylon scrubbing pad, which will not harm the surface. Minor scratches can usually be removed with very fine sandpaper followed by buffing with a nylon scrubbing pad to revive the finish.

11.0.0 ◆ FASTENERS USED FOR CABINET CONSTRUCTION AND INSTALLATION

A variety of fasteners are used in cabinet installation. Some were covered in earlier modules. This section covers some specialized fasteners commonly used with cabinets.

11.1.0 T-Nuts and Threaded Inserts

A T-nut is a threaded metal nut. Once a hole is drilled, it is driven in place so that its prongs penetrate the wood. Following this, a machine screw is installed and fastened from the opposite side (*Figure 23*). The threaded insert is screwed into a predrilled hole in the wood and becomes a receptacle for a machine screw. It is easier to conceal than T-nuts.

11.2.0 Tight-Joint Fastener

Tight-joint fasteners (*Figure 24*) are fasteners used at the joint of two mating pieces, such as between long countertop sections, the junction of 45-degree countertop miter joints, or other pieces that cannot be installed in one section. Tight-joint fasteners are installed so that they are concealed from the finished side and placed into a drilled hole in the two pieces to be joined. The fastener has a built-in bolt that is tightened with a wrench to pull the two pieces together to form a tight joint.

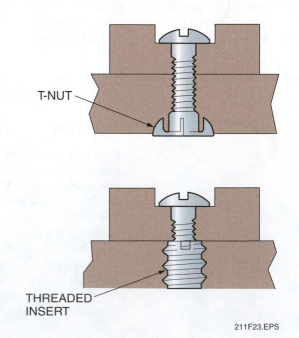

211F23.EPS

Figure 23 ◆ T-nut and threaded insert.

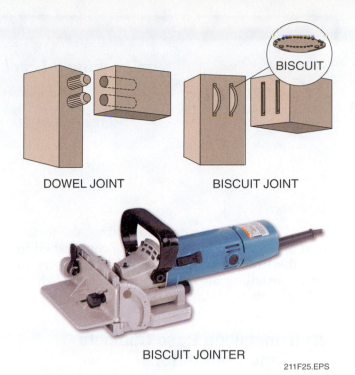

DOWEL JOINT   BISCUIT JOINT

BISCUIT JOINTER

211F25.EPS

Figure 24 ◆ Tight-joint fasteners.

Figure 25 ◆ Making joints using dowels and biscuits.

11.3.0 Dowels and Biscuit Fasteners

Dowels and biscuit fasteners are commonly used to strengthen butt, miter, and edge joints in cabinets. Fluted dowels are used to strengthen joints (*Figure 25*) by drilling matching (aligned) holes of the same depth in each piece to be joined. These holes should be drilled about $\frac{1}{16}$" deeper than the dowel length to contain the excess glue. After the holes are drilled, glue is squeezed into the holes in one of the pieces and the dowels are driven into the holes. Then, glue is squeezed into the holes in the opposite piece and the two pieces are fitted tightly together and clamped. They remain clamped until the glue is dry.

12.0.0 ◆ INSTALLING CABINETS

This section gives a general procedure for installing factory-built kitchen cabinets and countertops. Because kitchens normally require more cabinets than all other rooms combined, cabinet installation in general is best explained from the viewpoint of installing kitchen cabinets. The installations of other types of cabinets would be done using the same basic methods and procedures.

Cabinets are installed in accordance with the blueprints for the building and/or the design sketch or shop drawings that show their exact location and configuration within the room (*Figure 26*).

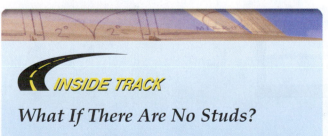

INSIDE TRACK

What If There Are No Studs?

In some homes and other buildings, wall studs are few and far between. Of course, if a wall is brick, there won't be any studs at all. For some installations, it may be necessary to build a new stud wall in front of the existing wall to accommodate the cabinets. For masonry that is in good condition, concrete anchors may be used.

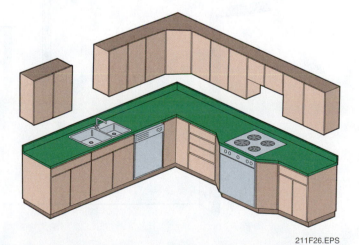

211F26.EPS

Figure 26 ◆ Simplified cabinet design sketch.

As with most carpentry procedures, the installation of kitchen cabinets can be accomplished in more than one way. Some carpenters prefer to start by installing the wall-mounted cabinets first; others choose to install the base units first.

12.1.0 Installing Base Cabinets and Countertops

The following general procedure for installing base cabinets and countertops assumes that the countertop used with the base cabinets is a manufactured unit that is ready to be installed.

Step 1   Make sure to thoroughly clean the area in which the units are to be placed. It is much easier to work in a room free of sawdust and scattered tools. If required,

remove any existing baseboard, window, or other trim from the walls in the work area.

Step 2   Check the wall surfaces with a straightedge for unevenness. Unevenness can cause cabinets to be misaligned, resulting in twisting of doors and drawer fronts. Wall high spots should be removed by shaving or sanding off excess plaster or drywall. Low spots can be shimmed later during cabinet installation.

Step 3   Locate and mark the position of all wall studs in the area where cabinets are to be installed (*Figure 27*). At each stud location, draw plumb lines on the wall. Locate the marks where they can easily be seen after the cabinets are in position.

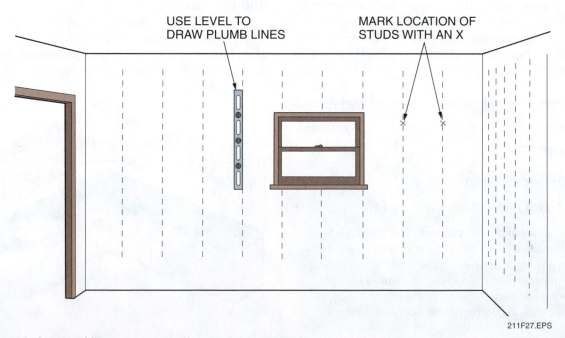

211F27.EPS

Figure 27 ◆ Marking stud locations on a wall.

Laying Out Cabinet Heights

Many carpenters use a water level, transit, or laser level to lay out the horizontal high point line. Some carpenters also use a story pole (layout stick) to mark the outlines for all the cabinets on the wall. A story pole is a narrow strip of wood that is cut to the exact length or height of a wall and marked with the different widths or heights of the cabinets to be installed on that wall. It is used to lay out and mark the locations of the cabinets on the wall by transferring the location marks from the story pole to the wall.

Step 4 Using a long straightedge and a 4' level, check the floor area where the base cabinets are to be installed for high spots. Find the highest point on the floor, then snap a level chalkline on the wall at this high point around the wall as far as the cabinets will extend (*Figure 28*). Using this line as a reference, measure up and mark the required distances for the tops of the base cabinets (normally 34½"), bottoms of the wall cabinets (normally 54"), and tops of the wall cabinets (normally 84"). Snap level chalklines around the walls at these heights.

Step 5 Remove the shipping skids, braces, and other packaging from the cabinets. Mark, then remove all the doors, drawers, and adjustable shelves and store them in a remote place where they cannot be damaged.

Step 6 Start the installation with a corner base unit. Slide it into place, then continue to slide the remaining base cabinets into position. Make sure that the types of cabinets and their placement match those shown on the building blueprints or cabinet layout sketch.

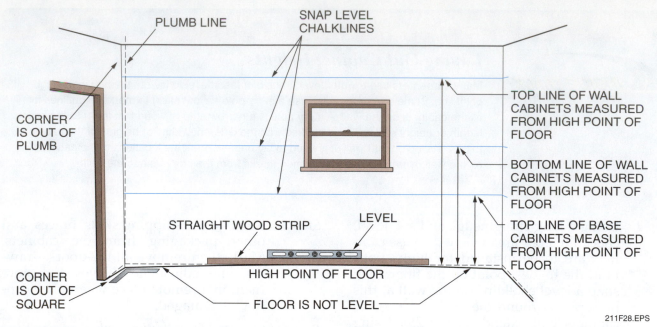

PLUMB LINE

SNAP LEVEL CHALKLINES

CORNER IS OUT OF PLUMB

TOP LINE OF WALL CABINETS MEASURED FROM HIGH POINT OF FLOOR

BOTTOM LINE OF WALL CABINETS MEASURED FROM HIGH POINT OF FLOOR

TOP LINE OF BASE CABINETS MEASURED FROM HIGH POINT OF FLOOR

STRAIGHT WOOD STRIP

LEVEL

CORNER IS OUT OF SQUARE

HIGH POINT OF FLOOR

FLOOR IS NOT LEVEL

211F28.EPS

Figure 28 ◆ Marking cabinet layout lines.

NOTE

Most frame-style cabinets have a ³⁄₁₆" or ¼" overlay on the stiles at the sides for applying finished sides. When joining two cabinets together, a gap of about ½" between the sides is created. A good practice is to shim and screw this gap to hold the cabinet square.

Step 7   When all of the base cabinets are in their correct positions, shim the corner cabinet so it is level and plumb and its top is aligned with the cabinet top level line marked on the wall. Fasten the cabinet to the wall by driving 2½" to 3" screws through the cabinet mounting rail into the studs or blocking in the wall.

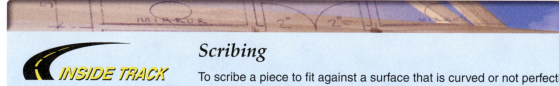

INSIDE TRACK

Scribing

To scribe a piece to fit against a surface that is curved or not perfectly straight and smooth, use the following procedure:

Step 1   Cut a piece of the appropriate material to length for a snug fit. Use a piece that is slightly wider than the space needs.

Step 2   Place the outside edge of the piece against the surface, so that the edge to be contoured is away from the surface.

Step 3   Use a piece of material that is the width you want the finished piece to be as a marking block. Slide the block along the surface, flat against the piece, with a pencil held against the block to mark the shape of the surface along the edge of the piece. Be sure to hold the point of the pencil firmly against the corner between the filler and the marking block, and keep the block firmly against the surface.

Step 4   Once you have the piece marked with the surface contour, use a belt sander, plane, hand planer, or, if the contour is not a straight line, a bandsaw or jig saw to duplicate the marked contour.

Step 8 Use suitable clamps to clamp the frame members of the corner and the adjoining cabinet together (*Figure 29*). Shim as necessary to make sure that the horizontal frame members form a level and straight line and that the frame faces are flush. After drilling countersunk pilot holes, fasten the cabinets together using screws or T-nuts and machine screws. One screw should be installed close to the top of the cabinet end stile and one close to the bottom. If using screws, lubricate the screws with wax or soap to prevent the screws from binding or snapping in hardwood.

Step 9 Continue to install the remaining base cabinets in the same manner as described above. Check the cabinet tops from front to back and across the front edges with a level (*Figure 30*). Shim as necessary between the wall and cabinet backs and the floor and cabinet bottoms until the cabinets are plumb and level and the tops are aligned with the mark on the wall.

Step 10 Fasten the cabinets to the wall by driving 2½" to 3" screws through the cabinet mounting rails into the studs in the wall. Use a chisel or utility knife to cut off any shims flush with the edges of the cabinet. If required, scribe and cut a filler strip to fit the remaining space between the cabinet and end wall. Clamp, then fasten the filler strip to the cabinet stile with countersunk screws. Use glue with thin filler strips.

Step 11 Place the countertop on top of the base cabinets and firmly against the wall. If the countertop backsplash does not fit tightly against the wall, scribe the backsplash to match the irregular wall surface. Following this, remove the countertop, place it on sawhorses, then plane or belt sand the backsplash to the scribed line. It is important to make sure the overhang is uniform before scribing the backsplash.

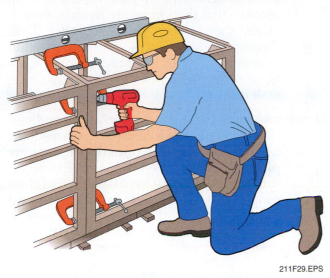

211F29.EPS

Figure 29 ◆ Joining base cabinet stiles.

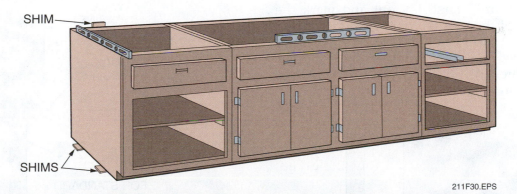

SHIM

SHIMS

211F30.EPS

Figure 30 ◆ Shimming cabinets so that they are plumb and level.

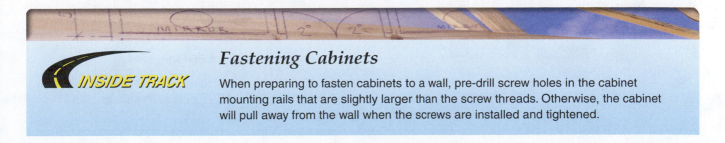

Fastening Cabinets

INSIDE TRACK

When preparing to fasten cabinets to a wall, pre-drill screw holes in the cabinet mounting rails that are slightly larger than the screw threads. Otherwise, the cabinet will pull away from the wall when the screws are installed and tightened.

Step 12 Reposition the countertop on top of the base cabinets and tightly against the wall and check the fit. Fasten the countertop from underneath by drilling and screwing through the corner blocks and/or diagonal bracing in the top of the base cabinets. Be careful not to drill through the countertop surface.

NOTE

The sink cutout will be cut in the countertop later in this procedure. Cover the countertop so that it is adequately protected from damage during the installation of the wall cabinets.

Most cabinet manufacturers offer filler strips. These strips can be attached at the end of a run, or at a corner, to provide drawer and door handle clearance, or to eliminate gaps between the cabinet and wall (*Figure 31*). While filler strips are usually not needed for custom cabinets, they are commonly needed for standard cabinets. The filler strip will need to be ripped to the required width and may need to be scribed to compensate for irregularities in the wall surface or to fit around baseboards.

12.2.0 Installing Wall Cabinets

To install wall cabinets, proceed as follows:

Step 1 Start the wall cabinet installation with a corner unit. First, lay out and mark the back of the cabinet for the stud locations, then predrill through the top and bottom cabinet mounting rails at these points. Countersink the drilled holes on the mounting rails.

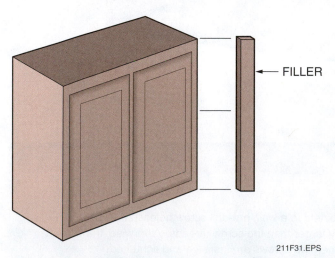

Figure 31 ◆ Filler strip.

Step 2 Position the corner cabinet in place with its bottom on the chalkline and temporarily hold it in place using a T-brace (*Figure 32*) or support it with a temporary cleat as shown in *Figure 32*. Make sure to get this cabinet installed plumb and level, using shims as necessary. Fasten the cabinet in place with screws of sufficient length to hold the cabinet securely against the wall, but do not fully tighten the screws.

Step 3 Position, brace, and fasten the adjoining wall cabinet so that leveling may be done without removing it. Align the adjoining stiles so that their faces are flush with each other, then clamp them together. Drill through the stiles and screw them together as was done with the base cabinets. Continue this procedure for the remaining wall cabinets.

Step 4 Use a level to check the horizontal and vertical cabinet surfaces. Shim between the cabinets and the wall until the cabinets are plumb and level (*Figure 33*). This is necessary if the doors are to fit, swing, and close properly. When the cabinets are level and plumb, tighten all mounting screws. If required, cut and install any filler strips.

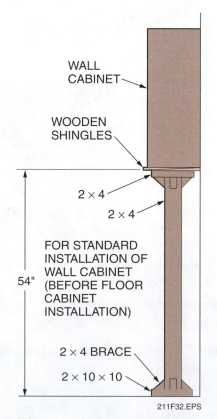

Figure 32 ◆ T-brace used to support wall cabinets.

TEMPORARY
CLEAT

211F33.EPS

Figure 33 ◆ Shimming wall cabinets.

12.3.0 Completing the Installation

To complete the installation, proceed as follows:

Step 1 Using the pattern supplied with the sink as a guide, drill starter holes in the corners of the pattern outline, then cut the sink cutout hole in the countertop using a saber saw and fine-toothed blade.

Step 2 Replace the doors, drawers, and shelves in their respective cabinets. Ensure that the doors open and close properly and that they are installed level and properly aligned with the doors on the other cabinets.

Summary

Cabinets and countertops are almost always made in a shop and installed somewhere else. The minor variations in angles and lines common to all trades work become critically visible in cabinet installation, and a test of the installers' skills. Knowing the ways the cabinets are constructed will allow the installer to fit them to the space better and to fasten them firmly. The work is difficult, and customers will appreciate good results. Exact measurements, precise layout and cutting, and careful assembly are required to produce a professional installation.

Notes

1. In terms of expense and quality, and starting from the lowest to the highest, cabinets are graded as _____.
   a. semi-custom, custom, and stock
   b. stock, semi-custom, and custom
   c. stock, custom, and semi-custom
   d. custom, semi-custom, and stock

2. The base cabinet part marked *A* in the figure below is the _____.
   a. countertop
   b. kickboard
   c. face frame
   d. bulkhead (interior standard)

3. The base cabinet part marked *B* in the figure below is the _____.
   a. end panel
   b. rail
   c. face frame
   d. bulkhead (interior standard)

4. The wall cabinet part marked *C* in the figure below is a _____.
   a. stile
   b. furr down
   c. door guide
   d. shelf support

5. All of the following natural woods are normally used in high-quality cabinets *except* _____.
   a. oak
   b. birch
   c. sweetgum
   d. cherry

6. A non-rabbeted cabinet door that mounts on the outside frame of the cabinet and conceals the cabinet opening is called a(n) _____ door.
   a. overlay
   b. lipped
   c. flush
   d. oversized

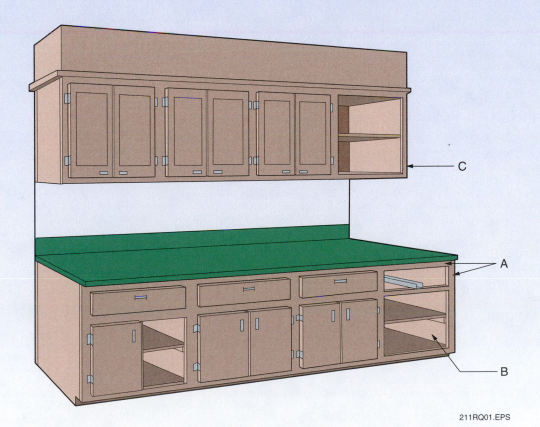

211RQ01.EPS

7. Most wooden cabinet shelves should be made of _____ stock.
   a. ¼"
   b. ⅜"
   c. ½"
   d. ¾"

8. One of the advantages of solid-surface countertops over laminate countertops is that solid-surface countertops are _____.
   a. less expensive
   b. more durable
   c. easier to fabricate
   d. available as prefabricated units

9. When measuring up from a level reference line marked at the high point of the floor, a line representing the tops of base cabinets would normally be drawn at _____ above the reference line.
   a. 25½"
   b. 32"
   c. 34½"
   d. 36"

10. Most frame-style cabinets have a small overlap on the stiles to _____.
    a. allow for application of finished sides on end cabinets
    b. allow for installation of a filler strip
    c. create a flush fit to a wall
    d. allow for the use of shims to square the cabinet

Trade Terms
Introduced in This Module

Backsplash: A piece that extends up the wall from a countertop.

Escutcheon plate: A decorative metal plate that is placed against a door or drawer face behind the pull.

Furr down: The enclosed section (which is usually decorated by paint or wallpaper) between the ceiling and the top of the wall cabinets; also called a soffit.

Veneer: A thin layer or sheet of wood intended to be overlaid on a surface to provide strength, stability, and/or an attractive finish. Thicknesses range between $\frac{1}{16}$" and $\frac{1}{8}$" for core plies and between $\frac{1}{125}$" and $\frac{1}{32}$" for decorative faces.

This module is intended to be a thorough resource for task training. The following reference works are suggested for further study. These are optional materials for continued education rather than for task training.

Cabinet Makers Association website, *www.cabinetmakers.org*.

Kitchen Cabinet Makers Association website, *www.kcma.org*.

Mill's Pride Cabinetry website, *www.millspride.com*.

NCCER CURRICULA — USER UPDATE

NCCER makes every effort to keep its textbooks up-to-date and free of technical errors. We appreciate your help in this process. If you find an error, a typographical mistake, or an inaccuracy in NCCER's curricula, please fill out this form (or a photocopy), or complete the online form at **www.nccer.org/olf**. Be sure to include the exact module ID number, page number, a detailed description, and your recommended correction. Your input will be brought to the attention of the Authoring Team. Thank you for your assistance.

Instructors – If you have an idea for improving this textbook, or have found that additional materials were necessary to teach this module effectively, please let us know so that we may present your suggestions to the Authoring Team.

NCCER Product Development and Revision

13614 Progress Blvd., Alachua, FL 32615

Email: curriculum@nccer.org
Online: www.nccer.org/olf

❑ Trainee Guide   ❑ AIG   ❑ Exam   ❑ PowerPoints   Other _____

Craft / Level: _____   Copyright Date: _____

Module ID Number / Title: _____

Section Number(s): _____

Description: _____

_____

_____

_____

Recommended Correction: _____

_____

_____

_____

_____

Your Name: _____

Address: _____

_____

Email: _____   Phone: _____

Cabinet Fabrication
27212-07

27212-07
Cabinet Fabrication

Topics to be presented in this module include:

Overview

Some students will realize during their training that they like building finished products such as cabinets. There are many small companies that build and install custom cabinets designed to fit into a specific area or to serve a particular need. Built-in entertainment centers and bookcases are examples of this work. Like trim carpentry, this is work that requires great precision, attention to detail, an eye for design, and the ability to use a variety of specialized tools that are unique to cabinet and furniture making.

Objectives

When you have completed this module, you will be able to do the following:

1. Recognize the common types of woods used to make cabinets.
2. Correctly and safely use stationary power tools.
3. Identify and cut the various types of joints used in cabinetmaking.
4. Build a cabinet from a set of drawings.
5. Install plastic laminate on a countertop core.

Trade Terms

Heartwood
Pawls

Sapwood

Required Trainee Materials

1. Pencil and paper
2. Appropriate personal protective equipment

212CMAP.EPS

Prerequisites

Before you begin this module, it is recommended that you successfully complete *Core Curriculum*; *Carpentry Fundamentals Level One*; and *Carpentry Framing and Finishing Level Two*, Modules 27201-07 through 27211-07.

This course map shows all of the modules in *Carpentry Framing and Finishing Level Two*. The suggested training order begins at the bottom and proceeds up. Skill levels increase as you advance on the course map. The local Training Program Sponsor may adjust the training order.

1.0.0 ◆ INTRODUCTION

One career option for carpentry trainees is cabinetmaking. A cabinetmaker uses a variety of stationary power tools to cut out, shape, and finish the pieces of a cabinet. The cabinetmaker also assembles cabinets using special joinery methods and fasteners. Like finish carpentry, the work of a cabinetmaker requires skill, patience, and attention to detail.

In the *Core Curriculum* and previous *Carpentry* modules, you were introduced to many types of woods, fasteners, and hand and power tools commonly used in carpentry work. This module expands on those foundations, examining in more detail some of the tools used in cabinet construction and the joining techniques used in cabinetry. Under the supervision of your instructor and/or supervisor, you will learn how to use these tools, perform joining techniques, and assemble typical cabinets.

2.0.0 ◆ WOODS AND MATERIALS USED IN CABINET CONSTRUCTION

High-quality cabinets are made from veneer and solid hardwoods, which are typically oak, cherry, birch, ash, or hickory. Hardboard, particleboard, plywood, or medium-density fiberboard (MDF) is commonly used for interior panels, for drawer bottoms, and as the base for plastic laminate countertops. Quality cabinets will incorporate the following features:

*   Face frame cabinets should have a ¾" hardwood face frame, ⅜" to ½" plywood or particleboard side frames, and a ¼" back panel.
*   Frameless cabinets should use ⅝" to ¾" particleboard for the entire cabinet. Wood veneer or high-pressure laminates are used for the exterior finish.
*   Drawers will slide smoothly with little play. They will close quietly and solidly. It is typical for metal ball bearing slides to be used. They can be regular or full extension drawers based on the specification. Look for a ½" to ¾" solid wood or 9-ply solid birch plywood drawer box with integral wood dovetail, shoulder, or dado joints. The box is typically attached to the drawer front with screws.
*   Door hinges and catch mechanisms will be well made and work without binding.
*   Rabbeted joints will be used where the top, bottom, back, and side pieces join.

2.1.0 Solid Woods

A wide variety of softwoods and hardwoods are used in making the visible exterior cabinet surfaces such as frames, doors, and drawers. This section provides an overview of the characteristics of various natural woods used for this purpose.

*   *Ash* – White ash is a very hard, dense, strong wood often used for tool handles that produces satin smooth finishes when machined properly. It is able to resist a great deal of shock and is well known for its bending abilities. White ash is light tan in color with light brown or red streaks.
*   *Beech* – Beech is hard, strong, and fairly dense. It can be bent and machined very easily and is often used in place of more expensive woods. Beech is pale tan to reddish-brown in the **heartwood**.
*   *Birch* – The two main species of birch lumber used in cabinetry are yellow and sweet birch, which is also called black or cherry birch. The wood of yellow and sweet birch is quite strong and hard. It machines well and finishes to a fine luster. The color is yellow to pinkish-white in the **sapwood** and is tan with a touch of red in the heartwood. Birch is considered one of the major cabinet woods and is often used to make doors and interior trim.
*   *Cedar* – Cedar is a semi-dense softwood with excellent machining qualities. It has a contrasting red heartwood and white sapwood. Aromatic cedar, because of its aroma and moth-repelling qualities, is most commonly used in building cedar chests and lining drawers and closets. Western red cedar and eastern white cedar do not have moth-repelling qualities. These types are used for siding, roofs, trim, and other outdoor applications.
*   *Cherry* – Black cherry (wild cherry) is a top-quality cabinet wood. It is part of the fruitwood group and is heavy, strong, and hard. This wood machines extremely well to produce a fine texture and luster when it is finished. The sapwood of the black cherry tree is yellow to pink; the heartwood is reddish-brown yet sometimes tinted with a bit of green. Cherry darkens to a deep red with exposure to the sun. The term used for this darkening is patina. An ultraviolet protective finish should be used where uneven sun exposure could cause uneven patina.
*   *Douglas fir* – Douglas fir is a moderately heavy and strong softwood. It is easy to work with but tends to splinter. Its color ranges from yellow to

yellow-red or reddish-brown. Douglas fir is one of the primary woods used for plywood veneer. It is also used for millwork and interior trim. Douglas fir is seldom used for an exposed surface in a cabinet; rather, it is used as the core material in the plywood used to make cabinets. It is extremely strong.

- *Genuine mahogany* – Genuine mahogany, also known as Honduras mahogany, is considered the finest cabinet wood in the world. It has been popular with furniture makers for many years. It is unsurpassed in machining and finishing qualities and can be polished to a high sheen. The color is an orange-brown. Genuine mahogany is used both as solid lumber and veneer facing on plywood in cabinets. It grows from the Everglades through the rain forests of Central America and into the Amazon.

- *African mahogany* – The qualities of African mahogany are considered equal to those of genuine mahogany. It has the same workability, stability, and durability. Its color is reddish-brown. It is popular as solid lumber and as a veneer facing on plywood in cabinets.

- *Philippine mahogany* – Philippine mahogany (lauan) is a cheaper grade of wood with a medium hardness and density. Lauan is normally orange-brown or reddish brown. It machines well but does not possess the same finishing qualities as the other types of mahogany.

- *Maple* – Hard maple comes from the sugar maple tree. It has a hard, dense, and strong composition. Hard maple is very resistant to wear and has an extremely fine texture and grain. The color varies from creamy-white to light tan. Hard maple has a very nice machining quality that makes it a prime candidate for making fine furniture. Hard maple is also used in flooring, especially in skating rinks, dance halls, and bowling alleys. It is the standard wood used for cutting boards and butcher block countertops. Soft maple has been used as a substitute for hard maple. It is not as hard or dense, but it has many of the visual and working characteristics of hard maple.

- *Oak* – Oak is probably the most well-known of the furniture woods. There are many different species of oak used in lumber production; however, white and red oak are the types most commonly sold. The differences between them are important in selecting the right wood for a cabinetmaking project. White oak has a tan or brownish color and is more water-resistant than red oak. Red oak is somewhat reddish and

does not stand up well against moisture when used outside. Both white oak and red oak are fine cabinet woods. Both bend quite well and are very stable. Oak also comes in a fine plywood form.

- *Pecan* – Pecan is a strong, durable wood with light brown heartwood and white sapwood. It is sometimes finished to simulate oak and walnut but has its own beauty if finished properly.

- *Pine* – There are 35 species of pine native to North America. About half are used in lumber production. There are notable differences among the species of pine in terms of color, texture, knottiness, moisture resistance, and stability. On the West Coast, the four main pine species used for lumber are the ponderosa, lodgepole, Western white (also known as Idaho white), and sugar pines. The heartwood in these species ranges from light yellow to light brown, and the sapwood is almost white. This wood is generally straight-grained, light, and easy to work with. Of the four, the Western white and sugar pines are the least apt to shrink or warp. On the East Coast, the principal species are the Eastern white and red (also known as Norway) pine. Both of these species are sold as white pine, but there are differences between them. The Eastern white pine's heartwood is light brown and darkens when exposed to air. It is generally straight-grained, easy to work, and highly stable wood. The red pine is slightly heavier and not as uniform in texture as the white pine. Its heartwood varies from pale red to reddish-brown. The Southern pines, which include the shortleaf, loblolly, slash, and longleaf species, are similar in appearance, with reddish-brown heartwood and yellowish-white to light orange sapwood.

- *Sweet gum* – Sweet gum is also called red gum because of the color of its autumn leaves. In the South, sweet gum is used in the furniture industry more than any other wood. The major reason for this is that it blends with the more expensive woods such as walnut. Sweet gum is fairly strong. It machines quite well. The sapwood is usually bland white, but the heartwood can be anything from mild tan to a heavy, dark brown. Sweet gum is rarely used for cabinets and millwork. It does not paint well and can be blotchy when stained. It also comes in plywood form.

- *Walnut* – Of the five species of North American walnut, black walnut is the principal lumber species. A dense, hard, and strong wood, it is one of the most popular furniture woods and is very expensive. It polishes to a deep, beautiful finish. The sapwood is white, while the heartwood ranges from light gray-brown to dark purple. Walnut is in very limited supply these days, which keeps the prices high. It is an excellent wood for carving and other machining techniques requiring crisp, good-looking joints and lines. Because of this property, it is often used for sculptured and contemporary furniture. Walnut is also used for cabinets, musical instrument cases, stereo cabinets, and gun stocks because of its exquisite beauty. A great deal of walnut is used as veneer and plywood. While beautiful, walnut is one of the most defective of all cabinet-grade woods; therefore, lengths over 8' can be hard to find.

- *Yellow poplar* – Poplar is a close-grained, light, and fairly soft wood. Its color ranges from yellow to brown and usually has a grayish tinge. This is a popular wood because it machines well with hand tools. In the furniture industry, it is primarily used for the more economical grades of furniture. It is the best hardwood for painting an opaque color. With proper care, it can also be stained. It is readily available in long, relatively defect-free lengths and as veneer.

2.2.0 Plywood

Plywood is an excellent building material and one of the most important modern cabinetmaking products. It has several advantages over lumber including availability in very large sheets, exceptional strength, and high resistance to warping. It is lightweight yet very strong in comparison to solid wood. It is also great for cabinets. Expensive wood can be used in plywood because it covers only the top ply; also, plywood does not split as easily as solid wood and is more stable across the grain. Thin wood layers (veneers) are glued together to form plywood sheets (*Figure 1*). Each veneer's grain runs at right angles to the next for strength and stability.

INSIDE TRACK

Book-Matched and Slip-Matched Veneers

During manufacturing, veneer patterns may be matched to produce a uniform, visually pleasing effect. Book-matched veneers are matched back to front to produce a matched design. Slip-matched veneers are joined side to side to create a repeating pattern.

Matched veneer panels can be ordered for custom cabinets made from the highest quality veneers. Matched panels are made using veneers produced from a single tree. The panels are numbered in sequence so that grain can flow from panel to panel.

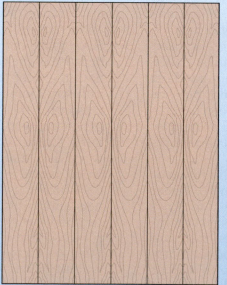

BOOK-MATCHED PANEL   SLIP-MATCHED PANEL

212SA01.EPS

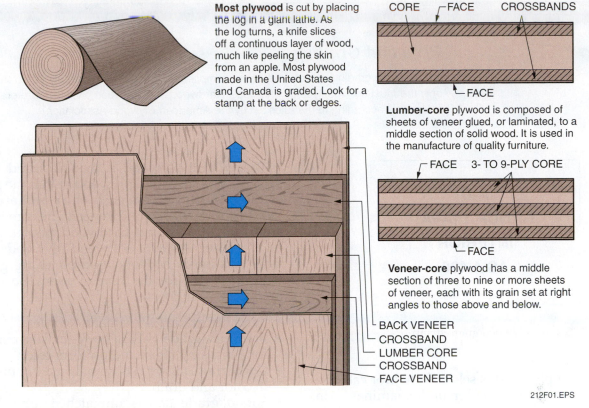

Most plywood is cut by placing the log in a giant lathe. As the log turns, a knife slices off a continuous layer of wood, much like peeling the skin from an apple. Most plywood made in the United States and Canada is graded. Look for a stamp at the back or edges.

Lumber-core plywood is composed of sheets of veneer glued, or laminated, to a middle section of solid wood. It is used in the manufacture of quality furniture.

Veneer-core plywood has a middle section of three to nine or more sheets of veneer, each with its grain set at right angles to those above and below.

CORE — FACE — CROSSBANDS — FACE

FACE — 3- TO 9-PLY CORE — FACE

BACK VENEER — CROSSBAND — LUMBER CORE — CROSSBAND — FACE VENEER

212F01.EPS

Figure 1 ◆ Veneer layers of plywood.

All plywood is divided into one of two categories: softwood or hardwood. The difference is in the species of wood covering the outer faces of a panel. Although both types are used in cabinetmaking, hardwood plywood is usually more expensive than softwood but is worth the extra money for its superior appearance.

2.2.1 Softwood Plywood

The most commonly used plywood in construction is softwood plywood, usually made of Douglas fir. Other softwood species include Western hemlock, redwood, cedar, Southern yellow pine, Western larch pine, and white fir.

All plywood comes in standard 4' × 8' sheets, although larger sizes are available. Nominal thicknesses of softwood plywood are ¼", ⅜", ½", ⅝", and ¾". Softwood plywood is graded according to the appearance of the front and back faces. Softwood grades (from finest to poorest) are designated N, A, B, C, and D.

- Grade N is defect-free with a uniform color and is used only when a flawless natural finish is desired.
- Grade A has neatly made repairs and is the type most commonly found in lumberyards.
- Grade B may have oval repair plugs and tight knots.

- Grade C can have knots slightly larger than 1" and is allowed to have small splits.
- Grade D, the poorest, has larger knots and wider splits. This grade should be used only where it will be hidden from view. It is not recommended for cabinet work.

Both faces of a panel graded A-A would be excellent when seen from both sides. A-C is fine when only one side will show and is the most common grade found in lumberyards. It is very good for built-ins. A-A grade is not stocked in some lumberyards, and its cost would rival some lower cost hardwood plywoods.

The core of most plywood is made from the same grade of veneer as used on the worst panel face, so you must judge the strength and suitability of a specific grade panel for the purpose intended by the quality standards of the core. Plywood is graded according to standards set by the American Plywood Association (APA). *Figure 2* shows an example of a grade stamp, which can be found on every sheet of plywood.

The first line of this stamp (A-D) indicates the grade of the front and back panels. The second line (GROUP 1) indicates the standards group followed in manufacture. The third line (INTERIOR) indicates the suitable application of the plywood. The last line (AS 4-430) indicates the product standards number and the mill number.

A-D

GROUP 1 (APA)

INTERIOR

AS 4-430

212F02.EPS

Figure 2 ◆ Sample plywood grade stamp.

Softwood plywood is seldom used for cabinet-making except for bases or under laminates. In such cases where the plywood would be visible at all, grades below B would not be suitable.

2.2.2 Hardwood Plywood

The main types of domestic hardwood veneers on plywood sheets are ash, birch, black walnut, oak, cherry, gum, maple, and poplar. Several imported woods such as lauan, mahogany, and teak are also available in specialized lumberyards or through mail order. In the hardwood plywood category, different fillers or inner cores are used. One type has a core of extra thick veneer, while another has a core made of particleboard (*Figure 3*).

In the particleboard-core plywood, the particleboard is sandwiched between two veneer faces. The inner core can be either conventional industrial 45-pound board or denser MDF board. It is strong and works well, but it chips a bit when sawing, so use caution.

Veneer core plywood is manufactured in the same way as softwood plywood in that the grains of the plies run across each other and are then faced with the hardwood veneer. The core is usually made of a tropical hardwood, poplar, aspen, or fir veneer. As with softwood plywoods, both faces of a hardwood plywood panel are graded without regard for the type of core. The veneer grades are as follows:

- Specialty grade (SP) is made to order.
- Premium grade (#1) is book- or slip-matched for a pleasing effect and uniform color.
- Good grade (also #1) is unmatched but uniform in color and grain.
- Sound grade (#2) is unmatched for color or grain, and small pin knots and repairs are allowed.
- Utility grade (#3) allows all natural defects, including open knots, wormholes, and splits, but it is fine for painting.
- Backing grade (#4) has unlimited defects. It should be used only in hidden areas.

A third type has a core of solid wood. This solid wood type is called lumber core plywood. The solid wood core consists of many strips of wood glued together much like the wood in a butcher block. Typically, the wood used in the core is a tropical hardwood or poplar. The use of this type of plywood for fine cabinet work was common in the past. Today it is rarely used for this purpose.

The plywood discussed here only includes the types used in cabinetmaking and built-ins covered in this module. There are many other types of plywood produced for specific situations and

Cutting Plywood

A table saw is the best tool for cutting plywood straight and is a must for a job site workshop. To avoid splintering the edge, work with the good edge up when cutting on the table saw and with the good edge down when using a circular saw. Use only a sharp plywood blade or a fine-toothed power saw blade. The saw blade should have very little set (a term that refers to the curve in the blade's teeth) and should project above (or below) the plywood surface just at the height of the teeth. Another method is to tape the plywood with masking tape, draw your line, and cut through the tape. The tape prevents the edges from splintering.

Make sure the panel is supported properly to keep it from sagging during the cutting and to prevent the blade from binding and jumping out of the cut. When cutting an angle with a circular saw, it is best to use a guide clamped to the piece so that a straight cut is achieved. Use extreme care when making the initial cuts so the project will fit together properly. The best working combination at a job site is a table saw, two sawhorses at the same height, and a full plywood sheet for a bench top.

Vertical panel power saws, like the one shown here, are used in commercial cabinetmaking operations.

212SA02.EPS

PARTICLE-
BOARD CORE

VENEER
CORE

212F03.EPS

Figure 3 ◆ Particle core and veneer core plywood.

exposures. Each type of plywood is graded according to how it is manufactured, what it is made of, what type of glue is used, and where it should be generally used.

Plywood is the best material for most job-built units because of its working ease and availability. In order to take advantage of these special characteristics, different techniques are needed to work with plywood. Most importantly, because plywood comes in large sheets, it is necessary to lay out the project on the wood or on paper before cutting. You must also remember to cut the pieces so the surface grain runs in the proper direction. For example, the grain of both the doors and the sides of cabinets should run in the same direction.

2.3.0 Particleboard

Particleboard is made from leftover materials from wood mills as well as trees that are not suitable for other purposes. Small chips of wood, sawdust, and even bent and twisted trees can be used. This creates a cheaper panel, which is a very important product in the cabinetmaking industry. There are two kinds of particleboard: mat-formed (flat) and extruded (thin, flat, or shaped). Most particleboard used in the cabinet industry is mat-formed. In either case, the basic raw materials are wood particles left over from planing mills, veneer plants, and pulp mills. The wood particles are first sorted for size by screening, dried, and

mixed with a resin glue. This material is then deposited on mats where it is heated and compressed at pressures up to 1,000 pounds per square inch (psi). The resulting particleboard mats are then cut to size (typically 4' × 8') and sanded smooth in huge sanders.

The particleboard used for most cabinets and built-ins will be standard medium density or cabinet grade (45-pound density). It is available in thicknesses from ⅜" to 1". Generally, ¾" particleboard is recommended for most cabinet projects. Extruded particleboard is commonly used by manufacturers to make countertops with preformed splash and drop edges. An advantage of this method is that plastic laminate can be bonded to the top as it is being made.

Machining can be carried out with ordinary saws, drills, and routers if they are sharp. Carbide is recommended. The panels are heavy (about 60 pounds for a ¾" × 4' × 8' panel), so be sure to provide adequate support when cutting them. Particleboard has no grain, so there is less chance of warping. It is ideal for covering with plastic laminates or wood veneer. Although particleboard can be fastened with nails, it does not hold them as well as plywood or solid wood. Cement-coated nails are best, but screws with pilot holes also provide good holding power. Almost any joint can be used with plywood or solid wood. However, the best holding power can be achieved with dado or rabbet joints.

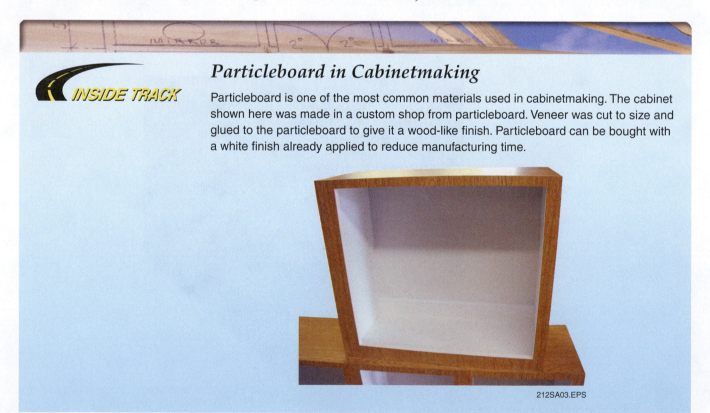

INSIDE TRACK

Particleboard in Cabinetmaking

Particleboard is one of the most common materials used in cabinetmaking. The cabinet shown here was made in a custom shop from particleboard. Veneer was cut to size and glued to the particleboard to give it a wood-like finish. Particleboard can be bought with a white finish already applied to reduce manufacturing time.

212SA03.EPS

Particleboard can be bought with a liner (usually white) on both sides. This saves time when building the main cabinet box frames. The inside of the cabinet can also be sprayed with sanding sealer to protect it from moisture. Particleboard will raise up and swell if allowed to get wet. Old material should be thoroughly inspected before use. If a cabinetmaker has a job to be covered with a plastic laminate, the best choice of material would be ¾" water-resistant, medium-density fiberboard (MDF) because it will not swell if it gets wet.

3.0.0 ◆ SHOP TOOLS USED IN CABINETMAKING

Cabinetmaking is best done with stationary tools designed for the purpose. Although it is possible to use handheld power tools, such as a circular saw, drill, router, and sander, the results will not be as satisfactory. This section covers the shop tools commonly used in cabinetmaking. Keep in mind that many of the tools can serve multiple purposes. For example, either a table saw or router table can be used to make a dado cut, and a router can do many of the tasks done on a shaper.

3.1.0 Table Saw

As its name implies, a table saw consists of a saw blade mounted in a level table. *Figure 4* shows a typical table saw. A circular saw blade is mounted below the surface of the table and driven by a motor. The blade may be raised or lowered to alter the depth of the cut. The angle of the blade may be adjusted for miter cuts (*Figure 5*). Note that the operator in *Figure 5* is using a push stick. The use of push sticks and push blocks (push boards) is very important. These tools can keep you from losing fingers or being injured by a kickback.

The key feature of the table saw is the table itself. The table is a machined metal surface with a slot through which the blade protrudes from below. Over the slot is a blade guard designed to protect the operator's hands from injury while operating the saw. On many table saws, the blade guard includes a kerf spreader to keep the cut pieces of wood separate from each other and prevent the blade from binding in the wood. The guard may also include an anti-kickback device. This device consists of sharp **pawls**, which keep the wood from being forced back toward the operator if the wood catches on the blade.

The table is grooved to hold a guide called a miter gauge, which provides a flat, perpendicular surface for pushing the stock through the blade. The miter gauge can be adjusted to any angle up to 60 degrees left or right for miter cuts.

A rip guide, or fence, can be attached to the table, parallel to the blade, for ripping lumber lengthwise. The fence clamps to the front and back of the table and has rails so that the distance between the fence and the blade can be adjusted.

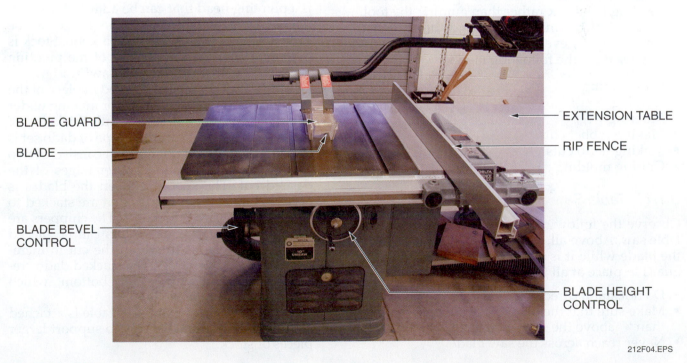

BLADE GUARD

BLADE

BLADE BEVEL CONTROL

EXTENSION TABLE

RIP FENCE

BLADE HEIGHT CONTROL

212F04.EPS

Figure 4 ◆ Table saw.

212F05.EPS

Figure 5 ◆ Using the table saw to make a beveled cut.

The table saw is best suited for any applications where a steady, straight cut is required. The stability of the table saw allows the operator to focus on feeding the stock rather than guiding the tool. Proper use of the miter gauge and the rip fence ensure accurate, even cuts. Table saws are commonly used for the following tasks:

- Crosscutting
- Ripping
- Mitering
- Making rabbet cuts
- Making dado cuts
- Cutting molding

3.1.1 Table Saw Safety

Observe the following safety rules when using a table saw. Above all, keep your hands away from the blade while it is running, and keep the blade guard in place at all times.

- Do not stand directly in line with the blade.
- Make sure that the blade does not project more than ⅛" above the stock being cut.
- Never reach across the saw blade.

- Use a push stick for ripping all stock less than 4" wide.
- Keep the guard over the blade while the saw is in use.
- Disconnect power when changing blades or performing maintenance.
- Never adjust the fence or other accessories until the saw has stopped.
- Enlist a helper or use a work support when cutting long or wide stock. The weight of the stock may cause the table to lean.
- Do not rip without a fence nor crosscut without a miter gauge.
- When the blade is tilted, be sure that the blade will clear the stock before turning on the machine.
- Be sure that the stock has a straight edge before ripping it.
- When using a dado or molding head, take extra care to hold the stock firmly.
- Never remove scraps from the saw table with your hands or while the saw blade is running.
- Use the proper blade for the job being done.

3.1.2 Table Saw Accessories

A wide variety of accessories are available for table saws. The primary accessories are described here:

- *Miter gauge* – The miter gauge (*Figure 6*) has a long strip that fits into a groove on the surface of the table saw. Attached to one end of the strip is a pivoting head that can be adjusted up to 60 degrees in either direction. The gauge can be locked at a desired angle with a knob. Stock is placed against the flat surface of the pivoting head and pushed through the saw.
- *Dado set* – A dado set can be used in place of the standard blade in a table saw for making wider cuts that don't penetrate the full thickness of the stock. The most common type of dado set is the stacked dado set (*Figure 7*). It consists of two blades that are set at the outer edges of the desired cut. The space between the blades is occupied by chipper blades that are stacked to provide the desired thickness. The chippers are typically ⅟₁₆" or ⅛" thick. Shims can be placed between the blades to widen the cut in increments as small as 0.004". The stacked dado creates a dado cut with a flat bottom, which provides a more stable joint.
- *Extension table* – An extension table is attached to either end of the table saw to support larger pieces of stock.

Figure 6 ◆ Miter gauge.

Figure 8 ◆ Radial arm saw.

Figure 7 ◆ Stacked dado set.

3.2.0 Radial Arm Saw

A radial arm saw (*Figure 8*) is a circular saw mounted on an overhead turret arm assembly. This keeps the saw on a steady track for making cuts while allowing it to be adjusted in any of several different directions.

The saw is mounted on a yoke that slides along the turret arm. In normal use, the yoke assembly is moved along the arm to make quick crosscuts. The yoke has a pivot built into it so that the blade can be turned up to 180 degrees to the left or right. This allows longer pieces of stock to be passed through the blade.

The arm of the radial arm saw can be raised or lowered to vary the depth of the cut. It can also be made to revolve around the supporting column. This makes the radial arm saw a very versatile tool. The radial arm saw is normally pulled toward the operator while cutting. A fence at the back of the table holds the stock in place.

Radial arm saws are used primarily for crosscutting lumber, which is a task that can be quickly and easily done by placing the stock against the fence on the table and pushing the saw through the stock. The flexible positioning allowed by the combination of the turret arm and the yoke allows miter cuts to be made.

The radial arm saw can be used for making dado cuts and, with the addition of specialized accessories, for cutting molding. Ripping can also be performed with a radial arm saw, but the table saw should be used for this purpose if there is one available. The blade must be set to be perpendicular to the turret arm and then locked in place. The stock is fed through as it would be done with a table saw.

The radial arm saw has many of the same accessories that are available for table saws, including stacked dado heads and molding heads.

3.2.1 Radial Arm Saw Safety

Observe the following safety practices when using a radial arm saw:

- Make sure the saw blade is set to the proper depth before turning it on.
- Use only a sharp and properly set blade in the saw.
- Do not stand directly in line with the blade.
- Allow the saw to reach maximum speed before starting the cut.
- Hold the stock firmly against the fence.
- Do not let the saw feed too fast when crosscutting heavy stock. The blade has a tendency to feed itself, which can result in jamming.
- Do not stop the blade by forcing a piece of scrap wood against the blade.

Stop Blocks

A stop block is used to set up a saw to cut several pieces of the same length. In this case, it is being used with a frame and trim saw. The stop block can be made from any convenient piece of material. It is clamped tightly to the fence at the desired distance from the blade. Note that a wedge is cut out of the stop block. Without the cutout, sawdust would pack into the space between the material and the stop block and would gradually shorten the lengths of the cut pieces.

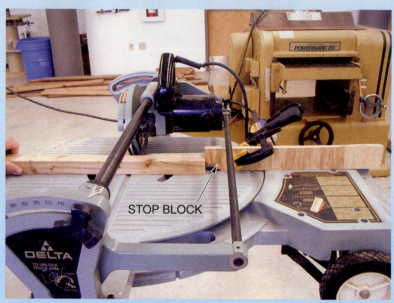

STOP BLOCK

212SA04.EPS

- Always lock the saw or hold it in place when starting it.
- Keep your hands and fingers away from the front of the blade. The blade has a tendency to run out into the material.
- Do not use a radial arm saw for ripping stock if a table saw is available.
- Always use an anti-kickback guard when ripping.
- Do not feed or pull the stock too fast when ripping. This will cause jamming.
- When ripping, feed the stock against the direction of blade rotation.

3.3.0 Compound Miter Saw

A compound miter saw (*Figure 9*) combines a miter box with a circular saw in a compact package that is easily taken to a job site. It is useful for either straight or miter cuts. The saw is attached at the back of the table and pivots so that miter cuts can be made. Unlike the basic power miter saw, the blade of the compound miter saw also tilts so that a combined miter and bevel cut, or compound cut, can be made. The compound miter saw shown is mounted on a set of rails so that the saw may be moved forward and back a short distance. This is necessary for making bevel cuts where the blade is set at an angle. For making straight cuts, the saw can be locked in place on the slide mechanism and brought straight down on the stock. For this reason, a compound miter saw is sometimes referred to as a chop saw.

The compound miter saw is especially useful for cutting small framing stock and moldings. The ability to set and keep a miter and/or bevel angle is particularly helpful when several identical cuts must be made.

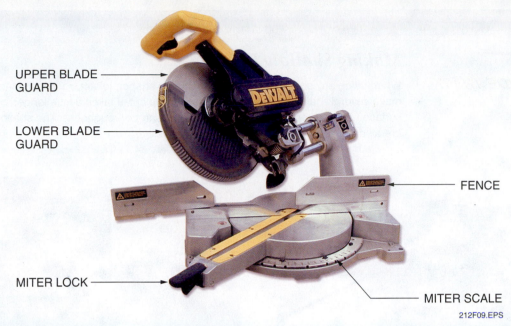

UPPER BLADE GUARD

LOWER BLADE GUARD

FENCE

MITER LOCK

MITER SCALE

212F09.EPS

Figure 9 ◆ Compound miter saw.

3.3.1 *Compound Miter Saw Safety*

Observe the following safety practices when using a compound miter saw:

- Always check the condition of the blade and make sure the blade is secure before starting the saw.
- Keep your fingers clear of the blade at all times.
- Before use, make sure the saw is sitting on a firm base. Fasten the saw securely to the base.
- Be sure that the saw is securely locked at the desired angle before starting the saw.
- Do not attempt to cut oversized material.
- Be sure that the blade guards are in place and working properly.
- Make sure that the stock is properly supported on either side, using either support stands or one or more helpers.
- Allow the saw blade to reach maximum speed before starting the cut.
- Hold the workpiece firmly against the fence when making the cut, and never attempt to feed stock into the blade.
- Never make adjustments of any kind while the saw is running.
- After the cut is made, turn the saw off.
- Never step away from the saw until the blade stops.

3.3.2 *Compound Miter Saw Accessories*

The following accessories are commonly used with a compound miter saw:

- *Dust collector* – A dust collector bag or hose can be connected to the back end of the compound miter saw to draw the dust away from the cutting area and make cleanup easier. The hose can be connected to a dust collection system or a stand-alone vacuum.
- *Hold-down clamp* – A hold-down clamp can be used on one side of the saw to hold the stock in place. The clamp attaches to the fence or to the saw base. It can be adjusted to apply enough pressure to the stock to keep it from slipping during cutting. The hold-down clamp is not a substitute for a firm grip on the stock; it is only meant to keep the stock from kicking away from the blade after the cut is made.

3.4.0 Jointer-Planer

A jointer-planer, sometimes referred to simply as a jointer, is used to true up the wavy or crowned edge of a board. The jointer-planer (*Figure 10*) consists of an infeed table, a cutter head composed of a set of spinning knives, an outfeed table, and a fence. The length of the knives determines the size

212F10.EPS

Figure 10 ◆ Jointer-planer.

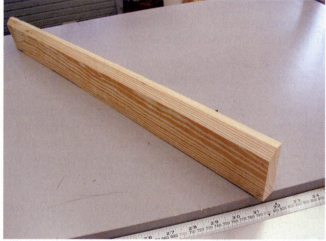

212F11.EPS

Figure 11 ◆ Making a chamfered edge on a jointer-planer.

of the jointer, with 6" being the common size. The knives rotate on a cylindrical head at about 4,500 rpm. The cutting head is covered by a spring-loaded guard, which is pushed aside by the stock as it is being fed through the jointer. The height of the infeed table can be adjusted to change the depth of the cut. The outfeed table must be adjusted to the maximum height of the cutting blades so the stock can be fed through without catching on the edge of the outfeed table. Make sure the infeed and outfeed tables are parallel; otherwise, the cut won't be straight.

The fence can be tilted toward or away from the cutting head, as shown in *Figure 11*. This forces the stock to be fed through the cutting head at an angle, creating a beveled or chamfered edge. Note that the operator is using push blocks to move the workpiece through the jointer.

3.4.1 Jointer-Planer Safety

The jointer-planer is one of the most dangerous woodworking tools. Most injuries result from hands or fingers coming into contact with the cutting head. Rules for the safe use of the jointer include the following:

- Don't let the knives extend more than ⅛" past the end of the cutting head.
- Keep the guard over the cutting head at all times.

- Adjust the depth of the cut before turning on the power.
- Lock the fence in place before turning on the power.
- Don't make cuts deeper than the tool is rated for, typically ⅛". Deeper cuts can cause the stock to kick back. If a deeper cut is needed, make multiple passes through the jointer-planer.
- Do not attempt to plane odd-shaped pieces that will not sit flat against the fence.
- Do not use stock that is less than 12" long, less than 1" wide, or less than ¼" thick.
- Cut with the grain of the wood to avoid kick-back.

Jointer Use

The jointer is not the ideal tool for making a chamfer. It is more common to use a table saw to cut the chamfer, and then use the jointer to smooth the chamfer.

- Allow the cutter head to come to full speed before feeding stock.
- Always use a push stick or push blocks for feeding stock into the jointer (*Figure 12*).
- Always use two people for any stock longer than the table. This operation can be very hazardous, and pressure is needed to hold the workpiece.
- Keep your hands on the top of the stock and never on the sides of the stock.
- Never allow your hands to pass directly over the cutting head. As the stock is fed through, position one hand over the stock on the outfeed table to keep the stock moving smoothly and steadily through the cutting head.
- Draw the stock all the way through the cutting head and allow the guard to return to the closed position before lifting the stock away from the machine.
- Make sure the cutter head comes to a complete stop before leaving the machine.

3.5.0 Shaper

A shaper (*Figure 13*), similar in function to a router table, consists of a worktable, an adjustable fence, and a motor-driven, adjustable-height spindle. The spindle holds a high-speed molding cutter and extends up through the worktable. The shaper may be used to carve profiles on the edge of the workpiece or to cut a groove partway through the workpiece. Such a groove may be smooth or may have a detailed shape depending on the cutting blade that is used.

The slower spindle speed and greater torque of the shaper, combined with a heavier base, make it able to drive larger cutters and make heavier cuts. Shapers use three-blade cutters instead of the two-blade cutters used on routers. Larger shapers have multiple heads to make large profile cuts with a single pass. As with router tables, however, cuts must be made at a slow and steady speed to prevent roughening. The finished side of the cut must be held firmly against the fence as the cutter approaches the corner to keep from rounding the corner.

Molding cutters typically have three blades and are available in a variety of designs, each of which produces a different edge shape. *Figure 14* shows a molding cutter. Irregularly shaped pieces of wood can be cut with the shaper. A depth collar is set on the cutting head to keep the work down against the table.

When shaping material, follow this sequence: one end grain first, then one long side, then the other end grain, then the other long side. This method will keep the wood from splitting.

212F12.EPS

Figure 12 ◆ Using push blocks to feed material through a jointer-planer.

212F13.EPS

Figure 13 ◆ Shaper.

212F14.EPS

Figure 14 ◆ Molding cutter.

3.5.1 Shaper Safety

 WARNING!

The shaper is one of the most dangerous shop tools because it has a tendency to kick back the workpiece. Use anti-kickback cutters with the shaper.

Rules for the safe operation of a shaper include the following:

- Use push sticks or push blocks to feed the work into the shaper.
- Be careful when handling cutting blades because they are extremely sharp.
- Do not use adjustable cutting heads.
- Install the cutter so that the bottom of the stock is shaped whenever possible. This allows the stock to cover most of the cutter and act as a guard.
- Make sure the cutter is locked securely to the spindle.
- Always position the fences so that they will support the work that has passed the cutters.
- Adjust the spindle to the correct height and lock it in position. Make sure it clears all guards and fences.
- Before placing the stock on the table for cutting, switch the power on and off and check the cutter to verify the direction of rotation.
- Do not attempt to cut stock with loose knots or stock that is split or cracked.
- When cutting straight stock, hold the stock down and against the fence with your hands on top of the stock, not against the side of the stock.
- Keep hands out of range of the cutters.
- Whenever appropriate, use proper guards, jigs, and clamping devices.

- Spring-loaded hold-down clips or clamps should not be set too tightly. They should hold the work against the fence without impeding movement.
- Use a depth collar to keep irregularly shaped pieces from riding up on the cutter.
- Use a guide pin to hold the work against the cutter.
- Never use the shaper on a piece of stock that is less than 10" long.
- Feed the thick end of the material first, and feed it against the cutting direction.
- Swing irregularly shaped pieces into the cutting head rather than sliding them in. Keep the piece in constant motion against the rotation of the cutter as it is being cut.
- If feeding more than one board through the shaper, feed only boards of the same size.

3.5.2 Shaper Accessories

A wide variety of cutting heads are available for the shaper, which makes it a very versatile tool. *Figure 15* shows some examples of cutting heads.

Because of the shaper's tendency to kick back work, the power feed shown in *Figure 16* is an important accessory for the shaper.

3.6.0 Routers

A router (*Figure 17*) is a handheld tool with a high-speed motor that drives a collet (collar) in which a wide variety of high-speed cutting and shaping bits can be mounted. The motor is mounted in a base that allows the depth of the cut to be adjusted. The cutting bits are made of high-speed steel or carbide-tipped steel. The base typically has a clear plastic panel for viewing and guiding the bit on the workpiece.

A cabinetmaking shop will commonly include a router table such as the one shown in *Figure 18*.

A router table allows a router to be mounted upside down under a table and is used much like a shaper. The router bit extends up through the table.

The router can be used for many different tasks because of the wide variety of tips and cutters that are available. Among the types of cuts that can be made with the router are the following:

- Shaping edges
- Rabbeting
- Beveling
- Chamfering
- Dovetailing
- Dadoing, fluting, and reeding
- Mortising

212F15.EPS

Figure 15 ◆ Shaper cutting heads.

212F17.EPS

Figure 17 ◆ Power router.

212F16.EPS

Figure 16 ◆ Power feed.

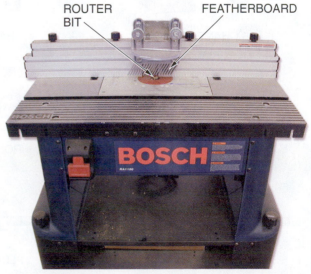

ROUTER BIT

FEATHERBOARD

212F18.EPS

Figure 18 ◆ Router table.

Figure 19 shows different router bits and the designs they make.

A router can be equipped with a plunge base, which is a spring-loaded base that is mounted under the router's own base. It is used for driving the router bit straight down into the workpiece.

3.6.1 Router Safety

Routers operate at speeds as high as 30,000 rpm. Therefore, it is important to observe the following safety precautions:

- Keep your hands clear of the router bits.
- If using a portable router, hold the tool firmly whenever the power is on.
- Be sure that the material is secured properly before routing.
- Approach the workpiece gently. Driving the tool quickly into the workpiece can cause the material to rip.
- Avoid making cuts that are too deep. Deep cuts can cause kickback or tearing.

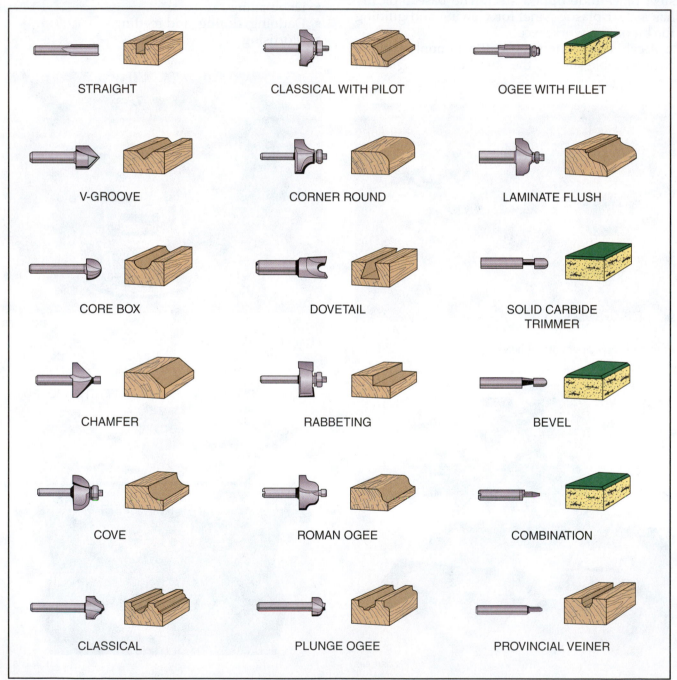

STRAIGHT

CLASSICAL WITH PILOT

OGEE WITH FILLET

V-GROOVE

CORNER ROUND

LAMINATE FLUSH

CORE BOX

DOVETAIL

SOLID CARBIDE TRIMMER

CHAMFER

RABBETING

BEVEL

COVE

ROMAN OGEE

COMBINATION

CLASSICAL

PLUNGE OGEE

PROVINCIAL VEINER

212F19.EPS

Figure 19 ◆ Router bits.

- As soon as the cut is finished, turn the tool off. Wait for the bit to come to a complete stop before removing the workpiece.
- Never rest a portable router on the bit. Raise the motor so the bit does not protrude from the base, remove the bit, or place the router on its top or side.

3.7.0 Sanders

Sanders come in a variety of styles, but the basic principle remains the same: Abrasive material is affixed to a tool that is used to move the abrasive over the workpiece rapidly to create a smooth surface. The most common sander in commercial cabinet shops is the stationary drum sander (*Figure 20*), which functions like a thickness planer. It is used to flatten and sand large surfaces and assemblies, such as doors.

A stationary belt-disc sander (*Figure 21*) combines a spinning abrasive disc and a moving abrasive belt in one machine. A typical belt-disc sander has a 6"-wide sanding belt and a 12"-diameter sanding disc. The belt portion can be tilted to operate in a vertical position for edge or end sanding; in the horizontal position for surface sanding; or at any angle in between. The disc table can be tilted 45 degrees up or down. Both the disc and the belt typically use aluminum oxide abrasive, which is available in a variety of grits.

Handheld sanders may also be used in cabinetmaking. These sanders fall into three categories: belt sanders, rotary disc sanders, and pad sanders.

212F20.EPS

Figure 20 ◆ Stationary drum sander.

A handheld belt sander, such as the one shown in *Figure 22*, operates on the same principle as the stationary belt sander. An endless belt coated with abrasive material travels through the device.

The sander is held to the material that is to be smoothed. A rotary disc sander has an abrasive

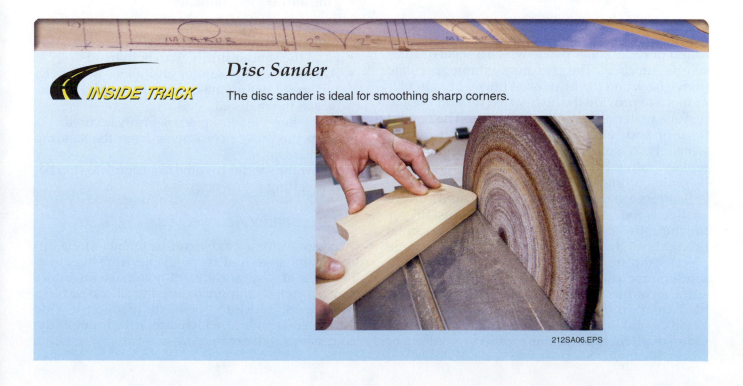

Disc Sander

INSIDE TRACK

The disc sander is ideal for smoothing sharp corners.

212SA06.EPS

Figure 21 ◆ Stationary belt-disc sander.

Figure 22 ◆ Handheld belt sander.

disc that spins at high velocity. A pad sander has a round or rectangular pad to which a sheet of abrasive is attached. The pad may vibrate, oscillate (move back and forth), or orbit (move in a circular motion) to provide the high-speed motion used to remove material from the workpiece. The newest form of pad sander is the random orbit sander, which both spins and orbits. It is coming to replace both vibrating and orbital sanders due to its rapid cutting and smooth results.

Sanders are used to provide a finished surface on the wood. A stationary sander is useful for finishing the surface of small pieces that can be brought to the sander and for running large pieces through before assembly. Handheld sanders are especially useful for finishing the surfaces of pieces that have already been assembled. In addition, sanders are useful for shaping wood by taking away enough material to provide small changes to the profile or contour of

the wood. Always sand in the direction of the wood grain. Sanding against the grain will scratch the wood.

3.7.1 Sander Safety

Safety precautions for belt-disc and spindle sanders include the following:

- Do not attempt to sand plastics, metals, or other materials for which the abrasive is not intended.
- Do not attempt to sharpen tools on a sander.
- Do not wear gloves while operating the sander. They can become caught in the rotating mechanism and cause injury.
- Never make adjustments while the sander is running.
- Keep fingers clear of the sanding surfaces.
- Hold the stock firmly with both hands while feeding it into the sander.
- Do not force the stock against the abrasive surfaces. Let the motion of the abrasive disc or belt do the work.
- Never leave a sander until it has stopped.

 WARNING!
Do not overload a spindle sander. Overloading can cause the workpiece to kick back.

When using a handheld sander, observe the following additional guidelines:

- Make sure the power switch is turned off before plugging in the sander.
- Hold the sander firmly with both hands.
- Always start the sander and allow it to reach full speed before making contact with the surface of the material.
- Make sure the workpiece is firmly secured.
- Keep the power cord away from the sanding mechanism.
- Allow the sander to come to a complete stop before setting it down.

3.7.2 Sander Accessories

Accessories for a sander consist mainly of the different abrasives that are available. Abrasives may be made of several different materials, including flint, garnet, aluminum oxide, and silicon carbide. The sheets, pads, and belts are made in a wide variety of grades, which are based upon the coarseness of the material.

Some sanders are equipped with dust bags. Others have connectors for attachment to a dust collection system.

3.8.0 Thickness Planer

A thickness planer is designed to shave the surface of a piece of lumber. Its function is to size stock to a desired thickness or to smooth and clean the surface of a piece of stock. *Figure 23* shows two types of thickness planers. The one in *Figure 23(A)* is used for trim pieces. This type will typically handle material from 4" to 12" wide and up to 6" thick. The one in *Figure 23(B)* is a shop tool designed to handle larger pieces of stock. Like hand planes, power planes are designed to shave off thin surface layers. It would typically be set to plane off a maximum of ⅛" depending on the machine size and horsepower. If more material needs to be removed, you would make multiple feeds, adjusting the machine for each new pass.

When using a surface planer, adjust the thickness, then place the stock on the planer table and allow the infeed rollers to draw it in. The rollers will feed the stock through the cutter head blades at the top of the machine and then onto the outfeed rollers.

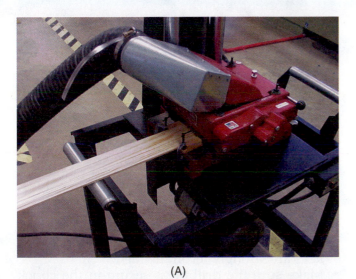

(A)

(B)

Figure 23 ◆ Thickness planers.

212F23.EPS

INSIDE TRACK

Using the Thickness Planer

The photo on the top shows a very rough piece of hardwood lumber. The photo on the bottom shows the same piece after it has been fed through the thickness planer.

212SA07.EPS

212SA08.EPS

3.8.1 Thickness Planer Safety

WARNING!

Setting the cut too thick can result in kickback. Make multiple passes to obtain the desired thickness.

When using a thickness planer, observe the following safety precautions:

- Keep your hands clear of the blades.
- Wear hearing protection.
- Check for the correct depth of cut before starting.
- Do not attempt to remove too much material with one cut.
- Make adjustments only after the power has been turned off. Make speed and depth adjustments according to the manufacturer's instructions.
- Do not use stock less than 12" long. The stock must be long enough to reach the outfeed rollers before it leaves the infeed rollers.
- Feed the thick end of the material first.
- Never bend over to look at the stock while it is being planed.
- Never push the material with your body.
- Allow the automatic feed mechanism to operate. Do not force the material.
- If more than one board is being fed, feed boards of the same length.

3.9.0 Drills

A drill is a device used to bore holes through a workpiece. Drills come in both handheld models and as a freestanding drill press. Both types can hold a variety of drill bits, which perform the actual cutting of the material. The drill press (*Figure 24*) holds the bit in a vertically mounted chuck that is mounted with the motor on a moving carriage. The drill press is moved downward, using a lever, to bring the bit into contact with the material. The press can be adjusted to compensate for the thickness of the material and the depth of the desired hole.

While the basic function of the drill is simple, the applications are widely varied. Drills are used for setting dowels, drilling starter holes for screws, boring holes for wiring and other uses, and cutting larger holes with hole saws. Portable drills can also be used for driving screws and other fasteners.

212F24.EPS

Figure 24 ◆ Drill press.

3.9.1 Drill and Drill Press Safety

When using a drill, observe the following:

- Always hold the tool firmly.
- Always remove the chuck key from the chuck before starting the drill. Spring-loaded chuck keys are strongly recommended.
- Make sure the drill bit is secure in the chuck before starting the drill.
- Never try to stop the drill by grabbing the chuck or pressing anything against the chuck or bit.
- Allow the spinning action of the drill to do the work. Forcing the drill into the material can cause the bit to break.
- Make sure the material is properly secured to the bed before drilling. Never point the drill at another person.
- Use the correct speed for the job being performed.

In addition, when using a drill press, take the following precautions:

- Do not wear gloves while using a drill press.
- Do not adjust the press while the drill is running.
- Never step away from a drill press until the chuck has come to a full stop.

Special-Purpose Shop Tool

This special press is designed to drill and insert Euro-hinges in one operation. The 7-bit drill attachment is designed to drill hole groups for a standard 32 mm system.

212SA09.EPS

4.0.0 ◆ JOINTS

Cabinetmakers use many specialized types of joints, but most are variations of a few simple joints. The following is a rough description of the basic types, and a few common variants:

- *Butt joint* – The butt joint (*Figure 25*) simply places an end of one piece of wood against the end or side of another, and uses nails, screws, or glue to hold them together. This is not a strong arrangement, especially in shear, so the addition of biscuits or dowels commonly makes a stronger connection between the two pieces. Cleats are also frequently used, with glue, screws, or pins, to strengthen the connected pieces.
- *Rabbet* – The rabbet (*Figure 25*) is strong especially in shear into the shoulder of the rabbet. It can be cut quickly and easily with a table saw or a router, and is commonly used for drawer edges and countertops where the joint must not be visible from one side.
- *Half lap joint* – The half lap joint (*Figure 26*) is a special case of the rabbet. Part of the edge is cut away from each piece, so that the two pieces make the corner together, and there is some glue penetration in each piece.
- *Miter* – The miter is an intersection in which the two edges are cut to half the angle they are to make.

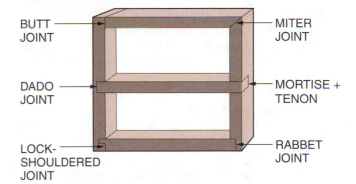

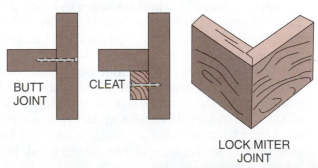

LOCK MITER JOINT

212F25.EPS

Figure 25 ◆ Common cabinet joints.

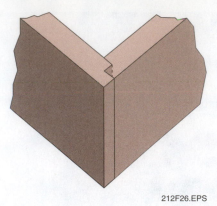

212F26.EPS

Figure 26 ◆ Half lap joint.

- *Lock miter joint* – The lock miter joint (*Figure 25*) is half rabbet and half miter, providing some locking of the joint.
- *Dado* – The dado (*Figure 27*) is a special case of the rabbet, in that the groove is moved away from the parallel edge, so that one piece of wood can be inserted into the other. The blind or stopped dado (*Figure 27*) is used to gain the strength of the joint without showing the joint. The joint can be made with a router or with a table saw and a dado kit. The dado is completed at the hidden end with a chisel to obtain the square corner needed for a precise fit. If the measurements and cuts are not very precise, the joint will have visible gaps and will be somewhat weakened.
- *Lock shouldered joint* – The lock shouldered joint (*Figure 28*) is a special case of the dado in that part of a side is cut thinner and is inserted and held in the groove near the edge, with the two pieces edge flush. This gives a locking action on the corner. An old variant is shown in *Figure 29*.
- *Dovetail joint* – The dovetail joint (*Figure 30*) is the strongest of the joints used for chests, boxes, or drawers. A properly made dovetail fitted together without glue will not pull apart, but must be pushed apart to the side. When glued properly, the dovetail will hold as long as the wood will. In fact, dovetail and dado joints with modern glues are said to break wood—to be able to hold under strains that would break the boards somewhere else.
- *Mortise and tenon* – Another joint that is functionally similar to the dado is the mortise and tenon. The mortise is a hole cut in one piece. The tenon is a tongue protruding from the other piece, so that it fits tightly into the mortise. There are many variations. Dowel pegs are a contemporary version of this joint.

DADO JOINT

BLIND DADO JOINT

212F27.EPS

Figure 27 ◆ Dado joint and blind dado joint.

212F28.EPS

Figure 28 ◆ Lock shouldered joint.

Figure 29 ◆ Variant of the lock shouldered joint.

212F29.EPS

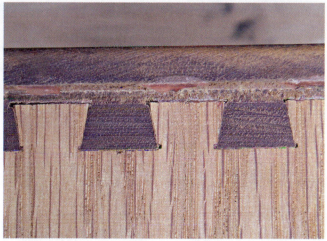

Figure 30 ◆ Dovetail joints.

212F30.EPS

4.1.0 Dado Joint

A dado joint is used to attach the edge of one piece to the surface of another piece. The dado joint requires only a single cut. A dado cut, or channel, is cut into one piece to be large enough for the end of the second piece to fit into. Attach the joint with screws or nails from the opposite side of the piece with the dado cut into it. For a cleaner-looking piece, glue the two pieces together.

Another form of the dado joint, called a stopped or blind dado joint, gives a more finished look. The dado is not cut all the way across the piece, and the other piece is notched. When they are joined, the dado groove is not visible.

4.2.0 Rabbet Joint

A rabbet joint is a corner joint in which the end of one piece is cut to allow the end of another piece to sit in a notch for added strength. One of the pieces has a rectangular section, called a rabbet, cut out of its length. The uncut piece fits into the cut piece, forming a smooth joint.

4.3.0 Lock Miter Joint

A lock miter joint is created by cutting both pieces to be joined with a specialized shaper or router bit, so they have matching profiles that fit together. A specially designed lock-miter bit allows both pieces to be cut with the same bit. The angled profile provides a much larger gluing surface for this joint, which makes for a very strong corner joint.

4.4.0 Half-Lap Joint

A half-lap joint is a corner joint where a rabbet cut is made in both pieces so they can be fastened together with their surfaces flush with each other.

The width of each cut is equal to the width of the other piece, and the depth of the cut is half of the piece's thickness.

The half-lap joint may be fastened with glue across the matching faces of the joint or with nails or screws through either outer face of the joint. If screws or nails are used, a minimum of two fasteners should be used for greater stability of the joint. Using a single fastener may allow the joint to shift or rock, causing the entire cabinet to shift out of square.

4.5.0 Dowels and Biscuit Fasteners

Dowels and biscuit fasteners are commonly used to strengthen butt, miter, and edge joints in cabinets. Fluted dowels (*Figure 31*) are used to strengthen joints by drilling matching (aligned) holes of the same depth into each piece to be joined. These holes should be drilled about 1/16" deeper than the dowel length to contain the excess glue. After the holes are drilled, glue is squeezed into the holes in one of the pieces, and the dowels are driven into the holes. Then, glue is squeezed into the holes in the opposite piece, and the two pieces are fitted tightly together and clamped. They remain clamped until the glue is dry.

Biscuits are used to strengthen joints in a similar manner. A biscuit is made of dried wood and kept in an airtight container to avoid contact with any moisture. The biscuit will swell up after glue is applied, and it is seated in the slots of the two pieces to be joined. This ensures a tight fit and a stronger joint than one might expect using a piece of wood only a few thousandths of an inch thick. The use of biscuits requires that properly sized grooves be cut into both of the pieces to be joined. The grooves are cut using a tool called a biscuit jointer or plate jointer (*Figure 32*). After the grooves are cut, the oval-shaped wooden biscuits are fit into the grooves and, with a little bit of glue, create a strong joint. Within reason, biscuit joints

can be aligned before the glue dries by laterally shifting the pieces. Once in position, the two pieces are clamped together until the glue dries. Biscuits are available in several sizes. Use the largest one that fits the joint.

4.6.0 Brad Gun

A brad gun (*Figure 33*) is similar to the pneumatic nail guns used in wood framing. The brad gun is smaller because it is designed to shoot a much

(A) BISCUIT JOINTER

(B) BLADE   (C) BISCUITS

212F32.EPS

Figure 32 ◆ Portable biscuit jointer, blade, and biscuits.

212F33.EPS

Figure 33 ◆ Brad gun and brads.

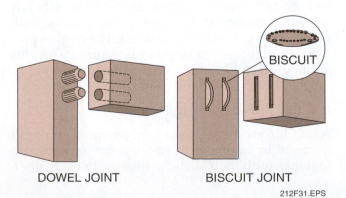

DOWEL JOINT   BISCUIT JOINT

212F31.EPS

Figure 31 ◆ Dowel and biscuit joints.

smaller nail. A brad is like a finish nail but is smaller and thinner. The brad gun has become a popular tool for assembling cabinets because it makes the work go much faster than previous assembly methods.

4.6.1 Brad Gun Safety

Brad guns are pneumatic guns that shoot very sharp nails. Be sure to observe the following safety precautions whenever you handle or use a brad gun:

- Never point a brad gun at yourself or another person.
- If you see somebody else using a brad gun improperly, report it to your instructor or supervisor.
- Be sure all safety devices are functioning properly.
- Only use the pressure specified by the manufacturer.
- Never disengage the firing safety mechanism.
- Always assume that fasteners are loaded in the brad gun.
- Be sure the brad gun is disconnected from the power source before making adjustments or repairs or before loading the tool.
- When attaching the brad gun to the air supply, keep it pointed in a safe direction. The fastener may discharge when attaching the air supply.
- Never leave a brad gun unattended while it is attached to the air supply.

5.0.0 ◆ ASSEMBLING THE CABINET

Cabinets can be assembled using brads, staples, nails, or screws. Many cabinetmakers prefer using glue because fasteners leave a hole that must be filled. Yellow wood glue (*Figure 34*) is commonly used when gluing the cabinet parts together. Clamps are used to hold the pieces together while the glue is drying.

Once the cabinet frame is built, wood veneer can be applied. In a custom cabinet shop, often both the surface of the cabinet and the surface of the wood veneer are sprayed with a light coating of contact cement. (A brush or roller can also be used to apply the contact cement.) When the cement has adequately dried, the wood veneer is carefully placed on the cabinet surface, and a roller is used to remove any air bubbles and to make a tight bond. Once the gluing process is done, the veneer is trimmed using a laminate trimmer. *Figure 35* shows a cabinet made of particleboard with a wood veneer surface. The veneer was cut from a 4 × 8 sheet using a table saw and was glued to the particleboard surface of the cab-

212F34.EPS

Figure 34 ◆ Wood glue.

212F35.EPS

Figure 35 ◆ Custom cabinet in process.

inet to simulate real wood. Matching rolls of edge band are also available from the same sources that supply sheets of wood veneer.

You might construct a cabinet following this series of steps:

Step 1   Use the table saw to cut out the sides, bottom, and back pieces of the cabinet and drawers.

Step 2   Use the thickness planer to smooth the surfaces and achieve the desired thickness.

Step 3   Use the jointer-planer to smooth any edges that will be visible.

Step 4   Use the radial arm or compound miter saw to cut the face frame.

Step 5 Use the drill press to drill the face frame to receive dowels.

Step 6 Assemble the face frame.

Step 7 Assemble the cabinet using a brad nailer and/or adhesive.

Step 8 Use a router to make dovetail joints for the drawer and a decorative drawer face, if desired.

Step 9 If a raised panel door is to be used, cut out the stiles, rails, and panel pieces using the table saw.

Step 10 Use the shaper or table saw to make the raised panel cuts and to shape any molding pieces that will be used for trim.

Step 11 Use a sander to smooth the wood for finishing.

Step 12 Apply the finish coat.

Step 13 Apply laminate to the top of the cabinet, if desired.

6.0.0 ◆ SANDING AND FINISHING

 WARNING!
To prevent eye and respiratory injury, always wear safety goggles and a dust mask when sanding.

The finish coat of your cabinet will only be as smooth as the surface to which it is applied. Bare wood should be sanded by starting with a grade of sandpaper that will do the best job to smooth the surface. Use progressively finer grades, finishing with fine or very fine sandpaper to achieve the smoothest finish. Dampening the wood will raise its grain slightly; sanding the wood after it dries will give a smoother finish. Always sand with the grain of the wood using straight, even strokes. Sanding against the grain will cause scratches that are difficult to remove. If you must sand end grain or across the grain, complete the job using sandpaper one grade finer than normal. This will allow subsequent finishes to apply more uniformly. If glue was used in the assembly of the object, make sure that all glue spots or drips are sanded off. Wood surfaces with glue on them will not accept stains and will leave a light spot. Check the surface of the wood frequently using an oblique light source to show imperfections. Use your fingers to check the finish because they are often more sensitive to flaws than your eyes.

When hand-sanding on flat surfaces, use a sanding block to prevent unevenness and rounded edges. For sanding curved or irregular surfaces, tear the sandpaper into strips and use it without a block in a back and forth shoeshine motion. Flex the sandpaper to make it softer. To flex the paper, pull the sheet (grain side up) over a sharp edge several times.

Each time sanding is done, thoroughly clean the surface using a vacuum cleaner and a tack cloth or rag dampened with the appropriate solvent. A tack cloth is a clean, soft, lint-free cloth treated with diluted varnish or similar material to make it tacky or sticky. Tack cloths can be made at the job site but are easily purchased.

6.1.0 Choosing Sandpaper

Two main considerations in choosing a sandpaper are its abrasive coating and its grit. Abrasive coatings include emery, garnet, aluminum oxide, and silicon carbide. Each type has different uses. The grit indicates how coarse or fine the sandpaper is. The grit of some sandpaper is described in words: very coarse, coarse, medium fine, very fine, extra fine, and super fine. Other types of sandpaper use numbers that indicate the grams of abrasive on a given amount of backing. The higher the number, the finer the grit. For example, 220 is very fine sandpaper; 36 is very coarse. The full range of grades goes from 12 to 1200. There are two densities of grit: closed coat and open coat. Closed coat means that the abrasive completely covers the paper. Open coat means that the abrasive covers only 50 percent. Open coat abrasives are normally used when loading or filling with the removed material may be a problem.

Aluminum oxide and silicon carbide sandpapers can handle all sanding jobs. They tend to be more durable and work faster than sand or garnet. Aluminum oxide, which is brown, is an ideal all-around sandpaper for use on bare wood. It is easy to use and produces a high-quality finish. For finishing varnish or lacquer coats, sandpaper coated with silicon carbide, a shiny black material, offers good results. Silicon carbide sandpaper comes in two forms: one for dry sanding and one for wet sanding with lubricants. When used with a light oil lubricant, fine-grit carbide gives lacquer a lustrous satin finish.

Unlike aluminum oxide and silicon carbide, which are both manufactured abrasives, emery, flint, and garnet are natural minerals mined from the ground. Because of its rounded grains, emery is relatively slow-cutting. This dull, black sandpaper is used primarily to clean or polish metal. Garnet sandpaper, which is reddish-brown, is commonly used for woodworking. It has medium-hard cutting edges, and it produces good results with a moderate amount of labor. The types of sandpapers and grits commonly used for different stages of wood finishing are shown in *Figure 36*.

Alternatives to sandpaper are steel wool and synthetic non-woven abrasives. They have the advantage of being non-clogging and are available in a range of grades from grade 4 (the coarsest) to 0000 (the finest). For fine finishing of wood, grades 0 through 0000 are typically used. Do not use steel wool on wood that will be finished with water-based stain or coatings because the residual steel wool particles embedded in the wood will rust.

6.2.0 Using Sealers

Sealers are used to prevent the wood from absorbing too much stain. They also stop resins in the wood from bleeding out and discoloring the finish. Depending on the job at hand, sealers are used at different points in the finishing process. For example, when finishing porous hardwoods, the sealer is typically applied after the surface has been filled and sanded. For nonporous wood, where the use of a filler is not needed, the sealer is usually applied after the stain. When finishing a softwood, such as pine, the wash coat or wood conditioner is typically applied before the stain to prevent overabsorption of the stain into the softer parts of the wood. For guidance as to when the sealer should be applied, always follow the manufacturer's recommendations for the particular coating system being used and the type of wood being finished.

The sealer used for any job must be compatible with the finish to be applied over the rest of the wood surface. Sealers for shellac, varnish, or lacquer finishes are often made on the job by thinning the shellac, varnish, or lacquer. The sealer used for a shellac finish typically is either orange or white shellac thinned to the proportion of about one part of shellac to about six parts of denatured alcohol. Orange shellac is normally used for dark-colored woods; white shellac is used for light-colored woods.

The sealer used for a varnish finish can be either shellac or a varnish sealer. A varnish sealer is typically made by thinning varnish with an equal quantity (one-to-one proportion) of turpentine or

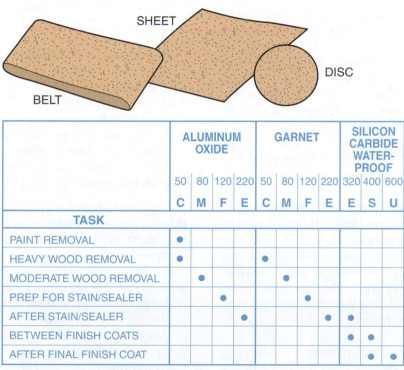

TASK	ALUMINUM OXIDE				GARNET				SILICON CARBIDE WATER-PROOF		
	50	80	120	220	50	80	120	220	320	400	600
	C	M	F	E	C	M	F	E	E	S	U
PAINT REMOVAL	●										
HEAVY WOOD REMOVAL	●				●						
MODERATE WOOD REMOVAL		●				●					
PREP FOR STAIN/SEALER			●				●				
AFTER STAIN/SEALER				●					●		
BETWEEN FINISH COATS									●	●	
AFTER FINAL FINISH COAT										●	●

C=COARSE, M=MEDIUM, F=FINE, E=EXTRA FINE, S=SUPER FINE, U=ULTRA FINE

NOTE: The grit value for extra-fine silicon carbide sandpaper (320) differs from the grit value of extra-fine aluminum oxide and garnet sandpaper (220).

212F36.EPS

Figure 36 ◆ Sandpaper grades and uses.

thinner as specified by the manufacturer of the varnish. A lacquer sealer generally consists of one part lacquer to six parts of the recommended lacquer thinner. Spraying lacquer sealer on provides better uniformity than using a brush. After a sealer coat has been applied and is thoroughly dry, it should be sanded lightly in the direction of the grain using 220-grit or finer sandpaper.

6.3.0 Using Wood Fillers

Fillers are used to fill the pores of open-grained wood in order to get a smooth surface for the finish coat. Using a wood filler begins with selecting the best filler for the job. When choosing a filler, it is important to keep in mind the type of finish you intend to use. Wood fillers come in a thin, cream-like consistency or a paste. Generally, paste fillers come in too thick a consistency to be used without thinning. When preparing a paste-type wood filler, first thin it using a solvent recommended by the filler manufacturer. It should have the consistency of thin cream.

Using a brush, burlap, or a coarse cloth, apply the filler liberally across the grain of the wood, making sure to fill all the pores in the wood. Do not work the filler in the direction of the grain because it may not hold in the pores. The filler should be allowed to cure per the manufacturer's instructions but not get hard. Timing is important. If the filler is wiped too soon, it will pull from the pores. If allowed to set too long, it will harden so much that it is difficult to remove. Typically, the surface can be wiped when the filler begins to lose its wet appearance. After the filler sets, the surface should be wiped clean with a rag or cloth to remove excess filler. When the filler is dry, the wood surface should be lightly rubbed with fine sandpaper to remove any small particles of filler that may remain on the surface.

When using both stain and a filler, it is best to apply the stain first. Sometimes the stain can be mixed with the filler material. When finishing with stain and filler in separate operations, the surface should be coated with a wash coat or wood conditioner after the stain has dried thoroughly and before applying the filler. Again, good practice is to follow the manufacturer's recommendations.

A wood's grain pattern may be emphasized by using a filler a shade darker than the color of the stained wood. The grain pattern may be brought out even more prominently by using a filler that is very dark. Two-tone effects can be obtained by rubbing a white or colored filler into a surface that has been stained a different color. The filler fills the pores of the stained wood, leaving them the color of the filler, while the rest of the surface remains the color of the stain.

6.4.0 Applying Stains

There are many kinds of stains and many ways to apply them. This section provides an overview. Until you learn what type of stain is right for a given situation, and the best way to apply it, you should ask an expert.

Stains should always be applied as recommended by the manufacturer's application instructions. Some general guidelines for staining are as follows:

- Wood must be dry, free of dirt, grease, wax, marks, and any old finish in poor condition.
- Sand wood to a smooth surface and remove sanding dust with a vacuum and tack cloth.
- Before applying a stain to the wood surface, always test a sample area on a concealed surface of the piece you are finishing before staining the entire area.
- A dark stain has a stronger effect on a light wood than on a dark one.
- Always keep a wet edge when applying stain to avoid color blotches.
- Plan your staining pattern before starting so that you maintain a uniform color.
- When trying to match colors, add stain gradually until an exact match is obtained. You can always make a surface darker with more stain, but it is impossible to lighten the color without resorting to bleach.
- Avoid stain runs by working on the underside of a piece first.
- Faults and end grain often absorb more stain than other areas, which makes them appear darker. Faults and end grain should be sealed with the proper sealer before staining.

No single type of applicator works for everything that needs staining. The brush is probably the most widely used applicator for staining interior woods because it is the most suitable for forcing the stain into the wood. Interior woods can be stained using pigmented, penetrating, or water stains. The coloring matter of pigmented stains consists of pigments instead of dyes. Pigmented stains are best suited for use on close-grained woods and are commonly used where it is desirable to hide the natural grain of the wood or to impart the color of the wood being imitated.

Pigmented stains should be stirred well before use and frequently during use because the pigments tend to settle to the bottom of the can. When the pigments have settled in the bottom of the can, pour off the liquid, and then add it back a little at a time while mixing to properly disperse the pigments in the liquid and achieve a uniform mixture.

Interior pigmented stains are usually applied with a brush but can be applied with a lint-free cloth. The stain should be applied liberally over the surface and then allowed to set for a short period of time (typically 5 to 15 minutes) as recommended by the manufacturer's instructions. The exact time depends on the product, the type of wood, and the color of the finish desired. Following this, the stain is wiped off to achieve a uniform tone. Wipe with the grain. If wiped too soon, the color may be lighter than desired. If wiped too late, it may become too dark or gummy.

Penetrating stains are usually preferred over pigmented stains when finishing open-grained woods. They have less pigmentation and require less stirring to keep the pigment in suspension.

Penetrating stains are normally applied with a brush using long, full strokes with the grain. Brush strokes should be started on unstained portions and worked into the stained portions without overlapping. When staining end grain, the stain should be wiped off quickly to prevent too much penetration or darkening of the wood. On flat surfaces, the stain should be wiped off using a soft cloth a few minutes after its application. An area that is too dark can be lightened using a cloth saturated with turpentine. Additional darkening or coloring of the wood may be obtained by applying a second coat of stain or by allowing the stain to remain on longer before wiping off the excess.

A water stain is simply a dye that is called an aniline. Water stains penetrate evenly and deeply into the pores of the wood and give clear, even tones. The dye is packaged in powder form and is mixed with boiling water. The stain is made up by adding the powdered dye to the water in different proportions depending on the color required. Add more dye to make the stain darker. The stain solution must be mixed thoroughly to make sure all the powder has dissolved.

Water stains can be applied using a spray gun or brush. They are slow drying and promote grain raising. To prevent this, dampen the surface of the wood, let it dry, and then sand it back before staining. When applying water stains with a brush, use as large a brush as practical and work with the grain, using long, straight strokes. The stain should be applied freely and spread as far and as quickly as possible. Avoid applying stain to the same spot twice as this can cause a darker brush lap. Be sure to maintain a wet edge; otherwise, laps and streaks will occur.

After an even coat of stain has been applied, the surface should be wiped to an even tone and allowed to dry.

Non-grain-raising (NGR) stains are made of synthetic dyes dissolved in a denatured alcohol or methanol vehicle. NGR stains dry much more quickly than water stains. They also require less sanding than is involved with water stains. NGR stains are best applied with a spray gun but can also be brushed on. When brushing, use a large brush and apply a wet coat of stain. Using a soft, lint-free cloth, immediately wipe off the excess while making the color uniform.

Once interior wood is stained, it normally is covered with a clear topcoat, such as varnish, polyurethane, or lacquer.

7.0.0 ◆ APPLYING PLASTIC LAMINATE TO A COUNTERTOP

This section provides guidelines and a general procedure for applying plastic laminate material to a countertop base. The construction of a typical countertop base was described earlier in this module.

Mark on plastic laminate with a pencil or grease pencil because it will clean off easily. Mark on the good side of the laminate.

Sheets of laminate are large, thin, and somewhat fragile. Handle them carefully; otherwise, they will bend and break. Large sheets of laminate can be cut to rough sizes needed for the countertop by using a table saw or handheld circular saw equipped with a fine-toothed blade. They can also be cut to rough size using a laminate cutter (*Figure 37*) or a router equipped with a solid-carbide flush (straight) trimmer bit. Some cabinetmakers prefer this method because

212F37.EPS

Figure 37 ◆ Laminate trimmer.

it is easier to run the trimmer across the laminate sheet than to run the laminate sheet across a table saw.

Most finish cutting and trimming of plastic laminates is done with either a laminate trimmer or router equipped with either a solid-carbide flush trimmer bit or a bevel trimmer bit. Use bits specifically designed for cutting plastic laminates (*Figure 38*). Some are made of solid carbide with integral pilots, while others are carbide-tipped with ball bearing pilots. Regardless of type, they are made so that the diameter of the pilot and the cutting circle of the blades are the same. As a result, the part being trimmed will be perfectly flush with the work edge. Thus, doing accurate trimming is a matter of moving the tool bit correctly along the work edge. The tool must be in the correct position throughout the pass because any accidental tilting will mar the work.

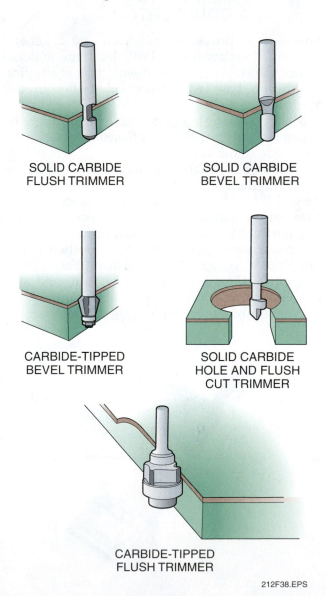

SOLID CARBIDE
FLUSH TRIMMER

SOLID CARBIDE
BEVEL TRIMMER

CARBIDE-TIPPED
BEVEL TRIMMER

SOLID CARBIDE
HOLE AND FLUSH
CUT TRIMMER

CARBIDE-TIPPED
FLUSH TRIMMER

212F38.EPS

Figure 38 ◆ Laminate trimmer bits.

7.1.0 Cutting the Laminate

To cut the laminate, proceed as follows:

Step 1 Measure and mark the cut lines on the laminate. Make sure to cut all the pieces slightly larger than their final size. Edge pieces should be cut about ¼" to ½" wider than needed; the top piece(s) should be about ⅜" to ½" larger in both length and width.

Step 2 Cut the pieces to rough size. If using a laminate trimmer or router for this task, clamp a straightedge to the sheet with C-clamps; then cut the pieces with the laminate trimmer or router and a flush trimmer bit by guiding the trimmer along the straightedge. Many cabinetmakers cut the edge pieces first because it's easier to cut a long, thin piece from a large piece.

Step 3 If it is necessary to make a seam or joint between two pieces of laminate for use on a long countertop, mirror-cut the joint. This can be done by clamping the two pieces of laminate in a straight line on some strips of ¾" stock with their ends butted together (*Figure 39*). Then, using one of the strips as a guide, run a laminate trimmer through the joint, cutting both pieces of laminate at the same time.

Step 4 Before unclamping the pieces, make a couple of pencil marks across them to use for alignment purposes later when gluing the pieces to the countertop.

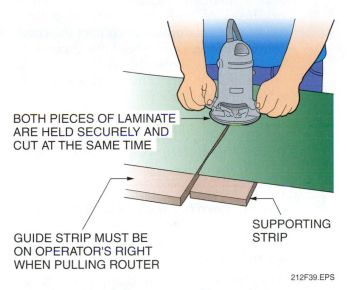

BOTH PIECES OF LAMINATE ARE HELD SECURELY AND CUT AT THE SAME TIME

GUIDE STRIP MUST BE ON OPERATOR'S RIGHT WHEN PULLING ROUTER

SUPPORTING STRIP

212F39.EPS

Figure 39 ◆ Cutting two adjoining sheets of laminate.

7.2.0 Applying the Laminate

Any surface to be covered with plastic laminate should be free of dirt, sawdust, and grease. It must also be flat. Any voids should be filled with wood filler and sanded flat and square. All edges should be smooth.

7.2.1 Laminating the Countertop Edges

To laminate the countertop edges, proceed as follows:

Step 1 Apply the edging one piece at a time, starting with the front edge. This produces a better appearance and prevents water entry. Use a narrow roller or paint brush to coat the back of the laminate strip and the wood or particleboard countertop edge with contact cement. Allow the cement to dry until it appears hazy and is not sticky. This typically takes 15 to 20 minutes.

Step 2 Carefully align the edge strip against the edge, allowing it to overlap both the bottom and top edges and at the ends. Allow the strip to make contact, press it firmly in place, then immediately roll it with a J-roller (*Figure 40*) or tap it into place with a smooth wood rubbing block and hammer to make sure no air is trapped between the two materials. Apply the laminate strips to the ends of the countertop base in the same manner as for the front piece. Make sure that the square ends of the sides butt up firmly against the back side of the overhanging front ends.

212F40.EPS

Figure 40 ◆ Rolling edges to eliminate air bubbles.

Step 3 Trim the overhanging ends and edges of the laminate. Start by trimming the overhanging ends of the front strip using the laminate trimmer and a bevel trimmer bit. Adjust the trimmer bit so that the trimmer cuts the laminate flush with the side piece but does not cut into the side piece.

Step 4 Trim the top and bottom overhang of the edging square using a flush trim bit (*Figure 41*). After trimming, go over the entire top edge with a belt sander or file to ensure a smooth and square edge. If using a belt sander, spray the material with a silicone lubricant or apply paraffin wax to prevent the edge from burning.

Step 5 If the edge of the countertop has rounded corners, the laminate can be bent. This is done by uniformly heating the laminate strip over the entire area of the bend with a suitable heat gun and then bending it until the desired radius is obtained. When doing this, make sure to keep your fingers away from the heated area because the laminate will retain the heat for some time.

 WARNING!
Contact cement is flammable. It can also be harmful or fatal if swallowed, and inhaling its vapors can be dangerous. Make sure to use contact cement in an open, well-ventilated area, and keep it away from open flames and lit cigarettes.

7.2.2 Laminating the Countertop

To laminate the countertop, proceed as follows:

Step 1 Begin by applying contact cement to both the laminate panel and the countertop using a roller or brush. Let the cement dry.

 CAUTION
When positioning the laminate to apply it to the countertop, take care that the two cemented surfaces do not touch. Once they touch, they are stuck and cannot be repositioned.

Step 2 Lay a series of wood strips or sheets of kraft paper on the countertop and set the laminate on top (*Figure 42*). Make sure that the laminate does not touch the countertop until you are satisfied with its alignment.

Step 3 Once the laminate is properly positioned, allow it to make contact at one end, then move along the length of the countertop and sequentially remove the wood strips or sheets of kraft paper one at a time, allowing the laminate to contact the countertop surface. Throughout this process, be sure to apply pressure on the laminate, working from the center out toward the edges to expel air bubbles.

Step 4 Roll the entire surface of the laminate vigorously to make sure it bonds tightly to the countertop surface. Work from the center toward the edges.

Step 5 If it is necessary to join two pieces, first cement down one piece of laminate. Then, at the seam, align the pencil marks and cement down the second piece, working away from the seam.

Step 6 After the top laminate is applied, trim the overhanging back edge with a flush trimmer bit. Trim the front edge and ends with a bevel trimmer bit (*Figure 43*); then use a flat file to smooth the trimmed edge.

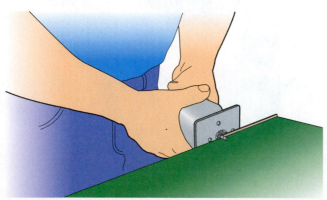
212F41.EPS

Figure 41 ◆ Trimming laminate edges.

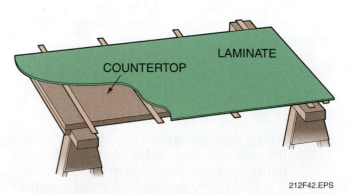
212F42.EPS

Figure 42 ◆ Using wood strips to separate the laminate from the countertop.

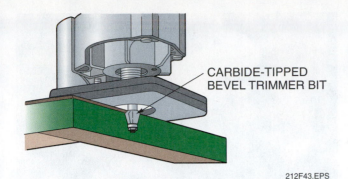

CARBIDE-TIPPED
BEVEL TRIMMER BIT

212F43.EPS

Figure 43 ◆ Trimming beveled edges.

8.0.0 ◆ INSTALLING SOLID-SURFACE COUNTERTOPS

For installation of solid-surface countertops, some manufacturers require that silicone adhesive be applied to the top of the base cabinet frames, then the countertop placed on top of it. Others require that 1" wood strips be attached to the cabinets with the countertop glued to the strips. Some guidelines for a quality installation of a solid-surface countertop are as follows:

- The counter should be leveled with shims rather than simply hiding any gaps with silicone.
- Any seams made at the job site should be inconspicuous and directly supported underneath.
- Inside corners should be cut on a radius and sanded to form a perfect curve rather than a right angle. This is because corners are stress points, and a radius withstands stress better than a sharp angle.
- The finish should be consistent over the entire surface. Check for consistency by examining the surface from several angles.

Solid-surface countertops should be maintained in accordance with the manufacturer's instructions. Most stains and spills can be removed with soap and water.

9.0.0 ◆ MASS-PRODUCTION CABINETMAKING

Local cabinet shops will design and build custom cabinets to suit the needs of individual customers. Other companies, more correctly called manufacturers than cabinetmakers, make the mass-market cabinets and cabinet systems you would find in building supply stores. These companies use huge computer-controlled machines to cut out, drill, and assemble cabinets. *Figures 44* through *48* show examples of these machines.

Figure 44 shows a table saw used to automatically cut sheets of plywood or particleboard into sizes and shapes needed for cabinet sides, bases, drawers, and shelves. The material comes prefinished.

Figure 45 shows a computer-controlled machine that automatically drills holes and makes dado cuts. *Figure 46* shows an example of a workpiece that has been through the machine.

The machine shown in *Figure 47* applies edge banding. Notice the roll of edge banding being fed into the machine. The cabinet manufacturer is able to buy large rolls of edge banding that match the finish on the pre-finished sheet material used to make the cabinets.

The boring machine in *Figure 48* is used to drill holes for dowels along the edges of cabinet sections.

212F44.EPS

Figure 44 ◆ Production table saw.

Figure 45 ◆ Computer-controlled boring and cutting machine.

Figure 47 ◆ Edge-banding machine.

Figure 46 ◆ Example of a workpiece after boring.

Figure 48 ◆ Computer-controlled boring machine for doweling.

Practice Exercises

Tool Safety Review

1. Brad guns do not require the same safety precautions as other pneumatic nail guns because they are not very dangerous.

   a. True
   b. False

2. When using a table saw, it is important to make sure your body is lined up with the blade.

   a. True
   b. False

3. You should always use a miter gauge when crosscutting on a table saw.

   a. True
   b. False

4. The radial arm saw is the best tool for ripping stock.

   a. True
   b. False

5. When using a compound miter saw, always wait to adjust the cutting angle until after the blade is running at full speed.

   a. True
   b. False

6. Jointer-planer guards can be left open while the machine is started.

   a. True
   b. False

7. When using a jointer-planer, you should lock the fence in place before turning on the power.

   a. True
   b. False

8. The shaper is considered one of the most dangerous tools because it has a tendency to kick back the workpiece.

   a. True
   b. False

9. Routers are generally not considered dangerous because they operate at low speeds.

   a. True
   b. False

10. The belt-disc sander is a good tool for smoothing metal.

    a. True
    b. False

Summary

Cabinetmaking, while similar to finish carpentry, requires many tools and techniques not used by carpenters. For example, cabinetmakers are more likely to use stationary tools. Also, cabinets use a variety of joining techniques not commonly used in finish carpentry.

Some cabinets, such as the ones you see in building supply stores, are mass-produced in manufacturing plants using special machines. This type of cabinet work does not require a cabinetmaker's skills. Custom cabinet shops usually serve their local area by designing and building cabinets to fit specific applications. This type of work requires skilled cabinetmakers.

Cabinets are often made of particleboard with a wood veneer applied to the surface because it is a relatively inexpensive method. Plywood, such as birch, is also used in finer cabinets. Furniture-quality cabinets are often made of hardwood lumber.

A wide variety of hinges and pulls are available. The majority of cabinets are made for use as kitchen cabinets and bathroom vanities. The top surface of such cabinets is usually covered with either a plastic laminate or a solid material, such as granite, marble, or a composite material.

Notes

1. A very hard, dense wood, very resistant to shock, light tan-colored with brown streaks is probably _____.
   a. ash
   b. cedar
   c. Douglas fir
   d. lauan

2. The red softwood used for lining drawers, chests, or closets is _____.
   a. mahogany
   b. cedar
   c. cherry
   d. ash

3. The wood often used as the core material in plywood or for millwork, but seldom for cabinet exteriors is _____.
   a. mahogany
   b. ash
   c. Douglas fir
   d. sweet birch

4. The wood traditionally used for cutting boards is _____.
   a. ash
   b. maple
   c. mahogany
   d. cherry

5. All the following solid woods are normally used in high-quality cabinets *except* _____.
   a. oak
   b. birch
   c. sweet gum
   d. cherry

6. The very expensive hardwood used for cabinets, musical instruments, and gun stocks is _____.
   a. yellow poplar
   b. yellow pine
   c. walnut
   d. Douglas fir

7. Veneer-core plywood is composed of sheets of _____ glued together.
   a. particleboard
   b. veneer
   c. HPL
   d. OSB

8. The plywood made from sheets of veneer attached to the outside of a piece of solid wood is called _____ core.
   a. particle
   b. solid
   c. lumber
   d. veneer

9. Which hardwood plywood veneer grade would you select if you needed uniform color and grain, but not book- or slip-matching?
   a. Specialty
   b. Good
   c. Premium
   d. Sound

10. Plywood should be cut to have the grain run _____.
    a. in the same direction
    b. opposite
    c. left to right
    d. right to left

11. The most common type of particleboard used for cabinets is _____.
    a. high-density
    b. 10-pound density
    c. 45-pound density
    d. 100-pound density

12. Table saws are not designed for making rabbet or miter cuts.
    a. True
    b. False

13. A stacked dado set used with a table saw creates a dado with beveled edges.
    a. True
    b. False

14. The primary use of a radial arm saw is _____.
    a. ripping lumber
    b. making compound cuts
    c. crosscutting lumber
    d. making radius cuts

15. A jointer can be used for making bevel or chamfer cuts.
    a. True
    b. False

16. A thickness planer must not be used on a board shorter than _____.
  a. 8"
  b. 12"
  c. 24"
  d. 36"

For Questions 17 through 21, match the name of the joint to its illustration in the figure below.

17. _____ Lock-shouldered

18. _____ Dovetail

19. _____ Rabbet

20. _____ Lock miter

21. _____ Stopped dado

22. A corner joint in which a rabbet cut is made in both pieces is known as a _____ joint.
  a. double rabbet
  b. half-lap
  c. blind dado
  d. lock miter

23. Wood veneer is typically applied using _____.
  a. wood glue
  b. brads
  c. contact cement
  d. construction adhesive

24. Aluminum oxide sandpaper is typically used to sand between finish coats.
  a. True
  b. False

25. When applying laminate to a countertop, _____.
  a. apply the front edge first
  b. apply the top edge first
  c. apply the side edges first
  d. it makes no difference which piece is applied first

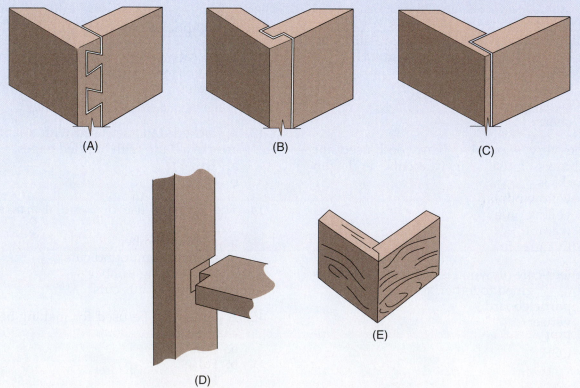

(A)   (B)   (C)

(D)

(E)

212RQ01.EPS

Trade Terms
Introduced in This Module

Heartwood: The part of the tree trunk between the pith and the sapwood that contains inactive (non-living) cells.

Pawls: Pivoting metal fingers used in power saws to prevent workpiece kickback.

Sapwood: The pale-colored living wood beneath the bark of the tree.

Practice Exercises

This appendix contains four exercises to complete under your instructor's supervision. They involve the cutting and joining skills you will need in order to build cabinets.

Practice Exercise A – Making Cabinet-Type Joints

This exercise will give you practice measuring and cutting wood, making rip cuts, and making joints commonly used in assembling cabinets.

 WARNING!
This work is to be performed under the supervision of your instructor and only after you have been checked out by your instructor on the safe and correct use of the tools involved.

TOOLS

- Table saw with stacked dado head
- Compound miter saw
- Router (optional for dado)
- Screw gun (if screws are used)
- Level
- Framing square
- Drill

MATERIALS

- 2 × 4 lumber
- Wood glue

PROCEDURE

Step 1 Cut four lengths of 2 × 4 using the table saw.

Step 2 Refer to *Figure A-1* and make each of the joints as indicated.

Step 3 Glue the pieces together and clamp them.

Practice Exercise B – Making a Covered Box

This project covers many of the same steps you would follow in making a cabinet drawer. In this case, however, you will be making something you can take home. This project will give you practice in making dado and dovetail joints.

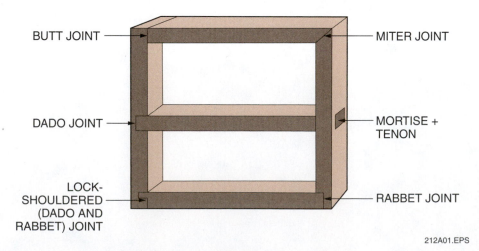

BUTT JOINT

MITER JOINT

DADO JOINT

MORTISE + TENON

LOCK-SHOULDERED (DADO AND RABBET) JOINT

RABBET JOINT

212A01.EPS

Figure A-1

TOOLS

- Table saw
- Router table and dovetail jig
- Brad nailer and brads
- Scroll saw or band saw

MATERIALS

- About 4' of 1 × 4 (or ripped plywood) for the sides and top
- A scrap piece of ¼" lauan plywood for the bottom
- Wood glue

PROCEDURE

Step 1 Rip the top, side, and end pieces from a 1 × 4 or plywood (see *Figure B-1*).

Step 2 Cut dadoes in the sides and back end piece as shown in the diagram. Use a stopped dado in the sides to compensate for the dovetails in these pieces.

Step 3 Cut dovetail joints for the drawer front and mating side pieces. Glue and assemble these pieces. Clamp the assembly until the glue dries.

Step 4 Cut the bottom piece to fit, and then slide it into place.

Step 5 Attach the back to the two side pieces using the brad nailer.

Step 6 Get creative: use the router table to cut out and decorate a top for the box, or you can also use a scroll saw or bandsaw to make a handle for the top.

Practice Exercise C – Making a 12" Raised Panel Cabinet Door

Cabinetmakers often purchase prefabricated raised-panel doors from companies that specialize in doors. However, if you want to custom-design a cabinet for a particular purpose, you may have to make the door(s) yourself. This exercise will give you practice in making a raised-panel door. The door consists of stiles (sides), rails (top and bottom), and the center panel, which is made by joining two pieces of 1 × 6. Most of the work can be done with a table saw. Before you start cutting, prepare a layout of the door.

TOOLS

- Table saw
- Jointer-planer
- Biscuit jointer
- Shaper (or table saw) for cutting raised panels
- Clamps

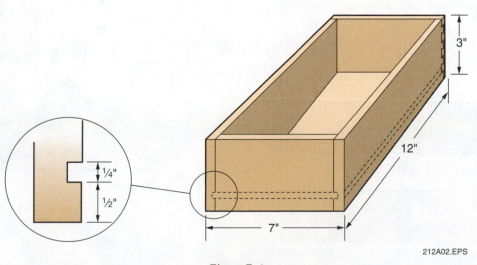

212A02.EPS

Figure B-1

MATERIALS

- 1 × 4 lumber for stiles and rails
- 1 × 6 lumber for panels
- Wood glue
- Biscuits

PROCEDURE

Step 1 Take two pieces of 1 × 4 (stiles and rails) and cut a dado in both sides of each piece (*Figure C-1*).

Step 2 Rip both 1 × 4s in half (*Figure C-2*). This will give you the stiles and rails. Cut them to the required lengths for your design.

Step 3 Use the jointer-planer to smooth the mating edges and surfaces of the 1 × 6 pieces.

Step 4 Use the biscuit jointer, biscuits, and glue to join the two pieces of 1 × 6 (*Figure C-3*). Clamp the assembly until the glue sets up.

Step 5 Cut a tenon on each end of each side, and then cut the dado about ¾" deeper on each side to create a mortise to receive the tenon (*Figure C-4*).

Step 6 Cut the 1 × 6 panel to size; note the depth of the dado to ensure a correct measurement.

Step 7 Raise the blade of the table saw and set the angle at 15 degrees.

Step 8 Stand the panel vertical and cut all four sides (*Figure C-5*).

Step 9 Cut a tongue on all four sides. Bring the blade back to zero degrees and lower it until the height of the blade equals the depth of the dado, as shown in *Figure C-6*.

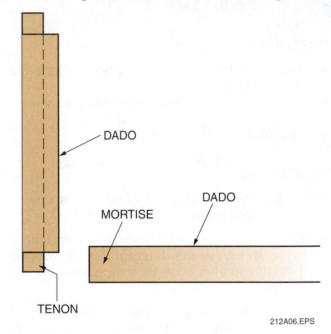

Figure C-4

Figure C-1

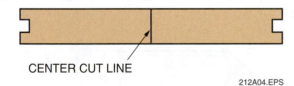

CENTER CUT LINE

Figure C-2

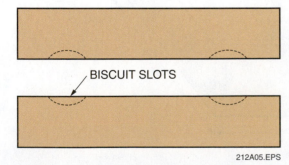

BISCUIT SLOTS

Figure C-3

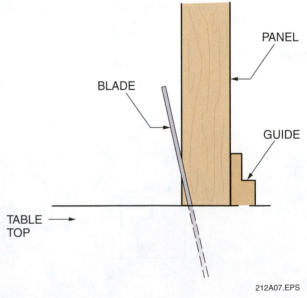

Figure C-5

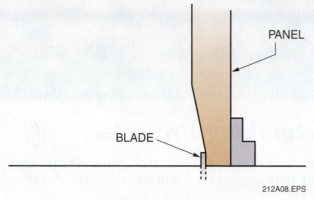

PANEL

BLADE

212A08.EPS

Figure C-6

Step 10 When cutting the back side of the panel, put a scrap piece of wood against the guide. If you do not use a dado blade, more than one pass will be needed to get the desired depth (*Figure C-7*).

Step 11 Glue the two sides to the bottom and slide the panel into the dado grooves (*Figure C-8*).

Step 12 Put on the top, and then glue and clamp.

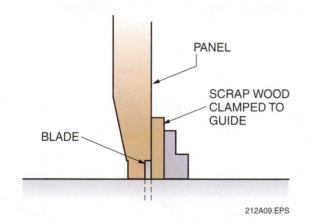

PANEL

SCRAP WOOD CLAMPED TO GUIDE

BLADE

212A09.EPS

Figure C-7

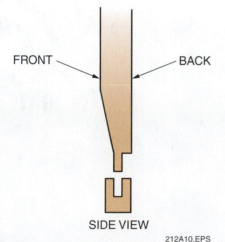

FRONT — BACK

SIDE VIEW

212A10.EPS

Figure C-8

Practice Exercise D – Applying Plastic Laminate to Plywood

The top surfaces of many cabinets are covered with plastic laminate. The same principles used in applying laminate are also used in applying wood veneers to cabinet surfaces. Therefore, this is a very important skill for a cabinetmaker. This exercise will allow you to practice that skill.

 WARNING!
This work is to be performed under the supervision of your instructor and only after you have been checked out by your instructor on the safe and correct use of the tools and materials involved. Be extremely careful with the contact cement. Check the MSDS for the contact cement you are using.

TOOLS

- Table saw to rip laminate
- Laminate trimmer
- Roller

MATERIALS

- Scrap countertop material* (use plywood as a substitute if necessary)
- Laminate material
- Contact cement
- Kraft paper or wood strips

*A local cabinet shop would probably be willing to donate scrap pieces of countertop material.

PROCEDURE

Step 1 Clean the top surface and one edge of the countertop.

Step 2 Apply the contact cement to the countertop surface and the laminate material.

Step 3 Apply laminate to the top, sides, and front of the countertop, following the procedure in Section 7.0.0 of this module.

Cabinetmaking Projects

This section contains two complete cabinet projects that you can build under your instructor's supervision. Both projects were used in SkillsUSA national competitions.

A number of organizations sponsor competitions where carpentry students and others can demonstrate and be recognized for their skills. SkillsUSA is the largest of these organizations. SkillsUSA is a student organization with more than 250,000 student and professional members. It is made up of 13,000 chapters in 4,100 high schools, career and technical education centers, and two-year colleges.

SkillsUSA conducts competitions at the local, district, regional, and state levels, leading to the national championships. In 2001, there were 10,000 local contests leading up to the national championships. More than 4,000 competitors, in 72 different career fields, competed in the nationals at Kansas City, Missouri. Prizes included medals, scholarships, tools, and other awards. Each year, a select few are chosen to represent the U.S. at the World Skills Competition. In 2002, that competition was held in St. Gallen, Switzerland.

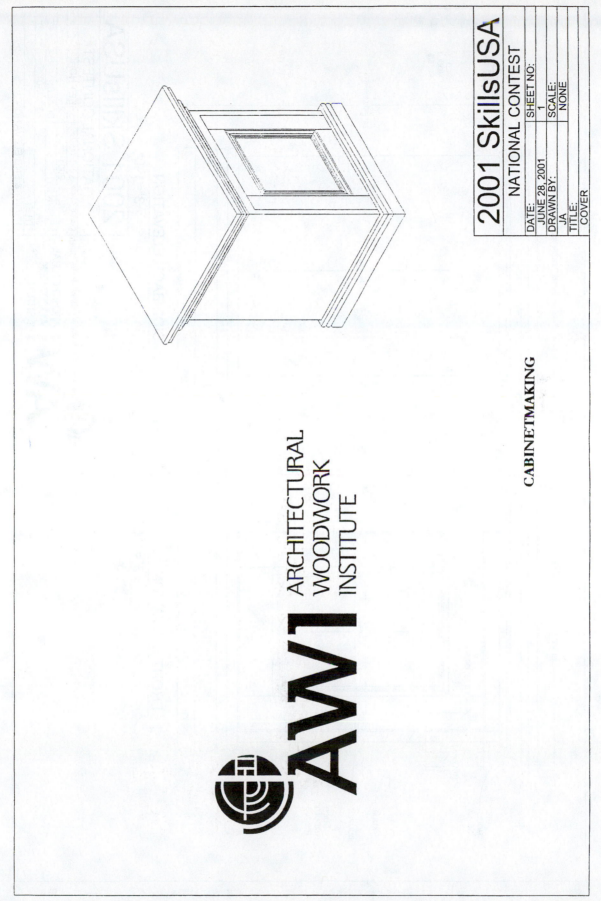

2001 SkillsUSA
NATIONAL CONTEST

DATE:	SHEET NO:
JUNE 28, 2001	1
DRAWN BY:	SCALE:
JA	NONE
TITLE:	
COVER	

CABINETMAKING

ARCHITECTURAL WOODWORK INSTITUTE

AWI

CM3F01.TIF

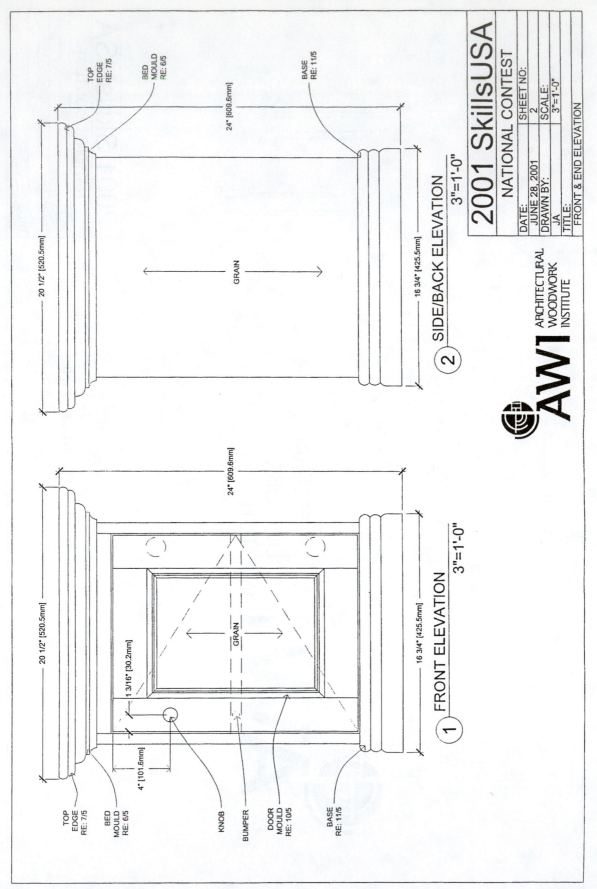

TOP EDGE RE: 7/5

BED MOULD RE: 6/5

BASE RE: 11/5

24" [609.6mm]

20 1/2" [520.5mm]

16 3/4" [425.5mm]

GRAIN

② SIDE/BACK ELEVATION
3"=1'-0"

24" [609.6mm]

20 1/2" [520.5mm]

16 3/4" [425.5mm]

GRAIN

1 3/16" [30.2mm]

4" [101.6mm]

① FRONT ELEVATION
3"=1'-0"

TOP EDGE RE: 7/5

BED MOULD RE: 6/5

KNOB

BUMPER

DOOR MOULD RE: 10/5

BASE RE: 11/5

2001 SkillsUSA
NATIONAL CONTEST

SHEET NO: 2
SCALE: 3"=1'-0"

DATE: JUNE 28, 2001
DRAWN BY: JA
TITLE: FRONT & END ELEVATION

AWI ARCHITECTURAL
WOODWORK
INSTITUTE

CM3F02.TIF

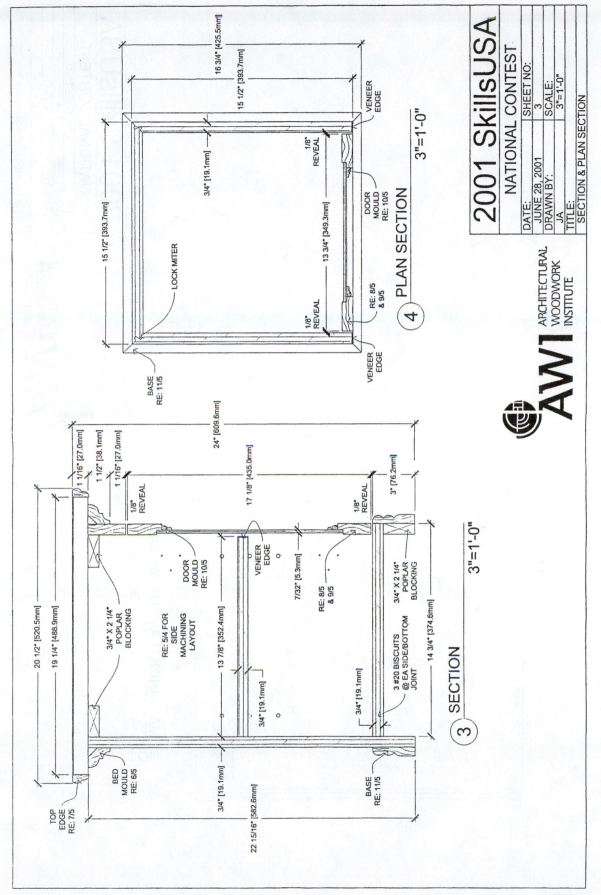

PLAN SECTION

16 3/4" [425.5mm]
15 1/2" [393.7mm]
13 3/4" [349.3mm]
15 1/2" [393.7mm]
3/4" [19.1mm]

VENEER EDGE
1/8" REVEAL
DOOR MOULD RE: 10/5
RE: 8/5 & 9/5
1/8" REVEAL
VENEER EDGE
LOCK MITER
BASE RE: 11/5

④ PLAN SECTION   3"=1'-0"

SECTION

24" [609.6mm]
1 1/16" [27.0mm]
1 1/2" [38.1mm]
1 1/16" [27.0mm]
17 1/8" [435.0mm]
3" [76.2mm]
20 1/2" [520.5mm]
19 1/4" [488.9mm]
13 7/8" [352.4mm]
14 3/4" [374.6mm]
22 15/16" [582.6mm]
3/4" [19.1mm]
3/4" [19.1mm]
3/4" [19.1mm]
7/32" [5.3mm]

1/8" REVEAL
DOOR MOULD RE: 10/5
VENEER EDGE
1/8" REVEAL
3/4" X 2 1/4" POPLAR BLOCKING
RE: 5/4 FOR SIDE MACHINING LAYOUT
RE: 8/5 & 9/5
3/4" X 2 1/4" POPLAR BLOCKING
3 #20 BISCUITS @ EA SIDE/BOTTOM JOINT
TOP EDGE RE: 7/5
BED MOULD RE: 6/5
BASE RE: 11/5

③ SECTION   3"=1'-0"

2001 SkillsUSA
NATIONAL CONTEST

DATE: JUNE 28, 2001
DRAWN BY: JA
TITLE: SECTION & PLAN SECTION
SHEET NO: 3
SCALE: 3"=1'-0"

AWI
ARCHITECTURAL
WOODWORK
INSTITUTE

CM3F03.TIF

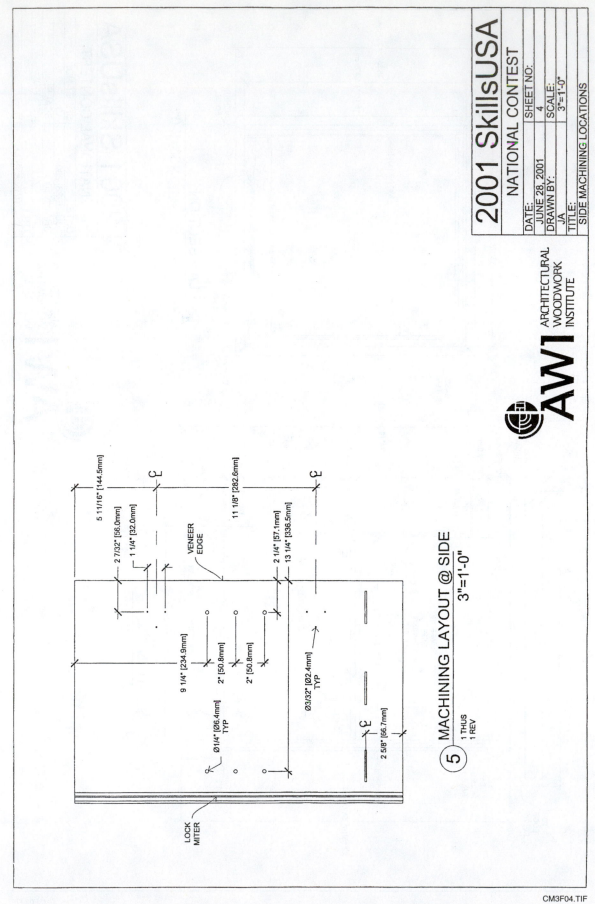

MACHINING LAYOUT @ SIDE
3"=1'-0"

5 1 THUS
   1 REV

5 11/16" [144.5mm]
2 7/32" [56.0mm]
1 1/4" [32.0mm]
VENEER EDGE
11 1/8" [282.6mm]
2 1/4" [57.1mm]
13 1/4" [336.5mm]
9 1/4" [234.9mm]
2" [50.8mm]
2" [50.8mm]
Ø3/32" [Ø2.4mm] TYP
Ø1/4" [Ø6.4mm] TYP
2 5/8" [66.7mm]
LOCK MITER

2001 SkillsUSA
NATIONAL CONTEST

SHEET NO:
4
SCALE:
3"=1'-0"

DATE:
JUNE 28, 2001
DRAWN BY:
JA
TITLE:
SIDE MACHINING LOCATIONS

AWI
ARCHITECTURAL
WOODWORK
INSTITUTE

CM3F04.TIF

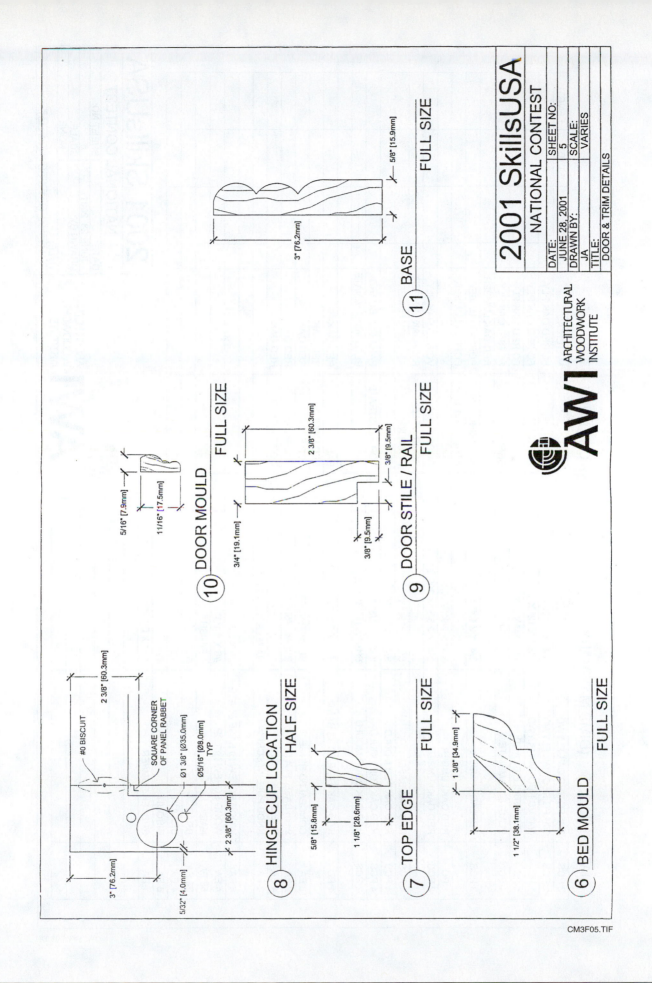

11 BASE
5/8" [15.9mm]
3" [76.2mm]
FULL SIZE

10 DOOR MOULD
FULL SIZE
5/16" [7.9mm]
1 1/16" [17.5mm]

9 DOOR STILE / RAIL
FULL SIZE
2 3/8" [60.3mm]
3/8" [9.5mm]
3/4" [19.1mm]
3/8" [9.5mm]

8 HINGE CUP LOCATION
HALF SIZE
#0 BISCUIT
SQUARE CORNER OF PANEL RABBET
Ø1 3/8" [Ø35.0mm]
Ø5/16" [Ø8.0mm] TYP
2 3/8" [60.3mm]
2 3/8" [60.3mm]
3" [76.2mm]
5/32" [4.0mm]

7 TOP EDGE
FULL SIZE
5/8" [15.8mm]
1 1/8" [28.6mm]

6 BED MOULD
FULL SIZE
1 3/8" [34.9mm]
1 1/2" [38.1mm]

2001 SkillsUSA
NATIONAL CONTEST

DATE:	
JUNE 28, 2001	SHEET NO:
DRAWN BY:	5
JA	SCALE:
TITLE:	VARIES
DOOR & TRIM DETAILS	

AWI
ARCHITECTURAL WOODWORK INSTITUTE

CM3F05.TIF

MATERIAL LIST   (DIMS MAY VARY)

3 PCS	3/4" FC RED OAK PLYWD	3/4" X 16" X 24"
2 PCS	3/4" FC RED OAK PLYWD	3/4" X 16" X 16"
1 PC	1/4" VC RED OAK PLYWD	7/32" X 12" X 16"
1 PC	1" PARTICLE BOARD	1" X 24" X 24"
1 PC	PLASTIC LAMINATE	24" X 24"
3 PCS	POPLAR LUMBER	3/4" X 2 1/4" X 16"
1 PC	RED OAK LUMBER	3/4" X 2 9/16" X 16"
2 PCS	RED OAK LUMBER	3/4" X 2 3/8" X 30"
4 PCS	RED OAK BASE	5/8" X 3" X 18"
4 PCS	RED OAK BED MOULD	3/4" X 2" X 20"
4 PCS	RED OAK TOP MOULD	5/8" X 1 1/8" X 23"
2 PCS	RED OAK DOOR MOULD	5/16" X 11/16" X 28"
1 PC	RED OAK TAPE EDGE BAND	.018 X 7/8 X 72
2 EA	CONCEALED HINGE	BLUM 75M5690.21
2 EA	9mm BASE PLATE	BLUM 175H7190
4 EA	SHELF SUPPORTS	
1 EA	KNOB PULL W/ SCREW	HAFELE 134.42.816
6 EA	WOOD SCREW @ BASE PLATE	#6 X 5/8"
4 EA	WOOD SCREW @ TOP	#8 X 1 1/4"
58 EA	NAIL	4 FINISH
24 EA	BRAD	5/8"
4 EA	BISQUIT	#0
6 EA	BISQUIT	#20
1 EA	RUBBER BUMPER	1/8" X Ø1/2"

CUTTING LIST

PART	MATERIAL	QTY	THICK	WIDTH	LENGTH
SIDE	FC RED OAK PLY				
BOTTOM	FC RED OAK PLY				
BACK	FC RED OAK PLY				
SHELF	FC RED OAK PLY				
RAIL	RED OAK LUMBER				
BLOCKING	POPLAR LUMBER				
STRETCHERS	POPLAR LUMBER				
DOOR PANEL	VC RED OAK PLY				
DOOR STILE	RED OAK LUMBER				
DOOR RAIL	RED OAK LUMBER				
TOP SUBSTRATE	PARTICLE BOARD				
TOP	PLASTIC LAMINATE				
BASE	RED OAK				
BED MOULD	RED OAK				
TOP EDGE	RED OAK				
DOOR MOULD	RED OAK				
DOOR MOULD	RED OAK				

AWI ARCHITECTURAL WOODWORK INSTITUTE

2001 SkillsUSA
NATIONAL CONTEST

DATE: JUNE 28, 2001	SHEET NO: 6
DRAWN BY: JA	SCALE: NONE
TITLE: MATERIAL LIST & CUTTING LIST	

CM3F06.TIF

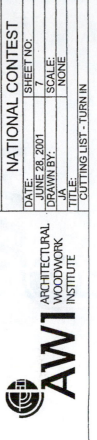

2001 SkillsUSA
NATIONAL CONTEST

DATE:	JUNE 28, 2001
DRAWN BY:	JA
TITLE:	CUTTING LIST - TURN IN

SHEET NO: 7
SCALE: NONE

ARCHITECTURAL WOODWORK INSTITUTE

CONTESTANT #

CUTTING LIST

PART	MATERIAL	QTY	THICK	WIDTH	LENGTH
SIDE	FC RED OAK PLY				
BOTTOM	FC RED OAK PLY				
BACK	FC RED OAK PLY				
SHELF	FC RED OAK PLY				
RAIL	RED OAK LUMBER				
BLOCKING	POPLAR LUMBER				
STRETCHERS	POPLAR LUMBER				
DOOR PANEL	VC RED OAK PLY				
DOOR STILE	RED OAK LUMBER				
DOOR RAIL	RED OAK LUMBER				
TOP SUBSTRATE	PARTICLE BOARD				
TOP	PLASTIC LAMINATE				
BASE	RED OAK				
BED MOULD	RED OAK				
TOP EDGE	RED OAK				
DOOR MOULD	RED OAK				
DOOR MOULD	RED OAK				

CM3F07.TIF

2002 Skills USA / VICA

National

Cabinetmaking

Championships

June 27, 2002

Sheet 1

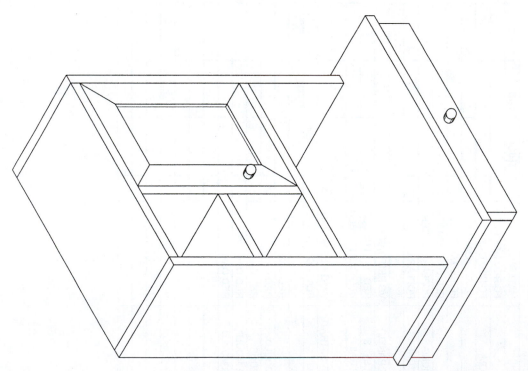

CM4F01.TIF

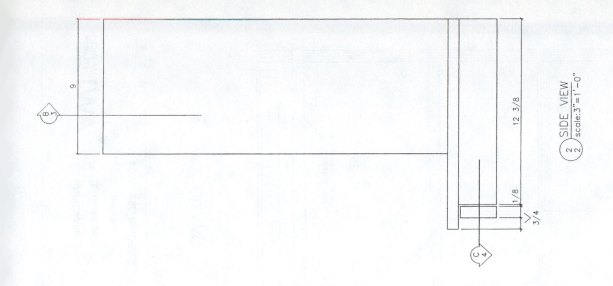

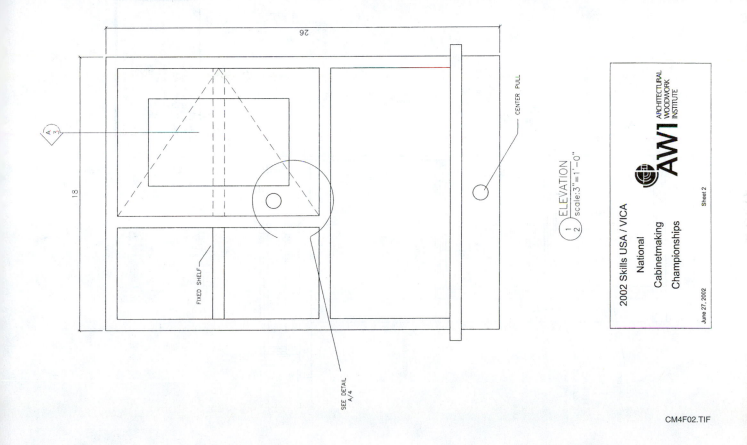

B 3

9

C 4

12 3/8

1/8

3/4

2 SIDE VIEW
2 scale:3"=1'-0"

A 3

26

18

FIXED SHELF

SEE DETAIL A/4

CENTER PULL

1 ELEVATION
2 scale:3"=1'-0"

AWI ARCHITECTURAL WOODWORK INSTITUTE

2002 Skills USA / VICA
National
Cabinetmaking
Championships

June 27, 2002   Sheet 2

CM4F02.TIF

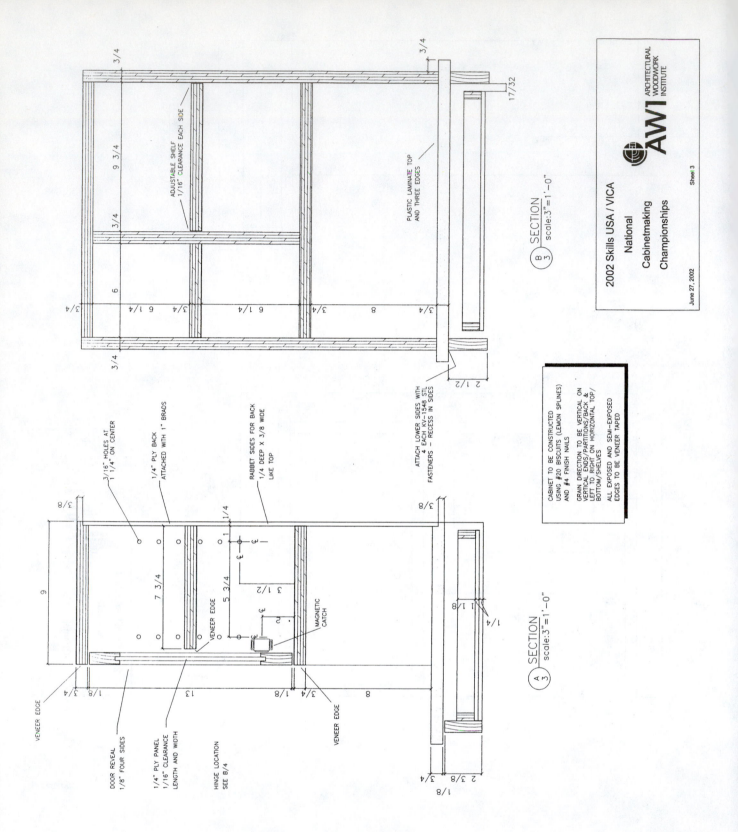

CM4F03.TIF

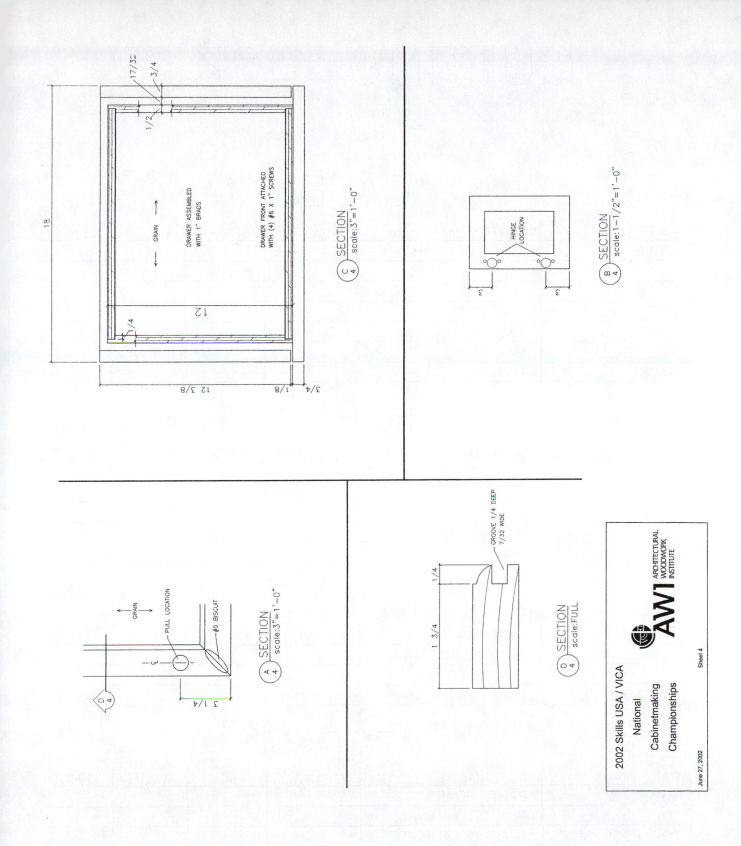

18

17/32 3/4

1/2

GRAIN →

← GRAIN

DRAWER ASSEMBLED
WITH 1" BRADS

DRAWER FRONT ATTACHED
WITH (4) #6 X 1" SCREWS

12

1/4

12 3/8

3/4   1/8

C SECTION
4 scale:3"=1'-0"

HINGE
LOCATION

3   3

B SECTION
4 scale:1-1/2"=1'-0"

GRAIN →

PULL LOCATION

#0 BISCUIT

3

3 1/4

D
4

A SECTION
4 scale:3"=1'-0"

GROOVE 1/4 DEEP
7/32 WIDE

1/4

1 3/4

D SECTION
4 scale:FULL

2002 Skills USA / VICA
National
Cabinetmaking
Championships

June 27, 2002   Sheet 4

AWI ARCHITECTURAL
     WOODWORK
     INSTITUTE

CM4F04.TIF

CUTTING TICKET

CASE

PART NAME	MATERIAL	QUANTITY	THICK	WIDTH	LENGTH
SIDES	OAK PLY				
TOP	OAK PLY				
FIXED SHELF	OAK PLY				
FIXED SHELF	OAK PLY				
PARTITION	OAK PLY				
ADJ. SHELF	OAK PLY				
BACK	OAK PLY				

DRAWER

PART NAME	MATERIAL	QUANTITY	THICK	WIDTH	LENGTH
CASE SIDES	OAK STOCK				
FRONT	OAK STOCK				
DRAWER SIDES	BALTIC BIRCH PLY				
DRAWER ENDS	BALTIC BIRCH PLY				
DRAWER BOTTOM	OAK PLY				

DOOR

PART NAME	MATERIAL	QUANTITY	THICK	WIDTH	LENGTH
DOOR STILE	OAK DOOR STOCK				
DOOR RAIL	OAK DOOR STOCK				
DOOR PANEL	OAK PLY				

TOP

PART NAME	MATERIAL	QUANTITY	THICK	WIDTH	LENGTH
TOP	PART. BOARD				
TOP	PLASTIC LAMINATE				
TOP EDGE SIDES	PLASTIC LAMINATE				
TOP EDGE FRONT	PLASTIC LAMINATE				

CM4F05.TIF

MATERIAL LIST

1 PC	PART. BOARD	3/4" x 16" x 24"
1 PC	PLAS. LAMINATE	16" X 24"
1 PC	OAK PLY	3/4" X 48" X 24"
1 PC	OAK PLY	3/4" X 24" X 24"
1 PC	OAK PLY	NOM 1/4" X 48" X 24"
1 PC	BALTIC BIRCH	1/2" X 6" X 20"
1 PC	OAK STOCK	3/4" x 3" x 28"
1 PC	OAK STOCK	3/4" x 3" x 20"
2 PCS	OAK DOOR STOCK	3/4" X 2" X 25"

HARDWARE LIST

2	BLUM HINGE PLATES W/SCREWS	# 175H7190
2	BLUM HINGE	# 75T1690
1 PR.	SLIDES W/SCREWS	# 230M3000
4	SHELF CLIPS	
2	KNOBS W/SCREWS	
1	MAGNETIC CATCH W/SCREWS	
4	DESK TOP FASTENERS	# KV1548 STL
2	DRAWER BUMPERS	

MISC. LIST

12 LF	VENEER TAPE
26	#20 BISQUITS
4	#0 BISQUITS
30	1 1/2" NAILS
40	1" BRADS
4	#6 X 1" SCREWS
8	#6 X 3/4" SCREWS

CM4F06.TIF

This module is intended to be a thorough resource for task training. The following reference works are suggested for further study. These are optional materials for continued education rather than for task training.

Cabinet Makers Association website, *www.cabinetmakers.org*.

NCCER, *Cabinetmaking Trainee Guide*, Contren Learning Series, Upper Saddle River: Pearson Education, 2007.

Kitchen Cabinet Makers Association website, *www.kcma.org*.

Mill's Pride Cabinetry website, *www.millspride.com*.

NCCER CURRICULA — USER UPDATE

NCCER makes every effort to keep its textbooks up-to-date and free of technical errors. We appreciate your help in this process. If you find an error, a typographical mistake, or an inaccuracy in NCCER's curricula, please fill out this form (or a photocopy), or complete the online form at **www.nccer.org/olf**. Be sure to include the exact module ID number, page number, a detailed description, and your recommended correction. Your input will be brought to the attention of the Authoring Team. Thank you for your assistance.

Instructors – If you have an idea for improving this textbook, or have found that additional materials were necessary to teach this module effectively, please let us know so that we may present your suggestions to the Authoring Team.

NCCER Product Development and Revision
13614 Progress Blvd., Alachua, FL 32615

Email: curriculum@nccer.org
Online: www.nccer.org/olf

❏ Trainee Guide   ❏ AIG   ❏ Exam   ❏ PowerPoints   Other _____

Craft / Level: _____   Copyright Date: _____

Module ID Number / Title: _____

Section Number(s): _____

Description: _____

_____

_____

_____

_____

Recommended Correction: _____

_____

_____

_____

_____

Your Name: _____

Address: _____

_____

Email: _____   Phone: _____

Glossary

Access: A passageway or a corridor between rooms.

Acoustical materials: Types of ceiling tile, plaster, and other materials that have high absorption characteristics for sound waves.

Acoustics: A science involving the production, transmission, reception, and effects of sound. In a room or other location, it refers to those characteristics that control reflections of sound waves and thus the sound reception in the area.

Apron: A piece of window trim that is used under the stool of a finished window frame. It is sometimes referred to as undersill trim.

Asphalt roofing cement: An adhesive that is used to seal down the free tabs of strip shingles. This plastic asphalt cement is mainly used in open valley construction and other flashing areas where necessary for protection against the weather.

Astragal: A piece of molding attached to the edge of an inactive door on a pair of double doors. It serves as a stop for the active door.

A-weighted decibel (dBA): A single number measurement based on the decibel but weighted to approximate the response of the human ear with respect to frequencies.

Backsplash: A piece that extends up the wall from a countertop.

Base flashing: The protective sealing material placed next to areas vulnerable to leaks, such as chimneys.

Beams: Loadbearing horizontal framing elements supported by walls or columns and girders.

Blocking: C-shaped track, break shape, or flat strap material attached to structural members, flat strap, or sheathing panels to transfer shear forces.

Board-and-batten: A type of vertical siding consisting of wide boards with the joint covered by narrow strips known as battens.

Brown coat: A coat of plaster with a rough face on which a finish coat will be placed.

Building paper: A heavy paper used for construction work. It assists in weatherproofing the walls and prevents wind infiltration. Building paper is made of various materials and is not a vapor barrier.

Bundle: A package containing a specified number of shingles or shakes. The number is related to square foot coverage and varies with the product.

Butt: Any kind of hinge, except a strap or T hinge.

Butt gauge: A marking gauge normally used to mark the depth and width of mortises for butts.

Callouts: Markings or identifying tags describing parts of a drawing; callouts may refer to detail drawings, schedules, or other drawings.

Cap flashing: The protective sealing material that overlaps the base and is embedded in the mortar joints of vulnerable areas of a roof, such as a chimney.

Casing: The exposed finish material around the edge of a door or window opening.

Catches: Spring bolts used to secure a door when shut.

Ceiling panels: Any lay-in acoustical board that is designed for use with an exposed grid mounting system. Ceiling panels normally do not have finished edges or precise dimensional tolerances because the exposed grid mounting system provides the trimout.

Ceiling tiles: Acoustical ceiling boards that are nailed, cemented, or suspended by a concealed grid mounting system. The edges are often kerfed and cut back.

Civil drawings: A drawing that shows the overall shape of the building site. It is also called a site plan.

Clip angle: An L-shaped piece of steel (normally with a 90-degree bend), typically used for connections.

Cold-formed steel: Sheet steel or strip steel that is manufactured by press braking blanks sheared from sheets or cut lengths of coils or plates, or by continuous roll forming of cold- or hot-rolled coils of sheet steel.

Compound cut: A combined bevel and miter cut.

Condensation: The process by which a vapor is converted to a liquid, such as the conversion of the moisture in air to water.

Contour lines: Imaginary lines on a site plan/plot plan that connect points of the same elevation. Contour lines never cross each other.

Convection: The movement of heat that either occurs naturally due to temperature differences or is forced by a fan or pump.

Coordinator: A device for use with exit features to hold the active door open until the inactive door is closed.

Coped joint: A joint made by cutting the end of a piece of molding to the shape it will fit against.

Corner bead: A metal or plastic angle used to protect outside corners where drywall panels meet.

Cornice: The construction under the eaves where the roof and side walls meet.

Course: One row of brick, block, or siding as it is placed in the wall.

C-shape: A cold-formed steel shape used for structural and non-structural framing members consisting of a web, two flanges, and two lips (edge stiffeners).

Curtain wall: A light, nonbearing exterior wall attached to the concrete or steel structure of the building.

Cylindrical lockset: A lock in which the keyhole and tumbler mechanism are in a cylinder that is separate from the lock case and can be removed to change the keying of the lock.

Deadbolt: A square-head bolt in a door lock that requires a key to move it in either direction.

Decibel (dB): A unit that is used to express differences in sound power. In acoustics, it is equal to ten times the logarithm of the ratio of one sound and a lower-intensity reference sound. One decibel indicates a difference of about 26% and is about the smallest change the ear can detect.

Dew point: The temperature at which air becomes oversaturated with moisture and the moisture condenses.

Diaphragm: A floor, ceiling, or roof assembly designed to resist in-plane forces such as wind or seismic loads.

Diffuser: An attachment for duct openings in air distribution systems that distributes the air in wide flow patterns. In lighting systems, it is an attachment used to redirect or scatter the light from a light source.

Diffusion: The movement, often contrary to gravity, of molecules of gas in all directions, causing them to intermingle.

Door closer: A mechanical device used to check a door and prevent it from slamming when it is being closed. It also ensures the closing of the door.

Door frame: The surrounding case of a door into which a door closes. It consists of two upright pieces called jambs and a horizontal top piece called the head.

Door jamb: See *door frame*.

Door stop: The strip against which a door closes on the inside face of the frame or jamb. It can also be a hardware device used to hold the door open to any desired position or a hardware device placed against the baseboard to keep the door from marring the wall.

Dry lines: A string line suspended from two points and used as a guideline when installing a suspended ceiling.

Dustproof strike: A metal piece used to receive the latch bolt when a door is closed. The piece has a spring action cover to keep particles out of the strike when the door is open.

Eave: The lower part of a roof, which projects over the side wall.

Elevation view: A drawing giving a view from the front, rear, or side of a structure.

Escutcheon plate: A decorative metal plate that is placed against a door or drawer face behind the pull.

Exposure: The distance (in inches) between the exposed edges of overlapping shingles.

Exterior insulation finish system (EIFS): A protective and decorative coating applied directly to insulation board.

Fascia: The exterior finish member of a cornice on which the rain gutter is usually hung.

Felt paper: An asphalt-impregnated paper.

Finger-jointed stock: Paint-grade moldings made in a mill from shorter lengths of wood joined together.

Finish coat: The final coat of plaster or paint.

Finish hardware: The exposed hardware in a building such as doorknobs, door hinges, door locks, door closers, window hardware, shelf and clothing storage hangers, and bathroom hardware.

Finishing sawhorse: A pair of trestles used to support lumber, molding, doors, stair parts, and other materials while they are being fitted and shaped for installation.

Fissured: A ceiling panel or ceiling tile surface design that has the appearance of splits or cracks.

Floating interior angle construction: A drywall installation technique in which no fasteners are used at the edge of the panel in order to allow for structural stresses.

Flush bolt: A sliding bolt mechanism that is mortised into a door at the top and bottom edge. It is used to hold an inactive door in a fixed position on a pair of double doors.

Flush door: A door of any size that has a totally flat surface.

Frequency: Cycles per unit of time, usually expressed in hertz (Hz).

Frieze board: A horizontal finish member connecting the top of the sidewall, usually abutting the soffit. Its bottom edge usually serves as a termination point for various types of siding materials.

Furr down: The enclosed section (which is usually decorated by paint or wallpaper) between the ceiling and the top of the wall cabinets; also called a soffit.

Girders: Large steel or wooden beams supporting a building, usually around the perimeter.

Gypsum board: A generic term for paper-covered gypsum core panels; also known as gypsum drywall.

Hanging stile: The door stile to which the hinges (butts) are fastened.

Hardware: Fastenings that allow for the movement and attachment of door components.

Head lap: The distance between the top of the bottom shingle and the bottom edge of the one covering it.

Head: The horizontal member at the top of a door or window opening.

Header: A horizontal structural framing member used over floor, roof, or wall openings to transfer loads around the opening to supporting structural framing members.

Heartwood: The part of the tree trunk between the pith and the sapwood that contains inactive (non-living) cells.

Hertz (Hz): A unit of frequency equal to one cycle per second.

Hinge: The hardware fastened to the edge of a door that allows the frame to pivot around a steel pin, permitting the operation of the door.

Isometric drawing: A three-dimensional type drawing in which the object is tilted so that all three faces are equally inclined to the picture plane.

Jamb: One of the vertical members on either side of a door or window opening.

Joint: A place where two pieces of material meet.

Joists: Horizontal members of wood or steel supported by beams and holding up the planks of floors or the lathes of ceilings. Joists are laid edgewise to form the floor support.

Kerf: A slot or cut made with a saw.

Knob lockset: A lock for a door, with the locking cylinder located in the center of the knob.

Knurled: A series of small ridges used to provide a better gripping surface on metal and plastic.

Landscape drawings: A drawing that shows proposed plantings and other landscape features.

Latch bolt: A spring-loaded bolt in a lock, with a beveled head that is retracted when hitting the strike.

Lateral: Running side to side; horizontal.

Ledger: A board to which the lookouts are attached and which is placed against the outside wall of the structure. It is also used as a nailing edge for the soffit material.

Lips: That part of a C-shape framing member that extends from the flange as a stiffening element that extends perpendicular to the flange. Also called edge stiffener.

Lock: A mechanical hardware device for securing a door in a closed position. This device is usually designed with a bolt operated by a key, combination, or electrical impulse.

Lockset: The entire lock unit, including locks, strike plate, and trim pieces.

Lookout: A member used to support the overhanging portion of a roof.

Louver: A slatted opening used for ventilation, usually in a gable end or a soffit.

Mil: A unit of measurement equal to $1/1,000"$.

Moldings: Decorative strips of wood or other material used for finishing purposes.

Mortise: A measured portion of wood removed to receive a piece of hardware such as a lock or butt.

Mortise lock: A rectangular metal box that houses a lock. It usually has a latch and deadbolt as part of the unit. Options consist of a locking cylinder and/or thumb turn by which it can be secured. It is used on residential entry doors and commercial doors.

Nail pop: The protrusion of a nail above the wallboard surface that is usually caused by shrinkage of the framing or by incorrect installation. Also applies to screws.

Overhang: The part that extends beyond the building line. The amount of overhang is always given as a projection from the building line on a horizontal plane.

Panelization: The process of assembling steel-framed walls, joists, or trusses before they are installed in a structure. Roll-formers normally cut studs to within ⅛" tolerance. This helps the framer to consistently create straight walls in the panel table that are easy to install in the field.

Panic hardware: Hardware that provides an emergency escape exit.

Pawls: Pivoting metal fingers used in power saws to prevent workpiece kickback.

Perm: The measure of water vapor permeability. It equals the number of grains squared of water vapor passing through a one square foot (sq ft) piece of material per hour, per inch of mercury difference in vapor pressure.

Permeability: The measure of a material's capacity to allow the passage of liquids or gases.

Permeable: Porous; having small openings that permit liquids or gases to seep through.

Permeance: The ratio of water vapor flow to the vapor pressure difference between two surfaces.

Pitch: The ratio of the rise to the span indicated as a fraction. For example, a roof with a 6' rise and a 24' span will have a ¼ pitch.

Plan view: A drawing that represents a view looking down on an object.

Plancier: The same as a soffit, but the member is usually fastened to the underside of a rafter rather than the lookout.

Plenum: A chamber, container, or other confined space for moving air under a slight pressure. In commercial construction, the area between the suspended ceiling and the floor or roof above is often used as the HVAC return air plenum.

Plumb: A true perpendicular/vertical position.

Prefinished: Material such as molding, doors, cabinets, and paneling that has been stained, varnished, or painted at the factory.

Prehung door: A door that is delivered to the job site from the mill already hung in the frame or jamb. In some instances, the trim may be applied on one side.

Rabbet: A groove cut in the edge of a board so as to receive another board.

Racking: Being forced out of plumb by wind or seismic forces.

Rail: A horizontal member of a door or window sash.

Rake: The slope or pitch of the cornice that parallels the roof rafters on the gable end.

Reveal: The amount of setback of the casing from the face side of window and door jambs or similar pieces.

Ridge: The horizontal line formed by the two rafters of a sloping roof that have been nailed together. The ridge is the highest point at the top of the roof where the roof slopes meet.

Rim track: A horizontal structural member that is connected to the end of a floor joist.

Riser diagram: A type of isometric drawing that depicts the layout, components, and connections of a piping system.

Roof rafter: A horizontal or sloped structural framing member that supports roof loads.

Roof sheathing: Usually 4 × 8 sheets of plywood, but can also be 1 × 8 or 1 × 12 roof boards, or other new products approved by local building codes. Also referred to as decking.

Rough opening: Any unfinished door or window opening in a building.

R-value: A numerical designation given to a material based on its insulating ability, such as R-19.

Saddle: An auxiliary roof deck that is built above the chimney to divert water to either side. It is a structure with a ridge sloping in two directions that is placed between the back side of a chimney and the roof sloping toward it. Also referred to as a cricket.

Sapwood: The pale-colored living wood beneath the bark of the tree.

Scarf joints: End joints made by overlapping two pieces of molding with 22.5- or 45-degree angle cuts.

Scratch coat: The first coat of cement plaster consisting of a fine aggregate that is applied through a diamond mesh reinforcement or on a masonry surface.

Scribe: To mark wood with a sharp knife or scriber.

Scrim: A loosely knit fabric.

Selvage: The section of a composition roofing roll or shingle that is not covered with an aggregate.

Shakes: Hand- or machine-split wood shingles.

Shear wall: A wall designed to resist lateral forces such as those caused by earthquakes or wind.

Shim: A thin, tapered piece of wood such as a wooden shingle used to fill in gaps and to level or plumb structural components.

Side lap: The distance between adjacent shingles that overlap, measured in inches.

Sill: The lowest member at the bottom of a door or window opening. It also refers to the lowest member of wood framing supporting the framing of a building.

Slope: The ratio of rise to run. The rise in inches is indicated for every foot of run.

Slurry: A thin mixture of water or other liquid with any of several substances such as cement, plaster, or clay.

Smoke gasket: A rubber strip that goes all the way around the door to keep smoke from penetrating an area in case of fire.

Soffit: The underside of a roof overhang.

Sound attenuation: The reduction of sound as it passes through a material.

Sound transmission class (STC): The rating by which the sound attenuation of a door is determined. The higher or greater the number, the better the sound reduction.

Square: The amount of shingles needed to cover 100 square feet of roof surface. For example, square means 10' square or 10' × 10'.

Square cuts: Cuts made at a right (90-degree) angle.

Stile: The vertical edge of a door.

Stool: The bottom horizontal trim piece of a window.

Striated: A ceiling panel or ceiling tile surface design that has the appearance of fine parallel grooves.

Strike or **strike plate:** A metal plate screwed to the jamb of a door so that when the door is closed, the bolt of the lock strikes against it. The bolt is then retracted and slides along the metal plate. When the door is fully closed, the bolt inserts itself into a hole in the plate to hold the door securely in place.

Substrate: The underlying material to which a finish is applied.

Sweep: A type of weatherstripping. A sweep is a felt or rubber flap mounted in a metal channel to seal door bottoms to prevent air infiltration.

Template: A thin piece of material such as plastic or heavy paper with a shape cut out of it, or the whole of it cut to a shape on its perimeter. A template is used to transfer that shape to another object by tracing it with a pencil or scribe.

Threshold: A piece of wood, metal, or stone that is set between the door jamb and the bottom of a door opening.

Top lap: The distance, measured in inches, between the lower edge of an overlapping shingle and the upper edge of the lapping shingle.

Track: A framing member consisting of only a web and two flanges. Track web depth measurements are taken to the inside of the flanges.

Transom: A panel above a door that lets light and/or air into a room or is used to fill the space above a door when the ceiling heights on both sides of the door opening allow.

Trim: Finish materials such as molding placed around doors and windows and at the top and bottom of a wall.

Underlayment: Asphalt-saturated felt protection for sheathing; 15-lb roofer's felt is commonly used. The roll size is 3' × 144' or a little over four squares.

Valley: The internal part of the angle formed by the meeting of two roofs.

Valley flashing: Watertight protection at a roof intersection. Various metals and asphalt products are used; however, materials vary based on local building codes.

Vapor barrier: A material used to retard the flow of vapor and moisture into walls and prevent condensation within them. The vapor barrier must be located on the warm side of the wall.

Vapor diffusion retarder (VDR): See *vapor barrier*.

Vapor retarder: See *vapor barrier*.

Veneer: A thin layer or sheet of wood intended to be overlaid on a surface to provide strength, stability, and/or an attractive finish. Thicknesses range between ¹⁄₁₆" and ¹⁄₈" for core plies and between ¹⁄₁₂₅" and ¹⁄₃₂" for decorative faces. Also, a brick face applied to the surface of a frame structure.

Vent: A small opening to allow the passage of air.

Vent stack flashing: Flanges that are used to tightly seal pipe projections through the roof. They are usually prefabricated.

Wainscoting: A wall finish, usually made of wood, stone, or ceramic tile, that is applied partway up the wall from the floor.

Wall flashing: A form of metal shingle that can be shaped into a protective seal interlacing where the roof line joins an exterior wall. Also referred to as step flashing.

Water stop: Thin sheets of rubber, plastic, or other material inserted in a construction joint to obstruct the seepage of water through the joint.

Water vapor: Water in a vapor (gas) form, especially when below the boiling point and diffused in the atmosphere.

Weatherstripping: Strips of metal or plastic used to keep out air or moisture that would otherwise enter through the spaces between the outer edges of doors and windows and the finish frames.

Web: That portion of a framing member that connects the flanges.

Module 27201-07

The Athenaeum of Philadelphia, Inc., 201SA02,
  Tom Crane, Photographer, 2000
Black and Veatch Corporation, Appendix B
Condor Rebar Consultants, Inc., Appendix A
The Construction Specifications Institute*,
  201F26
King's Material, Inc., 201F01 (school)
Dumond Chemicals, Inc., 201SA01
Ivey Mechanical Co., LLC, 201F16, 201F17
PERI USA, 201F01 (high-rise)
Tilt-Up Concrete Association, 201F01 (church,
  store)

Module 27102-07

Alum-a-Pole Corporation, 202SA05
ATAS International, Inc., 202SA11 (top)
Cedar Shake and Shingle Bureau, 202F05,
  202F90. Copyright © 2006. All rights reserved.
CertainTeed Corporation, 202SA01, 202F02,
  202F40. Copyright © 2006. Used with
  permission.
Cornell Corporation, 202SA14
DaVinci Roofscapes LLC, 202SA02, 202SA09.
  For more information about DaVinci's family
  of products, call 1-800-328-4624 or visit
  www.davinciroofscapes.com.
DBI/SALA & Protecta, 202F24
Follansbee Steel, 202F10
Hilltop Slate, Inc., 202F109
Johns Manville, 202F13

*The Numbers and Titles used in this textbook are from
MasterFormat™ 2004, published by The Construction Specifications Institute (CSI) and Construction Specifications
Canada (CSC), and used with permission from CSI. For those
interested in a more in-depth explanation of *MasterFormat™
2004* and its use in the construction industry, visit **www.
csinet.org/masterformat** or contact:

The Construction Specifications Institute (CSI)
99 Canal Center Plaza, Suite 300
Alexandria, VA 22314
800-689-2900; 703-684-0300
http://www.csinet.org

Maruhachi Ceramics of America, 202SA13
Reimann and Georger Corporation, 202SA06
The Stanley Works, 202F19 (hook knife)
Topaz Publications, Inc., 202F01, 202F04, 202F07,
  202F12, 202F19 (except hook knife), 202SA03,
  202F20, 202F44, 202F48, 202F51, 202SA07,
  202F66, 202F69 (photo), 202SA10, 202SA11
  (bottom), 202SA12, 202F130–202F137, 202F140,
  202F145, 202F148, 202F149 (photo), 202F150
  (photos), 202SA15

Module 27203-07

Amerimax Home Products, Inc., 203A04
Bemis Manufacturing Company, 203A06, 203A07
CertainTeed Corporation, 203SA01, 203F05,
  203F08. Copyright © 2006. Used with
  permission.
Dupont™ Building Innovations™, 203F31
  (center, bottom), 203F32, 203SA06
Engler, 203A02
International Code Council, 203F02.
  2006 International Energy Conservation Code™.
  Copyright © 2006. Falls Church, VA:
  International Code Council. Reproduced with
  permission. All rights reserved.
Johns Manville, 203F09, 203F31 (top)
Owens Corning, 203F15
Topaz Publications, Inc., 203F04, 203F07, 203SA05
US Greenfiber LLC, 203F06

Module 27204-07

Alum-a-Pole Corporation, 204SA03
Cedar Valley Shingle Systems, 204SA13
CertainTeed Corporation, 204SA15. Copyright ©
  2006. Used with permission.
Cummins Industrial Tools, 204SA06
James Hardie Building Products, 204SA14,
  204F45, 204F46
Milwaukee Electric Tool Corp., 204F44 (top)
Tapco Integrated Tool Systems, 204F50
Topaz Publications, Inc., 204SA01, 204SA02,
  204SA05, 204F26, 204SA08, 204SA09, 204SA11,
  204F57, 204F59
Zircon Corporation, 204F56

Module 27205-07

Erico, Inc., 205SA01
Klein Tools, Inc., 205F08
Senco Products, Inc., 205F02
Photo courtesy of Simpson Strong-Tie Company, Inc., 205F03
Steel Framing Alliance, 205F05, 205F06, 205F09, 205F11–205F13, 205F15–205F20, 205F21 (top), 205F22–205F26, 205F28–205F31, 205F36–205F41, 205F43–205F45, Table 1, Appendix B
Topaz Publications, Inc., 205F01, 205F04, 205F07, 205F21 (bottom), 205SA01, 205SA02, 205F32

Module 27206-07

AMICO, 206SA09 (left)
Crorey Builders, Inc., 206SA04
Gypsum Association, 206F01, 206F02, 206F03, 206F18–206F29, 206F31–206F35, 206F49–206F51, 206F53–206F55
John Hoerlein, 206F20, 206SA06, 206SA08
National Gypsum, 206SA10, 206SA11, 206SA12
RotoZip® by BOSCH, 206F10 (top)
Senco Products, Inc., 206SA07
The Stanley Works, 206F05, 206F06, 206F09, 206F10 (bottom), 206F11, 206F13, 206F14, 206SA03, 206SA05
Telpro, Inc., 206F16
Topaz Publications, Inc., 206F04, 206F07, 206F08, 206F15, 206F30 (top, middle), 206F43, 206SA09 (middle, right)
USG Corporation, 206SA01, 206SA02, 206F36, 206F37
Z-Pro International, 206F30 (bottom)

Module 27207-07

Ames Taping Tool Systems, Inc., 207F15, 207F18–207F21
Kraft Tool Company, 207F34
National Gypsum Company, 207F29
Porter-Cable Corporation, 207F22
The Stanley Works, 207F01, 207F03, 207F04 (lower), 207F05 (upper), 207F06, 207F09, 207F14, 207SA02, 207F47
Topaz Publications, Inc., 207F02, 207F04 (upper), 207F05 (lower), 207F07, 207F08, 207F10, 207F13, 207F16, 207F17, 207F23–207F28, 207SA01, 207F30, 207F31, 207F38–207F45, 207SA03, 207SA04, 207F48
USG Corporation, 207F33, 207F35, 207F54

Module 27208-07

Photo courtesy of Crane Revolving Doors, 208F133
Detex Corporation, 208SA15
Photo courtesy of DORMA Architectural Hardware, 208F97 (magnetic door holder)
HMMA Division of NAAMM, National Association of Architectural Metal Manufacturers, 208F08
Ingersoll-Rand Safety and Security, 208SA08, 208F81, 208SA09, 208F87, 208F90, 208F91, 208SA11, 208F97 (all except magnetic door holder), 208F98–208F102, 208F110–208F112, 208SA14
Marlite Decorative Wall Systems, 208F02
Pioneer Industries, 208F10–208F43, 208F127, 208F128
Porter-Cable Corporation, 208F67, 208F68, 208SA10, 208SA12
Raynor, 208F134
SDC Security Door Controls, 208F92–208F94, 208F95 (bottom), 208F96
Securitron Magnalock Corp., an ASSA ABLOY Group Co., 208SA13, 208F95 (top, middle)
Timely Prefinished Steel Door Frames, 208F09
Topaz Publications, Inc., 208SA01, 208SA02, 208SA03, 208F57
Yale Security Inc.[†], 208F79, 208F82–208F86, 208F88, 208F89, 208F103, 208F104.

Module 27209-07

Armstrong World Industries, Inc., 209F18, 209F19, 209F32
BPB Celotex—Capaul Series, 209F23
Ceilings Plus, 209F25, 209F35, 209F36
Chicago Metallic Corp., 209F33, 209F34
CST/berger, 209F04
Ritterbush-Ellig-Hulsing, P.C., Appendix A
USG Corporation, 209SA01, 209SA02
Topaz Publications, Inc., 209F02
Zircon Corporation, 209F03

[†]The drawings of Yale® locks used in these materials have been used with the consent of Yale Security Inc. Nothing in these materials is intended to describe or depict the actual performance of YALE locks or other products of Yale Security Inc.

Module 27210-07

Canamould Extrusions, Inc., 210SA02, 210F02 (photo), 210F05 (fancy crown molding). **www.canamould.com,** Trimroc interior mouldings by Canamould Extrusions, Inc., 2006

Collins Tool Company, 210SA04

Topaz Publications, Inc., 210F01, 210SA01, 201F03, 210F04, 210SA03, 210F05 (cove, crown, and bed moldings), 210F06–210F09, 210F12, 210F13, 210F15, 210SA05 (photo)

Module 27211-07

Blum, Inc., 211F10

Formica Corporation, 211SA01

Photo courtesy of **www.hdfiles.com,** 211F21

Makita Industrial Power Tools, 211F25 (photo)

Photo courtesy of SOSS Invisible Hinges, 211F11

Strasser Woodenworks, Inc., 211F03

Topaz Publications, Inc., 211SA02, 211F08, 211F15, 211F19, 211F20, 211SA03, 211F24

Module 27212-07

Makita Industrial Power Tools, 212F32 (portable biscuit jointer)

Robert Bosch Tool Corporation, 212F19

SkillsUSA, Appendix B

Topaz Publications, Inc., 212F03–212F18, 212SA02–212SA10, 212F21–212F24, 212F27–212F30, 212F32 (blade and biscuits), 212F33–212F35, 212F37, 212F44–212F48

WMH Tool Group, 212F20

Index